ENCYCLOPEDIA OF BIOTERRORISM DEFENSE

Edited by

Richard F. Pilch, M.D.

Center for Nonproliferation Studies
Monterey Institute of International Studies
Monterey, California

Raymond A. Zilinskas, Ph.D.

Center for Nonproliferation Studies
Monterey Institute of International Studies
Monterey, California

WILEY-LISS

A JOHN WILEY & SONS, INC., PUBLICATION

Published by John Wiley & Sons, Inc., Hoboken, New Jersey.
Published simultaneously in Canada.

For general information on our other products and services please contact our Customer Care Department within the U.S. at 877-762-2974, outside the U.S. at 317-572-3993 or fax 317-572-4002.

Wiley also publishes its books in a variety of electronic formats. Some content that appears in print, however, may not be available in electronic format.

Library of Congress Cataloging-in-Publication Data

ISBN 0-471-46717-0

Printed in the United States of America

10 9 8 7 6 5 4 3 2 1

CONTENTS

PREFACE

Despite the extreme danger [of chemical and biological weapons], we only became aware of them when the enemy drew our attention to them by repeatedly expressing concern that they can be produced simply.

Ayman al-Zawahiri, al-Qa'ida chief strategist under Usama Bin Ladin (Cullison and Higgins, 2001)

Since we first began assembling and editing the *Encyclopedia of Bioterrorism Defense* over a year ago, the above statement has been affixed on our office wall. It highlights what is perhaps the primary concern of those who publish in the national and international security sphere today: how is sensitive information to be published so that it optimally benefits policy makers and the public without generating added risks to our society? In the context of this volume, "sensitive" information is that which either is relevant to the development, production, formulation, or delivery of biological weapons, or reveals vulnerabilities of our society that ill-intending governments, groups, or individuals might exploit.

Certainly, the issue of how to manage sensitive but unclassified information is hotly debated both in the United States and other nations, particularly regarding openly published scientific information that contributes to the advancement of knowledge and often the betterment of mankind, but that also might be misdirected toward military, criminal, or terrorist purposes (Zilinskas and Tucker, 2002). The outcome of this debate cannot yet be discerned, though it appears increasingly likely that scientific publications containing sensitive information will face some restrictions in the near future, whether applied by publishers, professional societies, or, in the last instance, by executive or legislative government organs. But if the restriction of scientific publications poses such a challenge, what of reference books such as this, and other forms of media containing information that might be misused in the same way (an issue that is in fact addressed within, in an entry entitled "Media and Bioterrorism" by Joby Warrick of the *Washington Post*)? While we have made every effort to exclude information that might in some way serve illicit ends, most notably by referring materials to multiple experts in the field and deferring to their opinions on content, no doubt some will criticize the wisdom of a compilation detailing a topic as critical as the threat and defense of bioterrorism.

On the basis of nothing more than the opening statement above, the argument can be made that the disclosure of anything other than misinformation regarding the advantages of and vulnerabilities to biological weapons is inherently dangerous. But adopting such a stance outright overlooks what is a significant and deeply interdependent issue, and that is the need for the public and its representatives to have an accurate understanding not only of biological weapons themselves and what they can and cannot do but also of the host response to these weapons on both physical and psychological levels. Only by having this understanding is it possible for our society to plan and develop appropriate defenses against them and, should defense fail, to implement effective consequence management measures. As Dr. Paul Fildes, eminent bacteriologist and scientific leader of England's offensive biological warfare (BW) program during World War II, noted in the early stages of his work:

It is perhaps not generally understood that the basic problems of BW are basic problems of medicine. The applications only are different. Applied medicine is primarily concerned with defence, but defence cannot be arranged intelligently without study of offence. Similarly BW is concerned with the exploitation of offence. Thus both the BW workers and medical workers must know how the microbe carries out its offensive activities before the former can exploit them and the latter protect against them (Fildes, 1944).

Fildes's profound conviction serves to remind us that in these difficult times, when enemies armed with weapons of their choosing can come from any direction or within, the prevention of infectious diseases of whatever origin, be it nature or the laboratory, as well as the treatment of illnesses caused by pathogens or toxins depend to a large extent on the effective application of science. And scientific research is conducted most efficiently in an environment where ideas and information can flow freely. Thus, while information pertaining to, for example, the aerosol dispersal of a pathogen may be utilized by a terrorist group to develop and deploy a biological weapon, this information is also vital to those who develop personal and collective protective equipment, vaccines, detectors, and other defensive biologicals and materials. Further, on the policy level, such information is required to develop effective prevention and countermeasure strategies against the deployment of biological weapons against human, animal, or plant populations. Importantly, such advancement can be further directed as suggested toward the management of natural diseases, a factor of particular relevance in that natural outbreaks may well pose a greater health and economic risk at present than deliberate events, as has been demonstrated for example by the emergence of Severe Acute Respiratory Syndrome (SARS) as a global epidemic and the steadily increasing incidence of West Nile fever cases in the United States.

Apart from the physical threat, we should not underestimate the damages that the psychological response to biological weapons may cause, especially when weighing the disclosure of information against the need for education in the context of terrorism. After all, since it is generally recognized that one of the goals of terrorism is to generate fear, then the first step toward reducing the appeal and potential effects of bioterrorism is to inform people who might be targeted for attack about the possibilities and limitations of biological weapons. Since the limitations are truly severe, as proved by past incidents, an informed U.S. population protected by knowledgeable, well-trained first responders and medical professionals will be less likely to panic. Hospitals and emergency rooms will be less likely to be flooded by the so-called worried well population of unexposed individuals, allowing resources to be channeled toward those most in need.

It was with these thoughts in mind that we began this project in the fall of 2002 (only one year after the September 11, 2001, attacks and the subsequent "anthrax letters" delivered through the U.S. postal system), and these thoughts still remain in mind today. While we have long recognized the potential harm that an unscrupulous publication of the same scope might cause, we have also appreciated the vast benefits that would come as a result of work done right. Preparedness efforts aimed at reducing the threat of bioterrorism, and response efforts aimed at mitigating its consequences, rely upon a baseline understanding not only of the potential weapons and the potential terrorist motivations and capabilities for using them but also of the global web of bodies in place to address these considerations, from the local and state to national and even international level. This is what is meant by "bioterrorism defense," and this what we have sought to provide.

When we set out on this journey, we cast our net wide. The encyclopedia's initial statement of work read, "To provide complete A to Z coverage of the subject in order to offer a single authoritative resource for students, analysts, scientists, journalists, and policy makers worldwide." A year later, we have not strayed far from this initial objective, but perhaps like every academician we have allowed ourselves to wander, and linger on topics of interest and importance to us. What has thus evolved is a work in progress, a snapshot of the state of bioterrorism defense at the beginning of 2004, compiled by experts in the varying fields of policy and government; virology, bacteriology, and toxins; terrorism and threat reduction … the list goes on and on. And while we have aimed to be as comprehensive as possible, possibly the greatest lesson of our work is that, particularly in the (scientific) age of biotechnology and the (sociological) age of terrorism (at least as far as public perception is concerned), the realm of bioterrorism defense is without bounds. It would be entirely justifiable for John Wiley & Sons, Inc. to publish this volume in a loose leaf binder, so that readers may with some frequency update information as science evolves, new defenses are crafted, and terrorist motives and capabilities change. Of course, were there to be a bioterrorist event, it might spring surprises on us all, perhaps substantially altering the way we now think about how terrorists would use biological weapons and the best way to respond.

Conceding that changes pertaining to biodefense will surely come to pass, likely even between the time of this writing and the time of its publication, is not so difficult to deal with them on a personal level. What is more challenging, however, is to identify gaps and shortcomings in our own product as it stands today. This was an ambitious undertaking from the start, and admittedly in the back of our minds we were uncertain if all aspects of the subject matter could in fact be covered adequately in one volume. Simply put, and to our own dismay, we were right. In the end, we therefore have been forced to make certain difficult decisions regarding the content within. Some entries are not as well constructed as we would wish but are nevertheless included for the sake of completeness. Others were excluded for any of a variety of reasons, including unmet contributor obligations and unmet entry objectives.

Gaps include, for example, reviews of the U.S. and former Soviet biological weapons programs and lessons they might have for biodefense; discussions of the SARS and West Nile fever outbreaks; detailed studies of Centers for Disease Control and Prevention (CDC) Category B agents *Burkholderia pseudomallei*, *Coxiella burnetii*, and *Chlamydia psittaci*; analysis of plant and animal pathogens of importance with respect to agricultural terrorism; and overviews of the possibilities that macroparasites and fungi offer as biological weapons. In addition, it should be noted that from the outset this work was meant for an international audience, and while the preparedness efforts of a number of states and intergovernmental agencies are included, the encyclopedia is by no means comprehensive in this respect. We hope to improve on these deficiencies in a revised edition in the not-so-distant future.

Last, a word on technical accuracy. We have strived to be as accurate as possible when selecting terminology and information for inclusion in this work. However, since this volume is intended for a general audience, we cannot disregard certain popular parlance. For example, the terms "anthrax letters" and "anthrax mail attacks" are technically (and grammatically) inaccurate but are nevertheless commonly used to describe the delivery of envelopes containing *Bacillus anthracis* spores by the U.S. postal system in September and October 2001. Thus, these and other popular terms will be used throughout this volume for the sake of simplicity.

RICHARD F. PILCH
RAYMOND A. ZILINSKAS

Monterey, California

REFERENCES

Cullison, A. and Higgins, A., Computer in Kabul holds chilling memos, *Wall Street Journal*, December 31, (2001).

Fildes, P., *Post War Work on BW*, PRO, WO188/654, BIO/5198, August 22, 1944; referenced in Balmer, B., *Britain and Biological Warfare*, Palgrave, New York, 2001, p. 56.

Zilinskas, R.A. and Tucker, J.B., Limiting the Contribution of the Scientific Literature to the Biological Weapons Threat, *Journal of Homeland Security*, December 2002, http://www.homelandsecurity.org/journal/Articles/tucker.html.

CONTRIBUTORS

Praveen Abhayaratne, MS, Center for Nonproliferation Studies, Monterey Institute of International Studies

Gary Ackerman, MS, Center for Nonproliferation Studies, Monterey Institute of International Studies

Stephen S. Arnon, MD Chief, Infant Botulism Treatment and Prevention Program, California Department of Health Services

Jeffrey M. Bale, PhD, Senior Research Associate, Center for Nonproliferation Studies, Monterey Institute of International Studies

Michelle Baker, MS, Center for Nonproliferation Studies, Monterey Institute of International Studies

Stephen A. Berger, MD, Director, Geographic Medicine and Clinical Microbiology, Tel Aviv Medical Center, Tel Aviv, Professor of Medicine, Tel Aviv University School of Medicine

Anjali Bhattacharjee, MS, Center for Nonproliferation Studies, Monterey Institute of International Studies

Marcus Binder, MS, Center for Nonproliferation Studies, Monterey Institute of International Studies

Sara Renner Birkmire, U.S. Army Center for Health Promotion and Preventive Medicine, Aberdeen Proving Ground

Dallas A. Blanchard, Professor Emeritus, University of West Florida

W. Dickinson Burrows, U.S. Army Center for Health Promotion and Preventive Medicine, Aberdeen Proving Ground

Bridget K. Carr, DVM, MPH, DACVPM, Lieutenant Colonel, United States Air Force, Chief, Education and Training Department, Operational Medicine Division, United States Army Medical Research Institute of Infectious Disease (USAMRIID)

W. Seth Carus, Deputy Director, Center of Counterproliferation Research, National Defense University

Rocco Casagrande, PhD, Homeland Security Program, Abt Associates Inc., Cambridge, MA 02138

Andrea Cellino, Deputy for Policy Co-ordination to the Secretary-General of the NATO Parliamentary Assembly

Theodore J. Cieslak, MD, Colonel, MC, USA, Chief, Pediatrics Department, Brooke Army Medical Center

Dr. Rita R. Colwell, PhD, University of Maryland Biotechnology Institute

Norman M. Covert, Former Public Affairs Officer, Fort Detrick, Maryland

Skyler Cranmer, University of California, Davis

Malcolm Dando, PhD, Professor of International Security, Department of Peace Studies, University of Bradford

Greg Dasch, PhD, Viral and Rickettsial Zoonoses Branch, Centers for Disease Control and Prevention

Zygmunt F. Dembek, PhD, Assistant Clinical Professor, University of Connecticut Health Center, Masters in Public Health Program, Department of Community Medicine and Health Care

Adam Dolnik, MS, Research Associate, International Centre for Political Violence and Terrorism Research, Institute of Defence and Strategic Studies, Singapore

Edward Eitzen, MD, MPH, FACEP, FAAP, Colonel, United States Army, Senior Medical Advisor, Office of the Assistant Secretary for Public Health, Emergency Preparedness (OASPHEP), Department of Health and Human Services

Robert Emery, University of Texas Health Science Center—Houston, Environmental Health and Safety

Gerald Epstein, PhD, Senior Fellow for Science and Security, Homeland Security Program, Center for Strategic & International Studies

Marina Eremeeva, MD, PhD, Viral and Rickettsial Zoonoses Branch, Centers for Disease Control and Prevention

Daniel P. Estes, Sandia National Laboratories

Mats Forsman, Swedish Defence Research Agency, Umeå, Sweden

Pat Fitch, PhD, Lawrence Livermore National Laboratory

Odelia Funke, Office of Environmental Information, U.S. Environmental Protection Agency

Michael J. Glass, PhD, Chief, Special Programs Division, West Desert Test Center, U.S. Army Dugway Proving Ground

Gene D. Godbold, PhD, Research Scientist/Analyst, Battelle Memorial Institute, Charlottesville Operations

Erin Harbaugh, Center for Nonproliferation Studies, Monterey Institute of International Studies

C. Dix Harrell, DVM, MS, Area Epidemiology Officer, USDA-Animal and Plant Inspection Service

Anne Harrington, Deputy Director, Office of Proliferation Threat Reduction, U.S. Department of State

Lauren Harrison, Center for Nonproliferation Studies, Monterey Institute of International Studies

Richard Hartzell

Erika Holey, MS, Center for Nonproliferation Studies, Monterey Institute of International Studies

Bruce Hope, Vice President, Oregon Department of Environmental Quality

Nathaniel Hupert, MD, MPH, Department of Public Health, Weill Medical College of Cornell University

Anders Johansson, Swedish Defence Research Agency, Umeå, Sweden, Department of Clinical Microbiology, Infectious Diseases, Umeå University, Umeå, Sweden

Leslie Jacobs, University of the Pacific, McGeorge School of Law

Michelle Karch

Rebecca Katz, Woodrow Wilson School of International Affairs, Office of Population Research, Princeton University

Robyn Klein, Center for Nonproliferation Studies, Monterey Institute of International Studies

Wendy Kleinman, MPH, Arizona Department of Health Services, Office of Bioterrorism

Margaret E. Kosal, PhD, Center for Nonproliferation Studies, Monterey Institute of International Studies

Frank J. Lebeda, PhD, Chief, Department of Cell Biology & Biochemistry, Toxinology and Aerobiology Division, U.S. Army Medical Research Institute of Infectious Diseases

Michelle Longmire, Center for Nonproliferation Studies, Monterey Institute of International Studies

Stephen Marrin, Woodrow Wilson Department of Politics, University of Virginia

David McAdams, MS, Center for Nonproliferation Studies, Monterey Institute of International Studies

Kimberly McCloud, MS, Center for Nonproliferation Studies, Monterey Institute of International Studies

Dennis McGowan, Assistant Director, National Mass Fatalities Institute

Jennifer Mitchell, MS, Center for Nonproliferation Studies, Monterey Institute of International Studies

Alan Jeff Mohr, PhD, Chief, Life Sciences Division, U.S. Army Dugway Proving Ground

Stephen A. Morse, PhD, National Center for Infectious Diseases, Centers for Disease Control and Prevention

Donald L. Noah, DVM, MPH, DACVPM, Lt Colonel, USAF, BSC, Office of the Air Force Surgeon General

Dr. Gail H. Nelson, PhD, Association of Former Intelligence Officers

Andy Oppenheimer, Consultant, Jane's Information Group

Jason Pate, MS, Center for Nonproliferation Studies, Monterey Institute of International Studies

Elizabeth Rindskopf-Parker, University of the Pacific, McGeorge School of Law

Graham S. Pearson, Visiting Professor of International Security, Department of Peace Studies, University of Bradford (previously Director General & Chief Executive, Chemical and Biological Defence Establishment, Porton Down, Salisbury, Wiltshire, United Kingdom)

Richard F. Pilch, MD, Center for Nonproliferation Studies, Monterey Institute of International Studies

Dr. Daniel A. Pinkston, PhD, Senior Research Associate, Center for Nonproliferation Studies, Monterey Institute of International Studies

Mark A. Poli, PhD, DABT, Toxinology and Aerobiology Division, U.S. Army Medical Research Institute of Infectious Diseases

Serguei Popov, PhD, DSc, Director for Bacterial Pathogen Science, National Center for Biodefense, George Mason University

Yannick Pouliot, PhD, Consultant, Artificial Intelligence Center, SRI International

James A. Poupard, PhD, Pharma Institute of Philadelphia, Inc.

Don Prosnitz, Chief Science and Technology Advisor, U.S. Department of Justice

Gary Resnick, PhD, Associate Center Director for Chemical and Biological Defense, Los Alamos National Laboratory Center for Homeland Security

E.J. Richey, DVM, Professor Emeritis, Department of Large Animal Clinical Sciences, College of Veterinary Medicine, University of Florida

Silvana Rodriguez, Center for Nonproliferation Studies, Monterey Institute of International Studies

Roger Roffey, Swedish Defence Research Agency, Sweden

Jennifer Runyon, MS, Research Associate, Chemical & Biological Arms Control Institute

Joshua Sinai, PhD, Department of Homeland Security

Sammy Salama, MS, Center for Nonproliferation Studies, Monterey Institute of International Studies

Charlotte Savidge, Center for Nonproliferation Studies, Monterey Institute of International Studies

Reynolds M. Salerno, PhD, Sandia National Laboratories

Paul Setlak, Chicago College of Pharmacy

John Shaw

David W. Siegrist, Potomac Institute for Policy Studies

Maureen F. Sinclair, MPH, Epidemiology Program Officer, Centers for Disease Control and Prevention

Amy Smithson, PhD, Senior Fellow, International Security Program, Center for Strategic & International Studies

Jeremy Sobel, MD MPH, Foodborne and Diarrheal Diseases Branch, Centers for Disease Control and Prevention

Raymond A. Strikas, MD, National Immunization Program, Centers for Disease Control and Prevention

Masaaki Sugishima, School of Law, Asahi University, Gifu, Japan

Kathleen Thompson, MS, Center for Nonproliferation Studies, Monterey Institute of International Studies

Joseph Urrea, Arizona Department of Health Services

Sundara Vadlamudi, Center for Nonproliferation Studies, Monterey Institute of International Studies

Dr. Stephen Waring, DVM, PhD, Assistant Professor Epidemiology, Center for Biosecurity and Public Health Preparedness at UT-Houston, School of Public Health

Joby Warrick, Staff Writer, Washington Post

Fred Wehling, PhD, Center for Nonproliferation Studies, Monterey Institute of International Studies

Michael Woodford, Chair, Office International des Epizooties (OIE) Working Group on Wildlife Diseases

Angela Woodward, Legal Researcher, Verification Research, Training and Information Centre (VERTIC)

Alex Yabroff, Center for Nonproliferation Studies, Monterey Institute of International Studies

Alan P. Zelicoff, MD, Senior Scientific Consultant, ARES Corporation

Tyler Zerwekh, DrPH, Safety Specialist, Biosafety Division, University of Texas Health Science Center—Houston, Environmental Health and Safety

Raymond A. Zilinskas, PhD, Center for Nonproliferation Studies, Monterey Institute of International Studies

ACKNOWLEDGMENT

Planning, organizing, and editing this encyclopedia was perhaps the most difficult undertaking of our careers to date, particularly because it involves 99 contributors, intertwines the social sciences, security issues, weapons engineering, the natural sciences, and the physical sciences, and presents a moving target in that bioterrorism defense is a rapidly advancing and mutable field. The contributions of the experts who wrote the 136 entries it contains, and their forbearance during a long and arduous editing process, made our work less onerous than it otherwise would have been, so thank you so much. We are certainly grateful to Dr. William Potter, Director of the Center for Nonproliferation Studies (CNS), and Dr. Clay Moltz, Deputy Director of the CNS, who, despite their reservations about this project's feasibility, allowed us to carry on. And our special thanks go to the members of the CNS WMD Terrorism Database staff, who wrote many of the entries, helped with editing and other chores, and provided vital expertise in the Arab language. Since the membership of the WMD Terrorism Database staff is continually in flux (the team largely consists of graduate students who inevitably depart), we ask forgiveness if some of their names are missing from the following list: Gary Ackerman, Praveen Abhayaratne, Joel Baker, Jeffrey Bale, Adam Dolnik, Lauren Harrison, Jason Pate, Sammy Salama, Brian Traverso, and Sundara Vadlamudi.

Of course, this project would not have been possible at all had it not been for Ms. Luna Han, then a senior editor at Wiley and Sons, who inveigled us to take it on. Thanks a lot. . . After she departed Wiley, we are grateful to have received continuous support from other Wiley editors, including in particular Kristin Cooke Fasano, Thomas H. Moore, and Andrew Prince.

RICHARD F. PILCH
RAYMOND A. ZILINSKAS
April 8, 2005
Monterey, California

A

ABU SAYYAF GROUP

Jeffrey M. Bale

DEFINITION

A violence-prone Muslim separatist group in the southern Philippines that has carried out numerous acts of terrorism, including indiscriminate bombings in public places, murders of hapless Christian villagers, kidnappings-for-ransom of both Philippine nationals and foreigners, and ostentatious decapitations of captured soldiers and civilian hostages.

BACKGROUND

The origins of the Abu Sayyaf Group (ASG), which has also reportedly been referred to by its members as Al Harakatul Islamiya (the Islamic Movement), can be traced back to 1989. In its earliest incarnation, the organization was apparently known as the Mujahideen Commando Freedom Fighters (MCFF), but it was later renamed Jundullah (Soldiers of God) and thence, in the early 1990s, the ASG. Whichever moniker one uses, it is this relatively small but active terrorist organization that has since become the *bête noire* of both the Philippine government and the Bush administration.

The ASG was established by Abdurrajak Abubakar Janjalani, a member of the Tausug ethno-linguistic group, who was born on the southern island of Basilan in 1963. In 1981, he received a Saudi scholarship to study Islamic jurisprudence at 'Umm al-Qura` University in Mecca, and three years later, the charismatic young man returned home to preach Wahhabism—the puritanical doctrine formulated by Muhammad ibn `Abd al-Wahhab (1703–1792) and subsequently embraced by Saudi Arabia's ruling House of Sa`ud—in local mosques, where he soon built up a devoted following.

At some point during the 1980s, Janjalani also became an active member of Nur Misuari's Moro National Liberation Front (MNLF), which was then the largest and most important Muslim separatist group in the Philippines. Janjalani's uncompromising attitude, his eloquence as a speaker, and his personal charisma seemed to mark him as one of that organization's potential future leaders, but in 1986 he began openly questioning Misuari's policy of negotiating an autonomy agreement with the government, and eventually his outspoken anti-Misuari agitation led to a complete break with the MNLF.

In 1987, Janjalani went to Pakistan to join the anti-Soviet resistance movement. Upon arriving in Peshawar, he joined the so-called "Abu Sayyaf" Afghan guerilla group, the last of the seven foreign volunteer bands to be established there. This particular unit had been founded in 1986 by a Pashtun professor named Abdul Rasul Sayyaf, a rare Afghan follower of Wahhabism. 'Usama Bin Ladin himself was reportedly influenced by Abdul Sayyaf, and both men subsequently denounced the Saudi regime after it invited American troops into the kingdom during the months leading up to the 1991 Gulf War. This in turn seems to have prompted Abdul Sayyaf to make an ideological transition from Wahhabism proper to jihadist Salafism, and the very same doctrinal shift was thence apparently made by Janjalani, who was so inspired by his former Afghan trainer that he subsequently named his own armed Islamist group after him.

After completing his *mujahidin* training in Pakistan and returning home to Basilan, Janjalani joined together with seven other men who were likewise disenchanted with the MNLF's moderation. In 1989, this small group of Islamist militants broke away from the Philippine branch of the proselytizing, Pakistan-based Tabligh-i Jamaat (Association for the Propagation of the Faith) movement and formed a separate armed organization known as the MCFF. This group, the forerunner of the later ASG, originally consisted of a core of around 20 members, mainly Moro (Philippine Muslim) volunteers who had gone to Afghanistan to wage *jihad*, but it soon managed to recruit hundreds of men, including some provincial MNLF commanders, by persuading them that Misuari was not waging a real *jihad*.

During this formative period, and indeed up through the mid-1990s, al-Qa`ida provided considerable funding to the ASG through an array of "charitable" fronts set up by Muhammad Jamal Khalifa, one of Bin Ladin's brothers-in-law and a high-ranking member of the Ikhwan al-Muslimin (Muslim Brotherhood). For example, on January 29, 1992, the organization received 160,000 Philippine pesos from Khalifa, as well as large deliveries of weapons from Viktor Bout, a Russian arms dealer linked by international investigators to al-Qa`ida. During this same five-year period, the ASG was allegedly responsible for carrying out 67 terrorist attacks, more than half of which were indiscriminate bombings, thereby killing 58 people and wounding 398. According to the Philippine police, when convicted 1993 World Trade Center bomber Ramzi Yusuf moved to Manila in mid-1994 to set up an al-Qa`ida terrorist cell, he relied upon ASG contacts and intended to use ASG members to help him carry out his plans. Although the evidence for this claim is thin, especially since Yusuf soon apparently came to the conclusion that

Encyclopedia of Bioterrorism Defense, Edited by Richard F. Pilch and Raymond A. Zilinskas
ISBN 0-471-46717-0 Copyright © 2005 Wiley-Liss

the members of the group were too amateurish to rely upon, Janjalani's group nonetheless took credit for Yusuf's December 10, 1994, test bombing of a Philippine Airlines (PAL) flight from Cebu to Tokyo, which resulted in the death of a Japanese businessman. Later, when Yusuf's plot to assassinate the visiting Pope John Paul II was exposed in January 1995, the ASG also took credit for organizing that, ostensibly in the name of the "Islamic liberation struggle against the Manila government and the Catholic Church." The sudden breakup of the Yusuf cell after its terrorist plots were exposed in early 1995 had a profound impact on the ASG, which was quickly deprived of its principal source of external funding. It was this development, perhaps more than any other, which in the second half of the 1990s led to the devolution of Janjalani's organization from an ideologically committed Islamist group into one that increasingly resorted to criminal activities to finance itself.

On paper, the ASG was said to be loosely organized into a cellular structure under the direction of an Executive Committee headed by a "Caliph"—Janjalani himself—and consisting of eight other leaders. However, the ASG does not appear to have ever created an elaborate, well-defined organization with a clear chain of command, even though the Executive Committee may have established small subunits to handle specific functional responsibilities in addition to several elite strike forces and some lower-level, territorially based units. On the basis of various accounts, including those provided by hostages seized by the group, in its home islands the ASG seems to have been divided either into distinct bandit gangs or separate kin-based groupings resembling the minimal and medial Tausug alliance groups described by anthropologist Thomas Kiefer, but when groups of ASG fighters operated outside of their island bases in urban locales such as Metro Manila or Zamboanga City, they apparently organized themselves into standard terrorist-style cells and thereafter functioned in the clandestine and covert fashion typical of such cellular structures. There were two primary ASG groups, one on Basilan initially headed by Janjalani and one on the island of Jolo led by Galib Andang ("Commander Robot") and his deputy Mujib Susukan. The Jolo group was seemingly subordinate to the Basilan group, at least when the two groups collaborated operationally in joint actions, but under normal circumstances, both groups appear to have operated more or less independently on their home turf. In the wake of Janjalani's December 1998 death in a firefight with the Philippine military, the leadership of the Basilan group ended up in the hands of two men, his soft-spoken younger brother Kaddafy Janjalani—named after Libyan leader Mu`ammar al-Qadhdhafi—and the belligerent Aldam Tilao ("Abu Sabaya"). As of this writing, most of these original ASG leaders have either been killed or captured.

In the beginning, at least, the ASG espoused an Islamist agenda that was far more radical than that associated with the Moro Islamic Liberation Front (MILF), another breakaway faction of the MNLF that then rapidly developed into the largest armed Islamist separatist movement in the Philippines. Janjalani's primary objective was to unify "all sectors of the predominantly Muslim provinces in the South" and establish an Islamic state governed by the *shari`a* in that region, a state where "Muslims can follow Islam in its purest and strictest form as the only path to Allah." Moreover, he intended to accomplish this objective by means of armed struggle rather than through the gradual and peaceful process of proselytization (*da`wa*) embraced by Muslim evangelical groups such as the Tabligh. Thus, the ASG was noticeably uncompromising within its own milieu. Unlike the MNLF and the MILF, one of the ASG's specific goals was to rid the Sulu archipelago, Basilan, and parts of Mindanao of all Christians and non-Muslims, by force if necessary, since according to Janjalani, Islam permitted the killing of "our enemies" and "depriving them of their wealth." Indeed, anti-Christian animus seems to have been the principal motive underlying the group's initial wave of terrorist attacks, all of which specifically targeted "Crusaders," and such sentiments were thereafter consistently used as a rationale to justify its violent actions.

Nevertheless, it is generally acknowledged that the radical religio-political objectives promoted by the original ASG leadership cadre were gradually compromised and corrupted by material interests, and some observers have gone so far as to conclude that the ASG has transformed itself from an authentic rebel political group inspired by Islamist doctrines into a violent criminal gang that simply uses Islam as a convenient cover to conceal its mercenary aims. Yet, this was by no means an "either/or" process, and the most that one can say is that early on the ASG was more ideologically driven, even though it never eschewed criminal acts, but that as time progressed the group became more and more concerned with its own continued survival and material well-being than with the active pursuit of a regional or transnational holy war.

Since its creation, the ASG has carried out a wide variety of terrorist operations, including relatively small-scale terrorist bombings in public places, such as the August 23, 1992, bombing of the open-air Roman Catholic shrine at Fort Pilar in Zamboanga City, which killed five people, and the December 26, 1993, bombing of the San Pedro Cathedral in Davao City, which killed seven; small-scale raids and massacres; large-scale raids and massacres, the most important of which was the April 4, 1995, raid on the predominantly Christian town of Ipil, which resulted in the massacre of 53 people and the razing of the town's central market; small-scale kidnappings-for-ransom; and high-profile hostage seizures. In recent years, the group has gained most of its notoriety from a series of spectacular operations that fall into this last category. The most famous ASG hostage operations were two relatively recent seizures that victimized Western tourists, one on April 23, 2000, from the Sipadan Dive Resort on the tiny island of Sipadan off the coast of Sabah, Malaysia, and the other on May 28, 2001, from the Dos Palmas Arrecife Island Resort on Honda Bay in Puerto Princesa City, Palawan. In both cases, ASG gunmen traveled across local seas in motorized outriggers, suddenly disembarked at night along the beach, rounded up several stunned foreign tourists and resort workers, and whisked them away by sea to their strongholds on Basilan or Jolo. In the government's ensuing efforts to rescue the hostages, several ended up being executed by ASG members or accidentally killed in firefights, but in the end the surviving captives were released in exchange for the payment of a substantial ransom.

THE ASG AND WMD

As of early 2004, no one associated with the ASG has publicly expressed any interest in acquiring or employing chemical, biological, radiological, or nuclear (CBRN) weapons, and there is no evidence indicating that the group has made any tangible efforts to obtain or deploy them. There is also little reason to believe that this situation will change in the near future, since the ASG appears to lack both the motivational drive and the technical capacity to carry out a successful attack with such weapons.

The ASG has never had any moral qualms about murdering civilians or causing mass casualties in the process of carrying out conventional terrorist attacks. However, in spite of its occasionally overheated rhetoric about waging a worldwide *jihad* against unbelievers, most of the group's members seem to be fighting at most for the establishment of an independent "Islamic" state in Basilan and Sulu, if not for far more parochial causes or for naked material gain in one of the most impoverished regions of the Philippines. To the degree that they are focused on these practical and insular matters, they are arguably much less likely to resort to the use of weapons of mass destruction (WMD), especially within their home territory, since such an action would likely elicit even heavier crackdowns on the group by the forces of order.

Even if the ASG's leaders suddenly developed the urge to do so, it is unlikely that they could muster sufficient technological and operational sophistication to carry out a successful bioterrorism attack. Indeed, the ASG appears to have an unusually low level of technical competence in comparison with most other terrorist groups, be they Islamist or secular. The majority of the group's members are poor, uneducated, and illiterate, its religious leaders are at best knowledgeable about arcane matters such as Islamic law and theology, and only a few of its operational chiefs have had any advanced technical education. Hence, it is not surprising to find that firsthand observers have been singularly unimpressed with the group's operational methods, ranging from its fire discipline to its organizational security, command and control, logistical arrangements, and basic military tactics. One might suspect that their noticeable lack of a formal technical education could have been offset by the hands-on training ASG members received from professional terrorists in Afghanistan and Mindanao and the actual combat experience some of them subsequently gained, and to some extent this is true. Even so, the most damning verdict concerning their technical capacities was pronounced by terrorist professionals such as Ramzi Yusuf and other members of his Manila cell, who considered ASG personnel to be too incompetent and untrustworthy to be entrusted with serious operational matters and hence used them solely for logistical support.

CONCLUSION

Although the ASG is included in this encyclopedia because it appears on the U.S. government's list of Foreign Terrorist Organizations, as a collective entity it is very unlikely to resort to WMD terrorist attacks because of its increasingly criminal nature, its often parochial political objectives, and its generally low level of technical and operational capability. However, individual members of the ASG may well be drawn into the orbit of other organizations, including transnational Islamist terrorist groups and corrupt components of the Philippine security forces with which they have colluded in the past. These organizations, with or without the help of ASG fighters, may have both the technical capabilities and the motives to carry out some sort of CBRN attack in the Philippines, if only for psychological impact.

REFERENCES

Abuza, Z., *Militant Islam in Southeast Asia: Crucible of Terror*, Chapter 3, Lynne Rienner, Boulder, 2003.

Barreveld, D.J., *Terrorism in the Philippines: The Bloody Trial of the Abu Sayyaf, Bin Ladin's East Asian Connection*, Writer's Club, San Jose, CA, 2001.

Burnham, G. (with Dean Merrill), *In the Presence of My Enemies*, Tyndale House, Wheaton, IL, 2003.

Chalk, P., "Militant Islamic Extremism in the Southern Philippines," in J.F. Isaacson and C. Rubenstein, Eds., *Islam in Asia: Changing Political Realities*, New Brunswick, Transaction, 2002a, pp. 187–222.

Chalk, P., "Al-Qaeda and its Links to Terrorist Groups in Asia," in A. Tan and K. Ramakrishna, Eds., *The New Terrorism: Anatomy, Trends, and Counter-Strategies*, Eastern Universities Press, Singapore, 2002b, pp. 107–128.

Che Man, W.K., *Muslim Separatism: The Moros of Southern Philippines and the Malays of Southern Thailand*, Oxford University, Singapore and New York, 1990.

Gunaratna, R., *Jane's Intell. Rev.* (1 July 2001), at: www.janes.com/K2/docprint.jsp?K2DocKey=/content1/janesdata/mags/jir/history.

Gutierrez, E., "New Faces of Violence in Muslim Mindanao," in E. Gutierrez, et al. Eds., *Rebels, Warlords and Ulama: A Reader on Muslim Separatism and the War in the Southern Philippines*, Institute for Popular Democracy, Quezon City, 2000, pp. 349–362.

Kiefer, T.M., *The Tausug: Violence and Law in a Philippine Moslem Society*, Waveland, Prospect Heights, IL, 1972.

Ressa, M.A., *Seeds of Terror: An Eyewitness Account of Al-Qaeda's Newest Center of Operations in Southeast Asia*, Chapter 6, Free Press, New York, 2003.

Torres, Jr., J., *Into the Mountain: Hostaged by the Abu Sayyaf*, Claretian Publications, Quezon City, 2001.

Tucker, J., Ed., *Toxic Terror: Assessing Terrorist Use of Chemical and Biological Weapons*, M.I.T., Cambridge, MA, 2000.

Turner, M., *Contemp. Southeast Asia*, **17**(1), 1–19 (1995).

Vitug, M.D. and Glenda, M.G., *Under the Crescent Moon: Rebellion in Mindanao*, Part 3, Ateneo Center for Social Policy and Public Affairs/Institute for Popular Democracy, Quezon City, 2000.

AEROSOL (AEROBIOLOGY, AEROSOLS, BIOAEROSOLS, MICROBIAL AEROSOLS)

ALAN JEFF MOHR

INTRODUCTION

Aerobiology is the study of microbiological particles that have, either naturally or purposefully, been introduced into the air. The science of bioaerosols is extremely complex, and to fully understand the subject it is necessary to have an understanding of microbiology, meteorology, biochemistry, and aerosol physics. Bioaerosols are usually presented as polydispersed particles or droplets and range in size from 0.5 to 15.0 micrometers (μm). The optimal aerodynamic particle size range, which represents a hazard to the human respiratory tract, is between 1.0 to about 10.0 μm. Particles between 1.0 and 3.0 μm are typically the most hazardous because they can be inhaled into the deepest regions of the lungs (alveoli) and initiate respiratory disease.

Bioaerosols of significance to terrorism include both human and agricultural pathogens and toxins. These aerosols can be composed of spore-forming bacteria, vegetative bacteria, rickettsia, viruses, toxins, and fungi. Bioaerosols are attractive as terrorist weapons because they are silent, invisible, and odorless, and can be introduced clandestinely to strike without warning.

BACKGROUND

The controlled study of biological aerosols can be traced back to the early 1930s, when the Japanese military intentionally infected prisoners with bioaerosols to determine their application in war. In 1942, the United States initiated its offensive program to study the infectivity and aerosol stability of biological agents. Between 1949 and 1968, the United States studied the aerosol dispersion characteristics of biological simulants when released at various sites in the country. In 1998, it was revealed that for decades, the former Soviet Union had produced biological agents on a massive scale and had developed methods for weaponizing hemorrhagic fever viruses to enhance their BW potential.

Today, there are few authoritative texts dealing directly with aerobiology because it is a comparatively new discipline and has not been comprehensively studied. Results presented in peer-reviewed articles are often confusing, and contradictory results are common. These inconsistencies can be explained by the varied methodologies employed to study the subject. The results of aerosol stability studies are dependant on the method of microbiological sample preparation, sample storage, and sample quantification. Additionally, results are dependent on the type of aerosol aging drum used, the methods of aerosol generation and sampling, and the types of fluids used for suspension and collection.

TECHNICAL REVIEW

The fate and transport of bioaerosols is a complicated issue involving many physical and biochemical factors (Table 1). The transport of bioaerosols is primarily governed by hydrodynamic and kinetic factors, while their fate is dependent upon their specific chemical makeup and the meteorological parameters to which they are exposed. When a bioaerosol particle approaches a surface, the effects governed by the makeup of the biological agent's cell wall will influence deposition. Specific surface-surface interactions then dictate the release of the particle (Lighthart and Mohr, 1994).

The vast majority of airborne microorganisms are immediately inactivated upon release because of environmental stresses (e.g., desiccation, temperature, and oxygen) that act to alter the makeup of the outer surface of the microorganism. The most significant environmental factors influencing viability are relative humidity (RH), temperature, and oxygen. Additional influences are exerted through air ions, solar irradiance, and open air factors (OAF).

Bioaerosol particles can be either solid or liquid and can come from a number of natural and anthropogenic sources. Although few generalities can be made concerning the aerosol stability or fate of microorganisms, bacteria tend to behave differently compared to viruses, which behave differently from molds and fungus. Some microorganisms have built-in mechanisms that act to repair damage inflicted during the aerosolization and transport phases.

Encyclopedia of Bioterrorism Defense,   Edited by Richard F. Pilch and Raymond A. Zilinskas
ISBN 0-471-46717-0   Copyright © 2005 Wiley-Liss

Table 1. Factors Affecting the Viability of Airborne Microorganisms

Primary Factors	References
Humidity	Cox et al., 1974; Cox and Goldberg, 1972; Sattar et al., 1984; Theunissen et al., 1993
Temperature	Dimmock, 1967; Theunissen et al., 1993
Radiation	Donaldson, 1972; Jensen, 1964
OAF	de Mik and de Groot, 1977; Donaldson and Ferris, 1975; Druett, 1973

Secondary Factors	References
Method of aerosol generation	Heidelberg et al., 1997; Marthi et al., 1990
Composition of generation fluid	Cox, 1987
Sampling method	Marthi and Lighthart, 1990; Warren et al., 1969
Collection medium	Elazhary and Derbyshire, 1977

PHYSICAL PARAMETERS

Water Content

The state of water and the water content associated with bioaerosols are fundamental factors influencing the fate or viability of microorganisms. As the RH decreases, so does the water available to the exterior environment of the microorganism. Loss of water can cause dehydration, resulting in inactivation of many microorganisms. Of all the measurable meteorological parameters, RH is the most important with respect to aerosol stability (Cox, 1987). Israeli et al. (1994) studied freeze-dried microorganisms and showed the importance of water content to the viability of microorganisms. They concluded that biomembranes, as phospholipid bilayers, undergo conformation changes from crystalline to gel phases as a result of water loss. These transformations induce changes to cell proteins, which in turn result in a loss of viability. The targets of RH-induced inactivation appear to be membrane phospholipids and proteins.

Temperature

The vapor pressure and therefore the RH of a system are dependent on the temperature. This relationship makes it very difficult to separate the effects of temperature and RH. Studies to determine the effect of temperature on aerosol stability have generally shown that increases in temperature tend to decrease the viability of airborne microorganisms (Dimmock, 1967). Additionally, frozen cells tend to lose cellular proteins (namely permease) that enhance aerosol inactivation rates. As with RH, the targets of temperature-induced inactivation appear to be membrane phospholipids and proteins.

Oxygen Concentration

Oxygen concentration can have an important effect on the aerosol stability and infectivity of some bacteria (Cox

et al., 1973, 1974). Free radicals of oxygen have been suggested as a cause of inactivation. Some investigators have observed negative relationships between oxygen concentration and viability. It has been shown that oxygen susceptibility increases with dehydration, increased oxygen concentration, and time of exposure. The targets of oxygen-induced inactivation also appear to be phospholipids and proteins.

Electromagnetic Radiation

Aerosol inactivation caused by electromagnetic radiation has been shown to be dependent on the wavelength and, hence, the intensity of the radiation. Shorter wavelengths contain more energy and are generally more deleterious to aerosolized microorganisms. The targets of radiation-induced inactivation appear to be phospholipids, proteins, and nucleic acids. The negative impact of solar irradiance on the viability of bioaerosols cannot be overemphasized.

BIOLOGICAL PARAMETERS

The type, species, or strain of a microorganism will affect its airborne survival. The atmospheric environment is hostile to all microorganisms—no species has evolved to fill the niche of the open atmosphere.

Bacteria

There is some evidence that suggests that some bacteria form new cells in the airborne state (Dimmick et al., 1979). Gram-negative bacteria contain more phospholipids than gram-positive bacteria. For this reason, they are more susceptible to inactivation through such mechanisms as Maillard reactions (reactions between proteins and carbohydrates), dehydration, and osmotic shock. Bacteria grown on enriched media display widely varying aerosol stability rates, as do bacteria in different phases of development. The physiological age of a bacterial population also influences its stability, because old cells die off at higher rates than do newly grown samples. Further, cells grown during the log phase may be more active metabolically, and some systems may be more vulnerable to stresses, than those in the stationary phase of growth.

Table 2 shows some of the parameters that have been shown to influence the viability of selected bacteria. Differences are not only due to dissimilar test procedures and data representation but also to the greater structural and metabolic complexity of bacteria (cell walls, membranes, and metabolism). Some generalities can be inferred concerning the aerosol stability of bacteria:

- Loss of aerosol viability is caused by desiccation and oxygen toxicity.
- Some gram-negative bacteria are inactivated by oxygen after desiccation.
- Many bacteria have complicated RH-inactivation profiles in inert atmospheres.
- Bacterial survival can be greatly increased by the addition of stabilizers (carbohydrates, for example).

Table 2. Aerosol Stability Parameters for Selected Bacteria

Bacteria	Stability Parameters	References
Bacillus subtilis	Death rate	Webb, 1959, 1960
Bacillus patchiness	Ambient temperature, bacterial viability	Webb, 1959, 1960
Bacillus violaceous	Ambient temperature, bacterial viability	Webb, 1959, 1960
Chlamydia pneumoniae	RH, temperature	Warren et al., 1969
Erwinia herbicola	RH, temperature	Marthi, 1994
Escherichia coli	RH, temperature, O_2, wet, dry	Cox, 1966, 1968
Mycoplasma pneumoniae	RH, temperature, solar	Israeli et al., 1994
Pasteurella tularensis	RH, wet and dry generation, solar	Cox and Goldberg, 1972
Serratia marcescens	RH, O_2, freeze, time	Cox et al., 1973, 1974
Staphylococcus albus	Time, ambient temperature	Webb, 1959, 1960
Staphylococcus aureus	RH, temperature	–

Table 3. Aerosol Stability Parameters for Selected Viruses

Virus	Stability Parameters	References
Foot and mouth viruses	Radiation, RH, temperature, weather factor	Barlow, 1972
Influenza viruses	RH	Loosli et al., 1943; Miller and Artenstein, 1966
Newcastle's disease virus	RH, temperature	Songer, 1966
Pigeon pox	RH, inositol	Webb et al., 1963
Rouse sarcoma virus	RH, inositol	Webb et al., 1963
St. Louis encephalitis virus	RH, temperature	Rabey et al., 1969
Venezuelan equine encephalitis virus	RH, temperature, solar	Dimmick et al., 1979
Vesicular stomatitis virus	RH, temperature, O_3	Donaldson, 1972; Songer, 1966
Yellow-fever virus	RH, temperature	Marthi and Lighthart, 1990

- Aerosolized spores of many bacteria are resistant to inactivation by oxygen concentration, RH, and temperature.

Viruses

Table 3 shows some of the parameters that have been shown to influence the viability of selected viruses. Viruses are normally very resistant to inactivation by oxygen. Because of this characteristic, and because of the relative simplicity of their structure, the results of aerosol inactivation studies are more consistent for viruses than for bacteria. Some generalities can be made about aerosol inactivation rates for viruses:

- Viruses with lipids in their outer coat or capsid are more stable at low RHs than at high RHs.
- Viruses without lipids are more stable at high RHs than at low RHs.
- When viable viruses can no longer be detected after aerosol collection, nucleic acids can be isolated and are still active (this evidence suggests that aerosol inactivation of viruses is not caused by nucleic acid inactivation but by denaturation of coat proteins).
- Prehumidification during sampling increases recovery of viruses that lack lipids in their outer coat.

It is also interesting to note that some viruses exhibit higher survival rates at mid-range RH, regardless of temperature, while some display better survival at low and high RH and the lowest survival at mid-range RH. However, survival may depend on the temperature and oxygen content of the test atmosphere at the time of testing. There are some viruses that are stable in the airborne state over broad temperature and RH ranges.

Fungi

Fungi and their spores seem to be resistant to desiccation, but little work has been completed in assessing their survival rates in aerosols. Interest has intensified recently with the phenomenon termed "sick building syndrome" (SBS). It is believed that fungi and bacteria are responsible for most SBS cases, even though most of the 100,000 known species of fungi do not cause disease in healthy people. More common effects of the presence of fungi in buildings are asthma, allergenic-type responses, and sometimes significant respiratory distress. Many fungi produce toxins that, when inhaled in high concentrations (as can occur with farm workers or people working in granaries), can result in significant health consequences.

Toxins

There are very few references addressing the stability of aerosolized toxins. Because toxins are chemicals, their inactivation during travel downwind is less affected by meteorological factors (desiccation due to low RH, temperature, and solar irradiance) than viable bacteria and viruses. For these reasons, the aerosol activity of toxins presumably falls between stable bacterial spore formers (*Bacillus anthracis*) and fragile vegetative bacteria (such as *Yersinia pestis*, *Francisella tularensis*, and so forth).

INACTIVATION MECHANISMS

Bioaerosols are subject to inactivation during storage, during dissemination, and especially during aerosol transport. Desiccation of the droplet is the main factor responsible for inactivation (Wells, 1934). Applicable particle sizes range from 0.5 to 15 μm. (Particles smaller than 5.0 μm act as vapors and follow the streamlines of the local airstreams.) Some relationships between aerosol stability and the biological composition of microorganisms have been identified. Bacterial aerosol stability is considerably more complex than that observed for viruses. For bacteria aerosolized into inert atmospheres at mid to high RH, the biological membrane constituents become destabilized through loss of water molecules. Additives that supersaturate, such as polyhydroxyl compounds, can stabilize these structures. The polyhydroxyl compounds, by binding to sites on proteins, cause conformational changes and thereby stabilize the proteins, making them less susceptible to denaturation. This is convincing evidence that the state of proteins on the outer membrane of some microorganisms is critical to the resultant stability profile.

Cox's (1987) explanation of the sequence of events is as follows. There is little doubt that during the desiccation process polyhydroxyl compounds and amino acids react together, causing conformational changes that strengthen the overall protein structure. The presence of sugar additives causes conformational changes in the coat proteins, and in the new configuration, these coat proteins do not react (or react more slowly) with the polyhydroxyl coat moieties. In the absence of sugar additives and free molecules, the coat proteins may react irreversibly, through Maillard reactions, with the polyhydroxyl coat moieties and cause loss of viability. In addition, the sugar additives could compete with the polyhydroxyl coat moieties for the reaction sites of the coat proteins or physically hinder those reactions' molecular collisions. The result in each case would be more aerostable microorganisms.

CURRENT ISSUES

Dissemination

The physical properties of bioaerosols are primarily dependent upon the generation method applied. Two of the most important variables, particle size distribution and concentration, are directly related to the aerosol generator. The most efficient sized particles for deposition in the human respiratory tract lie between 1.0 and 10.0 μm.

Biological aerosols are usually produced from a *liquid* suspension. Most bioaerosols generated from liquids are polydispersed in nature. The energy required to produce small particles can come from pressurized air, electricity, centrifugal forces, impaction, or heat. Many of these forces are so violent that inactivation of microorganisms, especially vegetative bacterial cells, can result. Fluid associated with newly aerosolized particles will instantaneously start to evaporate, and equilibrium

with the surrounding water vapor will be established, resulting in a dehydrated solid particle.

The energy applied to aerosolize *dry* particles may come in the form of pressurized air, scraping a dried cake and then applying air, or employing an explosive device. Bioaerosols that are initially disseminated from dry preparations are often very electrostatic; that is, electrically charged particles will be attracted to one another and thus form large clumps.

Several significant issues must be overcome to successfully disseminate infectious bioaerosols. It is true that unsophisticated methods (agricultural sprayers, commercial aerosol devices, etc.) can be used to produce bioaerosols, but the results will almost certainly be just as basic as the method of delivery. To efficiently and consistently produce infectious bioaerosols in the 1.0- to 10.0-μm size range, a committed testing program would be required to verify infectivity and pathogenicity as well as particle size range. Problems that would have to be overcome include inactivation caused by the dissemination system, a thorough knowledge of appropriate meteorological parameters, biological particle stabilizers, and clogging of nozzles. The use of spores would preclude many of the above concerns.

Sampling

A significant part of the science of aerobiology has been directed toward the development of samplers that collect a representative parcel of air while striving to minimize the stresses that cause inactivation of the target. Aerosolized biological particles can be collected on agar surfaces (slit-to-agar samplers) or into buffered liquid medias (impinger, cyclone, etc.). Biological particles collected onto filters are often inactivated owing to the severe drying affects of the procedure. Liquid cyclone samplers are among the best collectors because they are efficient and minimize biological inactivation. Currently there is no biological sampler available that is highly efficient, collects isokinetically, and maintains a high degree of viability.

FUTURE OUTLOOK

Techniques employed to produce, concentrate, purify, and store microorganisms have changed very little since the 1950s. Additionally, there has been little recent advancement concerning the procedures used to dry, mill, weaponize, and disseminate biological agents. It is probable that biotechnology and genetic engineering could perhaps be applied to affect the aerosol stability of microorganisms and toxins, but research required to accomplish this would be substantial and time consuming. Advancements in polymer science and microencapsulation technology could significantly decrease desiccation in an aerosol particle, which would act to prolong the aerosol stability and hence the infective nature the biological particle.

Biological aerosols can be produced utilizing sophisticated or crude methods. Techniques required to produce dry spore preparations are quite simple, but years of work were required to develop those procedures. Liquid suspensions of vegetative or spore-forming bacteria would be the most straightforward to disseminate. If an unsuitable technique were employed for dissemination, the size range of the ensuing particles would likely be too large (and quickly settle out) and the process would likely inactivate most of the biological particles. To optimize fully the effectiveness of a biological aerosol effort, it would be necessary to have a sophisticated testing program—including production and concentration, animals for infectivity, and aerosol characterization for particle size and viability—to evaluate the processes.

REFERENCES

Barlow, D.F., *J. Gen. Virol.*, **17**, 281–288 (1972).

Cox, C.S., *J. Gen. Microbiol.*, **43**, 383–399 (1966).

Cox, C.S., *J. Gen. Microbiol.*, **50**, 139–147 (1968).

Cox, C.S., *The Aerobiological Pathway of Microorganisms*, John Wiley & Sons, Chichester, New York, Brisbane, Toronto, Singapore, 1987.

Cox, C.S. and Goldberg, L.J., *Appl. Microbiol.*, **23**, 1–3 (1972).

Cox, C.S., Baxter, J., and Maidment, B.J., *J. Gen. Microbiol.*, **75**, 179–185 (1973).

Cox, C.S., Gagen, S.J., and Baxter, J., *Can. J. Microbiol.*, **20**, 1529–1534 (1974).

de Mik, G. and de Groot, I., *J. Hyg.*, **78**, 175–180 (1977).

Dimmick, R.L., Wolochow, H., and Chatigny, M.A., *Appl. Environ. Microbiol.*, **37**, 924–927 (1979).

Dimmock, N.L., *Virology*, **31**, 338–353 (1967).

Donaldson, A.I., *Vet. Bull.*, **48**, 83–94 (1972).

Donaldson, A.I. and Ferris, N.P., *J. Hyg. (Camb.)*, **74**, 409–415 (1975).

Druett, H.A., "The Open Air Factor," in J.F.Ph. Hers and K.C. Winkler, Eds., *Airborne Transmission and Infection*, John Wiley & Sons, New York, Toronto, 1973, pp. 141–151.

Elazhary, M.A.S.Y. and Derbyshire, J.B., *Can. J. Comp. Med.*, **43**, 158–167 (1977).

Heidelberg, J.F., Shahamat, M., Levin, M., Rahman, I., Stelma, G., and Colwell, R.R., *Appl. Environ. Microbiol.*, **63**, 3585–3588 (1997).

Israeli, E., Gitelman, J., and Lighthart, B., "Death Mechanisms in Bioaerosols," in B. Lighthart and A.J. Mohr, Eds., *Atmospheric Microbial Aerosols*, Chapman & Hall, New York, 1994, pp. 166–191.

Jensen, M.M., *Appl. Microbiol.*, **12**, 412–418 (1964).

Lighthart, B. and Mohr, A.J., *Atmospheric Microbial Aerosols. Theory and Applications*, Chapman & Hall, New York, 1994.

Loosli, C.G., Lemon, H.M., Robertson, O.H., and Appel, E., *Proc. Soc. Exp. Biol. Med.*, **53**, 205–206 (1943).

Marthi B. "Rescitation of Microbial Bioaerosols," in B. Lighthart and A.J. Mohr, Eds., *Atmospheric Microbial Aerosols*, Chapman & Hall, New York, 1994, pp. 192–225.

Marthi, B., Fieland, V.P., Walter, M., and Seidler, R.J., *Appl. Environ. Microbiol.*, **56**, 3436–3467 (1990).

Marthi, B. and Lighthart, B., *Appl. Environ. Microbiol.*, **56**, 1286–1289 (1990).

Miller, W.S. and Artenstein, M.S., *Proc. Soc. Exp. Biol. Med.*, **123**, 222–227 (1966).

Rabey, F., Janssen, R.J., Kelley, L.M., *Appl. Environ. Microbiol.*, **18**, 880–882 (1969).

Sattar, S.A., Ijaz, M.K., Johnson-Lussenburg, C.M., and Springthorpe, V.S., *Appl. Environ. Microbiol.*, **47**, 879–881 (1984).

Songer, J.R. *Appl. Microbiol.*, **15**, 1–16 (1966).

Theunissen, H.J., Lemmens-den Toom, N.A., Burggraaf, A., Stolz, E., and Michel, M.F., *Appl. Environ. Microbiol.*, **59**, 2589–2593 (1993).

Warren J.C., Akers, T.G., and Dubovi, E.J., *Appl. Microbiol.*, **18**, 893–896 (1969).

Webb, S.J., *Can. J. Microbiol.*, **5**, 649–669 (1959).

Webb, S.J., *Can. J. Microbiol.*, **6**, 89–105 (1960).

Webb, S.J., Bather, R., and R.W., *Can. J. Microbiol.*, **9**, 87–94 (1963).

Wells, W.F., *Am. J. Hyg.*, **20**, 611–627 (1934).

FURTHER READING

Cox, C.S., *The Aerobiological Pathway of Microorganisms*, John Wiley & Sons, Chichester, New York, Brisbane, Toronto, Singapore, 1987.

Dimmick, R.L. and Akers, A.B., *An Introduction to Experimental Aerobiology*, Wiley-Interscience, New York, 1969.

Fuchs N.A., *The Mechanics of Aerosols*, Dover Publications, New York, 1964.

Gregory, P.H., *The Microbiology of the Atmosphere*, 2nd ed., John Wiley & Sons, New York, 1973.

Lighthart, B. and Mohr, A.J., *Atmospheric Microbial Aerosols. Theory and Applications*, Chapman & Hall, New York, 1994.

Willeke, K. and Barons, P.A., *Aerosol Measurement*, Van Nostrand Reinhold, New York, 1993.

AGRICULTURAL BIOTERRORISM (AGRICULTURAL BIOSECURITY, AGROTERRORISM)

ROCCO CASAGRANDE

INTRODUCTION

In this chapter, agricultural bioterrorism is defined as the intentional spread of the pathogens of livestock or crops in order to cause economic harm. This definition is used here to distinguish agricultural bioterrorism from bioterrorist attacks on the food supply (the intentional spread of pathogens in the food supply to cause human casualties and economic disruption). In practice, the distinction between attacks on the food supply and on agriculture may be blurry. For example, if the pathogen spread in the attack is zoonotic (i.e., capable of infecting humans and lower animals), an attack on livestock could also potentially cause human casualties through the ingestion of tainted products (threatening food security) or through contact with contagious animals. Nevertheless, the distinction, however artificial, is valuable because biological attacks targeted at agriculture differ markedly from attacks on the food supply or direct biological attacks on civilians in the technical sophistication required to perpetrate the attacks, the means used to attack the target, and the motives of the terrorists executing the attack.

Unlike other types of biological attacks, agricultural bioterrorism has received relatively little attention until recently. According to a recent study by the National Academy of Sciences:

At the time the study was initiated in early 2001, concerns about possible terrorist attacks on US agriculture were not prevalent. Terrorist attacks during the fall of 2001 dramatically changed the prevailing attitude. Economic terrorism in the form of biological attacks on agriculture is now widely perceived as a threat to the nation. There is general recognition of the need to defend the nation against such threats, and additional resources have been allocated to the USDA [United States Department of Agriculture] and other departments to do so (National Academy of Sciences, 2003).

Additionally, only after September 11, 2001, was agriculture formally incorporated into the national counterterrorist strategy of the United States (Chalk, 2003). These statements may cause one to think that biological warfare (BW) targeted at agriculture is a new idea and that before this the United States and other major powers did not consider agriculture as a suitable target for a biological attack. This suggestion could not be further from the truth.

HISTORY OF STATE PROGRAMS IN BIOLOGICAL WARFARE TARGETED AT AGRICULTURE

The vulnerability of agriculture to a biological attack was realized long ago, and was at the heart of most states' biological weapons programs from their inception. This history has been thoroughly analyzed by others (Whitby, 2002; Geissler and Moon, 1999; Davis, 2003) and will only be summarized here.

The first use of BW in modern times was not against soldiers or civilians but animals (Wheelis, 1999): During World War I, German agents infected draft animals in several countries, including the United States, with *Burkholderia mallei* and *Bacillus anthracis* (Wheelis, 1999). The German program also researched the use of the wheat pathogen *Pucinia graminis* (Whitby, 2002). Between the two world wars, France and Germany researched agricultural pathogens such as rinderpest, late blight, wheat rust, and several beetle pests of crops in order to establish a capability to destroy the livestock and the staple crops of their enemies (Geissler and Moon, 1999).

Encyclopedia of Bioterrorism Defense,   Edited by Richard F. Pilch and Raymond A. Zilinskas
ISBN 0-471-46717-0   Copyright © 2005 Wiley-Liss

The British produced and weaponized BW agents during World War II; their program focused mostly on the production of *B. anthracis* for attacks on livestock. In fact, the first biological munition produced by the British was the "cattle cake," an edible linseed cake contaminated with *B. anthracis* spores that was to be dropped over Axis pastureland in quantity, infecting enemy livestock upon consumption (Balmer, 2001; Davis, 2003).

Although the United States began its BW program later than the other major participants in the world wars, by the 1960s U.S. scientists had researched and weaponized several animal and plant pathogens. To attack the staple crops of USSR and China, the United States weaponized and produced the spores of wheat rust and rice blast (Whitby, 2002). To attack livestock, the United States weaponized and produced the causative agents of hog cholera and Newcastle disease (Davis, 2003). To disseminate these agents, U.S. scientists developed bomblets containing feathers coated with agent to be released high above animal pens and fields; upon detonation of the bomblet, the feathers would be released into the air, float down to the target, and produce several foci of infections (Whitby, 2002).

After the renunciation of biological weapons by the United States, United Kingdom, France and other countries, the Soviet Union stepped up its offensive BW efforts in earnest. According to Ken Alibek, the Soviet program included research into several agricultural pathogens alongside its immense anti-personnel program under the code name "Ecology" (Alibek, 1999). Ecology scientists studied the causative agents of such diseases as wheat rust, rice blast, African swine fever (ASF), food and mouth disease (FMD), and rinderpest.

Today, it is estimated that approximately a dozen countries possess active offensive BW programs. It is unclear how many of these include programs in agricultural warfare, but if the historical record offers these bioliogical warfare programs likely also have a robust and dangerous anti-agriculture component.

QUALITIES OF U.S. AGRICULTURE THAT MAKE IT VULNERABLE TO BIOLOGICAL ATTACK

Agriculture is a Vital Sector of the U.S. Economy

While state-level programs have historically sought to undermine the fighting ability of the enemy by decreasing its ability to feed troops and factory workers, the food supply of the United States is too diverse and plentiful for even a series of well-planned and well-executed biological attacks to cause food shortages. However, agricultural incidents still could produce economic hardships, even in our extremely industrialized economy. Agriculture and industries based on agricultural production (such as textile manufacture, food processing, and transportation of agricultural products) account for an economic output of $1.5 trillion a year, roughly 16 percent of the gross domestic product (Casagrande, 2000). Although farming itself only employs less than three percent of the American population, related industries employ approximately 17 percent of the work force (Chalk, 2003, USDA, 2001). Any

incident that disrupts agriculture could severely reduce the purchasing power of this 17 percent, generating ripple effects in all sectors of the economy that depend on it. Furthermore, the export of agricultural goods, worth roughly $50 billion in the year 2000, contributes significantly to the positive balance of trade (USDA-NASS, 2001).

The Federal Response to an Agricultural Disease Outbreak

Outbreaks Among Livestock. Partially because of the unrelenting efforts of the USDA, U.S. agriculture has been untouched by disease outbreaks that have devastated the agricultural economies of other industrialized nations, such as the "mad cow" disease outbreak in the United Kingdom, the classical swine fever outbreak in Holland, and the FMD outbreaks in the United Kingdom, Denmark, and Taiwan. However, the actions of the USDA are designed to prevent the introduction and spread of diseases introduced accidentally, and may prove to be insufficient to protect against the deliberate spread of disease.

When a livestock producer notices strange symptoms in a herd, he or she may call a local veterinarian. If that veterinarian suspects that the symptoms are caused by a foreign animal disease, the case is reported to the state veterinarian, who may then call on USDA resources to control and contain the outbreak. First, within 24 hours of notification, the USDA will dispatch a Foreign Animal Disease Diagnostician to confirm that the disease is occurring. If the Foreign Animal Disease Diagnostician deems it necessary, an Early Response Team, which typically consists of a few epidemiologists and veterinary specialists may be assembled and called to respond. If the disease outbreak is large or growing, one of two Regional Emergency Animal Disease Eradication Organizations (READEOs) is deployed to assist in disinfection, vaccine administration, and culling (USDA-APHIS, 2001). These READEOs can call on approximately 40,000 private-practice veterinarians accredited by the USDA's Animal and Plant Health Inspection Service (APHIS) to respond to animal disease outbreaks, but of these only 450 veterinarians across the 50 states have received special training in foreign animal disease control (the diseases of most concern in terms of terrorism) (National Academy of Sciences, 2003). Some states have created special training courses (usually through the local land-grant university) for large animal veterinarians on the diagnosis and control of foreign animal diseases, imparting experience in the control of diseases with which these veterinarians would otherwise have no familiarity. However, veterinarians with any practical experience working with animals infected with FMD, ASF, or other high-consequence animal diseases are still rare.

Plant Disease. The response to outbreaks of crop disease is far less defined (National Academy of Sciences, 2003). If a grower thinks that his plants may be infected by a foreign plant pest, he or she may send a sample to a local university for diagnosis. If the plant pathologist at this university diagnoses the pest as a high-consequence foreign plant pest, APHIS and the state plant regulatory official are immediately notified. Outbreaks are responded to by Plant Protection and Quarantine Rapid Response Teams (RRTs), a cadre of specialists trained in the eradication and control of plant diseases who provide support and direction to local workers (National Academy of Sciences, 2003). This informal reporting network (that includes often overworked academic laboratories) is partially to blame for the fact that plant disease outbreaks are typically discovered after several cycles of infection have occurred and disease incidence is too high to consider eradication (National Academy of Sciences, 2003).

Assessment. The current resources of the USDA are often strained by naturally occurring outbreaks and appear insufficient to respond to or to control a large or multifocal disease outbreak of the type likely caused by the intentional introduction of pathogens by terrorists (National Academy of Sciences, 2003). The recent outbreak of exotic Newcastle disease in the southwest United States illustrates the problem. This outbreak, likely caused and spread by the illegal importation of fighting cocks from Mexico, was first discovered in December 2002 but was not eradicated until September 2003 (USDA, 2003). One reason for the supposed difficulty in disease control was the spread of the virus by the illegal movement of fighting cocks.

Some of the deficiencies have been realized, and efforts are under way to improve the way agricultural diseases are diagnosed and reported. The system is advancing: in 2002, funding was appropriated to establish the national animal and plant diagnostic laboratory network to improve communication, reporting, and analysis in disease detection, and five regional plant diagnostic centers have been established to reduce the time needed to return test results in outbreaks and enhance diagnostic capability (Lambert, 2003). Recently, the response to an outbreak of livestock or crop disease has been harmonized to resemble the response to terrorist incidents or natural disasters through the adoption of the National Interagency Incident Management System by the USDA and an emergency operations center now coordinates the response to agricultural disease outbreaks in the same way that responses to terrorist attacks are coordinated (Lambert, 2003). These seemingly minor changes will greatly facilitate the interaction among responders in a case of agroterrorism, in which law enforcement, transportation, animal or plant health and other communities will have to collaborate closely. One may wonder why such a robust response is required to help protect U.S. agriculture.

The Livestock Industry

There are three qualities of animal agriculture in the United States that make it vulnerable to a biological attack: it is highly concentrated, animals are moved throughout the country, and animals are farmed in a way that makes them susceptible to disease.

Concentration of Animal Agriculture. The fact that animal agriculture is highly concentrated is illustrated by the following examples. The four largest poultry producers grow and mill the feed, hatch the chicks, and dress the carcasses of about half the chickens produced

in the United States (Bastian et al., 1994). The largest 30 feedlots fattened almost 5 million head of cattle in 1998 (Heffernan et al., 1999). In 1997, 63 percent of all cattle slaughter took place in companies that slaughter a minimum of 1 million head of cattle each year (Davis, 2003), and the largest five beef packers processed 83 percent of all beef in the United States (Heffernan et al., 1999). This concentration enables a terrorist to affect a large proportion of the target with one attack.

Movement of Animals. Animals and their products travel between otherwise isolated farms and regions of the country, greatly facilitating the spread of disease and complicating disease-control efforts. Animals today are often born on a breeding farm, transported to a facility in another state to be fattened, and then transported to another region where slaughter and processing occurs (Lautner, 1999). One study estimates that meat travels roughly 1000 miles before reaching the dinner plate (Wilson et al., 2000). In fact, it was the movement of people, animals, and equipment between farms that was the primary driver of the spread of classical swine fever in the Netherlands in 1997 (Stegman et al., 1999).

Modern Farming Practices. Modern agricultural practices also include regular sterilization, dehorning, branding, and overcrowded conditions that elevate the stress levels of the animals, conditions which are claimed to make the animals more susceptible to infection (Chalk, 2003). Furthermore, only the most productive varieties of livestock are raised in the United States. This lack of genetic diversity increases the chances that any particular pathogen that establishes itself in a herd will be able to spread in an uncontrolled fashion.

Case Study. Illustrating the above points, as part of a U.S. government study, undercover agents posed as buyers and successfully infiltrated sale barns where livestock from several producers were shown and auctioned. These barns are a central location where animals gather from across the region, mingle, and then disperse back across the region. Thus, a contagious disease introduced at one of these barns could potentially affect animals that reside in hundreds of otherwise isolated facilities. The agents smuggled spray bottles filled with water, meant to simulate a suspension of a livestock virus, in with them, and succeeded in spraying several animals and holding areas without bystanders asking any questions (Wilson et al., 2000). If the water had been a suspension of a contagious animal pathogen, many animals would have been infected and would have then traveled all over the country while incubating and later spreading the disease.

Crops are Grown over Wide Areas and are Impossible to Secure

The commercial cultivation of crops is not as geographically concentrated as the livestock industry, and therefore, it is problematic to contaminate a significant portion of any particular crop. However, the fact that crops are grown over wide areas makes them virtually impossible to secure. A terrorist could cause multiple foci of infection simply by throwing a contaminated product into a handful of suscep-tible fields while driving along the highway (Casagrande, 2000). Alternatively, a terrorist could infect several sites by contaminating products that are stored in bulk and then shipped across the country (such as seed, agricul-tural marking foam, or some types of fertilizer). Either of these modes of attack could cause multiple foci of infec-tions in geographically dispersed fields, possibly hitting many of the important growing areas for that crop in the United States. Facilitating the spread of disease from one field to another is the fact that, similarly to livestock, only the most productive cultivars are raised in our fields.

QUALITIES OF AGRICULTURAL PATHOGENS THAT ENHANCE THEIR CAPACITY AS WEAPONS

Agricultural pathogens can take advantage of the characteristics of agriculture to spread rapidly. But, for a terrorist to use an agricultural pathogen as a weapon, he or she must acquire the pathogen, magnify it, weaponize it, and disperse it.

Agricultural disease agents are in fact readily acces-sible. Before September 11, 2001, there were few con-trols regulating the possession or transfer of agricultural pathogens from laboratory collections (Casagrande, 2000). And while the Public Health Security and Bioterrorism Preparedness Response Act of 2002 now requires that any entity that possesses, uses, or transfers select plant and animal pathogens register with the USDA (Bioterrorism Act, 2002), reducing the ability of terrorists to obtain agricultural biological agents from a laboratory collection, agricultural pathogens are relatively easy to acquire from the environment (further aiding the perpetrator in that he or she need not chance discovery by attempting to obtain pathogens from laboratories and other traceable sources).

To obtain agricultural pathogens, a terrorist must first determine where an outbreak is currently occurring. This is not complicated, however, because many of the most dangerous agricultural pathogens are endemic to several areas in the developing world (such as FMD to most of Africa and Asia, soybean rust to South America and elsewhere, and rinderpest to parts of Africa), such that a terrorist traveling to these areas needs only to recognize the disease in the field and purchase the infected plants or animals to obtain a sample.

Simplifying the magnification of agricultural pathogens is the fact that relatively few cause zoonotic disease (i.e., the majority do not infect people). Therefore, a perpetrator need not obtain personal protective equipment or take other precautions to prevent self-infection during production.

Weaponization and dispersal of agricultural pathogens are simplified by the fact that many are highly communicable from animal to animal. There is thus no need to create high-quality aerosols of agent to reach thousands of animals or plants, greatly reducing the technical barriers to a successful attack. Therefore, to reach thousands of animals or acres of plants, a perpetrator may only need to cause a few foci of infection, after which the disease will spread by itself (facilitated by the movement of agricultural products) to potentially thousands of other organisms. For example, in the 1997 outbreak of FMD in Taiwan, the disease spread from one farm to 28 in one week and to 717 in the next (USDA-FAS, 1997). The recent FMD outbreak in the United Kingdom also supposedly started on one farm in Heddon-on-Wall but caused about 2000 confirmed cases, the culling of 4 million animals, and direct and indirect economic losses between $6 and $30 billion (Feguson et al., 2001; Becker, 2001).

There are several agricultural diseases that possess the above characteristics. The Office International des Epizooties (OIE), an international agency that monitors the health of livestock, maintains a list of transmissible diseases ("List A diseases") that have "the potential for very serious and rapid spread, irrespective of national borders, which are of serious socio-economic or public health consequence and which are of major importance in the international trade of animals and animal products" (Table 1; for more information, please refer to the entry on the Office International des Epizooties elsewhere in this volume) (OIE, 2003). Although this list is maintained primarily to guide inspection and export controls to contain naturally occurring diseases, it is representative of the wide variety of agents available for a bioterrorist to attack livestock. The majority of diseases in list A are caused by highly contagious viruses.

The highest consequence plant pathogens are primarily fungal parasites. These fungi produce spores that can be spread by the wind or water droplets to spread from plant to plant and field to field. Some high-consequence

Table 1. OIE List A Livestock Diseases (OIE, 2003)

Disease (OIE, 2003)	Species Affected (ISU, 2003)
African horse sickness	Horses, donkeys, mules, zebras, and camels
African swine fever	Swine (wild and domestic)
Bluetongue	Sheep, cattle, goats, deer, and other ruminants
Classical swine fever	Swine (wild and domestic)
Contagious bovine pleuropneumonia	Cattle
Foot and mouth disease	Cattle, swine, sheep, goats, deer, and other cloven-hoofed ruminants
Highly pathogenic avian influenza	Poultry and a wide variety of domestic and wild birds
Lumpy skin disease	Cattle
Newcastle disease	Poultry and many species of wild and domestic birds
Peste des petits ruminants	Goats and sheep
Rift valley fever	Many wild and domestic animals (including humans, cattle, sheep, and dogs—horses and pigs are resistant)
Rinderpest	Most cloven-hoofed animals (esp. cattle)
Sheep pox and goat pox	Sheep and goats
Swine vesicular disease	Swine
Vesicular stomatitis	Cattle, horses, and swine (to a lesser extent, humans and other primates, and wild carnivores and ruminants)

Table 2. Selected Important Plant Pathogens (Whitby, 2002)

Pathogen	Disease Caused and Species Affected
Erwinia amyovora	Fire blight of apple, pear, quince, and related species
Pseudomona solanacearum	Wild of potato, tomato, and tobacco
Pyricularia orzae	Rice blast
Ustilago maydis	Corn smut
Xanthomonas albilineans	Leaf scald of sugarcane (and, to a lesser extent, corn and other grasses)
Xanthomonas compestris pv. Oryzae	Bacterial blight of rice
Tilletia tritici (or *caries*)	Smut or bunt of wheat
Sclerotinia sclerotorium	White mold of vegetables (also attacks soy beans)
Puccinia graminis	Wheat
Ohytoohthora infestans	Potatoes and tomatoes

plant pathogens are bacteria or viruses that are spread by insect vectors. For example, citrus canker, a viral disease spread by insects, caused the destruction of over 1.5 million citrus trees at a cost of over $200 million in a recent outbreak (Schubert et al., 2001). Furthermore, crops can be attacked by insects intentionally introduced into the country. Foreign insect pests may attack a target directly, like the Colorado beetle attacks potato crops, for example, or may simply spread endemic viruses in a more efficient manner than domestic species of insects. Table 2 demonstrates some of the diversity of fungal and bacterial plant pathogens.

It should be noted that it is difficult to attribute an outbreak of agricultural disease to a malicious actor, adding to the relative appeal of such diseases in this context. The lengthy incubation periods for many agricultural diseases provide the perpetrator with much time to flee a crime scene and go underground. If caught, a terrorist using his own crops or animals as vectors (by bringing infected animals to trade shows, for example) could simply claim to be an innocent victim. And the presence of natural outbreaks lends a level of camouflage to any illicit release. For example, rabbit calicivirus appeared in a farm in central Iowa in March 2000, then in Utah in 2001, and then in New York; in all cases, it was never determined how the animals came in contact with the virus, which is not native to the United States (Davis, 2003).

CONSEQUENCES OF AN AGRICULTURAL DISEASE OUTBREAK

As stated above, an agricultural attack will not cause food shortages in the United States; the primary impact will be economic. The most obvious economic costs of an outbreak are the direct losses caused by the destruction of saleable animals or crops and the costs incurred in attempting to control and contain the disease. These costs in themselves can be enormous, as illustrated by the $2.5 billion paid to farmers in the United Kingdom to compensate for culled

animals during the "mad cow" disease outbreak (National Academy of Sciences, 2003).

Besides these direct losses, more severe losses may be associated with efforts to prevent the spread of disease. To prevent spread within the country, transportation of agricultural products can be severely restricted (in fact, transportation in its entirety can be halted within a perimeter around an affected premises), and the work of industries that depend on receiving the agricultural products can be brought to a halt (such as meat packers, food processing plants, textile mills, and bottling plants). A California study modeling eight hypothetical FMD outbreaks estimated that each day the disease was uncontrolled resulted in a loss of $1 billion to the state (Chalk, 2003). Similarly, the Wisconsin Department of Agriculture, Trade and Consumer Protection (DATCP) calculated that any disease outbreak that disrupts the transportation of dairy products will cost the state $10 million a day (Washburn, 2003).

To prevent the spread of disease to trading partners, export restrictions may be enacted, preventing the sale of affected agricultural products overseas (in 2000, overseas agriculture sales by the United States amounted to some $50 billion) (USDA-NASS, 2001). Importing countries would then be required to turn to other producers to satisfy their import requirements, and might remain with these new relationships after export restrictions are lifted. This case is illustrated by successive outbreaks of FMD in pork-exporting countries. Prior to 1982, Denmark was the leading exporter of pork to the Japanese market (the largest pork importer in the world). In 1982, Denmark suspended its pork exports because of an outbreak of FMD, at which time Taiwan took over the Japanese pork market. Denmark never recovered it. Then in 1997, the United States took over the Japanese pork market when Taiwan suspended its pork exports because of an outbreak of FMD (USDA-FAS, 1997), and to this day the United States remains Japan's largest pork supplier.

Because of the disruption to trade and transportation enacted to control disease spread, even limited outbreaks can have severe economic consequences. A recent outbreak of karnal bunt (a pathogen of wheat) was limited to four Texas counties, but cost $27 million in direct and indirect losses (Bevers et al., 2001). A similar but larger outbreak in the southwest United States in 1996 took longer to control and cost $250 million in lost exports. And the discovery of a single "mad cow" in Canada currently is estimated to have cost $2.5 billion in losses to the beef industry (Yovich, 2003). Because limited disease outbreaks can cause severe economic consequences, a terrorist need not cause a massive, uncontrolled infection to substantially affect its target.

SUMMARY OF VULNERABILITIES OF AGRICULTURE TO A BIOLOGICAL ATTACK

- Agriculture is an important and vital sector of the U.S. economy.
- The federal response to disease outbreaks is geared toward the prevention and control of naturally introduced diseases, not terrorist attacks.

- The livestock industry is highly concentrated, enabling a terrorist to strike at a large proportion of his target with one attack.
- Susceptible animals are moved between otherwise isolated locations, complicating control of disease and facilitating spread.
- Crops are grown over large areas and are impossible to secure.

Agricultural disease agents:

- Are simple to find in the environment.
- Are not dangerous to manipulate because the highest consequence pathogens are nonzoonotic.
- Spread rapidly from organism to organism and farm to farm, eliminating the need to make special preparations of agent or construct mechanically complicated munitions.
- Disease outbreaks cause severe economic losses due to lost sales, control efforts, and export restrictions.
- Costs of a disease outbreak can be great even though the disease outbreak is limited.

TERRORISTS WITH THE MOTIVES TO ATTACK AGRICULTURE

Because the technical barriers to the carrying out of a biological attack on agriculture are minimal, there is a great diversity of terrorist groups with the means to successfully execute this type of attack. As discussed more fully elsewhere (Casagrande, 2000) and summarized here, a number of terrorist groups possess the motivation to carry out attacks on agriculture.

Terrorists with political motives may turn to agricultural attacks against an adversary. These terrorists depend on support from their audience for finance, logistics, and possibly political change. An agricultural attack is less likely to alienate supporters than a mass-casualty attack while still holding the potential for substantial damage. Furthermore, a terrorist group is likely to get the attention it seeks if it can become the first group to successfully execute a large-scale biological terrorist attack on U.S. soil. Apolitical terrorists who only seek to destroy their adversary (such as al-Qa'ida), may turn to an agricultural attack to destroy the economy of their target as part of a larger strategy.

Because the primary consequence of a biological attack on agriculture is economic, those who may execute this type of attack may have economics as their motivating force, using biological attacks or the threat of an attack to gain economic advantage. Blackmailers may threaten to release a livestock pathogen in the United States, the greedy may attempt to manipulate futures markets or stock prices of agrochemical firms, and exporting concerns may attempt to capture foreign markets from competitors.

For other groups, an agricultural attack could directly satisfy their goals rather than just being a means to an end. Such might be the case for extremist animal rights and environmental groups. Some of these groups have a history of terrorism: for example, the FBI estimates that the Earth Liberation Front and Animal Liberation Front combined have committed nearly 600 criminal acts, resulting in over $45 million in damages since 1996 (Davis, 2003). It is not inconceivable that this type of organization may turn to an agricultural attack to directly harm those who keep animals in a "concentration camp-like" existence or "defile the earth" with genetically modified organisms.

THE FUTURE

As discussed above, the United States has realized that it is vulnerable to a biological attack on agriculture and is taking steps to address this vulnerability. For example, the Department of Homeland Security has deemed the risk to agriculture so critical that food and agriculture threat response represents two of the four biological defense portfolios that drive research in the Science and Technology Directorate (Albright, 2003).

Although the possibility of terrorism is discussed at length here, as far as is known there has not been a successful, large-scale bioterrorist attack on agriculture to date. Nor has a state-level attack of this kind ever occurred, despite the fact that states have known about and attempted to exploit the vulnerability of agriculture for approximately 100 years. It is uncertain whether an agricultural attack will ever be realized; however, it is certain that accidental introductions of disease, such as those noted in this chapter to have caused billions of dollars in damages, will continue to regularly occur. Because much of the effort directed toward the prevention, mitigation, and management of agroterrorism applies to natural disease outbreaks as well, the cost of these measures is not wasted should an attack never come: The next time an agricultural disease outbreak occurs, whether natural or deliberate, the United States will be better prepared to respond.

Two efforts to decrease vulnerability to agricultural disease are much discussed but have yet to be implemented in the United States. First, over the past few years, there has been significant effort to improve the tracking of livestock. If the recent movements of a diseased animal were immediately known upon discovery of the animal's illness, control efforts could immediately be directed to all locations where that animal and its contacts have traveled. It is estimated that at least 25 countries are in the process of implementing individual identification systems on a national level, some tracking individual animals from birth to slaughter (Davis, 2003). Certain industries, such as dairy, have had some success in the United States with animal tracking, and headway is being made in the registration of agricultural premises. However, many livestock producers are concerned that such measures will provide the government with access to their "private" business information, and have thus balked at further efforts to implement tracking of all livestock.

The second effort is to supply the USDA with the resources required to respond to large or multifocal disease outbreaks. These resources may come in the form of additional READEOs or RRTs to give the surge capacity required to direct the control and eradication of disease at multiple sites. While not responding to outbreaks of disease, these teams could enforce normal biosanitary guidelines on farms, provide extra security at auction barns, and educate producers in the recognition of foreign agricultural diseases. Additionally, these teams could spend time responding to disease outbreaks in the developing world. This role would have several benefits. Since most disease outbreaks in the industrialized world originate in the developing world, the control of these diseases overseas is a proactive measure to prevent the accidental introduction of the disease into the United States. In addition to an obvious humanitarian benefit, limiting the extent of the disease in nature increases the difficulty for terrorists to intentionally obtain the diseases. The overseas activity will give hands-on experience in the control of foreign diseases that may not have occurred in the United States in the lifetime of the team members, enhancing their ability to respond to similar disease outbreaks in the United States.

REFERENCES

Albright, P., Statement for the Records, Testimony presented before the Government Affairs Committee of the US Senate on November 19, 2003.

Alibek, K. and Handleman, S., *Biohazard: The Chilling True Story of the Largest Covert Biological Weapons Program in the World–Told From Inside by the Man Who Ran It*, Random House Press, New York, 1999.

Balmer, B., *Britain and Biological Warfare: Expert Advice and Science Policy, 1930-65*, Palgrave Publishers, New York, 2001.

Bastian, C., Bailey, D., Menkhaus, D., and Glover, T., Today's Changing Meat Industry and Tomorrow's Beef Sector, *Wyoming Stockman-Farmer*, October 12–13, 1994.

Becker, S., Presentation to the Committee on *Biological Threats to Agricultural Plants and Animals*, November 15, 2001 as cited in National Academy of Sciences, 2003.

Bevers, S., McAlavy, T., and Baughman, T., *Texas A&M News Release*, November 12, 2001.

Casagrande, R., Bioliogical Terrorism Targeted at Agriculture: The Threat to US National Security, *Nonproliferation Review*, **7**, 92–105 (2000).

Chalk, P., *The Bio-Terrorist Threat to Agricultural Livestock and Produce*, Testimony presented before the Government Affairs Committee of the US Senate on November 19, 2003.

Davis, R., "Agroterrorism: Overview of a Sleeping Target," manuscript provided to author, which formed the basis of: *Agroterrorism: Need for Awareness, Perspectives in World Food and Agriculture*, Iowa State University Press, Ames, Iowa, 2003.

Ferguson, N., Donnelly, C., and Anderson, R., *Nature*, **413**, 542–548 (2001).

Geissler, E. and Moon, J., Eds., "Biological and Toxin Weapons: Research, Development and Use from the Middle Ages to 1945," *SIPRI Chemical and Biological Warfare Studies*, No 18, Oxford University Press, Oxford, 1999.

Heffernan, W., Hendrickson, M., and Gronski, R., *Consolidation in the Food and Agriculture System*, Report to the National Farmers Union, Aurora, CO, February 1999.

ISU (Iowa State University), *Emerging & Exotic Diseases of Animals*, Searchable Database, 2003.

Lambert, C., *Agroterrorism: The Threat to America's Breadbasket*, Testimony presented before the Government Affairs Committee of the US Senate on November 19, 2003.

Lautner, B., *Industry Concerns and Partnerships to Address Emerging Issues, Annals of the New York Academy of Sciences*, issue 894, 1999.

National Academy of Sciences, *Countering Agricultural Bioterrorism*, National Academy Press, Washington, DC, 2003.

OIE, Terrestrial Animal Health Code, Section 1.1, 2003.

Public Health Security and Bioterrorism Preparedness Response Act of 2002, signed into Law by President G.W. Bush on June 12.

Schubert, T., Rizvi, S., Sun, X., Gottwlad, T., Graham, J., and Dixon, W., *Plant Dis.*, **85**, 340–356 (2001).

Stegman, A., Elbers, R., Smak, J., and de Jong, M., *Prev. Vet. Med.*, **42**, 219–234 (1999).

USDA, *Food and Agricultural Policy: Taking Stock for the New Century*, 2001, available at http://www.usda.gov/news/pubs/farmpolicy01/fpindex.htm.

USDA, *USDA Lifts Quarantine Restrictions For Exotic Newcastle Disease*, In California, Press Release, September 16, 2003, available at http://www.usda.gov/news/releases/2003/09/0321.htm.

USDA-APHIS, *Animal Health Programs: Combining Surveillance, Detection, and Response*, Miscellaneous Publication No. 1573, Issued September 2001, available at http://www.aphis.usda.gov/oa/pubs/brotradc.pdf.

USDA-ARS, Agricultural Research Service Homepage, 2003, available at http://www.ars.usda.gov.

USDA-FAS (Foreign Agricultural Service), Foot and Mouth Disease Spreads Chaos in Pork Markets, FASonline, *Livestock and Poultry: World Markets and Trade Circular Archivest*, October 1997, available at www.fas.usda.gov/DLP2/circular/1997/97-10LP/taiwanfmd.htm.

USDA-NASS (National Agriculture Statistics Service), *Agricultural Statistics 2001*, U.S. Government Printing Office, Washington DC, 2001.

Washburn, C., *Dairy Farms Affect Main Street, Too*, Advertisement, Trail's Media Group, Spring 2003.

Wheelis, M. "Biological Sabotage in World War I, Biological and Toxin Weapons: Research, Development and Use from the Middle Ages to 1945," in E. Geissler and J. Moon, Eds., *SIPRI Chemical and Biological Warfare Studies*, No 18., Oxford University Press, Oxford, 1999, pp. 35–63.

Whitby, S., *Biological Warfare Against Crops*, Palgrave Publishers, New York, 2002.

Wilson, T., Logan-Henfrey, L., and Weller, R., *Agroterrorism, Biological Crimes and Biological Warfare Targeting Animal Agriculture, Emerging Diseases of Animals*, ASM Press, Washington, DC, 2000.

Yovich, D., Single BSE-infected cow costs Canada $2.5 billion, Meatingplace.com, Published on November 19, 2003, quoting a Canadian Government Study, available at http://www.meatingplace.com/DailyNews/forward.asp?iArticle=11510.

See also HOMELAND DEFENSE; OFFICE INTERNATIONAL DES EPIZOOTIES: WORLD ORGANIZATION FOR ANIMAL HEALTH; and SCIENTISTS, SOCIETIES, AND BIOTERRORISM DEFENSE IN THE UNITED STATES.

ALIENS OF AMERICA: A CASE STUDY

MONTEREY WMD-TERRORISM DATABASE STAFF

INTRODUCTION

The Aliens of America appears to have been the creation of Muharem Kurbegovic, better known as the Alphabet Bomber. "Aliens of America" first surfaced when Kurbegovic, under the pseudonym Isaiak Rasim, declared himself the "Chief Military Officer" of the group in a taped message to the *Los Angeles Times* in the summer of 1974. Kurbegovic, a legal Yugoslavian immigrant who came to the United States in 1967, had begun terrorizing various Los Angeles police commissioners and a Los Angeles judge starting in the fall of 1973 in response to their denial of his application to open a dance club. The Aliens of America tape bragged of the group's possession of nerve gas and its desire to liberate itself from "the audacity and terror of the U.S. government."

SELECTED CHRONOLOGY

- *June 15, 1974.* Kurbegovic warned that he had hidden "nerve gas" under the postage stamps on nine postcards mailed to the nine U.S. Supreme Court justices. When the postcards were received on June 16, 1974, inspection of the metal vials found under the postage stamps confirmed that the incident was a hoax.
- *August 6, 1974.* An 11-pound bomb placed in a locker at Los Angeles International Airport exploded, killing three people and injuring 35. The same evening, Rasim (Kurbegovic) called a Los Angeles newspaper, representing the Aliens of America, to claim responsibility for the bombing. In a taped message on August 9, 1974, Rasim explained that the airport bombing stood for the letter "A" and that subsequent bombs would represent the rest of the letters in the name "Aliens of America," earning him the moniker "the Alphabet Bomber."
- *August 15, 1974.* A tape sent by Rasim to a Los Angeles newspaper threatened a "nerve gas" attack on Capitol Hill in the following months, as well as the introduction of "nerve gas" into the air conditioning systems of unspecified Los Angeles skyscrapers.
- *August 16, 1974.* Rasim, in another taped message, warned of a second bomb placed in the locker of a Los Angeles bus terminal. This bomb, which symbolized "L" (for "locker"), weighed 25 pounds and was for the time one of the largest explosive devices uncovered in the history of Los Angeles crime.
- *August 20, 1974.* Muharem Kurbegovic, a.k.a. Isaiak Rasim, was identified and apprehended after placing a taped message at a Carl's, Jr. restaurant in Los Angeles, California.
- *July 1979.* A "member" of the Aliens of America threatened to use *Legionella pneumophila*, the causative agent of Legionnaires' disease, to contaminate the U.S. Consulate in Munich. Kurbegovic was involved in competency hearings for his 1980 trial at the time of this threat, and thus the perpetrator may have been an admirer trying to imitate him.

CASE RESOLUTION

A police search of Kurbegovic's home at the time of his arrest uncovered pipe bombs; bomb components; two U.S. Army gas masks; chemical and laboratory equipment catalogues; literature on chemical, biological, and conventional weapons as well as law enforcement operations, equipment, and strategies; 25 pounds of sodium cyanide; a bottle of nitric acid; and supplies of carbon tetrachloride, phosphoric acid, sodium chlorite, ammonium nitrate, and chloroform. Because many of the named agents may serve as precursors to chemical weapons, a legitimate intention to stage chemical attacks was suspected. However, despite the fact that literature regarding biological weapons was found among Kurbegovic's possessions, there was no evidence of any acquisition or production of biological weapons. Further, Kurbegovic's taped statements made no reference to biological agents. Kurbegovic is currently serving a life sentence at Pelican Bay State Prison in Crescent City, California.

REFERENCES

Monterey WMD-Terrorism Database, Center for Nonproliferation Studies, 1998–2003.

Encyclopedia of Bioterrorism Defense,   Edited by Richard F. Pilch and Raymond A. Zilinskas
ISBN 0-471-46717-0   Copyright © 2005 Wiley-Liss

AL-QA'IDA (AL-QAIDA, AL-QAEDA, AL-QA'IDAH, ISLAMIC ARMY FOR THE LIBERATION OF THE HOLY PLACES, WORLD ISLAMIC FRONT FOR THE JIHAD AGAINST JEWS AND CRUSADERS, ISLAMIC SALVATION FOUNDATION (A CHARITABLE FOUNDATION SET UP TO RAISE FUNDS FOR THE ORGANIZATION), THE 'USAMA BIN LADIN NETWORK)

KIMBERLY McCLOUD

INTRODUCTION

Al-Qa'ida (Arabic for "the Base") is a worldwide network of organizations and individuals dedicated to *jihad*, or "holy war," allegedly for the cause of Islam. According to various estimates, there are anywhere from 5000 to 15,000 individuals active in al-Qa'ida cells in as many as 60 countries of Asia, Europe, the Middle East, North Africa, North America, and South America. Terrorist and insurgency groups that have been closely linked with al-Qa'ida include the Egyptian al-Jihad al-Islami and al-Jama'a al-Islamiyya, Pakistan's Harakat ul-Mujahidin, Algeria's Groupe Islamique Armée, the Abu Sayyaf Group in the Philippines, and the Islamic Movement of Uzbekistan. Most of the individual groups associated with al-Qa'ida strive to establish Islamic governments that would enforce *shari'a* (Islamic law) in their countries of origin. In many cases, they have carried out terrorist attacks on carefully selected targets in these countries in order to achieve this end. The broader network of terrorist cells, however, is responsible for acts of violence on an international scale, primarily targeting U.S. assets and allies.

BACKGROUND

Much of the attention on al-Qa'ida has centered on its founder and leader, 'Usama Bin Ladin. Information regarding Bin Ladin's involvement in international terrorism dates back as early as 1993, when his associate Ramzi Ahmad Yusuf was implicated in the first terrorist attack on New York City's World Trade Center towers. The U.S. government began to publicly identify Bin Ladin as an international terrorist in the mid-1990s after revelations surfaced concerning his connection to attacks on U.S. military personnel and assets in Somalia (1992) and Saudi Arabia (1995–1996). Bin Ladin was linked to several unsuccessful plots to commit terrorist attacks as well, including the assassination of the Pope during his trip to the Philippines in 1994 and that of U.S. President Bill Clinton, also in the Philippines in 1995.

On August 20, 1998, in the wake of the near-simultaneous August 7 U.S. Embassy bombings in Kenya and Tanzania, President Clinton amended Executive Order 12,947 to include al-Qa'ida as a Foreign Terrorist Organization (FTO). Also on August 20, Clinton commanded air strikes against a Bin Ladin camp in Khost, Afghanistan, as well as what was believed to be an al-Qa'ida chemical weapons facility in Sudan (even though the actual nature of the facility still remains a matter of substantial debate). On June 7, 1999, Bin Ladin was added to the Federal Bureau of Investigation's (FBI) "Ten Most Wanted List," and a $5 million reward was offered for his capture.

Although several al-Qa'ida suspects were arrested and tried for the various terrorist attacks and plots both in the United States and abroad throughout the 1990s, Bin Ladin himself evaded capture and continued his campaign of terror from Afghanistan as a guest of the radical Taliban regime. In November 2000, the USS Cole was bombed in the harbor of Aden, Yemen, killing 19 U.S. servicemen and servicewomen. And of course on September 11, 2001, 19 al-Qa'ida members hijacked four airliners and perpetrated suicide attacks on both the World Trade Center and the Pentagon, collectively killing over 3000 people in New York, Washington, DC, and Pennsylvania.

HISTORY

'Usama Bin Ladin was one *mujahid* ("holy warrior") among the thousands that participated and fought in the decade-long resistance to the Soviet occupation of Afghanistan (1979–1989). From a wealthy and prominent Saudi family of Yemeni origin, Bin Ladin brought not only himself to the cause but also significant financial support. While the radical Palestinian scholar Abdullah Azzam, the main organizer of Arab support for the Afghans, focused on the primary task of driving the Soviet forces out of Afghanistan, Bin Ladin's plans extended internationally, a fact that would eventually lead him to split with Azzam. Bin Ladin believed that defeating the Soviet Union would be only the first step in a worldwide jihad campaign

Encyclopedia of Bioterrorism Defense, Edited by Richard F. Pilch and Raymond A. Zilinskas
ISBN 0-471-46717-0 Copyright © 2005 Wiley-Liss

to eradicate the oppression of Muslims by repressive regimes everywhere.

In Afghanistan, Bin Ladin's early jihadist supporters included individuals from the radical militant Egyptian group al-Jihad al-Islami, which was implicated in the assassination of Egypt's President Anwar al-Sadat in 1981. One of al-Jihad's prominent leaders, Ayman al-Zawahiri, would later form part of al-Qa'ida's leadership with Bin Ladin and commit one of his lieutenants, Muhammad Atif, as a military commander to the organization (Atif is believed to have died while fighting against U.S. troops in Afghanistan in early 2002). Al-Jihad favored terrorism and violence as the means to wage international *jihad*, a characteristic that influenced Bin Ladin's own development. Bin Ladin eventually named his group the World Islamic Front for the Jihad Against Jews and Crusaders, which later evolved into al-Qa'ida, a reference to the base camp where *mujahidin* fighters registered in Afghanistan when they arrived to join the *jihad*.

After the Afghan war, young Muslim men were recruited from the Middle East; North America; European countries such as Belgium, France, Germany, Italy, Russia (particularly Chechnya), the United Kingdom, and Yugoslavia; and Central, East, and South Asian states such as China (particularly Xinjiang province), Indonesia, Malaysia, Pakistan, the Philippines, Singapore, and Uzbekistan. Many flocked to Bin Ladin from poverty-stricken, disenfranchised, and repressed strata of Arab societies whose discontent had long been ignored by autocratic governments. Its principal leaders, however, were well-educated, middle- to upper-class individuals raised in a variety of different societies and cultures, including the West. These intellectuals had at one time represented all extremes of the political and ideological spectrum, from right-wing nationalism to left-wing communism.

IDEOLOGY

There are competing assessments regarding the motivations and ideology of al-Qa'ida as a group, as well as that of its individual members.

Some analysts argue that al-Qa'ida is a political organization, and that its leaders and members have specific political goals. Repeatedly outlined by Bin Ladin himself in various videotaped statements created for media consumption, for example, Bin Ladin himself repeatedly outlined his political goals, including (1) that the United States should withdraw its military forces from Saudi Arabia; (2) that the United States should stop causing the suffering of innocent Muslims in Iraq through its economic sanctions; (3) that the United States should stop supporting Israel at the expense of the suffering of Palestinians; and (4) that Muslims should reestablish the "Islamic nation," or Caliphate, stretching from Eastern Europe to Western Asia.

In 2003, the United States indeed began a complete withdrawal of its military forces from Saudi Arabia, and economic sanctions on Iraq were lifted after the United States and United Kingdom toppled the regime of Saddam Husayn and took over the administration of the state. While the first of these events could potentially lead to a decline in al-Qa'ida attacks on the United States, the second development could lead to an increase in attacks in response to U.S. occupation of a Muslim country. In fact, news reports in mid- to late 2003 indicate that foreign fighters with suspected al-Qa'ida ties have indeed increased their presence in Iraq and contributed to the near daily violence and terror attacks on selected U.S., coalition, and international targets.

The alternative argument is that al-Qa'ida is an organization devoted to "terrorism for terrorism's sake," whose members are so radicalized that they are unlikely to ever return to peaceful, nonviolent ways. Thus, even if their proclaimed political goals are accomplished, the group will continue to find reasons to propagate violence. Some analysts argue that the lofty, unlikely goal of reestablishing the Caliphate will in any event remain a justification for continuing terrorist attacks without a specific cause. If this analysis proves accurate, it may indicate an increased willingness to pursue weapons of mass destruction (WMD).

CURRENT STATUS

Since the attacks on the World Trade Center and the Pentagon on September 11, 2001, many new developments with regard to 'Usama Bin Ladin and al-Qa'ida have taken place. On October 9, 2001, nearly a month after the attacks, U.S. President George Bush directed military strikes against al-Qa'ida camps and other targets in Afghanistan. As a result of this intensive international military campaign, dubbed "Operation Anaconda" by the U.S. Department of Defense, al-Qa'ida was in large part forced out of its safe havens in Afghanistan and the Taliban regime was removed from power. Aerial bombing destroyed virtually all al-Qa'ida camps in Afghanistan. Hundreds of al-Qa'ida fighters were killed, and hundreds were detained and incarcerated by the United States at its military base in Guantanamo, Cuba. The United States believes that one of the primary al-Qa'ida leaders, Muhammad Atif, was killed in the mountainous Tora Bora region of Afghanistan. Other leaders, however, including Bin Ladin and al-Zawahiri, remain missing.

Nonetheless, given the multitude of terrorist attacks that have occurred since September 11 in places such as Tunisia, Yemen, Pakistan, Indonesia, Kenya, Morocco, and Iraq, concern about this terrorist organization remains high. Although the post-September 11 developments have damaged the cohesion and operational capabilities of al-Qa'ida, several factors make the organization difficult to eradicate entirely, including the unknown fate of al-Qa'ida's leaders; the loose but internationally pervasive network character of the organization, which complicates tracking its members' identities and whereabouts; the self-sufficiency and secrecy of individual cells worldwide; the international financial assets at the disposal of al-Qa'ida members and cells; and the continuing political and social conditions—especially the continuing conflict between Israelis and Palestinians—that generate anger and resentment toward the United States and its allies.

SELECTED CHRONOLOGY

February 26, 1993
The first bombing of the World Trade Center in New York City, killing six and injuring 1042 people.

October 3–4, 1993
Eighteen U.S. servicemen were killed by al-Qa'ida-trained fighters in a firefight in Mogadishu, Somalia.

November 13, 1995
Bombing outside Saudi Arabia's National Guard Communications Centre in Riyadh, Saudi Arabia, killing two Indians and five U.S. servicemen.

June 25, 1996
Bombing of the U.S. military housing complex, Khubar Towers, in Dhahran, Saudi Arabia, killing 19 U.S. servicemen.

August 7, 1998
Near-simultaneous bombing of the U.S. embassies in Nairobi, Kenya, and Dar al-Salaam, Tanzania, killing 224 people, including 12 Americans and 38 Foreign Service Nationals, and injuring more than 4585 people.

February 16, 1999
A letter threatening to kill U.S. and British citizens living in Yemen with *Bacillus anthracis* was faxed to the London office of an Arabic newspaper. The authors instructed the Westerners, whom they referred to as "diplomats, oil and gas thieves, missionary zealots, journalists, tourists, visitors, military officers, academic brain-washers, arrogant CIA and FBI employees, espionage recruiters, atheists, doctors, filthy businessmen and any other filth you came with," to "[p]ack up and go home. Get out of this land which has nothing to offer you but hatred and violence. And when you go, take your filthy husbands with you. Or else their rotten corpses will be shipped out to you in coffins… if you stay, then you've chosen death." The letter went on to say, "You have a 276 hour ultimatum to leave this country starting at 4 a.m. February 16, 1999." The anthrax threat was stated as follows: "We have obtained weapons of what's called mass destruction… if your demons make you stay, then await Jihad anthrax attacks." The letter was signed by the "Army of Suicidals Group 66. Bin Ladin Militant Wing. Chemical and Biological Apparatus." This group was not known to Yemeni law enforcement officials, and it remains unclear whether the authors of the letter actually had ties to 'Usama Bin Ladin (Monterey WMD-Terrorism Database, 2003).

October 12, 2000
Bombing of the *USS Cole* in the port of Aden, Yemen, killing 17 U.S. servicemen.

September 11, 2001
Four suicide airline hijackings in the United States. Two planes were flown into New York City's World Trade Center, causing all seven towers to collapse entirely. The third was flown into the Pentagon in Arlington, Virginia. The fourth, believed to be heading for Washington, DC, crashed in Pennsylvania because of a passenger revolt on board. Including those who died in the airplanes, an estimated 3000 people were killed, with thousands more injured.

May 8, 2002
A vehicle bomb exploded next to a U.S. Navy shuttle bus in Karachi, Pakistan, killing 10 French nationals and two Pakistanis. Nineteen others were wounded.

14 June, 2002
A car bomb exploded on the main road near the U.S. Consulate and Marriott Hotel in Karachi, Pakistan. Eleven persons were killed and 51 injured.

October 6, 2002
Bombing of the French oil tanker *Limburg*, anchored approximately five miles from the port of al-Dhabbah, Yemen. One person was killed and four others wounded.

October 8, 2002
U.S. soldiers conducting a military exercise were attacked by gunmen on Faylaka Island, Kuwait. One U.S. Marine was killed and another wounded.

October 12, 2002
A car bomb exploded outside the Sari Club Discotheque in Bali, Indonesia. At least 187 international tourists were killed, with some 300 others injured.

November 28, 2002
Two antiaircraft missiles were launched at an airplane taking off from Mombasa, Kenya, en route to Israel. The missiles missed their target.

May 12, 2003
Simultaneous car bombs were detonated at three different housing complexes and an office in Riyadh, Saudi Arabia. At least 34 people, including seven Americans, were killed, and over 200 others were injured.

May 16, 2003
Five simultaneous suicide bomb attacks occurred in downtown Casablanca, Morocco. At least 39 people were killed, with approximately 100 injured.

August 5, 2003
A bomb exploded outside the JW Marriott Hotel in Jakarta, Indonesia. Thirteen people were killed and 149 others injured.

August 19, 2003
Truck bombing of UN Headquarters in Baghdad, Iraq. Twenty-three people were killed, including the head of the mission, Sergio Vieira de Mello. At least 100 were wounded.

ASSESSMENT OF BIOLOGICAL WEAPONS AND OTHER WMD CAPABILITY

Despite a decade of organized al-Qa'ida terrorist activity, as well as the potentially irrelevant hoax elucidated above, at the moment, the network does not appear to possess any serious WMD capabilities. In the al-Qa'ida training manual, there is no discussion of how to conduct chemical, biological, or radiological attacks (http://www.justice.gov/ag/trainingmanual.htm, accessed 3/25/03), but the extraction process of the toxins ricin and abrin for assassination purposes is described in some detail, as is the production of botulinum toxin (although it is never named specifically, the preparation and effects of botulinum toxin are covered in a segment entitled "Poisoning from Eating Spoiled Food"). In addition, a host of news reports, allegations, and rumors have described interest in and attempts to acquire and/or develop WMD by al-Qa'ida. Such reports have included details of secret meetings between al-Qa'ida members and Iraqi and Sudanese intelligence agents, as well as of purchases of chemical, biological, and radiological agents and materials from Eastern Europe, including the Czech Republic and Yugoslavia.

One al-Qa'ida member, Jamal Ahmad al-Fadhl, testified during the trial following the 1998 U.S. embassy bombings that he had been directly involved in the initial phase of an attempt to purchase uranium canisters for Bin Ladin in Sudan in late 1993 or early 1994. Significantly, he was not the one to complete the purchase and thus could not tell investigators whether or not it ever actually occurred.

A Bin Ladin lieutenant on trial in Egypt in October 2001, Ahmad Ibrahim al-Najjar, maintained that Bin Ladin had purchased *B. anthracis* spores as early as 1998 through the Filipino Islamist separatist group the Moro Islamic Liberation Font, along with other biological warfare (BW) agents and equipment from Eastern Europe, specifically the Czech Republic. However, no evidence surfaced to support al-Najjar's claims.

By the beginning of 2002, after U.S. military forces had destroyed most al-Qa'ida and Taliban installments in Afghanistan, more reports began to surface describing al-Qa'ida's attempts to pursue a WMD capability. Journalists claimed to have found plans for the construction of nuclear bombs as well as chemical and biological weapons, and two journalists even claimed to have found vials of an alleged biological agent (these vials never physically surfaced). The U.S. military spokesperson in Afghanistan, General Richard Meyers, reported in March 2002 that U.S. forces had discovered what was thought to be the beginnings of a BW laboratory near Kandahar. He also reported that trace amounts of *B. anthracis* and ricin were detected in several hideouts used by al-Qa'ida in Afghanistan, but that the amounts were so minimal that they did not provide sufficient evidence to verify the existence of a BW program. In the end, the U.S. military, according to Defense Secretary Donald Rumsfeld, held that while al-Qa'ida appeared to have an "appetite" for WMD, no conclusive evidence of any such capability surfaced during the campaign in Afghanistan.

Still, claims by al-Qa'ida members continued. In April 2002, Abu Zubayda, an al-Qa'ida cell leader captured in the United States, claimed that the group had the know-how to produce a radiological weapon, and that it actually possessed such a weapon in the United States. In May 2002, another al-Qa'ida member arrested in Chicago, Abdullah al-Muhajir (formerly Jose Padilla), claimed that he was involved in planning a radiological bomb attack in the United States. No such attack occurred, and no

evidence was ever disclosed publicly that served to verify such a capability.

Potentially, the most significant piece of evidence supporting the claim that al-Qa'ida was in fact pursuing WMD came in the form of a video released by CNN on August 18, 2002—which had allegedly been found at a house where Bin Ladin had stayed in Afghanistan and was then provided to the news network by an anonymous source—showing what appeared to be the testing of a crude chemical weapon, most likely hydrogen cyanide, on three dogs, in either Afghanistan or Pakistan.

Yet another alarming discovery was that of a makeshift laboratory in a London area home in January 2003, where trace amounts of the toxin ricin were found. Six Algerians were arrested as a result of the discovery. While some suspected that these men were connected to al-Qa'ida, neither the United States nor any other government has confirmed this connection to date.

INTENT TO ACQUIRE AND USE WMD

In his public statements, 'Usama Bin Ladin has both implied and directly stated that al-Qa'ida is both interested in and has worked to acquire a WMD capability. The most recent and specific of these statements, in November 2001, alluded to chemical and nuclear but not biological weapons. In an interview with a Pakistani journalist, Bin Ladin stated that "[w]e have chemical and nuclear weapons as a deterrent and if America used them against us we reserve the right to use them." The lack of evidence supporting this claim suggests that Bin Ladin may have simply been using the WMD "buzz words" as a scare tactic.

Al-Qa'ida's loosely organized nature complicates the issue of intent, because while one cell may not be interested in WMD another may deeply desire such weapons. The six individuals arrested north of London in January 2003, for example, may have been operating in accordance with instructions from al-Qa'ida superiors, but may just as easily have planned to produce ricin on their own. On the other hand, the loose nature of the organization, as well as its loss of its camps and infrastructure in Afghanistan, may have made it more difficult for al-Qa'ida to pursue biological weapons or any other WMD. While cells might indeed be capable of producing a crude chemical or biological weapon for small-scale use, it is unlikely that the group would be able to support any larger capability without the necessary facilities and financial support.

In the final analysis, conventional terrorist methods—however creative and innovative—have been and are likely to remain more cost-effective for al-Qa'ida, and therefore more likely to be pursued and perpetrated. Nonetheless, there may be individuals or cells more inclined to experiment with and attempt to use biological agents or other WMD in the future.

IMPLICATIONS

As of the end of 2003, it has not been confirmed that al-Qa'ida possesses biological weapons or any other WMD capability. Nor has the group ever perpetrated a terrorist attack employing WMD. While weakened and dislocated both by the U.S. military during the ousting of the Taliban regime from Afghanistan in 2001–2002 as well as by massive international intelligence and policing efforts, al-Qa'ida remains a formidable terrorist organization, with its leaders at large and its members scattered throughout the world. Conventional terrorist attacks carried out by individuals with apparent links to al-Qa'ida have continued after the September 11 attacks in the United States, but have generally been localized in their nature and intent.

Al-Qa'ida's ability to reconstitute its tactical capabilities and reestablish training camps or a central base of operations will likely determine whether or not the group can continue its concerted and sophisticated international campaign of violence. In this context, it is necessary that the United States and the international community continue to aid lawless and failing states in their recovery, and further to ensure the creation of viable and lawful governments in such countries as Afghanistan and Iraq so that al-Qa'ida cannot take root there again. Similar troubled states and regions such as Chechnya, Somalia, and Sudan may serve as safe havens for al-Qa'ida or its offshoot organizations in the future.

While al-Qa'ida members have reportedly had contact with various governmental officials from countries such as Iraq, Pakistan, Sudan, and possibly even Iran, there is no conclusive evidence of any state-level WMD assistance to al-Qa'ida at this time. Nonetheless, states seeking to counter U.S. global hegemony (or that of Israel in the Middle East region, India in South Asia, etc.) might turn to terrorism, and support terrorists, in the future, especially given the relative success of al-Qa'ida in this arena. In the context of a state that poses a WMD capability, such a scenario could present a real and significant WMD threat, whether through al-Qa'ida, another terrorist organization, or the state itself.

REFERENCES

Al-Qa'ida Training Manual, excerpts in translation. Released by U.S. Department of Justice, December 6, 2001. Available online at http://www.justice.gov/ag/trainingmanual.htm, (cited March 25, 2002).

Bodanksy, Y., *Bin Ladin: The Man Who Declared War on America*, Prima Publishing, Rocklin, CA, 1999.

Corbin, J., *Al-Qa'ida: In Search of the Terror Network that Threatens the World*, Thunder's Mouth Press, New York, 2003.

Gunaratna, R., *Inside Al-Qa'ida: Global Network of Terror*, Berkley Publishing, New York, 2003.

Kepel, G., *Jihad: The Trail of Political Islam*, Belknap Press, Cambridge, MA, 2002.

Mir, H., Osama Claims He Has Nukes, *Dawn* (Pakistan), November 10, 2001.

Monterey WMD-Terrorism Database, Center for Nonproliferation Studies, 1998–2003.

Reeve, S., *The New Jackals*, Northeastern University Press, Boston, MA, 1999.

Roy, O., Bin Ladin: An Apocalyptic Sect Severed from Political Islam, *Internationale Politik*, December 2001.

Venzke, B. and Aimee, I., *The Al-Qa'ida Threat*, Tempest Publishing, Alexandria, VA, 2003.

ANIMAL AID ASSOCIATION: A CASE STUDY

Monterey WMD-Terrorism Database Staff

On January 15, 1992, British Columbia Television (BCTV) received a letter from the Animal Aid Association (AAA) stating that the group had injected "Cold Buster" bars with HIV in protest of perceived animal cruelty on the part of the product's manufacturer. Only weeks earlier, a group calling itself the Animal Rights Militia claimed that it had contaminated this same product, suggesting a copycat element to the incident. The letter from the AAA read as follows: "Effective now, AAA will start random injecting with AIDS of the Cold Buster Bars to protest Asiatic treatment of pets and diversion of funds from AIDS research" (Monterey WMD Terrorism Database, 2003). While authorities did subsequently find evidence of product tampering, affected bars proved to have been contaminated only with innocuous saline solution. Thus, the incident has been classified as a hoax.

REFERENCES

Monterey WMD Terrorism Database, Case Number 279, http://cns.miis.edu/dbinfo/about.htm, accessed on 9/25/03.

ANTHRAX (*BACILLUS ANTHRACIS*)

GENE GODBOLD

INTRODUCTION

Bacillus anthracis is the causative agent of anthrax. It can occur in three forms: cutaneous, gastrointestinal, and inhalational. The disease is named for the coal-black lesions that result from its cutaneous form (Greek *anthrakis* = coal). *Bacillus cereus* and *Bacillus thuringiensis* are species closely related to *B. anthracis*. These bacilli have very few genetic differences between them outside of the virulence plasmids of the anthrax bacterium, denoted pXO1 and pXO2 (Ivanova et al., 2003; Radnedge et al., 2003; Read et al., 2003).

There are nearly 100 species of *Bacillus*. They are commonly found in the soil and, less frequently, in water all over the world. Anthrax is principally a disease of herbivores—humans are just an incidental host. People in occupations exposed to quantities of untreated animal hair or hides are especially prone to infection; thus, anthrax has become known as "Wool sorter's" and "Rag picker's" disease. While inhalational anthrax is the rarest form of the disease, 18 naturally occurring cases have been reported in the United States since 1900. Prior to October 2001, the last recorded case of inhalational anthrax was in 1976 (Lew, 2000).

B. anthracis is ubiquitous in the soil and can be present on the hair or hides of some animals, particularly livestock. In areas where livestock vaccination is limited, the vegetative form can be isolated from afflicted or recently deceased animals. *B. anthracis* cells are nonmotile, gram-positive rods ranging from 1 to 10 μm in length and from 1 to 1.5 μm in width. These so-called vegetative cells are what cause infection in hosts. The vegetative cells can transform themselves into spores, ranging in diameter from 1 to 4 μm. Sporulation can be triggered by any of several environmental factors, particularly by a shortage of moisture and vital nutrients. Spores are extremely hardy, being able to withstand dryness, heat, ultraviolet radiation, and many disinfectants. Germination of spores into vegetative bacteria follows contact with water or the availability of previously missing vital nutrients.

VIRULENCE PROPERTIES

The vegetative form of *B. anthracis* is protected by a capsule that heightens its resistance to phagocytosis by cells of the host immune system. The capsule consists of a polymer of γ-D-glutamic acid. There are at least three enzymes responsible for capsule formation and these are encoded on the pXO2 plasmid (Mock and Fouet, 2001).

The anthrax bacterium secretes three proteinaceous toxins that are essential for virulence: lethal factor (LF), edema factor (EF), and protective antigen (PA). The genes for these proteins are encoded on the pXO1 plasmid. The presence of PA is a necessary requirement for cellular entry for both LF and EF; without it, they cannot exert their toxic effects. PA binds to one of two receptors on the target cells of the host (Bradley et al., 2001; Scobie et al., 2003), undergoes proteolytic cleavage, and oligomerizes into a pore-like structure of seven PA subunits within the membrane, allowing molecules of LF or EF to bind. The cells then internalize the PA/LF or PA/EF complexes into endosomes. Endosomal acidification provokes a conformational change in PA such that enzymatically active LF and EF are released into the cell cytosol.

Each of the three toxins exerts its unique action on host cells. EF is a calmodulin-dependent adenylate cyclase that converts ATP to cyclic AMP. LF is a Zn^{2+} protease that is active against nearly all forms of mitogen-activated protein kinase kinase (MAPKK), a molecule essential for the proper transmission of a wide variety of cellular signals—including the cytokine signaling that enables proper coordination of the host immune response. Its proteolytic inactivation by LF is a critical factor in the subsequent death of the cell (Brossier and Mock, 2001).

PATHOGENESIS, CLINICAL COURSE, AND PROGNOSIS

As noted above, anthrax infections in humans can take three forms: cutaneous, gastrointestinal, and inhalational.

Encyclopedia of Bioterrorism Defense,   Edited by Richard F. Pilch and Raymond A. Zilinskas
ISBN 0-471-46717-0   Copyright © 2005 Wiley-Liss

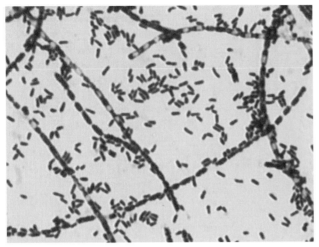

Figure 1. Gram stain showing both vegetative and spore forms of *B. anthracis*.

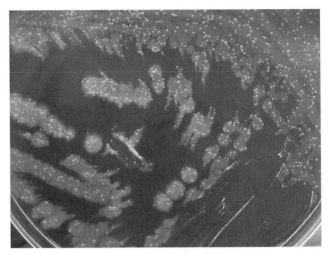

Figure 3. *B. anthracis* growing on SBA agar about 36 h after streaking; the characteristic 'comet-like' projections (aka 'medusa-head') can be seen at the colony margins.

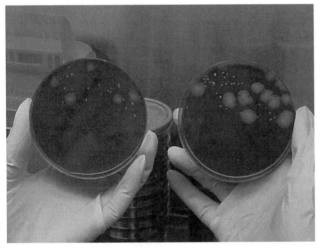

Figure 2. *B. anthracis* growing on SBA and CHOC agar; the large, gray, flat colonies have an irregular margin due to outgrowths of long filamentous projections of the bacteria.

Figure 4. Culture plate from a nasal swab at 36 h; colonies show characteristic 'whipping' when pulled from media.

In each form of infection, spores are taken up by macrophages and transported to regional lymph nodes that drain the site of inoculation. Once in the lymph node, the spores germinate into cells, rendering the organism capable of causing infection. After germination, the vegetative cells produce and release PA, EF, and LF, with subsequent disease progressing rapidly.

How the spores taken up by the macrophages survive the antimicrobial environment of the phagolysosome to germinate is unknown. How the vegetative forms of *B. anthracis* escape the macrophage is also uncertain (Mock and Fouet, 2001). The precise cause of death from anthrax infection remains unclear, and is an ongoing focus of research (Moayeri et al., 2003).

Cutaneous Anthrax

Most (95 percent) anthrax infections in developed countries occur when the bacterium enters a cut or abrasion on the skin during the handling of contaminated wool, hides, leather, or hair products from infected animals. Skin infection typically begins as an itchy bump that resembles an insect bite. Within approximately two days, the bump develops into an ulcer surrounded by vesicles, followed by formation of a painless eschar. This eschar is usually 1 to 3 cm in diameter, with a characteristic coal-black area in the center where the skin cells undergo necrotic death. Lymph nodes that drain the affected area may swell, as may the tissue immediately surrounding the ulcer. Some systemic symptoms such as fever, malaise, and headache may accompany the cutaneous infection. After one to two weeks, the eschar dries and eventually separates, usually without leaving a permanent scar (Lew, 2000). About 20 percent of untreated cases of cutaneous anthrax result in death.

Gastrointestinal Anthrax

Epidemic cases of gastrointestinal anthrax are occasionally reported in the developing world. Infection follows the consumption of contaminated meat. This form of anthrax is characterized by an acute inflammation of the intestinal tract. The incubation period is typically three to seven days. Initial signs of nausea, loss of appetite, vomiting, and fever are followed by abdominal pain, vomiting of blood, and severe diarrhea. Later, toxemia develops with shock, cyanosis, and death. Fatalities following intestinal anthrax range from 25% to 60% of untreated cases. The time from onset of symptoms to death can vary from two to five days. The oropharyngeal variant of gastrointestinal anthrax involves swelling and tissue necrosis in the mouth and/or throat. Sometimes a lesion resembling the cutaneous eschar will form in the oral cavity. Victims present with sore throat, dysphagia, fever, and swollen lymph nodes in the neck (Lew, 2000).

Inhalational Anthrax

After spores are inhaled, the length of time before symptoms of disease appear can vary, but usually within nine days (with a range of 2–43 days) (Jernigan et al., 2002). However, spores lodged in the lungs can have an incubation period of longer than a month. Patients with inhalational anthrax progress through two distinct phases. Initially, nonspecific symptoms such as malaise, fever, chills, muscleaches, and fatigue occur. During this stage, upper respiratory symptoms such as nasal congestion, inflammation, and sneezing are typically absent. A dry cough may be present early or late in disease. A subjective feeling of substernal pressure may be described. Two characteristics of inhalational anthrax are (1) hemorrhagic necrotizing lymphadenitis, involving the chest lymph nodes and observed as a widened mediastinum on chest X-ray, and (2) an accumulation of fluid around the lungs (pleural effusion). The patient may show signs of improvement after two to four days. These initial symptoms are then followed by sudden onset of severe respiratory distress with labored breathing, profuse sweating, and a bluish skin pallor resulting from lack of oxygen. Swelling of the chest wall may be observed. Other symptoms may include headache, abdominal pain, nausea, and/or vomiting. Physical findings are nonspecific except for a wheezing (whistling) sound that may be heard in the lungs upon auscultation. Titers of bacteria higher than 10 million organisms per milliliter of blood are not uncommon in mature inhalational anthrax infections (Dixon et al., 1999). Without treatment, shock and death usually follow within 24 to 36 h of onset of severe symptoms (Lew, 2000).

DETECTION AND DIAGNOSIS

Detection of *B. anthracis* in the environment depends on assays that can distinguish the *B. anthracis* bacterium from closely related species. Several assays have been developed as part of the biodefense effort in the United States. These assays are reportedly quite robust and have as their targets several unique bacterial determinants (Slezak, 2003). They have been automated as part of a nationwide monitoring program and sited in over two dozen cities across the United States (Mintz, 2003). Some assays also exist as handheld units for testing of specifically identified areas of contamination concern (e.g., a postal facility).

A high index of suspicion followed by a positive blood culture, which generally shows growth in 6 to 24 h, provides the working diagnosis for *B. anthracis* infection. However, physician recognition of the (pathognomonic) widened mediastinum and/or pleural effusions expected in cases of inhalational anthrax should prompt immediate therapy when there is reason to believe exposure to *B. anthracis* has occurred, for example, when large numbers of people complaining of similar symptoms of flu-like illness present to the hospital at approximately the same time (Inglesby et al., 2002). Confirmatory testing, for example via enzyme-linked immunosorbent assay (ELISA) for PA and polymerase chain reaction (PCR), can be performed at national reference laboratories such as the Centers for Disease Control and Prevention (CDC) but is not necessary for the initiation of treatment (Inglesby et al., 2002).

PREVENTION AND TREATMENT

Naturally occurring *B. anthracis* is susceptible to a variety of antibiotics. Penicillin G given intravenously is the treatment of choice. In cases in which optimization of the organism by genetic manipulation is suspected (as in a terrorist attack) as well as in patients with penicillin allergy, ciprofloxacin or doxycycline is recommended (Lew, 2000). This treatment should continue for at least 60 days (Inglesby et al., 2002). Levofloxacin may eventually be a component of a treatment regimen since it appears to be at least as effective as penicillin (Friedlander et al., 1993; Bush et al., 2003).

After infection has progressed to a certain point, treatment with antibiotics by itself can no longer rescue the patient. The toxins that have already been secreted are presumably sufficient to cause irreversible systemic damage and lead to death (Smith and Keppie, 1954). At this time, one treatment option is to neutralize existing toxins with antibodies. For example, passive immunization with (rabbit) polyclonal antibodies raised against the *B. anthracis* protective antigen (PA) provides dose-dependent protection in guinea pigs when administered 24 h after an inhaled dose of *B. anthracis* spores 25 times greater than the LD_{50} (median lethal dose). Increasing the time between inoculation and administration of antiserum decreases the survival rate (Kobiler et al., 2002). Recent work also has used directed evolution to optimize monoclonal antibodies to PA such that they can provide adequate protection for rats intoxicated with LF and PA (Maynard et al., 2002).

The discovery of receptors allowing the entry of anthrax toxins into host cells has precipitated the development of new antitoxin treatments based on inhibition of the binding event between PA and the receptors (Bradley and Young, 2003). Soluble forms of the anthrax toxin receptor can prevent toxin entry in cultured cells (Bradley et al.,

2001; Scobie et al., 2003). Another promising therapy involves treating patients with a "dominant negative" form of PA. These mutant PA molecules, when mixed with normal PA, inhibit the entry of toxin into cells (Sellman et al., 2001). The hope is that administration of this mutant PA will (at least) delay progression of the disease. Similarly, a peptide has been developed that competes with LF and EF binding to the processed, heptameric PA but does not interact with free PA. When many of these peptides are chemically linked to a single polymer, the multivalent complex prevents cell death following toxin entry in cell culture (Mourez et al., 2001). Molecules that prevent cleavage of the MAPKK substrate might someday be developed to combat LF that enters cells (Chaudry et al., 2002). Within a decade, patients afflicted with anthrax might have several treatment options available to them. It is expected that each therapeutic option will target a discrete step along the continuum of pathogenesis and will complement the existing antibiotic and antibody-based antiserum therapies.

Other potential options for treatment of anthrax infection include development of inhibitors to capsule formation by the *B. anthracis*. If the three enzymes responsible for assembly of the protein coat could be inhibited, they would render the bacterium less resistant to phagocytic cells of the host immune system. Research in this area is in its infancy. A more immediately available approach is to use a specific viral predator of the bacillus known as a "bacteriophage," or one of the phage's enzymatic tools, to combat the invading bacteria (Schuch et al., 2002). The enzyme PlyG lysin is a product of the γ bacteriophage of *B. anthracis* and consists of two portions, a binding domain and an enzymatic domain. The binding domain specifically adheres to the bacillus cell wall while the enzymatic domain catalyzes the hydrolysis of that wall, thereby killing the bacterium and enabling the assembled bacteriophage to escape. Incubation of PlyG lysin with a culture of *B. anthracis* results in the rapid disintegration of the bacteria. Besides *B. anthracis*, only one very closely related *B. cereus* strain is affected, while other *B. cereus* strains and all *B. thuringiensis* strains appear immune. A reassuring feature of PlyG lysin action is that heretofore, *B. anthracis* colonies resistant to its activity have not been generated, even using powerful chemical mutagens (Schuch et al., 2002).

PA is the key component of the currently licensed (Biothrax®) anthrax vaccine in the United States. It is also the principal antigen in the next generation of vaccines. Numerous studies in animals have demonstrated the capacity of PA to protect against aerosol challenge with *B. anthracis* spores (reviewed in Welkos et al., 2001). The vaccine is administered at 0, 2, and 4 weeks and 6, 12, and 18 months with subsequent annual boosters. It is fairly well tolerated; the most common side effect is a local reaction with soreness and swelling at the site of injection. Current research is focused on improving this vaccine, as well as discovery of other anthrax proteins (antigens) that may be incorporated into newer generation vaccines to help boost the host immune response to the vaccine.

INFECTION CONTROL AND DECONTAMINATION

Various methods to sterilize areas contaminated with anthrax and related bacilli have recently been reviewed (Whitney et al., 2003). These include treatment with hypochlorite (bleach), hydrogen peroxide, peracetic acid, radiation (ultraviolet and gamma), formaldehyde, ethylene dioxide, chlorine dioxide, methylene bromide, propylene oxide, ozone, and heat. After the 2001 anthrax attacks in the United States, undelivered mail was successfully decontaminated with a radiation dose of at least two megarads from a cobalt source (Whitney et al., 2003).

ANTHRAX AS A BIOLOGICAL WEAPON

B. anthracis has been developed for weapons use by the United Kingdom, the United States, Iraq, Japan, South Africa, and the Soviet Union. Other countries suspected of having weaponized *B. anthracis* include Iran, North Korea, and Syria (Tenet, 1997). The *B. anthracis* spore has several features that render it amenable to weaponization. Chief among these is its lethality when inhaled. Estimates of the number of fatalities that would result from a nocturnal attack in which 50 to 100 kg of spores are aerosolized over a major urban area start at 100,000 and range into the millions. The LD_{50} for untreated inhalational anthrax in humans is thought to range between 2500 and 55,000 spores (Inglesby et al., 2002). Second, the spore is relatively durable and can withstand being dispersed with explosives (Alibek and Handelman, 1999). Third, the progression of resulting disease is usually rapid. Finally, the fact that humans are dead-end hosts provides some assurance to those using *B. anthracis* as a weapon that the resulting disease will not spread to their own populations (a margin of safety that disappears when contagious agents such as the smallpox virus are considered).

It was not long after the discovery of *B. anthracis* as the causative agent of anthrax by Robert Koch and Louis Pasteur at the end of the nineteenth century that it was first used as a weapon. During World War I, German agents waged a campaign of biological warfare (BW) using both *B. anthracis* and *Burkholderia mallei*. Beginning in 1915, they targeted beasts of burden shipped from neutral countries (Romania, Spain, Norway, the United States, and Argentina) to the Allies (Wheelis, 1998).

Perhaps as much as 1 g of *B. anthracis* spores was accidentally released on April 2, 1979, from a military facility involved in biological weapons research in the Soviet city of Sverdlovsk. The release killed over 65 people and caused illness in at least a dozen more (Meselson et al., 1994), though some analyses have suggested that hundreds of people were infected as a result of the mishap (Inglesby et al., 2002). Pathology samples taken from those who died showed that the fatalities had resulted from infection by at least four different *B. anthracis* strains (Jackson et al., 1998), suggesting that the scientists at Sverdlovsk may have been working with manipulated cultures of the organism. The Soviets were not deterred by the mishap, however, and by the end of the 1980s were reportedly capable of producing nearly

5000 tons of *B. anthracis* spores annually (Alibek and Handelman, 1999).

The Japanese doomsday cult Aum Shinrikyo sprayed a culture of *B. anthracis* over a suburb of Tokyo in June of 1993 (and multiple other times both before and after this date as well). The aerosolization attempt was crude and no one was infected, though reports of a horrible smell prompted an investigation that provided material for later analysis. The *B. anthracis* culture used by the cult lacked the pXO2 virulence plasmid, making it consistent with the Sterne strain commonly used for livestock vaccinations (Keim et al., 2001). It is therefore presumed that the cult members procured their strain from a veterinary source without realizing that it was unsuitable for weaponization.

In October of 2001, the first inhalational anthrax victim of the United States "anthrax mail attacks" was identified and reported by the CDC. In all, 22 people were infected by *B. anthracis* spores carried in at least four envelopes mailed from Trenton, New Jersey, with half of the victims suffering from inhalational and half from cutaneous anthrax. Of the cases of inhalational anthrax, five ended in death (Inglesby et al., 2002). Among these were postal workers exposed to aerosolized spores that presumably escaped from the envelopes during handling (Dewan et al., 2002). As of this writing, the perpetrator(s) of this act of bioterrorism remains unknown (Gugliotta and Matsumoto, 2002).

IMPORTANT ASPECTS OF WEAPONIZATION

Weaponization of *B. anthracis* spores requires preparing them for aerosol delivery. In order to permeate the alveoli of the lungs, a product that is consistently no more than a few micrometers in diameter must be generated. In the process of weaponization, spores are typically treated to minimize physicochemical interactions with surrounding spores and other materials. Providing each bacterium with an electrostatic charge also contributes to efficient dispersion (Matsumoto, 2003). If the spores are thus manipulated, the ease with which they become airborne and the length of time that they can remain in such a state are maximized.

PRESENT AND FUTURE IMPLICATIONS

China, Iran, Libya, North Korea, and Syria are believed to have (at a minimum) a research and development program for biological weapons (Tenet, 1997). Of these, at least Iran, North Korea, and Syria are thought to have pursued *B. anthracis* in a production capacity (Center for Nonproliferation Studies, 2002). While Iraq is known to have pursued *B. anthracis* in the past, it is unlikely that this effort has persisted in view of the second Gulf War.

In 1992, Russia's president Boris Yeltsin issued a decree banning all work related to biological weapons, and Russian stockpiles were destroyed. Most of the facilities of the former program have been decommissioned, but four facilities under the Ministry of Defense (Ekaterinburg, Sergiev Posad, Vyatka, and the Military Medical Academy, St. Petersburg) remain operational and closed to outsiders (Hoffman, 1998). Vyatka, given its focus on bacteriological defense, is the most likely of the three to be involved in anthrax research if any such research is taking place. With the myriad economic and social problems faced by Russia, it seems improbable that a large-scale offensive program could or would be funded, but current political conditions in that region probably favor continued low-level research and development.

Future research will be directed at further characterizing the interplay of virulence factors of the bacterium at the molecular level. How the anthrax toxins individually and collectively assist in the pathogenic process will should remain a focus of study. A great deal remains unclear as to how the bacterium kills the host, especially in primates, and this issue continues to be explored. Development of anthrax therapeutic agents appears very promising from a biodefense/public health standpoint, although the economic benefit of such work is liable to be slight for the developers.

Acknowledgments

The author would like to thank Drs. Roy Barnewall, Bruce McClelland, Jason Mott, David M. Robinson, and James V. Rogers of the Battelle Memorial Institute and Dr. Jeffrey Tessier of the University of Virginia for helpful discussions and assistance with this article. He is especially grateful for the contributions of Dr. Molly A. Hughes of the University of Virginia. Thanks are also due to Drs. Greg Martin and Robert Paolucci of the Infectious Diseases Service at the Naval Medical Research Institute for kindly providing the photographs.

REFERENCES

Alibek, K. and Handelman, S. *Biohazard: The Chilling True Story of the Largest Covert Biological Weapons Program in the World—Told from the Inside by the Man Who Ran it*, Random House, New York, 1999.

Bradley, K.A., Mogridge, J., Mourez, M., Collier, R.J., and Young, J.A.T., *Nature*, **414**, 225–229 (2001).

Bradley, K.A. and Young, J.A.T., *Biochem. Pharmacol.*, **65**, 309–314 (2003).

Bush, K., Kao, M., Barnewall, R., Estep, J., Hemeryck, A., and Kelley, M.F., Efficacy of Levofloxacin in the Inhalational Anthrax (post-exposure) Rhesus Monkey Model, Poster Session 28-Animal Models of Gram-Positive Infection: Pathogenesis and Treatment—*43rd ICAAC Annual Meeting*, Chicago, IL, 2003.

Center for Nonproliferation Studies, *Chemical and Biological Weapons: Possession and Programs Past and Present*, 2002, http://cns.miis.edu/research/cbw/possess.htm.

Chaucry, G.J., Moayeri, M., Liu, S., and Leppla, S.H., *Trends Microbiol.*, **10**, 58–62 (2002).

Dewan, P.K., Fry, A.M., Laserson, K., Tierney, B.C., Quinn, C.P., Hayslett, J.A., Broyles, L.N., Shane, A., Winthrop, K.L., Walks, I., Siegel, L., Hales, T., Semenova, V.A., Romero-Steiner, S., Elie, C., Khabbaz, R., Khan, A.S., Hajjeh, R.A., and Schuchat, A., and members of the Washington, DC anthrax response team. Inhalational anthrax outbreak among postal workers, Washington, D.C., 2001, *Emerg. Infect. Dis.*, **8**, 1066–1072 (2002).

Dixon, T., Meselson, M., Guillemin, J., and Hanna, P., *N. Engl. J. Med.*, **341**, 815–826 (1999).

Friedlander, A.M., Welkos, S.L., Pitt, M.L., Ezzell, J.W., Worsham, P.L., Rose, K.J., Ivins, B.E., Lowe, J.R., Howe, G.B., Mikesell, P., Lawrence, W.B., *J. Infect. Dis.*, **167**, 1239–1243 (1993).

Gugliotta, G. and Matsumoto, G., FBI's Theory on Anthrax is Doubted, *Washington Post*, October 28, 2002, p. A01.

Hoffman, D., A Puzzle of Epidemic Proportions, Source of 1979 Anthrax Outbreak in Russia Still Clouded, *Washington Post Foreign Service*, Wednesday, December 16, 1998, p. A01, (http://www.central.edu/homepages/weihep/bio101/USSR.htm).

Inglesby, T.V., O'Toole, T., Henderson, D.A., Bartlett, J.G., Ascher, M.S., Eitzen, E., Friedlander, A.M., Gerberding, J., Hauer, J., Hughes, J., McDade, J., Osterholm, M.T., Parker, G., Perl, T.M., Russell, P.K., Tonat, K., and Working Group on Civilian Biodefense, *J. Am. Med. Assoc.*, **287**, 2236–2252 (2002).

Ivanova, N., Sorokin, A., Anderson, I., Galleron, N., Candelon, B., Kapatral, V., Bhattacharyya, A., Reznik, G., Mikhailova, N., Lapidus, A., Chu, L., Mazur, M., Goltsman, E., Larsen, N., D'Souza, M., Walunas, T., Grechkin, Y., Pusch, G., Haselkorn, R., Fonstein, M., Ehrlich, S.D., Overbeek, R., and Kyrpides, N., *Nature*, **423**, 87–91 (2003).

Jackson, P.J., Hugh-Jones, M.E., Adair, D.M., Green, G., Hill, K.K., Kuske, C.R., Grinberg, L.M., Abramova, F.A., and Keim, P., *Proc. Natl. Acad. Sci. U.S.A.*, **95**, 1224–1229 (1998).

Jernigan, D.B., Raghunathan, P.L., Bell, B.P., Brechner, R., Bresnitz, E.A., Butler, J.C., Cetron, M., Cohen, M., Doyle, T., Fischer, M., Greene, C., Griffith, K.S., Guarner, J., Hadler, J.L., Hayslett, J.A., Meyer, R., Petersen, L.R., Phillips, M., Pinner, R., Popovic, T., Quinn, C.P., Reefhuis, J., Reissman, D., Rosenstein, N., Schuchat, A., Shieh, W.J., Siegal, L., Swerdlow, D.L., Tenover, F.C., Traeger, M., Ward, J.W., Weisfuse, I., Wiersma, S., Yeskey, K., Zaki, S., Ashford, D.A., Perkins, B.A., Ostroff, S., Hughes, J., Fleming, D., Koplan, J.P., Gerberding, J.L., and National Anthrax Epidemiologic Investigation Team (2002), *Emerg. Infect. Dis.*, **8**, 1019–1028 (1998).

Keim, P., Smith, K.L., Keys, C., Takahashi, H., Kurata, T., and Kaufmann, A., *J. Clin. Microbiol.*, **39**, 4566–4567 (2001).

Kobiler, D., Gozes, Y., Rosenberg, H., Marcus, D., Reuveny, S., and Altboum, Z., *Infect. Immun.*, **70**, 544–560 (2002).

Lew, D.P., "*Bacillus anthracis* (Anthrax)," in G.L. Mandell, J.E. Bennett, and R. Dolin, Eds., *Mandell, Douglas, and Bennett's Principles and Practice of Infectious Diseases*, 5th ed., Churchill Livingstone, Philadelphia, PA, 2000, pp. 2215–2226.

Matsumoto, G., *Science*, **302**, 1492–1497 (2003).

Maynard, J.A., Maassen, C.B., Leppla, S.H., Brasky, K., Patterson, J.L., Iverson, B.L., and Georgiou, G., *Nat. Biotechnol.*, **20**, 597–601 (2002).

Meselson, M., Guillemin, J., Hugh-Jones, M., Langmuir, A., Popova, I., Shelokov, A., Yampolskya, O., *Science*, **266**, 1202–1208 (1994).

Mintz, J., U.S. provides a peek at air sensor program, *Washington Post*, Saturday November 15, 2003, p. A03.

Moayeri, M., Haines, D., Young, H.A., and Leppla, S.H., *J. Clin. Invest.*, **112**, 670–682 (2003).

Mock, M. and Fouet, A., *Annu. Rev. Microbiol.*, **55**, 647–671 (2001).

Mourez, M., Kane, R.S., Mogridge, J., Metallo, S., Deschatelets, P., Sellman, B.R., Whitesides, G.M., and Collier, R.J., *Nat. Biotechnol.*, **19**, 958–961 (2001).

Radnedge, L., Agron, P.G., Hill, K.K., Jackson, P.J., Ticknor, L.O., Keim, P., and Andersen, G.L., *Appl. Environ. Microbiol.*, **69**(5), 2755–2764 (2003).

Read, T.D., Peterson, S.N., Tourasse, N., Baillie, L.W., Paulsen, I.T., Nelson, K.E., Tettelin, H., Fouts, D.E., Eisen, J.A., Gill, S.R., Holtzapple, E.K., Okstad, O.A., Helgason, E., Rilstone, J., Wu, M., Kolonay, J.F., Beanan, M.J., Dodson, R.J., Brinkac, L.M., Gwinn, M., DeBoy, R.T., Madpu, R., Daugherty, S.C., Durkin, A.S., Haft, D.H., Nelson, W.C., Peterson, J.D., Pop, M., Khouri, H.M., Radune, D., Benton, J.L., Mahamoud, Y., Jiang, L., Hance, I.R., Weidman, J.F., Berry, K.J., Plaut, R.D., Wolf, A.M., Watkins, K.L., Nierman, W.C., Hazen, A., Cline, R., Redmond, C., Thwaite, J.E., White, O., Salzberg, S.L., Thomason, B., Friedlander, A.M., Koehler, T.M., Hanna, P.C., Kolsto, A.B., and Fraser, C.M., *Nature*, **423**, 81–86 (2003).

Schuch, R., Nelson, D., and Fischetti, V.A., *Nature*, **418**, 884–889 (2002).

Scobie, H.M., Rainey, G.J., Bradley, K.A., and Young, J.A., *Proc. Natl. Acad. Sci. U.S.A.*, **100**(9), 5170–5174 (2003).

Sellman, B.R., Mourez, M., and Collier, R.J., *Science*, **292**, 695–697 (2001).

Slezak, T., Computational analysis of biothreat agents: from algorithms to assays, *6th Annual Computational Genomics Conference*, Cambridge, MA, October 8–11, 2003.

Smith, H. and Keppie, J., *Nature*, **173**, 869–870 (1954).

Tenet, G., *The Worldwide Threat in 1997, Global Realities of Our National Security*, Statement before the Senate Select Committee on Intelligence, February 27, 1997.

Welkos, S., Little, S., Friedlander, A., Fritz, D., and Fellows, P., *Microbiology*, **147**, 1677–1685 (2001).

Wheelis, M., *Nature*, **395**, 213 (1998).

Whitney, E.A.S., Beatty, M.E., Taylor, T.H., Weyant, R., Sobel, J., Arduino, M.J., and Ashford, D.A., *Emerg. Infect. Dis.*, **9**, 623–627 (2003).

ANTHRAX HOAXES: CASE STUDIES AND DISCUSSION

Monterey WMD-Terrorism Database Staff

INTRODUCTION

This entry reviews four well-described cases involving the threat of *Bacillus anthracis* use, as documented in the Monterey Institute's Weapons of Mass Destruction (WMD) Terrorism Database. The cases are presented in chronological order: the Counter Holocaust Lobbyists of Hillel (1997), Brothers for the Freedom of Americans (1998), unknown perpetrator (1999), and Secret Operating Base (2001).

CASE ONE: COUNTER HOLOCAUST LOBBYISTS OF HILLEL

On April 23, 1997, a mail clerk at the B'nai B'rith headquarters in Washington, DC, discovered a package from the "Counter Holocaust Lobbyists of Hillel" containing a petri dish labeled "anthracis." After a thorough examination, authorities found that the dish had not been inoculated with *B. anthracis* but rather with its close relative *Bacillus cereus*, a common organism that is not particularly pathogenic (Carus, 1999). There were no casualties in the incident.

Timeline of the Incident

When an envelope addressed to the headquarters of the Jewish organization B'nai B'rith but providing no other identification was received by the organization's mail clerk, it was set aside until the next morning without much thought. Upon returning to the envelope, however, the clerk found it to be leaking a red, gelatinous substance and emitting an ammonia-like odor (Horwitz, 1997; Dorf, 1997; Powell and Lengel, 1997). The envelope was placed in a trashcan, and authorities were notified.

The FBI's Joint Terrorism Task Force, fire and police departments, a bomb squad, and the Centers for Disease Control and Prevention (CDC) responded en masse to the scene (Horwitz, 1997; Powell and Lengel, 1997). All buildings and streets within a wide radius were shut down. Authorities opened the package and inside found a letter and a broken petri dish. As noted, the dish was labeled "anthracis" and the letter was signed by the "Counter Holocaust Lobbyists of Hillel." The letter stated that the dish contained an agent of "chemical warfare" (Horwitz, 1997).

Over 100 employees were quarantined for eight hours while the material was sent to the Naval Medical Research Institute (NMRI) in Maryland for testing. During that time, the mail clerk and a security guard exposed to the package complained of headache and shortness of breath (Powell and Lengel, 1997). In all, four employees and 12 firefighters underwent decontamination procedures outside the building. After testing, the substance was determined to be *B. cereus*.

Afterwards, some reports expressed concern about how poorly the Washington, DC, fire and police departments reacted to the incident. For example, the local emergency response units were not equipped with decontamination tents or other specialized equipment. Further, Ray Sneed, president of the International Association of Firefighters Local 36, informed reporters that only a small number of firefighters were trained to manage chemical and biological events (Horwitz, 1997). In the years since, significant steps have been taken to correct these deficiencies.

The Group and the Letter

Anti-Defamation League of B'nai B'rith's Regional Director David C. Friedman stated on April 25, 1997, that he "had never heard of the 'Counter Holocaust Lobbyists of Hillel' (Horwitz, 1997)." The group's two-page letter contained references to both Nazis and Hillel, the Jewish college organization, as well as statements attacking Jewish liberalism and declaring that "the only good is Jew is an Orthodox Jew" (Dorf, 1997).

The Counter Holocaust Lobbyists of Hillel were never heard from again. However, in the year 2000, a hoax letter containing a suspicious white powder that proved innocuous was delivered to the Hillel organization of University of Pennsylvania (Margulies, 2000; Associated Press, 2000). Both hoax incidents ultimately proved successful in creating a sense of fear among members of the target group, and the Jewish community at large.

CASE TWO: BROTHERS FOR THE FREEDOM OF AMERICANS

On August 17, 1998, an employee of the Joan Finney State Office Building in Wichita, Kansas, discovered a package

Encyclopedia of Bioterrorism Defense, Edited by Richard F. Pilch and Raymond A. Zilinskas
ISBN 0-471-46717-0 Copyright © 2005 Wiley-Liss

containing a white powder identified in an enclosed note as "anthrax." The building was evacuated, but authorities soon concluded that the powder was not *B. anthracis* but a common household product (FDHC Federal Department and Agency Documents, 1998). The following day, a local television station received a message identifying the "Brothers for the Freedom of Americans" as the perpetrators (Lessner, 1998). There were no casualties in the incident.

Timeline of the Incident

Initially discovered by an employee in the building's stairwell, the package was carried by hand to the employee's supervisor, who immediately dialed 911. Ten specialists in hazardous materials from nearby McConnell Air Force Base joined the local fire department's team of first responders in responding to the call (FDHC Federal Department and Agency Documents, 1998). Though 20 to 25 people were exposed to the unknown substance, no injuries or symptoms were reported (AP Online, 1998). As a precaution, however, 15 people were given decontamination showers on-site. Nearly four city blocks were evacuated and cordoned off for approximately 12 h (FDHC Federal Department and Agency Documents, 1998; Kansas City Star, 1998). Preliminary tests revealed that the powder did not contain *B. anthracis*. According to Special Agent Jeff Lanza of the FBI, the substance was a nontoxic "common household product" (Lessner, 1998). Authorities also found traces of the substance on the control panels of three elevators located in the building.

The next day, local television station KWCH Channel 12 and Nevada-based radio talk show host Art Bell each received an identical 11-page letter signed by the "Brothers for the Freedom of Americans" claiming responsibility for the incident. Opening with the statement, "Congratulations—you've been infected," the self-proclaimed splinter organization asserted that it intended to scare government officials and would ultimately "defeat the corrupt government" (Lessner, 1998; Laviana, 1998). The group also claimed to be working in conjunction with both the Christian Identity Church and individuals from Ireland who maintained connections to the Irish Republican Army (IRA). According to Bell, the letter contained very specific threats (Laviana, 1998). Ultimately, however, it was never ascertained whether the Brothers for the Freedom of Americans in fact existed as anything other than a fictitious name.

CASE THREE: UNKNOWN PERPETRATOR

On March 5, 1999, an employee of the Southern Baptist Convention Executive Committee office in Nashville, Tennessee, opened a letter that threatened contamination with *B. anthracis* but contained no visible substance (http://www.biblicalrecorder.org/news/3_12_99/sbc.html, accessed on 5/21/99). The letter was postmarked from the west coast of the United States. The FBI, Metro Health Department, Metro Police Department, Nashville Fire Department, and Metro Office of Emergency Management responded to the scene. Authorities sent the letter to a laboratory for further analysis, and the building was closed and remained off limits through March 8, 1999. Four employees and a fire chief showered with a bleach solution and were given antibiotics as a precaution against exposure. On March 19, 1999, the Metro Health Center announced that lab results confirmed that the letter was not contaminated with *B. anthracis* or any other substance (Klausnitzer, 1999).

As a result of this incident, the Southern Baptist Convention began a training program for employees who handled mail in order to better prepare them for similar incidents in the future (Baptist Press, 1999).

CASE FOUR: SECRET OPERATING BASE

Between October 16 and 17, 2001, two envelopes containing suspicious white powder were received in Seoul, South Korea. The first letter was received at the Seoul office of Woonigjin Coway, a manufacturer of home utilities such as air purifiers and rice cookers. The second was discovered at a local post office; its final destination has not been reported.

Enclosed with the powder in each was a message from the "Secret Operating Base, South Korean Section of Al-Qaeda" reading: "The moment you open this letter, you will be blessed by Almighty God Allah." (Hwa-sop, 2001).

The Secret Operating Base had not been heard from before, and has not resurfaced since. The powder in the letters was determined not to be *B. anthracis* (Yonhap News, 2001), leading to the assumption that the attacks were hoaxes inspired by the legitimate "anthrax letters" in the United States during the same period. Indeed, during October 2001, some 11 similar hoax incidents were reported in the Seoul area alone (Yonhap News, 2001).

REFERENCES

Anthrax Ruled Out in Scare: FBI Runs More Tests on Powder Found in Wichita State Building, *Kansas City Star*, August 20, 1998; Internet, available from www.kcstar.com, accessed on 8/26/98.

AP Online, Anthrax Note Evacuates Kansas City Office, August 18, 1998; Internet, available from www.nytimes.com/aponline/, accessed on 8/18/98.

Baptist Press, *No Anthrax in Baptist Attack*, March 24, 1999; Internet, available from www.churchnet.org.uk/newsm, accessed on 5/21/99.

Carus, S., *Bioterrorism and Biocrimes: The Illicit Use of Biological Agents in the 20th Century*, Working Paper, Center for Counterproliferation Research, National Defense University, July 1999 revision.

Dorf, M., *In Wake of B'nai B'rith Chemical Scare, Questions Remain*, Jewish News of Greater Phoenix, May 2, 1997; Internet, available from www.jewishaz.com/jewishnews/970502/bnai.shtml.

FDHC Federal Department and Agency Documents, *McConnell Responds to Anthrax Threat*, August 19, 1998.

Horwitz, S., FBI Sends Alert to Jewish Groups, Package in D.C. Incident Called Threatening But Not Toxic, *Washington Post*, April 26, 1997, p. C1.

Hwa-sop, L., Powder Pranks Plague Police, *Yonhap News*, October 22, 2001.

Klausnitzer, D., *Anthrax Note at Southern Baptist Convention was Hoax*, *The Tennessean*, March 20, 1999, Internet, available at http://www.tennessean.com/sii/99/03/20/anthrax20.shtml, accessed on 5/21/99.

Lessner, L., Group Takes Credit for Prank at Building: FBI Says the White Powder that Forced Evacuation of Downtown Building as Nontoxic, Common Household Product, *Wichita Eagle*, August 21, 1998, p. 15A.

Letter with Powder inside Believed to be Hate Threat, *Associated Press*, 2000.

Margulies, J., *U. Penn Hillel Receives Death Threat in Hate Mail*, Daily Pennsylvanian, April 24, 2000; Internet, available at http://news.excite.com/news/uw, accessed on 4/26/00.

Powell, M. and Lengel, A., Chemical Alert Traps Workers in Buildings, Discovery of Leaking Package at B'nai B'irth Turns Into Nine-Hour Ordeal in D.C., *Washington Post*, April 25, 1997, p. A1.

SBC Building Evacuated After Anthrax Scare, *Biblical Recorder: Journal of the Baptist State Convention of North Carolina*, March 12, 1999; Internet, available at http://www.biblical recorder.org/news/3_12_99/sbc.html, accessed on 5/21/99.

ANTI-MATERIAL AGENTS

Margaret E. Kosal

INTRODUCTION

At the core of antimaterial biological weapons is the concept of focused exploitation of biological species. Through the evolution process, some microorganisms have gained capabilities to metabolize or digest materials that we utilize for infrastructure and daily purposes, such as plastics and metal, which thus can be directed toward legitimate environmental needs. Some of these microbes have further demonstrated extraordinary environmental robustness (e.g., microbes used in the remediation of radioactively contaminated areas), enhancing their utility all the more (Brim et al., 2000). However, the biocatalytic abilities of such agents can also be developed to serve as defensive or offensive antimaterial weapons, the prospect of which is discussed in detail below (Knoth, 1994).

APPLICATIONS

From the iron-utilizing bacteria of the North Atlantic ocean that biodegrade the hull of the *Titanic* to fungi that transform hard plastics into soft gels, microorganisms target a wide range of materials in a variety of different ways.

Metals

Microbiologically influenced corrosion (MIC), or biocorrosion, has been documented for all common metals and alloys, with the possible exception of titanium (Little and Wagner, 1996). Biocorrosion destroys materials via the facilitation of pitting on the metal surface. Pitting generally occurs via biofilms, which promote the formation of microsized electrochemical cells, leading to anodic and cathodic sites on the metal surface. Oxidation reactions occur at the anode and reduction reactions at the cathode, which for biofilms is in the surrounding aqueous environment or some separation from the initial pitting. In the absence of an antimaterial agent, a thin layer of surface oxidation from atmospheric oxygen passivates the exposed metal surface, effectively creating a protective layer from further oxidation. With the presence of an anti-material biofilm, rather than just having the thin, passive protective layer, the anode enables the oxidation-reduction (redox) reaction to continue. Biofilms—really, the responsible microorganism constituting biofilm flora—will often make available ionic species to the system as one way of supporting the redox reaction. Rather than occurring lamellarly, the oxidation is observed to be pronounced at a single discrete site (or multiple discrete sites), hence pitting and ultimately the destruction of the targeted material.

The aforementioned iron-utilizing bacteria, a group of aerobic bacteria that oxidize iron(II) or manganese(II), accelerate corrosion and the weakening of critical infrastructure. Typical examples of biocorrosive iron-utilizing bacteria are found in the *Gallionella, Clonothrix, Leptothrix*, and *Sphaerotilus* genera. Iron, as the major elemental component of steel (approximately 80 percent), can be targeted for attack via corrosion of pipes and fittings, for example. Railroads would also be susceptible. In such cases, pitting is frequently concentrated at weld seams. Microbiologically induced corrosion can also lead to the blockage of pipes and therefore affect the potability of water. This biocorrosion is very hard to control because the formation of iron salts around the bacteria and inside the cell interferes with antibacterial control agents (Cullimore and McCann, 1978).

Sulfate-reducing bacteria (SRB) are most commonly found in seawater, for example, *Desulfonema* spp. and *Desulfovibrio* spp. While not directly interacting with metal, these biofilm-producing bacteria generate reactive hexaphosphate compounds, which are destructive to iron-containing alloys. The metal surface is degraded, leading to the precipitation of black ferric sulfides and ferric phosphides (Iverson, 2001). Small black specks indicate chemically triggered damage to the metal surface that occurs in the presence of the sulfate-reducing bacteria.

Plastics

"Polyurethane" is a general term applied to polymeric plastics formed from the condensation of alcohols and isocyanates. Increasingly popular due to their physical properties and low cost, polyurethanes are found in construction materials, coatings (often to protect against exposure to inclement weather), and adhesives, along with myriad household and consumer goods. While highly resistant to oxidation, backbone polyurethane bonds are susceptible to hydrolases and esterases produced by microorganisms, which are able to hydrolyze the essential ester linkages (Howard, 2002). The most susceptible to biodegradation are thin, flexible coatings, which provide a large accessible surface area. Rigid, bulk samples, with their increased crystallinity and extensive hydrogen bonding and cross-linking, are more resistant to biodegradation.

The fungi *Chaetomium globosum* and *Aspergillus terreus* are among the microorganisms capable of degrading polyurethanes. The application of these fungi suffers from a major drawback, however, in that their growth is inhibited in the absence of additional carbon- and nitrogen-containing media. A number of bacteria have also demonstrated the ability to reduce the tensile strength and percentage of elongation of polyurethane after three days growth in the presence of supplementary nutrient-containing broths, for example, *Corynebacterium* spp., *Pseudomonas* spp. (Kay et al., 1993), and *Bacillus* spp. (Blake et al., 1998). However, none of the microbes shows any enzymatic activity when applied to polyurethane alone. In addition, one specific strain of *Comamonas acidovorans* has exhibited the ability to completely degrade a solid cube of polyurethane in seven days when a supplemental nitrogen source is provided (Nakajima-Kambe et al., 1995).

Petroleum Products

All kerosene-based fuels (including jet aviation fuels) contain antimicrobial chemicals to inhibit the growth of *Aspergillus flavus* fungi (Lopes and Gaylarde, 1996) and bacteria that would otherwise propagate in long chain hydrocarbon-based fuels. If these added antimicrobials were overwhelmed, however, the viscosity of the fuel could substantially increase, potentially changing its characteristics in such a way that combustion is affected or mechanical failure results. In this way, a terrorist might target the fuel used for long transcontinental or international flights, for example.

Alternatively, hydrocarbon-degrading bacteria such as *Rhodococcus erythropoliso*, *Pseudomonas* spp., and *Sphingomonas* spp. have been reported to consume up to 83 percent of a total petroleum source over 65 days when supplemental nitrogen and phosphorous sources are available, and thus could be illicitly applied to reduce this type of resource in a military or terrorism setting (Thamassin-Lacroix et al., 2001, 2002).

Petroleum-degrading bacteria can also affect roads and runways as embrittlement agents, emulsifying the hydrocarbons in asphalt via the secretion of biosurfactants (Pendrys, 1989). Controlled studies using *Acinetobacter* sp. have shown that after three weeks of growth with no additional nutrients, asphalt surfaces became brittle and flaky (although depth of penetration was limited). Certain fungi have also been found to degrade hydrocarbon-based lubricants (Little et al., 2001).

Concrete

Microbiologically induced corrosion of concrete has primarily been studied with respect to sewage collection systems. Biocorrosion of cementious material often results in a white pasty surface. The most damaging agents, *Thiobacillus* spp., are sulfur-oxidizing bacteria that are unable to grow in media deficient in thiosulfate or elemental sulfur (Nica et al., 2000); therefore, a bioterrorist could not just pour a bacteria-containing broth on a concrete runway and anticipate it dissolving over time. Nor are the bacteria viable when thiosulfate or elemental sulfur is the sole source of energy, such that spiking a surface with sulfur or thiosulfate alone would not prove sufficient to a bioterrorist in possession of a cement-degrading bacterial culture. In order to promote biocorrosion of a concrete surface, a complex nutrient broth must be available to the *Thiobacillus* spp., as is found for example in sewer pipes and wastewater.

Other Materials

Fiberglass, or fiber-reinforced polymeric composites, is also susceptible to biodegradation by sulfate-reducing bacteria and hydrogen-producing bacteria (Wagner et al., 1996). Loss of strength on fiber surface, separation of fiber layers, and erosion of bonding resin are among the various ways that bacteria can affect the integrity of fiber-reinforced polymeric composites. Most at risk are those materials exposed to seawater, such as ship hulls.

Primarily of interest as environmentally benign polymers, polyhydroxyalkanoates (PHAs) are plastic-like materials that are secreted by bacteria during the fermentation of sugars and lipids and that have the potential to clog filters and convert lubricants into gums or abrasives (Anderson and Dawes, 1990; Luzier, 1992, Poirier et al., 1995). Because PHA characteristics are species particular and must be carefully provided the proper nutrients in order to express the appropriate physical characteristics, for example, rigid stiffness versus rubbery malleability, the bacteria must be carefully selected for application. They also tend to be susceptible to thermodegradation; if the temperature is too hot, the PHAs melt.

LIMITATIONS

The over 85 years that the *Titanic* has remained intact on the floor of the North Atlantic demonstrates the primary limitation to the use of naturally occurring antimaterial microbes by a bioterrorist: the long time required for significant degradation to take place. The estimated rate of localized corrosion of a steel container by sulfate-reducing bacteria is 1.8 mm over one thousand years (Little and Wagner, 1996). Bioremediation—the safe and beneficial use of antimaterial biological agents—commonly attempts to accelerate such rates by modifying the target material through aeration or deliberate nutrient feeding, usually in the form of fertilizers. This need to feed the bacteria was demonstrated during the cleanup operation following the *Exxon Valdez* disaster, when even under assisted conditions the seeded and supplemented biodegradation of the oil spilled into Prince William Sound required a three to five day lag period for bacterial growth, and only reached significant levels after twenty plus days (Atlas, 1995).

The formation of biofilms responsible for microbiologically influenced corrosion requires an aqueous environment and supplemental nutrients. Thus, metals in warm climates that are in contact with soil, seawater, or lubricants lacking antimicrobial additives are the most susceptible targets for an antimetal bioagent.

Nonetheless, even in perfect conditions, corrosion rates are on the order of millimeters per year. Additionally, most sulfate-reducing bacteria are extremely difficult to grow in vitro because of the requirements for anaerobic conditions, unique growth media, and interdependence with other microorganisms (Santegoeds et al., 1998). A bioterrorist would therefore have to have access to a fairly sophisticated laboratory and have substantial experience culturing microorganisms in order to possess the technical ability to exploit them for destructive purposes.

While genetic engineering hypothetically can provide a means to accelerate the kinetics of antimicrobial agents, there has not been a great deal of success to show for efforts directed toward this end (Zwillich, 2000). Elevated activity levels have not been reported despite the successful cloning of antimaterial proteins in unrelated bacteria (Howard, 2002; Vega et al., 1999; Nakajima-Kambe et al., 1997) and the more recent success in increasing the resistance of antimaterial agents to other toxic chemicals (e.g., radioactive waste remediator *Deinococcus radiodurans* was modified to withstand ionic mercury (Hg^{II})) (Brim et al., 2000). Other recombinant DNA work has focused on identifying and removing the portion of the genome in a particular bacterium coding for production of biocorrosive enzymes (Dubiel et al., 2002).

Another hindrance to the use of antimaterial agents is the necessity for a specific environment. For example, many of the aqueous iron-degrading bacteria require specific pH ranges to grow (Cullimore and McCann, 1978). Most iron-utilizing bacteria grow preferably in reduced oxygen atmospheres, for example, 6 percent O_2 for *Gallionella* sp., and for some bacteria the presence of any oxygen serves as an inhibitor to growth, for example *Gallionella ferruginca*. Thus, if a potential terrorist were to try to use an antiiron agent, it would be critical that the proper species or, in some instances, specific strain be selected.

CONCLUSION

The use of an antimaterial bioterrorist agent requires sophisticated knowledge of microbiology. While research programs involving antimaterial agents have been conducted by the U.S. Army and Navy, details remain classified (Knoth, 1994; Naval Studies Board, 2003). For most applications, a biological antimaterial agent also requires an intervening human for contamination of the target, thereby reinforcing the critical need for intelligence and security. If such stumbling blocks were overcome by a prospective bioterrorist, however, the consequences could be severe: A 2001 report jointly commissioned by the U.S. Federal Highway Administration and NACE International—the Corrosion Society determined direct corrosion costs due to normal chemical and biological effects to be $276 billion, or 3.1 percent of the U.S. GDP (Federal Highway Administration, 2001).

REFERENCES

Anderson, A.J. and Dawes, E.A., *Microbiol. Rev.*, **54**, 450–472 (1990).

Atlas, R.M., *Int. Biodeterior. Biodegrad.*, **35**, 317–327 (1995).

Blake, R.C., Norton, W.N., and Howard, G.T., *Int. Biodeterior. Biodegrad.*, **42**, 63–73 (1998).

Brim, H., McFarlan, S.C., Frederickson, J.K., Minton, K.W., Zhai, M., Wackett, L.P., and Daly, M.J., *Nat. Biotechnol.*, **18**, 85–90 (2000).

Cullimore, D.R. and McCann, A.E., "The Identification, Cultivation and Control of Iron Bacteria in Ground Water," in F.A. Skinner and M.J. Shewan, Eds., *Aquatic Microbiology*, Academic Press, 219–261, 1978.

Dubiel, M., Hsu, C.H., Chien, C.C., Mansfeld, F., and Newman, D.K., *Appl. Environ. Microbiol.*, **68**, 1440–1445 (2002).

Federal Highway Administration, Corrosion Costs and Preventive Strategies in the United States, Report FHWA-RD-01-156, Office of Infrastructure Research and Development, Washington, DC, September 2001.

Howard, G.T., *Int. Biodeterior. Biodegrad.*, **49**, 245–252 (2002).

Iverson, W.P., *Int. Biodeterior. Biodegrad.*, **47**, 63–70 (2001).

Kay, M.J., McCabe, R.W., and Morton, L.H.G., *Int. Biodeterior. Biodegrad.*, **31**, 209–225 (1993).

Knoth, A. *Int. Defense Rev.*, **7**, 33–39 (1994).

Little, B. and Wagner, P., *Can. J. Microbiol.*, **42**, 367–374 (1996).

Little, B., Staehle, R., and Davis, R., *Int. Biodeterior. Biodegrad.*, **47**, 71–77 (2001).

Lopes, P.T.C. and Gaylarde, C., *Int. Biodeterior. Biodegrad.*, **37**, 37–40 (1996).

Luzier, W.D., *Proc. Natl. Acad. Sci. U.S.A.*, **89**, 839–842 (1992).

Nakajima-Kambe, T., Onuma, F., Akutsu, Y., and Nakahara, T., *J. Ferment. Bioeng.*, **83**, 456–460 (1997).

Naval Studies Board, *An Assessment of Non-Lethal Weapons Science and Technology*, National Academy Press, Washington, DC, 2003, p. 64.

Nica, D., Davis, J.L., Kirby, L., Zuo, G., and Roberts, D.J., *Int. Biodeterior. Biodegrad.*, **46**, 61–68 (2000).

Pendrys, J.P., *Appl. Environ. Microbiol.*, **55**, 1357–1362 (1989).

Poirier, Y., Nawrath, C., and Sommerville, C., *Biotechnology*, **13**, 142–150 (1995).

Santegoeds, C.M., Ferderlman, T.G., Muyzer, G., and De Beer, D., *Appl. Environ. Microbiol.*, **64**, 3731–3739 (1998).

Thomassin-Lacroix, E.J.M., Yu, Z., Eriksson, M., Reimer, K.J., and Mohn, W.W., *Can. J. Microbiol.*, **47**, 1107–1115 (2001).

Thomassin-Lacroix, E.J.M., Yu, Z., Eriksson, M., Reimer, K.J., and Mohn, W.W., *Appl. Microbiol. Biotechnol.*, **59**, 551–556 (2002).

Vega, R., Main, T., and Howard, G.T., *Int. Biodeterior. Biodegrad.*, **43**, 49–55 (1999).

Wagner, P.A., Little, B.J., Hart, K.R., and Ray, R.I., *Int. Biodeterior. Biodegrad.*, **38**, 125–132 (1996).

Zwillich, T. *Science*, **289**, 2266–2267 (2000).

FURTHER READING

Borenstein, S.W., *Microbiologically Influenced Corrosion Handbook*, Cambridge Ed., Woodhead, London, 1994.

ARMED ISLAMIC GROUP: A CASE STUDY

MONTEREY WMD-TERRORISM DATABASE STAFF

BACKGROUND

The Armed Islamic Group (GIA) is an Algerian Islamic fundamentalist group with antigovernment, anti-intellectual, antisecular, and anti-West policies. Its stated goal is to establish an Islamic state in Algeria. The precise numbers of the group are unknown, probably fewer than 100.

The GIA began its violent activity in 1992 after Algiers voided the victory of the Islamic Salvation Front—the largest Islamic opposition party—in the first round of legislative elections in December 1991. Since 1992, the group has conducted a terrorist campaign of civilian massacres, sometimes wiping out entire villages in its area of operation (though over time, the group's dwindling numbers have caused a decrease in the number of attacks). Since announcing its campaign against foreigners living in Algeria in 1993, the GIA has killed more than 100 expatriate men and women, mostly Europeans. The group routinely carries out assassinations and bombings, including car bombs, and has a reputation for slitting the throats of kidnap victims. The GIA also hijacked an Air France flight to Algiers in December 1994. In 2002, a French court sentenced two GIA members to life in prison for conducting a series of 1995 bombings in France.

CASE DETAILS

On 5 March 1998, the Belgian police raided two GIA safe houses in the Ixelles district of Brussels, Belgium. Initially, it was reported that police found isolation boxes containing botulinum toxin and an address book containing the location of a laboratory in Schaarbeek, Brussels. Three months later, on 26 May 1998, safe houses in Belgium and throughout Europe were searched as well, along with the laboratory listed in the GIA address book. On 27 May 1998, Brussels public prosecutor and the Gendarmerie-Royale denied that botulinum toxin had been found during the 5 March 1998 raids and emphasized that the medical laboratory in Schaarbeek was searched as a precaution, not because the laboratory was suspected of supplying GIA with botulinum toxin. Ultimately, it was revealed that only instructions for producing botulinum toxin, and not the toxin or its causative microorganism, were found during the March raids (the searches also uncovered 500 g of Securex, a dynamite derivative).

As a result of the 5 March 1998 raids, the French police criticized Belgian authorities for committing operational blunders that prevented the capture of several GIA members. Belgian police had planned to raid several GIA safe houses on 5 March simultaneously, but personnel shortages had prevented this from happening. Regardless, some key GIA members were arrested, including Farid Marouk, Ali Hajjaji, Omar Marouf, and Bakhti Raho Moussa.

REFERENCES

Center for Nonproliferation Studies, Monterey WMD-Terrorism Database, 1998–2003.

Herbert, S., Arrests Sparked by Fears of World Cup Terrorism, *Calgary Herald*, May 27, 1998.

La Libre Belgique, *Belgium: Reports on Botulism Find in GIA Laboratory Said False*, May 27, 1998, available from FBIS, document identification number fbtot05271998000868.

La Une Radio Network (Brussels), *Belgium: Police Find Biological Cultures Linked to Botulism*, May 26, 1998, available from FBIS, document identification number fbtot05261998000815.

Story Behind a Belgium Police Raid, *Intelligence Newsletter*, April 23, 1998.

United States Department of State, *Patterns of Global Terrorism, 2002*, April 2003.

WEB RESOURCES

Armed Islamic Group Under Djamel Zitouni (Algeria) Role Profile, Role Profiles Archive, March 31, 1996, Internet, available from www.link.lanic.utexas.edu, accessed on 8/3/98.

ARMY OF GOD

DALLAS A. BLANCHARD

BACKGROUND

There is little agreement about the true nature of the Army of God. Is it simply a term of convenience used by unconnected individuals, a "state of mind" shared by members of an activist milieu (Mark Potok, in Nifong, 1998), a loose confederation of occasionally interacting individuals, or a formal organization (National Abortion Federation, Undated)?

The term "Army of God" has a long history, and it has been applied to or adopted by widely diverse groups. Although traditional Judeo-Christian scriptures do not use the term in a literal sense, the concept is there in the warfare imagery of the Battle of Armageddon, the notion of the Lord of Hosts, and other passages. It is also implied in the Manichaean doctrine of the warfare between God and Satan, a dualism found in contemporary fundamentalisms.

Paul, in Ephesians 6:1017, mentions that the people of God are engaged in spiritual warfare and should don the armor of God. That tradition was maintained and reappeared as a theme during the Crusades, beginning with the First Crusade in 1095. Christian hymnody followed centuries later with "Onward, Christian soldiers."

Other martial imagery in the Christian milieu can be seen even in the Salvation Army, complete with its military ranks and uniforms, as well as to the reference to youthful Mormon missionaries as "God's Army" in a 2000 movie with that title.

An Internet search will soon yield a large number of church web pages, sermons, and church programs referring to the Church as the Army of God. Such usages generally are comparable to the concept of the "invisible" Church of the "People of God," which exists beyond denomination and does not necessarily include only church members but rather all those considered faithful.

However, similar terms are also widely employed in other religions as well, especially by Islamic and Jewish fundamentalists, most frequently when they seek to justify the use of violence against those perceived as "enemies" of God. (Some incorrectly translate "Hizb'allah" as Army of God, although most do not.)

Since 1982, the use of the appellation in North America has been almost completely limited to antiabortion activists, especially by those using violence. This is also true in the public mind, especially since the USA

Network (Levin and Pinkerson, 2001) television broadcast of "Soldiers in the Army of God," a made-for-TV movie featuring Neal Horsley and other prominent persons assumed to be members of the Army of God.

The first usage of the name "Army of God" in abortion-related violence was by Don Benny Anderson in the 1982 kidnapping of Dr. Hector Zevallos and his wife and held them captive for more than a week under threat of death unless Dr. Zevallos agreed to stop performing abortions. During the captivity, Anderson posted letters to federal agents in filling station restrooms and signed them "Army of God."

Later, Michael Bray, Thomas Spinks, and Kenneth Shields, who in 1984 bombed eight or more clinics in the Washington, DC, area, attributed at least one of those bombings to the Army of God. Also in 1984, Supreme Court Justice Harry Blackmun received a death threat letter from someone attributing it to the Army of God.

In 1996, Eric Rudolph reportedly exploded a bomb at the Olympic Park in Atlanta, killing a woman; then, in 1997, he allegedly bombed a gay bar and an abortion clinic in the Atlanta area; and finally, in 1998, he was charged with the bombing of a Birmingham, Alabama, clinic, which killed an off-duty police officer and maimed a clinic nurse (Department of Justice Press Release, 1998). At the time of this writing, Rudolph is in custody awaiting trial for the Birmingham incident, in which the bomb was reportedly set off by remote control as the officer leaned over to examine it. The evidence against him includes letters sent to the *Atlanta Journal-Constitution* in the name of the Army of God. (Mapping of incidents against abortion providers currently attributed by offenders to the Army of God will be found in Figure 3.)

In 2001, Clayton Waagner sent more than 550 packages threatening contamination with *Bacillus anthracis* to abortion clinics in the Midwest and East, and signed most of his threats "the Army of God." He also threatened to kill 42 clinic workers on the Army of God website. While the testing of each of the packages eventually revealed that they did not contain *B. anthracis* spores, the substance enclosed, *Bacillus thuringiensis*, did yield an initial positive result due to its close resemblance to *B. anthracis*, necessitating more extensive security precautions and testing.

Encyclopedia of Bioterrorism Defense,   Edited by Richard F. Pilch and Raymond A. Zilinskas
ISBN 0-471-46717-0   Copyright © 2005 Wiley-Liss

Figure 1. Atlanta bomb squad division bumper sticker. *Source:* www.christiangallery.com/aogpics.html.

Figure 2. Pensacola hunt club bumper sticker. *Source:* www.christiangallery.com/aogpics.html.

FORMAL ORGANIZATION OR UNIQUE ENTITIES?

All of these aforementioned occurrences could conceivably be isolated, unconnected events. They certainly do not provide evidence of a formal organization, nor even of a loose confederacy of interacting persons. There is no evidence of any interaction at all between Anderson, the Bray "gang," Rudolph, and Waagner. The only indicators of possible interactions are Donald Spitz's website, http://www.armyofgod.com (Spitz, undated), and the Army of God Manual (see Figure 4), which was first discovered buried in the backyard of Rachelle "Shelley" Shannon. In 1993, Shannon was convicted for the attempted murder Dr. George Tiller in Wichita.

The Website

Spitz, who makes his headquarters in the Norfolk, VA, area, signed Paul Hill's "Justifiable Homicide" statement; was a central figure in the American Coalition of Life Activists, which supported justifiable homicide; and continues to post a wide assortment of inflammatory items on his website. John Salvi, who murdered clinic workers

in Boston and subsequently was arrested after firing shots at a clinic in Norfolk, had a piece of paper with Spitz's telephone number on it in his possession when arrested.

The Manual

The manual suggests more definitively the existence of an organization. Self-reportedly in its third edition, this manual is a blueprint for violence that was based in part on both the CIA's Nicaraguan "Contra" Manual and *The Anarchist's Cookbook*, with additional materials such as instructions for the manufacture and use of butyric acid (Bustilloz, Undated). Butyric acid has been the agent of choice in chemical attacks carried out against abortion clinics. It has an extremely nauseating and pervasive odor, which in even small amounts permeates carpeting, furniture, and walls, thus frequently requiring the almost total remodeling of clinics following release. The substance has been placed in the return air ducts of heating and air conditioning systems during the operating hours of clinics, while abortion procedures are being performed, thereby causing uncontrollable vomiting in patients, physicians, and personnel. According to the National Abortion Federation ("Butyric Acid Attacks, " Undated), there have been more than 100 butyric acid attacks on clinics in the United States and Canada (some of which were committed by Shelley Shannon).

The Manual's introduction dedicates itself to more than 50 antiabortion activists through aliases or anonyms. Indications are that some of those aliases represent James Kopp (Atomic Dog), Shelley Shannon (Shaggy West), John Burt, Paul Hill, John Cavanaugh-O'Keefe (the "father" of clinic blockades), David Trosch (the initiator of the justifiable homicide position), Joan Andrews, John Brockhoeft, Father Norman Weslin (organizer of the Lambs of Christ), Joseph Scheidler, Jayne Bray (Michael Bray's wife), Dennis Malvasi, Loretta Marra, and others.

THE ARMY OF GOD NETWORK

A number of these persons, and other activists interacting with them regularly, have established informal networks through the years, especially while they were placed in jail together in Atlanta during the Democratic National Convention in 1988 and in Buffalo during other massive blockades; at events sponsored and organized by Joseph Scheidler in Chicago (Scheidler has attended the trials and sentencings of a large number of persons convicted of serious crimes against clinics and their personnel, has reportedly visited every person convicted in prison, holds award ceremonies for many of them after their release, and sponsors regular gatherings for activists); and at the White Rose Banquets held each January by Michael Bray to benefit the families of those in prison. (Bray, Spitz, Horsley, and others have regularly solicited donations to provide at least temporary support for jailed perpetrators, including those convicted of murder). Major trials have also become a basis for interactions. For example, at the federal trial for the Pensacola Four in 1985, the judge opened a

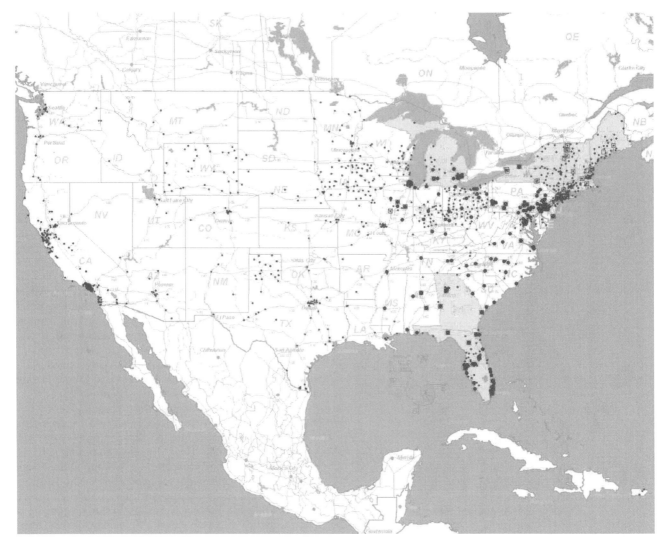

Figure 3. Army of God: currently mapped incidents. *Credit*: Jim Skiles, Counterterrorism Research, University of West Florida.

second, larger courtroom with closed circuit television to accommodate the large number of antiabortion activists who came for the eight-day trial from across the nation.

ASSERTIONS ABOUT THE ARMY OF GOD

A number of unsubstantiated assertions about the Army of God have been made from time to time:

- "Army of God leader Michael Bray..." (National Abortion Federation, Undated). This could be true, but there is no definitive evidence of it. While Bray assumes a leadership role under the rubric of some "Army of God" events, the National Abortion Federation (NAF) statement implies that he is *the* leader and that the Army of God is a formal organization.

- "It has metamorphosized into a more modern kind of hate group that includes anti-government rhetoric,

anti gays and lesbians..." (Jamieson, 1997). There was no metamorphosis. Such issues, and others associated with them, had always been the agenda of these persons. All of them appear to have a Christian Reconstructionist agenda, their goal being a theocratic society.

- "For some time, we have thought a group of people have been acting in concert to terrorize abortion clinics" (Jamieson, 1997). Despite this statement by Eleanor Smeal, and despite clear evidence of small groups interacting in the planning and execution of violence against clinics and their personnel (such as the Bray "gang" in Bowie, the Anderson group in Granite City, IL, the Pensacola Four, and a number of others), the only clear evidence of wider interaction lies in concerted efforts to avoid detection *after* events, as in the case of Kopp's utilization of networks to avoid capture.

- "Best known for its terror campaign against abortion providers, the militant Army of God has lately displayed a virulent animus in recent postings on its website" (Clarkson, 2002). This assertion assumes that Spitz's website represents a group that is formally organized. However, it is unclear whether (1) the group is in fact a formal organization, and whether (2) Spitz's site belongs to the group.

- "Waagner and others like him do not act alone. Now is the time to target the organization that enables Waagner, and men like him, to carry out violent acts against innocent people" (Kieschnick, 2001). This statement assumes that Waagner was interacting with other major players in the violence, which has never been established, and, in fact, has been denied by Waagner himself (Horsley, 2003).
- "Based out of Virginia and headed by Reverend Donald Spitz, The Army of God..." (Zackem, Undated). This statement assumes that there is a formal organization, that it is based in Virginia, and that Spitz is the leader, all of which is speculation.
- "Don Benny Anderson is likely the leader of the Army of God" (Jamieson, 1997). Anderson has been in prison since 1983.

"The Army of God is an evanescent, amorphous, autonomous and spontaneous eruption of individuals who are responding to precisely the dynamic I describe in the introduction of my book posted on the Internet at http://www.christiangallery.com/book.html." (Horsley, 2003) Horsley adds, "... it is a network where the people where the people do not communicate with one another, have never met each other, and are unaware of each other's soldier status." (Horsley, 2003).

CONCLUSIONS

There is no evidence that the anonymous persons anonymed in the Army of God Manual interacted with Anderson, Waagner, Bray, and Rudolph *prior to their offenses*. The overwhelming evidence is that the majority of offenders began seriously interacting only after their offenses. Therefore, it appears likely that the early cases in which acts of violence were identified as the responsibility of the Army of God were attributed as such only in the most generic sense of the term, in the same way as phrases such as the "People of God." The use of this name was probably intended to create an aura of threat, an impression that there was a much larger group behind the action. If this was in fact the intention, it was successful. Successive perpetrators (and potential perpetrators) may soon have discerned the utility of enhancing and exacerbating such perceptions by attributing their actions to the Army of God as the party responsible, suggesting that today's ambivalent and evasive statements by those perceived to be in the Army of God may be intentionally designed to further spread those perceptions. Some have even published bumper stickers to lend credence to the perceptions of an organized group (see Figures 1 and 2).

In addition, it is useful to law enforcement and pro-choice organizations to attribute anti-abortion violence

Figure 4. Army of God manual cover. *Source*: www.courttv. vom/archive/onair/shows/mugshots/indepth/shannon/aog1.html.

to an actual, formally organized group, especially given the national concern about international and national terrorism since September 11, 2003. Giving the name formal organizational status heightens public fear and support for prosecution of anti-abortion violence.

Thus, the designation appears to have evolved through time from (1) an appellation egotistically assumed by individuals or small groups; to (2) associations of acquaintances who assisted one another in avoiding detection or apprehension and occasionally providing resources prior to carrying out individual acts of violence; to (3) a loose confederation of persons frequently interacting and performing in a fashion similar to the "leaderless resistance" concept advocated by certain U.S. militia groups.

In the end, contemporary Americans calling themselves the Army of God have exploited and appropriated to themselves many of the implied and attributed qualities of past usages. This assessment is supported by Neal Horsley (1997 and 2003), an ardent anti-abortion activist. In that case, they have now metamorphosized into a loose confederation of occasionally interacting individuals who now interact more and more frequently on an organized basis, such as through the American Coalition of Life Activists.

REFERENCES

Bustilloz, L., *Re: Army of God?* http://www.skepticfiles.org/misc3/aog-book.htm, n.d., accessed on 6/19/04.

Clarkson, F., Brand New War for the Army of God? *Salon*, July 17, 2002.

Department of Justice Press Release, *Rudolph Charge in Centennial Olympic Park Bombing*, October 14, 1998.

Horsley, Neal, *Exploding the Myth of The Army of God* http://www.christiangallery.com/ExplodingArmyofGodMyth.htm. November 14, 2003, acccessed on June 19, 2004.

Horsley, Neal, *Understanding The Army of God* http://www.christiangallery.com/aog.html.n.d., 1997, acccessed on June 19, 2004.

Jamieson, R., *Crime Tape Removed Near Nightclub*, CNN, February 25, 1997.

Kieschnick, M., Treat the Army of God as a Terrorist Organization, *Act for Change*, December 14, 2001.

Levin, M. and Pinkerson, D. Directors/Producers, *Soldiers in the Army of God*, USA Network, 2001.

National Abortion Federation, *The Facts About Butyric Acid*, 1999.

National Abortion Federation, *The Army of God and Justifiable Homicide*, May 2003.

Nifong, C., Anti-Abortion Violence Defines Army of God, *Christian Science Monitor*, February 4, 1998.

Spitz, D., http://armyofgod.com/, accessed on 6/19/04.

Zackem, A., *Army of God*, http://www.ithaca.edu/buzzsaw/. n.d., accessed on 6/19/04.

ARMY TECHNICAL ESCORT UNIT

JENNIFER RUNYON

INTRODUCTION

The U.S. Army created the Technical Escort Unit (TEU, initially called the Guard and Security Division of the Chemical Warfare Service) on January 20, 1943, to safely transfer chemical munitions. TEU began escorting radioactive material in 1949, and by 1969 had expanded its mission to include a biological component.

ACTIVITIES

TEU is a quick response unit designed to prevent or preempt an attack from occurring. The unit's robust peacetime mission involves working daily with live agents in dangerous areas. TEU's mission has expanded from its founding purpose of escorting materials, an activity on which it currently spends 10 to 15 percent of its time, to providing specialized support to combatant commanders, the homeland, and civil authorities (Fig. 1).

TEU has supported combatant commanders with various chemical and biological operational capabilities during operations such as Desert Shield/Storm in 1990–1991. These capabilities include field verification, sampling (following NATO standard protocol), packaging, escorting, disposing, and remediation of suspected chemical or biological devices and munitions. The unit also possesses limited decontamination capabilities designed for its soldiers, cargo, and equipment. The unit was also deployed to Afghanistan in October 2001, as part of Operation Enduring Freedom and was also involved in Operation Iraqi Freedom in 2003. When TEU deploys, it does so in small teams of 3 to 35 people and not as a company. The teams are tailored to specific missions, and vary by number of people, equipment, and skill sets.

Under its homeland mission, TEU involves predeploys to special events and responds to suspected chemical or biological material. TEU has forward-deployed to numerous events, including Presidential Inaugurations and the Superbowl. Examples of TEU's response efforts are the removal of chemical munitions from the Spring Valley section of Washington, DC, in 1993 and the environmental cleanup of former military sites used for testing or training of chemical and biological materials. The unit also provides the United States Northern Command, the military unit with responsibility for North America, with support similar to what it provides to combatant commanders. After the 9/11 and anthrax terrorist attacks, TEU sampled various suspected contaminated sites around Washington, DC, including the Pentagon and several mail rooms.

TEU has a mission to support civil authorities as well, including first responders and the National Guard Weapons of Mass Destruction Civil Support Teams. In this capacity, TEU provides additional assistance if first responders and subsequently the Civil Support Teams are overwhelmed by a situation.

FUTURE OUTLOOK

The U.S. Army is determined to consolidate its capabilities to respond to chemical, biological, radiological, nuclear, and explosive ordnance incidents. To do so, it created the Guardian Brigade on October 9, 2003 (U.S. Army Research, Development, and Engineering Command, 2003) that then became the CBRNE Command on October 16, 2004 (FORSCOM News Service, 2004). As the Army reorganizes, TEU continues to be a valued asset.

REFERENCES

FORSCOM News Service, *Army Activates CBRNE Command*, http://www.forscom.army.mil/pao/news/1004/CBRNEactivated.htm, October 16, 2004.

Multiple interviews with Cathryn L. Kropp, US Army Technical Escort Unit Public Affairs Officer, 2003.

U.S. Army Research, Development and Engineering Command (Provisional) News Release, *New organization to be recognized during ceremony*, September 29, 2003.

FURTHER READING

Chemical and Biological Arms Control Institute, *Bioterrorism in the United States: Threat, Preparedness and Response*, November 2000.

Figure 1. U.S. Army Technical Escort Unit, SBCCOM: http://www2.sbccom.army.mil/teu/factsheet.htm.

Chemical and Biological Arms Control Institute, *Fighting Bioterrorism: Tracking and Assessing U.S. Government Programs*, forthcoming.

Fein, Geoff S., *Army Sets Up 'One-Stop Shop' for Chem-Bio Response*, National Defense Magazine online, http://www.nationaldefensemagazine.org/article.cfm?Id=1404, April 2004.

Scott, Gourley, Guardian Brigade, *Military Medical Technology* online edition, **8**(5), 2004.

See also MARINE CORPS CHEMICAL AND BIOLOGICAL INCIDENT RESPONSE FORCE.

ASSASSINATIONS

Erika Holey

INTRODUCTION

While generally employing conventional methods (i.e., guns and bombs), a small number of assassinations have been attempted, some successful, using biological or toxin weapons in the past. In particular, intelligence agencies of the former Soviet Union and United States have been accused of incorporating these complicated and unpredictable weapons into Cold War assassination plots, though physical evidence to support such allegations is lacking in many instances, making conclusions difficult to draw.

WORLD WAR II

Historically, there has been speculation that a biological agent played a role in at least one high-profile event during World War II: the assassination of Reinhard Heydrich, Reich Protector of Bohemia-Moravia, by two Czech Patriots, Joseph Gabchik and Jan Kubish. Heydrich was the target of a grenade attack on May 27, 1942, as he was being driven through the streets of Prague, Czechoslovakia. The grenade's blast injured his lung, diaphragm, and spleen, and caused a significant amount of blood loss, yet upon receiving immediate intensive care for his injuries, Heydrich appeared to be well on his way to recovery. A few days after the attack, however, his condition suddenly destabilized, and twenty-four hours later—on June 4, one week after first receiving treatment—Heydrich died. Doctors conducting the autopsy were unable to determine the cause of death.

Several interpretations about the true cause of Heydrich's death have evolved over the years, including the possible delivery of botulinum toxin via fragments of the grenade. One of the few certainties in the case is that Czech Patriots trained and equipped by the British did in fact carry out the attack. The United Kingdom was involved in offensive biological warfare (BW) research at the time, a temporal association that has led some historians to speculate that the bomb used by the Czech assassins could have been contaminated with an agent provided by the British. Evidence to support this theory includes reports that the bomb was modified in a way that would have allowed its contents to be contaminated before use, and a comment made by a respected British scientist suggesting that he had played a part in the assassination. Nevertheless, the exact cause of Heyrich's death remains unresolved.

THE CIA

The U.S. Central Intelligence Agency (CIA) has been accused of plotting to use unconventional methods in assassinations, namely in the targeting of prime minister of Congo Patrice Lumumba and Cuban president Fidel Castro.

Patrice Lumumba

In May 1960, Patrice Lumumba was elected the first Prime Minister of newly independent Congo. After Congo formally separated from Belgium in the ensuing weeks, however, public services faltered and social and political unrest broke out. Lumumba asked the Soviet Union for aid in resolving the situation, arousing fears in the West that a new breeding ground for communism was taking shape. To counter this development, CIA scientific advisor Sid Gottlieb was dispatched to Leopoldville, the capital of the Congo, with orders to assassinate Lumumba. Gottlieb brought with him a container of botulinum toxin for this purpose, but by the time Gottlieb arrived in the Congo, Lumumba had already been taken into protective custody by UN peacekeepers and was being held incommunicado. After some days of waiting, Gottlieb aborted the mission, discarded the toxin into the Congo river, and returned to the United States. Eventually Lumumba left UN custody to join his supporters in Stanleyville, but was hunted down by a group led by Moises Tshombe, the secessionist leader of Katanga, and was jailed for some months. On January 17, 1961, Lumumba was delivered to the Katanga secessionist regime, which immediately executed him and two of his supporters.

Fidel Castro

In the 1960s, the CIA developed several creative plans to remove Cuban leader Fidel Castro from power. Some of these plots were meant to discredit Castro through nonlethal means, for example by exposing him to a hallucinogen prior to a public speaking engagement and by using thallium salts to make his beard fall out (which U.S. officials associated with Castro's power, hence the nickname "The Beard"). Lethal schemes were contrived as well, such as providing Castro with cigars and food laced with botulinum toxin and a scuba diving suit contaminated with *Mycobacterium tuberculosis*. These particular plans never made it beyond the initial planning stages.

Encyclopedia of Bioterrorism Defense,   Edited by Richard F. Pilch and Raymond A. Zilinskas
ISBN 0-471-46717-0   Copyright © 2005 Wiley-Liss

GEORGI MARKOV

The most widely known example of an assassination employing a BW agent is the case of Georgi Markov, a Bulgarian dissident who was injected with a steel pellet containing the toxin ricin while awaiting a bus on central London's Waterloo Bridge. Markov had defected to the West in 1969 and had remained critical of the Bulgarian communist regime, in particular the leadership of Todor Zhivkov, while working as an author and broadcaster for the BBC. His broadcasts were thought to have been a source of inspiration for the Bulgarian dissident movement.

On September 7, 1978, Markov was grazed by the tip of a passerby's umbrella. Unbeknown to him, the umbrella contained a pneumatic device that had covertly discharged the loaded pellet into his leg, and though he felt pain almost immediately he had no reason to suspect foul play, and thus the encounter was disregarded as unimportant for a time. Within 36 hours, however, Markov was admitted to the hospital with symptoms of fever, swollen lymph nodes, and inflammation near the area of injection. He died two days later, leaving physicians and staff at the hospital confused as to what had caused his sudden, unexpected end. The question was finally answered when dissection of Markov's affected leg at autopsy revealed the responsible pellet, made of a platinum–iridium alloy and indented on two sides with 0.016-in diameter pits that had presumably contained the toxin prior to injection.

As medical investigators were still struggling to identify Markov's cause of death, they learned of another prominent Bulgarian dissident who had experienced circumstances similar to those of Markov. Vladimir Kostov, a broadcaster for Radio Free Europe, noticed a sharp pain in his back as he exited a Paris metro station 10 days before the attack on Markov. When he turned around, Kostov saw a man rushing away from him, umbrella in hand. He developed a fever soon afterward, which resolved over the next few days. On an X ray taken of his chest, a steel pellet, also flanked on either side by a small hole, was identified in the subcutaneous tissue of his back. The pits were plugged with wax designed to melt at body temperature, but which had remained intact because the pellet had not been inserted deeply enough into Kostov's back to fully reach body temperature. Thus, only a small portion of the pellet's toxin payload had been released into the blood stream, a quantity that was insufficient to kill Kostov.

The British opened an investigation into the Markov case in October 1990, but it remains open to this day. Despite the lack of official attribution, most analysts believe that Markov (and Kostov) was attacked by the Bulgarian intelligence service with technical support from the KGB. Former KGB General Oleg Kalugin has publicly accused the former Bulgarian Minister of the Interior, Dimitur Stoyanov, of requesting Soviet assistance in the assassination of Markov, and Kalugin has also named Vladimir Todorov, former Bulgarian intelligence chief, as being the primary organizer of the crime. Both Stoyanov and Todorov have denied these accusations. Regardless, in 1992, Todorov was sentenced to 16 months in jail for destroying files relating to Markov's death.

REFERENCES

Bulgarian News Agency, Former Bulgarian Interior Minister denies approaching KGB over Georgi Markov, March 29, 1991.

Bulgarian News Agency, KGB General Blames Todor Zhivkov for Georgi Markov, March 5, 1991.

Davis, R., *Surgery, Gynecol. Obstet.* 304–317.

Franz, D. and Jaax, N., "Ricin Toxin," in R. Zajtchuk, Eds., *Textbook of Military Medicine: Medical Aspects of Chemical and Biological Warfare*, Office of the Surgeon General, Department of the Army, USA, 1971.

Franz, D. and Middlebrook, J., "Botulinum Toxin," in R. Zajtchuk, Ed., *Textbook of Military Medicine: Medical Aspects of Chemical and Biological Warfare*, Office of the Surgeon General, Department of the Army, USA.

Gup, T., The Coldest, *Washington Post*, December 16, 2001, p. W09.

Regis, E., *The Biology of Doom: The History of America's Secret Germ Warfare Project*, Henry Holt and Company, New York, 1999.

AUM SHINRIKYO AND THE ALEPH

MASAAKI SUGISHIMA

Article Contents

INTRODUCTION

The Aum Shinrikyo is perhaps best known for the so-called Tokyo sarin incident in which the deadly nerve agent sarin was released in the Tokyo subway system, killing 12 people and injuring over 1000. But this was not the cult's only attempt to indiscriminately cause a large number of deaths; Aum Shinrikyo also several times attempted to use biological weapons against the Japanese population and U.S. servicemen.

This section describes Aum Shinrikyo's acquisition and use of biological weapons. First, a brief historical background of the cult is provided, and then the efforts involved in developing each bioliogical warfare agent are described. Finally, the cult's motives for using biological weapons are discussed and lessons that can be learned from their activities are analyzed.

BACKGROUND

In early 1984, Shoko Asahara (born Chizuo Matsumoto) established the religious cult *Aum Shinsen No Kai*, a name he changed to *Aum Shinrikyo* in 1987 (Fig. 1). The doctrine of the Aum was largely based on yoga and Buddhism. Followers were asked to worship Asahara as the ultimate existence and devote themselves entirely to realizing his will.

The Aum obtained religious corporation status from the Tokyo Metropolitan Government in August 1989. According to the Public Security Examination Commission, an independent administrative board of the Ministry of Justice, the cult had only 15 followers in 1985 but in excess of 11,000 by 1994. By that time, the number of branches and related facilities established in Japan had reached twenty-four (Kaplan 2000).

During the same time, the Aum managed to establish a close relationship with officials of the Russian government, including the former secretary of the Russian Security Council, Oleg Lobov. Prosecutors later noted that Aum followers had obtained AK-74 submachine guns in Russia and had planned to manufacture replicas in Japan. For this purpose, a Russian scientist was said to have furnished detailed information on each part of the gun. Yoshihiro Inoue, former intelligence minister of the Aum,

Figure 1. Shoko Asahara, founder and leader of the Aum Shinrikyo. *Source*: Shoko Asahara, *Hi Izuru Kuni Wazawai Chikashi (Disaster is approaching toward the country of the rising sun)*, Aum Publication 1995.

testified at the Tokyo District Court in 1997 that the cult had received blueprints for its sarin production plant from a Russian source and had given Lobov about 10 million yen the next year as payment. The Aum also purchased a Mil-17 helicopter from Russia in 1994 to disseminate chemical and biological agents.

Encyclopedia of Bioterrorism Defense, Edited by Richard F. Pilch and Raymond A. Zilinskas
ISBN 0-471-46717-0 Copyright © 2005 Wiley-Liss

The Aum is believed to have sought outside help in developing its biological weapons. Thus, there are reports that Aum followers identified and approached at least one Russian scientist with specialized knowledge in the field; Dr. Anatoly Vorobyev, a former deputy director of Biopreparat, the ostensibly civilian organization with responsibility for developing Soviet biological weapons. Whether anything meaningful came out of this action is not known.

After the Tokyo sarin incident in March 1995, Asahara and most of his senior followers were arrested and prosecuted for murder, kidnapping, theft, and other various crimes under relevant provisions of Japanese criminal law. Further, the Japanese government requested the Public Security Examination Commission to issue an order for the disbandment of the Aum in accordance with procedures set up by the Subversive Activities Prevention Law of July 1952. In January 1997, however, the commission turned down this request as it concluded that the Aum had become too weak organizationally to carry out subversive acts in future. Not being satisfied with this action, the Japanese government enacted the law on the Regulation of the Group Committing the Act of Indiscriminate Murder (the so-called Anti-Aum Law) in December 1999, and the commission instructed the Public Security Investigation Agency to put the cult under close watch. The law authorized the agency and the police department both to conduct on-site inspections of Aum (later Aleph) facilities whenever it was deemed necessary and obligated Aum to provide authorities with any information on demand.

In January 2000, the Aum changed its name to the *Aleph* (the Hebrew letter "A"), and Fumihiro Joyu, the Public Relations Minister of the Aum, took over management operations (he had been released from jail the previous year). In May 2001, the Aleph reported to the Public Security Investigation Agency that it had 1141 followers and 10 facilities in Japan. However, the agency and other organizations have estimated these figures to be higher: 1600 followers in Japan and 300 in Russia, with 26 facilities in Japan and two in Russia.

THE AUM'S ATTEMPTS TO DEVELOP BIOLOGICAL WEAPONS

Clostridium botulinum

The Aum began to develop its biological weapons after its across-the-board defeat in the Japanese Diet election campaign of February 1990. The Aum's political party, *Shinrito* (the Supreme Truth Party), had been formed specifically for the campaign. After all of its candidates—including Asahara himself—were defeated, Asahara ordered his senior followers to cultivate a sample of *Clostridium botulinum* and dispatched a team to Hokkaido (the northern island of Japan) to collect samples from soil believed to contain this organism. Beginning in March 1990, work began in the Aum laboratory to isolate and cultivate *C. botulinum*. Asahara intended to extract botulinum toxin from the culture and, possibly inspired by the balloon bomb operation of the Imperial Japanese Army during World War II, disseminate it from

balloons to inflict indiscriminate casualties throughout Japan. The Aum also purchased two small vessels in order to conduct dissemination operations from the sea. A *C. botulinum* production plant was built at the Aum's compound in Kamiku-Isshiki village, at the foot of Mt. Fuji in the Yamanashi prefecture. The first attack was planned for mid-April 1990, but the Aum's scientists could not overcome the technical hurdles associated with isolating and cultivating the microorganism, so the necessary preparations for the operation were not completed by that date.

Next, the Aum made plans to disseminate botulinum toxin from vehicles. Followers constructed a customized mobile spraying device for this operation and targeted the U.S. naval base in Yokosuka, the Imperial Palace in Tokyo, and the headquarters of the Soka Gakkai Institute (a rival religious group), among other locations. The operation was conducted in May 1990. According to the Judgment on Murder and Other Cases of Minister of Home Affairs Tomomitsu Niimi, issued on June 26, 2002, by the Tokyo District Court, Aum followers disseminated "*Botulinus kin yo no mono*" (*C. botulinum*-like substance—presumably the liquid solution that Aum followers believed to contain botulinum toxin) from vehicles, but to no effect.

In July 1990, followers attempted to contaminate a water purification plant located in a certain mountain with this same substance. The police intercepted the operation, however, and confiscated the case containing the substance. Later, court proceedings did not make clear whether *C. botulinum* was successfully cultivated, but the Public Security Examination Commission held in 1997 that the Aum had failed to isolate a toxigenic strain of the pathogen. In addition, it was later learned that an Aum medical doctor had informed Niimi that the cult was unsuccessful in cultivating the pathogen.

After a series of failed plots, the Aum terminated operation of its *C. botulinum* production plant and dismantled it. The cult did not give up on the substance entirely, however. The Tokyo District Court pointed out in the above judgment that in November 1993, the Aum again tried to disseminate the "*Clostridium botulinum*-like substance" together with saline around the headquarters of the Soka Gakkai Institute. Moreover, immediately prior to the March 1995 sarin attack in Tokyo, followers placed three portable disseminators contained within separate briefcases in the Kasumigaseki subway station in Tokyo. A substance believed by the perpetrators to be botulinum toxin was released by a timing mechanism from each, but no harm came from the attack. Though it has been reported that an Aum member in a crisis of conscience had secretly filled the disseminators with water, Sei-ichi Endo, the microbiologist responsible for the development of the Aum's biological weapons admitted at the court in January 2002 that, after all, botulinum toxin had never been produced successfully (Fig. 2).

Bacillus anthracis

Since the series of plots utilizing *C. botulinum* was not successful, sometime in 1992, Asahara asked Endo about other potential biological warfare agents. Endo suggested *Bacillus anthracis*, and Asahara ordered him to acquire a

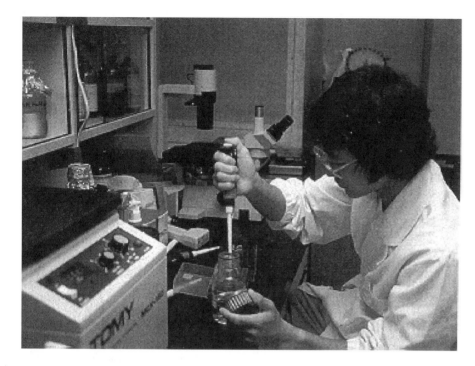

Figure 2. Sei-ichi Endo, the microbiologist responsible for the development of the Aum's biological weapons. *Source*: Shoko Asahara, *Hi Izuru Kuni Wazawai Chikashi (Disaster is approaching toward the country of the rising sun)*, Aum publication 1995.

seed stock of the pathogen. Endo subsequently obtained a vaccine strain of *B. anthracis* from another follower and cultivated it at the New Tokyo General Headquarters located in Kameido, Koto Ward. Asahara chose the May 9, 1993, wedding parade of the Crown Prince Hironomiya as the initial target, but followers could not finish the necessary preparations in time. Then, in late May of the same year, Asahara ordered the dissemination of *B. anthracis* from the Kameido Headquarters building into the open environment. Fumihiro Joyu planned the operation. The Aum created two disseminators for this task based on a round-shaped cooling tower widely available on the open market. The device, dubbed the *Water Mach*, was set up on the rooftop of the building, and Asahara himself pushed the starter switch for the first dissemination.

Between June 29 and July 2, 1993, Kameido residents complained of a foul odor and, on July 1, mist generated by the device was witnessed as well. Having been notified of the problem by a local resident, the Koto Ward Office requested the Aum for permission to inspect the building. The Aum agreed, but removed all equipment related to the dissemination from the building prior to the inspection. Much later, in 1999, the National Institute of Infectious Disease, with the help of the U.S. Centers for Disease Control and Prevention (CDC), analyzed a jelly-like substance that had been collected around the building during the Koto Ward Office inspection. The sample tested positive for *B. anthracis*, but on further analysis it was determined to be the avirulent Sterne strain (the Sterne strain is commonly used as a component of anthrax vaccine). To date, this is the only case of the Aum's use of biological weapons that has been confirmed by hard evidence (Keim et al., 2001; Takahashi et al., 2004).

The Aum continued to cultivate *B. anthracis* in their Kamiku-isshiki compound, and sprayed slurry of the organism from three customized vehicles in the summer of *1993*. Again, Joyu supervised the operation (as an aside, court proceedings later revealed that the slurry had splashed on Joyu while he was checking the spraying device, with no ill effect save a bad odor). The Kangawa Prefectural Office and the Imperial Palace were targeted in this operation. However, the nozzles of the spraying devices clogged during the attack and the operation therefore failed.

Coxiella burnetii

The Aum reportedly also attempted to obtain a culture of *Coxiella burnetii*, the causative pathogen of Q fever. Followers initially tried to steal research materials on Q fever from a laboratory at Gifu University but failed. Later, after the sarin incident, Asahara sent a videotaped message to the Japanese media that he and his followers had been attacked by biological weapons, and specifically had suffered from Q fever. Reportedly, in early January of 1995, the Aum's Treatment Minister (medical doctor) Ikuo Hayashi found that an unknown infectious disease was spreading among followers living in the Aum's Kamiku-isshiki compound. Soon afterwards, Endo told Hayashi that the symptoms were identical to those of Q fever as described in a medical journal. Endo tested samples with a diagnostic kit imported from Australia and found that almost half of the samples were positive. Hayashi asked the Shizuoka Institute of Environment and Hygiene to analyze the same blood samples (Shizuoka prefecture is adjacent to Yamanashi prefecture where the Aum's compound was located), but the test results were uniformly negative. Asahara and Endo told Hayashi that they believed their own test result and that the institute might have falsified its findings. Recently, however, it has become clear that Q fever is endemic in Japan; therefore, even if the strange disease had in fact been Q fever, it may well have been of natural origin (Hayashi later thought that the respiratory syndrome observed among followers might alternatively have been caused by the Aum's sarin). Neither prosecutors nor the court held that the Aum ever cultivated *C. burnetii*.

Ebola Virus

The Aum is said to have dispatched a medical mission to Zaire in 1992 to obtain blood samples of patients suffering from Ebola fever. Further, in the Aum's publications, Endo and other followers discussed the use of hemorrhagic fever viruses as potential biological weapons agents and analyzed defensive measures against them. There is, however, no evidence that the Aum was successful in obtaining such viruses.

MOTIVATIONAL FACTORS FOR BIOLOGICAL WEAPONS USE

Prosecutors and the Public Security Examination Commission pointed out that the defeat in the election campaign of 1990 was a turning point for the Aum. Asahara told his senior followers that the result of the campaign clearly showed that it was not possible to save Japan by legitimate means. Accordingly, Asahara came to rely on the doctrine of *Tantra Vajrayana* in Tibetan Buddhism to justify destruction and murder as a means of salvation. In Asahara's view, killing people was a righteous act to keep them from ending up in hell, and he thus asked his followers to join him in "saving" individuals in this way. In addition, Asahara increasingly made mention of Armageddon and a looming third world war in his speeches, prophecies that the Public Security Examination Commission later concluded had been made in order to cause social disturbance.

Asahara testified in 1999 at his followers' court trial that the cult had not disseminated *C. botulinum* around Tokyo in May 1990, but instead had sprayed solutions containing either *Escherichia coli* or *Staphylococcus* sp. In 2000, Joyu described the operation in Kameido as an exercise to simulate the coming Armageddon predicted by Asahara, noting that at the time Asahara had called the operation a kind of religious exercise. While such an explanation hardly seems credible to many of us, it should be remembered that for Aum followers, Asahara's commands were final and absolute, and were to be obeyed unconditionally—casting doubt on any of his orders was considered a grave sin. Even when followers were assigned tasks by Asahara that seemed strange and were beyond their ability to comprehend, they considered them "breakthrough practice" (*Mahamudra*) and thus willingly performed them. In autumn 2001, the Aleph publicly admitted on its Internet homepage that Aum followers had in fact disseminated a harmless strain of *B. anthracis* from its New Tokyo Headquarters building.

DISCUSSION

In 1997, the Public Security Examination Commission unanimously concluded that there were insufficient grounds to believe that the Aum would repeat its acts of violence and thus could pose a danger to society in the future because almost all of Aum's scientists who had developed and produced biological weapons had been arrested (Table 1), the police department had confiscated all equipment and documents related to bioterrorism, and the Aum's biological laboratories had been completely destroyed. However, the cult was, and is, still viewed with considerable suspicion by the Japanese society, and thus remains under surveillance by the Japanese government.

The Aum' bioterrorism efforts were the first of their kind: The cult clearly attempted to use pathogens and toxins to indiscriminately generate a large number of casualties. Though the Rajneeshees sickened 751 local residents in The Dalles, Oregon, in 1984, the pathogen it employed (*Salmonella typhimurium*) was selected for its incapacitating rather than killing effects and was disseminated by contaminated food. Conversely, the Aum chose deadly pathogens and toxins for its biological weapons. Moreover, the Aum developed its biological weapons with aerosol dissemination in mind, which is the most certain method for causing mass casualties. If the Aum's scientists had possessed the expertise sufficient to acquiring a virulent strain of *B. anthracis* and toxigenic strain of *C. botulinum* and had overcome the technical difficulties related to disseminating aerosols, the Kameido incident and other attempts at aerosol dispersal might well have ended up as tragedies on a monumental scale.

The Aum incident clearly showed the existence of at least two kinds of technical hurdles in producing effective biological weapons. The first one relates to the development of the means of delivery and the second one relates to the acquisition of deadly pathogens or toxins. The Aum could not surmount these hurdles—they failed not only in disseminating *B. anthracis* in aerosolized form but in obtaining a virulent strain. Similarly, the Aum was unable to secure a toxigenic strain of *C. botulinum*, so the attacks where botulinum toxin supposedly was used could not cause harm. These technical hurdles can be expected to be present should terrorist groups of the future attempt to acquire biological weapons (Sugishima 2003).

With regard to the expertise necessary for producing biological weapons, we should bear in mind the rapid advancement of biotechnology, pharmaceutics, and other bioscientific disciplines. Though various opinions exist as to whether such advancement has in fact lowered the technical hurdles mentioned above for would-be bioterrorists, governmental agencies responsible for the counterterrorism should pay attention to dual-use technologies (e.g., peaceful technologies having possible application for the development of biological weapons) and take measures to prevent such malicious usage. In regards to the availability of dangerous pathogens and toxins, at the time of Kameari anthrax incident, the Japanese government was a member of the Australia Group and had adopted laws to regulate the export of dangerous microorganisms; however, no such law existed for their domestic transfer. As demonstrated by the Aum probably having obtained the *B. anthracis* Sterne strain from a

Table 1. List of Senior Aum Shinrikyo Members who Participated in Bioterrorism Plots

Hideo Murai	"Minister of Science and Technology." Right-hand man of Asahara. Supervised almost all activities of the Aum's armament program including the development of weapons of mass destruction. Killed in April 1995.
Sei-ichi Endo	"Minister of Health." The microbiologist who led the Aum's activity related to the development of biological weapons. Tried to weaponize botulinum toxin and *Bacillus anthracis*. Sentenced to death by a local court.
Fumihiro Joyu	Aum's spokesman after the Tokyo sarin incident. Suggested the balloon bomb attack with botulinum toxin to Asahara in 1990. Headed the New Tokyo Headquarters at the time of Kameari anthrax incident. Sentenced to a three-year term of imprisonment for perjury and other crimes. Now the spiritual leader of the Aleph after having served his jail sentence.
Kiyohide Hayakawa	"Minister of Construction." Tried to contact Anatoly Vorobyev in Russia. Sentenced to death by a local court.
Tomomasa Nakagawa	"Director of Household Agency." A medical doctor who taught the toxic property of botulinum toxin to Asahara and participated in a series of bioterrorism plots. Sentenced to death by a local court.
Tomomitsu Niimi	"Minister of Home Affairs." Participated in a series of failed bioterrorism attempts by the Aum. Sentenced to death by a local court.
Yoshihiro Inoue	"Minister of Intelligence." Participated in the cultivation of *Bacillus anthracis* used in the Kameari anthrax incident. Set up portable disseminators designed to release botulinum toxin at Kasumigaseki Station in March 1995. Sentenced to life terms imprisonment by a local court.
Kozo Fujinaga	"Vice Minister of Science and Technology." Designed spraying device mountable on the vehicle that was used to disseminate *Bacillus anthracis* in the summer of 1993. Sentenced to 10-years term of imprisonment at the court of appeal.
Toru Toyoda	Senior member of the "Ministry of Science and Technology." Designed and manipulated the spraying device used in the Kameari anthrax incident. Sentenced to death by a local court.
Ken-ichi Hirose	Senior member of the "Ministry of Science and Technology." Participated in the extraction of botulinum toxin in 1990. Sentenced to death by a local court.
Shigeo Sugimoto	Senior member of the "Ministry of Home Affairs." Drove the vehicle equipped with spraying device to the U.S. Navy base in Yokosuka, and so on in May 1990. Sentenced to life term imprisonment by a local court.

Note: No Aum member, including Asahara, has been prosecuted for bioterrorism-related activities. The senior members listed above (except Murai) were prosecuted for their involvement in the sarin incidents and other crimes. Most of Aum court trials are still in progress at the time of this writing (October 2004).

domestic source, there is a need for nations to institute effective control mechanisms for controlling the transfer of microorganisms domestically as well as internationally.

REFERENCES

Global Proliferation of Weapons of Mass Destruction: A Case Study on the Aum Shinrikyo. Global Proliferation of Weapons of Mass Destruction, Hearings before the Permanent Subcommittee on Investigations of the Committee on Governmental Affairs, United States Senate, 104th Congress, 1st Session, Part 1, October 31 and November 1, 1995.

Kaplan, D.E., "Aum Shinrikyo (1995)," in J.B. Tucker, Ed., *Toxic Terror*, MIT Press, Cambridge, MA, 2000, pp. 207–226.

Keim, P.K., Smith, K.L., Keys, C., Takahashi, H., Kurata, T., and Kaufmann, A., *J. Clin. Microbiol.*, **39**(12), 4566–4567 (2001).

Sugishima, M., "Biocrimes in Japan," *A Comprehensive Study on the Bioterrorism*, Asahi University Legal Research Institute Monograph #6, Asahi University, Mizuho, Gifu, Japan, 2003, pp. 86–109.

Takahashi, H., Keim, P., Kaufmann, A.F., Keys, C., Smith, K.L., Taniguchi, K., et al., *Emerg. Infect. Dis.*, **10**(1), 117–120, (2004).

WEB RESOURCE

The Aleph's homepage (in English), http://english.aleph.to.

B

BAADER-MEINHOF GANG (BAADER-MEINHOF GROUP, ROTE ARMEE FRAKTION)

DAVID MCADAMS

OVERVIEW

The Baader-Meinhof Gang (BMG) was a West German left-wing terrorist group that strove to create a Marxist–Leninist state by using urban guerrilla tactics against what it considered to be the representatives of imperialist exploitation and oppression. The BMG went through several incarnations during the group's active life span. Of particular relevance to this discussion is that the first generation "Baader-Meinhof Gang" spawned what eventually became known as the Red Army Faction (RAF), which included those original BMG members who had not been imprisoned or forced to flee Germany in the 1970s, along with members of other left-wing terrorists groups such as the German Socialist Patient's Collective. Over the course of its existence, the RAF consisted of some 30 dedicated members (including leaders such as Andreas Baader, Ulrike Meinhof, Gundrun Ensslin, and Jan-Carl Raspe) and up to 200 supporters (including such prominent individuals as politicians Horst Mahler and Silke Maier-Witt) (Huffman, 2002).

BACKGROUND

The BMG's origin can be traced to the era of student protests during the 1960s. Several leftist student groups developed during this period, many of which drew inspiration from the 1967 shooting of a protester at a rally against the visiting Shah of Iran. The BMG surfaced after a small group of students led by the charismatic Andreas Baader firebombed two department stores. When Ulrike Meinhof led a public escape that freed Baader from arrest, the BMG name was bestowed upon the group (Pearlstein, 1984). The BMG soon became a notorious but popular bank-robbing outfit, and at one point reportedly enjoyed the support of up to 25 percent of West Germans under the age of thirty (Wright, 1990). This support quickly waned, however, when the group began to incorporate lethal means in its attacks.

In 1970, Baader, Meinhof, and several other members of the group fled West Germany and traveled to the Jordanian desert to train with the Popular Front for the Liberation of Palestine (PFLP). The group lasted two months in the desert before being thrown out because of the strains they placed on their Palestinian trainers. Yet this proved to be a significant "milestone" for terrorism,

since it was purportedly the first time that one terrorist group had ever successfully trained another (Hoffman, 1998).

Upon returning to Germany, the BMG proceeded to rob banks and plant bombs targeting prominent individuals and industries that they viewed as capitalist exploiters. A number of U.S. military installations were targeted as well, for example, the 1972 pipe bombing of a U.S. Army installation in response to the U.S. mining of North Vietnamese harbors (Hoffman, 1998). Leaders Baader, Meinhof, Ensslin, and Raspe were arrested in 1972, after which the group's efforts shifted toward trying to free its imprisoned comrades. These efforts included several murders, kidnappings, hijackings, and hostage seizures. The BMG also capitalized on a report of missing mustard agent, a well-known chemical warfare (CW) agent, by threatening the German government with the missing substance in an attempt to negotiate the release of its leaders (Claridge, 2001). Finally, after numerous attempts failed to bring about their release, the group's leaders apparently committed multiple suicide in Stammheim prison in October 1977 (Huffman, 2001). The terrorist activities of successive generations of RAF members continued until 1998, when the RAF sent a formal communiqué to the Reuters wire service declaring that the group had officially disbanded.

SELECTED CHRONOLOGY

- *June 2, 1967.* Student protester Benno Ohnesorg is shot at a Berlin demonstration against the Shah of Iran, leading many of the members of the BMG to formally meet for the first time and agree to take up arms against "imperialism" (Huffman, 2001).
- *April 2, 1968.* Kaufhaus Schneider Department Store is firebombed by Baader, Ensslin, and others (Huffman, 2001).
- *May 14, 1970.* Baader escapes from a guarded library while researching a book with Meinhof. The media then dubs the group the "Baader-Meinhof Gang" (Huffman, 2001).
- *February 2, 1972.* The group bombs the West Berlin British Yacht Club, killing one (Huffman, 2001).
- *May 11, 1972.* Three pipes bombs placed by group members at the headquarters of the U.S. Army in Frankfurt am Main, killing one U.S. officer and

Encyclopedia of Bioterrorism Defense, Edited by Richard F. Pilch and Raymond A. Zilinskas
ISBN 0-471-46717-0 Copyright © 2005 Wiley-Liss

injuring several others. The bombs were placed to protest the mining of the North Vietnamese harbors by the United States (Hoffman, 1998).

- *May 19, 1972*. RAF members place six bombs in the Springer Press office. Three explode, injuring 17 people (Huffman, 2001).
- *May 24, 1972*. Car bombs placed by the group kill three U.S. soldiers at the Campbell Barracks in Heidelberg (Huffman, 2001).
- *June 1972*. Baader, Raspe, Ensslin, and Meinhof are arrested by German authorities in a string of raids (Huffman, 2001).
- *November 10, 1974*. Günter von Denkmann, president of Germany's Superior Court of Justice, is killed in a botched kidnapping by 2nd of June Movement, a BMG breakaway faction, in response to Holger Meins' death in prison (Huffman, 2001).
- *February–March 1975*. 2nd of June Movement members kidnap Peter Lorenz, the Christian Democratic candidate for mayor of West Berlin, and demand the release of several RAF and 2nd of June members imprisoned for offenses other than murder. Several are released, including Peter Lorenz (Huffman, 2001).
- *April 24, 1975*. Six members of the RAF take over the West German Embassy in Stockholm, seizing hostages and rigging the building to explode. After their demands for the release of the BMG's leaders are not met, the group executes two embassy attachés before the bombs in the basement prematurely detonate, killing two of the terrorists. The rest surrender without incident (Huffman, 2001).
- *May 1975*. Reports begin to circulate that the RAF has stolen mustard agent from a U.S./British military storage facility, and that it is planning to use the gas against several German cities if the RAF leaders were not released. It was later learned that the group had been trying to capitalize on a British news release describing the missing mustard canisters. The canisters were eventually found (Claridge, 2001).
- *May 9, 1976*. Ulrike Meinhof hangs herself in her cell in Stammheim prison (Huffman, 2001).
- *October 1977*. Palestinians hijack a Frankfurt-bound 737 and demand the release of the BMG's leaders. The plane lands in Mogadishu and is stormed by a German GSG-9 team, which kills all but one of the terrorists. After hearing of the failed attempt, Baader, Raspe, and Ensslin commit suicide on 17 October, 1977. Irmgard Müller, another group member who was imprisoned with its leaders, stabs herself multiple times but survives (Huffman, 2001).
- *April 1998*. Members of the RAF send Reuters a formal communiqué announcing its disbandment (Huffman, 2001).

THE BMG AND BIOLOGICAL WEAPONS

In the early 1980s, Germany's *Neue Illustrierte Revue* and the French paper *Le Figaro* reported that in 1980 French and German police had raided an RAF safe house in Paris, where they found an improvised laboratory with flasks of botulinum toxin. Later, a detailed account of this purported makeshift laboratory appeared in *America the Vulnerable* (Douglass and Livingston, 1987):

> The sixth-floor apartment contained typed sheets on bacterial pathology. Marginal notes were identified by graphologists as being the handwriting of Silke Maier-Witt, a medical assistant by profession, terrorist by night. Other items included medical publications dealing with the struggle against bacterial infection…. In the bathroom, the French authorities found a bathtub filled with flasks containing cultures of *Clostridium botulinum*.

However, the Public Prosecutor's Office in Karlsruhe eventually concluded that the account from *Le Figaro* "did not correspond with the facts and drew false conclusions," and further noted that there was "no evidence whatsoever that the members of the 'RAF' had planned or prepared an attack using biological agents" (Taylor and Trevan, 2001). The terrorists purportedly involved in the incident, Maier-Witt in particular, were never charged with any crimes related to the reports (Taylor and Trevan, 2001).

CONCLUSION

In retrospect, the use of mass casualty weapons falls outside the BMG's modus operandi, which focused largely on high-profile West German political and economic targets. RAF members never issued any public statements regarding the acquisition of biological agents, and no verified evidence of biological weapons production by the group exists. While it is possible that the publicity received by the RAF's 1975 report regarding stolen mustard agent encouraged the group to explore (if not produce) biological and chemical weapons due to their obvious blackmail potential, such efforts have never been confirmed.

REFERENCES

Claridge, D., "The Baader-Meinhof Gang 1975," in *Toxic Terror: Assessing Terrorist Use of Chemical and Biological Weapons*, 4th Print, Chapter 6, MIT Press, Cambridge, MA, 2001, pp. 95–106.

Douglass Jr., J. and Livingston, N., *America the Vulnerable: The Threat of Chemical and Biological Warfare*, Lexington Books, Lexington, MA, 1987, p. 29.

Hoffman, B., *Inside Terrorism*, Columbia University Press, New York, NY, 1998, pp. 82.

Huffman, R., This is Baader-Meinhof, Seattle, WA, June 2, 2002, http://www.baader-meinhof.com.

Noe, D., The Baader-Meinhof Gang; Meinhof: Journalist to Terrorist, *The Crime Library*, http://www.crimelibrary.com/terrorists_spies/terrorists/meinhof/1.html?sect=22.

Pearlstein, R.M., *The Mind of a Political Terrorist*, SR Books, Wilmington, Del, 1984.

Taylor, T. and Trevan, T., "The Red Army Faction," in *Toxic Terror: Assessing Terrorist Use of Chemical and Biological Weapons*, MIT Press, Cambridge, MA, 2001, pp. 106–113.

Wright, J., *Terrorist Propaganda—The Red Army Faction and the Provisional IRA, 1968-86*, St. Martin's Press, New York, 1990.

BIOLOGICAL SIMULANTS

Margaret E. Kosal

INTRODUCTION

Simulants are less lethal or nonlethal substitutes for biological warfare (BW) agents. Most simulants are themselves microorganisms, the vast majority being nonpathogenic "nearest neighbor" species that exhibit properties similar to BW agents but are less dangerous. In addition to nonvirulent microbes, other compounds can also serve as simulants, such as isolated proteins or chemical compounds. Simulants have historically been employed in both offensive and defensive biological weapons programs.

BACTERIAL SIMULANTS

Serratia marcescens

The use of BW simulants was first reported in 1933, when covert tests by the Germans using *Serratia marcescens* allegedly were performed in Paris subways and at other French locations (Smart, 1997). *Serratia marcescens* was initially considered to be a harmless organism, a prime qualification in simulant selection, and further was easily grown and manipulated. Perhaps most appealing to testing programs was that the organism was known to produce a characteristic red-pink color that made it easily identifiable when cultured, making the determination of its successful spread simple and straightforward.

Between September 1950 and February 1951, aerosolized *S. marcescens* (SM in U.S. military code) was intentionally released from offshore Navy vessels in the San Francisco Bay and spread over the inland San Francisco area in order to test the effectiveness of novel dispersal methods (Christopher et al., 1997). The simulant was successfully disseminated but later was implicated in the death of one man and the hospitalization of 10 other men and women (Carlton, 2001). Though a dired connection with this particular outbreak has not been determined, *S. marcescens* is now known to be a human pathogen responsible for a significant percentage of nosocomial (i.e., hospital-acquired) infections (Yu, 1979). The bacterium is most commonly observed to cause urinary tract infections, wound infections, and pneumonia, and is highly opportunistic in cystic fibrosis patients. Many *Serratia* spp. are currently of concern to medical authorities due to the increasing incidence of associated disease, their escalating virulence, and their emergent resistance to antibiotics (Choi et al., 2002).

Bacillus globigii

Bacillus globigii (BG), also known as *Bacillus subtilis* var. *niger*, has frequently been used as a surrogate species for *Bacillus anthracis* because of physical similarities between the two organisms. BG is a ubiquitous, naturally occurring, saprophytic (i.e., feeding on decaying matter) bacterium that is commonly recovered from soil, water, air, and decomposing plant material. Under most conditions, it exists in spore form and is not biologically active. BG is not known to be a human pathogen but does produce the proteolytic enzyme subtilisin, which has been implicated in cases of allergic asthma, hypersensitive skin reactions, and pulmonary inflammation upon repeated exposure (Pepys, 1992; Tripathi and Grammer, 2001; Hendrick, 2002).

BG was one of the first simulants produced by the U.S. Army and has since seen widespread use as a BW agent stimulant (Smart, 1997). For example, in 1966, BG was released into the New York City subway system to model the dispersal of *B. anthracis* spores in an enclosed system with significant airflow (Smart, 1997; Cole, 1997). The test was alarmingly successful in its results in terms of demonstrating the vulnerability of subways to biological attack, but no injuries or illnesses were reported to have resulted from the test.

VIRAL SIMULANTS

Comparatively, simulants for viral agents have received little attention. Perhaps most prominently, after the 1991 Gulf War, it was learned that scientists of the Iraqi offensive BW program had researched the camelpox virus, likely as a simulant for the related orthopoxvirus Variola major, the causative agent of smallpox (please refer to the entry entitled "Camelpox" for more information). Bacteriophages (viruses that infect bacterial cells) have often been used as viral BW agent surrogates, of which the Levivirus MS2 coliphage is a typical model (Parker, 2002; O'Connell et al., 2002).

Encyclopedia of Bioterrorism Defense,   Edited by Richard F. Pilch and Raymond A. Zilinskas
ISBN 0-471-46717-0   Copyright © 2005 Wiley-Liss

TOXIN SIMULANTS

For testing and development purposes, ovalbumin, a protein typically isolated from chicken eggs, is commonly used as a simulant for botulinum toxin (Parker, 2002). Ovalbumin is similar to human serum albumin. Trichothecene mycotoxins, while potential bioterrorist agents themselves, have been used as toxin surrogates as well.

NONSPECIFIC SIMULANTS

Chemical compounds that can be easily monitored and tracked from a distance have been used as simulants for biological agents. For example, zinc cadmium sulfide (ZnCdS) particles were used extensively by the pre-1969 U.S. BW program to model aerosol dispersion of BW agents over large portions of continental United States during the 1950s and 1960s (National Research Council, 1997; Smart, 1997; Cole, 1990). At the time, these particles could be commercially obtained in a size that resembled aerosolized BW agents, and were known to be stable in the environment and to disperse similarly to BW agents, hence their appeal. Not considered, however, were the long-term deleterious physiological effects of cadmium, particularly when inhaled.

CURRENT APPLICATIONS: BIOTERRORISM OFFENSE

There are a number of means by which a simulant might be employed in the context of bioterrorism. If the perpetrator's intention were to engender panic rather than to generate lethality, the release of an aerosolized simulant would suffice. This has in fact occurred in the United States with the delivery of *Bacillus thuringiensis*, which can be considered a stimulant of *B. anthracis*, in a letter from the Army of God (please refer to the entry entitled "Army of God" for more information). Repeated false incidents of this type, and particularly on a larger scale with a more advanced delivery device, have the potential to both overwhelm emergency response systems and paradoxically lull the public into complacency or disregard. A simulant release might also serve as a diversionary tactic. One problem caused by simulants is the inability of many BW agent detectors to differentiate the lethal from the nonlethal species, such that false positive results generated would lead to an alarmed public and potentially overwhelm emergency response infrastructure. It is therefore imperative to improve differentiation technologies, which will also be of value for forensics of actual BW agents.

CURRENT APPLICATIONS: BIOTERRORISM DEFENSE

Simulants have a number of applications in BW and bioterrorism defense. The military employs simulants in research, development, and testing of protective clothing and masks. Testing laboratories employ simulants in the development of new detection and identification techniques and devices (Parker, 2002; O'Connell et al., 2002; McBride et al., 2003). Basic research related to select agents is frequently conducted with more easily manipulated, nonlethal simulants before final testing with virulent agents. In these and other capacities, simulants are presently, and will remain in the future, of unquestionable value.

REFERENCES

Carlton, J., Of microbes and mock attacks—51 years ago, the military sprayed germs on U.S. Cities, *Wall Street J.*, October 26, 2001.

Choi, S.H., Kim, Y.S., Chung, J.W., Kim, T.H., Choo, E.J., Kim, M.N., Kim, B.N., Kim, N.J., Woo, J.H., and Ryu, J., *Infect.Control Hosp. Epidemiol.*, **12**, 740–747 (2002).

Christopher, G.W., Cieslak, T.J., Pavlin, J.A., and Eitzen, E.M., *J. Am. Med. Assoc.*, **278**, 412–417 (1997).

Cole, L.A. *Clouds of Secrecy: The Army's Germ Warfare Tests Over Populated Areas*, Ch.7 and 8, Rowman and Littlefield, Savage, MD, 1990.

Cole, L.A., *The Eleventh Plague: The Politics of Biological and Chemical Warfare*, W.H. Freeman and Company, New York, 1997, pp. 18, 160.

Hendrick, D.J., *Occup. Med.*, **52**, 56–63 (2002).

McBride, M.T., Gammon, S., Pitesky, M., O'Brien, T.W., Smith, T., Aldrich, J., Langlois, R.G., Colston, B., and Venkateswaran, K.S., *Anal. Chem.*, **75**, 1924–1930 (2003).

National Research Council, *Toxicologic Assessment of the Army's Zinc Cadmium Sulfide Dispersion Tests*, National Academy Press, Washington, DC, 1997.

O'Connell, K.P., Khan, A.S., Bucher, J.R., Anderson, P.E., Cao, C.J., Gostomski, M.V., and Valdes, J.J., Application of real-time fluorogenic polymerase chain reaction (PCR) to nucleic acid-based detection of simulants and biothreat agents, in *Proceedings of the 23rd Army Science Conference*, December 2–5, 2002, Orlando, FL, http://www.asc2002.com/oral_summaries/G/GO-02.PDF.

Parker, A., Rapid field detection of biological agents. *Sci. Technol. Rev.*, January/February 24–26, (2002), http://www.llnl.gov/str/JanFeb02/pdfs/01_02.4.pdf.

Pepys, J., *Am. J. Ind. Med.*, **21**, 587–593 (1992).

Smart, J.K., "History of Chemical and Biological Warfare: An American Perspective," in F.R. Sidell, E.T. Takafuji, and D.R. Franz, Eds., *Medical Aspects of Chemical and Biological Warfare*, TMM Publications, Washington, DC, 1997, pp. 32 and 43.

Tripathi, A. and Grammer, L.C., *Ann. Allergy Asthma Immunol.*, **4**, 425–427 (2001).

Yu, V.L., *N. Engl. J. Med.*, **300**, 887–893 (1979).

BIOREGULATORS

Richard F. Pilch

INTRODUCTION

In 1989, the results of Project Bonfire, an ambitious undertaking of the former Soviet Union's biological warfare (BW) program, were presented at a review conference attended by Ken Alibek, then deputy director of the civilian component of the Soviet program known as Biopreparat. Alibek later summarized the project as follows: "A new class of weapons had been found. For the first time, we would be capable of producing weapons based on chemical structures produced naturally in the body... [that] could damage the nervous system, alter moods, trigger psychological changes, and even kill. Our heart is regulated by peptides. If present in unusually high doses, these peptides will lead to heart palpitations and, in rare cases, death" (Alibek and Handelman, 1999).

Project Bonfire demonstrated for the first time that deliberate human action could turn a substance produced normally by body cells so it harmed the body itself. For Project Bonfire, the substance was "myelin toxin," named as such by Soviet scientists because of its ability to damage the myelin sheaths of nerve cells and thus disrupt nerve impulse transmission throughout the body. Soviet scientists had identified the gene coding for the production of the toxin and, with the help of novel genetic engineering techniques, successfully inserted it into the bacterium *Yersinia pseudotuberculosis*. When tested, the genetically engineered bacterium acted both as the pathogen and the myelin toxin, which caused paralysis.

Myelin toxin is a bioregulatory peptide, or "bioregulator," a genetically coded chain of amino acids produced naturally in the human body that is essential for normal physiological functioning. In the body, the actions of bioregulators can range from the mediation of sensations such as pain and fear to the regulation of blood pressure, heart rate, and respiration. When present at an abnormally elevated level, however, a given bioregulator can overwhelm the body's compensatory mechanisms, causing its effects to go unchecked—the desired end result with respect to its use as a biological weapon. Though science has only begun to elucidate the role of most bioregulators in controlling biological processes, the deliberate introduction into the body of those that are known can lead to such incapacitating or deadly effects as altered sensorium, delusional behavior, pain, loss of consciousness, elevated or depressed blood pressure (hyper- or hypotension), and death under certain circumstances.

HISTORICAL AND FUTURE CONSIDERATIONS FOR WEAPONS DEVELOPMENT OF BIOREGULATORS

A bioregulator can conceivably be delivered in two ways. In one approach, the gene that codes for it is inserted into a microorganism using well-established recombinant DNA techniques, and the microorganism is then introduced into the body via ingestion, injection, or inhalation. As the genetically engineered microorganism colonizes the body tissue, it secretes the bioregulator in large quantities, producing the effects described above. In the second approach, the bioregulator itself is synthesized chemically or enzymatically in a laboratory and then, at least in theory, delivered like any other pathogen, toxin, or pharmaceutical agent, in other words, via injection, ingestion, or inhalation. Bioregulators in fact lend themselves to fast and cost-efficient synthesis due to the abbreviated nature of their constituent amino acid chains. Scientific and commercial developments have allowed for the inexpensive production of large quantities of similar peptides, as demonstrated for example by the 4 million kilograms of NutraSweet manufactured per year in the late 1980s (Dando, 2001). Further, much like is being done in drug design, chemical synthesis can be programmed to produce analogs of the original bioregulator that can be more potent, specific, and toxic than the naturally occurring substance. Thus, it is wise to assume that in the future the illicit production—and, as prudence dictates, use—of these chemicals and their enhanced analogs will occur (Dando, 2001).

While it is known that Soviet scientists performed basic research on multiple bioregulators and successfully inserted a number of them into various pathogens, it is unclear whether any of the results of their research were applied by the Soviet BW program, or how much developmental work on peptides progressed before the collapse of the Soviet Union in December 1991. It has been reported, however, that Soviet scientists did in fact successfully transfer myelin toxin into *Yersinia pestis*, the causative agent of plague and close relative of *Y. pseudotuberculosis*, but that no chimeric weapon was ever produced in quantity (Alibek and Handelman, 1999). Whether a successor program to Project Bonfire exists now within the closed Russian military biological laboratories is not known.

Military research of bioregulators probably was not limited to the former Soviet Union: Project Coast, the apartheid-era chemical and biological weapons program of South Africa, may have looked into the development of bioregulators as well. Program leader Wouter Basson testified during hearings of the post-apartheid South Africa's Truth and Reconciliation Committee that such work had in fact been performed at two separate facilities, the laboratory complex of the large-scale chemical weapons production plant Delta G Scientific and a covert research laboratory concealed within the Special Forces headquarters. According to Basson, scientists at the Special Forces laboratory successfully synthesized an unspecified thymic peptide (different peptides are produced by the thymus, for example alpha-thymosin, which acts as an immune enhancer, and beta-thymosin, which has been linked to both tumorigenesis and metastasis), as well as growth hormone and other unnamed peptides produced by the pituitary gland. While Basson's suggestion that bioregulators were in fact considered for military research and development has been corroborated, his claims of success have not been verified. Further, Basson had reason to offer false testimony in this regard in order to substantiate his own claim that he had allocated substantial government funds, rather than embezzling them, toward the purchase of a peptide synthesizer that later could not be accounted for. To what degree Project Coast activities actually focused on bioregulators thus remains uncertain (Bale, 2003).

While an unknown but undoubtedly large number of bioregulators exist, only a few have been considered in the context of biological weapons. Security analysts tend to focus on substances that influence those functions of the body most vital for survival. The barriers to successful production serve to further limit the pool of candidate agents, especially in the context of terrorism, given the perceived limitations of terrorist technical capabilities.

And terrorist motivations may in fact reduce this pool to zero at the present time: While the threat or success of a biological event is presumed to carry with it a significant psychological impact, no greater impact would be expected from the use of a bioregulator than a more well-known BW agent such as *Bacillus anthracis*, which is far easier to acquire, produce, and deliver effectively.

Examples of bioregulators include endothelin, a potent vasoconstrictor produced in the lining of blood vessels and capable of causing rapid and severe hypertension; endorphins and enkephalins, opioids similar to morphine that are produced in the brain and whose use lead to analgesia, respiratory depression, and hypotension; vasopressin, also known as antidiuretic hormone (ADH), a hormone produced in the pituitary gland that causes water retention by the kidneys and vasoconstriction, both of which act to elevate blood pressure; and Substance P, a neuropeptide responsible for the transmission of the sensation of pain along nerve fibers.

REFERENCES

Alibek, K. and Handelman, S., *Biohazard: The Chilling True Story of the Largest Covert Biological Weapons Program in the World—Told From Inside by the Man Who Ran It*, Random House, Inc, New York, 1999.

Bale, J., South Africa Country Profile, www.nti.org, accessed on 2003.

Dando, M., *The New Biological Weapons: Threat, Proliferation, and Control*, Lynne Rienner Publishers, Boulder, CO, 2001.

FURTHER READING

Spertzel, R. et al., *Technical Ramifications of Inclusion of Toxins in the Chemical Weapons Convention (CWC)*, Supplement, Defense Nuclear Agency, Alexandria, VA, 1993.

See also BIOTECHNOLOGY AND BIOTERRORISM.

BIOSECURITY: PROTECTING HIGH CONSEQUENCE PATHOGENS AND TOXINS AGAINST THEFT AND DIVERSION

REYNOLDS M. SALERNO

DANIEL P. ESTES

INTRODUCTION

The tragic events of September 2001 and the subsequent dissemination of *Bacillus anthracis* spores through the U.S. postal system underscored the dangers to national and international security posed by terrorist attacks, especially those involving pathogenic microorganisms and toxins. Since the time of these incidents, state-sponsored biological warfare (BW) programs and terrorists believed to have pursued biological weapons have attracted increased attention and concern. In addition, the general public has acquired an unprecedented fascination with, and fear of, the powers of bioscience and bioterrorism.

Many different strategies are now being applied to combat the proliferation and use of biological weapons. Most strategies—such as increasing the effectiveness and availability of vaccines and antibiotics, improving disease surveillance and detection, building public health capacities, and developing biosensor technologies—are reactionary in nature, focusing on improving the ability to detect and respond to a bioterrorist event after it has occurred. But the international community has also begun employing preventive strategies, and one of the principal strategies in this category is biosecurity, the protection of dangerous pathogens and toxins from unauthorized access. Biosecurity aims to stop proliferation before it starts by protecting dangerous pathogens and toxins, the basic building blocks of biological weapons, against theft or malicious diversion from bioscience institutions. By preventing potential bioterrorists or proliferant states from acquiring certain dangerous biological materials, biosecurity provides the first line of defense against both state-based biological weapons proliferation and bioterrorism.

Thousands of bioscience facilities around the world conduct critical research on pathogens and toxins that could be used as biological weapons (Carlson, 2003). Yet the academic and private biological research communities—in which the majority of this research takes place—have not been accustomed to operating in a security conscious environment. In fact, security applied to microbiology laboratories has often been perceived to be ineffective, intrusive, expensive, and likely to obstruct or jeopardize vital biomedical and bioscience research (Kwik et al., 2003; Steinbruner et al., 2003).

It is evident that the increased biological weapons and bioterrorist threat justifies improving control and oversight over certain biological material that could be misused in this fashion. It is now essential and appropriate to establish biosecurity systems, practices, and procedures that deter and detect the malicious diversion of these biological materials. However, it is critically important to strike an appropriate balance between protection of biological material that could be used in a biological weapon and preservation of an environment that promotes legitimate and lifesaving microbiological research (Salerno et al., 2003).

Balancing security and research at biomedical and bioscience facilities is no trivial matter, especially because few microbiologists are knowledgeable about modern security systems, and few security experts have any familiarity with microbiology. Moreover, the concept of biosecurity remains in its relative infancy. In fact, the international microbiological community has yet to even reach a consensus on a clear definition of the term "biosecurity" itself.

SECURITY FUNDAMENTALS

There are at least two fundamental truisms about security. First, a security system cannot protect every asset against every conceivable threat, and second, security resources are not infinite. A degree of risk will always exist; therefore, it is important to understand and document what risks the facility management is prepared to accept. These are the risks that the security system cannot protect against, which in turn define the issues that the incident response planning must address. Designing a security system compels institutional managers to make important decisions about how limited security resources should be allocated. To make these decisions with

Encyclopedia of Bioterrorism Defense,   Edited by Richard F. Pilch and Raymond A. Zilinskas
ISBN 0-471-46717-0   Copyright © 2005 Wiley-Liss

confidence, facility managers must be able to articulate and defend the purpose and scope of their security systems.

Security systems should be based on the assets or materials that require protection. Institutional administrators must be cognizant of what materials they possess, the nature and locations of these materials, and the people who have access to them. Those materials that could cause a national or international security incident if diverted and misused must receive greater protection than other assets. In addition, managers and administrators must evaluate how an adversary would attempt to divert, steal, destroy, or release those assets. Defining which assets are critical to protect and which methods would be employed to harm them establishes the security system's objectives and scope.

In addition to appreciating the nature of the assets that require protection, security systems must uniquely apply to the operations of the specific environment. System designers must understand the characteristics and purposes of all other critical operating systems that will have to interact with the security system. In a biological research environment, one of the most important operational considerations is *biosafety*. Biosafety aims to reduce or eliminate exposure of laboratory workers or other persons and the outside environment to potentially hazardous agents involved in microbiological or biomedical research. Biosafety is achieved by implementing various degrees of laboratory "containment," or safe methods of managing infectious materials in a laboratory setting (National Institutes of Health and Centers for Disease Control and Prevention, 1999; World Health Organization, 2003).

Biosafety and biosecurity systems should be complementary, even though they have different objectives and strategies. As noted, the objective of biosecurity is to protect dangerous pathogens and toxins, as well as critical related information, against theft or diversion by those who intend to pursue biological weapons proliferation or bioterrorism. Thus, simply stated, biosafety aims to protect people from dangerous pathogens, while biosecurity aims to protect pathogens from dangerous people. Both methodologies are now critical to the operation of a modern bioscience institution.

Similar in many ways to the practice of biosafety, biosecurity depends on the implementation of comprehensive policies and procedures that affect the facility's operations only to the extent that is required. Ideally, security measures taken should not hinder a researcher's ability to perform experiments in a timely manner, delay or prevent authorized access to materials, or impede communication between associates and peers. To the extent possible, the security system should be transparent to those who are required to use it. In other words, the emphasis of biosecurity should be on creating and sustaining a "security culture" at the bioscience facility—a culture in which individuals understand the rationale and support the need for systems to protect dangerous pathogens and toxins from theft and diversion.

CHALLENGES ASSOCIATED WITH PROTECTING PATHOGENS AND TOXINS

There are several unique challenges posed by microorganisms that differentiate biosecurity from other forms of high security (National Research Council of the National Academies, 2003). First, although certain biological agents have the potential to cause serious harm to the health and economy of a population if misused, many of them have legitimate uses for medical, commercial, and defensive application. The possession of any one of these inherently "dual-use" materials does not necessarily signal an intention to use that material as a weapon. Generally speaking, this characteristic sets biological weapons apart from other materials used to develop weapons of mass destruction (WMD), such as certain nuclear and chemical materials that have no peaceful uses.

Second, biological agents are widespread. They exist in nature and are globally distributed in research laboratories, collection centers, and clinical facilities. By contrast, special nuclear materials are much less widely available and, thus, are much more difficult for terrorists to acquire than biological agents.

Third, biological agents are living, self-reproducing organisms, the volumes of which continually change throughout legitimate research activities. They can be found in a number of locations within a facility, including freezers, incubators, and infected animals and their waste. Quantification of actual amounts of materials is further encumbered because the amounts of a biological agent required for effective use or as a basis for growth are typically small, involving microgram to gram-sized quantities.

And fourth, although aerosol clouds that may or may not contain viable biological agents can currently be detected in a nondiscriminatory fashion, microorganisms cannot be speciated in real time with current standoff detection systems.

Despite these challenges, protecting certain pathogens and toxins is an essential component of both national and international biological weapons nonproliferation strategies. In order to design an appropriate security system that considers all of these unique challenges, system designers should employ a biosecurity methodology that aims to establish clear objectives for the biosecurity system. In other words, system designers should employ a risk management approach that establishes which assets should be protected against which threats (US General Accounting Office, 2001; US General Accounting Office, 2003). To accomplish this task, system designers and institutional managers should identify and prioritize the facility's assets, identify the adversaries who would likely attempt to divert or steal those assets, develop scenarios of undesirable events involving those assets and threats, and conduct a security risk assessment incorporating these scenarios.

ASSET IDENTIFICATION AND PRIORITIZATION

The purpose of the security system must be clear to the organization's management and staff. A fundamental step in achieving this clarity is defining and prioritizing the assets that the security system is designed to protect. Assets should be divided into separate categories based on the consequences of their loss (e.g., low, medium, and high). In this manner, the security system can be designed to have graded levels, with the highest consequence assets receiving the highest level of protection, and lower consequence assets receiving appropriately lower levels of protection.

Primary Assets

High-consequence or primary assets are those materials the loss of which could result in an event that would have national or international security consequences. The primary assets that most biosecurity systems should aim to protect are generally limited to *dangerous pathogens and toxins*. However, not all pathogens are equally likely to be diverted for purposes of biological weapons proliferation and, thus, not all pathogens should receive the same level of security. Agents should be evaluated on the basis of their attractiveness to an adversary. In other words, how easy would it be to deploy the agent as a weapon, and how significant would the consequences be should that agent be used as a weapon?

Those agents that are in a form that would be easiest for an adversary to deploy and most likely to yield a high-consequence event are the most attractive to an adversary and are therefore primary assets that require the highest level of security. In contrast, there are many characteristics that make certain agents less attractive to adversaries. For instance, if an agent requires processing to improve dissemination or virulence, is environmentally fragile, requires technical equipment or materials to amplify, causes an easily recognizable and treatable infectious disease, or is readily defeated by high degrees of local immunity, it is presumably less attractive to adversaries than other high-consequence assets.

Those agents that require the highest level of protection are defined as High Consequence Pathogens and Toxins (HCPTs), which are microorganisms or their by-products that are capable, through their use as a weapon, of severely affecting national or international public health, safety, economy, and security. In other words, HCPTs are those agents that have properties and attributes that would make them effective weapons material, and are therefore the agents most likely to be targeted for diversion from a legitimate biological research laboratory for the purposes of bioterrorism or biological weapons proliferation. (Three U.S. Codes of Federal Regulations (42 CFR 72, 9 CFR 121, and 7 CFR 331) have defined the "select agents" that must be secured at bioscience facilities in the United States. In the opinion of the authors, all of these select agents are not necessarily HCPTs, and all do not require the same level of protection against theft or diversion. For the U.S. regulations, as well as additional articles and documents on biosecurity, see www.biosecurity.sandia.gov.)

Determining which pathogens and toxins are HCPTs, and therefore primary assets, requires an assessment of their infectious disease risk and the risk that they could be developed into or used as weapons. The important characteristics to consider in this assessment are (U.S.

Congress Office of Technology Assessment, 1993; U.S. General Accounting Office, 1999; Patrick, 1994; Zilinskas and Carus, 2002):

- Infectious disease risk
  - Infectivity (ability to invade a host organism)
  - Pathogenicity (ability to cause disease in a host organism)
  - Lethality (ability to cause death in a host organism)
  - Transmissibility (ability to spread disease from host to host)
- Weaponization risk
  - Availability (number of facilities that house the pathogen or toxin)
  - Ease of amplification (rate of growth, nature of growth media, level of technical equipment and expertise required, etc.)
  - Ease of processing (including ease of aerosolization and increased inhalational characteristics)
  - Environmental hardiness (viability in a broad range of temperatures, hydration levels, light sensitivity, etc.)
  - Lack of availability of countermeasures/immunity (pharmacotherapies or prophylaxis)
  - Ability to be camouflaged as an endemic or common disease.

Secondary Assets

Medium-consequence or secondary assets are those materials whose loss could result in an event of somewhat lesser magnitude than the loss of a primary asset, or whose loss could assist an adversary in achieving an event of national or international consequence. Secondary assets could not only be pathogens and toxins that are not as dangerous as those identified as primary assets, but could also be information related to a given security system, newly discovered attributes of dangerous pathogens or toxins, or techniques that could be exploited by a terrorist.

Examples of secondary assets in a biological research environment include the following:

- Pathogens or toxins whose loss would be significant, but somewhat less than the consequences associated with losing an HCPT.
- Information related to HCPTs.
  - Agent databases containing information on which agents are stored at a given facility, where they are stored, and who has access to them.
  - Nonpublic critical information related to the maintenance or manipulation of HCPTs.
  - HCPTs shipping and receiving information.
  - Information regarding new techniques or discoveries that may aid a terrorist in developing a more effective biological weapon.
- Human resources records that reflect personal information of those individuals who work with or otherwise have access to HCPTs, security systems, or computer systems.

- Information related to the security system that protects the dangerous pathogens and toxins (e.g., facility blueprints).
- Information related to mission critical systems (e.g., the control centers that manage the security systems, the computer network, and containment-related environmental controls).

Tertiary Assets

Low-consequence or tertiary assets are those materials whose loss could result in an event of somewhat lesser magnitude than the loss of a secondary asset, or whose loss could assist an adversary in gaining access to a secondary asset. Tertiary assets could be pathogens and toxins or types of information that are not as dangerous as those identified as secondary assets, but are most often assets associated with the operations of a facility. While any operational asset that could be destroyed to cause a medium- or high-level consequence should be redundantly installed, operating systems such as electrical power, air-handling equipment, and laboratory equipment present a specific vulnerability and are thus considered tertiary assets.

THREAT IDENTIFICATION AND SECURITY RISK ASSESSMENT

After defining which assets the security system should protect, the institution's management should establish threat scenarios (who, what, when, where, how) and evaluate them on the basis of the relative likelihood of each threat materializing and the associated consequences. This threat identification and security risk assessment process should not be a description of all possible malevolent actions that could befall a facility. Instead, this exercise should evaluate the relative risk of reasonable threat scenarios. Then, the institution's management should decide which of those risks to protect against and which of them warrant incident response plans.

The threat identification step should define the characteristics, motivations, and capabilities of the adversaries who may attempt to steal or disperse the target assets. What kind of adversary would target this facility? What would the adversary know about the facility? What tools or skills would the adversary have? How might the adversary attack the facility? The objective of this review should be a list of scenarios of undesirable events based on the defined assets and the defined adversaries. For instance, which adversaries would attempt to steal biological agents, which would attempt to steal information, and which would attempt to destroy or deface the facility.

The security risk assessment step should evaluate all of these developed scenarios according to their probability and consequences. The highest risk scenarios are those judged to be more likely than others to occur, and with consequences that could result in a national or international security incident. The relative risk of each additional scenario then declines as either the probability or the consequences of the event decreases. The following

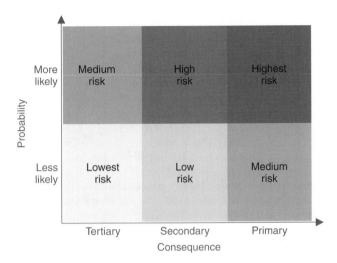

Figure 1. The risk prioritization process.

chart depicts the risk prioritization process, which should aim to position each scenario into at least one of the risk levels (Fig. 1).

After all of the developed scenarios have been assigned relative risk levels, the institution's management must decide which risks the security system must protect against. In addition, the risk assessment helps management define which possible but unlikely scenarios the security system should not be required to protect against; these are the risks that the management accepts, and develops incident response plans to address. This final step in the security risk assessment reflects the management's level of risk tolerance and/or risk aversion. The more risk tolerant the management is, the fewer resources it will need to invest in security. And conversely, the more risk averse the management is, the more resources it will have to invest in security. Thus, the risk assessment is a critical "resource allocation" step because it helps ensure that funds are expended primarily to prevent the high-consequence and high-probability events.

In general, the security risk assessment of a bioscience facility would be expected to reveal that adversaries would not likely conduct an overt external assault to steal agents. First, these agents are not unique materials: They can be isolated in nature and exist in laboratories throughout the world. And second, an overt attack using force would signal authorities to respond with medical and/or agricultural countermeasures that could mitigate the consequences of a bioterrorist attack with the targeted or acquired agent(s).

Bioscience facilities should therefore first concern themselves with defending against an insider who has approved access. An insider who is willing to divert a primary asset may be a disgruntled employee or one who is financially desperate, personally threatened, psychologically unstable, or motivated by any number of other reasons. Insiders are familiar with the protocols of the institution, and have knowledge of, and access to, the asset.

Bioscience facilities should also concern themselves with outsiders who would attempt to steal a biological agent covertly. This type of adversary would endeavor to avoid detection, and would probably abort the operation if discovery or capture appeared imminent. Such covert outsiders could include visiting scientists, students, or short-term maintenance workers.

Insiders and covert outsiders do not comprise the traditional threat group against which high-security systems have been designed to protect. For this reason, it is necessary for the bioscience community, in collaboration with security experts who are knowledgeable in the field of microbiology, to develop, publish, and employ uniquely tailored biosecurity standards that can guide facility managers responsible for implementing biosecurity systems.

ACHIEVING BIOSECURITY

These evaluations—asset identification and prioritization, threat identification, and security risk assessment—determine the objectives of the biosecurity system. The analyses provide the information necessary for an institution's management to define the assets and threat scenarios that must be protected against, and those that incident response planning must address. In this manner, the institution's management sets the design parameters and performance objectives of the biosecurity system.

With these objectives as their parameters, biosecurity experts should conduct a vulnerability assessment that identifies those vulnerabilities of the facility that would allow a high-risk scenario to occur. The security system should be designed to mitigate only those identified vulnerabilities that are associated with the risks the institution has decided to mitigate.

It is important to recognize that a security system can effectively protect the defined assets against the defined threats without mitigating every conceivable facility vulnerability. For instance, a facility's security system may be unable to protect a building from a large-scale physical assault, but may prevent a visiting scientist from stealing a primary or secondary asset. The institution's management can then address the risk of a large-scale assault—in this example, a risk that management has decided to accept—through incident response planning.

An effective biosecurity system includes many different components and should not rely on physical security and technologies alone. In fact, the most important aspects of a biosecurity system are procedural and cultural—elements that do not require large expenditures of resources. For example, a biosecurity system should physically consolidate, to the extent possible, all dangerous pathogens and toxins. Access to those biological materials should then be controlled by a combination of door locks or access controls limiting the number of authorized personnel.

The personnel who receive permission to access these areas should provide evidence that they have a legitimate need to handle, use, or transport the dangerous pathogens or toxins (or other assets) within, and that they have completed specific biosafety and biosecurity training. In addition, these personnel should be subject to a level of background screening that demonstrates their honesty and reliability, as would be performed as part of a personnel reliability program. Procedures should also be established for escorting visitors and support personnel who only need occasional access to areas where dangerous pathogens and toxins are located.

A biosecurity system should establish control and accountability of dangerous pathogens and toxins by documenting exactly what materials exist at the facility, where in the facility they are located, who has access to them, and who is responsible for them. Material control and accountability procedures should avoid trying to apply quantitative material-balance inventory accounting principles, which are impossible to achieve in a biological environment. Because dangerous pathogens and toxins are often transferred between facilities and shared among researchers, it is important for a biosecurity system to implement procedures to document, account for, and control both internal and external transfers of that particular material as well. Ideally, the procedures would demonstrate continuous custody of dangerous pathogens and toxins during both internal and external transfers.

All of the components of the biosecurity system should be documented in a biosecurity plan, which should be regularly reviewed and revised. In addition, an incident response plan should be written as well as regularly reviewed and revised. These core texts of the biosecurity system, as well as the many biosecurity policies and procedures, indicate that there is also a genuine need for information control and oversight. Biosecurity systems should include procedures for handling, using, and storing certain sensitive information related to dangerous pathogens and toxins and the various methods for accessing and protecting them.

Perhaps most importantly, a biosecurity system should include a security program management infrastructure that develops and maintains the biosecurity plan and incident response plan and conducts regular security training for the institution's staff. Creating and sustaining a biosecurity culture is the responsibility of the security program management staff.

CONCLUSION

Although biosecurity cannot in and of itself prevent biological weapons proliferation or bioterrorism, it is appropriate to take steps that reduce the probability that high-consequence pathogens and toxins will be stolen from a given bioscience research laboratory. The increased biological weapons and bioterrorist threat justifies improving control and oversight over those biological materials that could be used to cause a devastating or highly disruptive event. However, a balance between security and research must be achieved in order to protect critical assets as well as allow vital bioscience to advance.

Achieving this balance requires a comprehensive knowledge of the assets, threats, risks, and vulnerabilities associated with bioscience research. The security system, policies, and procedures should be designed specifically to address these unique bioscience characteristics. To achieve this goal, biosecurity objectives must be clearly defined and articulated to the research community. Simply applying the security standards that currently protect other high-value or high-consequence assets could result in inadequate protection of certain biological agents, inefficient use of limited resources, and the potential jeopardy of biomedical research.

REFERENCES

Carlson, R., *Biosecur. Bioterror. Biodefense Strategy Sci. Pract.*, **1**(3), 203–214 (2003).

Kwik, G., et al., *Biosecur. Bioterror. Biodefense Strategy, Sci., Pract.*, **1**(1), 27–36 (2003).

National Institutes of Health and Centers for Disease Control and Prevention, *Biosafety in Microbiological and Biomedical Laboratories*, 4th ed., May 1999, http://bmbl.od.nih.gov/contents.htm.

National Research Council of the National Academies, *Biotechnology Research in an Age of Terrorism: Confronting the Dual Use Dilemma*, Washington, DC, October 2003, http://www.nap.edu/books/0309089778/html/.

Patrick, W.C., III, "Biological Warfare: An Overview," in K. Bailey, Ed., *Proliferation*, Lawrence Livermore National Laboratory, Livermore, 1994.

Salerno, R.M., et al., Balancing Security and Research at Biomedical and Bioscience Laboratories, *BTR 2003: Unified Science and Technology for Reducing Biological Threats and Countering Terrorism—Proceedings*, Albuquerque, NM, March 2003.

Steinbruner, J., et al., *Controlling Dangerous Pathogens: A Prototype Protective Oversight System*, CISSM Working Paper, February 5, 2003, http://www.puaf.umd.edu/CISSM/Publications/AMCS/finalmonograph.pdf.

U.S. Congress Office of Technology Assessment, *Technologies Underlying Weapons of Mass Destruction. OTA-BP-ISC-115*, US Government Printing Office, Washington, DC, 1993.

U.S. General Accounting Office, *Combating Terrorism: Need for Comprehensive Threat and Risk Assessments of Chemical and Biological Attacks*, Washington, DC, September 1999.

U.S. General Accounting Office, *Homeland Security: A Risk Management Approach Can Guide Preparedness Efforts*, GAO-02-208T, Washington, DC, October 2001.

U.S. General Accounting Office, *Combating Bioterrorism: Actions Needed to Improve Security at Plum Island Animal Disease Center*, GAO-03-847, Washington, DC, September 2003.

World Health Organization, *Laboratory Biosafety Manual*, 2nd ed., (revised), 2003, http://www.who.int/csr/resources/publications/biosafety/who_cds_csr_lyo$_{20034}$/en/.

Zilinskas, R. and Carus, W., *Possible Terrorist Use of Modern Biotechnology Techniques*, Chemical and Biological Defense Information Analysis Center, April 2002 (For Official Use Only).

FURTHER READING

Barletta, M., *Biosecurity Measures for Preventing Bioterrorism*, Center for Nonproliferation Studies, November 2002, http://cns.miis.edu/research/cbw/biosec/pdfs/biosec.pdf.

Royse, C. and Johnson, B., "Security Considerations for Microbiological and Biomedical Facilities," in J. Richmond, Ed., *Anthology of Biosafety: V. BSL-4 Laboratories*, ABSA, Mundelein, IL, 2002.

Tucker, J.B., *Biosecurity: Limiting Terrorist Access to Deadly Pathogens*, US Institute of Peace, November 2003, http://www.usip.org/pubs/peaceworks/pwks52.html.

WEB RESOURCES

Sandia National Laboratory's dedicated biosecurity website: www.biosecurity.sandia.gov.

BIOTECHNOLOGY AND BIOTERRORISM

Raymond A. Zilinskas
Malcolm Dando

Article Contents

INTRODUCTION

Political leaders in major democratic states have often warned of a grave new security threat from terrorists willing to kill large numbers of people and with access to biological weapons. Different factors—social, political, economic, and scientific—will influence the evolution of this threat. Here we consider how the impact of science and technology, specifically, how advances in modern biotechnology might affect the biological weapons acquisition process and deployment methods. For the purpose of this section, we accept the U.S. Office of Technology Assessment's definition of biotechnology (Office of Technology Assessment, 1991):

> Any technique that uses living organisms or substances from those organisms to make or modify a product, to improve plants or animals, or to develop microorganisms for specific uses. These techniques include the use of novel technologies such as recombinant DNA, cell fusion, and other bioprocesses.

Of course, advances in the life sciences will have multiple impacts. Since biotechnology is predominantly dedicated to benign peaceful purposes, we expect it to deliver, amongst other benefits, better means of detecting pathogens and toxins, enhanced methods for diagnosing illnesses, and improved forms of prophylaxis (especially vaccines) and treatments. However, our main concern here is with the advanced biotechnology techniques that terrorists may apply to attack civil societies in new and more dangerous ways. While this problem has only recently come to public attention, it should be understood that it has long been a concern of defense organizations (Norqvist, 1993).

Technologies other than biotechnology could have a drastic impact on how future biological weapons are developed and used. Terrorists could adapt advances in the technology of biocontrol—of weeds or drug crops for instance—to improve their capabilities to mount effective attacks on staple crops. Similarly, recently developed techniques to administer vaccines to animals by aerosol dispersal might be adapted to effectively disperse living pathogens over a target population, and sophisticated meteorological maps of major urban centers commonly available on the Internet could guide terrorists in dispersing aerosols for maximum effect. Furthermore, discussion of the letters containing *Bacillus anthracis* spores sent to public figures in the United States during September/October 2001 served to focus attention on the key problems of drying, milling, and formulating agents so that they are of the right particle size when dispersed and are able to survive environmental stresses. Anyone with access to the Internet will have little difficulty discovering the enormous effort within commercial companies to find more effective technologies for the dry aerosolized delivery of drugs and for delivery of aerosolized vaccines, as well as the much-improved large-scale spraying systems used in agricultural biocontrol. It can be seen that the bioscience component of biological weapons research, development, production, and deployment is only a part of the process whereby biological weapons are acquired and used, albeit a very important one.

Biological weapons could be targeted against plants and animals as well as against human beings, and as the advances in biotechnology affect our understanding of all aspects of biology, human biology is not the only one affected by ongoing developments. While we concentrate here on the impact of developments as they affect humans, this broader context should be borne in mind. Additionally, while we stress here the threat to human beings from microorganisms and toxins, it should be understood that threats could also arise from our increasing understanding of the bioregulators that control normal body functions. If these were used by an attacker in unusual forms or quantities, great harm could be caused. We touch on this lesser-known threat at the end of this analysis.

In 1996, the U.S. Department of Defense (DoD) listed several possibilities that the techniques of genetic engineering might provide (US Department of Defense, 1996). Since then, several security analysts have expanded on the DoD's list of possibilities to theorize that genetic and molecular biology techniques will be used to substantially enhance the capabilities of those who may apply them in research and development (R&D) aiming to weaponize pathogens and nonpathogens in the future (Dando, 1996; Dando, 2001; Kadlec and Zelicoff, 1999; Poste, 2002, 2003).

The concerns of these analysts have a solid basis. The Soviet Union used sophisticated biotechnology techniques to research and develop genetically engineered microorganisms for biological warfare (BW) purposes. Dr. Ken Alibek, a former deputy director of the ostensibly civilian component of the Soviet Union's BW program, Biopreparat, claims that Soviet weapons scientists genetically engineered hybrid pathogenic viruses as part of their research (Alibek and Handelman, 1999). Others who once worked for Biopreparat have confirmed this allegation. However, it appears unlikely that the Soviet program ever actually weaponized genetically engineered BW agents or deployed them in weapons systems (*Leitenberg and Zilinskas, forthcoming*). Although President Boris Yeltsin ordered the termination of the Soviet BW program in 1992, it might continue in some form in Russian military biological laboratories today (Miller et al., 2002). Certainly, genetic engineering is currently being employed in peacefully directed research involving both bacteria and viruses at several institutes that belong to the Biopreparat system, which now has been privatized and calls itself a "joint venture" company.

Fears have been expressed that well-funded, technically capable nonstate groups also might exploit advanced biotechnology techniques for their purposes. The Aum Shinrikyo cult, which released the nerve agent sarin in the Tokyo subway system in 1995, is known to have had molecular biologists and medical doctors among its members, some of whom operated a biological weapons R&D program (Farley, 1995; Kaplan, 2000). Although there are no verified reports that Aum Shinrikyo or any other terrorist group has ever successfully developed genetically engineered microorganisms (Leitenberg, 1999), the rapid spread of expertise in the use of the new technologies opens the possibility that such uses could emerge in the future.

The assumption must be that terrorists will in the not-too-distant future solve the problems involved in using the classical biological weapons agents in the most effective way to cause mass casualties. After all, past state-offensive programs selected classical agents for weaponization because they had demonstrated particular advantages in biological attacks. Further, we assume that governments have anticipated this conclusion and are focusing defense measures (in particular, detection, vaccination, and treatment) against such agents. This situation will change in short order as findings of advancing bioscience are applied to produce ever more sophisticated molecular tools that are accessible to an ever-growing population of microbiologists and biotechnologists. This being so, the questions of interest to us are: Can the techniques of modern biotechnology be used by terrorists to wholly or partially overcome defenses erected by governments against biological events? And if so, which tools and developments are we most likely to see in the immediate future? In this section, we seek to answer these questions.

The section has five parts. First, we describe some of the advanced biotechnology techniques that we believe are applicable to the weaponization of BW agents. Second, we list characteristics of pathogens and toxins that could be deliberately affected using advanced biotechnology techniques and analyze whether this might be done within the next five years. Third, we briefly discuss the human resources needed by terrorists to apply biotechnology for weaponization. Fourth, we briefly describe and discuss some of the wider implications of biotechnological applications. Last, we give consideration to the threat from the misuse of bioregulators and how that is being impacted by the advances in biotechnology.

ADVANCED BIOTECHNOLOGIES OF POSSIBLE CONCERN

Four advanced biotechnology techniques in particular appear applicable to R&D, aiming to weaponize pathogens and toxins: genetic engineering, DNA technologies, protein engineering, and cell and tissue culture.

Genetic Engineering

The oldest, most commonly used, and best-known genetic engineering technique is gene cloning (or splicing), which produces recombinant DNA (rDNA). Simply put, rDNA techniques allow scientists to isolate a gene from the many genes that constitute one organism's genome (the donor) and emplace it in the genome of another organism (the host). The host is said to have been transformed after it has received the foreign gene and expression is achieved.

One of the more important steps to achieve successful genetic engineering is to transfer the desired gene across the host cell's membrane without causing it injury. There are many techniques available to scientists who wish to do so, including transfection, transduction, transformation, biolistics, electroporation, and microinjection. All of these techniques have been extensively described in the literature and therefore are well known to bioscientists everywhere (Ausubel et al., 2002). Knowledge about and use of genetic engineering is now so pervasive that it can be assumed that most scientists who might be sought by terrorists could use genetic engineering in R&D aimed to weaponize pathogens. Genetic engineering can be used to enhance the characteristics of both bacteria and viruses for weapons use by making them hardier, antibiotic resistant, host specific, and so on. However, manipulating microorganisms in this way might also bring about pleiotropic effects, which are unwanted new characteristics that would make the altered microorganism less suitable for production or weaponization (pleiotropic effects are discussed below).

Genetic engineering of cells is becoming easier to do because of the rapid development of many types of kits. Without a kit, persons attempting a genetic engineering project would have to develop a protocol for their research, make the reagents themselves, and proceed to follow the many steps set forth in the protocol to achieve success. This process is much more difficult than if the scientist is able to purchase a kit for the same protocol. In fact, today, anyone can buy cloning kits. Kits are quality controlled, have complete instructions including troubleshooting tips, and are fast and cheap. Granted, the kit operator still has to understand how to apply the technology, but the kit makes it easier than ever for scientists to accomplish the transformation of many types of microorganisms.

There are also certain possibilities inherent in specific genetically engineered viruses that have been developed for purposes of gene therapy, including retroviruses, adenoviruses, and baculoviruses, which scientists may employ to transfer genetic material designed to harm the host. Of the available virus vectors, the adenoviruses appear to offer the best prospects to weapons scientists because they can handle large transgenes, efficiently infect a large variety of both dividing and nondividing cells, and may be produced at high titer. Further, several kits now available on the market allow a researcher to create customized recombinant adenoviruses relatively easily (Stull, 2000).

Security experts have feared that naked DNA might be useful for BW for some time (Zimmerman, 1984). Naked DNA is already used in research as gene delivery vectors in gene therapy (Mountain, 2000) and as vaccines (Davis and McCluskie, 1997). To wit, it might be possible for scientists to design and produce small DNA segments that act as either toxins or powerful antigens. The naked DNA could be produced in quantity by appropriately genetically engineered bacteria or yeast, or in cell culture (Glaser, 1999). Possible applications of naked DNA for BW may also be generated by findings from experiments with DNA immunization.

Although the development of genetically engineered microorganisms is no longer technically difficult to do, this does not guarantee that the altered strains will be better suited for weapons use than their natural relatives. The reason is that no one can know at the onset of research whether or not the newly genetically engineered strain, for whatever reason it is done, will evidence undesired pleiotropic effects; for example, the newly engineered strains will possess not only the desired characteristic of antibiotic resistance, it also will manifest additional but negative characteristics that will make it unsuitable for weapons purposes. For example, at the same time as having acquired enhanced antibiotic resistance, the newly engineered strain might also have become less virulent or hardy than the parent strain (or both). Pleiotropy has been a common problem with genetically engineered organisms.

DNA Technologies

DNA technologies encompass three techniques: gene machines, sequence banks for proteins and nucleic acids, and functional genomics. Each is considered in turn, after which the important issues of protein engineering and cell and tissue cultures are discussed.

Gene Machines. As noted above, genetic engineering involves the transfer of genes between different organisms for the purpose of endowing the recipient organism with the ability to produce a new protein, increase yields of a protein, or adapt the organism to different environments. Genes required for genetic engineering may be obtained by cloning them in cells programmed for that purpose, assembling them from cloned fragments of DNA, or synthesizing them using a gene machine (also

called a DNA synthesizer). Thus, a gene machine is a device, usually computerized, that combines nucleotides in a specified order to produce the gene. A scientist working for terrorists might use a gene machine to assemble genes that code for the production of proteins useful when weaponizing pathogens such as toxins and virulence factors.

Since viruses consist primarily of genetic material, the synthesis of a virus using a gene machine is possible. In fact, an article published in summer 2002 reported on the synthesis of the poliovirus by a scientific team that purchased the necessary gene sequences from readily available commercial supply houses and stitched them together using a common synthesizer (Cello et al., 2002). Articles immediately appeared in the popular press hypothesizing that similar methods could be used to synthesize highly pathogenic viruses such as the Ebola virus (Westphal, 2002). It also appears possible for scientists to use gene machines to construct synthetic bacteria. With the gene sequences of *Mycoplasma genitalium*, constituted by 541 genes, and *Mycoplasma pneumoniae*, which has 677 genes, having been published in the open literature, the possibility arises that a synthetic pathogenic organism possessing between 300 and 500 genes might be constructed in the laboratory. A weapons scientist might eventually be able to design a synthetic bacterium that contains virulence factors, including toxins, thereby making it a human pathogen.

Sequence Banks for Proteins and Nucleic Acid. Bioinformatics is the use and organization of information of biological interest. Much of bioinformatics is concerned with organizing databases that contain this information and making that information available to those who need it. Research has generated an enormous amount of data on DNA sequences, protein sequences, the human genome, and enzymes, that is stored in databases maintained by organizations such as National Center for Biotechnology Information, the DNA Data Bank of Japan, the Genome Database (GDB) of the Human Genome Project (HGP), and the European Molecular Biology Laboratory (see Web Resources). Programmers have designed a large number of computer software programs to help scientists utilize the enormous amount of data available in these databases for such purposes as designing macromolecules, including toxins (http://www.netsci.org/Resources/Software/Modeling/CADD/top.html, accessed 8/29/03). Moreover, new techniques are being developed to mine such data in novel ways, which may have implications for combining previously unrelated literatures in ways that could be misused by terrorists (Kostoff, 1999).

Functional Genomics. Microbial sequencing initiatives have given rise to a new scientific field called genomic information technology, more commonly known as "functional genomics." Functional genomics attempts to correlate the activity of a gene with specific activities such as protein production, disease processes, signaling between body cells, and many others. It has been aptly stated that "[t]he fundamental strategy in a functional genomics approach is to expand the scope of biological investigation from studying single genes or proteins to studying all genes or proteins at once in a systematic fashion" (Hieter and Boguski, 1997). Using functional genomics, scientists are beginning to clarify how genes interact with one another and to understand complex interactions between genes, virulence factors, and the environment. Scientists working for or at the behest of terrorists can, as with scientists involved in peacefully directed research, easily access the GDB. Conceivably, they might use HGP data for such purposes as identifying:

- The few genes among the mass of genes constituting the human genome that would be special targets for pathogens and toxins. In particular, informatics and proteomics can be used to identify the structures of receptors, which would enable the reverse engineering of novel toxins and synthetic viruses or bacteria.

- Genetic variations or differences between populations that explain varying disease susceptibility and individual responses to pathogens and toxins. These differences might be utilized to design more effective biological weapons based on pathogens and toxins to which a target population is especially susceptible.

At the time of this writing, the genomes of 95 bacteria, including 39 human pathogens, have been sequenced, with another approximately 100 close to completion (Fraser, 2003). There is no doubt that scientists will sequence hundreds of genomes of bacterial, fungal, and viral genomes within a few years. These efforts provide information on the factors that when working together result in a microorganism being a pathogen. For example, when two strains, one pathogenic and one nonpathogenic, of the bacterial species *Helicobacter pylori* were sequenced, a comparison between the two genomes revealed which genes were involved in pathogenic processes, such as acidic survival, toxin production, motility, and attachment to epithelial cells (Ge and Taylor, 1999). Information derived from genome sequences of pathogens will undoubtedly accelerate scientists' understanding of pathogens, an understanding that can lead to the development of new therapeutic strategies for controlling these pathogens (Relman and Strauss, 2002). However, ill-willed scientists could also use this information to develop more infective, pathogenic, and virulent bacterial and viral pathogens.

Protein Engineering

Protein engineering is the modification of the chemical structure of a naturally occurring protein. Therefore, protein engineering is done using existing natural proteins as a starting point. This procedure might be done for such purposes as making the molecule more stable, altering the pharmacological properties of the parent protein, or, if the protein is an enzyme, changing its substrate specificity. Further, scientists could apply protein engineering in order to produce a new type of protein, one that is not found in nature, called a fusion protein.

A scientist might use protein engineering to develop various toxins for weapons use. Genes for a sizeable number of proteinaceous toxins have been cloned, the regulation of the expression of these genes is well understood, and the three-dimensional structures of many of these toxins have been clarified (Del Giudice and Rappuoli, 1999). While this information is being used in the pharmaceutical industry to develop new vaccines and toxoids, weapons scientists could also use it to deliberately alter the chemical structure of toxin molecules. This may be done for the purpose of, for example, developing more stable toxin molecules that do not dissociate if placed in water or resist destruction by cooking. Another possibility is to change a toxin so it becomes "activatable"; that is, the altered toxin is made more toxic than its unaltered predecessor by chemicals present in the targeted host. It is not difficult for a toxinologist to change one form into the other by biochemical means (O'Brien, 2001).

Holding possibly even more promise for terrorism than modified toxins are fusion toxins (Arnon, 2003). During the last 10 years, reports on how to construct various kinds of fusion proteins have appeared with some frequency in the scientific literature. Fusion toxins are a type of fusion protein. Briefly, the chemical structure of most protein toxins is composed of a light chain and a heavy chain. In 1992, a fusion toxin was constructed that combined one chain from the diphtheria toxin and one from a *Pseudomonas aeruginosa* toxin. The *Pseudomonas* chain made it possible for the extremely toxic diphtheria toxin to enter cells that it otherwise could not, and once inside, to exert its cytotoxic effects (Guidi-Rontani, 1992). Another fusion toxin that has been developed in the past combines tetanus and anthrax toxins. Tetanus toxin is nature's second most powerful toxin (after botulinum toxin). Since it is a neurotoxin, it is capable of attacking only neuronal cells. Conversely, the anthrax protective antigen is capable of easily translocating foreign proteins into a nonneural body cell's cytoplasm. In an experiment, the tetanus toxin was fused to an appropriate segment of anthrax toxin so that it was then able to enter and destroy nonneuronal cells, such as the secretory cells of the pancreas. The fusion toxin therefore proved to be more toxic and have a wider scope of action than either of its parents (Arora et al., 1994).

The foregoing research was performed in order to elucidate how toxins affect physiological functions and to develop methods for effectively attacking cells that must be destroyed for therapeutic purposes, such as cancer cells. However, there seems to be little doubt that with technologies available today fusion toxins could be developed within a short time for purposes of terrorism. Even in cases when unaltered toxins are used to harm and kill persons, investigators have great difficulty in determining the etiology of the illness because toxins are difficult to detect and their actions are subtle. These difficulties would be compounded if a perpetrator were to use a fusion toxin to attack individuals or populations. Further, its effects could be more deadly than naturally occurring toxins. The possible combinations of fusion toxins are nearly endless, potentially opening a Pandora's box to terrorists.

Cell and Tissue Culture

Cell culture is the use of isolated animal, insect, or plant cells to manufacture products in bioreactors. Tissue culture is the use of multicell assemblies to manufacture products in bioreactors. These technologies differ from classical fermentation, which uses whole bacterial or fungal cells to propagate biomass. The major advantage of cell culture and tissue culture over classical fermentation is that they are much better at producing correctly modified and immunologically active proteins from eukaryotic genes. There are advantages and disadvantages inherent in each type of cell culture system. Animal cells can produce complex substances, but are expensive to grow in bulk and require exacting growth conditions. Insect cells can express many proteins, but some are not well suited to therapeutic use. Insect cells are inexpensive to use because they grow at room temperature and do not require CO_2. Further, efficient methods have been perfected for transferring genes into insect cells using baculovirus as a vector. Plant cells are able to produce secretory antibodies, something neither of the other two systems can do, and plant cells are inexpensive to grow. The disadvantages, generally, of animal and cell culture is that the cells grown in culture are considerably more fragile than bacterial cells and more fastidious in terms of nutrient requirements. Owing to the difficulties inherent to using cell and tissue culture systems, their successful operation requires the active involvement of biochemical engineers.

However, because chemical supply houses are beginning to offer new kits for optimized cell culture production of proteins, it takes less and less training for users to operate these kits. Kits being marketed in 2003 allow operators to both increase yields of proteins and broaden the range of the types of proteins they are able to produce (Dutton, 2003; Glaser, 2003). Over 80 types of therapeutic proteins are being produced via different cell culture systems today, and this number is certain to double within a year. There is every reason to believe that should a person wish to produce proteins for illicit purposes, he or she would easily be able to purchase the appropriate kit on the open market with no questions asked.

While weapons scientists are likely, for the foreseeable future, to continue using classical fermentation techniques to produce BW agents and toxins on a large scale, cell and tissue culture systems provide them with substantial added production capabilities. In particular, they could use these culture systems to produce certain complex proteins, including fusion and chemically altered toxins, which could not be done through the use of classic fermentation techniques. It is estimated that in the next five years, many commercial cell culture production systems in the 100 to 300 L range will be marketed, making it likely that there will be extensive commercialization of recombinant proteins (Wrotnowski, 2000). If this estimate is realized, it would be possible for scientists employed by terrorists to purchase systems allowing them to manufacture relatively large quantities of toxins of the terrorist's choice, including fusion toxins. Further, if new life forms are created in the laboratory (such as the synthetic viruses and bacteria discussed above), they are likely to be first propagated in appropriate cell culture systems.

POSSIBLE USES OF BIOTECHNOLOGY TO WEAPONIZE PATHOGENS AND TOXINS

In the 1950s, Australian scientists used myxoma virus as a biological weapon against rabbits (Fenner and Ratcliffe, 1965). This virus causes myxomatosis, a deadly disease of four species of leporids, including the common rabbit. After the Australian scientists deliberately introduced myxomatosis into their continent, within two years it killed over 90 percent of the rabbit population. To this day, myxomatosis exerts a powerful population-control effect on rabbits in Australia. Lessons from Australia's experience with myxomatosis as a successful biological weapon allow us to identify five attributes that characterize a "perfect" military BW agent (Zilinskas, 1986). They are as follows:

- High ability to infect and harm the targeted host
- High degree of controllability
- High degree of resistance to adverse environmental forces
- Lack of timely countermeasures available to the attacked population
- Ability to camouflage the BW agent with relative ease.

These attributes serve as a useful starting point for consideration of the scientific objectives scientists working for terrorists may have when applying modern biotechnology to weaponize microorganisms. Thus, terrorists might try to apply scientific advances to enhance any or all of six characteristics or traits of microorganisms considered important for weaponization—hardiness, resistance, infectivity, pathogenicity, specificity, and detection avoidance. Each will be discussed in turn, with each discussion consisting of three parts: (1) the characteristic or trait is described and related to those of the "perfect" biological agent listed above; (2) the possible applications of modern biotechnology techniques to enhance the characteristic or trait under consideration are presented; and (3) conclusions as to the feasibility for such research are stated.

Hardiness

Hardiness refers to the ability of a microorganism or a bacterial or fungal spore to survive being enclosed in a storage container or munition and, after release onto the target, endure physical and chemical stresses encountered in the open environment. Referring to the five attributes of a "perfect" BW agent above, hardiness correlates with a high degree of resistance to adverse environmental forces.

A scientist might attempt to enhance the hardiness of bacteria, fungi, and viruses in one of two ways. First, he could try to enhance the organism's ability to resist desiccation, withstand UV radiation, and survive decontamination procedures. In this case, the hardiness of bacterial cells depends mostly on the bacterium's repair mechanism; that is, the quickness and thoroughness with which the bacterium's internal organelles are able to repair damage caused by stressors to the cell wall, chromosome, and other structures. If successful, the BW agent would survive longer after release, thereby increasing its potential for causing casualties. However, because of inadequate scientific knowledge about the genetic control over bacterial repair mechanisms and limits to scientists' ability to transfer multigene constructs from one organism to another, it is unlikely that success will be achieved in genetically increasing the hardiness of a bacterial species within the next five years.

Second, a scientist may attempt to stabilize genetically determined traits, such as virulence, in the weaponized agent. If crowned with success, the pathogen constituting the payload of a given biological weapon would have a longer shelf life, thus lessening the need for weaponeers to continually exchange "expired" pathogens with freshly propagated ones. However, the research required to stabilize traits can be a time-consuming endeavor since various approaches, such as attaching different promoters to genes controlling these traits, have to be attempted with a low rate of success. Therefore, scientists working for terrorists are not likely to pursue this avenue of biotechnological advance in the foreseeable future.

Some viruses, such as the smallpox virus, are naturally exceedingly hardy, being able to withstand desiccation for many hours. But most viruses die within minutes of release into the open environment because of desiccation. It appears that the hardiness of viruses depends mostly on the chemical structure of their outer coat. In any case, while it is possible to attempt to alter the outer coat of some viruses to change their antigenic properties (see below), there today is insufficient knowledge on how to do so for augmenting the hardiness of viral pathogens.

Resistance

Resistance refers to the ability of a microorganism to defeat the actions of therapeutic drugs and preventives such as vaccines. Referring to the five attributes possessed by a "perfect" BW agent, resistance relates to the attacked population lacking appropriate countermeasures.

The means by which different microorganisms are able to resist drugs and preventives vary considerably from type to type. With respect to bacteria, a scientist might attempt to develop a strain that is resistant to antibiotics used by the target population; if a virus, the aim could be to develop a viral strain that is unaffected by the enemy's antiviral therapeutic drugs; or if a fungus, an effort could be made to develop a strain that resists fungicides and antifungals. The advantage to the bioterrorist of using highly resistant strains in an attack is the potential for greater casualty generation and higher lethality among those attacked.

It is no longer a substantial technical challenge for scientists to imbue a bacterial strain with antibiotic resistance. Many plasmids carrying resistance genes against various antibiotics are available for purchase from biological supply houses or can be procured from research laboratories on a courtesy basis. Resistance genes may be transferred to new hosts using classical or genetic engineering techniques. During the last 15 years, it has become common in both classical microbiology and genetic engineering experiments for scientists to insert genes coding for antibiotic resistance as markers. By doing so, it is easy to differentiate between cells that have been successfully genetically manipulated and cells that remain in the original state.

Russian scientists have published findings from a research project during which they imbued an avirulent strain of *B. anthracis* with resistance to multiple antibiotics (Pomerantsev et al., 1995). Such a strain has value to the Russians because they use a live vaccine, in addition to antibiotics, as part of the treatment accorded to persons exposed to *B. anthracis*. However, the techniques used by the Russian scientists in this research can be applied equally well to increase the ability of a weapons strain of *B. anthracis* to resist antibiotics. This research, which clarifies the methodology required to imbue bacterial species of many types with an increased ability to resist antibiotics, has been published in the open literature. Presumably, scientists working for terrorists could access this information and do this kind of development today.

Much less is known about how viruses and fungi resist respectively antiviral and antifungal compounds than bacteria. This being so, we would not expect scientists working for terrorists to attempt R&D aiming to enhance resistance of viruses and fungi for the foreseeable future.

Infectivity

Infection is the process whereby microorganisms attach themselves to certain cells of the host, enter the cells by penetrating their walls and membranes, and colonize the tissues of the host. Referring to the list of attributes of a "perfect" BW agent, infectivity relates to high ability to infect and harm the targeted host.

The bioterrorist scientist can attempt to enhance the invasive abilities of microorganisms being developed for BW in several ways. Since body surfaces usually are the host's first and primary defense against invading pathogens, increasing a pathogen's ability to penetrate these barriers could be the primary objective of a weapons scientist. To enhance a pathogen's ability to penetrate a cell wall or membrane, a scientist could try to develop a microorganism that secretes two kinds of powerful enzymes: proteinases that break up peptide bonds, and phospholipases that hydrolyze phospholipids. An example of such work has been published; scientists attempting to enhance the insecticidal activity of baculovirus inserted genes that code for the production of three proteases into the virus. As a result, insect larvae infected with the genetically engineered baculovirus were killed off at a 30 percent higher rate than when attacked by wild-type baculoviruses (Harrison and Bonning, 2001).

Another approach would be for a scientist to attempt to enhance the ability of bacterial cells to adhere to the walls of the respiratory or intestinal tracts. In immunocompetent hosts, these tracts are continuously flushed by fluids and are lined by cells capable of secreting protective substances such as mucus and antibodies. To overcome these defenses, pathogenic bacteria produce special proteins, adhesins, which bind specifically to receptors (technically speaking, proteins on both interacting cells may be called "receptors") located on host cells. Since a substantial amount of information is available in the scientific literature about these substances and how they are produced by pathogens, it is possible that scientists could use this information to design projects aiming to imbue pathogens that normally do not produce adhesins with the capability to do so, or enable pathogens to secrete viscous substances, such as an alginate capsule or polysaccharide slime, thereby increasing their ability to adhere to host cells.

There also might be possibilities for increasing the infective capabilities of viruses. Before being able to initiate infection, viruses must attach to an appropriate receptor on the prospective host's body cells. For example, the human immunodeficiency virus (HIV) produces a special protein (gp120) that attaches to receptors on T lymphocytes, thus allowing the virus to enter these cells whose normal function is to destroy invading microorganisms. Using information that has been published about viral pathogens, scientists can attempt research that aims to alter the genetic makeup of a virus so that it can attach more efficiently to traditional host cell receptors or alternatively to receptors that it normally could not bind. Findings from recently reported research indicate that a very small change in a virus's genome can result in a shift in receptor usage for virus entry (Baranowski et al., 2001).

Despite mentioning these possibilities for enhancing infectivity, there is as yet little information about how microorganisms penetrate mammal skin. Therefore, it is unlikely that anyone will be in a position for some years to enhance this particular attribute in a pathogen or transfer the gene (or genes) that controls it from one organism to another. Although more is known about adhesins, and their genetic control, than about membrane penetration mechanisms, it is not known whether the gene controlling adhesion in one microorganism would be expressed in another microorganism. Further, even if such a gene was expressed after transfer, it is possible that the transfer would result in pleiotropic effects. For these reasons, we believe that terrorists probably would not find research in this area worthwhile for the foreseeable future.

Pathogenicity

Pathogenicity refers to the ability of the pathogen, once it has successfully colonized the host, to traverse the bloodstream or lymphatics, evade the intrinsic defenses of the host, enter target tissues, and obtain nutrients for itself. Referring to the five attributes possessed by a "perfect" BW agent, pathogenicity relates to high ability to infect and harm the targeted host.

The successful invader's virulence (ability to produce morbidity and mortality in a host) is a complex process. For example, it has been estimated that over 200 genes of *Salmonella typhimurium*'s genome, which is approximately 4 percent of the organism's total number of genes, have functions related to virulence (Marcus et al., 2000). Among the most important determinants of virulence are virulence factors, many of which often act

in unison to destroy the host's defenses and bring about symptoms of disease.

A weapon scientist wishing to add or modify virulence factors could select from a large menu of candidates. To illustrate the array of virulence factors possessed by pathogens, to date eight virulence factors have been found in the fungal pathogen *Cryptococcus neoformans* (Buchanan and Murphy, 1998). For the opportunistic bacterial pathogen *P. aeruginosa*, the following virulence factors have been identified: pili, flagella, liposaccharides, proteases, quorum sensing, exotoxin A, and exoenzymes secreted by the type III secretion system (Lyczak et al., 2000). It is reasonable to assume that each bacterial and fungal pathogen possesses its own array or complement of virulence factors.

Among all types of virulence factors that exist, toxins may be of most immediate interest. Many bacterial pathogens are able to secrete toxins; over 200 bacterial toxins have been identified as of the date of this writing. It would appear that the genes controlling the production of some of these proteins would not be difficult to identify and transfer to microorganisms being developed for BW purposes. For example, a gene coding for the production of one of *Staphylococcus aureus*' most common superantigens (a type of toxin), SEG, has been identified and cloned (Abe et al., 2000). Also, it was recently reported that researchers have developed methodologies for the transfer and efficient expression of the genes encoding botulinum and tetanus toxins to *Escherichia coli*, the ubiquitous and generally harmless resident bacterium of the normal human intestinal flora (Zdanovsky and Zdanovskaia, 2000). It is probable that scientists attempting to weaponize bacteria and fungi would have a plethora of choices as to which virulence factors they could use.

Again, while it is not technically difficult for an appropriately trained scientist to transfer a gene coding for a virulence factor from one bacterium to another, the person doing the transferring cannot be certain whether the newly transformed bacterium will exhibit pleiotropic effects that will render it less suitable for weapons purposes than the original. If this occurs, a sizeable, long-term R&D effort would have to be undertaken to rid the transformed bacterium of pleiotropic effects while retaining the newly acquired virulence factors intact.

Specificity

Specificity refers to a pathogen's propensity for a specific host. Referring to the five attributes possessed by the "perfect" BW agent, specificity relates to high host specificity.

A scientist working for bioterrorists might find it useful to either increase a pathogen's preference for a specified target population or decrease the pathogen's ability to attack populations other than the target population. By doing so, the probability of a biological weapon causing collateral damage is decreased, thus increasing the bioterrorist's ability to control the weapon. When considering biological weapons against humans, the ultimate manifestation of specificity would be an ethnic weapon. Ethnic weapons are discussed elsewhere in this volume.

Host preferences among pathogens vary widely. At the one end of the scale, some species of viruses (e.g., poxviruses) and bacteria (e.g., *Mycobacterium lepri*) tend to be species specific. At the other end of the scale, there are many bacterial and fungal species that attack more than one animal or plant species. For example, there are subspecies of *P. aeruginosa*, a bacterium that is ubiquitous in the environment, which can cause disease in every kind of animal and in virtually all tissues (Lyczak et al., 2000).

The biological relationships between hosts and pathogens are exceedingly complex, having evolved over thousands or more years. While research on the genetic basis governing some host-pathogen relationships is beginning to produce findings, knowledge about these relationships is still rudimentary. Therefore, it is unlikely that even the most qualified scientist will be able to enhance the specificity of any type or species of pathogenic microorganism in the next five years.

Detection Avoidance

There are two types of detection avoidance. First, detection avoidance could involve the deliberate altering of properties possessed by well-characterized BW agents, such as engineering it to express surface antigens it normally would not express. If so, the target population, using existing methods, would have problems with detecting and identifying the modified form of pathogen. Second, an organism could be deliberately altered to defeat immunological defense systems present in a target population. Regardless of the type being discussed, referring to the five attributes possessed by the "perfect" BW agent, detection avoidance relates to lack of timely countermeasures by the attacked population or ability to camouflage the BW agent.

In reference to the first type of detection avoidance, all known biological threat agents have been characterized to the point that were one of them to be used in an attack, it would be identified within a short time, enabling the exposed populations to quickly receive appropriate treatment. To defeat these defensive measures, a bioterrorist scientist might endeavor to make it difficult for defenders to quickly and correctly identify an organism causing an outbreak. Thus, a scientist working for terrorists might alter a specified organism's antigenic properties, thereby making it difficult for defenders to identify the BW agent using existing detection methods.

To develop a bacterial strain that defeats detection by clinical methods, a scientist could attempt to manipulate one or a few genes that control bacterial metabolism or the production of proteins constituting the bacterium's cell wall. By altering a bacterium's metabolic properties, the work of the clinical laboratory to identify the bacterium is made more difficult. In order to make a definitive identification of a bacterium, clinical microbiologists test the bacterium's ability to ferment a variety of sugars (such as glucose, lactose, sucrose, etc.) and reactions with chemicals to produce indicator chemicals (such as hydrogen sulfide gas, indole, urea, etc.). By comparing the pattern of the unknown bacterium's fermentation and chemical reactions with the patterns of known bacterial species, identification can usually be made 48 to 72 h after the specimen containing the bacterium is first inoculated on growth media. However, when an unknown bacterium exhibits a fermentation pattern that does not fit known patterns, the clinical microbiologist is forced to perform additional tests or send the unknown bacterium to a reference laboratory that is better equipped to handle difficult identifications. In either case, at least an additional 48 h are necessary before the unknown bacterial pathogen can be identified. With regard to altering the bacterial cell wall, if this were done, the modified organism's antigenic properties would be sufficiently changed to confuse detection methods usually employed in the clinical laboratory, such as polymerase chain reaction, to identify organisms to the level of species. Similarly, the modified organism might avoid detection by field investigators employing array kits designed to quickly identify any of a number of biological threat agents.

The second type of detection avoidance refers to circumventing primed immunodefense systems of the target population. Both human and animal populations in industrialized nations are vaccinated against many common diseases. Vaccinated individuals develop antibodies that most often are able to defeat the pathogens against which vaccines have been developed and administered. To defeat this type of defense, a scientist working for terrorists could attempt to genetically engineer a classical threat agent so that its modified form is antigenically different from the parent. If he were successful, the antibodies constituting part of the target population's immunodefenses would not recognize the new antigenic properties, leaving the host vulnerable to infection by the modified form. In bacteria, this could be done by altering the cell wall as described above. With viral species, the scientist could attempt to change the viral coat. Many viruses, especially RNA viruses such as influenza viruses, mutate frequently in nature, in the process changing their antigenic properties. Research has been, and is being, conducted for clarifying how viruses accomplish this; some findings of this research have been published.

From a technical standpoint, it would not be difficult for an appropriately trained scientist to alter the antigenic properties of some bacterial and viral pathogens (as opposed to fungal pathogens, which are genetically more complex). For example, Russian scientists were able to genetically engineer a strain of *B. anthracis* so that some of its antigenic properties were quite different than the original strain (Stepanov et al., 1996). Methods similar to those used by the Russians could be employed to alter the genetic properties of other bacterial species. With viruses, genes coding for the production of the viral coat are amenable to manipulation. In either case, the modified bacterium or virus probably would be antigenically different from the original strains and thus able to avoid detection by both clinical laboratories and human immunodefenses.

Of the two types of detection avoidance, the first type, altering a BW agent's antigenic properties, could be done relatively easily by a well-trained scientist. However, if genetic manipulations were done on an organism for the purpose of, for example, altering a cell wall or viral coat, it is almost certain that the manipulated organisms would exhibit pleiotropic effects, such as having a weakened external structure or lowering the organism's ability to withstand defensive actions by the host, such as phagocytosis. Therefore, it hardly would be worthwhile for a terrorist organization to pursue it.

As to the second type of avoidance detection, research to accomplish this probably would not be successful in the next five years. There are two main reasons why we draw this conclusion. First, research performed by scientists of the former Soviet Union's BW program to alter the smallpox virus was conducted for many years but was nevertheless unsuccessful, demonstrating the difficulty of the task. Second, before such research could be undertaken, the immunological status of the target population would have to be known. Studies for this purpose would take a long time to accomplish, and in the end might not generate sufficient data to identify weaknesses or defects in the target population's immunological defenses.

HUMAN RESOURCE REQUIREMENTS FOR WEAPONIZATION R&D

When regarding human resource requirements for biological weapons acquisition, two elements must be considered: (1) levels of education and training that may be possessed by those who seek to develop biological weapons, and (2) the human resources required to carry out the six types of R&D described in the previous section. These will be discussed in turn.

Education and Training for Biological Weapons Acquisition

One of six levels of education and training may be possessed by those who might be retained by terrorists seeking to acquire biological weapons: (1) high school graduate, (2) technician, (3) research assistant, (4) junior scientist, (5) scientist, or (6) a scientific team.

High School Graduate. Many U.S. high schools have advanced programs providing gifted students with experience as interns or assistants in university biology laboratories. Cloning experiments are being performed in some high school biology classes. As a result, tens of thousands of high school students have been exposed to advanced biotechnology techniques. Nevertheless, the most that can be expected from high school students, even if gifted, is the ability to follow instructions cookbook style. Thus, a gifted high school biology student could culture and propagate common bacteria and assist someone with more advanced training by performing routine tasks, but would be unable to apply advanced biotechnology techniques to reach even the most simple research objectives in microbiology. Therefore, excepting the rare prodigy, high school graduates are not likely to have had sufficient training to take on substantial bioweapons-related R&D.

Technician. Some community colleges in the United States offer programs that produce microbiology technicians in about two years. Students in these programs

learn to prepare complex culture media, culture most of the common aerobic and anaerobic gram-positive and gram-negative bacteria, culture fungi, culture viruses in egg embryos and cell culture, and perform basic clinical testing to identify cultured organisms. They might receive training in some advanced biotechnology techniques. After graduation, a technician can find work as an assistant to junior scientists or scientists, performing routine tasks such as inoculation, culturing, and diagnostic testing. With experience, he or she should be able to independently perform simple experiments, including cloning and genetic engineering of typical laboratory microbial strains such as *E. coli* and *Bacillus subtilis*. After five or more years of experience, a competent technician would have gained approximately as much expertise as a recent graduate with a bachelor's degree.

Research Assistant. A graduate from a four-year college biology program with an emphasis on microbiology should have good knowledge of all aspects of basic bacteriology, including the metabolism and growth characteristics of common bacteria, host-parasite interactions, and endotoxin and exotoxin production. He or she will have been exposed to virology and mycology, will have learned about many advanced biotechnology techniques, and will have practiced them in laboratory work. Thus, he or she can follow directions in published protocols that provide detailed and frequently updated instructions of basic methods in DNA preparation and isolation, sequencing, genetic manipulation, and so on (Ausubel et al., 2002). As a result, he or she can carry out complex experiments under the supervision of a junior scientist or scientist and can design and undertake simple microbiological experiments. After graduation, he or she is qualified to work in a research laboratory as a research assistant to a principal investigator or enter into training in a clinical laboratory in preparation for becoming a licensed clinical microbiologist. After five or more years of work experience, the research assistant would have approximately the same level of expertise as the recent graduate with a master's degree.

Junior Scientist. A graduate with a master's degree will have the same qualifications as the bachelor's degree graduate, but in addition will have had extensive course work and hands-on experience in a specialized field of microbiology. In his or her area of specialization, the master's graduate can design complex experiments that might result in producing new findings. After graduation, the junior scientist is qualified for employment as the head of a microbiology section in a clinical laboratory (but working under a laboratory director who has an advanced academic or professional degree), in the biotechnology or pharmaceutical industry as a junior scientist, and in an academic laboratory as a senior research assistant to a scientist. In view of the many thousands of persons who graduate every year with a degree in the biosciences, terrorist groups would not find it difficult to recruit a willing junior scientist to work for them.

Scientist. A graduate who has earned a doctorate will possess the experience and expertise necessary to conceptualize hypotheses and to design and carry out complex experiments to prove or disprove those hypotheses. In doing so, he or she is likely to generate original findings that increase science's knowledge of scientific phenomena or lead to new applications in agriculture or industry. In a BW program, scientists often have key roles because they are best equipped to (1) search the literature for information generated by peacefully directed research but applicable to the weaponization of microorganisms and toxins; (2) use that information to design applied weaponization research on microorganisms or toxins of interest; and (3) carry out the research successfully, which means learning from mistakes as they occur and avoiding less promising or dead-end research approaches. As the Aum Shinrikyo experience demonstrates, terrorist groups might be successful in recruiting the rare scientist to their cause.

Scientific Team. Advanced national R&D programs—such as that of the former Soviet Union—that aim to undertake sophisticated R&D to weaponize microorganisms for military uses typically employ teams constituted by scientists trained in various disciplines, including microbiologists, biochemists, fermentation engineers, immunologists, molecular biologists, and others. If the work involves viruses, the service of a virologist is required as well. To develop aerosols, the program requires an aerobiologist or appropriately trained physicist. The contributions of weapons engineers are needed if the biological agent were to be used in bombs, shells, and missiles. A terrorist group would find it very difficult to assemble such a team.

Impact of Level of Expertise on Capabilities

If the bioterrorist wanted to conduct a low-technology attack, such as contaminating food with a foodborne pathogen, the needed level of expertise would be low and requirements for such an attack would be uncomplicated: A small quantity of a foodborne pathogen, such as a *S. typhimurium*, *Shigella dysenteriae*, *Listeria monocytogenes*, or *Bacillus cereus* (if the aim is casualty generation) or *B. anthracis* (if the aim is lethality) suspended in the culture media in which the organisms were propagated. A person having the training of a technician or above and who works in a hospital or clinic should easily be able to obtain a foodborne pathogen suitable for this type of criminal act and would have no problem propagating it at home in sufficient quantity to cause hundreds of casualties (but with low mortality).

If the terrorist's aim was to cause mass casualties by mounting an airborne attack, the pathogen of choice would be *B. anthracis* (Friedlander, 1997). Procuring a virulent strain of this pathogen would be the most difficult aspect of the acquisition process. However, on the assumption that the terrorist had obtained a seed culture of virulent *B. anthracis*, it is an easy bacterium to culture and propagate because it grows readily in commonly available nutrient media at 37 °C. It is technically more demanding to convert germinating cells to spores, but the information on how to accomplish this is available

in the open literature. Someone possessing the training equivalent to that of a technician or higher would be qualified to produce *B. anthracis* spores in quantities needed for bioterrorist acts. However, it would require a fairly high level of expertise, at the level of junior scientist or scientist, for someone to be able to develop an efficient formulation for aerosol delivery. A substantial amount of experimentation performed over some months would be required before this junior scientist or scientist would be able to perfect a delivery system to efficiently and reliably dispense a *B. anthracis* formulation over a target population. An adequate knowledge of local meteorology would be required as well.

If a terrorist group desired to develop a genetically engineered BW pathogen, perhaps through the enhancement of one or more of the characteristics described above, the scientific/technical expertise required to research, develop, test, and produce the requisite genetically modified pathogen would be high. If the pathogen of choice were a bacterial species, the planning and implementation of the developmental project would require the services of a scientist with years of experience in bacteriology, assisted by other scientists with expertise in genetics, biophysics, and aerobiology. If the pathogen were to be a viral species, the lead scientist would have to be a virologist, again assisted by a geneticist, biophysicist, and aerobiologist.

For some of the more complicated R&D to, for example, alter the genetic presentation of a bacterium or virus, it would require a substantial effort by a team of highly qualified scientists to successfully undertake such a project. We estimate that it would require scientists possessing expertise in bacteriology, bioinformatics, biophysics, gene therapy, human genetics, molecular biology, and virology to staff such a team. As the project progresses, there probably would be a need for additional expertise.

THE WIDER IMPACT OF MODERN BIOTECHNOLOGY

Modern biotechnology is founded on our increased understanding of the structure and function of the genome. This understanding is at the center of modern biology and is changing it from a descriptive to a predictive science (Kitano, 2001). As such, this profound change affects the whole of biology. Thus, while most concern about misuse is appropriately now centered on pathogens and toxins, some serious analysts have pointed out that in the longer term we must think about possibilities "beyond bugs" (Poste, 2003).

Disease is, in effect, a disruption of the normal functioning of the affected host. Normal functioning of body cells is coordinated through three major systems and signaling molecules that are collectively known as bioregulators: the immune system (cytokines), the endocrine system (hormones), and the nervous system (neurotransmitters). Rather than use a pathogen for the purpose of disrupting bioregulators, an approach that could be taken would therefore be to attempt to disrupt one of these systems directly. Clearly, for example, if the immune system defenses are disrupted effectively, the victim will be open to infection by a wide array of opportunistic pathogens, and it appears that bioregulators of this type could be easily dispersed over very wide areas (Kagan, 2001). Bioregulators are discussed in more detail elsewhere in this volume.

As will have been noted throughout this entry, there is an increasing awareness of the importance of receptor molecules in cellular function. Signaling bioregulators are only effective because cells have specific receptor proteins that are designed to respond to the appropriate signals. One of the key impacts of modern biotechnology has been to facilitate the elucidation of the structure and function of such crucial receptor molecules. We see the potential impact of this advance clearly in relation to efforts to design toxins and bioregulators to attack the nervous system. During the Cold War period, both eastern and western nations attempted to move beyond their understanding of how lethal nerve agents interfered with neurotransmitters in order to design agents that would incapacitate humans for various times in different ways. One such "incapacitating agent" was used in the attempt to break the Moscow hostage siege (Dando, 2003), and this event demonstrated the problem with that technology because it was not possible for the Russian operators to separate the sedative and respiratory depression effects of the agent they used to subdue attackers and victims alike.

In short, during the Cold War period it was known that neurotransmitters operated through different types of receptors, but the amazing diversity of subtypes of receptors was as yet unknown. Now scientists understand this diversity, and the brain circuits within which the particular subtypes are located are increasingly being elucidated. Thus, military organizations are taking an active interest in finding new incapacitating agents, but this has to be seen as only the first stage in the possibilities for malign manipulation of signaling systems.

CONCLUSION

In the last 18 months, there have been several publications reporting on findings from bioscientific research that could be applied by persons intent on developing biological weapons in proliferant countries or for terrorist purposes. For example:

- A study aiming to determine how human immunodefenses are broken down by the smallpox virus but not by the vaccinia virus clarified the small genetic differences between the two (Rosengard et al., 2002), differences that might be bridged through the correct application of genetic engineering.
- A live pathogenic virus, the polio virus, was for the first time synthesized by scientists, who purchased the required genetic sequences from commercial sources and assembled them in their own laboratory (Cello et al., 2002); similar techniques might be used to assemble viruses of interest to bioweaponeers.

- A sophisticated method for encapsulating microorganisms was published in a large-circulation scientific journal (Loscertales et al., 2002); this method could be adapted by ill-willed persons to protect microorganisms of interest to bioweaponeers from environmental stresses.

These three examples follow one of the most frightening developments ever to be reported openly in the scientific literature, namely, the use of genetic engineering to convert a low-order pathogenic virus to an exceedingly deadly virus (Jackson et al., 2001). In this case, Australian scientists added the gene for the cytokine IL-4 to the mousepox virus, a relative of the smallpox virus, in an attempt to produce an enhanced immune response in the host. They expected to see an enhanced production of antibody, but they did not expect to find that the other arm of the immune response—which specifically kills virus-infected cells—would be shut down at the same time. Alarmingly, the genetically altered virus killed resistant strains of mice even if they had been vaccinated. Although this occurred by accident, it is probable that the same technique could be applied deliberately to weaponize other nonpathogenic viruses or to enhance the pathogenicity of low-order pathogens.

Moreover, we are continually reminded that nature is full of surprises, such as mechanisms for interfering with the operation of RNA (interfering RNA), which may be used to "shut off" genes, and means of greatly "speeding up" evolution such as the technique of DNA shuffling. In short, it must be assumed that many more biological mechanisms will become available to terrorists in the coming decades and be directed toward illicit ends. Indeed, if the current problem is to stop the spread of state-offensive biological weapons programs (and the potential "leakage" from such programs), in the years ahead, there will be a quite different and much more complex problem—that of dealing with substate groups, or even deranged individuals, capable of carrying out mass casualty attacks through the use of biological weapons.

This analysis raises many questions that should concern the biomedical community and wider public alike. For instance, which are the main areas of science and technology that could be of concern? How fast are dangerous capabilities likely to arise and spread in these areas? What might best be done to halt or at least slow such developments and prevent the spread of capabilities into the hands of those with malign intent? Is it possible for more advanced countermeasures to be devised to ensure that defense authorities continue to have the advantage over those wishing to cause harm with biological weapons? While we do not as yet have answers to these questions, and while some may never be satisfactorily answered, what is certain is that ever-developing biotechnology will be the basis of a worldwide industry sooner rather than later in this century, and that the problem of preventing criminal and terrorist use of biotechnological capabilities will become increasingly important to civil society.

REFERENCES

Abe, J., Ito, Y., Onimaru, M., Kohsaka, T., and Takeda, T., *Microbiol. Immun.*, **44**(2), 79–88 (2000).

Alibek, K. and Handelman, S., *Biohazard: The Chilling True Story of the Largest Covert Biological Weapons Program in the World—Told From Inside by the Man Who Ran it*, Random House, New York, 1999.

Arnon, S.S., Personal communication with R. A. Zilinskas, February 8, 2003.

Arora, N., Williamson, L.C., Leppla, S.H., and Halpern, J.L., *J. Biol. Chem.*, **269**(42), 26165–26171 (1994).

Ausubel, F.M., Brent, R., Kingston, R.E., Moore, D.D., Seidman, J.G., Smith, J.A., and Struhl, K., Eds., *Current Protocols in Molecular Biology*, John Wiley & Sons, New York, 2002.

Baranowski, E., Ruiz-Jarabo, C.M., and Domingo, E., *Science*, **292**, 1102–1105 (2001).

Buchanan, K.L. and Murphy, J.W., *Emerg. Infect. Dis.*, **4**(1) (1998), http://www.cdc.gov/ncidod/EID/vol4no1/buchanan.htm.

Cello, J., Paul, A., and Wimmer, E., Chemical Synthesis of Poliovirus cDNA: Generation of Infectious Virus in the Absence of Natural Template, *Science Online*, July 11, 2002, http://www.aaas.org/scienceonline.

Dando, M.R., "New Developments in Biotechnology and Their Impacts on Biological Warfare," in O. Thränert, Ed., *Enhancing the Biological Weapons Convention*, Verlag J.H.W. Dietz Nachfolger, Bonn, Germany, 1996, pp. 21–56.

Dando, M.R., *The New Biological Weapons: Threat, Proliferation, and Control*, Lynne Rienner Publishers, Boulder, CO, 2001.

Dando, M.R., *The Danger to the Chemical Weapons Convention from Incapacitating Chemicals*, First CWC Review Conference Paper No. 4, University of Bradford, 2003. Available at http://www.brad.ac.uk/acad/scwc.

Davis, H.L. and McCluskie, M.J., *Microbes Infect.*, **7**(21), 7–21 (1997).

Del Giudice, G. and Rappuoli, R., *Vaccine*, **17**, S44–S52 (1999).

Dutton, G., *Genet. Eng. News*, **23**(1), 18–19 (2003).

European Molecular Biology Laboratory, http://www.embl-heidelberg.de, accessed on 8/29/03.

Farley, M., Japanese police focus on elite scientists in sect, *Los Angeles Times*, March 30, 1995, p. A6.

Fenner, F. and Ratcliffe, F.N., *Myxomatosis*, Cambridge University Press, Cambridge, UK, 1965.

Fraser, C., Bioterrorism and Genomics, Paper presented at the meeting *Scientific Openness and National Security*, Washington, DC, January 9, 2003.

Friedlander, A.M., "Anthrax," in F.R. Sidell, E.T. Takafuji, and D.R. Franz, Eds., *Medical Aspects of Chemical and Biological Warfare*, Office of the Surgeon General, Washington, DC, 1997, pp. 467–478.

Ge, Z. and Taylor, D.E., *Annu. Rev. Microbiol.*, **53**, 353–387 (1999).

Glaser, V., *Genet. Eng. News*, **19**(20), 10 (1999).

Glaser, V., *Genet. Eng. News*, **23**(1), 14–15 (2003).

Guidi-Rontani, C., *Mol. Microbiol.*, **6**(10), 1281–1287 (1992).

Harrison, R.L. and Bonning, B.C., *Biol. Control*, **20**(3), 199–209 (2001).

Hieter, P. and Boguski, M., *Science*, **278**, 601–602 (1997).

Jackson, R.J., Ramsay, A.J., Christianson, C.D., Beaton, S., Hall, D., and Ramshaw, I., *Journal of Virology*, **75**, 1205–1210 (2001).

Kadlec, R.P. and Zelicoff, A.P., "Implications of the Biotechnology Revolution for Weapons Development and Arms Control," in R.A. Zilinskas, Ed., *Biological Warfare: Modern Offense and Defense*, Lynne Rienner Publishers, Boulder, CO, 1999, pp. 11–26.

Kagan, E., *Clin. Lab. Med.*, **21**(3), 607–618 (2001).

Kaplan, D.E., "Aum Shinrikyo (1995)," in J.B. Tucker, Ed., *Toxic Terror: Assessing the Terrorist Use of Chemical and Biological Weapons*, MIT Press, Cambridge, MA, 2000, pp. 207–228.

Kitano, H., Ed., *Foundations of Systems Biology*, MIT Press, Cambridge, MA, 2001.

Kostoff, R.N., *Technovation*, **19**, 593–604 (1999).

Leitenberg, M., *Terror. Polit. Violence*, **11**(4), 149–158 (1999).

Leitenberg, M. and Zilinskas, R.A., *Doing the Devil's Work: The Soviet Union's Biological Warfare Program* (in preparation).

Loscertales, I.G., Barrero, A., Guerrero, I., Cortijo, R., Marquez, M., and Gañán-Calvo, A.M., *Science*, **295**, 1695–1698 (2002).

Lyczak, J.B., Cannon, C.L., and Pier, G.B., *Microbes Infect.*, **2**(9), 1051–1060 (2000).

Marcus, S.L., Brumell, J.H., Pfeifer, C.G., and Brett Finlay, B., *Microbes Infect.*, **2**, 145–156 (2000).

Miller, J., Engelberg, S., and Broad, W., *Germs*, Simon & Schuster, New York, 2002.

Mountain, A., *Trends Biotechnol.*, **18**, 119–128 (2000).

Network Science: Welcome to NetSci's List of CADD, CAMD Software. Network Science Corporated, http://www.netsci.org/Resources/Software/Modeling/CADD/top.html, accessed on 8/29/03.

Norqvist, A., *Biotechnology*, Swedish National Defence Research Establishment, Stockholm, 1993.

O'Brien, A.D., Pathogenesis of *E. coli* O157:H7, Paper presented at the symposium *Epidemiology, Pathogenesis, and Risk Assessment of E. coli O157:H7*, Annual meeting of the American Society for Microbiology, Orlando, FL, May 21, 2001.

Office of Technology Assessment, *Biotechnology in a Global Economy*, US Government Printing Office, Washington, DC, 1991.

Pomerantsev, A.P., Mockov, Yu.V., Marinin, L.I., Stepanov, A.V., and Podinova, L.G., *Anthrax Prophylaxis by Antibiotic Resistant Strain STI-AR in Combination with Urgent Antibiotic Therapy*, Centre for Applied Microbiology & Research, Porton Down, England, 1995.

Poste, G., Biotechnology and Terrorism, *Prospect*, April 25, 2002.

Poste, G., Knowledge Production in the Life Sciences: Assessing the Threat, Paper presented at the meeting *Scientific Openness and National Security*, Washington, DC, January 9, 2003.

Relman, D.A. and Strauss, E., *Microbial Genomes: Blueprints for Life*, American Academy of Microbiology, Washington, DC, 2002.

Rosengard, A.M., Liu, Y., Nie, Z., and Jimenez, R., *Proc. Natl. Acad. Sci. U.S.A.*, **99**(13), 8808–8813 (2002).

Stepanov, A.V., Marinin, L.I., Pomerantsev, A.P., and Staritsin, N.A., *J. Biotechnol.*, **44**, 155–160 (1996).

Stull, D.L., *Scientist*, **14**(24), 30, 33 (2000).

US Department of Defense, *Biotechnology and Genetic Engineering: Implications for the Development of New Warfare Agents*, US Department of Defense, Washington, DC, 1996.

Westphal, S.P., Ebola virus could be synthesized, *New Scientist*, July 17, 2002, http://www.newscientist.com/news/print.jsp?id=ns99992555, accessed on 6/28/04.

Wrotnowski, C., *Genet. Eng. News*, **20**(8), 8–9 (2000).

Zdanovsky, A.G. and Zdanovskaia, M.V., *Appl. Environ. Micro-biol.*, **66**(8), 3166–3173 (2000).

Zilinskas, R.A., "Recombinant DNA Research and Biological Warfare," in R.A. Zilinskas and B.K. Zimmerman, Eds., *The Gene-Splicing Wars: Reflections on the Recombinant DNA Controversy*, Macmillan Publishers, New York, 1986, pp. 167–203.

Zimmerman, B.K., *Polit. Life Sci.*, **2**(2), 188–191 (1984).

WEB RESOURCES

DNA Data Bank of Japan, Welcome to DDBJ!, http://www.ddbj.nig.ac.jp/Welcome-e.html, accessed on 8/29/03.

The Genome Database, http://www.gdb.org/, accessed on 8/29/03.

National Center for Biotechnology Information, http://www.ncbi.nlm.nih.gov, accessed on 8/29/03.

BIOTERRORIST ATTACK, STAGES, AND AFTERMATH

RAYMOND A. ZILINSKAS

INTRODUCTION

Three assumptions underlie this consideration of the stages of a bioterrorist attack and its aftermath, namely that the perpetrator(s) will: (1) use one or more types of bacterial, fungal, or viral pathogens to achieve their objectives; (2) endeavor to disseminate a sufficient quantity of the pathogen over or onto the target population to cause illness or death to most of its members; and (3) use such means of delivering the pathogens as aerosol dissemination, contamination of food or water, direct application to a targeted individual, or contagion by previously infected individuals. Although a biological attack could be mounted against human, animal, or plant populations, in this section only terrorist attacks against humans are considered. Further, the emphasis throughout is on biological events that generate mass casualties; that is, when more than 1000 persons are affected or, alternatively, "an incident in which the medical system is overwhelmed and the balance between resources and demands is undermined" (Shapira et al., 2003).

On the basis of an analysis of three past incidents in which criminals or terrorists deliberately deployed biological agents or chemicals in attacks (the Rajneeshees, Aum Shinrikyo, and the "anthrax letters" of September/October 2001 by an unknown perpetrator or perpetrators), as well as several exercises that modeled or simulated the aftermaths of bioterrorist attacks, a biological emergency essentially comprises three overlapping stages: (1) preattack stage, (2) attack stage, and (3) attack management stage (which has either four or five substages, depending on which biological agent has been released and in what manner). A detailed discussion of each stage follows.

PREATTACK STAGE

The preattack stage is that period leading up to an attack when a terrorist acts to acquire a biological weapon with the intent of using it against a target population in order to gain some perceived benefit.

Acquiring an effective biological weapon and carrying out a successful biological attack requires the criminal to take four vital steps: (1) secure a culture of a suitable pathogen or a quantity of toxin; (2) develop an appropriate formulation; that is, a combination of the pathogen or toxin and the substrate in which it is suspended or dissolved; (3) obtain an appropriate container to safely store and transport the formulation; and (4) apply an efficient mechanism to disperse the pathogen or toxin over or onto the target population. In addition, if the biological warfare (BW) agent is to be delivered by aerosol, a fifth factor is essential, namely, favorable meteorological conditions for the act of dispersion. If these four or five vital elements are in place, the likelihood of a criminal being able to mount a successful attack is high. The preattack stage ends with the deployment of the biological weapon to the site of the planned attack.

Now to consider the other side of the coin, the defenders. Defenders can be divided into two groups. The first consists of police, security, and intelligence agencies whose responsibility it is to prevent illicit biological activity and, should this fail, to track down and apprehend the perpetrator. For the purposes of this abbreviated discussion, it is sufficient to note that no police, security, or intelligence agency appears to have had forewarning of any of the three cases of biological criminality or terrorism noted above. It is probable that unless a police, security, or intelligence agent has been able to penetrate a criminal group intent on using a biological weapon, which is an unlikely possibility, future biological attacks are going to be as unexpected as these attacks were in the past.

Physicians, emergency medical personnel, public health officials, and others involved with delivering health care and securing public health constitute the second group. Obviously, these people cannot prevent or deter a biological event, but they can prepare themselves so that they are capable of dealing with the aftermath of a biological attack to the greatest extent possible. The issue of preparedness is addressed in another entry in this volume.

ATTACK STAGE

During the attack stage, the terrorist activates the deployed biological weapon. The biological weapon could be armed with either a noncontagious or contagious agent.

With noncontagious agents, if the terrorist's aim is to generate mass casualties, the pathogen or toxin must be dispersed either as an aerosol over a city or in food or beverages that will be consumed by large numbers of people. The various approaches taken by the Aum Shinrikyo to carry out attacks with aerosolized noncontagious pathogens and toxins are described elsewhere in this volume, but the underlying lesson is that these attacks were unsuccessful because of

Encyclopedia of Bioterrorism Defense,   Edited by Richard F. Pilch and Raymond A. Zilinskas
ISBN 0-471-46717-0   Copyright © 2005 Wiley-Liss

technical problems. Whether other terrorists will learn from Aum's mistakes is as yet an open question. If a competent terrorist group were intent on acquiring biological weapons, it is best to assume that members of this group would either possess the requisite expertise to overcome such technical hurdles or would act expediently to gain it. Further, the Aum experience might be studied by potential perpetrators for the sake of avoiding the same mistakes, and by doing so the perpetrators probably would be in a good position to carry out a successful aerosol attack. In this case, success would be measured in the number of casualties caused among the target population.

Also as described in this volume, the Rajneeshees used a foodborne pathogen in its attack. Although the cult did not generate mass casualties according to the first definition set forth above, it did come very close. It is not farfetched to believe that if this attack had been carried out with more rigor and had employed more operatives, several thousands of Oregon inhabitants would have been stricken with debilitating gastrointestinal illness. Further, had the cult used a more deadly pathogen for foodborne dispersal, such as a virulent strain of *Bacillus anthracis*, the mortality rate among the sick would have been high.

If a perpetrator chooses to employ a contagious pathogen in an attack, he or she would most likely select one of the pathogens designated by the Centers for Disease Control and Prevention (CDC) as Category A agents (CDC Strategic Planning Workgroup, 2000). Contagious viruses on the Category A list include smallpox virus, several filoviruses, and Lassa fever virus. The only contagious bacterial species on the list is *Yersinia pestis*, the causative agent of plague. Since it is not a very contagious pathogen and is relatively fragile (i.e., does not survive long in the open environment), it would be difficult for terrorists to initiate a large-scale pneumonic plague outbreak. This being the case, a Category A virus would be more likely to be utilized in a terrorism scenario. While terrorists would find it nearly impossible to secure some viral strains (such as smallpox virus) and very difficult to secure others (such as filoviruses), if such an agent were acquired, initiating an epidemic within the target population would not be difficult since most populations are susceptible to these pathogens. The easiest method probably would be for the attacker to use the biological equivalent of a suicide bomber; that is, a person who has been deliberately infected with a contagious agent and dispatched to the target population before disease symptoms appear (during the prodromal phase of the illness). In many viral diseases, including influenza, the infected individual is more contagious in the prodromal state than after disease symptoms have appeared. With other viral diseases, the person must be showing signs of the disease before he or she is infective. For example, a person afflicted with smallpox is essentially not contagious until rash appears.

Regardless of the approach taken by a criminal to carry out an attack with a contagious agent, once the agent is acquired less technical preparation is required for dissemination of a contagious agent than a noncontagious one. Thus, there is little or no need to prepare a large quantity of the contagious agent (a single infected individual might be sufficient), there is no need to develop a formulation for the agent of choice, and there is no need to disseminate the agent via mechanical dispersal.

There are, however, several disadvantages to using contagious agents for terrorist purposes. To begin, contagious agents are much more hazardous to laboratory workers than noncontagious agents. If scientific workers who develop and deploy contagious pathogens do not take extreme care, they could be infected by the pathogens on which they are working or pathogens could escape the laboratory and production facilities that house them and come to infect people working and living nearby and, through secondary spread, those who associate with these primary victims. But most importantly, once the contagious agent has been dispersed and an epidemic has commenced, depending on the effectiveness of medical and public health interventions, it may spread far beyond the target population to cause disease in neutral populations, in populations friendly to the attacker, or even in the attacker's own population.

The role of defenders is limited during the attack stage. If police, security, and intelligence agencies have not been able to penetrate a criminal group about to deploy a biological weapon, they will not be in a position to know that a biological attack is about to take place or has occurred. Conversely, if a terrorist group planning to stage a biological attack is penetrated, the assumption is that this information would be transmitted to the authorities, which would then act so as to not allow the attack to take place.

As to health providers and public health professionals, these individuals would not know that an attack is taking place, and would only become so aware when victims begin presenting to primary and acute care facilities after the incubation period of the affecting agent. At this point, the third stage of the attack commences.

ATTACK MANAGEMENT STAGE

During the attack management stage, a complicated series of either four or five activities takes place: casualty presentation, quarantine (if the pathogen in question is contagious), triage, treatment, and disposition, resolution, epidemiological investigation, and police investigation.

Casualty Presentation

The first sign that a biological attack has occurred will probably be that many sick people present to physicians' offices and emergency medical facilities. This may occur proximal to the site of the attack or, if the attack was mounted against an underground railroad station or airport, may be widely dispersed, with patients presenting to health providers in many separate locations. In addition, the longer the incubation period of the pathogen used in the attack, the more dispersed will be its victims.

There have been several attempts in the United States to model the sequelae of biological attacks in which different pathogens are used. In 1999, an analysis was done of a theoretical attack against a football stadium filled with people in Baltimore, Maryland; the pathogen used was *B. anthracis* (Inglesby et al., 1999). During an exercise in 2000 called "TOPOFF," the city of Denver was attacked with a weapon utilizing *Y. pestis* (Inglesby et al., 2000b) A mathematical model was constructed in 2001 to describe the spread of disease following the deliberate release of the smallpox virus (Meltzer et al., 2001). During June 2001, the exercise "Dark Winter" was carried out to simulate the reaction of government at various levels to the deliberate release of the smallpox virus; this exercise ended with more than 16,000 persons contracting smallpox in 25 states and 10 countries, of which some 1000 died (O'Toole and Inglesby, 2001). Lessons have also been drawn from the responses of hospitals and emergency medical facilities to the pressures created by severe outbreaks of natural diseases such as influenza (Schoch-Spana, 2000), Nipah virus infection, West Nile fever, and Severe Acute Respiratory Syndrome (SARS) (Gerberding, 2003; United States General Accounting Office, 2003a).

The main finding of these exercises and studies is disconcerting: the U.S. health care delivery system and public health system are ill equipped for many reasons to adequately manage the consequences of biological attacks that generate mass casualties (United States General Accounting Office, 2003b). Further, there would be great confusion and inefficiencies among the local, state, and federal agencies responsible for assisting or coordinating assistance between local health providers and other first responders directly involved in managing the aftermath of a biological attack. Specific problem areas and deficiencies include the following:

- Supplies of vaccine are far less than needed to alleviate a national biological catastrophe involving most viruses (except smallpox).

- If the causative agent is contagious, quarantine facilities and equipment and supplies required to isolate stricken persons may quickly prove inadequate.

- If the causative agent of the catastrophe is a bacterial species, local antibiotic supplies may be inadequate to fulfill the need and therefore world run out quickly.

- Riots uncontrollable by local authorities may erupt as desperate people fight over limited supplies of antibiotics or vaccines.

- Large numbers of the affected city's critical workforce may quickly disappear, including police, firefighters, transportation operators, sewage treatment workers, utilities staff, and food market operators. Similarly, health workers at all levels may desert their posts as medical supplies run out and safety equipment proves inadequate.

- Communications within the affected city may quickly break down or become so clogged as to be unusable. The lack of dependable information may aggravate the sense of panic likely to be present among the affected city's inhabitants.

- Beyond a certain point, health care facilities may become overtaxed and cease diagnosing and treating casualties. As a result, desperate persons may try to enter health care facilities by force.

At the local level, an important issue for health care facilities is how to provide adequate shelter to victims of a mass casualty event. In the Unites States, in order to become accredited and maintain their accredited status, hospitals and emergency medical departments must prepare a plan for meeting the exigencies of a disaster, and are enjoined to stage exercises on a regular basis that serve to familiarize staff with the plan and its operational provisions. These plans typically include provisions to be followed in case a disaster occurs that generates a large number of casualties who present to the hospital. Some provisions spell out how a large number of casualties are to be accommodated, procedures for triage, apportioning resources, and so on. Most, if not all, accredited hospitals can be assumed to have developed plans for dealing with large numbers of injured and ill persons, although how well these plans will work in an actual emergency cannot be known until one occurs.

Undoubtedly, each hospital has a breaking point; that is, the point at which a hospital is no longer able to accommodate all persons who present themselves for treatment. It is reasonable to believe that the breaking point for a small community hospital will be lower than a large county or city hospital. Whatever the specific case, a hospital's disaster plan should make clear what needs to be done when the situation approaches or reaches the breaking point. Were a biological emergency to occur, a hospital approaching or at its breaking point must be able to ask for assistance from higher levels of government; in the first instance, a state government. In the United States, hospital preparedness and response plans have instructions on how to do this, including telephone numbers of responsible persons and agencies.

Quarantine

Quarantine is considered whenever health providers deal with a patient infected, or believed to be infected, with a contagious pathogen. If a contagious pathogen were used in a biological attack, it would bring about a more complicated and dangerous situation for first responders, emergency medical facilities, health providers, and public health professionals compared to a noncontagious pathogen. The difficulties would accrue from the very first moment of the disease outbreak. When the first ill persons present for care, those performing the initial medical evaluations and administrative functions would be unknowingly exposed to the causative pathogen. If initial physicians have a high index of suspicion and recognize the illness to be potentially contagious early on, the question that must immediately be answered is what to do with the ill persons. Generally, even the largest and best equipped emergency medical facilities in the United States have only one or at the most two rooms in which a patient can be completely isolated; that is, a room that can be sealed off from the rest of the emergency medical facility, has negative air pressure (air from the isolation room does not flow to the open environment without first having been filtered), and has an air handling apparatus equipped with High Efficiency Particulate Air (HEPA) filters. Of course, the parent hospitals of emergency medical facilities have regular rooms that can be quickly adapted for isolation use, but when this is done the specialized equipment required becomes rapidly depleted. In view of these universally limited resources, were a sizeable outbreak to take place in the United States, isolation rooms at the local level would be filled, and then overfilled, almost immediately.

Similar to solving problems related to receiving and housing an overwhelming number of casualties, the directors of many hospitals and emergency medical facilities have recognized the limited ability of their institutions and departments to receive and process persons afflicted with contagious diseases that require isolation. Disaster plans of these facilities usually specify the conversion of a certain section of the hospital to an isolation ward; the rapid assembly of temporary housing, such as communal tents; or the transport to the hospital's site of large trailers in which patients may be sheltered in isolation from the general population. Even if a hospital does not have a specific plan to meet the challenges posed by a biological event, as noted above, all accredited hospitals have disaster plans that include provisions for sheltering large numbers of casualties. These shelters could be designated as isolation facilities where sick persons are kept apart from the healthy if necessary.

Triage, Treatment, and Disposition

Hospital disaster plans always have provisions that address triage. Triage is defined as "... the sorting of and allocation of treatment to patients and especially battle and disaster victims according to a system of priorities designed to maximize the number of survivors" (Merriam-Webster, 2004). Emergency physicians in particular must be prepared to perform triage in cases in which large numbers of sick persons present to the emergency department (ED) and resources to treat them are limited. Generally, the physician who performs the initial evaluation must decide on one out of three courses of action. If the person being evaluated is minimally affected, treatment is delayed and/or he or she is rapidly evacuated. If the person is moribund, he or she will be made comfortable but denied scarce resources. And a person who is severely injured or very sick but has a good chance of survival if accorded rapid treatment is given first call on medicines and services.

The physician responsible for performing triage has a very challenging role (Burkle and Frederick, 2002). To do it right, he or she must be an astute clinician and be aware of the extent of available human health-care resources, the patient carrying capacity of permanent and temporary hospital facilities, and the extent of stocks of medicines and other expendable supplies. He or she must also be able to rapidly recalculate options and possibilities as a situation improves (for example, outside assistance becomes available) or worsens (for example, more than the expected number of sick persons appear). The efficiency and effectiveness of triage depends, in the first instance, on the quality of the physician or physicians responsible for performing the triage, and also on how well hospital disaster plans are developed, how clearly protocols are written, and the smoothness with which triage procedures are performed (best honed by repeated practical exercises). If these elements are in place, triage will work well; if not,

patients will suffer as confusion reigns and treatment is either misapplied or needlessly denied.

Before treatment is given to a victim of a biological event, health providers must ask themselves whether it is first necessary to decontaminate that person. Unlike victims of chemical events, in most cases of biological exposure no decontamination is necessary beyond disrobing the victim and washing exposed areas of the body with soap and water, because most pathogens and toxins (with the exception of the rare mycotoxin) are not dermally active (Macintyre et al., 2000). Also, since victims of biological attacks will likely have changed clothing during the intervening time between exposure and the development of disease symptoms and signs (i.e., during the incubation period of the disease), and will likely have taken showers or baths as well, in all probability no residual contamination will remain in need of decontamination. In the field, however, decontamination would become a serious and immediate issue. For example, referring to September/October 2001, once the letters carrying *B. anthracis* spores were opened, the need for the decontamination of affected rooms and persons became obvious and immediate.

The ability of local health providers to effectively manage the aftermath of a biological event depends primarily on the number of patients that present for treatment and the timeframe in which they do so. There are very few hospitals and emergency medical facilities in the United States that are in a position to receive, treat, and dispose of over a few hundred victims of a biological attack presenting within a period of one or two days. If the victims were to number more than 1000, probably no single health-care facility anywhere would be able to adequately handle the load: there simply would not be a sufficient supply of human resources, antibiotics, antidotes, and critical care equipment, such as ventilators and respirators, on hand to treat so many persons on such short notice. This being the situation, most victims would have to be either transferred to nearby health care facilities, thereby spreading the treatment burden, or housed in temporary facilities in wait of new medical supplies and equipment en route from stockpiles elsewhere. The efficient implementation of either approach demands a high degree of coordination between local, state, and federal authorities (see below).

Were a mass casualty event to occur, the civilian authorities would have to coordinate activities with the military. In the United States, states have National Guard units that can be activated by the state government executive, usually the governor. Such activations are not rare; they take place every year with varying degrees of frequency when nature unleashes tornadoes, hurricanes, earthquakes, fires, floods, and other destructive forces that cause widespread damage. National Guard units can be assembled relatively quickly and usually have access to stores of medical equipment, motor parks filled with vehicles, and other supplies that would be useful in efforts to manage the aftermath of a biological event. In addition, they have the manpower to control crowds, enforce quarantine, and so forth.

Were the National Guard to prove inadequate, state governments could seek assistance from the U.S. Department of Defense (DoD) and the Veterans Administration. If it were clear to the federal government that a disaster was in the making and help was desperately needed, the president would order the DoD to provide the requested assistance. The military would have all of the manpower, supplies, and transportation needed for most large casualty events, but, as U.S. law makes clear, the local civil authorities would remain in charge upon the arrival of federal military aid. In addition, the Veterans Administration, which operates a large number of large hospitals and controls significant stockpiles of pharmaceutical and other medical supplies, can be expected to make its hospitals available for patients that cannot be accommodated in civilian hospitals.

Regarding treatment, in the United States there are stockpiles of antibiotics and other medicines strategically located throughout the country. In particular, the CDC controls the Strategic National Stockpile (SNS), formerly the National Pharmaceutical Stockpile (NPS). These medicines, vaccinations, and other assets can be distributed upon formal request by the governor of an affected state to the Director of the CDC. Once requested, the CDC Director has the authority, in consultation with the Surgeon General and the Secretary of Health and Human Services, to order the deployment of the materials. If deployed, the materials would be accompanied by CDC technicians, who would serve in an advisory role to state or local authorities to ensure that SNS assets were put to prompt and effective use. The CDC also has over 80 million doses of smallpox vaccine on hand that could be sent to any locality affected by this disease. In addition to the SNS, the Veterans Administration has stores of broad-spectrum antibiotics on hand, such as Ciprofloxin, while private medical and surgical supply houses store rather large quantities of expendable supplies in major cities, all of which could be accessed fairly quickly. All in all, there is a substantial capability resident in the U.S. health delivery system to quickly deliver to local health providers the means for treating a large number of sick persons. However, nobody knows what the carrying capacity of this system is. Would it be able to supply sufficient antibiotics within 12 and 18 hours to treat 1000 victims of an aerosol attack in which spores of *B. anthracis* were used? 5000 victims? What about if a future attack were to utilize a contagious virus, such as one of the hemorrhagic fever viruses, against which there are no vaccines?

Indications of answers to these questions can be had from results of TOPOFF, the exercise mention above (Hoffman and Norton, 2000). In this exercise, terrorists released an aerosol containing *Y. pestis* at a center of performing arts in Denver, Colorado. Three days later, the first sick persons presented to health providers and emergency medical facilities. By the end of the third day, 3700 cases had been diagnosed, and of these 950 died. By this time, available resources were depleted. Isolation was not possible because patients and healthy but worried persons overwhelmed both the ability of health care providers to conduct interviews and the capacity of the hospital isolation rooms. Therefore, an emergency declaration was issued ordering all citizens of Denver to remain at home. This home quarantine of some two million persons could not be maintained as food ran out and panicked persons tried to depart from the city, however. Because of limited supplies, antibiotics were provided only to health care workers, first responders, public safety workers, and their families. The Colorado state epidemiologist wrote that the single most important lesson learned from the exercise was that "it became clear that unless controlling the spread of the disease and triage and treatment of ill persons in hospitals received equal effort, the demand for health-care services will not diminish" (Hoffman and Norton, 2000).

In practical terms, it is difficult to evaluate the worth of exercises. It is possible that persons from agencies at all levels of government who usually do not interact now have a better appreciation of the need to cooperate with one another. It might be that decision-making, which proved inefficient during the TOPOFF exercise, will be improved so that when a real event occurs, orders for distribution of supplies and manpower will be given more quickly and followed more surely. And there may be a more efficient management strategy for sick persons as a result of this exercise as well. One would hope that these benefits are accruing over time, but in the end only a real biological event will show the worth of such exercises.

Resolution Stage

During the resolution stage, the biological event terminates or is controlled. Thus, victims of a biological accident or attack have either succumbed to disease or recovered, health care providers and public health professionals resume their normal duties, medicines and expendable supplies are replenished, and society as a whole heals from the injuries and insults that the event brought about. For the purposes of this entry, the epidemiological investigation that inevitably will be carried out during this period is of primary importance. If this investigation indicates or demonstrates that the biological event was deliberately caused, a police investigation would ensue.

Some biological events may not be resolvable given today's resources and knowledge (such as the presently under way AIDS pandemic), others may take a long time to resolve (such as the legacy of the field testing of *B. anthracis* spores on Gruinard Island (Manchee and Steward, 1988) in 1943), and still others may be resolved with dispatch (such as three of the biological events noted above). We do not know what type of biological event will next affect our society, but if it is a deliberately caused event, such as a biological attack, its resolution will happen in one of three ways. First, if it is a food- or beverage-borne attack, either of which is generally self limiting, its resolution will depend on the removal or complete consumption of the contaminated food or beverage. This might be done in the course of normal activities; for example, the contaminated food is consumed so that none remains to infect persons other than those unlucky enough to have partaken of the original potion, or the dishes containing food remnants are washed and their contents flushed away. As nothing remains of the contaminant, no new illnesses occur and the outbreak ends. Public health professionals will probably deem an outbreak that ends in this way as having had a natural etiology. Second, a contaminated food or beverage might be diluted to such an extent that its pathogen or toxin load is no longer sufficient to infect persons, or alternatively the contaminant may be broken down by natural forces so that it is no longer viable or toxic. In this situation, the outbreak will end as described above, and its etiology probably will also be deemed to have been natural. Third, an epidemiological investigation generates findings that reveal the etiology of the outbreak, and this knowledge is used to destroy or decontaminate the source of the infectious or toxic material. Once this is done, the outbreak ends as described above. An outbreak that ends in this way probably will be investigated further, with a high probability that its deliberate nature is recognized.

Epidemiological and Police Investigations

In the United States, local and state health departments have the responsibility for carrying out epidemiological investigations. Most states employ a state epidemiologist to direct such efforts; epidemiologists who staff county health departments assist these directors. If problems arise that are too complex or difficult for state and local health departments to handle, or if they would like to confirm the validity of their findings, they can call on help from the federal CDC. The CDC has a special unit set up for the purpose of performing field investigations as requested by both local U.S. authorities and foreign governments. This unit, which was established in the early 1950s (Langmuir, 1980), is called the Epidemiological Intelligence Service (EIS). It has highly trained epidemiologists on call at all times, ready to travel to the furthest places of the world on short order, and. has an enviable record for rapidly uncovering the etiology of disease outbreaks in which the responsible pathogen is unfamiliar or unknown to the medical community before the outbreak. For example, in the 1970s, the EIS cleared up the mystery surrounding an outbreak of what proved to be Legionnaires diseases in Philadelphia (Winn, 1988), and in 1993 it clarified the etiology of an outbreak of atypical pneumonia in Four Corners, New Mexico, which turned out to be caused by a previously unknown strain of hantavirus (Nichol et al., 1993). It is reasonable to believe that were a deliberately caused outbreak to occur in the future, its etiology would be fairly quickly determined by epidemiologists from local or state health departments assisted as necessary by the CDC.

There is, however, a caveat to the foregoing: it could be very difficult to know whether a biological attack has taken place if the attacker uses a foodborne pathogen. The difficulty is that the incidence of disease caused by foodborne pathogens is already a significant problem and one that has received considerable attention in recent years. According to an estimate prepared by the CDC, foodborne disease is probably responsible for 76 million illnesses, 325,000 hospitalizations, and 5000 deaths every year in the United States (Anonymous, 2001). More than 200 pathogens are associated with foodborne illness, including at least 20 identified only in the past two decades, and more than 80 percent of all illnesses and 64 percent of deaths are attributed to pathogens that have not yet been identified (Mead et al., 1999).

It can be seen that public health professionals investigating an outbreak caused by a deliberately dispersed foodborne pathogen or toxin may not know how it came about and therefore would tend to believe that its etiology was a natural source. The history of deliberately caused biological events is indicative. Although there have been very few such incidents, the largest in terms of causing casualties resulted from the deliberate contamination in 1984 of salad bars by members of the Rajneeshee cult (Carus, 2000). The cult's attack caused 751 persons to contract gastrointestinal disease. However, the initial investigation performed by the local health department determined that the rash of illnesses had a natural etiology. It was not until some months after this investigation had been concluded that a Rajneeshee member had crisis of conscience and confessed to his involvement in the biological attack. If he had not done so, the outbreak in The Dalles, Oregon, probably would still be recorded as a natural event today. The lesson here is that despite the United States possessing very sophisticated capabilities in reference to the performance of epidemiological investigations at the local and federal levels, a biological attack may go undetected because its manifestations cannot be distinguished from outbreaks having a natural origin, or because the outbreak resulting from a biological attack is never fully investigated due to a bias from the beginning on the part of investigators, who decide without further ado that it has a natural etiology.

Although mistakes can be made, it should be made clear that the United States has a powerful epidemiological investigative capability at local and federal levels. The lynchpin of this capability is the CDC, which has an extensive and thorough training program for producing many of the epidemiologists that staff county, state, and federal health departments and agencies. It is reasonable to believe that this system will detect biological attacks that manifest overtly suspicious characteristics and clarify their etiologies within a reasonable time. Biological attacks that utilize common pathogens to sabotage food or beverages are more likely to escape detection, as would bethose caused by contagious pathogens in which a limited number (one or only a handful) of individuals are infected at the onset. However, were there to be a biological attack in the future, it would seem that the perpetrator would wish this to be known and widely publicized in order to satisfy the likely goal of the attack, whether that is to generate fear, draw attention to a specific cause or ideology, or both. If so, the perpetrator would be certain to make known what had occurred, or he or she would stage the attack in such a way that its etiology was obvious. In any case, a thorough epidemiological investigation would have to be done in order to gather clues and evidence that could be used by police and the FBI to track down and convict the perpetrator.

CONCLUSION

What would happen if a real biological event generated such a large number of casualties that no single hospital or collection of hospitals in a municipality would be able to care for them all? Would the health care delivery system in the United States be able to adequately process and treat such a high number of victims? Would public health authorities be able to contain the outbreak and determine its etiology?

Several large-scale exercises have been carried out to try to answer these questions. In these exercises, pathogens such as *B. anthracis*, *Y. pestis*, and smallpox virus have been dispersed by virtual terrorists, and the effects of the resulting outbreaks on health care delivery systems, public health systems, and governance have been observed and analyzed. As is discussed above, the findings from these exercises are worrisome because they indicate that a massive breakdown of all three systems could occur. As a result, there is a wide recognition among policymakers of the need to improve preparedness in the United States to meet the challenges posed by large-scale biological events (Independent Task Force, 2003).

Actually, the U.S. government's concern about biological events, especially bioterrorist events, precedes these exercises. After the 1995 Aum Shinrikyo attack, in which the chemical nerve agent sarin was released in the Tokyo subway, the Clinton administration recognized that terrorists not only posed a threat to the United States, but that this threat included the possible employment of chemical, biological, radiological, or nuclear (CBRN) weapons against Americans. As a result, a program was designed, funded, and implemented to improve preparedness for such events at the local, state, and federal levels. The total cost of this effort is approximately $11 billion per year; of this, about $1 billion is being spent for biological and chemical preparedness and defenses. The Bush administration's thinking on this subject appears to resemble that of the Clinton administration, such that preparedness efforts are likely to continue for the foreseeable future. To illustrate, the Bush administration is reportedly seeking $11 billion over two years (2003–2004) to protect the nation against biological terrorism (Miller, 2002). Further, in his State of the Union address on January 28, 2003, President Bush announced the implementation of Project Bioshield, a massive effort to strengthen the U.S. homeland security strategy (Fauci, 2003). Bioshield was authorized by Congress in spring 2003. It bears mentioning that there now is an enormous amount of literature on the U.S. preparedness program in which authors debate the program's intent, appropriateness, provisions, size, and so on. For the purposes of this entry, it is not necessary to discuss the pros and cons of this program; the program exists and, as noted, is likely to continue and expand. Whether it will work effectively cannot be known until a biological mass casualty event occurs.

Acknowledgments
This entry is an abridged version of the following article: Zilinskas, Raymond A. 2003. "Preparedness for biological events in Japan and the United States: A comparative study of hospitals, emergency medical departments, and governmental agencies," in A Comprehensive Study of Bioterrorism (English Part) by Masaaki Sugishima (ed.), Legal Research Institute, Asahi University, Japan, pp. 48–84.

REFERENCES

Anonymous, *MMWR*, **50**(13), 241–246 (2001).

Burkle, F.M., A Systems Approach to Triage and Management of a Large-Scale Bioterrorism Event, Paper presented at the *6th Asia-Pacific Conference on Disaster Medicine*, Fukuoka, Japan, February 21, 2002.

Carus, W.S., "The Rajneeshees (1984)," in J.B. Tucker, Ed., *Toxic Terror: Assessing Terrorist Use of Chemical and Biological Weapons*, MIT Press, Cambridge, MA, 2000, pp. 115–138.

CDC Strategic Planning Workgroup, *MMWR*, **49**(RR04), 1–14 (2000).

Fauci, A.S., The power of biomedical research, *Washington Times*, July 9, 2003.

Gerberding, J.L., *N. Engl. J. Med.*, **348**(2), 2030–2031 (2003).

Hoffman, R.E., and Norton, J.E., *Emerg. Infect. Dis.*, **6**(6) (2000), http://www.cdc.gov/nciDoD/eid/vo6no6/hoffman.htm.

Independent Task Force, Council on Foreign Relations, *America Still Unprepared—America Still in Danger*, Council of Foreign Relations, Washington, DC, 2003.

Inglesby, T.V., and O'Toole, T., *Biosecur. Bioterror. Biodef. Strat. Pract., Sci.*, **1**(2), 97–110 (2003).

Inglesby, T.V., Grossman, R., and O'Toole, T., *Biodefense Q.*, **2**(2), 1–10 (2000b).

Inglesby, T.V., Henderson, D.A., Bartlett, J.G., Ascher, M.S., Eitzen, Jr., E.M., Friedlander, A.M., Hauer, J., McDade, J., Osterholm, M.T., O'Toole, T.P.G., Perl, T.M., Russell, P.K., Tonat, K, Working Group, Civilian Biodefense. *J. Am. Med. Assoc.*, **281**(18), 1735–1745 (1999).

Langmuir, A.D., *Public Health Rep.*, **95**(5), 470–477 (1980).

Macintyre, A.G., Christopher, G.W., Eitzen, Jr., E.M., Gum, R., Weir, S., DeAtley, C., Tonat, K., Barbera, J.A., *J. Am. Med. Assoc.*, **283**(2), 242–249 (2000).

Manchee, R.J. and Stewart, W.D.P., *Chem. Britain*, **24**(7), 690–691 (1988).

Mead, P.S., Slutaker, L., Dietz, V., McCaig, L.F., Bresee, J.S., Shapiro, C., Griffin, P.M., and Tauxe, R.V., *Emerg. Infect. Dis.*, **5**(5), 607–625 (1999).

Meltzer, M.I., Damon, I., Leduc, J.W., and Millar, J.D., *Emer. Infect. Dis.*, **7**(6) (2001), http://www.cdc.gov/nciDoD/eid/vol7no6/meltzer.htm.

Merriam-Webster Online Dictionary. Merriam-Webster Incorporated 2004: http://www.m-w.com/.

Miller, J., Bush to request big spending push on bioterrorism, *New York Times*, February 4, 2002, 1.

Nichol, S.T., Spiropoulo, C.F., Morzunov, S., Rollin, P.E., Ksiazek, T.G., Feldmann, H., Sanchez, A., Childs, J., Zaki, S., and Peters, C.J., *Science*, **262**, 914–917 (1993).

O'Toole, T. and Inglesby, T., *Biodefense Q.*, **3**(2), 1–3 (2001).

Schoch-Spana, M., *Biodefense Q.*, **1**(4), 1, 2, 8 (2000).

Shapira, S.C. and Shemer, J., *Isr. Med. Assoc. J.*, **4**, 489–492 (2003).

United States General Accounting Office, *Infectious Disease Outbreaks: Bioterrorism Preparedness Efforts have Improved Public Health Capacity, but Gaps Remain (Statement by Janet*

Heinrich, Director, Health Care-Public Health Issues), Report GAO-03-654T, 2003a.

United States General Accounting Office, *Bioterrorism: Preparedness Varied across State and Local Jurisdictions*, Report GAO-03-373, 2003.

United States General Accounting Office, *Severe Acute Respiratory Syndrome: Established Infectious Disease Control Measures*

Helped Contain Spread, But a Large-Scale Resurgence May Pose Challenges (Statement by Marjorie E. Kanof, Director, Health Care—Clinical and Military Health Care Issues), Report GAO-03-1058T, United States General Accounting Office, Washington, DC, 2003b.

Winn, Jr., W.C., *Clin. Microbiol. Rev.*, **1**(1), 60–81 (1988).

BOTULINUM TOXIN (BOTULINUM NEUROTOXIN, BOTULIN (ARCHAIC), BONT/A, BONT/B (SCIENTIFIC NOMENCLATURE FOR BOTULINUM TOXIN TYPE A, B, ETC.), BOTOX®, DYSPORT®, MYOBLOC®, NEUROBLOC® (WHEN USED AS A THERAPEUTIC FOR CONDITIONS CHARACTERIZED BY EXCESSIVE MUSCLE CONTRACTION))

STEPHEN S. ARNON

INTRODUCTION

Botulinum toxin causes the illness called botulism. There are three natural forms of human botulism; foodborne, wound, and intestinal (infant), and two man-made forms, inhalational (from aerosolized toxin) and iatrogenic (from excess therapeutic toxin). Foodborne botulism was recognized as a distinct illness early in the nineteenth century in Germany and termed sausage poisoning, after which it eventually became known as botulism (from the Latin word for sausage, *botulus*). While botulism occurs worldwide, it is rare in the United States, causing approximately 25 foodborne cases per year, 25 wound cases per year, and 100 intestinal infant cases per year. Intestinal botulism, more commonly known as "infant" botulism, can also occur very rarely in older children and adults. People of all ages and both genders are susceptible to botulinum toxin.

Botulinum toxin is produced by the obligate anaerobic bacterium *Clostridium botulinum* and by rare species of *C. baratii* and *C. butyricum* that have naturally acquired the toxin gene (Fig. 1). Botulinum toxin is produced in one or more of seven "toxin types" that have been arbitrarily assigned the letters A–G and that are distinguished from each other by slight variations in their shapes. Thus, neutralizing antibody ("antitoxin") against type A toxin will neutralize type A toxin but will not neutralize toxins B–G, and so forth. Hence, to protect fully against potential botulinum toxin weapons, antitoxins against all seven toxin types are needed. In addition, in recent years, the genes for botulinum toxin have been spliced into various nonclostridial bacteria, for example, *Escherichia coli*, which thus become potential biological weapons as well (*E. coli* is a common colon bacterium present in all humans).

Botulinum toxin is one of just six "Class A" (maximum threat) potential biowarfare/bioterrorism agents identified by the Centers for Disease Control and Prevention (CDC) and is the only one that is nonreplicating.

BACKGROUND

Botulinum toxin has been weaponized by a number of nations. In the 1930s, the Japanese Army's biological warfare (BW) program Unit 731 killed Chinese prisoners in Manchuria by feeding them live cultures of *C. botulinum.* During World War II, the United States produced botulinum toxin and also made more than one million doses of botulinum toxoid vaccine for Allied troops preparing to invade Normandy on D day because of concerns that Germany had weaponized botulinum toxin. Research on botulinum toxin as a weapon was also performed by the former Soviet Union during the Cold War. After the 1991 Gulf War, Iraq admitted to the United Nations that it had loaded 10,000 L of botulinum toxin (theoretically enough to kill twice the entire world's population by inhalation) into specially designed artillery bombs and SCUD missiles. Some of the weaponized botulinum toxin known to have been produced in Iraq and the former Soviet Union may still exist. Other nations classified by the U.S. Department of State as "State sponsors of terrorism" (Iran, North Korea, Syria) have developed, or are believed to be developing, botulinum toxin as a weapon. The Japanese cult Aum Shinrikyo attempted the release of aerosols of botulinum toxin on civilian and military targets in Japan at least three times between 1990 and 1995 but failed to produce casualties for technical reasons: The cult's *C. botulinum* seed cultures, which it had obtained from soil in northern Japan, were nontoxigenic (Arnon et al., 2001).

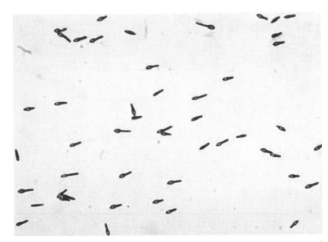

Figure 1. A photomicrograph of *Clostridium botulinum* type A viewed using a Gram stain technique. Public Health Images Library (PHIL) id# 2131; Credit: CDC/Dr. George Lombard.

Encyclopedia of Bioterrorism Defense,   Edited by Richard F. Pilch and Raymond A. Zilinskas
ISBN 0-471-46717-0 Copyright © 2005 Wiley-Liss

TECHNICAL SUMMARY

Botulinum toxin is a dichain protein that consists of an approximately 100,000 dalton "heavy" chain and an approximately 50,000 dalton "light" chain joined by single disulfide bond. It can be inactivated simply by heating it, or solutions and foods containing it, to 85 °C for 5 min. The heavy chain contains the binding and cell-internalization domains, while the light chain is an enzyme that cleaves key "fusion" proteins that enable the motor neuron to release acetylcholine and cause the muscle fibers to contract. Botulinum toxin can be disseminated as an aerosol or, more easily, in foods or beverages that will not be adequately heated before being consumed. Foodborne botulism and botulinum toxin are not contagious and thus cannot be transmitted from person to person. Although botulinum toxin is temporarily stable in untreated water and some beverages, no instance of waterborne botulism has ever been reported.

Botulinum toxin is the most poisonous substance known to science. The lethal dose of botulinum toxin for humans has been estimated from primate studies: For a human weighing 70 kg (154 pounds), the lethal amount of crystalline type A toxin would be approximately 0.09 to 0.15 µg intravenously or intramuscularly, 0.70 to 0.90 µg inhalationally, and 70 µg orally (Arnon et al., 2001). The toxin molecule produces flaccid paralysis by enzymatically blocking the release of neurotransmitter (acetylcholine) from peripheral motor neurons. Botulinum toxin does not affect sensory nerves or the central nervous system. Hence, patients with botulism are afebrile and have a clear sensorium but appear lethargic and even comatose because of droopy eyelids, flaccid facial expression, and difficulty in speaking. Death results from paralysis of airway and respiratory muscles and consequent suffocation.

Symptoms of foodborne botulism may begin as early as 2 h or as long as eight days after ingestion of toxin; patients typically present 12 to 72 h after the implicated meal. The diagnosis of botulism is established by identification of botulinum toxin in serum, feces, gastric washings, wound debris, or an epidemiologically implicated food, and the presence of *C. botulinum* organisms in these materials may also contribute to diagnostic certainty.

CURRENT STATUS

A licensed equine-derived trivalent (ABE) antitoxin is made available for civilian use by CDC; an equine-derived heptavalent (A–G) antitoxin developed by the U.S. Army is under evaluation as an Investigational New Drug (IND). The license application for human-derived botulinum antitoxin for the treatment of infant botulism is under review by the U.S. Food and Drug Administration (FDA). Despite equine antitoxin treatment, most adult patients with botulism have a lengthy (weeks to months) hospital stay. Treatment of infant botulism patients with human-derived antitoxin within three days of hospital admission shortens mean hospital stay from approximately six weeks to approximately two weeks and reduces mean hospital costs by approximately $100,000. Treatment of patients with botulism is quite expensive because of their frequent need for intensive care, mechanical ventilatory support, and tube or parenteral feeding.

Botulinum toxin is a particularly hazardous terrorist threat agent because of its ease of production, transport, and dissemination in foods or beverages. In addition, because it can be aerosolized, the toxin presents a plausible albeit unlikely threat of large-scale dissemination. Economic modeling of a bioterrorism attack with aerosolized botulinum toxin in a suburban Canadian setting estimated that 50,000 persons would become paralyzed and the associated cost burden in U.S. dollars would be $5.6 billion (St. John et al., 2001).

THE FUTURE

Efforts are underway to develop bioengineered human neutralizing antibodies against all seven serotypes of botulinum toxin; such human bioengineered antitoxin could be given prophylactically as well as therapeutically. An abundant, locally pre-positioned supply of this product would effectively deprive terrorists of botulinum toxin as a mass casualty weapon and would thereby "deweaponize" botulinum toxin as a Class A agent (Arnon et al., 2001). Efforts are also underway to develop recombinant vaccines against all seven toxin serotypes using a fragment of the heavy chain; such vaccines would replace the current IND pentavalent (A–E) toxoid. When available, the heptavalent recombinant vaccines will likely be restricted to military or other high-risk populations because immunization with these vaccines will deprive the recipient from ever benefiting from therapeutic botulinum toxin. Rapid, ultrasensitive, and specific detection and diagnostic tests for botulinum toxin, now under development, are very much needed.

REFERENCES

Arnon, S.S. et al., *J. Am. Med. Assoc.*, **285**(8) (2001); http://jama. ama-assn.org/cgi/content/full/285/8/1059#ACK, December 7, 2003.

St. John, R., Finlay, B., and Blair, C., *Can. J. Infect. Dis.*, **12**, 275–284 (2001).

WEB RESOURCES

CDC website:

Botulism: Technical Information, http://www.cdc.gov/ncidod/dbmd/diseaseinfo/botulism_t.htm.

Botulism in the United States, 1899–1996; Handbook for Epidemiologists, Clinicians, and Laboratory Workers, http://www.cdc.gov/ncidod/dbmd/diseaseinfo/botulism.pdf

Facts about Botulism, http://www.bt.cdc.gov/agent/botulism/factsheet.pdf.

See also TOXINS: OVERVIEW AND GENERAL PRINCIPLES.

BREEDERS: A CASE STUDY

Monterey WMD-Terrorism Database Staff

INTRODUCTION

The Breeders are a single issue environmental group that in 1989 claimed responsibility for a Mediterranean fruit fly (or Medfly) infestation in Southern California, with the professed motive of protesting pesticide spraying in the area (Chavez and Simon, 1989). The group claimed responsibility for the infestation through letters sent to a variety of locations, including the offices of Los Angeles Mayor Tom Bradley, the Department of Agriculture, and various media outlets. As these letters are the only known evidence of the group's existence, information on its membership, structure, and ideology is unavailable. No contact with the group has been made since 1989, leading many to assume that the letters were a hoax; however, the infestation did display some characteristics suggesting that its origin may in fact have been sabotage.

CHARACTERISTICS OF THE INFESTATION

The infestation began in Whittier, California, on September 26, 1989 (Johnson, 1989). As the infestation progressed, officials noticed the following characteristics:

- Far fewer Medfly larvae were found during the infestation than would normally have been expected.
- Over two-thirds of the captured flies were inexplicably female.
- The size of the affected region continued to grow despite the vigorous employment of pesticides, with new flies often discovered directly outside recently sprayed areas (Johnson, 1990).

The irregular pattern of infestation led some officials to consider the possibility of a deliberate event, and letters claiming responsibility on the part of the Breeders only contributed to the speculation. Finally, on December 6, 1989, a scientific panel was convened to review the case found. While the panel had no proof that the recently discovered letters were legitimate, it was unable to provide answers for the mysteries surrounding the infestation (Johnson, 1989).

CHARACTERISTICS OF THE LETTERS

In their letters, the Breeders stated that they had imported Medfly larvae to Southern California and had "started breeding them there" (Chavez and Simon, 1989). The attacks were described in the letters as follows: "Every time the copters go up to spray, we'll go into virgin territory or old Medfly problem areas and release a minimum of several thousand blue-eyed Medflies" (Dunn, 1990). The letters directed those wishing to contact the group to the classified section of the *Los Angeles Times*, but nevertheless stated that "we are under no obligation to reply" (Dunn, 1990). Ultimately, the U.S. Department of Agriculture placed multiple ads in the *Los Angeles Times* without reply, and the Breeders were never heard from again (Dunn, 1990).

REFERENCES

Chavez, S. and Simon, R., Mystery letter puts a strange twist on latest medfly crisis, *Los Angeles Times*, December 3, 1989.

Dunn, A., *Los Angeles Times*, February 10, 1990.

Johnson, J., Invasion of pesky medfly defies logic, *Los Angeles Times*, December 30, 1989.

Johnson, J., Female medfly found in sun valley close to area targeted earlier, *Los Angeles Times*, January 4, 1990.

Monterey WMD-Terrorism Database, Center for Nonproliferation Studies, 1998–2003.

BRUCELLOSIS (*BRUCELLA* SPP.)

E.J. Richey

C. Dix Harrell

INTRODUCTION

Brucellosis in humans can be caused by *Brucella abortus* (generally a disease of cattle), *Brucella melitensis* (generally a disease of goats), *Brucella suis* (generally a disease of pigs), and *Brucella canis* (generally a disease of dogs). At least the first three species have been considered in the context of biological warfare (BW). Iraq's former biological weapons program, for example, may have developed *B. melitensis* in an offensive capacity (Leitenberg, 2000), while the pre-1969 U.S. biological weapons program is known to have weaponized *B. suis*. This entry focuses on *B. abortus* as a representative cause of brucellosis due to its pertinence in the realm of agricultural biosecurity as well as human and animal bioterrorism. Other species of *Brucella* can be expected to act and be managed similarly; however, not all of the information below applies to all forms of the disease.

BACKGROUND

Brucellosis is an infectious and contagious bacterial disease of animals primarily caused by *B. abortus* (Fig. 1). This form of the disease primarily affects cattle and the American buffalo. Brucellosis sometimes infects horses, dogs, swine, and humans; however, brucellosis is considered to be a "dead-end disease" in these species, meaning it will not be transmitted to other animals. In the United States, the disease has been a major economic problem for the beef and dairy industries, their consumers and related agricultural industries. Brucellosis once affected more than 10 percent of the cattle population and perhaps as many as 30 percent of the cattle herds in the United States. Annual production losses to U.S. dairy producers alone were estimated at $499 million annually in 1951 (1993 dollars) before an eradication program began. After investing over $3.5 billion in state, federal, and industry funds since 1951 in joint state-federal cooperative eradication programs, the United States has reduced the brucellosis-infected herd count from 124,000 in 1957 to 40 as of November 30, 1996. If there were not a brucellosis eradication program, it is estimated that

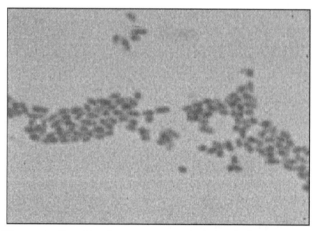

Figure 1. *Brucella abortus*, gram stain (x3200), CDC website: http://www.bt.cdc.gov/documents/PPTResponse/agentsglenfinal.pdf.

current losses due to reduced supplies of meat and milk would be in excess of $800 million annually.

In Cattle

In cattle, brucellosis is primarily a disease of the female, the cow. Bulls can be infected but they do not readily spread the disease. The brucellosis organism localizes in the testicles of the bull and produces an orchitis (inflammation of the testicles), whereas in the female, the organism localizes in the udder, uterus, and lymph nodes adjacent to the uterus. The infected cows exhibit symptoms that may include abortion during the last third of pregnancy, retained afterbirth, and weak calves at birth. Infected cows usually abort only once. Subsequent calves may be born weak or healthy and normal. Some infected cows will not exhibit any clinical symptoms of the disease and give birth to normal calves. The brucellosis organism is shed by the millions in the afterbirth and fluids associated with calving and aborting. The disease is spread when cattle ingest contaminated forages or lick calves or aborted fetuses from infected cattle. Outside the animal, the afterbirth, and aborted calves, the brucella bacteria are easily killed by sunlight, high temperatures, and drying;

Encyclopedia of Bioterrorism Defense, Edited by Richard F. Pilch and Raymond A. Zilinskas
ISBN 0-471-46717-0 Copyright © 2005 Wiley-Liss

however, the brucella organisms are difficult to control while they are in the animal. There is no economical cure for a brucellosis-infected animal.

In Humans

In humans, *B. abortus* causes undulant fever, a disease characterized by intermittent fever, headaches, fatigue, joint and bone pain, psychotic disturbances, and other symptoms. It is contracted through exposure to *B. abortus*-contaminated milk and infected organs from infected animals. Livestock and slaughter industry workers, and consumers of unpasteurized milk and other dairy products made from unpasteurized milk are at the greatest risk of contacting undulant fever. Transmission occurs through contact with the tissues of infected animals at slaughter or ingestion of unpasteurized milk derived from infected cattle. (Note that currently only one dairy herd in the United States is known to be infected with brucellosis).

CONTROL AND ERADICATION

The brucellosis control and eradication program has been and continues to be multifaceted; the program uses surveillance testing at the farm, at the stock markets, and at slaughter facilities; quarantine and herd depopulation with indemnity payments; herd management; and vaccination. Any bovine that is known to be infected with the field strain of *B. abortus* is required to be placed under quarantine until slaughtered.

Surveillance Testing

Surveillance testing consists of conducting a serological test or a series of serological tests to detect the presence of *B. abortus* antibodies in the animal. Animals that test negative to the initial official screening test are classified as "negative" and are considered NOT to be infected with brucellosis. Animals that are positive on the initial screening test are classified as "positive," and further testing is required to assist in a more definitive diagnosis of brucellosis. Serological screening tests for *B. abortus* antibodies include the Buffered Acidified Plate Antigen (BAPA) test, which is used solely in the laboratory, and the "Card Agglutination" test, which is an official test conducted at the livestock markets or in the laboratory to classify animals as either positive or negative. Samples that test positive on the BAPA test are subsequently subjected to the "Card Agglutination" test in the laboratory. Screen-test positive samples could come from either an animal infected with field *B. abortus* organisms or from an animal that has persistent antibodies to a brucellosis vaccination.

To determine if a screen-test positive animal is either infected with the disease or if it still has antibodies that were caused by vaccination, its blood samples are subjected to official supplemental serological testing. The Rivanol, the Complement Fixation (CF), and the Particle Concentration Fluorescent Immuno Assay (PCFIA) tests are official supplemental serological tests used in the laboratory. The results of the official supplemental testing are subsequently interpreted by a trained brucellosis epidemiologist who makes the final determination on the screen-test positive animal; it is either designated as "suspect" and will require further testing, or it is classified as a "reactor" and must be sent to slaughter under permit.

Suspects, if they have not been slaughtered, are required to be retested in 30 to 60 days. Quarantining the suspect animal or the herd from which the suspect originated will be determined by the trained epidemiologist.

Reactors, those animals at highest risk of being infected with field strain brucellosis, are required to be appropriately tagged and branded within 15 days of the owner receiving official notification of the reactor status. The herd from which the reactor originated is placed under immediate quarantine. All reactors are "B" branded on the left tail-head; the reactor animal(s) must be slaughtered at an approved slaughter facility within 15 days of tagging or branding.

Quarantine

When a herd has been officially quarantined because of brucellosis, any movement of nonneutered cattle into and out of the herd is restricted (neutered animals are steers and spayed heifers). Movement of nonneutered cattle out of quarantined herds is allowed to approved destinations only and only with a written permit (USDA Form VS-127) obtained from state or federal animal health personnel or a veterinarian accredited by United States Department of Agriculture (USDA). Brands are required on all nonneutered quarantined animals prior to movement; reactors are "B" branded on the left tail-head and all exposed or known suspect cattle are "S" branded on the left tail-head. Approved destinations include quarantined pastures, quarantined feedlots, and approved slaughter facilities.

Release from Quarantine

Herds that are classified as "Quarantined Reactor Herds" and are subsequently found NOT to be infected with field stain *B. abortus* are released from quarantine without any further restrictions. Quarantined infected herds can be released following a minimum of two negative whole-herd tests. One additional negative whole-herd test is required following quarantine release for the herd to remain released from quarantine. A whole-herd test must include all animals in the herd over 6 months of age except steers, spayed heifers, brucellosis-vaccinated heifers less than 24 months of age, and bull calves less than 18 months of age.

The first negative whole-herd test must be conducted no less than 30 days following removal of the last reactor cattle from the herd and the second negative whole-herd test must be conducted no less then 180 days following the removal of the last reactor animal from the herd. For the herd to remain released from quarantine, a postquarantine whole-herd test must be conducted no less than 6 months following release of the herd from quarantine. Failure to conduct the postquarantine whole-herd test can result in the herd being placed back under quarantine until the required testing is completed.

HERD DEPOPULATION

Quarantined herds that are determined to be infected with field strain *B. abortus* will remain under quarantine. A quarantined herd may be considered for depopulation if the owner agrees and the herd is approved for depopulation by state and federal authorities. If an infected herd is selected to be depopulated, all nonneutered cattle will be slaughtered and the owner will receive an indemnity for each animal included in the depopulation. Neutered cattle (steers and spayed heifers) may be allowed to remain on the premises for additional growth before marketing; however, spayed heifers will be required to be appropriately branded with an official "spay" brand.

HERD MANAGEMENT

Cattle herds in states or areas that are not free of brucellosis are at risk of contracting the disease. Factors that increase a herd's risk of being infected with brucellosis in endemic areas are listed as follows:

- *Replacement Cattle*. Purchasing replacement cattle is a common method of herd-to-herd transmission of brucellosis. The more replacement animals a herd owner purchases the higher the risk of contracting brucellosis.
- *Source of Additions*. The source of purchased herd additions also affects an owner's risk of purchasing brucellosis-infected cattle.
  - Buying replacement cattle from special auction sales, herd dispersal sales, and/or consignment sales does not appear to significantly increase a herd's risk of contracting brucellosis.
  - Purchasing cattle from private treaty sales and livestock dealers/order buyers increases the risk of buying brucellosis-infected cattle. Cattle moving through these marketing channels often bypass change of ownership testing prior to the sale. In addition, some owners may be selling their cattle by these methods in order to avoid their cattle being tested and the potential that their herd may be placed under quarantine.
  - Herds consisting of replacements bought from regular auction sales have a higher risk of contracting brucellosis. Some cattle moving through these sales can test negative but be in the incubation stage of the disease. During the incubation stage of the disease, the animal is infected but as yet has not developed clinical signs of brucellosis nor has it had sufficient time to produce antibodies against the organism; hence the test is negative. More likely, however, brucellosis-infected cattle that are marketed through these sales are untested (but infected) brucellosis-vaccinated heifers that are considered "too young to test." In this way, infected heifers may move to other farms without a test, spreading brucellosis in the process.
- *Distance between Herds*. The distance between a noninfected herd and a brucellosis-infected herd affects the risk of contacting brucellosis. Herds located less than half a mile away from brucellosis-infected herds are more than four times as likely to become infected as other herds. Herds half a mile to one mile away are still at risk of brucellosis spreading through the cattle community. However, the spread of the disease into a herd is much less likely to occur when it is located over one mile from the brucellosis-infected herd.
- *Stray Cattle*. Herds reporting stray cattle are more than twice as likely to have brucellosis as herds not reporting stray cattle intrusion into the herd.
- *Contact with Animals*. In areas where brucellosis exists, herds that have contact with foxes, stray dogs, and coyotes are at higher risk of contacting brucellosis. Domestic and wild canids can spread brucellosis by dragging dead/aborted calves and afterbirth (placenta) between neighboring herds. Coyotes have been shown to shed the brucellosis organism in their feces for several weeks following ingestion of infected material.
- *Multiple Herds*. Ownership of multiple herds potentially increases the risk of a herd being infected with brucellosis. The more herds a rancher owns the higher is their probability of being exposed to brucellosis.
- *Culling Practices*. Culling practices are important factors when considering a herd's risk of having brucellosis. Owners of small cattle herds usually cull less than 10 percent of their herd each year. In general, it requires 3 to 5 years (or longer) to detect an infected cow through livestock marketing or slaughter testing if the owners cull less than 10 percent of their cow herd each year. If the industry solely relies upon market and slaughter testing to detect brucellosis-infected herds, many small brucellosis-infected herds will remain undetected just because of the culling practices used on the herd.

VACCINATION OF FEMALE CATTLE AGAINST BRUCELLOSIS

Strain-19 Vaccine

Until recently, "Strain-19 vaccine" was the only brucellosis vaccine used in the brucellosis control programs for cattle in the United States. Strain-19 vaccine was and still is an effective tool in brucellosis control. However, as with any tool, using Strain-19 vaccine has its advantages and disadvantages. Strain-19 is a live vaccine that stimulates the immune system of the vaccinated animal to resist a brucellosis disease challenge, produce antibodies against the disease organisms, and kill off the vaccine organisms. Normally, a vaccinated animal will retain the resistance to the disease for an extended period of time (years) but the detectable antibodies will disappear in a few months.

Unfortunately, the serological tests used to detect brucellosis-infected cattle cannot differentiate between antibodies produced against the Strain-19 Vaccine and antibodies produced against the brucellosis disease organism; hence, if a vaccinated animal is tested too soon following vaccination or if the vaccinated animal retains the antibodies stimulated against the vaccine for an extended period, the vaccinated animal would test positive. In addition, some animals vaccinated with Strain-19 vaccine (such as those whose uteruses and/or mammary glands have begun to develop) will become permanently infected with the vaccine organism, constantly producing antibodies against it and thus continuing to test positive.

Calves born to Strain-19-vaccinated cows will acquire anti-brucellosis antibodies from the cow through the colostrum (first-milk) immediately after birth. These acquired antibodies will normally be circulating in the calves' blood system for 4 to 6 months, and can neutralize or kill the live vaccine organisms if the calf is vaccinated during the time it still possesses the antibodies. Hence, it is required that no heifer calf be vaccinated before 4 months of age. Because some calves mature early and become permanently infected with the live vaccine organism, and because older calves can produce excessive antibodies when vaccinated, heifer calves should be vaccinated before they are 10 months of age. The routine vaccination of cattle/herds against brucellosis is restricted to heifers between 4 and 10 months of age.

RB-51

In 1996, the USDA officially recognized and began using a new brucellosis vaccine for vaccinating cattle against the disease. Like Strain-19, the new vaccine, "RB-51," is a live vaccine derived from the cattle brucellosis bacteria, *B. abortus*. Unlike Strain-19, the RB-51 vaccine does not stimulate antibodies that are detected by the standard brucellosis serological tests. Thus, the problem of having some brucellosis-vaccinated cattle testing positive has been alleviated. As with Strain-19, the RB-51 vaccine is to be administered only by state and federal brucellosis program personnel and USDA-accredited veterinarians. The right age for vaccinating heifers is between 4 and 10 months, and proper permanent identification is required for all vaccinated cattle. Permanent identification consists of applying an official USDA brucellosis ear tag and an official ear tattoo to each vaccinated animal.

Who Should Vaccinate?

Cattle owners have inquired about the feasibility of continuing to vaccinate heifers against brucellosis. The answer is not a simple one; it has to be the decision of the herd owner. But the herd owners must realize that brucellosis has not been eradicated from the United States, and if their herd is located in an area of brucellosis infection, then vaccinate. If their herd is at risk of contracting brucellosis (described above), then vaccinate. If they are selling replacement heifers and their clients will only purchase vaccinated heifers, then vaccinate.

ROUTINE HERD TESTING

If a herd has had at least two consecutive negative whole-herd blood tests between 10 and 14 months apart, the herd is eligible for "certification." Certification is not automatic; it requires, in addition to the negative tests, that the herd owner apply for the "Brucellosis Certified Free" status at their respective State Veterinarian's Office. Once certified, a herd must have an annual negative whole-herd test between 10 and 14 months after the certification anniversary date to be eligible for recertification. Herd owners also frequently ask if they should continue to test their herds annually to retain a "Brucellosis Certified Free Herd" status. Basically, it is an economic decision depending upon the type of operation and the location of the herd(s). If the cattle operation is selling breeding stock or replacement heifers throughout the year, a certified herd keeps them ready to sell at any time without any additional brucellosis test requirements. In addition, the certified status provides the buyers with some degree of confidence that they are not "buying" brucellosis. If the herd is located in an area where brucellosis-infected herds exist, an annual test will allow the owner to detect infection faster than waiting for a market/slaughter test to find it. Whole-herd testing every 10 to 14 months provides an excellent monitoring tool to insure that brucellosis has not been introduced into a brucellosis-free herd. In reality, the owner should consider having a "Brucellosis Certified Free" status a bonus from using an important disease-monitoring tool.

BRUCELLOSIS IN THE UNITED STATES

As of December 1, 1996, there were 37 brucellosis-free states and 14 Class A states in the United States. The brucellosis Class Free classification is based on a state carrying out all requirements of the state-federal cooperative eradication program and finding no cases of brucellosis in cattle or bison during a 12-month period. To be eligible for Class A status, a state must have a herd infection rate of less than 2.5 herds per 1000 for a 12-month period (0.25 percent). Several of the current Class A states have not detected a brucellosis-infected herd during the last year and are on the 12-month countdown to Class Free. However, experience indicates that if a state has no infected herds for six years, it has eradicated the disease and no more brucellosis remains in the cattle population waiting to cause a problem; many states fall within this category. All-in-all, the brucellosis eradication program has made major advances in controlling this disease; however, if you maintain cattle in areas that are still at risk, please consult your private veterinary practitioner to discuss whole-herd testing of your cattle and whether brucellosis vaccination of heifers should be continued as a part of your herd health program.

As of September 16, 1998, the following states are brucellosis-free: Alabama, Alaska, Arizona, Arkansas, California, Connecticut, Colorado, Delaware, Georgia, Hawaii, Idaho, Indiana, Illinois, Iowa, Kentucky, Maine, Maryland, Massachusetts, Michigan, Minnesota, Montana, Nebraska, Nevada, New Hampshire, New

Jersey, New Mexico, New York, North Carolina, North Dakota, Ohio, Oregon, Pennsylvania, Rhode Island, South Carolina, Tennessee, Utah, Vermont, Virginia, Washington, Wisconsin, West Virginia, Wyoming, Canada, and Puerto Rico.

Class A states are Florida, Kansas, Louisiana, Mississippi, Missouri, Oklahoma, South Dakota, and Texas.

Acknowledgment

This is a reprint of document VM-100, one of a series of the Department of Large Animal Clinical Sciences, College of Veterinary Medicine, Florida Cooperative Extension Service, Institute of Food and Agricultural Sciences, University of Florida.

It was first printed in March 1997. Updates on numbers are provided by the Colorado Department of Agriculture State Veterinarian's office. Only the introduction was added for this volume.

REFERENCES

Leitenberg, M., *Asian Perspect.*, **24**(1), (2000).

WEB RESOURCES

For an online version of this article, please visit, http://www.ag. state.co.us/animals/livestock_disease/bruc.html.

C

CAMELPOX

Wendy Kleinman

BACKGROUND

The camelpox virus is the closest relative of the variola virus, the cause of smallpox (Gubser and Smith, 2002). In recent years, it has been speculated that camelpox virus could be utilized as a biological weapon because of the genetic similarities it shares with variola, either by the deliberate manipulation of its genome or natural evolution to fill the ecological niche created when smallpox was eradicated over two decades ago. Both camelpox virus and variola are members of the *Orthopoxvirus* genus, which consists of some of the largest and most complex viruses known to science (Henderson et al., 1999). Orthopoxviruses are endemic in many animal species, causing such diseases as cowpox and monkeypox. Researchers suspect that the camelpox virus and variola stem from a common ancestor, most likely a rodent virus that existed some 7000 years ago (Gubser and Smith, 2002). While camelpox is widespread among the world's camel populations, the camelpox virus has rarely, if ever, infected humans (Fenner et al., 1988).

POTENTIAL FOR FUTURE HUMAN PATHOGENIC EFFECTS

The similarities in genetic structure of the camelpox virus to variola raise some serious concerns about the natural progression of the disease. Recent research suggests that camelpox virus may possess some of the genes necessary to become a human pathogen (Gubser and Smith, 2002). Further, the possibility of the camelpox virus, or other orthopoxviruses, emerging as a threat to human health increases as the proportion of the world's immunologically naïve population grows. Routine smallpox vaccination ceased years ago, and recent efforts to resume vaccinations have stalled. Therefore, much of the world's population carries no immunity to smallpox and other orthopoxviruses. In addition, the increasing number of immunocompromised individuals, such as those with AIDS, cancer, and organ transplants, increases the probability of successful invasion of a human host (Gubser and Smith, 2002).

BIOLOGICAL WEAPONS APPLICATION

There is concern among security analysts that the camelpox virus may be used for biological warfare (BW) purposes. For example, in 1995, Iraq admitted to United Nations Special Commission inspectors that it had researched camelpox virus within its offensive BW program. It has been suggested that this virus was being investigated with the idea that as a result of frequent contact with camels, the Iraqi populations would be immune to camelpox but foreign troops would not. However, most analysts believe that the virus was being studied as simulant for the smallpox virus (Coghlan and Mackenzie, 2002). It is unclear how many countries, if any, are currently pursuing research on variola or genetically similar virus for BW purposes.

CONCLUSION

The camelpox virus, and other orthopoxviruses, is a significant threat to all human populations (note the monkeypox outbreak in the United States in June 2003). There exists a need for continued research into these complex viruses in order to prevent and prepare for epidemics they may cause, whether of natural or laboratory etiology.

REFERENCES

Coghlan, A. and Mackenzie, D., Fear over camelpox as bioweapon. *New Scientist*, published online April 17, 2002, www.newscientist.com.

Fenner, F., Henderson, D.A., Arita, I., Jezek, Z., and Ladnyi, I.D., *Smallpox and its Eradication*, World Health Organization, Geneva, 1988.

Gubser, C. and Smith, G.L., *J. Gen. Virol.*, **83**(4), 855–872 (2002).

Henderson, D.A., Inglesby, T.V., Bartlett, J.G., Cher, M.S., Edward, E., Jahrling, P.B., Jerome, H., Marcelle, L., Joseph, Mc.D., Osterholm, M.T., Tara, O., Gerald, P., Trish, P., Russell, P.K., and Kevin, T., *J. Am. Med. Assoc.*, **281**, 2127–2137 (1999).

WEB RESOURCES

UNSCOM Reports to the Security Council, January 25, 1999, http://www.fas.org/news/un/iraq/s/990125/.

Encyclopedia of Bioterrorism Defense, Edited by Richard F. Pilch and Raymond A. Zilinskas
ISBN 0-471-46717-0 Copyright © 2005 Wiley-Liss

CDC CATEGORY C AGENTS

MICHELLE LONGMIRE

INTRODUCTION

In the year 2000, the Centers for Disease Control and Prevention's (CDC) Strategic Planning Workgroup devised a list of three categories of critical biological agents based on the following characteristics: "(1) public health impact based on illness and death; (2) delivery potential to large populations based on stability of the agent, ability to mass produce and distribute a virulent agent, and potential for person-to-person transmission of the agent; (3) public perception as related to public fear and potential civil disruption; and (4) special public health preparedness needs based on stockpile requirements, enhanced surveillance, or diagnostic needs" (Rotz et al., 2002). Threat categories are denoted as A, B, and C, with Category A agents deemed to be the greatest national security threat. Category C consists of agents not presently considered public health threats in terms of bioterrorism but which could emerge in this capacity as the scientific understanding of each grows in the future. Category C agents Nipah virus, the hantaviruses, and multiple drug-resistant (MDR) *Mycobacterium tuberculosis* are discussed in turn below.

NIPAH VIRUS

History

Named after the town where it was first discovered in Malaysia, Nipah virus emerged in 1998 as a newly recognized zoonotic virus. Although the outbreak began in September of that year, it remained unnoticed by health authorities until June of 1999. During this period of time, 265 cases of encephalitis and 105 deaths were reported. Since the initial outbreak, Nipah virus has continued to afflict Malaysia and Singapore. Like other viruses of the family *Paramyxoviridae*, Nipah virus warrants concern because of its broad host range and high level of mortality (Chua et al., 2000).

Biological Weapons Application

There have been no known applications of this virus in a weapons context.

Natural Host

It is believed that fruit bats serve as the primary host for Nipah virus. Fruit bats in the east and southeast

regions of Australia, Indonesia, Malaysia, the Philippines, and some Pacific Islands carry the virus but show no sign of illness and remain largely unaffected by Nipah infection.

Transmission

The exact mode of transmission is unclear, but it appears that intimate contact with contaminated tissue or bodily fluids is required to transmit the Nipah virus. It is unknown if bat-to-human transmission is possible. Evidence suggests that pigs are the primary source of infection in human cases, but other domestic animals have not been ruled out (World Health Organization, 2003). There are no documented cases of human-to-human transmission.

Pathogenesis

Blood vessels are a primary target of Nipah virus. Endothelial ulceration is the most common type of vascular damage observed at autopsy, although lesions caused by multinucleated syncytium in the endothelium have also been reported. Autopsies have also revealed extensive vascular lesions in the brain of Nipah virus victims (Wong et al., 2002).

Clinical Course

The incubation period of the virus ranges from 4 to 18 days. After incubation, many cases remain mild or subclinical. However, some cases advance to flu-like symptoms, high fever, and myalgia. A proportion of these cases can progress to encephalitis, causing symptoms such as disorientation, convulsions, and coma. Fifty percent of symptomatic cases result in death, and encephalitis can for reasons unknown become a relapsing condition in survivors.

Treatment

Early treatment with the antiviral drug ribavirin has been shown to reduce the duration and severity of illness (Wong et al., 2002). However, because of the limited availability of this drug, treatment primarily relies on intensive supportive care.

Encyclopedia of Bioterrorism Defense,   Edited by Richard F. Pilch and Raymond A. Zilinskas
ISBN 0-471-46717-0   Copyright © 2005 Wiley-Liss

HANTAVIRUSES

Overview

Hantaviruses are members of the family *Bunyaviridae*, which consists of 5 genera and some 250 species. Infection can cause hantavirus pulmonary syndrome (HPS) or hemorrhagic fever with renal syndrome (HFRS): HPS is primarily found in North and South America, while HFRS occurs most frequently in the Eastern Hemisphere (CDC, 2003).

History

Hantavirus-related illnesses were first reported in Russia in 1913, but remained unnoticed by Western physicians until the early 1950s, when a large portion of UN forces in Korea fell ill with HFRS. In 1993, a previously unknown group of hantaviruses emerged in the Four Corners region of the United States (New Mexico, Colorado, Arizona, and Utah), causing the acute respiratory disease later named HPS (Schmaljohn and Hjelle, 1997). Since the emergence of HPS, intense research has led to the discovery of several new hantaviruses.

Biological Weapons Application

The hantavirus causing Korean hemorrhagic fever was researched by the United States' former biological weapons program. Its potential utility as a biological weapon became apparent to the United States when its soldiers became ill with the disease while stationed in Korea in the early 1950s. However, as far as is known, hantavirus was never weaponized.

Transmission

In a favorable environment, hantaviruses are hardy, and able to live outside the host for prolonged periods, during which they may become aerosolized and incidentally infect humans. The primary route of transmission to humans is inhalation of aerosolized rodent excreta, especially urine. However, transmission can also occur from rodent bites or other direct contact of infectious particles with broken skin or mucous membranes.

Natural exposure has been linked with outdoor activities such as heavy farm work, threshing, sleeping on the ground, and military exercises. Indoor exposure is correlated with activities that cause rodent excreta to become aerosolized, for example, sweeping out old sheds or interior home renovation work.

Clinical Course

The clinical courses of both HFRS and HPS vary from mild to moderate to severe, depending on which virus is responsible for the disease.

HFRS. Approximately 150,000 to 200,000 cases of hospitalization due to HFRS are reported worldwide annually. As noted, HFRS collectively refers to a group of hantavirus-related illnesses that primarily affect the Eurasian landmass and adjoining areas, and is a potential sequelae of such diseases as Korean hemorrhagic fever, epidemic hemorrhagic fever, and nephropathia epidemica.

The clinical course of HFRS has five overlapping phases: febrile, hypotensive, oliguric, diuretic, and convalescent. Symptoms typically appear suddenly and include headache, backache, fever, and chills. Early indications of hemorrhage include facial flushing and injection of the conjunctiva and mucous membranes during the febrile phase of illness.

HPS. The full clinical spectrum of HPS is undefined. HPS is generally limited to a mild pulmonary disease, though some cases may involve a renal component. The incubation time for HPS is unknown but is thought to be anywhere between one and five weeks. Common early symptoms include fatigue, fever, and myalgia. About half of all patients experience headaches, dizziness, chills, and/or abdominal problems such as nausea, vomiting, diarrhea, and abdominal pain. Late symptoms occur four to ten days after the onset of initial symptoms and include coughing and severe shortness of breath. Although the disease is similar to HFRS, in cases of HPS, the associated capillary leakage is localized exclusively in the lungs rather than in the retroperitoneal space; the kidneys are generally not affected. Death results from shock and cardiac complications.

Most cases of HPS in the United States and Canada have been caused by the so-called Sin Nombre virus. However, a newly recognized group of hantaviruses—which includes Bayou, Black Creek Canal, and Andes viruses—has also been found to cause HPS.

Treatment

Judicious application of the antiviral agent ribavirin shortens the length of illness and improves survival in HFRS. An open label trial of ribavirin in 1993–1994 did not find the drug to be effective against HPS (most patients who received the drug were noted to be critically ill at the time, however). Early intensive care is necessary in HPS, including administration of inotropic drugs (dobutamine), mechanical ventilation, and careful monitoring of oxygenation, blood pressure, and fluid balance.

MDR-MYCOBACTERIUM TUBERCULOSIS

History

While tuberculosis (TB) has afflicted humans since ancient times, only after the development of antitubercular drugs in the 1970s did strains of antibiotic-resistant TB appear. MDR-TB, defined as the disease caused by TB bacilli resistant to isoniazid (INH) and rifampin (the two most powerful anti-TB drugs), usually results from incomplete or poorly supervised TB treatment. Cases are seen throughout the world but are especially prevalent in areas with large numbers of immunocompromised persons or inadequate healthcare systems, such as Africa and Eastern Europe (WHO, 2003).

Biological Weapons Application

M. tuberculosis is thought to have been researched by several countries for potential application as a biological weapon. The two primary countries believed to have engaged in such research are World War II Japan and North Korea; however, the details of these programs are not well known.

Transmission

Tuberculosis is a contagious, airborne disease but requires fairly close and prolonged contact for transmission to occur (American Lung Association, 2003). As such, MDR-TB is transmitted by inhalation of airborne infectious bacterial particles. Patients with MDR-TB remain infectious for longer periods of time than those with drug-susceptible TB because of the lack of therapeutic efficacy.

Treatment

MDR-TB is treatable with extensive chemotherapy. However, the treatment can take up to two years and is much more expensive than drug-susceptible TB treatment. Treatment first requires determining what drugs are effective against the isolate. Patients with INH-resistant tuberculosis can be treated with rifampin, pyrazinamide, and ethambutol for six to nine months. Patients with rifampin-resistant tuberculosis should be treated with INH, pyrazinamide, and ethambutol for at least nine months. If an isolate is resistant to both INH and rifampin, treatment with second-line agents is necessary and should be extended to at least 18 months (Johns Hopkins University, 2003).

REFERENCES

American Lung Association, *Tuberculosis*, http://www.lungusa. org/diseases/lungtb.html#how, accessed on 9/25/03.

Centers for Disease Control and Prevention, *All About Hantavirus*, Special Pathogens Branch, http://www.cdc.gov/ncidod/ diseases/hanta/hps/noframes/hpsslideset/index.htm#slide%20 show, accessed on 9/25/03.

Chua, K.B., Bellini, W.J., Rota, P.A., Harcourt, B.H., Tamin, A., Lam, S.K., Ksiazek, T.G., Rollin, P.E., Zaki, S.R., Shieh, W.-J., Goldsmath, C.S., Gubler, D.J., Roehrig, J.T., Eaton, B., Gould, A.R., Olson, J., Field, H., Daniels, P., Ling, A.E., Peters, C.J., Anderson, L.J., and Mahy, B.W.J., *Science*, **288**, 1432–1435 (2000).

John Hopkins University, *Multi-Drug Resistant Tuberculosis*, http://hopkins-id.edu/diseases/tb/tb_treat.html, accessed on 9/25/03.

Rotz, L.D., Khan, A.S., Lillibridge, S.R., Ostroff, S.M., and Hughes, J.M., *Emerg. Infect. Dis.*, **8**, 2 (2002).

Schmaljohn, C. and Hjelle, B., *Emerg. Infect. Dis.*, **3**, 2 (1997).

Wong, K.T., Shieh, W.J., Zaki, S.R., and Tan, C.T., *Springer Semin. Immunopathol.*, **24**, 215–228 (2002).

World Health Organization, *Nipah Virus Factsheet*, http://www. who.int/inf-fs/en/fact262.html, accessed on 9/25/03.

World Health Organization, *Tuberculosis Factsheet*, http://www.who.int/mediacentre/factsheet/who104/en/print. html, accessed on 9/25/03.

WEB RESOURCES

World Heath Organization, *Nipah Virus Factsheet*, http://www.who.int/inf-fs/en/fact262.html.

CDC, *All About Hantaviruses*, http://www.cdc.gov/ncidod/ diseases/hanta/hps/.

CDC, *Questions and Answers About TB*, http://www.cdc.gov/ nchstp/tb/faqs/qa.htm.

CENTERS FOR DISEASE CONTROL AND PREVENTION'S BIOTERRORISM PREPAREDNESS PROGRAM

RAYMOND A. STRIKAS
MAUREEN F. SINCLAIR
STEPHEN A. MORSE

INTRODUCTION

This chapter describes the Centers for Disease Control and Prevention/Agency for Toxic Substances and Disease Registry's (CDC/ATSDR) bioterrorism preparedness activities and programs as of September 2003. It includes descriptions of (1) the bioterrorism agents of primary concern, (2) the CDC's cooperative agreement with state and local health departments to foster bioterrorism preparedness, (3) the CDC's Laboratory Response Network (LRN), (4) the Strategic National Stockpile, (5) the CDC's Emergency Operations Center, (6) the national smallpox vaccination program, (7) international collaborations to prevent biological warfare (BW), and (8) the benefit these programs have offered in combating other infectious disease threats, in particular Severe Acute Respiratory Syndrome (SARS).

BIOTERRORISM AGENTS

Many biological agents can cause illness in humans, but not all are capable of affecting public health and medical infrastructures on a large scale (Rotz et al., 2002). Thus, in order to bring focus to public health preparedness activities, the CDC convened a meeting of national experts in 1999 to review the criteria for selecting the biological agents that posed the greatest threat to civilians and to help develop a prioritized list of agents (Rotz et al., 2002). This list of "Critical Agents" (Table 1) was prioritized on the basis of considerations such as the ability of the agent to cause mass casualties; the ability of the agent to be widely disseminated either by aerosol or other means; the ability of the agent to be transmitted from person to person; the public's perception, correct or incorrect, associated with the intentional release of the agent; and special public health preparedness needs (e.g., vaccines, therapeutics, enhanced surveillance, and diagnostics).

As currently defined, Category A agents, which include some of the classic BW agents, are high-priority organisms and toxins that are most likely to cause mass casualties if deliberately disseminated and require broad-based public health preparedness efforts. Natural infections caused by agents in Category A are uncommon or nonexistent in the United States. For example, prior to the bioterrorist attacks with letters carrying *Bacillus anthracis* spores in 2001, the last case of inhalational anthrax in the United States was in 1976 (CDC, 1976). Furthermore, the World Health Organization (WHO) declared smallpox, also a Category A agent, eradicated in 1977 (WHO, 1980). Category A agents also have some potential for large-scale dissemination, but generally cause less illness and death than those in Category A. Many of the Category B agents have been weaponized in the past, or are being considered as weapons by some state-sponsored programs (Miller et al., 2001). Some Category B agents could be used to contaminate food or water sources. In addition, many of these agents are relatively easy to obtain and thus are more likely to be used in the setting of a biological crime or bioterrorism (Kolavic et al., 1997). Biological agents that are not currently believed to present a high bioterrorism risk to public health, but which could emerge as future threats, were placed in Category C. Some of these agents are associated with emerging infections or have characteristics that terrorists could exploit for deliberate dissemination. The Critical Agent list has been used by the National Institutes of Health (NIH) in establishing priorities for bioterrorism-related research (NIAID, 2002).

In the United States, there is limited clinical and laboratory experience with Category A agents, as well as many of the Category B agents, which complicates efforts to recognize and confirm the diseases they cause. The low numbers of human infections in the United States caused by Category A agents (as well as for many of those in Category B) has been suggested as one reason why there has historically been a general lack of interest by the commercial sector in spending money for the development, manufacture, and Food and Drug Administration (FDA) approval of diagnostic tests for these agents. This situation has therefore created the

Encyclopedia of Bioterrorism Defense, Edited by Richard F. Pilch and Raymond A. Zilinskas
ISBN 0-471-46717-0 Copyright © 2005 Wiley-Liss

Table 1. Critical Biological Agents for Public Health Preparedness (Modified from Khan, A.S., Morse, S., and Lillibridge, S., *Lancet*, 356, 1179–1182 (2000))

Agent	Disease
Category A	
Variola major	Smallpox
Filoviruses (e.g., Ebola and Marburg)	Hemorrhagic fever
Arenaviruses (e.g., Lassa and Junin)	Lassa fever, Argentine hemorrhagic fever
Bacillus anthracis	Anthrax
Yersinia pestis	Plague
Francisella tularensis	Tularemia
Clostridium botulinum neurotoxins	Botulism
Category B	
Alphaviruses (e.g., Venezuelan, Eastern, and Western encephalomyelitis viruses)	Encephalomyelitis
Coxiella burnetii	Q fever
Brucella spp.	Brucellosis
Burkholderia mallei	Glanders
Burkholderia pseudomallei	Melioidosis
Staphylococcal enterotoxin B	Staphylococcal food poisoning
Ricin from *Ricinus communis*	Ricin intoxication
Clostridium perfringens epsilon toxin	–
Food- and waterborne agents, including (but not limited to)	
Salmonella sp.	Salmonellosis
Shigella dysenteriae	Bacillary dysentery
Escherichia coli O157:H7	Hemolytic uremic syndrome
Vibrio cholerae	Cholera
Cryptosporidium parvum	Cryptosporidiosis
Category C	
Multidrug-resistant *Mycobacterium tuberculosis*	Tuberculosis
Yellow-fever virus	Yellow fever
Tickborne encephalitis complex (flavi) viruses	Encephalitis
Tickborne hemorrhagic fever viruses	Hemorrhagic fever
Nipah and Hendra Complex viruses	
Hantaviruses	Hantavirus pulmonary syndrome

need for the government to support the development and limited distribution of biodetection assays and specialized reagents that would not otherwise be available to support the public health infrastructure and national security interests of the United States (Meyer and Morse, 2002).

COOPERATIVE AGREEMENT

The CDC's strategic plan for biological and chemical terrorism preparedness and response was implemented in 1999 through the funding of cooperative agreements with states and several large municipalities (CDC, 2000). Awards were based on up to five focus areas, with each area integrating training and research. These focus areas were (1) preparedness and planning; (2) detection and surveillance; (3) diagnosis and characterization of biological and chemical agents; (4) response; and (5) communication. After three years of funding, there have been substantial improvements (see the discussion below of the LRN); however, additional preparedness efforts were needed.

In the aftermath of the incident involving the dissemination of *B. anthracis* spores via the mail (CDC, 2001; Jernigan et al., 2002), an unprecedented increase in resources for all state health departments and select major metropolitan cities and territories for bioterrorism preparedness occurred. These additional resources continue to support state and local health departments, allowing them to greatly enhance their capacities for surveillance, epidemiology, and laboratory investigations of possible bioterrorism, natural infectious disease outbreaks, and other public health threats and emergencies. The CDC is working to integrate the surveillance of illnesses resulting from bioterrorism into a strengthened U.S. disease surveillance system. As part of this program, there is increased accountability on the part of the awardees to ensure that specific goals are met, such as increasing the number of epidemiologists in a given population area and strengthening ties with the public health laboratory. Currently, the CDC is developing a comprehensive yet flexible set of "preparedness indicators" that will serve as a way to measure movement toward universal goals and objectives regarding public health preparedness. The CDC cooperative agreement is paired with a similar agreement from the Health Resources Services Administration to assure increased capacity to identify and treat victims of terrorism through the nation's health care and hospital systems.

LABORATORY RESPONSE NETWORK

Because there is only a small window of opportunity during which prophylaxis or other control measures can be implemented to reduce the morbidity and mortality associated with a bioterrorism event, the public health response to such an event must be rapid to be effective (Kaufman et al., 1997). In order to facilitate the rapid identification of threat agents, the Laboratory Response Network (LRN) was created in August 1999. This national system is designed to link state and local public health laboratories with other advanced capacity clinical, military, veterinary, agricultural, and water- and food-testing laboratories, including those at the federal level, building upon the existing interaction of nationwide public health laboratories and their complementary disease surveillance activities (Gilchrist, 2000). The LRN is a critical component of the CDC's public health mission, enhancing readiness to detect and respond to bioterrorism at the local, state, and federal levels.

The LRN was established by the CDC, in concert with the Association of Public Health Laboratories (APHL) and with collaboration from the Federal Bureau of

Investigation (FBI) and the United States Army Medical Research Institute of Infectious Diseases (USAMRIID), to address the problem of extremely limited existing national infrastructure of diagnostic testing laboratories competent to deal with biological (or chemical) terrorism. It is the first example of a public health-law enforcement partnership. Importantly, the LRN has a dual function in that it has the ability to detect and respond not only to agents released intentionally but also to those that occur naturally, a capacity that warrants emphasis because it will generally not be known at the time of detection whether the outbreak is intentional or natural. This network is, therefore, by its very nature of great benefit to public health whether or not a terrorist incident ever occurs.

The LRN collaborative partnership assigns member laboratories to operate in either a sentinel or reference capacity, with the latter characterized by progressively stringent safety, containment, and technical proficiency capabilities. A member laboratory provides its own agent-specific self-designation, meaning that a particular laboratory may have a different capacity for testing for *B. anthracis* than testing for botulinum toxin.

Sentinel laboratories are, for the most part, hospital and other community clinical laboratories. It is likely that in the aftermath of a covert bioterrorism attack, patients will seek care at many widely dispersed hospitals, many of which would house such laboratories (Gilchrist, 2000). These laboratories participate in the LRN by ruling out or referring critical agents (Table 1) encountered in their routine work to nearby LRN reference laboratories. To make this process as rapid as possible, the CDC, in collaboration with the American Society for Microbiology (ASM), has developed protocols and algorithms for clinical laboratories. These algorithms can be found on the Internet at either organization's website: www.asmusa.org or www.bt.cdc.gov.

Reference laboratories that perform confirmatory testing are primarily local and state public health laboratories, employing both biosafety level 2 (BSL-2) facilities where BSL-3 practices are observed and public health laboratories with full BSL-3 facilities, as well as laboratories with certified animal facilities necessary for performing the requisite mouse toxicity assay for the detection of botulinum toxin. Some of these reference laboratories can perform additional tests requiring BSL-3 containment, including handling powders suspected of containing *B. anthracis* spores. There currently are 120 LRN reference laboratories in the United States and Canada. All 50 states now have laboratories with the capacity to perform presumptive and confirmatory identification of *B. anthracis*, *Yersinia pestis*, and *Francisella tularensis*, as well as having access to specialized LRN laboratories for confirmation of *Brucella* spp., *C. botulinum* neurotoxin, and Variola major.

There are two federal laboratories (CDC and USAMRIID) with BSL-4 facilities that can handle extremely dangerous pathogens, such as the Ebola and Variola major viruses, for which other laboratories have insufficient safety facilities and/or unvaccinated staff. The federal laboratories identify agents in specimens submitted by reference laboratories, and can also identify recombinant microorganisms (e.g., chimeras) that may not be recognizable by conventional isolation and identification methods. These federal laboratories also maintain extensive culture collections of critical agents against which the isolate(s) from a bioterrorist event may be compared using molecular methods in order to determine its likely origin.

The LRN played a critical and successful role in the nation's response to the bioterrorism-related anthrax letter incidents during October and November 2001. The LRN concept was validated when on October 3, 2001, the LRN laboratory in Jacksonville, Florida, rapidly confirmed that the gram-positive rod isolated from the cerebrospinal fluid and blood of the index case in Palm Beach County was in fact *B. anthracis* (Jernigan et al., 2001). Although the bioterrorism-related anthrax cases were limited to four states (Florida, New Jersey, New York, and Connecticut) and the District of Columbia (Jernigan et al., 2002), the impact was felt nationwide. During this period, LRN laboratories tested over 125,000 clinical specimens and environmental samples, involving approximately 1 million assays. The majority of these environmental samples were either part of hoaxes or specimens provided by frightened or concerned individuals; none turned out to be positive for *B. anthracis*.

Improvements made in every state as a result of the funds provided by the U.S. Congress have strengthened the national public health infrastructure. Among the improvements have been the construction of a number of BSL-3 laboratories, the hiring of additional personnel, and the purchase of state-of-the-art equipment for performing real-time nucleic acid amplification and time-resolved fluorescence antigen detection assays.

The LRN now can be described as a unified network of integrated laboratories functioning through a single operational plan and represented by laboratory first responders (sentinel laboratories) and confirmatory testing laboratories (reference laboratories) representing several disciplines. The maturity of the LRN structure is commensurate with a parallel increase in technical capacity and response ability of the public health system.

Confirmatory laboratories use standard protocols and reagents for the identification and confirmation of threat agents. Because bioterrorism is a criminal act and specimens or cultures collected during the investigation of such an event will be evidence in criminal and judicial proceedings, the protocols also have information concerning chain of custody requirements. The protocols were written by subject matter experts at CDC, USAMRIID, and the FBI, and reviewed for accuracy and ease of use by laboratory workers representing the LRN. The protocols, which are available to LRN members on a secure website currently managed by the CDC, contain the information for ordering the necessary reagents and control strains for performing the tests. The CDC currently develops, produces, validates, packages, and ships all reagents used in the rapid screening and confirmatory tests developed by CDC laboratories and its federal partners for these agents.

THE STRATEGIC NATIONAL STOCKPILE

In 1999, Congress charged the Department of Health and Human Services (HHS) and the CDC with the establishment of the National Pharmaceutical Stockpile (NPS). The mission was to establish a capability to provide a resupply of large quantities of essential medical materiel to states and communities during an emergency within 12 h of the federal decision to deploy, thus supplementing state and local public health agencies anywhere within the United States or its territories at times of crisis. The national repository contained (and continues to contain) antibiotics, chemical antidotes, antitoxins, life-support medications, intravenous administration kits, airway maintenance supplies, and medical/surgical items for this purpose.

The Homeland Security Act of 2002 charged the Department of Homeland Security (DHS) with defining the goals and performance requirements of the NPS, as well as managing the actual deployment of assets. Effective on March 1, 2003, the NPS became the Strategic National Stockpile (SNS), managed jointly by the DHS and the HHS (CDC, 2003a). The SNS program works with governmental and nongovernmental partners to upgrade the nation's public health capacity to respond to a national emergency. Critical to the success of this initiative is ensuring that a capacity is developed at federal, state, and local levels to receive, stage, and dispense SNS assets. A significant factor for the success of deployment to a state or local agency is the ability of the public health agency to stage, receive, and distribute stockpile assets when needed in an emergency. Thus, all state and selected local health departments currently are receiving funds through the CDC cooperative agreement to prepare the logistics and operating procedures needed to ensure the successful transfer of federal assets to the state and local level.

Composition and Quality

To determine and review the composition of the SNS Program assets, the DHS, DHHS, and CDC jointly consider many factors, such as current biological and/or chemical threats, the availability of medical materiel, and the ease of dissemination of pharmaceuticals. One of the most significant factors in determining SNS composition, however, is the medical vulnerability of the U.S. civilian population. The SNS program ensures that the medical materiel stock is rotated and kept within potency shelf-life limits, which involves quarterly quality assurance/quality control checks on all push packages, annual 100 percent inventory of all package items, and inspections of environmental conditions, security, and overall package maintenance.

Decision to Deploy

The decision to deploy SNS assets may be based on evidence showing the overt release of an agent that might adversely affect public health, but it is more likely that subtle indicators such as unusual morbidity and/or mortality identified through the nation's disease outbreak surveillance and epidemiology network will alert health officials to the possibility (and confirmation) of a biological or chemical incident or a national emergency. To receive SNS assets, the affected state's governor's office will directly request the deployment of the SNS assets from

the CDC, DHS. DHS, or DHHS, and other federal officials will evaluate the situation and determine a prompt course of action.

Implementation

The SNS is organized for flexible response. The first line of support is an immediate response involving 12-h push packages. These are caches of pharmaceuticals, antidotes, and medical supplies designed to provide rapid delivery of a broad spectrum of assets for an ill-defined threat in the early hours of an event. The push packages are positioned in strategically located, secure warehouses ready for immediate deployment to a designated site within 12 h of the federal decision to deploy SNS assets. Exact locations are not publicly disclosed for security reasons. The 12-h push packages have been configured to be immediately loaded onto either trucks or commercial cargo aircraft for the most rapid transportation. Concurrent with SNS transport, the SNS Program will deploy its Technical Advisory Response Unit (TARU) to coordinate with state and local officials so that the SNS assets can be efficiently received and distributed upon arrival at the site.

If an incident requires additional pharmaceuticals and/or medical supplies, follow-on vendor managed inventory (VMI) is shipped to arrive within 24 to 36 h. If the threat is well defined, VMI can be tailored to provide pharmaceuticals, supplies, and/or products specific to the suspected or confirmed agent(s) employed. In this case, the VMI can act as the first option for immediate response from the SNS. Thus, in the event of a national emergency, state and local first responders and health officials can use the SNS to bolster their response with a 12-h push package, VMI, or a combination of both (depending on the situation), but the SNS is not a first response tool in and of itself.

DHS will transfer authority for the SNS materiel to the state and local authorities once it arrives at the designated receiving and storage site. State and local authorities will then open the 12-h push package and apportion its contents. SNS-TARU members will remain on site in order to assist and advise state and local officials in putting the SNS assets to prompt, effective use.

Preparedness Training and Education

The SNS Program is part of a nationwide preparedness training and education program for state and local health care providers, first responders, and governments (to include federal officials, governors' offices, state and local health departments, and emergency management agencies). This training not only explains the SNS Program's mission and operations but also alerts state and local emergency response officials to the important issues that they must plan for in order to receive, secure, and distribute SNS assets. To conduct this outreach and training, CDC and SNS Program staff are currently working with DHS, DHHS agencies, Regional Emergency Response Coordinators at all of the U.S. Public Health Service regional offices, state and local health departments, state emergency management offices, the Metropolitan Medical Response System cities, the

Department of Veterans' Affairs, and the Department of Defense.

DIRECTOR'S EMERGENCY OPERATIONS CENTER

The CDC/ATSDR Director's Emergency Operations Center (DEOC) serves as the agency's central public health incident management center for coordinating and supporting staff, information, and other assets. The DEOC coordinates and supports the operations and logistics associated with CDC/ATSDR's preparedness and response to public health emergencies, disease outbreaks, and investigations. The DEOC provides a secure location for CDC leaders, technical experts, and emergency coordinators to transmit and receive vital information during a public health event. The CDC/ATSDR Director determines the scope of DEOC activities, which may extend across all CDC/ATSDR CIOs (Centers/Institutes/Offices) and may involve coordination with multiple local, state, and federal government agencies, as well as private industry. These activities generally involve six distinct functions: receiving information, distributing information, analyzing information, submitting recommendations, integrating resources, and synchronizing resources. The DEOC also serves as a central point for monitoring and tracking CDC/ATSDR's worldwide public health commitments, and provides a single access point for the full range of CDC/ATSDR assets and capabilities, which include specialized laboratory support, specialized technical consultation and expertise, the SNS, CDC's public health information resources (Health Alert Network, *Morbidity Mortality Weekly Report*, and EPI-X secure communications), and all agency centers, institutes, and offices.

NATIONAL SMALLPOX PREPAREDNESS PROGRAM

Smallpox has been identified as one of the two highest priority biological agents in relation to bioterrorism (Henderson, 1999). As a result of this high-priority perception, the United States began a national smallpox preparedness program in 2002. The CDC requested that grantees receiving funds through the bioterrorism cooperative agreement develop plans for managing a smallpox outbreak and also vaccinate designated civilian public health and hospital heath care smallpox response team members as a baseline preparatory measure. As of September 2003, over 38,000 response team members had received the smallpox vaccine. In addition, over 14,000 persons had been trained as vaccinators by state and local public health staff. By the time of this writing, CDC estimated that over 140,000 persons will have received training about smallpox and smallpox preparedness via in-person courses, satellite broadcasts, and CD-ROM trainings, and over 1.6 million had accessed Internet informational materials. The vaccination program was conducted safely, with no reports of those severe adverse events historically associated with the smallpox vaccine, including progressive vaccinia, eczema vaccinatum, or contact transmission (CDC, 2003b). While rare cases of myopericarditis was reported and assessed to be likely causally related to the smallpox vaccine (CDC, 2003c), as of September 2003, all those

with this reported complication had recovered from their illness. Reports of ischemic heart disease (angina and myocardial infarction) and dilated cardiomyopathy following vaccination are still under investigation, but to date the association between these events and vaccination is unclear (CDC, 2003c). The CDC is developing detailed performance measures for both the smallpox preparedness program and the overall bioterrorism cooperative agreement, to be available by the end of 2003, which will assist greatly in assuring critical levels of preparedness for smallpox and other bioterrorism threats.

INTERNATIONAL COLLABORATIONS

The CDC's international efforts against bioterrorism include collaborations with national ministries of health and international organizations to strengthen epidemiologic, surveillance, and laboratory capacity, technical support through the DEOC international team, research projects on biological threat agents with scientists in the former Soviet Union, and training programs and materials through the Public Health Training Network.

Collaborative Efforts to Strengthen Epidemiologic, Surveillance, and Laboratory Capability

While the majority of CDC's international efforts in epidemiology and surveillance are not directed specifically at bioterrorism defense, it has been recognized that strengthening the public health surveillance and response capacity for naturally occurring infectious diseases is an efficient and effective means for detecting and responding to potential bioterrorism-related threats (Polyak et al., 2002; Sandhu et al., 2002). The CDC's international collaborations to strengthen epidemiologic, surveillance, and laboratory capacity are concentrated in three program areas: applied epidemiology and training programs (AETPs), international emerging infections programs (IEIPs), and surveillance and health information systems development.

Applied Epidemiology and Training Programs. AETPs are long-term in-service training programs, based on CDC's Epidemic Intelligence Service (EIS), designed to build epidemiological capacity of host nations by providing training in applied epidemiology and other public health competencies in the context of health service delivery systems (White et al., 2001). Currently, there are approximately 30 AETP programs in existence worldwide. CDC-supported AETPs are implemented through CDC's Division of International Health in collaboration with Ministries of Health and other international partners, such as WHO, World Bank, and USAID. AETPs are responsible for detecting, investigating, and responding to acute health events in many Ministries of Health, placing them in the first line of response to bioterrorism in their countries. Epidemiologists trained in these programs contributed significantly to the global response to SARS, as well as to the investigation of possible anthrax cases and other suspected bioterrorist activities following the anthrax attacks in the United States in October–November 2001 (Sandhu et al., 2002).

International Emerging Infections Programs. IEIPs are laboratory centers of excellence that integrate disease surveillance and applied research, prevention, and control activities. IEIPs are implemented through a partnership between the CDC's National Center for Infectious Diseases and host government ministries of health. Additional partners include local universities, medical research institutes, AETPs, and U.S. military laboratories. IEIP sites strengthen national public health capacity and provide hands-on training in laboratory science, epidemiologic science, and public health administration. The first of several IEIPs that are planned worldwide was established in Thailand in 2002. The second IEIP will be located in Kenya and is scheduled to begin operating in 2004. CDC's ultimate goal is to link all IEIPs in a global laboratory network.

Surveillance and Health Information Systems. Over the past 10 years, CDC has consulted on and developed a number of international surveillance and health information systems. Most recently, CDC staff in the Division of Public Health Surveillance and Informatics designed, developed, and deployed the National Egyptian Disease Surveillance System, an electronic communicable disease surveillance information system, in collaboration with the Egyptian Ministry of Health and Population, the WHO, and the U.S. Naval Medical Research Unit No. 3. The next version of this system will support outbreak detection and identification of clusters of unusual events; provide secure, multilevel access to health information; support ad hoc data analysis and reporting; and enable linkage to statistical packages and laboratory software. This system strengthens the national disease surveillance system by providing information for early detection of and prompt response to public health threats, including bioterrorism. The system has been designed with an interface that supports data entry and manipulation in different languages and could therefore be adapted for use in other countries. Currently, CDC is negotiating with partners to support the adaptation of this system to countries in the former Soviet Union.

Technical Support Through the DEOC

In October 2001, the CDC established an international team in its Emergency Operations Center to respond to inquiries from other countries regarding anthrax and bioterrorism. The international team provided varying forms of support in response to such inquiries: consultation regarding laboratory methods for isolation and identification of *B. anthracis*, clinical and epidemiologic support, and policy and preparedness guidance (Polyak et al., 2002). The team also prepared and disseminated information about anthrax and bioterrorism to CDC international assignees and global partners, and actively collaborated with WHO.

An international team was activated to respond to the recent SARS crisis as well. In addition to providing technical support, the international team coordinated the deployment of several CDC staff members to international locations to provide direct assistance to ministries of health in affected countries. The international team

will be activated during any international public health emergency in which the DEOC is involved.

Collaborative Research with Scientists of the Former Soviet Union

The CDC is involved in collaborative research projects with scientists in countries of the former Soviet Union through the DHHS' Biotechnology Engagement Program (BTEP). The BTEP program is one of several U.S. government-sponsored programs focused on preventing the proliferation of biological weapons through redirection of biotechnical scientists (Cook and Woolf, Undated). One of the CDC's most active BTEP research collaborations involves scientists from the CDC's Division of Vector-Borne Infectious Diseases (DVBID) and the Kazakh Scientific Center for Quarantine and Zoonotic Diseases (KSCQZD) in Almaty, Kazakhstan. This collaboration focuses on comparing the enzootic nature of plague in Kazakhstan and in the United States, and determining the molecular characteristics of hundreds of *Y. pestis* isolates archived at the KSCQZD. The DVBID has recently initiated a second, multiyear project with KSCQZD on identifying molecular characteristics of *F. tularensis*. In addition, DVBID members have visited and assessed the scientific research capabilities of various Kazakh institutes, including the Scientific Research Agricultural Institute in Otar and the Institute of Microbiology and Virology in Almaty.

The Public Health Training Network

The Public Health Training Network (PHTN) was established in 1993 to provide an effective system for education of the public health workforce. It is centrally led and staffed by the CDC's Public Health Practice Program Office. The PHTN uses a mixture of instructional media to reach learners, including satellite, web, CD-ROM, videotape/DVD and other audio resources, on-site courses and conferences, and print. This mixed-media approach makes PHTN products accessible to both national and international audiences. A large portion of the products produced by PHTN has focused on terrorism and emergency response programming. Training products pertaining to anthrax, smallpox, bioterrorism preparedness, and SARS have been produced for delivery to global audiences with as little as 48 h lead time.

CDC RESPONSE TO SEVERE ACUTE RESPIRATORY SYNDROME

During November 1, 2002, to July 11, 2003, an epidemic of severe acute respiratory disease caused by a hitherto unknown coronavirus occurred. The outbreak was first recognized in Asia, but cases were ultimately reported from 29 countries overall. This disease was later designated as SARS. As of July 2003, 8427 probable cases and 813 deaths had occurred worldwide, of which 418 cases and 0 deaths were documented in the United States (CDC, 2003d). The CDC's initial response to SARS included the formation of internal teams to work with states and local areas to develop, implement, and coordinate disease control recommendations, with particular emphasis on the isolation of suspected and known SARS cases and their contacts. State health departments credited their collaboration with CDC on smallpox preparedness and other bioterrorism planning efforts as having prepared them to deal with SARS more rapidly and effectively than would otherwise have occurred (Selecky, 2003).

REFERENCES

CDC, *Emergency Preparedness and Response*, Strategic National Stockpile, 2003a, http://www.bt.cdc.gov/stockpile/index.asp accessed September 20.

Centers for Disease Control, *MMWR*, **25**, 33–34 (1976).

Centers for Disease Control and Prevention, *MMWR*, **49**(RR04), 1–14 (2000).

Centers for Disease Control and Prevention, *MMWR*, **50**, 893–897 (2001).

Centers for Disease Control and Prevention, *MMWR*, **52**, 819–820 (2003b).

Centers for Disease Control and Prevention, *MMWR*, **52**, 492–496 (2003c).

Centers for Disease Control and Prevention, *MMWR*, **52**, 664–665 (2003d).

Cook, M.S. and Woolf, A.F. *Preventing Proliferation of Biological Weapons: US Assistance to the Former Soviet States*, CRS Report for Congress, Document code RL 31368, April 10, 2002.

Gilchrist, M.J.R., *Milit. Med.*, **165**(Suppl. 2), 28–31 (2000).

Henderson, D.A., *Science*, **283**, 1279–1282 (1999).

Jernigan, D.B., Raghunathan, P.L., Bell, B.P., Brechner, R., Bresnitz, E.A., Butler, J.C., Cetron, M., Cohen, M., Doyle, T., Fischer, M., Greene, C., Griffith, K.S., Guarner, J., Hadler, J.L., Hayslett, J.A., Meyer, R., Petersen, L.R., Phillips, M., Pinner, R., Popovic, T., Quinn, C.P., Reefhuis, J., Reissman, D., Rosenstein, N., Schuchat, A., Shieh, W.-J., Siegal, L., Swerdlow, D.L., Tenover, F.C., Traeger, M., Ward, J.W., Weisfuse, I., Wiersma, S., Yeskey, K., Zaki, S., Ashford, D.A., Perkins, B.A., Ostroff, S., Hughes, J., Fleming, D., Koplan, J.P., and Gerberding, J.L., and the National Anthrax Epidemiologic Investigation Team, *Emerg. Infect. Dis.*, **8**, 1019–1028 (2002).

Jernigan, J.A., Stephens, D.S., Ashford, D.A., Omenaca, C., Topiel, M.S., Galbraith, M., Tapper, M., Fisk, T.L., Zaki, S., Popovic, T., Meyer, R.F., Quinn, C.P., Harper, S.A., Fridkin, S.K., Sejvar, J.J., Shepard, C.W., McConnell, M., Guarner, J., Shieh, W.-J., Malecki, J.M., Gerberding, J.L., Hughes, J.M.,

and Perkins, B.A., and members of the Anthrax Bioterrorism Investigation Team, *Emerg. Infect. Dis.*, **7**, 933–944 (2001).

Kaufman, A.F., Meltzer, M.I., and Schmid, G.P., *Emerg. Infect. Dis.*, **3**, 83–94 (1997).

Khan, A.S., Morse, S., and Lillibridge, S., *Lancet*, **356**, 1179–1182 (2000).

Kolavic, S.A., Kimura, A., Simons, S.L., Slutsker, L., Barth, S., and Haley, C.E., *J. Am. Med. Assoc.*, **278**, 396–398 (1997).

Meyer, R.F. and Morse, S.A., *Mayo Clin. Proc.*, **77**, 619–621 (2002).

Miller, J., Engelberg, S., and Broad, W., *Germs. Biological Weapons and America's Secret War*, Simon & Schuster, New York, 2001.

National Institute of Allergy and Infectious Diseases, The Counter-Bioterrorism Research Agenda of the National Institute of Allergy and Infectious Diseases (NIAID) for CDC Category A agents, National Institutes of Health, Bethesda, MD, 2002.

Polyak, C.S., Macy, J.T., Irizarry-De La Cruz, M., Lai, J.E., McAullife, J.F., Popovic, T., Pillai, S.P., and Mintz, E.D., and the Emergency Operations Center International Team, *Emerg. Infect. Dis.*, **8**(10), 1056–1059 (2002).

Rotz, L.D., Khan, A.S., Lillibridge, S.R., Ostroff, S.M., and Hughes, J.M., *Emerg. Infect. Dis.*, **8**, 225–230 (2002).

Sandhu, H.S., Thomas, C., and Nsubuga, P. et al., Assessment of bioterrorism response capacity of applied epidemiology and training programs worldwide—2001, *Proceedings of International Conference on Emerging Infectious Diseases*, Atlanta, GA, March 24, 2002.

Selecky, M., Presentation to the Institute of Medicine's Committee on Smallpox Vaccination Program Implementation, Washington, D.C., May 1, 2003.

White, M.E., McDonnell, S.M., and Werker, D.H. et al., *Am. J. Epidemiol.*, **154**(11), 993–999 (2001).

World Health Organization, *The Global Eradication of Smallpox: Final Report of the Global Commission for the Certification of Smallpox Eradication*, World Health Organization, Geneva, Switzerland, 1980.

WEB RESOURCES

CDC Bioterrorism website, 09/07/03 http://www.bt.cdc.gov/.

CDC Severe Acute Respiratory Syndrome (SARS) website, http://www.cdc.gov/ncidod/sars/index.htm.

CDC National Laboratory Training Network Factsheet, http://www.cdc.gov/od/oc/media/presskit/training.htm.

CENTRAL INTELLIGENCE AGENCY

MICHELLE BAKER

BACKGROUND

The U.S. Central Intelligence Agency (CIA) is the most recognized entity of the U.S. Intelligence Community (Fig. 1). The creation of the CIA coincided with the implementation of the National Security Act under President Harry S. Truman in 1947, which drastically restructured both foreign policy and the military in the United States. During World War II, the Office of Strategic Services (OSS) served as the United States' main intelligence body, yet the merger of the OSS with other smaller agencies to form the CIA allowed for greater coordination in intelligence gathering and dissemination as political leaders sought to prevent another intelligence failure similar to Pearl Harbor (http://www.state.gov/r/pa/ho/time/cwr/17603.htm, accessed on 8/29/03).

As a facet of this governmental reorganization, the Director of Central Intelligence (DCI), currently George J. Tenet, became the president's primary advisor on intelligence matters affecting national security, this in addition to his position as head of the CIA and the U.S. Intelligence Community. Although the CIA is directly responsible to the U.S. President through the DCI, it remains an independent agency directed by the National Security Council (NSC) (http://www.cia.gov/cia/information/info.html, accessed on 8/29/03). This independence, however, does not negate the CIA's intrinsic involvement in operations undertaken by other sectors of the government or military. In 1992, Congress amended the National Security Act to allow the CIA greater freedom in collecting foreign intelligence and providing guidance to other intelligence agencies. Coinciding with the passage of this amendment, the CIA was given the mandate of providing Congress "substantive analysis" of intelligence (http://www.access.gpo.gov/su_docs/dpos/epubs/int/index.html, accessed on 8/29/03).

INTELLIGENCE GATHERING IN THE UNITED STATES

The CIA, although only one entity of the U.S. Intelligence Community, is perceived by many Americans to be the sole source of U.S. intelligence. This view is not entirely inaccurate, as the DCI focuses the majority of his efforts on overseeing the CIA, but other members of the Intelligence Community play a substantial role as well (http://www.intelligence.gov/1-members.shtml, accessed on 8/29/03). These include the following:

- Defense Intelligence Agency (DIA)
- National Security Agency (NSA)
- National Reconnaissance Office
- National Imagery and Mapping Agency
- Military Intelligence Agencies (Army, Navy, Air Force, and Marines)
- Department of Homeland Security (DHS)
- Department of State
- Department of Energy
- Department of the Treasury
- Federal Bureau of Investigation (FBI)
- U.S. Coast Guard

Budget figures for intelligence agencies within the Intelligence Community are not readily available; however, in 1998, the aggregate budget for all intelligence agencies was $26.7 billion, of which the CIA accounts for approximately one-eighth(http://www.cia.gov/cia/public_affairs/faq.html, accessed on 8/29/03; http://www.access.gpo.gov/su_docs/dpos/epubs/int/index.html, accessed on 8/29/03).

According to the Special Assistant to the DCI, collection and analysis of information concerning biological weapons proliferation involving both states and nonstate actors is one of the greatest challenges currently facing the Intelligence Community. The U.S. Intelligence Community

Figure 1. The CIA seal. Source: www.cia.gov/.

Encyclopedia of Bioterrorism Defense,   Edited by Richard F. Pilch and Raymond A. Zilinskas
ISBN 0-471-46717-0   Copyright © 2005 Wiley-Liss

focuses on three major elements in combating the threat of biological weapons proliferation (http://usembassy-australia.state.gov/hyper/WF981120/epf508.htm, accessed on 8/29/03):

- Assessment and warning
- Deterrence, disruption, and protection
- Monitoring of arms control regimes

The CIA is increasingly concerned about the ability of terrorists and other lone actors to acquire and use pathogens and toxins, particularly given that many of these agents are readily available and may be weaponized through the use of sophisticated biotechnologies. Detection of weaponization programs is becoming progressively more difficult, as facilities engaged in legitimate research can simultaneously carry out clandestine biological weapons-related activity without a change in their appearance. Dual-use concerns are compounded for the CIA as they attempt to monitor the status of former Soviet biological weapons facilities and scientists, which could conceivably supply both rogue states and terrorist groups with technology and expertise relating to biological warfare (BW). Determining the threat of proliferation requires the Intelligence Community to maintain awareness not only of the biological capabilities of states and other actors but also the "intentions" of those involved (http://www.cia.gov/cia/public_affairs/speeches/1999/lauder_speech_030399.html, accessed on 8/29/03).

EFFORTS TO COMBAT BIOLOGICAL WEAPONS DEVELOPMENT AND USE

Currently, the CIA is engaged in a broad range of efforts to combat biological weapons development and use, particularly in relation to threats involving nonstate actors and terrorists. Reports stress the lack of constraint on nonstate actors regarding biological agent acquisition, and the significant gaps in intelligence concerning terrorist acquisition of potentially deadly agents. Understanding the threat of biological weapons proliferation involves examination of both supply and demand of BW agents, which the CIA attempts to accomplish through careful analysis of current and prior country and group capabilities. The CIA is required to produce an unclassified biannual report on current dual-use or weapons of mass destruction (WMD) proliferation threats, including those involving biological weapons, which seeks to inform both Congress and the public of those issues encountered by the Intelligence Community. This report, combined with both classified and unclassified findings presented to Congress throughout the year, help to fulfill the obligation of the CIA to keep the legislative branch informed of security developments. Intelligence interest in biological weapons continues to strengthen, as the CIA monitors legitimate biotechnology programs along with potential biological weapons facilities (http://www.fas.org/irp/news/1998/11/98112003_nlt.html, accessed on 8/29/03).

COLLABORATION WITHIN THE INTELLIGENCE COMMUNITY

As the CIA has moved to counter biological weapons proliferation in the past decade, its reliance on integration with other members of the Intelligence Community has become more evident. In the previous decade, the DCI restructured the "nonproliferation intelligence community" in conjunction with the expansion of its Nonproliferation Center as an attempt to respond to the difficulties of obtaining intelligence concerning the WMD proliferation (http://www.cia.gov/cia/public_affairs/speeches/1999/lauder_speech_030399.html, accessed on 8/29/03). The Nonproliferation Center, established in 1992, is largely comprised of staff from the CIA and is located at CIA headquarters in Langley, Virginia. Increasing cooperation with other intelligence agencies, such as the DIA, FBI, and NSA, along with the military, has allowed for experientially diverse representation within the Nonproliferation Center, despite the fact that other agencies often continue to view the Center as simply a department within the CIA (http://www.fas.org/irp/eprint/snyder/proliferation.htm, accessed on 8/29/03). In addition to the appointment of a Special Assistant to the DCI for Nonproliferation, who serves as Director of the DCI Nonproliferation Center, one of two deputies in the Center is to be drawn from the Department of Defense (http://www.fas.org/irp/offdocs/dcid7-2.htm, accessed on 8/29/03).

In 2001, George Tenet formed a new unit within the CIA Directorate of Intelligence: the DCI Center for Weapons Intelligence, Nonproliferation and Arms Control (WINPAC). This unit was created to integrate the existing Nonproliferation Center, the Arms Control Intelligence Staff, and Office of Transnational Issues' Weapons Intelligence Staff into a broader, more varied center of expertise (http://www.ifa-usapray.org/ARCHIVE_TERRORISM/CIA%20Steps%20Up%20Weapons%20Monitoring.html, accessed on 10/27/2003). According to the U.S. Intelligence Community website, WINPAC is one of the intelligence community's National Centers, designed to address a "specific, nontraditional threat," engaging personnel from different sectors of the Intelligence Community (http://www.intelligence.gov/2-community_centers.shtml, accessed on 10/27/2003). Weapons Intelligence, Nonproliferation and Arms Control has therefore replaced the DCI Nonproliferation Center as the prominent organization within the CIA addressing current biological weapons threat while still drawing upon the experience of the DCI Nonproliferation Center as it secures its place in the CIA Directorate of Intelligence.

Members of the Intelligence Community also provide information on various biological agents to the Department of Defense in order to prepare the military to defend troops against various pathogens in the event of a BW threat or attack (http://usembassy-australia.state.gov/hyper/WF981120/epf508.htm, accessed on 8/29/03). Such collaboration has enhanced communication within the Intelligence Community regarding issues of proliferation and intensified efforts to increase the number of BW and CW experts involved in information collection and analysis (http://www.cia.gov/cia/public_affairs/speeches/1999/lauder_speech_030399.html, accessed on 8/29/03).

In addition to these recent internal steps taken by the Intelligence Community, the DCI has attempted to collaborate with outside organizations to increase the breadth of information available to analysts in the field. The DCI has employed a top U.S. virologist as the Senior Science and Technology Advisor for Nonproliferation; this advisor has assembled a senior scientific panel to further counsel the DCI on scientific matters. However, leaders within the Nonproliferation Center stress that efforts to combat proliferation emerge not only from within one center, but through interaction between many diverse agencies. These measures allow the Intelligence Community to enhance its information-gathering capability in the field of nonproliferation and allow for a more coherent strategy to combat biological weapons proliferation (http://www.cia.gov/cia/public_affairs/speeches/1999/lauder_speech_030399.html, accessed on 8/29/03).

REFERENCES

Central Intelligence Agency, *About the CIA*, 2003, http://www.cia.gov/cia/information/info.html.

The Central Intelligence Agency, *Preparing the 21st Century: An Appraisal of US Intelligence*, Commission on the Roles and Capabilities of the United States Intelligence Community, Permanent Select Committee on Intelligence, House of Representatives, March 1, 1996.

Central Intelligence Agency, *CIA Frequently Asked Questions*, http://www.cia.gov/cia/public_affairs/faq.html.

CIA Official Discusses BCW Threat at Hoover Institution, USIS Washington File, November 20, 1998.

Dougherty, J., *CIA Steps up Weapons Monitoring*, 2001, WorldNetDaily.com, http://www.ifa-usapray.org/ARCHIVE_TERRORISM/CIA%20Steps%20Up%20Weapons%20Monitoring.html.

Gannon, J.C., *Global Challenges for the 21st Century: Nonproliferation and Arms Control*, National Intelligence Council to the Defense Threat Reduction Agency, May 31, 2000.

National Security Act of 1947, US Department of State, http://www.state.gov/r/pa/ho/time/cwr/17603.html.

Oversight of the US Intelligence Community's Efforts to Combat the Proliferation of Weapons of Mass Destruction and their Means of Delivery, Director of Central Intelligence Directive 7/2, May 7, 1999.

Siegel, J., *Controlling Nuclear Anarchy: The Role of the United States Intelligence Community in Monitoring Nuclear Nonproliferation in the Post-Cold War Era*, January 6, 1997, http://www.fas.org/irp/eprint/snyder/proliferation.html.

Statement by Special Assistant to the DCI for Nonproliferation John A Lauder on the Worldwide Biological Warfare Threat to the House Permanent Select Committee on Intelligence, (prepared statement), Central Intelligence Agency, March 3, 1999.

The National Centers, United States Intelligence Community, http://www.intelligence.gov/2-community_centers.shtml.

See also DIRECTOR OF CENTRAL INTELLIGENCE COUNTERTERRORIST CENTER and INTELLIGENCE COLLECTION AND ANALYSIS.

CHARACTERISTICS OF FUTURE BIOTERRORISTS

Adam Dolnik

OVERVIEW

In order to adequately address the threat of mass-fatality bioterrorism, it is necessary not only to understand the level of this threat but also to identify the potential perpetrators of such an event. This chapter attempts to outline the probable characteristics of future bioterrorists in order to provide a profile by which groups that may attempt to procure and even use biological agents can be identified and preempted.

TERRORISM: OLD VERSUS "NEW"

At the most basic level, future bioterrorists must possess both the ability to acquire and successfully weaponize lethal agents and the motivation to inflict indiscriminate mass casualties. Despite the fact that terrorism typically involves killing and destruction, most terrorists practice a level of restraint regarding these outcomes. Traditionally, terrorists have not necessarily been interested in killing a lot of people but rather in spreading fear among the general population by killing only the necessary few. In this respect, perhaps the best definition of terrorism is the ancient Chinese proverb "kill one, frighten ten thousand," or perhaps Brian Jenkins' observation that "terrorists want a lot of people watching, not a lot of people dead" (Jenkins, 1975). Possibly for this very reason, terrorists have traditionally not been interested in chemical, biological, radiological, and nuclear (CBRN) weapons because such weapons generate outcomes that are generally large-scale and potentially problematic. Massive destruction is likely to be counterproductive for terrorists, who typically strive to attract some measure of popular support in order to bring about political change, such as the creation of a homeland or the implementation of social justice norms within the targeted state. Mass killing would likely hinder such support rather than attract it. Moreover, a large-scale attack might also strengthen the affected government's resolve to track down and punish the terrorists, and may thus jeopardize the group's very existence.

While this traditional interpretation of terrorism has been the consensus for decades, many authors have observed that over the past 20 years the phenomenon has displayed disturbing new trends. These include the rise of violent activities motivated by a religious imperative, as opposed to the still lethal but arguably more comprehensible motives of ethnic nationalism and revolutionary ideologies. Some authors have claimed that religious terrorists are not constrained by traditional political concerns such as their popular image or the reactions of their potential constituency or the targeted state. Rather, since they base their justifications for using violence on the sanction of a supernatural authority whose will is absolute, these "new" terrorists are less rational and therefore more prone to perpetrate acts of indiscriminate mass-casualty violence (Hoffman, 1998). While this logical and widely accepted interpretation of recent trends in terrorism makes intuitive sense, a grave danger lies in its mechanical application to threat assessment without further inquiry into the nature of a given organization's belief system.

It is true that terrorist attacks have over the last 20 years become more violent, as documented by the decreasing annual number of incidents but a simultaneously rising level of overall casualties in those attacks (Hoffman, 1998). Yet although it is also true that in the wake of the Cold War religion has tended to replace secular ideologies as the ostensible philosophical basis for terrorism, the underlying motives in the belief systems of the majority of today's terrorists have arguably not changed. Even the religious fanatic typically perceives his violent activities as essentially altruistic acts of self-defense. It is still the perception of victimization and injustice that drives the so-called religious terrorist, not solely a perceived command from God. The use of holy rhetoric by most of the groups that are commonly labeled "religious" serves much more as a unifying and morale-boosting tool than as a universal justification for acts of unrestrained violence. As a result, the general belief that these "new terrorists" are religious fanatics who do not seek to forge a constituency and whose violent actions are not a means to an end but rather a self-serving end in and of themselves does not necessarily apply to the majority of today's terrorists. Implicitly, many of the organizations included in the statistics that seem to indicate the rise of indiscriminate, divinely sanctioned violence do not fall into this narrowly defined category, rendering the alarmist interpretation of such statistics much less useful than is generally believed.

Encyclopedia of Bioterrorism Defense,   Edited by Richard F. Pilch and Raymond A. Zilinskas
ISBN 0-471-46717-0   Copyright © 2005 Wiley-Liss

For these reasons, the tendency to base assessments of future unconventional terrorism solely on such factors as the frequency of the use of the word "God" in a given organization's statements does not adequately reflect the CBRN threat level posed by that group. A more productive approach would be to focus on the individual characteristics of potential mass-casualty terrorists in order to assess the threat.

MOTIVATIONAL CHARACTERISTICS OF FUTURE BIOTERRORISTS

As noted above, it is clear that successful mass-casualty bioterrorists will have to possess both (1) the capability to acquire and deliver biological agents and (2) the motivation to kill thousands of people indiscriminately.

Contrary to popular belief, only organizations possessing a rather unique combination of very specific characteristics are likely to satisfy the requirements for mass-casualty terrorism using CBRN weapons (Stern, 1999). Of greatest concern on the motivational level are cult-like groups that are completely isolated from mainstream society and are driven by an apocalyptic ideology that could be described as destroying the world in order to save it. Religious and other cult-like organizations that share the worldview that our planet could use a radical makeover are not in short supply, but most such organizations have yet to resort to outward violence. If such a turn of events were to occur, however, the potential ability of apocalyptic organizations to justify killing people as a benevolent gesture by sending them to a better place makes such groups particularly dangerous. As in most terrorist attacks, the use of violence in this scenario would again be perceived by the terrorists to be altruistic, with the critical difference being that their constituency in this case would be the victims themselves. The victims would not necessarily be "enemies" that one kills out of hatred or for their symbolic value, but rather as unfortunate beings who are going to benefit from being killed. Under these circumstances, killing thousands of people indiscriminately would be psychologically much easier than carrying out such an action as part of a political strategy or for the purposes of revenge. For instance, the Thuggees, an Indian cult of Kali worshippers that according to some killed over a million people in acts of sacrificial violence between the seventh and mid-nineteenth centuries (Hoffman, 1998), displayed some of these motivational elements. According to David Rapoport, the Thuggees believed that if they did not shed blood, their victims would go to paradise. Allegedly for this reason, the cult used strangulation as its main operational method. Assuming that this group actually existed and that the number of murders attributed to them is correct, which many historians now contest, the Thuggees' average killing rate of 800 people per year would makes them the deadliest terrorist group in history.

The Thuggees are just one historical example of a terrorist group's "altruistic" desire to bring about Armageddon. And while it cannot be definitively asserted that an act of mass-fatality bioterrorism will never occur in the absence of such ideology, it is clear that similar belief systems should be a warning sign in this regard. Another key point to emphasize is that a terrorist group does not necessarily have to be religious in nature in order to reach an apocalyptic stage. Fundamentalist environmental or animal rights groups, as well as ethnic-based violent movements, might under certain circumstances also reach this juncture.

Another characteristic likely to be demonstrated by mass-casualty terrorists using CBRN weapons is a strong sense of paranoia among the group's members. Such paranoia enhances the polarization of the terrorists' perception of the world into an "us versus them" mentality and consequently increases their willingness to victimize nonmembers of the organization in an indiscriminate way. The greater the presence of paranoia in a group's perception, the greater the sense of urgency among that group's members to unite into a cohesive unit and eliminate dissenters. This is especially critical, as the utility of mass-casualty violence tends to be a topic of disagreement within most terrorist organizations, which might well lead to the creation of undesirable schisms within the organization. If a given group completely eliminates internal dissent, the potentially restraining nature of a vibrant intragroup debate concerning the utility of using weapons of mass destruction will be lost.

Another potential characteristic of future bioterrorists relates to the expressive value attached to a particular mode of attack, in this case perhaps a desire to kill without shedding blood, or a divine fascination with poisons and plagues as God's tools. An example of this is the frequent reference to biblical plagues by various radical Christian groups. Aum Shinrikyo leader Shoko Asahara's fascination was so great that he wrote poems about sarin. Alternatively, environmentalist cults may interpret diseases as natural tools used by Mother Nature to eliminate the human race, which through technological advances and an inconsiderate use of natural resources has caused a natural imbalance that can only be restored by eliminating the world's most destructive species.

A self-perception characterized by grandiosity and ideological uniqueness is another important element that might indicate a given terrorist group's potential for biological pursuits. And while it is true that most terrorist organizations believe in their exceptionality, which helps to explain why most armed struggles usually involve not one but several rival terrorist organizations with virtually identical goals, very few groups define their uniqueness on the basis of such narrow distinctions as weapons selection. The most significant differences among terrorist groups with a common cause and enemy involves their overall strategy of violence as a part of the revolutionary process, their leaders' personalities and ambitions, their allegiance to a particular state or nonstate sponsor, and their views concerning the appropriateness and desired frequency of individual acts of violence, the legitimacy of targeting civilians, and other similar factors. Future bioterrorists, however, are likely to attach extreme importance to the use of biological agents as a distinct feature of the group.

If an organization possessing the above characteristics is led by an uncontested, charismatic, and violence-prone leader who has the ability to convince his followers that his instructions are direct orders from a supernatural authority, the deadly combination of motivational attributes needed to indiscriminately kill masses of people with biological weapons will likely be established.

ORGANIZATION AND CAPABILITY

At the organizational level, a group that is successful in conducting a biological attack will likely be structured into either a tight hierarchical formation or into a number of small independent cells in order to limit the potential for the infiltration of the group and the forestalling of its grandiose plans. Furthermore, powerful mechanisms of social control such as heavy indoctrination, complete isolation, and intimidation will likely be in place in order to prevent internal defections that could jeopardize attack preparations. With respect to technical capability, given the difficulty of weaponizing a given biological agent in a way that can produce mass fatalities, a successful terrorist group will require significant financial, logistical, and human resources. Very few groups possess such resources, although the assistance of state sponsors has the potential to significantly alter that situation. Alternatively, unemployed scientists from state-level bioweapons programs might be recruited into the group through the use of incentives such as money, the opportunity to conduct high-level research, or the "scientifically cosmic" nature of the given organization's ideology.

DESIRE AND ABILITY TO INNOVATE

Another attribute that future mass-casualty CBRN terrorists must possess is the desire and ability to innovate on both technological and tactical level. Most terrorist groups to date have been rather conservative, innovating only when forced to do so by antiterrorist countermeasures such as the deployment of barometric pressure chambers, metal detectors, X-rays, and vapor detectors at airports. As a result, most of the innovation that has taken place in the realm of terrorist groups has been geared toward improving their existing methods of weapons concealment and delivery, as opposed to the adoption of new types of weaponry per se (Hoffman, 1993). This is quite logical considering that terrorist groups fear failure—an attack that fails wastes resources and leaves clues, and most importantly can have a negative effect on both the outward image of the organization and on the self-esteem of the group's members. Most organizations therefore stick to the methods that have proven to be successful in the past until such means become ineffective because of the defensive countermeasures put in place by their adversaries, or unless some other factors create the perception of a need for a tactical or technological shift. On the tactical level, terrorist innovation has historically had a more or less cyclical, multiplying character, utilizing proven traditional tactics in a combined and synchronized fashion. An example of this phenomenon is the increasingly frequent use of secondary explosive devices that are designed to target first responders or bystanders that gather around to watch the impact of the primary explosion. This method

has proven quite effective in generating a high body count in many terrorist bombings.

Overall, the successful progression to biological weapons requires a much more significant level of innovation than the vast majority of terrorist groups have demonstrated thus far. In order to undergo such a long and demanding process, an organization must first possess a combination of several important attributes (Jackson, 2001). First, the decision to innovate requires a high level of technological awareness, something that most organizations that are completely isolated from the rest of the world may find difficult to maintain. Next, the group has to be open to new ideas, so that the organization's members are not afraid to put forward their proposals for adopting new methods. Most cult-like organizations that fulfill the motivational "mass-casualty nonconventional terrorism" characteristics identified above do not possess this attribute—their members are highly controlled, dissent is not tolerated, and individuality is suppressed. In order for such a group to pursue innovative means, its leader must be fascinated with biological weapons or the process of innovation itself. Such an inclination on the part of the leader is likely to be heavily reflected in the group's ideology as well. Highly innovative organizations will have to further demonstrate a positive attitude toward risk-taking, with respect both to the risk of failure and the physical risks associated with handling lethal biological agents.

Once a group makes the decision to innovate, other important factors influencing the successful adoption of new technology will emerge. Most important is the nature of the technology and the difficulties associated with its acquisition and successful use. Agents that can be delivered via direct personal contact are much easier to apply than pathogens or toxins that require aerosolization. The assistance of a state sponsor can be a valuable asset when attempting to adopt high-level technology, and organizations that have received such assistance have historically been significantly more deadly than the groups that receive no such support (Hoffman, 2001). However, states have traditionally stayed away from providing high-level technology to proxies, which can never be fully controlled and whose affection toward the sponsoring state may only be short-lived. Furthermore, it is even less likely that a state, no matter how "rogue," would provide a lethal bioliogical agent to an untested, highly volatile, and apocalyptic cult, whose ideological foundation does not even remotely resemble that of the state. Consequently, most organizations that satisfy the aforementioned motivational characteristics of potential mass-casualty bioterrorists cannot hope for state support and are left to their own materials and abilities.

Besides financial or material resources, an organization requires personnel with the necessary expertise and sufficient time to devote their full attention to acquiring and weaponizing biological agents. Organizations whose members are only part-time terrorists that hold daily jobs, or groups that are involved in reciprocal battles in the field, can hardly devote a significant level of their human resources to this type of activity. At the same time, groups that perpetrate terrorist operations infrequently, and therefore do have the time to devote to discovering new technologies, are likely to encounter difficulty when attempting to learn how to use such technology effectively, precisely because of the absence of experience resulting from the infrequent nature of their attacks (Jackson, 2001).

CONCLUSION

The trends in terrorism are ominous. The rising frequency of spectacular attacks along with the existence of global terrorist networks seems to support the hypothesis that the ever-escalating levels of terrorism are likely to yield a mass-casualty terrorist incident with nonconventional weapons at some point in the future. Advances in communications and weapons technologies, as well as the questionable security of CBRN facilities in the former Soviet Union, also seemingly provide more violent and reckless "new" terrorists with the tools necessary to perpetrate such an attack.

However, the technological hurdles of perpetrating a mass-fatality bioterrorist incident are still significant and should not be overlooked. Even Aum Shinrikyo, the infamous Japanese cult that possessed an estimated $1 billion in assets, some 20 university-trained scientists working in top-notch research facilities, and the freedom to conduct unlimited experiments, completely failed in all 10 attempts to conduct a biological attack (Center for Nonproliferation Studies, 2001). The conditions and resources that were available to Aum Shinrikyo are unparalleled by even the deadliest terrorist organizations today, including al-Qa'ida.

Moreover, most organizations do not seek weapons of mass destruction initially; low- to medium-level violence, which helps the terrorists acclimate themselves to indiscriminate killing, usually precedes their escalation to mass-fatality attacks. It is therefore highly unlikely that organizations possessing both the motivational and the capability characteristics described above will be able to stay off the radar screen of intelligence agencies for long. At the same time, it should be noted that the low-level violent activity practiced by cult-like organizations may not be immediately obvious, as it is likely to take the form of violence within the group designed to eliminate dissent. Nevertheless, terrorist organizations usually follow an escalatory pattern of violence, and most organizations simply do not last long enough to progress all the way to nonconventional weapons. It is estimated that only one out of 10 groups survives the first year of its operation, and only half of the groups that do make it through the first year survive for a decade (Rapoport, 1998). On the other hand, the organizations that do manage to survive for a long enough period to be able to attain a biological weapons capability tend to develop support networks and constituencies over time, which generally serve to create or reinforce rational strategic calculation among the group's leadership. This means that even organizations that rise to the spotlight by perpetrating exceptionally unrestrained high-fatality attacks are usually forced to adjust their strategies and to scale down the level of violence over time in order to maintain levels of popular support that have, sometimes unwittingly, been accumulated. In the absence of such an adjustment, the given organization's credibility as an alternative to the existing world order fades, lessening the chances of the group's long-term survival.

As a final point, even groups that do overcome all of the motivational constraints against indiscriminate mass-fatality violence have yet to exploit the full killing potential of their current conventional capabilities. If their desire really is to kill as many people as possible, why not just attack more often, at more locations, or on a greater scale with weapons that are already available and have proven to be effective? Why invest massive amounts of precious resources in a new technology that very few may know how to use and that could potentially end up killing the perpetrators themselves—all without any guarantee of success? Why risk a negative public reaction and a possibly devastating retaliation as a result of the use of nonconventional weapons? Today's terrorist organizations may have faced such questions at some point, and have either decided that nonconventional weapons were not worth pursuing or made limited and unsuccessful attempts to explore this avenue. Groups that have decided that such weapons are an attractive option, or groups that will do so in the future, are likely to possess a mixture of unique and rare characteristics. They are likely to be apocalyptic cults with violence-prone charismatic leaders who are fascinated with diseases and poisons and are not afraid to fail or get killed in their attempts to pursue such technology. The greatest overall danger is posed by religious cults combining such apocalyptic visions with outward-oriented violence and suicidal tendencies. However, most "suicide cults" tend to direct their violence only inward, committing collective suicide without attacking outsiders. Apocalyptic cults that do kill nonmembers, on the other hand, somewhat surprisingly tend to be oriented toward their continued survival.

Also, the groups that are particularly dangerous with respect to their motivation to inflict mass casualties may be in a more difficult position to breach the technological hurdles necessary to carry out a biological weapons attack. Acquiring the necessary financial, logistical, and human resources is challenging for isolated cults with an obscure ideology. In essence, the more extreme the organization, the less likely it is to attract mass support. For an extremely radical group, attracting state-level assistance and finding a safe haven in which it can conduct research and low-level violent activities while remaining undetected by intelligence agencies may be particularly difficult. Moreover, the total suppression of individuality in such cults, along with their isolation from mainstream society, does not provide for the organizational dynamics that would be favorable for successful adoption of new technology.

As a result, the same inverse relationship between the motivation to produce mass fatalities and the ability to do so that was described by Post on the individual level seems to apply to organized formations as well. (Post has argued that individuals who want to kill masses of people indiscriminately are likely to suffer from significant psychological idiosyncrasies. For individuals suffering from such idiosyncrasies, it is nearly impossible to function in groups—even though one need not operate in a group [Post, 2000]). For this reason, the likelihood of a successful mass-casualty nonconventional terrorist attack remains relatively low. That being said, it must be conceded that many conventional terrorist organizations have likely noticed the enormous fear of chemical, biological, and nuclear weapons among the general public. It therefore seems likely that some will attempt to exploit this fear. Such attempts are likely to take the form of threats and an expressed desire to use such weapons, attacks involving a small amount of a crudely delivered chemical or pathogen, or the inclusion of some chemical, biological, or radiological agent in a conventional bomb. Such attempts should be understood as psychological operations aimed at creating disproportionate fear, and not necessarily as representative of a terrifying shift to catastrophic terrorism.

REFERENCES

Center for Nonproliferation Studies, *Chronology of Aum Shinrikyo's CBW Activities*, March 2001, Internet, available at http://cns.miis.edu/pubs/reports/aum_chrn.htm, accessed on 12/12/02.

Hoffman, B., "Terrorist Targeting: Tactics, Trends, and Potentialities," in P. Wilkinson, Ed., *Technology and Terrorism*, Frank Cass, London, 1993, p. 12.

Hoffman, B., *Inside Terrorism*, Orion Publishing Co., New York, 1998, p. 89.

Jackson, B., *Stud. Conflict Terror.*, **24**, 189–213 (2001).

Jenkins, B., *Will Terrorists Go Nuclear?* RAND Paper P-5541, 1975, p. 4.

Post, J., "Psychological and Motivational Factors in Terrorist Decision-Making: Implications for CBW Terrorism," in J. Tucker, Ed., *Toxic Terror*, MIT Press, London, 2000, pp. 271–289.

Rapoport, David, C., Terrroism and the Weapons of the Apocalypse. *National Security Studies Quartely*, **6**(3), 58, (1999).

Stern, J., *The Ultimate Terrorists*, Harvard University, London, 1999, p. 70.

CHECHEN SEPARATISTS

JENNIFER MITCHELL

OVERVIEW

Following the collapse of the Soviet Union in 1991, a group of Chechen "freedom fighters" took up the cause of establishing an independent Republic of Chechnya, sometimes referred to as the Republic of Ickheria (Sikevich, 2002). What began as a mission to free the Chechen region from Russian rule evolved with a shift in group leadership to become, at least in part, ideologically motivated as well. The majority of Chechens are Muslims. Today, some 200 of several thousand Chechen separatists are foreign militants (Council on Foreign Relations, 2003), and this group is believed to maintain ties with al-Qaʻida and certain Chinese separatist groups. In addition, the Chechen movement receives funding from the Global Relief Foundation, an Islamic "charity," and the Benevolence International Foundation, a group also known to finance al-Qaʻida (Council on Foreign Relations, 2003; Associated Press, 2003).

SELECTED CHRONOLOGY

- *May 2002.* Bombing during a military parade in Kaspiisk, killing 41 people (Council on Foreign Relations, 2003).

- *October 2002.* Takeover of the Dubrovka Theater in Moscow. Seven hundred people were held hostage until Russian Special Forces attempted to overcome the hostage-takers, killing approximately 120 people in the process (Council on Foreign Relations, 2003).

- *December 2002.* Suicide bombing at a Chechen government building in Grozny, the capital of Chechnya, which killed 83 people. It has been alleged that the Chechens received support from international terrorist groups in perpetrating this act (Council on Foreign Relations, 2003).

- *May 2003.* Suicide truck bombing of a government complex near Grozny. This attack left some 50 dead and was followed less than 48 h later by a suicide attack in Iliskhan-Yurt at a religious parade, in which 20 more were killed (*Radio Australia*, 2003).

BIOLOGICAL WEAPONS CAPABILITY

While the Chechens have established a highly effective conventional weapons capability and have apparently acquired or threatened to acquire chemical and radiological weapons, at least to a limited extent, their pursuit of biological weapons appears to be quite limited. Ricin seems to be the group's biological agent of choice, if any. In January 2003, a raid on an apartment in London revealed the presence of trace amounts of ricin. Some of the suspects arrested in connection with the raid were linked to Chechen separatists and may also have trained in Chechnya (Al-Shafi'i, 2003). A week later, a dead Chechen fighter was found with instructions for making ricin. And according to Russian authorities, a raid on Chechen field commander Rizvan Chitigov's home led to the discovery of instructions on how to produce ricin (Dougherty, 2003). Previous phone conversations between Chitigov and Chechen field commander Hizir Alhazurov disclosed Chitigov's request for instructions on the "homemade production of poison" (Shashkov, 2001).

Russian sources have also reported that the Chechen separatists have attempted to produce botulinum toxin, *Bacillus anthracis*, and even the smallpox virus: According to Russian officials, hundreds of ampoules of botulinum toxin have been discovered in Chechen territory, and Chechen gunmen are sometimes armed with *B. anthracis* and the smallpox virus (Zolotaykina, 2002). The validity of such claims is highly questionable given the political motivations behind them, namely, the desire to draw support for the Russian cause.

Yet Chechen rebel leaders have made numerous claims that they already possess biological weapons, will soon attain them, or will actually use them against Russia. Here are some examples:

- *October 1999.* Guerrilla fighter Amir Khattab, an Arab jihadist commander who fought in Chechnya prior to his death, allegedly awaited test tubes containing different cultures of highly dangerous pathogens from Arab countries (*Moscow Segodnya*, 1999).

- *October 1999.* Russian Interior Minister Vladmimir Rushailo announced that instructions for unspecified biological weapons use had been found on dead Chechen rebels earlier in the month (Shytov, 1999).

- *April 2000.* A member of the Dagestani Interior Ministry reported that Chechen rebels had been supplied with four containers of an unnamed biological agent by a foreign source. However, a Kremlin spokesman disputed the claim as "misinformation" (*Interfax Russian News*, 2000).

Encyclopedia of Bioterrorism Defense,   Edited by Richard F. Pilch and Raymond A. Zilinskas
ISBN 0-471-46717-0   Copyright © 2005 Wiley-Liss

Like the claims of their adversaries, the validity of Chechen claims regarding biological weapons is highly questionable given the motivations underlying them, specifically the psychological and deterrent effects that the threat of such weapons present.

ASSESSMENT

Regardless of the continued effectiveness of their conventional attacks in their fight for independence, the Chechens' reported ability and desire to obtain chemical and radiological weapons suggests that they are at least potentially willing to cross the weapons of mass destruction (WMD) threshold. Some Chechen leaders have already threatened to use biological weapons, and together with some of their men others have allegedly been found to possess biological weapons–related materials. The connections of some factions within the Chechen resistance movement to al-Qa'ida are well documented, suggesting a potential source of income and perhaps even material that could be applied toward a biological weapons capability. Ultimately, it appears that the Chechen movement warrants some concern from an international community bent on the prevention of the spread of biological weapons and all WMD use in the near and long term.

REFERENCES

Al-Shafi'i, M., Al-Qa'ida Leader Reportedly Detained in Raid on Finsbury Park Mosque, *Al-Sharq al-Awsat*, January 22, 2003.

Council on Foreign Relations, "Chechnya-based Terrorists," Terrorism: Q&A, accessed at www.terrorismanswers.com, 5/22/03.

Dougherty, J., Moscow: Ricin Recipe Found on Chechen Fighter, CNN, accessed at www.cnn.com, 1/13/03.

Khattab promised 'biocultures' from Arab country, *Moscow Segodnya*, October 23, 1999.

Rebels may have biological weapons—Dagestani official, *Interfax Russian News*, April 28, 2000.

Radio Australia, Second Suicide Bombing in Chechnya in 48 Hours, accessed at www.abc.net.au, 5/14/03.

Shashkov, A., Militants' Poisoning Plans Disclosed by FSB, *Itar-Tass*, August 15, 2001.

Shytov, A., Bacteriological Warfare Directions Found on Chechens, *Itar-Tass*, October 30, 1999.

Sikevich, Z., Who are the Chechens? Rosbalt University, accessed at www.rosbaltnews.com/print/print?cn=60225, 11/6/02.

Zolotaykina, M., Moscow Daily Views Possible Chechen Chemical, Bacteriological Threat, *Moscow Nezavisimaya Gazeta*, December 23, 2002.

WEB RESOURCES

Chechen Republic Online, http://www.amina.com/

Chechnya-based Terrorists, http://www.terrorismanswers.com

Kavkaz Center, http://www.kavkazcenter.info/

War in Chechnya, http://www.kafkasya.com/thewar.html.

CHRISTIAN IDENTITY

Jeffrey M. Bale

DEFINITION

A theologically idiosyncratic and sectarian Christian doctrine characterized by virulent anti-Semitism, racism, millenarianism, and conspiratorial thinking.

BACKGROUND

The importance of Christian Identity (CI) in the context of bioterrorism is that it has been openly embraced by certain U.S. right-wing "militia" and terrorist cells whose members have expressed an interest in acquiring or utilizing pathogens and toxic chemical agents—and in some cases have actually acquired and planned to use them—as weapons against their opponents, including representatives of the "Zionist Occupation Government" (ZOG) that they feel is controlled by "satanic" Jews.

The principal ideological and organizational foundations of the American CI movement were laid by an earlier movement that originated in England and was known as Anglo-Israelism or British-Israelism (BI). The essence of BI, doctrinally speaking, is the idea that the British people are the lineal descendants of the "ten lost tribes" of the Kingdom of Israel, who according to the Old Testament were forced out of their homeland and into exile by Assyrian armies in 722–721 B.C.E. As early as the seventeenth century C.E., certain Puritan sectarians were already arguing that Britain had a key role to play in the prophesied return of the Jews to Palestine, an event that had to transpire before the rest of the projected millennial scenario could unfold. However, it was an eccentric named Richard Brothers (1757–1824) who first argued that there was a direct biological link between the English and these biblical tribes by postulating the existence of a "Hidden Israel" that erroneously believed itself to be Gentile. He was convinced that most real Jews were actually hidden amidst the existing northern European, primarily British peoples, even though they were blithely unaware of their own supposedly exalted biblical heritage.

The real founder of the BI movement was John Wilson (d. 1871), who sought to prove on the basis of various false linguistic analogies that the biblical "lost tribes" had migrated to northwestern Europe. His basic thesis was that it was tribes from the kingdom of Israel, the Israelites, who had migrated to Europe and thereby constituted the "Jews" of prophecy, whereas the actual Jewish tribes from the kingdom of Judah had remained where they were and interbred over the centuries with spiritually inferior "Gentiles." Although these real Jews were also destined to become a part of the prophesied "All-Israel," and thence had to be protected by the European "Israelites" and aided in returning to their homeland, in his scheme they were nevertheless portrayed, as Michael Barkun justly notes, as "erring brothers who needed to be shown the true path of salvation by the spiritually more advanced Israel/Britain, now made aware of its true Identity." Therein lay the basis of the patronizing and paternalistic but still vaguely philo-Semitic doctrine of BI, which CI theorists were later to transform into peculiar forms of overt anti-Semitism. Wilson also claimed that other northern European peoples, above all the Germans and Scandinavians, were descendants of the Israelite tribes, a variant known as "Teutonism" that was increasingly rejected by most English BI adherents as conflicts between Britain and Germany became more acute during the late nineteenth and early twentieth centuries.

Wilson's principal disciple was the "anti-Teutonist" Edward Hines (1825–1891), who was primarily responsible for propagating BI ideas and consolidating the movement in both England and the United States. He believed that the remnants of the 10 "lost tribes" had all ended up in the British Isles, but his principal doctrinal innovation was the idea that the "thirteenth" tribe of Manasseh had found its way to the United States, so that "all of Israel that was not Jewish" could be found in those two countries. The other two tribes, those of Judah and Levi, were made up of actual Jews, and all 13 of these biblical tribes had to be resettled in Palestine to fulfill the millennial prophecies. However, Hines's interpretations never really amounted to a new orthodoxy, since throughout its existence BI was characterized by both doctrinal permissiveness and a lack of organizational centralization. The BI movement in Britain, which peaked in 1920 at around 5000 members, mainly appealed to individuals from the middle and upper classes, most of whom adopted patriotic and pro-government attitudes.

The first American to publish a BI work was Joseph Wild, whose 1879 pamphlet proclaimed that the United States would play the role of the thirteenth tribe of Manasseh, but Hines's key follower in the United States was former Army artillery officer C. A. L. Totten. Together, they helped to spread BI ideas to influential evangelists and key Pentecostal leaders such as Charles Fox Parham, and in the process managed to establish three loci of BI activity in the United States, one in the Northeast, one in the Midwest, and one in the West. Among those people who embraced BI concepts was an Oregon clergyman named Reuben H. Sawyer, who in the early 1920s was also an active member of his state's Ku Klux Klan faction. Sawyer's career demonstrated that BI views could be combined with far right political activism despite some inherent tension between BI's philo-Semitism and the latter's anti-Semitism, and in that sense it foreshadowed the careers of many later CI leaders.

A more secure organizational basis for the BI movement in the United States was laid by Howard B. Rand (1899–1991), a Massachusetts lawyer, who created a truly national, Detroit-based organization known as the Anglo-Saxon Federation of America (ASFA) to help bring together the existing congeries of often competing local and regional BI groups. It was during this period that Rand began collaborating actively with William J. Cameron (1878–1955), Henry Ford's personal media assistant and the author of numerous anti-Semitic articles in the newspaper he then edited, the *Dearborn Independent*; articles that were later gathered together in a notorious four-volume work attributed to the authorship of Ford himself, *The International Jew*. It is therefore hardly surprising that in 1933, Cameron claimed that the Bible was really the story of the "Anglo-Saxon race" that had descended from the ancient Israelites, and that the "anti-Israel" Jews had instead been corrupted by their intermarriage to the Edomites, Esau's supposed descendants. Thus the Jews, being members of the tribe of Judah, were neither God's "chosen people" nor members of the same racial group as Jesus. Rand's BI activities peaked during the 1930s, and the institutional framework of the ASFA declined precipitously after World War II.

The latent anti-Semitism of certain segments of the BI movement became increasingly overt and virulent when Jews in the Holy Land began opposing British rule by force, actions that were viewed by their members as a repudiation of the idea that Anglo-Americans constituted the "true Israel." It was a Canadian BI group originating in Vancouver, the British Israel Association (BIA), which not only created an extensive West Coast network but also introduced more conspiratorial and openly anti-Semitic themes into American BI. Indeed, a novel published by the BIA in 1944, H. Ben Judah's *When?*, may have been the first genuine CI text, since it depicted the Jews as the spawn of Satan. Other BIA-published works soon followed that likewise incorporated anti-Semitic themes that later became characteristic of certain currents of CI, such as the identification of Cain as the founder of the "synagogue of Satan," the idea that the blood of fallen angels resided in the Jews, and the acceptance of the historicity of the *Protocols of the Elders of Zion*, a notorious anti-Semitic treatise forged by the Tsarist secret police.

However, the figure who best epitomized the shift from BI's original philo-Semitism to CI's subsequent rabid anti-Semitism was the Canadian C. F. Parker. According to Parker, the blood of Esau and his descendants, the Edomites, had contaminated and corrupted the "All-Israel" blood of the Jews, who had then become the carriers of many subversive and revolutionary doctrines, including communism and Zionism. This latter doctrine was particularly pernicious because it advocated the expulsion of the Holy Land's "rightful owners—Israel-Britain." The BIA would play a prominent role, together with certain other BI groups and Pentecostal sects, in introducing BI themes into the variegated religious milieu of Southern California, the very region from whence the early leaders of the CI movement were later to emerge.

The individuals who played the most significant role in the creation of CI, either as a doctrine or as a movement, were already—or were soon to become—crucial figures in the development of the postwar American radical right. Among them was Gerald L. K. Smith (1898–1976), the infamous anti-Semitic radio broadcaster, who sought to mobilize CI adherents in support of a broad range of his favored right-wing political causes and thence, by assiduously cultivating leading CI figures, helped to solidify a fledgling CI network on the West Coast; Wesley Swift (1913–1970), an evangelical Protestant preacher who founded the Church of Jesus Christ Christian, and was perhaps the person who was most responsible for popularizing CI in far right circles, essentially by combining BI, political radicalism, and virulent anti-Semitism; Colonel William Potter Gale (1916–1988), a former guerrilla organizer on General Douglas MacArthur's staff in the Philippines during World War II and a postwar Swift protégé who later broke with the latter, after which he established both a far right paramilitary group by the early 1960s known as the Christian Defense League (CDL) and, two decades later, an armed tax protest group known as the Committee of the States; Richard Girnt Butler, an enthusiastic Gale and Swift follower who was appointed as the CDL's first president and national director and later went on to found a new Church of Jesus Christ Christian and its political wing, the Aryan Nations, in Hayden Lake, Idaho; and James K. Warner, a former member of the "neo-Nazi" National Socialist White People's Party and ex-Odinist who assumed the leadership of the CDL in 1973, after Butler had left California. Today, the most influential CI theologian is probably Dan Gayman, a former Mormon, who in 1972 founded his own evangelical congregation known as the Church of Israel in Schell City, Missouri.

IDEOLOGICAL CHARACTERISTICS OF CHRISTIAN IDENTITY

In this particular context, the most distinctive aspects of CI are (1) its peculiar apocalyptic millenarian views, specifically its rejection of the standard Protestant Fundamentalist belief that select Christian believers will be spared from terrible suffering before the establishment of Christ's reign on Earth; and (2) its virulent and uncompromising anti-Semitism. Both of these characteristics predispose CI adherents to prepare for and be ready to wage, on the earthly plane, the looming cosmic struggle between the "Aryan" forces of Light and the "satanic" Jewish forces of Darkness.

Like BI, CI is rooted in a millenarian theology, which means that its adherents are expectantly awaiting a collective, this-worldly, and imminent process of salvation that will be brought on by a recognized supernatural agency. By definition, such salvation doctrines promise a forthcoming rescue from evil, if only for a small exclusive community of believers, at some point during the approaching "Last Days" or "End Times." Yet, CI displays certain unique features within the context of Christian millenarianism.

CI adherents are convinced that the Antichrist is about to appear and inaugurate a terrible period of Tribulation prior to the anticipated return of Jesus Christ and the initiation of His reign before the final cataclysmic battle and the Last Judgment, but unlike most other Christian millenarians, they do not believe that God's anointed will be "raptured" or lifted up into the heavens by Jesus and thereby spared from suffering during this Tribulation era. On the contrary, those who are ultimately destined to be saved will have to ensure their own survival throughout this difficult period by establishing true Christian communities that are isolated from the corrupt wider society, communities where good people can take refuge and from which they can organize and carry on an imminent life-or-death struggle against the forces of Evil operating here on Earth. Actual examples of the formation of such microcosmic CI communities are James Ellison's Covenant, the Sword, and the Arm of the Lord (CSA) in Arkansas, Robert Millar's Elohim City in Oklahoma, and the aforementioned Aryan Nations in Idaho, all of which included members who have been linked to serious acts of violence.

CI doctrine holds that the agents of that Evil on Earth are none other than the Jews, along with their minions drawn from the ranks of Caucasian "race traitors" and members of various "pre-Adamic" races, inferior nonwhite "mud people" that God supposedly created before eventually perfecting humankind with His creation of Adam and Eve. There are several different variants of CI anti-Semitism, the most extreme being the so-called two-seed theory based on the idea that Jews are the literal offspring of Satan. According to this "seedliner" theory, after being tempted in the Garden of Eden, Eve had copulated with Satan or one of his underlings (who had assumed the form of a serpent) and given birth to Cain, thereby becoming the progenitor of the despoiled Canaanites, and ultimately of the entire Jewish race. Meanwhile, Adam and Eve gave birth to Abel (who was later murdered by Cain) and Seth, and it was this latter who founded Noah's line and became, via Sham, Abraham, and Isaac, the ancestor of the Aryan "Israelites" whose history the Bible allegedly recounts. Whether or not they know it, today's Anglo-Saxons are none other than the ancestors of the "ten lost tribes" of those very Israelites, who had all along been engaged in a bitter conflict with the Jewish "Sons of Satan." Hence those Aryans who had recently discovered their true "identity" as the descendants of the biblical "Israelites" must continue to wage this eons-old struggle against Satanic Jewry and eventually resettle in the Holy Land in order to prepare the way for Christ's prophesied return.

CHRISTIAN IDENTITY AND CBRN USE

Given their apocalyptic visions and utopian millenarian goals, as well as their burning desire to overthrow the existing "Zionist Occupation Government" in the United States, it is hardly surprising that certain CI-inspired terrorist groups have displayed few if any qualms about carrying out mass casualty attacks or that they have already engaged in periodic efforts to acquire so-called weapons of mass destruction (WMD), including biological weapons. Since these groups are unable to overthrow the government by employing conventional military tactics, they are necessarily forced to adopt an asymmetrical strategy to achieve their objectives. Apart from advocating "leaderless resistance" as an organizing principle and the carrying out of acts of violence by committed "lone wolves," another objective of their strategy may be to strike a devastating blow against the symbols and representatives of ZOG or otherwise smite their "satanic" Jewish and "mud people" enemies, whether by means of spectacular conventional terrorist attacks or unconventional attacks using chemical, biological, radiological, or nuclear (CBRN) weapons.

A few examples should suffice to highlight this last potential danger. In 1994 and 1995, four men belonging to antigovernment militia group called the Minnesota Patriots Council were arrested under the 1989 Biological Weapons Anti-Terrorism Act for possessing ricin, which they apparently intended to deploy to kill local police and Internal Revenue Service agents. In February 1998, Larry Wayne Harris, a microbiologist and former member of the Aryan Nations, and another man were arrested in Nevada for possessing a container of what later turned out to be a nonlethal form of anthrax bacteria, which the two had reportedly threatened to disseminate in the New York subway system. Furthermore, certain CI spokespersons have repeatedly threatened to employ biological weapons, and a few CI adherents have perpetrated biological warfare (BW) hoaxes by mailing threatening letters containing powder to their enemies. There have also been several cases in which CI believers have been arrested while in possession of harmful chemical agents that they seem to have been planning to use to kill or injure designated enemies. Although these efforts have so far turned out to be amateurish and ineffective, at some point such threats, hoaxes, and forestalled plots may give way to the successful deployment of biological agents.

CONCLUSION

Given their sectarian religious worldviews, violence-prone CI groups appear to be more willing—and therefore likely—to violate traditional moral and ethical taboos by employing mass-casualty terrorism and making use of CBRN than most of their nationalist/separatist or secular ideological (Marxist, anarchist, neo-fascist) terrorist counterparts. After all, to the extent that violent extremist groups are absolutely convinced that they are doing God's bidding, virtually any action that they decide to undertake can be justified, no matter how heinous, since the "divine" ends are thought to justify the means.

However, there is not necessarily any direct correlation between religious extremism and a particular terrorist group's decision to employ WMD. Many other factors are also undoubtedly involved, so the most that can be said is that under certain circumstances religious extremism can be a very important contributory factor in permitting a group to rationalize its development and use of BW agents. Whether CI adherents prove to be more willing to cross the WMD threshold than other types of religious militants remains to be seen. Although the previous efforts of a few disgruntled CI believers to deploy BW and chemical warfare agents suggest that this could be the case, such people are probably less likely to carry out a successful attack of this type than Islamist terrorists, mainly because of the relatively low level of their organizational, operational, and technological capabilities.

REFERENCES

Aho, J.A., *The Politics of Righteousness: Idaho Christian Patriotism*, University of Washington, Seattle, WA and London, 1990.

Barkun, M., *Religion and the Racist Right: The Origins of the Christian Identity Movement*, University of North Carolina, Chapel Hill, NC, 1996.

Daniels, T., "Introduction," in T. Daniels, Ed., *A Doomsday Reader: Prophets, Predictors, and Hucksters of Salvation*, New York University, New York, 1999, 1–18.

De Armond, P., "Right Wing Terrorism and Weapons of Mass Destruction: Motives, Strategies, and Movements," in B. Roberts, Ed., *Hype or Reality?: The "New Terrorism" and Mass Casualty Attacks*, Chemical and Biological Arms Control Institute, Alexandria, VA, 2000.

Levitas, D., *The Terrorist Next Door: The Militia Movement and the Radical Right*, St. Martin's, New York, 2004.

Neiwert, D.A., *In God's Country: The Patriot Movement and the Pacific Northwest*, Washington State University, Pullman, WA, 1999.

Noble, K., *Tabernacle of Hate: Why They Bombed Oklahoma City*, Voyager, Prescott, Ontario, 1998.

Seymour, C., *Committee of the States: Inside the Radical Right*, Camden Place Communications, Mariposa, CA, 1991.

Smith, B.L., *Terrorism in America: Pipe Bombs and Pipe Dreams*, Ch. 4–5, SUNY, Albany, NY, 1995.

Swain, C.M. and Nieli, R., Eds., *Contemporary Voices of White Nationalism in America*, Cambridge University, Cambridge, 2003 (see Chapter 7 by Gayman).

Tucker, J., Ed., *Toxic Terror: Assessing Terrorist Use of Chemical and Biological Weapons*, M.I.T., Cambridge, 2000 (see Chapters on the CSA, the Minnesota Patriots Council, and Larry Wayne Harris).

CONSEQUENCE MANAGEMENT

ZYGMUNT F. DEMBEK

THEODORE J. CIESLAK

INTRODUCTION

The response to a bioterrorism attack, similar to the response to any disaster whether natural or man-made, can be divided into two phases—crisis management and consequence management. Crisis management includes the immediate and coordinated actions necessary to resolve a potential public health disaster. Consequence management activities surrounding a bioterrorism event include the sustained measures necessary to protect public health, maintain essential government and health care services, and provide emergency relief to those agencies, businesses, and individuals in need. Of necessity, bioterrorism consequence management includes a partnership of many state and federal agencies, and involves first responders, law enforcement officials, emergency medical, hospital, and public health participants, and relevant volunteer organizations. All of these groups and individuals need to work closely together to reestablish normalcy in the community.

Consequence management activities evolve from the initial crisis management phase of response to a bioterrorism event. The bioterrorism consequence management team probably needs to function without having a complete knowledge of the unfolding scenario. In fact, a rapid response under such difficult circumstances may often be necessary in order to protect the health of the community. For example, with the anthrax letters of fall 2001, some of the deaths attributed to inhalational anthrax could not be linked to a mail-related exposure (Barakat et al., 2002; Mina et al., 2002). Initially, public health and law enforcement officials were unaware of the cause of the inhalational anthrax deaths (Lipton and Johnson, 2001). It was not until the mail source of the anthrax infections was realized that mitigation efforts for both the potentially exposed (the use of ciprofloxacin antibiotic prophylaxis) and for the potentially contaminated facilities (physical decontamination) could be optimized.

The challenges encountered while both conducting mass prophylaxis for postal workers potentially exposed to anthrax (Partridge et al., 2003) and performing decontamination of the Hart Office building (Weis, 2002) demonstrate the degree of difficulty that can be encountered in our complex society while attempting to restore normalcy through consequence management. These challenges are amplified in the response to a contagious agent (i.e., one that is transmissible from person to person), such as the smallpox virus. In this case, a rapid epidemiological investigation would be required on the part of public health authorities in order to identify both individuals with the disease and those who have come into close contact with them. The investigative efforts of law enforcement officials might be utilized to complement the public health response in this setting (USDOJ, 2003).

This entry discusses consequence management as it pertains to bioterrorism, including mitigation, response, and recovery, and the psychosocial challenges posed by such an event. It also discusses the efforts necessary by multiple federal, state, private, and volunteer agencies as they work together in an integrated response network.

EMERGENCY ACTIVITIES OVERVIEW

Emergency activities are traditionally divided into four phases: preparedness, mitigation, response, and recovery (FEMA, 2003). Consequence (and at times crisis) management generally relates to the latter three phases, discussed in turn below. Psychosocial factors may influence all phases of consequence management, and therefore present an additional point for discussion.

Mitigation

Disaster mitigation is an important component of consequence management. Typical disaster mitigation considerations following a natural disaster include hazard analysis, comparison assessment and prioritization of risk, and hazard mitigation (FEMA, 1998). These same principles can be used to mitigate a bioterrorism incident. In this case, a significant concern is the potential for exhaustion of resources. For example, following an aerosol release of a biological agent upwind of a heavily populated area, hundreds of thousands of patients may require antibiotic prophylaxis, straining both human resources and materiel stockpiles. Preparedness is a key in addressing this and other possibilities.

Response and Recovery

While "response" is generally synonymous with crisis management, which includes activities taken to save lives and reduce damage, it involves consequence management as well; for example, rebuilding a community so that individuals, businesses, and government regain self-function. After such efforts are realized, the community enters the recovery phase, which largely consists of protecting against reoccurrence. The timeline for recovery is unique to the event and the community, and is dependent upon the nature of the damage caused.

Encyclopedia of Bioterrorism Defense,   Edited by Richard F. Pilch and Raymond A. Zilinskas
ISBN 0-471-46717-0 Copyright © 2005 Wiley-Liss

The ultimate goal in all cases is to restore both personal lives and community livelihood. Because of the nature of a bioterrorism event and the public misconceptions of biological organisms, full recovery may be an extended process.

Psychosocial Challenges

A large infectious disease outbreak, whether natural or deliberate, could present complex psychosocial challenges for the health care system. Related consequence management problems could arise subsequent to a bioterrorism event, such as inordinate numbers of individuals presenting at health care facilities with psychological concerns (Lord, 2001), an increase in problematic behaviors (DiGiovanni, 2001), and issues related to the hoarding of antibiotics and other vital medicines and materials. Potential psychiatric disturbances and disorders could result in long-term effects for those afflicted (Norwood et al., 2001).

RESPONDING TO AN EVENT

Local and State Chain of Command

The response to any disaster begins, of necessity, at the local level. Since federal and state law enforcement and public health authorities are unlikely to be on site at the beginning of a disaster, the immediate presence of well-trained local personnel are critical to good crisis response. When local resources are overwhelmed, local authorities may seek help at the state level through the State Coordinating Officer (SCO). The SCO can then advise the governor to activate extensive emergency management resources available at the state level. These resources might include the law enforcement capabilities of the State Police and National Guard, which remains a state asset unless federalized. Many state National Guards now include military Weapons of Mass Destruction Civil Support Teams that can offer expert advice and liaison to more robust military assets at the federal level. Moreover, most National Guard units can provide public works assistance and mobile field hospitalization capability.

All state governors have authority and responsibility for issuing state or area emergency declarations, initiating state response actions (e.g., personnel or materials), activating emergency contingency funds and/or reallocating regular budgets for emergency activities, overseeing emergency management for all four phases, and applying for and monitoring federal assistance. The State Emergency Management Agency carries out and coordinates statewide emergency management activities and provides financial assistance on a supplemental basis through a process of application and review. All health-based decisions would likely be made in consultation with the state health officer and the state epidemiologist.

Federal Involvement

When response requirements exceed the capabilities available at state level, the SCO may contact the Federal Coordinating Officer (FCO). The FCO may then activate a federal response under the auspices of the Federal Response Plan (FRP) (Federal Emergency Management Agency, 1992), which implements Public Law 93–288 (the Robert T. Stafford Disaster Relief and Emergency Assistance Act) to provide the authority for federal disaster response. When the FRP is activated, the federal government assumes an active role in coordinating the response and can provide state and local authorities with personnel, technical expertise, equipment, and other resources.

As of March 1, 2003, the Federal Emergency Management Agency (FEMA) and the National Domestic Preparedness Office of the Federal Bureau of Investigation (FBI)—along with other key counterterrorism entities such as the Strategic National Stockpile (SNS) and the National Disaster Medical System (NDMS) from Department of Health and Human Service (HHS), the Nuclear Incident Response Team from the Department of Energy, and the Domestic Emergency Support Teams from the Department of Justice—were placed under the control of the new federal Department of Homeland Security. However, the responsibility of dealing with a bioterrorism attack will fall on multiple federal, state, and municipal agencies and on the civilian health community.

Resources available under the FRP are grouped into 12 emergency support functions (ESFs), and are provided by various federal agencies and nongovernmental organizations such as the American Red Cross. Each ESF is headed by a primary agency, with other agencies providing support as necessary. Among the relevant ESFs to a bioterrorism event are ESF 6 (Mass Care: American Red Cross) and ESF 8 (Health and Medical Services: HHS and the U.S. Public Health Service).

Disaster Medical Assistance Teams (DMATs) of HHS/NDMS can augment existing medical resources. DMATs consist of 35 or more medical and support personnel who can deploy to disaster sites and support themselves for a period of 72 h while providing medical care. In October 2001, five DMATs were deployed to New York City to evaluate and offer antibiotic prophylaxis to over 7000 U.S. Postal Service employees (Partridge et al., 2003).

Emergency Operations Planning

An Emergency Operations Plan (EOP) describes the mechanisms by which government agencies respond to an emergency. Each state's bioterrorism EOP is a key component of its emergency operations program: The EOP establishes the overall authority, roles, functions, and resources to be used during an event. The bioterrorism EOP is activated once a bioterrorism event has occurred in order to guide the emergency response and recovery, and only those parts of the EOP required for the statewide response are activated. The EOP enables the state to be prepared to take immediate action.

Fatality Management

Consequence management following a bioterrorism event may also include the need for fatality management. Out of necessity, this would include local coroners and state medical examiners, who by law must investigate cases of sudden, violent, suspicious, or unexplained death (Nolte, 2002). These individuals may recognize unusual deaths before other health care providers, because patients who die of infectious disease often die at home and generally require postmortem examination (Nolte et al., 2000). Local coroners, state medical examiners, and the state funeral directors association would together manage a mass-fatality event, as was the case during the 1918 influenza pandemic in which more than 12,000 people died during a single month in Philadelphia alone, adding a tremendous burden to the normal avenues for the storage and handling of bodies (Association of State and Territorial Health Officials, 2003). Should a similar event occur today, the state medical examiner will have the lead in fatality management activities.

Other fatality management considerations during a bioterrorism event include maintaining a mortuary registry of similar deaths; managing familial visits; using morgues to provide central processing; establishing long-term fatality storage facilities; determining final disposition for corpses; establishing family assistance centers; implementing mass cremation if necessary; coordinating release of remains to families; and establishing temporary internment options as appropriate (US Department of Justice, 2003).

Incident Command System

FEMA, with its historical lead role in consequence management, endorses the Incident Command System (ICS) as a mechanism by which to achieve a coordinated disaster response effort. The ICS permits a unity of command in an emergency situation that has proved useful with fires, natural disasters, and hazardous materials events. Without a clear designation of authority during an emergency situation, confusion of participants and diversion of resources from critical needs could readily occur. In this respect, the response to a bioterrorism incident should not differ from that to any other emergency situation. Disparate groups that are perhaps accustomed to responding to an emergency event in their own manner (e.g., the public health response to a disease outbreak) must learn to work in a coordinated method with others that may have jurisdiction (i.e., the FBI) or political control (i.e., the governor) over a bioterrorism response.

It is equally important for health care facilities to incorporate ICS principles into their emergency preparedness plans. Rapid casualty care is of utmost importance immediately subsequent to a mass-casualty incident, and the use of ICS will enable a given health care facility's staff to fully integrate their activities with community emergency response assets (MacIntyre et al., 2000). Should emergency medical services become overwhelmed, auxiliary replacement of casualty transport may occur through the use of state, military, or other resources. A fail-safe statewide communications network should be in place in order to provide advanced notification and coordinate casualty transport (SBCCOM, 2000). Although the initial hospital response may be centered on patient treatment, other services have vital roles, such as patient and traffic control by security

personnel. Hospital emergency, critical care services, plant operations, supply services, infectious disease, respiratory therapy, and toxicology departments, as well as hospital pharmacies and laboratories, should be fully trained in ICS principles (MacIntyre et al., 2000).

Additional Considerations

Other elements of consequence management are somewhat unique to the problems associated with a bioterrorism incident. First, a national network of laboratories, the Laboratory Response Network, has been established to expedite the diagnosis of infectious pathogens. The Centers for Disease Control and Prevention (CDC) and the United States Army Medical Research Institute of Infectious Diseases provide Level D reference laboratories capable of sophisticated biological threat agent analysis. These laboratories, which are capable of banking microbial strains, probing for genetic manipulations, and operating at Biosafety Level 4 (BSL-4, the highest level of laboratory protection), provide backup to Level C labs at state health departments (CDC, 2000), which in turn support Level A (hospital clinical laboratories, for example, which have responsibility for ruling out certain pathogens) and Level B (public health laboratories, for example, which are responsible for "ruling in" or including pathogens) labs at local and regional levels. Epidemiological consultation is also available from the CDC, as are critical drugs and vaccines necessary to combat a large bioterrorism attack. These pharmaceuticals are stockpiled at several locations throughout the country, available via CDC's SNS program for rapid deployment to an affected area.

CONCLUSION

As is evident, the myriad tasks associated with consequence management of a bioterrorism attack are beyond the response capabilities of a single agency. It is only by multiple federal, state, private, and volunteer agencies working together in a fully integrated response network, guided by well-developed EOPs, organized through the ICS, and supported at the federal level by multiple agencies under the rubric of the FRP, that the effects of a bioterrorist attack can be minimized and mitigated.

REFERENCES

Association of State and Territorial Health Officials, *Preparedness Planning for State Health Officials: Nature's Terrorist Attack—Pandemic Influenza*, Washington, D.C., 2003, p. 15.

Barakat, L.A., Quentzel, H.L., Jernigan, J.A., Kirschke, D.L., Griffith, K., Spear, S.M., Kelley, K., Barden, D., Mayo, D., Stephens, D.S., Popovic, T., Marston, C., Zaki, S.R., Guarner J., Shieh W. J., Carver, H.W., 2nd, Meyer, R.F., Swerdlow, D.L., Mast, E.E. and Hadler, J.L., *J. Am. Med. Assoc.*, **287**, 863–868 (2002).

CDC, *Morbid. Mortal. Weekly Rev.*, **49**(RR-04), 1–14 (2000).

DiGiovanni, C., *Milit. Med.*, **166**(Suppl. 2), 59–60 (2001).

Federal Emergency Management Agency, *The Federal Response Plan for Public Law 93–288*, as amended, April 1992.

FEMA Emergency Management Institute, *Principles of Emergency Management*, Independent study course, FEMA IS230, March 2003.

FEMA Emergency Management Institute, *Introduction to Mitigation*, Independent study course, FEMA IS393, April 1998.

Howard, R.D. (2000) "The National Security Act of 1947 and Biological and Chemical Weapons," Chapter 9.2, in R.D. Howard, and R.L. Sawyer, Eds., *Terrorism and Counterterrorism: Understanding the New Security Environment*, McGraw-Hill/Dushkin, 2003, Dubuque, IA, pp. 442–471.

Lipton, E. and Johnson, K., "A nation challenged: the anthrax trail; tracking bioterror's tangled course," *New York Times*, December 26, 2001.

Lord, E.J., *Milit. Med.*, **166**(Suppl. 2), 34–35 (2001).

MacIntyre, A.G., Christopher, G.W., Eitzen Jr., E., Gum, R., Weir, S., DeAtley, C., Tonat, K. and Barbera, J.A., *J. Am. Med. Assoc.*, **283**, 242–249 (2000).

Mina, B., Dym, J.P., Kuepper, F., Tso, R., Arrastia, C., Kaplounova, I., Faraj, H., Kwapniewski, A., Krol, C.M., Grosser, M., Glick, J., Fochios, S., Remolina, A., Vasovic, L., Moses, J., Robin, T., DeVita, M. and Tapper, M.L., *J. Am. Med. Assoc.*, **287**, 858–862 (2002).

Nolte, K.B., *J. Am. Med. Assoc.*, **287**, 984–985 (2002).

Nolte, K.B., Yoon, S.S., and Pertowski, C., *Emerg. Infect. Dis.*, **6**, 559–560 (2000).

Norwood, A.E., Holloway, H.C., and Ursano, R.J., *Milit. Med.*, **166**(Suppl. 2), 27–28 (2001).

Partridge, R., Alexander, J., Lawrence, T. and Suner, S., *Ann. Emerg. Med.*, **41**, 441–446 (2003).

Soldier and Biological Chemical Command, CDC/DoD Smallpox Workshop, DTIC/SBCCOM SPO700-00-D-3180, Battelle Edgewood Operations, Bel Air, MD, April 17–19, 2000.

US Department of Justice, *Criminal and Epidemiological Investigation Handbook*, US Department of Justice Office of Justice Programs and the US Soldier and Biological Chemical Command, 2003; http://hld.sbccom.army.mil/ip/ceih_download.htm.

Weis, C.P., *J. Am. Med. Assoc.*, **288**, 2853–2858 (2002).

FURTHER READING

Howard, R.D. and Sawyer, R.L., Eds., *Terrorism and Counterterrorism: Understanding the New Security Environment*, McGraw-Hill/Dushkin, New York, 2003.

Maniscalco, P.M. and Christen, H.T., *Understanding Terrorism and Managing the Consequences*, Prentice Hall, Upper Saddle River, NJ, 2001.

WEB RESOURCES

Agencies Offer Tips for Consumers Eyeing Online Anthrax Cures: FTC Says Fraudsters Prey on Consumers' Fears, Federal Trade Commission, Washington, DC, November 11, 2001, http://www.ftc.gov/opa/2001/11/alert.htm, accessed on 6/27/03.

Criminal and Epidemiological Investigation Handbook, http://hld.sbccom.army.mil/ip/ceih_download.htm, accessed on 6/27/03.

Public Health Service Disaster Medical Assistance Team (PHS-1 DMAT) website, http://oep.osophs.dhhs.gov/dmat/, accessed on 6/27/03.

US Department of Homeland Security Office of Domestic Preparedness website, http://www.ojp.usdoj.gov/odp, accessed on 6/27/03.

See also CRISIS MANAGEMENT; FATALITY MANAGEMENT; and LABORATORY RESPONSE TO BIOTERRORISM

COST-EFFECTIVENESS OF BIOLOGICAL WEAPONS

DAVID W. SIEGRIST

INTRODUCTION

One of the most cited figures on biological weapons' cost-effectiveness is the following:

"A group of [chemical and biological weapons] experts, appearing before a UN panel in 1969, estimated 'for a large scale operation against a civilian population, casualties might cost about two thousand dollars per square kilometer with conventional weapons, eight hundred dollars with nuclear weapons, six hundred dollars with nerve gas weapons, and one dollar with biological weapons" (Douglas and Livingstone, 1987; Danzig, 1996). This statement does not further define what level of destruction would be caused by such weapons, and appears to be simply measuring the marginal cost of materials to attack a square kilometer of territory, not including the delivery or development cost of such weapons. Indeed, the experts did not have their widely cited opinion published in the actual UN report. Rather, the report emphasized the inherent uncertainty of chemical and biological weapons to achieve any particular effect (United Nations, 1970).

A subsequent, more complete report by this World Health Organization (WHO) panel of experts also did not publish particular cost estimates; however, it may have provided insight into the methodology used to generate the figures (World Health Organization, 1970). The group of WHO consultants, including among other notables Professors Matthew Meselson and Joshua Lederberg, stated that this report was similar to the report of the United Nations but more technical. It discussed inflicting 50 percent casualties per square kilometer from an aircraft under ideal attack conditions against an unprotected population, but did not tie those to costs in the written document. It justified its use of such assumptions favorable to the offense in terms of its charter to illustrate the potential risk of such weapons; hence it presented worst plausible case scenarios. However, those assumptions also exerted downward pressure on the cost estimates, which may be why they did not put them in the published document.

COMPARATIVE COST ANALYSIS

There have been many changes that have affected weapons' costs since the WHO analysis in 1970. One of the most dominant is the fact that chemical and biological weapons have now been outlawed. Even if a regime that has not signed the Biological and Toxin Weapons Convention (BWC) or Chemical Weapons Convention (CWC) is considering developing such weapons, it would presumably do so in secret, against the background of a strong international norm against the development and use of such weapons. Therefore, in considering weapon costs, a realistic estimate must include the development effort for such weapons or the premium expected to be paid for them on the clandestine weapons market. In addition, weapons for use by terrorists or rogue regimes have differing requirements that also affect their cost. For instance, rogue regimes might be expected to pay for full "weaponization" and delivery systems for such agents, while terrorist use would likely be improvised, reducing thier costs. A further complexity is that compared to terrorist groups, rogue regimes often have more technical capability, more time, and secure locations in which to experiment and produce such weapons. Despite those production advantages, nation states have the additional practical disincentive to choose to produce chemical and biological weapons because they are critically dependent on environmental factors and striking relatively unprotected target populations for their effective use, and are thus unreliable as military weapons. In addition, while weapons of mass destruction (WMD) may be thought by some to provide a deterrent against attack, Iraq recently provided the spectacle of a rogue regime that claimed not to have such weapons but which was invaded precisely because it was thought to possess them (anti-deterrence). Hence, their value for deterrence may be said to be ambiguous at best.

In the assessments that follow, costs will be compared on inflicting 50 percent casualties on a square kilometer of territory.

Conventional Weapons

The cheapest conventional explosive in widespread use is Ammonium Nitrate and Fuel Oil (ANFO). Ammonium nitrate is used as fertilizer and sold by the ton. Pure, thoroughly mixed with fuel oil, and detonated by dynamite, it can provide a powerful blast, as demonstrated by the Oklahoma City bombing. Despite its horrendous effect, however, the blast radius in this event was limited to approximately 150 ft for structural damage to surrounding buildings (Hoffman, 1998). In reference to one square kilometer, a loose extrapolation of this figure suggests that four such truck bombs would be required. The FBI estimates that it cost $5000 to make the Oklahoma City bomb (up from an earlier $1000 estimate) (Hoffman, 1998). Therefore, approximately $20,000 might be required to inflict 50 percent casualties on a square kilometer with this type of conventional explosive, while reflecting as well the

Encyclopedia of Bioterrorism Defense, Edited by Richard F. Pilch and Raymond A. Zilinskas
ISBN 0-471-46717-0 Copyright © 2005 Wiley-Liss

costs of development, if required, and the relevant delivery system.

ANFO explosives probably do not appeal to rogue regimes, because they do not generate the blast of military explosives, even when boosted with aluminum particles. The fuel oil settles rapidly, necessitating that the explosives be prepared shortly before use. They are also extremely bulky and dangerous to make. Therefore, to lay waste to a square kilometer with conventional explosives, a state might instead choose a traditional military weapon system such as the Russian BM-21 122-mm rocket launcher, a proven technology that dates back to the "Stalin's Organ" of World War II. Further, using rockets does not involve the expense of acquiring an aircraft, generating air superiority over the target, training (and loss, whether in action or due to retirement) of pilots, or other encumbrances. The attacker can also "stand off" some kilometers from the target and attack during virtually any kind of weather with high confidence of the target being hit. The BM-21 launcher can be purchased for about $45,000 on the world market (http://www.pmulcahy.com/self-propelled_artillery/chinese_mrls.html, accessed on 8/13/03) and is not covered by particularly onerous weapon treaties. One hundred rockets for the launcher, more than enough for the task, are available similarly for $350 apiece (http://www.pmulcahy.com/rocket_launchers/russian_rocket_launchers.htm, accessed on 8/13/03). This makes the total cost of the conventional option for rogue states approximately $80,000. Note that the development cost of the rocket system is intentionally not covered in the estimate, since the weapon system is relatively freely available on the world commercial market.

Chemical Weapons

Chemical weapons are made from caustic chemicals and are inherently more expensive to produce than conventional or biological weapons. The Australia Group lists the following typical equipment as requisite for their production, virtually all of which must be glass lined or have a high content of fairly exotic alloys (http://www.australiagroup.net/control_list/dual_chemicals.htm, accessed on 8/13/03):

- Reaction vessels, reactors, or agitators
- Storage tanks, containers, or receivers
- Heat exchangers or condensers
- Distillation or absorption columns
- Remotely operated filling equipment
- Specialized valves
- Multiwalled piping
- Multiwalled piping incorporating a leak detection port
- Pumps: multiple-seal, canned drive, magnetic drive, bellows, or diaphragm
- High alloy incinerators
- Toxic gas monitoring systems and detectors

Aum Shinrikyo spent over $30 million on its weapons production plant for sarin and other weapons, a price tag that included the purchase of precision machine tools for conventional weapon manufacturing as well. However, the glass-lined reactor vessel required to make precursor chemicals for sarin costs $200,000 alone (Kaplan and Marshall, 1996). Other estimates claim that the cost of producing 1000 kg of sarin on small laboratory purchases of raw materials, would be in the neighborhood of $200,000 (Purver, 1995). It would take 300 kg (over 650 pounds) of pure sarin to inflict 50 percent casualties in a 0.8 square kilometer target area (US Congress, 1993). The production of such a large quantity of material appears unrealistic in a terrorism scenario, though sarin could still be used effectively to inflict mass casualties on an enclosed environment such as a building, stadium, mall, or subway. A relatively small attack that required only 280 grams of sarin would cost only $131, according to one researcher who ordered the precursor chemicals through the mail (Musser, 2001). More toxic weapons such as VX have the required effectiveness but are much more technically demanding to produce. For instance, the United States converted a heavy water plant and implemented a four-step "transesterification" process to produce VX (*Jane's Chemical and Biological Defense Guidebook*, 1999). Aum Shinrikyo experienced significant difficulty producing the less-demanding sarin, including probable fatal accidents. Iraq also had difficulty with sarin, overfluorinating it so that its stability was sufficient only for rapid use (within months of production). To reduce costs further, terrorists or states may consider binary weapons, that is, chemical weapons that are finally synthesized at the point of attack. The formulas for such weapons are available, and have been produced for artillery shells by the United States and Iraq, among others. Using binary agents eases production and storage, eliminating the need for specialized equipment such as negative pressure production facilities and some specially coated tubing that will not corrode and contaminate the facility. Nevertheless, the terrorist chemical threat may find an even less expensive and demanding chemical weapon, by using toxic industrial chemicals. The U.S. Environmental Protection Agency (EPA) claims that 123 industrial sites in the United States threaten over 1 million persons in a worst-case scenario chemical release. Such dire warnings may be overstated. For instance, in January 2002, seven tank cars with chlorine gas ruptured in Minot, North Dakota, creating a toxic cloud over a residential neighborhood (Baumann, 2002). Although over 1300 victims received medical care, only one person died. Nevertheless, terrorists seeking to devastate square kilometers might more readily choose to attack a previously identified hazardous chemical plant with conventional explosives or projectiles rather than pursue the expensive, lengthy, hazardous, insecure, and unsure route to military-type chemical weapons. The cost (and planning) of an operation would be similar to that of a standard terrorist operation utilizing high explosives. As a precedent for this type of attack, perpetrators in Bridgeport, Texas, planned in 1997 to attack a massive natural gas refinery, releasing a toxic cloud, merely as a diversion while they robbed an armored car 9 miles away (Kaplan and Tharp, 1997). It would appear that these perpetrators considered the costs of attacking the refinery small relative to the money they would gain from the armored car robbery. For rogue regimes that may wish military-type chemical weapons, it is estimated that a sophisticated sarin-type nerve agent plant would cost about $30–50 million, or as little as $20 million if one were willing to forego waste-handling facilities (*Jane's Chemical and Biological Defense Guidebook*, 1999). Cost of materials would be limited, because precursor chemicals are often available as bulk commodities. Cost of weaponization for military delivery could be significant, because point detonation bursting charges typically consume 20 to 30 percent of chemical toxic payload with missiles. However, even improvised aerosol chemical dispensers would be adequate to achieve the goals of a limited attack.

Biological Weapons

The Australia Group lists the following equipment as potentially applicable to the production of biological weapons (http://www.australiagroup.net/control_list/bio_equip.htm, accessed on 8/13/03):

- Complete containment facilities at P3 or P4 containment level (though the Japanese World War II–era BW program Unit 731, among others, has historically been able to make do without expensive specialized biocontainment facilities)
- Fermenters
- Centrifugal separators
- Cross (tangential) flow filtration equipment capable of continuous separation of pathogenic microorganisms, viruses, toxins, and cell cultures, as well as *in situ* sterilization
- Freeze-drying equipment
- Aerosol inhalation chambers for test animals
- Protective and containment equipment
  - Protective full suits, half suits, and hoods dependent upon a tethered external air supply and operating under positive pressure
  - Class III biological safety cabinets or isolators with similar performance standards (e.g., flexible isolators, dry boxes, anaerobic chambers, glove boxes, or laminar flow hoods (closed with vertical flow)).

This equipment is all inherently dual-use, with legitimate commercial purposes, with the possible exception of the aerosol test chambers, which nevertheless are not critical to terrorist weapon production. There have been various estimates for how much a sufficient set of such equipment would cost. Enviro Control, Inc performed a study for the military that estimated the cost to produce 14 liters (~2.1 kg) dry anthrax spores to be $250,000, updated to 2003 dollars (Enviro Control, Inc, 1976). A lower informed estimate has been made by Kathleen Bailey, former assistant director of the U.S. Arms Control and Disarmament Agency (Cole, 1997). She puts the figure at $10,000, more than an order of magnitude lower than several other estimates. However, the FBI has projected the

cost of producing the *Bacillus anthracis* spores from the fall 2001 anthrax mailings in the United States, and projected the cost to produce it as "in the few thousands" (Shane, 2003). Regardless of how it was made, the *B. anthracis* powder had high viable spore density and buoyancy. Since 5 g were delivered in each of the last two letters, it is reasonable to assume that enough of the agent could have been made by the same techniques to supply an infectious dose to affect half the population in a square kilometer—an amount of about 100 g (Wilkening, 1999). The additional consumables to produce that amount, including typical growth media and most additives, would be rather inexpensive (bentonite, sold as pond clay, might literally be "dirt cheap"). The original seed stock would likely be acquired from nature, stolen, or diverted from a legitimate source at a nominal cost. Although the cost of requisite labware and materials might only come to some $15,000, the perpetrator(s) would still require significant expertise to use them effectively. Thus, they would need not only technical education but also practical experience working with and dispersing living microbes. If perpetrators could be found who possess this expertise and are willing to direct it toward such illicit ends, the cash outlay to generate such weapon agents might be relatively small as described.

State BW programs, however, tend to be significantly more expensive. Much of this increased cost and complexity flows from weaponization requirements. While terrorists may count on improvised procedures, rogue regimes would need to implement more standardized methods. Regardless, both the Aum Shinrikyo and Iraq spent a significant amount of time and effort on mastering dissemination techniques with comparatively negligible results.

Nuclear and Radiological Weapons

Nuclear weapons are much more difficult and expensive to make than any of the other weapons described. However, their effects are devastating. Fabricating a nuclear weapon requires a team with knowledge of "the physical, chemical and metallurgical properties of the various materials to be used, as well as characteristics affecting their fabrication; neutronic properties; radiation effects, both nuclear and biological; technology concerning high explosives and/or chemical propellants; some hydrodynamics; electrical circuitry; and others" (Mark et al., 2002). This range of relevant expertise makes the organizational "footprint" quite large and increases the probability that development efforts will be detected.

A central problem in nuclear weapons development is gaining the special nuclear materials. The simplest weapon design is the "gun" type. However, this design requires highly enriched uranium rather than the comparatively more available plutonium, which can be chemically separated from spent reactor rods. Plutonium weapons, on the other hand, require complex implosion-type designs that make them unlikely terrorist choices for development. The fissionable uranium isotope differs from its much more prominent stable form by the weight of only three neutrons. Techniques to separate the two may run over 4000 stages (gaseous diffusion) and require fluorinating the uranium into a highly volatile and reactive gas. This would be very difficult for a terrorist group to accomplish, although it might be possible for a rogue regime. Cost estimates to refine the special nuclear materials and build a nuclear weapon run to the millions of dollars.

On the other hand, radiation dispersal devices (RDDs), the so-called dirty bombs, are much easier to make and much less effective than nuclear devices. RDDs require some preprocessing of fresh ("hot") nuclear materials, and their immediate effects are typically dominated by their blast impact. Terrorists drawn toward this type of attack might add symbolic amounts of radioactive materials to conventional explosives, or alternatively attack nuclear reactors, including their cooling ponds, to try to spread contamination. Costs to do the latter would be comparable to high end conventional terrorist attacks.

CONCLUSION

Cost will probably not be the driving factor of terrorist groups or rogue states seeking to develop unconventional weapons. Rather, these weapons will be sought to accomplish certain objectives, namely, to instill public fear, gain 'status' or potentially cause massive casualties. Generally speaking, states may have the means to create unconventional weapons, but are likely to be deterred from using them, while terrorist groups are less likely to be able to make them successfully but may also be less likely to be deterred from using them if they do get them. Biological weapons have a comparatively low production cost, but the success of their development is determined by the expertise of the developer. And their effective use is highly scenario dependent. However, as the biotechnology revolution continues, more people are becoming familiar with advanced biological techniques that can be used for either good or ill, increasing the pool of potential perpetrators of biological attack. Biological weapons also hold the potential to inflict large numbers of casualties if used effectively, so increasing preparedness against their use is extremely important.

REFERENCES

Baumann, J., *Protecting Our Hometowns*, U.S. Public Interest Research Group, Washington, 2002, p. 10.

Cole, L., *The 11ᵗʰ Plague*, Freeman and Co., New York, 1997, p. 4.

Danzig, R., *Biological Warfare: A Nation at Risk–A Time to Act*, National Defense University Institute of Strategic Studies Strategic Forum, Number 58, January 1996.

Douglass, J. and Livingstone, N., *America the Vulnerable: The Threat of Chemical and Biological Warfare*, Lexington Books, Lexington, MA, 1987, p. 16.

Enviro Control, Inc, *Minimum Resource for Biological Weapons Capability* (U), 1976. Secret. Cited in *Sample Technical Directive for a Family of Systems (FoS) Design*, Found at http://www.smdc.army.mil/Contracts/Guardian/W9113M_04_R_0005/Sample%20Tech%20Directive%20-%20Final%20Version%2018%20Dec.doc, accessed on 6/24/04.

Hoffman, D., *The Oklahoma City Bombing and the Politics of Terror*, Feral House, Venice, CA, 1998, pp. 423–452.

http://www.australiagroup.net/control_list/bio_equip.htm, accessed on 8/13/03.

http://www.australiagroup.net/control_list/dual_chemicals.htm, accessed on 8/13/03.

http://www.constitution.org/ocbpt/ocbpt_01.htm, accessed on 8/13/03.

http://www.pmulcahy.com/rocket_launchers/russian_rocket_launchers.htm, accessed on 8/13/03.

http://www.pmulcahy.com/self-propelled_artillery/chinese_mrls.html, accessed on 8/13/03.

Jane's Chemical and Biological Defense Guidebook, Jane's Information Group, Alexandria, VA, 1999, pp. 29–32.

Kaplan, D. and Marshall, A., *The Cult at the End of the World*, Crown Publishers, New York, 1996, p. 103.

Kaplan, D. and Tharp, M., Terrorism Threats at Home. *US News and World Report*, December 29, 1997.

Mark, T., *Can Terrorists Build Nuclear Weapons?*, Nuclear Control Institute, Washington, DC. http://www.nci.org/k-m/makeab.htm, accessed on 6/24/04.

Musser, G. *Scientific American*, **285**(6), 20–21, November 5 (2001).

Purver, R. "Chemical Terrorism," in *Chemical and Biological Terrorism: The Threat According to the Open Literature*, Canadian Security Intelligence Service, 1995. Found at http://www.csis-scrs.gc.ca/eng/miscdocs/chemter e.html#r70, accessed on 6/24/04.

Shane, S., Tests point to domestic source behind anthrax letter attacks, *Baltimore Sun*, April 11, 2003, p. 1.

United Nations, *Chemical and Bacteriological (Biological) Weapons and the Effects of Their Possible Use*, 1970.

US Congress, Office of Technology Assessment, *Proliferation of Weapons of Mass Destruction*, U.S. Government Printing Office, Washington, DC, 1993, pp. 53–54.

Wilkening, D., "BCW Attack Scenarios" S.D. Drell, A.D. Sofaer and G.D. Wilson, Eds., *The New Terror: Facing the Threat of Biological and Chemical Weapons*, Hoover Press, Stanford, CA, 1999, p. 111.

World Health Organization, *Health Aspects of Chemical and Biological Weapons*, 1st ed., WHO, Geneva, 1970, pp. 84–100.

CRISIS MANAGEMENT

Zygmunt F. Dembek
Theodore J. Cieslak

INTRODUCTION

The response to a biological event, like any incident, begins at the local level, where immediate actions taken before state and federal resources are likely to become available will prove critical in mitigating that event's effects (Garrett et al., 2000). When response requirements exceed those available at the local level, state and federal assistance can be sought. Response at the federal level is provided for, and guided by, the Federal Response Plan (FRP) (FEMA, 1992). Under the provisions of the FRP, response is divided into two phases: a crisis management phase and a consequence management phase.

The crisis management phase includes the immediate and coordinated actions necessary to resolve a potential public health disaster. These actions may be supported by consequence management activities, which should operate concurrently (consequence management is addressed elsewhere in this volume). This entry discusses crisis management as it pertains to bioterrorism and the collaborative efforts required by first responders, law enforcement and emergency medical personnel, and public health participants as they work together in an established integrated response network.

BIOTERRORISM: SPECIAL CONSIDERATIONS

An act of bioterrorism differs from other forms of terrorism in that its recognition may be greatly delayed. This is because of the incubation periods that are inherent in diseases caused by biological agents (Rotz et al., 2002). Most infectious diseases have incubation periods of days to weeks. Such a time lag between the release of an agent and the presentation of sick individuals can be expected to delay the coordinated response necessary for optimal crisis management, and might also enable a perpetrator to avoid attribution (as of this writing, almost two years after the anthrax mailings of fall 2001, the perpetrator of that act has not yet been identified by law enforcement authorities). Moreover, this time lag could permit the wide geographic dispersion of victims, which is of particular concern in a contagious disease scenario.

The occurrence of a bioterrorist attack will likely be detected clinically, through the recognition of disease symptoms by "frontline physicians" (e.g., physicians in emergency departments and acute care centers, as well as primary care physicians). Cases that present may

be far removed in time and space from the point of exposure, and a lengthy delay therefore may occur before the epidemic itself, let alone the sinister nature of the outbreak, is recognized (Ashford et al., 2003). Because diseases caused by the majority of potential bioterrorism agents have low natural incidence rates among humans, the lack of clinician experience with these diseases can impede their rapid diagnosis and reporting to public health authorities (Chang et al., 2003). Currently, the public health disease surveillance network must rely on disease reporting from community or hospital-based clinicians in order to detect a disease outbreak, although various state and city hospital-based syndromic surveillance systems are being developed and implemented (Lazarus, 2002; Begier et al., 2003). The National Notifiable Diseases Surveillance System, whereby laboratory findings are reported through the states to the Centers for Disease Control and Prevention (CDC), could be too slow to lead to an effective bioterrorism response (Chang, 2003), though future technological developments are expected to improve this system.

CRISIS RESPONSE TO A BIOTERRORIST EVENT

The multijurisdictional crisis response to a bioterrorist event would be expected to proceed as follows:

- A municipal or county health department is notified of an unusual case(s) of, for example, respiratory or cutaneous illness that has presented to a health care provider or local hospital.
- This local health agency initiates an epidemiological investigation and proceeds to notify and/or request assistance from the state health public health agency.
- The state health department in turn contacts the CDC in order to provide notification of a nationally significant event and to receive further public health assistance. In doing so, the state health department initiates the state public health crisis response (CDC, 2001).

At each level of response (i.e., municipal, county, state, or federal), a coordinated response network must be established that incorporates the abilities and assets of all essential emergency crisis response partners. Such a network is generally coordinated at the city, county, or state office of emergency management, with input from participants to a central manager under the direction of the chief elected official.

Law Enforcement

Suspicion of a bioterrorism-related disease outbreak should prompt the notification of law enforcement authorities. The law enforcement community would then conduct its own investigation, beginning with local and state agencies. The Federal Bureau of Investigation, which under Presidential Decision Directive 39 has lead responsibility for the operational response to any terrorist threat or incident including a biological attack, would become rapidly involved. Because any act or threat of bioterrorism constitutes a federal felony, additional federal law enforcement agencies may be called upon as well (USDOJ, 2003).

First Responders

The term "first responders" refers to those individuals likely to be the first to arrive at the scene of an incident, usually emergency medical technicians, police officers, or firemen. A bioterrorism event will likely pose a significant challenge to such individuals who are relatively undertrained in managing bioterrorism as opposed to fires, natural disasters, and hazardous chemical emergencies. Traditional state and municipal hazardous materials, or hazmat, training scenarios do not adequately train a crisis response network for a bioterrorism incident (Waeckerle, 2000). A locally based medical asset that should enhance the medical response component within the first 48 h of a bioterrorism event is the Metropolitan Medical Response System (MMRS). Overseen by the Department of Homeland Security, the MMRS is a nationwide effort to develop or improve existing emergency preparedness systems to effectively respond to a public health crisis, especially a weapons of mass destruction (WMD) event.

Incident Command System

The incident command system (ICS) is a national standard emergency management concept designed to permit a coordinated effort by disparate first response participants. It is an extremely flexible system, capable of expanding or contracting to meet the needs of the incident (FEMA, 1998). The ICS was developed to be utilized in response to any traditional crisis (e.g., fire, hazmat incident, natural disaster), but is considered an essential format for crisis response to a bioterrorism event, in which more than one agency will likely share crisis management responsibility. Since many emergency management functions remain similar in various types of disasters (e.g., providing food and coordinating volunteers), those functions and the methodologies for their provision are familiar to those using the ICS (PERI, 2003). Moreover, the unified command proscribed by the ICS allows all involved agencies to contribute to the command process by determining overall objectives, providing for joint planning while conducting integrated operations, and maximizing the use of available resources (FEMA, 1998).

Under the ICS, on-scene crisis management is placed under control of the incident commander, traditionally the local fire chief or a designee (FEMA, 1998). This role is initially assumed by the first responder on arrival, then reassigned as feasible to an on-site senior official, with the agency primarily responsible for management of the incident. The organizational structure of the ICS develops in a modular fashion, with the establishment of four separate functional areas as the need arises: planning, operations, logistics, and finance/administration. Within these functional areas, several additional branches may also be established as necessary (FEMA, 1998).

The ICS allows health care personnel providing relevant expertise to the specific scenario (hospital, emergency, and public health responders) to do their job (e.g., patient care or epidemiological investigation) while other experts are delegated to crisis management, logistical support, and other critical tasks. Thus, the system allows for an orderly response to any multiagency or multijurisdictional event, preventing further injury and damage and avoiding prolongation of the incident. It enables a coordinated response effort in the event that a fully developed and vetted bioterrorism response plan is not available.

Emergency Operations Center

Once a statewide crisis is identified (as with the occurrence of a bioterrorism event), the statewide Emergency Operations Center (EOC) is activated. The state EOC includes representatives of state agencies (e.g., the state health department, state police, and state National Guard units), private agencies (e.g., the utility and banking industry), and volunteer organizations (e.g., the American Red Cross and various religious groups). The EOC serves to coordinate these organizations' responses to an emergency event. An EOC manager, reporting to the governor or his representative, coordinates the work of the EOC.

Dissemination of Information to the Public

The information officer under the ICS reports directly to the incident commander. Similarly, the public information officer (PIO) for a statewide crisis (e.g., a bioterrorism event) would report directly to the state EOC manager. It is of vital importance during any crisis to manage the flow and content of crisis-relevant information to the media. Recent history is rife with examples of various media sources containing conflicting or inaccurate information during the occurrence of an incident. This serves to heighten concern and induce panic, as exemplified by media coverage of the events of September 11, 2001, and the subsequent anthrax mailings. Public health authorities and government officials attempted to allay the fears of the public, but messages were inconsistent, did not occur in real time, and were sometimes contradictory (Altman and Kolata, 2002). Hoarding of the antibiotic ciprofloxacin reached crisis levels, with medication being offered for sale on the Internet and from out-of-country sources (Federal Trade Commission, 2001). This illustrates how important it is that those who generate or transmit news be intimately familiar with and use sound principles of risk communication.

Given the intense media coverage devoted to bioterrorism issues, it is advisable that communication during a bioterrorism crisis be highly coordinated by a designated media spokesperson to minimize misunderstandings and prevent rumor generation among participants and the media. Such coordination may be accomplished by designating a centralized PIO, or media spokesperson, who acts as a sole source of information to media conduits throughout a crisis (ASTDHPPE, 2001). Rumor control is critical to controlling misattribution of symptoms and promoting appropriate behavior, and risk communication through the media will play a central role in determining how the public responds.

The Role of the Hospital

Crisis management at the hospital level consists of a fully coordinated partnership of both prehospital personnel (emergency medical services, or EMS) and the hospital staff who would be involved in responding to a bioterrorism incident. In fact, the Joint Commission on Accreditation of Healthcare Organizations (JCAHO), in its Environment of Care Standards (EC.1.4), requires hospitals to conduct a hazard vulnerability analysis, develop an emergency management plan, and evaluate this plan annually (JCAHO, 2003). Moreover, JCAHO advocates an "all-hazards" approach to disaster planning and management. Hospital personnel from administration, security, public relations, laboratory, pharmacy, and facilities management should be familiar with the plan, know when and how to activate it and the hospital command center, and understand their roles in the crisis response (Schultz et al., 2002).

Community hospitals are a vital partner in the identification, triage, and treatment of those affected by bioterrorism. Rapid casualty care is of utmost importance immediately subsequent to a mass-casualty incident. Patient transportation needs will increase greatly, and the local EMS may become overwhelmed. A bioterrorism event may necessitate auxiliary replacement of casualty transport through the use of state military or other resources. A fail-safe statewide communications network should be in place in order to provide advance notification and coordinate movement to hospitals of incoming casualties (SBCCOM, 2000).

In preparation for a bioterrorism event, hospitals need to complete a number of pertinent tasks (USDOJ, 2003). During the event, hospitals must conduct patient screening and triage. Subsequent release of asymptomatic individuals, based on potential for exposure and illness, may be necessary in order to avoid overwhelming treatment facilities. Bona fide casualties will likely require admission until maximum capacity is reached. Hospitals must then be able to redirect noncritical admissions, identify a backup or overflow emergency evaluation and triage facility (for example, an alternative health care facility of the ICS), and maintain constant and direct communication with public health authorities.

The Role of the Military

Unlike some other civil disasters, there may be a state or federal military component to a bioterrorism incident. Specific roles for the state military department may include transportation of critical medical supplies such as the distribution of the federal pharmaceutical strategic national stockpile, the detection of a WMD agent, or communications linkages. The National Guard, through its state components, has been authorized to develop 32 weapons of mass destruction civil support teams (WMD-CSTs) throughout the country. These teams are trained to deploy to an incident in support of the ICS incident commander to assess a known or suspected chemical, biological, or radiological incident, advise the incident commander and first responders of appropriate actions, and facilitate requests for assistance from further state or federal assets. A given WMD-CST has a minimal response time of 4 h, and possesses state-of-the-art detection equipment and capabilities for sample collection, analysis, and shipment, as well as a response vehicle that offers a secure unified command suite with the ability to communicate via every available communication network (HF, UHF, VHF, cell phone, the Internet, and worldwide satellite systems).

Psychological Consequences and Considerations

An important component of bioterrorism crisis response is psychological stress management and support, since it is probable that for every person seeking hospital care for physical injuries or infection following a biological or chemical terrorist incident, at least 6 to 10 will present with psychological concerns (Lord, 2001). The release of a biological agent, in particular, presents complex psychosocial challenges for a civilian population (Norwood, 2001). Features of bioterrorism may make group panic more likely (DiGiovanni, 2001). A disorganized, ineffective response to a bioterrorism event will heighten the public's fear and break down trust in public institutions (DiGiovanni, 2001). For example, anxiety engendered by media coverage of the fall 2001 anthrax mailings led some employees of the U.S. Postal System to question the safety and health recommendations of the CDC (Lipton and Johnson, 2001).

A large infectious disease outbreak could present complex psychosocial challenges for the health care system. Although the vast majority of traumatized individuals can be expected to recover over time, sequelae including posttraumatic stress disorder, major depression, substance abuse, dissociation, somatization disorder, recurrence of endemic psychiatric problems, and psychological factors affecting physical conditions will likely occur (Norwood et al., 2001). Although true panic is rare and preventable, it could occur through a combination of the public recognition of a bioterrorism event and information mismanagement of that event. Public health authorities need to be involved in providing accurate health risk communication to the public throughout such an event (Glass and Schoch-Spana, 2002).

CONCLUSION

Successful crisis management of a bioterrorism event must coordinate disparate entities to respond to a public health crisis in a rapid manner to save lives and reduce and ameliorate harm to the community. These organizations must work well with each other and with other first responders using the ICS method and the state and federal emergency response system. The institutional response to a bioterrorist attack, like the institutional response to any disaster, can be divided into a phase of crisis management and a phase of consequence management. To some degree, this distinction is artificial and, in most cases, consequence management activities must begin before crisis management is complete. Nonetheless, meticulous planning and attention to crisis management may diminish the consequences of an attack or disaster, and subsequently, the need for extensive consequence management.

REFERENCES

Altman, L.K. and Kolata, G., Anthrax Missteps Offer Guide to Fight Next Bioterror Battle, *New York Times*, January 6, 2002.

Ashford, D.A., Kaiser, R.M., Bales, M.E., Shutt, K., Patrawall, A., McShan, A., Tappero, J.W., Perkins, B.A. and Dannenberg, A.L., *Emerg. Infect. Dis.*, 9, 515–519 (2003).

Association of State and Territorial Directors of Health Promotion and Public Education, *Model Emergency Response Communications Planning for Infectious Disease Outbreaks and Bioterrorist Events*, Washington, DC, 2001, p. 164.

Begier, E.M., Sockwell, D., Branch, L.M., Davies-Cole, J.O., Jones, L.H., Edwards, L., Casani, J.A. and Blythe, D., *Emerg. Infect. Dis.*, 9, 393–396 (2003).

CDC, *Morb. Mortal. Wkly. Rev.*, 50, 877 (2001).

Chang, M.-H., Glynn, M.K., and Groseclose, S.L., *Emerg. Infect. Dis.*, 9, 556–564 (2003).

DiGiovanni, C., *Mil. Med.*, 166(Suppl. 2), 59–60, 2001.

Federal Emergency Management Agency, *The Federal Response Plan for Public Law 93–288*, as amended, April 1992.

Federal Trade Commission, *Agencies Offer Tips for Consumers Eyeing Online Anthrax Cures: FTC Says Fraudsters Prey on Consumers' Fears*, Federal Trade Commission, Washington, DC, November 11, 2001.

FEMA Emergency Management Institute, *Incident Command System*, FEMA Independent study course IS195, January 1998.

Garrett, L.C., Magruder, C., and Molgard, C.A., *J. Public Health Manag. Pract.*, 6, 1–7 (2000).

Glass, T.A. and Schoch-Spana, M., *Clin. Infect. Dis.*, 34, 217–223 (2002).

Jernigan, J.A., Stephens, D.S., Ashford, D.A., Omenaca, C., Topiel, M.S., Galbraith, M., Tapper, M., Fisk, T.L., Zaki, S., Popovic, T., Meyer, R.F., Quinn, C.P., Harper, S.A., Fridkin, S.K., Sejvar, J.J., Shepard, C.W., McConnell, M., Guarner, J., Shieh, W.J., Malecki, J.M., Gerberding, J.L., Hughes, J.M., Perkins, B.A., and the Anthrax Bioterrorism Investigation Team, *Emerg. Infect. Dis.*, 7, 933–944 (2001).

Joint Commission on Accreditation of Healthcare Organizations, *2003 Hospital Accreditation Standards*, JCAHO, Oakbrook Terrace, IL, 2003, pp. 221–224.

Lazarus, R., *Emerg. Infect. Dis.*, 8, 753–760 (2002).

Lipton, E. and Johnson, K., A nation challenged: the anthrax trail; tracking bioterror's tangled course, *New York Times*, December 26, 2001.

Lord, E.J., *Mil. Med.*, 166(Suppl.2), 34–35 (2001).

Norwood, A.E., Holloway, H.C., and Ursano, R.J., *Mil. Med.*, 166(Suppl.2), 27–28 (2001).

Norwood, A.N., *Mil. Med.*, 166(Suppl.2), 27–28 (2001).

Public Entity Risk Institute, *Characteristics of Effective Emergency Management Organizational Structures*, Fairfax, VA, 2003.

Rotz, L.D., Khan, A.S., Lillibridge, S.R., Ostroff, S.M. and Hughes, J.M., *Emerg. Infect. Dis.*, 8, 225–230 (2002).

Schultz, C.H., Mothershead, J.L., and Field, M., *Emerg. Med. Clin. N. Am.*, 20, 437–455 (2002).

Soldier and Biological Chemical Command. CDC/DoD Smallpox Workshop. DTIC/SBCCOM SPO700-00-D-3180. Battelle Edgewood Operations, Bel Air, MD, April 17–19, 2000.

US Department of Justice, *Criminal and Epidemiological Investigation Handbook*, US Department of Justice Office of Justice Programs and US Soldier and Biological Chemical Command, 2003.

Waeckerle, J.F., *J. Am. Med. Assoc.*, 283, 252–254 (2000).

FURTHER READING

Landesman, L.Y., *Public Health Management of Disasters*, American Public Health Association, Washington, DC, 2001.

Novick, L.F. and Marr, J.S., *Public Health Issues in Disaster Preparedness: Focus on Bioterrorism*, Aspen Publishers, New York, 2001.

WEB RESOURCES

Criminal and Epidemiological Investigation Handbook, http://hld.sbccom.army.mil/ip/ceih_download.htm, accessed on 6/27/03.

Federal Trade Commission, *Agencies Offer Tips for Consumers Eyeing Online Anthrax Cures: FTC Says Fraudsters Prey on Consumers' Fears*, Washington, DC, November 11, 2001, http://www.ftc.gov/opa/2001/11/alert.htm, accessed on 6/27/03.

Fight Against Bioterrorism, http://health.cwftx.net/terror16.htm, accessed on 6/27/03.

Global Security.org, http://www.globalsecurity.org/military/agency/army/wmd-cst.htm, accessed on 6/27/03.

Metropolitan Medical Response System, http://www.mmrs.hhs.gov/, accessed on 6/27/03.

Public Entity Risk Institute, http://www.riskinstitute.org, accessed on 6/27/03.

Weapons of Mass Destruction Civil Support Teams, http://www.defenselink.mil/specials/destruction/, accessed on 6/27/03.

See also CONSEQUENCE MANAGEMENT and HOMELAND DEFENSE.

CUBA

Raymond A. Zilinskas
Silvana Rodriguez
Michelle Baker

INTRODUCTION

This entry has two sections. The first considers Cuba's history of terrorist support and examines the United States's accusations regarding current terrorist relations. The second reviews Cuba's biotechnological capability and how it might relate to the potential development of biological weapons.

HISTORY OF CUBAN SUPPORT OF TERRORISM AND NATIONS OF PROLIFERANT CONCERN

From the Communist revolution in 1959 through the 1980s, the Cuban government openly supported leftist guerilla movements in Latin America and Africa (Fig. 1). Over three decades, Cuba's foreign policy in Latin America included support for the Sandinistas in Nicaragua (Marx, 2002), the Leftist Revolutionary Movement (MIR) in Chile, the Montoneros and the People's Revolutionary Army (ERP) in Argentina, the Tupamaros in Uruguay (US State Department, 1981), the Farabundo Marti National Liberation Front (FMLN) in El Salvador (Landau and Smith, 2002), the Armed Forces of National Liberation (FALN) in Venezuela (Center for Free Cuba, 2003), and the National Liberation Army (ELN) in Bolivia (SpecialOperations.Com, 2001). The Cuban government also lent support to independence movements in Algeria, Angola, Ethiopia, Guinea-Bissau, and the Congo in the 1960s and early 1970s (Landau and Smith, 2002).

Currently, Cuba is accused of permitting up to 20 members of Basque Fatherland and Liberty (ETA) to reside in Cuba, providing "some degree of safe haven and support" to members of the Colombian Revolutionary Armed Forces of Colombia (FARC) and ELN groups, and serving as the residence of Sinn Fein representative Niall Connolly (US Department of State, 2002). Further, the U.S. government claims that Castro has "vacillated" on the war against terrorism and has "continued to view terrorism as a legitimate revolutionary tactic" (US Department of State, 2002). However, the United States has published no evidence that any of the groups named above have active terrorist camps in Cuba or are using Cuba as a base for activities elsewhere.

After the collapse of the Soviet Union, whose subsidies to Cuba might have made assistance to terrorists possible, Fidel Castro publicly announced that Cuba would no longer provide material support to revolutionary groups trying to overthrow established governments. Although the U.S. government maintains that Cuba continues to support terrorist groups today, citing examples of how mostly old leaders of terrorist groups are being provided refuge in Cuba, the evidence for ongoing assistance is questionable and largely unrecognized by the international community. Over the past decade, the Cuban government has publicly condemned "all acts, methods, and practices of terrorism" (Olson, 2003; Ministry of Foreign Affairs of Cuba, 2003). Further, Cuban officials often have sought to portray their nation as being the victim of terrorist attacks carried out by Cuban exiles (*Daily Trust*, 2002).

Only hours after the September 11, 2001, terrorist attacks, Fidel Castro publicly stated that the Cuban government "vigorously rejects and condemns the attacks" and that the Cuban people "express their solidarity with the US people and their total willingness to cooperate to the extent of their modest possibilities, with the health institutions and any other medical or humanitarian institution of that country" (*British Broadcasting Corporation*, 2001). Further, Cuba is party to all 12 international counterterrorism conventions and protocols. Cuba ratified the Biological and Toxin Weapons Convention (BWC) in 1976. The Cuban government offered to sign a bilateral agreement with the U.S. government in early 2002 providing for joint efforts against terrorism, but it was turned down by the United States (Smith, 2002; Ministry of Foreign Affairs of Cuba, 2003.)

A cause for concern for the U.S. government is Cuba's recent efforts to cultivate relations with Iran, Syria, and Libya, countries known to support terrorists. In May 2001, Fidel Castro visited the three countries to bolster Cuba's political and economic ties in the region. While the bulk of the known cooperation among the countries centers on trade issues and public health exchanges, there is little or no public information on the three countries' dealings with Cuban intelligence or other ties to alleged Cuba-sponsored terrorism (Council on Foreign Relations, 2003).

BIOTECHNOLOGY AND BIOLOGICAL WEAPONS

In 2002, the U.S. government lodged the first of what was to become a series of accusations that Cuba had

Encyclopedia of Bioterrorism Defense,   Edited by Richard F. Pilch and Raymond A. Zilinskas
ISBN 0-471-46717-0   Copyright © 2005 Wiley-Liss

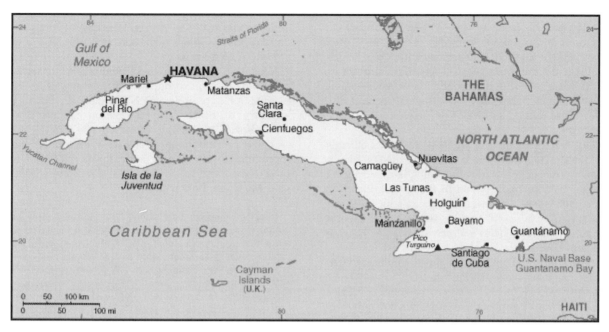

Figure 1. Map of Cuba, *CIA World Factbook*, CIA website: http://www.cia.gov/cia/publications/factbook/geos/cu.html.

both fostered the international proliferation of biological weapons and possessed a national biological warfare (BW) capability. As this is being written, we are not in a position to either corroborate U.S. contentions or disprove them. What we can do is provide information and options for the consideration of the issue whether Cuba is indeed misusing its substantial capabilities in biotechnology. Thus, this section has three parts. First, we describe Cuba's accomplishments in biotechnology. Second, we discuss Cuba and BW. Third, we discuss the likelihood of Cuba having misused its biotechnological capabilities.

Accomplishments in the Biosciences and Biotechnology

From the beginning of his rule, Fidel Castro has stated his belief that biotechnology is important to Cuba's economic development (Limonta, 1989; Colwell, 1998; Kaiser, 1998; Ramkisson, 1999; Lage, 2000; Satz, 2000; Reid-Henry, 2002; Baker, 2003). As a result of according biotechnology a high-priority area for development, and having spent the funds necessary to make Cuban biotechnology reach world-class status, Cuba had by the middle 1990s created one of the most technologically advanced biological industries in the world, able to compete favorably with many industrialized nations. Thus, in the past decade, Cuba has successfully developed a meningitis B vaccine, hepatitis B vaccine, cattle tick "vaccine," and monoclonal antibodies for kidney transplants. Currently, Cuba is conducting trials involving epidermal growth factor, cancer vaccines, AIDS vaccine, and hepatitis C vaccine, as well as with pest-resistant sugar cane. It is claimed that a transgenic fish, tilapia, is already being sold in Cuban markets; an estimated 30 tons of transgenic tilapia was produced already in 1998 (*Nature*, 1999). These activities clearly demonstrate Cuba's versatility in biotechnology research and production (Kaiser, 1998).

With the expansion of Cuban biotechnology industry, it became a major source of both medicine and scientific technology to the developing world. Fidel Castro has stated that Cuba began to export biotechnology products in 1990, with the value of exports increasing every year since then (Castro, 2002). Cuba currently has technology trade agreements with 14 countries, with negotiations for trade underway with several other states. In the past decade, Iran, China, India, Algeria, Brazil, and Venezuela have become the main recipients of Cuban technology (Castro, 2002). Cuba has also helped to initiate joint biotechnology enterprises within other developing countries, especially Iran, China, and India, transferring technology from several different scientific institutions, including the Center for Genetic Engineering and Biotechnology (CIGB) and the Center for Molecular Immunology (*Prensa Latina*, 2002). There are, however, claims that Cuba's biotechnology capability has been in a sharp decline since about 1991, when the Soviet Union dissolved and its aid to Cuba stopped (de la Fuente, 2001).

Cuba Vis-À-Vis Biological Warfare

Cuban government officials have often asserted that the United States has used biological weapons to attack its population and agriculture. During 1962 to 1997, these attacks allegedly triggered diseases affecting humans (dengue fever, dengue hemorrhagic fever, acute hemorrhagic conjunctivitis, and optic and peripheral neuropathy), animals (African swine fever and Newcastle disease), and plants (tobacco blue mold disease and sugar cane rust), as well as an infestation of crops by a destructive flying insect called *Thrips palmi*. A thorough examination of biological, entomological, epidemiological, and meteorological data has been made, the conclusion of which was that each outbreak or infestation either had a natural cause or was accidentally introduced into

Cuba (Zilinskas, 1999). None resulted from biological attack. However, it is clear that almost all outbreaks resulted from or were made worse by policies instituted by Cuban leaders that led to deficiencies in public health and diminished its ability to control agricultural diseases and infestations. For this reason, the Cuban government might have found it politically worthwhile to shift the blame for these outbreaks away from itself and onto the United States.

Turning the table on the Cubans, in 2002, officials of the Bush administration began what looks like a campaign to inform the world that Cuba is misusing its capabilities in biotechnology and in doing so is violating the BWC. But before discussing the actions by the Bush administration, it is useful to first consider past reports by U.S. government agencies and private persons bearing on Cuba and BW.

In 1998, the Defense Intelligence Agency (DIA) and other agencies constituting the U.S. intelligence community reported: "Cuba's current scientific facilities and expertise could support an offensive BW program in at least the research and development stage. Cuba's biotechnology industry is one of the most advanced in emerging countries and would be capable of producing BW agents" (Defense Intelligence Agency, 1998). Other than stating the obvious, namely, that Cuba had the potential to acquire biological weapons should its leadership so decide, no mention is made of Cuba actually being involved in activities that might violate the BWC or be of concern to the U.S. government. In subsequent reports issued by the U.S. Department of State in 1999 and 2002, seven states were identified as sponsors of international terrorism, including Iran, Iraq, Syria, Libya, Cuba, Sudan, and North Korea (US Department of State, 2000, 2002). Analysts have found it alarming that several of these states are believed to also have a BW capacity, although Cuba has not been specifically pointed out in this regard.

In 1999, two events occurred that is of relevance to this entry. First, Dr. Ken Alibek, an important defector from the former Soviet Unions BW program, noted in his book that although he did not have firsthand knowledge of Cuba's programs, his boss, Major General Yuri Kalinin, after a visit to Cuba in 1990, became convinced that the Havana government was deeply involved in a BW research effort (Alibek and Handelman, 1999). Since no other evidence buttresses this secondhand information, it has not been universally accepted as correct (Tamayo, 1999). Also in 1999, another private person, Manuel Cereijo, published his assessment of the threats that Cuba posed to U.S. security, including the supposed biological weapons threat (Cereijo, 1999). (Manuel Cereijo claims that he is an engineer with numerous years of experience in control systems and communications as well as being a professor of the subject at Florida International University. However, he is not listed as a faculty member on the Florida International University website.) While not stating directly that Cuba actually possesses biological weapons, he claimed: "The arsenal of Cuba could include weapons based on tularemia, anthrax, epidemic typhus, smallpox, dengue fever, Marburg, Ebola. It could also extend to neurological agents, based on chemical substances produced naturally in the human body" (Cereijo, 1999). Cereijo's claims have not been supported by anything beyond his own words.

Allegation about supposed Cuban misdeeds articulated by senior U.S. government officials began to appear in 2002. As a prelude, on November 19, 2001, Under Secretary of State John R. Bolton named Iraq and North Korea as having breached the BWC, but also warned of possible violations by other, unnamed countries. Approximately six months later, in May 2002, he named three of them: Libya, Syria, and Cuba (Bolton, 2002a). In regard to Cuba, Bolton stated:

Here is what we now know: The United States believes that Cuba has at least a limited offensive biological warfare research and development effort. Cuba has provided dual-use biotechnology to other rogue states. We are concerned that such technology could support BW programs in those states. We call on Cuba to cease all BW-applicable cooperation with rogue states and to fully comply with all of its obligations under the Biological Weapons Convention (Bolton, 2002a).

Bolton's assertion more than implied that Cuba was violating the BWC because it was undertaking an offensive BW development effort (research does not fall under the purview of the BWC) and assisting Iran to acquire its own BW capability. It received extensive, intense press coverage and stimulated much discussion in Washington. Perhaps in an attempt to desensitize the situation, Secretary of State Colin Powell issued a clarification:

As Under Secretary Bolton said recently, we do believe that Cuba has a biological offensive research capability. We didn't say that it actually had such weapons, but it has the capacity the capability to conduct such research. This is not a new statement… So Under Secretary Bolton's speech which got attention on this issue again wasn't breaking new ground as far as the United States' position on this issue goes (Powell, 2002).

Several senior officials subsequently followed Powell's lead in expressing U.S. concerns about Cuba possibly misusing its biotechnological research capabilities, including Assistant Secretary of State for Intelligence and Research Carl W. Ford Jr. (Ford, 2002) and Bolton himself (Bolton, 2002b).

The Cuban government strongly denied the allegations that it has or is supporting illicit BW activities. After Bolton's statements, President Castro himself took the stage to rebut the charges, calling them "Olympic-size lies" (*Cubavision*, 2002). He noted that Cuba has The Law against Terrorism Acts that spells out "The person who manufactures, facilitates, sells, transports, sends, introduces in the country or keeps in his or her possession, under any form or in any place chemical or biological agents… is liable to sanctions of 10 to 30 years of imprisonment, life sentence or capital punishment" (*Cubavision*, 2002). Castro also asserted:

No one has ever produced a single piece of evidence that any program for developing nuclear, chemical or biological weapons has been set up in our country… it would be utterly stupid to behave in any other way... Any such program would lead the economy of any small country to bankruptcy. Cuba would never have been able to transport such weapons. Moreover, it would be a mistake to use them in battle against an enemy that has a thousand times more of those weapons and that would be only to [sic] happy to find an excuse to use them (*Cubavision*, 2002).

Shortly after Castro's speech, the Center for Defense Information in Washington, DC, was invited by the Cuban government to organize a visit by a group of experts to biological facilities of their choice. The Center accepted the offer, assembled a multidisciplinary group, and departed for Cuba. After being greeted by Castro personally, the expert team visited nine facilities it deemed most significant in Cuban biotechnology: the CIGB, Center for Molecular Immunology, National Center for Agricultural and Livestock Health, Laboratorios DAVIH, Pharmaceutical Biological Laboratories, Center for Marine Bioactive Substances, Special Processing Plant "La Fabriquita," Carlos J. Finlay Institute, and Pedro Kouri Institute of Tropical Medicine. Soon after returning from its visit, the team wrote about its impressions (Baker, 2003). Included were observations made by each team member. None had observed anything suspicious. The statement by team member Terrence Taylor, a former UN Special Commission biological inspector in Iraq, appears to reflect the opinion of most if not all team members:

Neither I nor any member of the group found any indications that Cuba was involved in other than legitimate biotechnical activities. We did not expect that we would uncover anything to the contrary. The consensus of the group is that while Cuba certainly has the capability to develop and produce chemical and biological weapons, nothing that we saw or heard led us to the conclusion that they are proceeding on this path.

Between the issuance of the Center for Defense Information's report and the writing of this entry, no new substantial developments have occurred bearing on alleged Cuban BW-related activity. We then assume that the situation remains about the same as after Colin Powell's press briefing of May 13, 2002; that is, the Bush administration remains suspicious about the intent of Cuba's biotechnological research and development activities at home and whether dual-use capabilities are exported to other nations, while the Cuban government holds that it is doing nothing that contravenes the BWC, including its Article 3 (which forbids state parties from transferring BW-related knowledge and materials to other nations or subnational entities).

Discussion

Without doubt, Cuba possesses a powerful capability in biotechnology, one that could be used for biological weapons acquisition should its leaders so decide. The Bush administration claims that such a decision has been made and this capability is being misused, possibly to develop biological weapons and help Iran institute a BW program. If the U.S. government indeed possesses high-quality intelligence strongly indicative of Cuba being

involved in activities that contravene the BWC, it could seek redress under international law. Since both Cuba and the United States are BWC State Parties, either can call for a consultative meeting to consider the other's activities that allegedly contravene the BWC. Or if this approach is, from past experience, perceived as being insufficient to satisfactorily settle the matter, either can invoke the BWC's Article 6, which specifies that the UN Security Council investigate the claim. Yet a third possibility exists; either side can ask the U.N. Secretary General to investigate alleged breeches of the BWC. If either of the two latter approaches was to be invoked, the accuser would have to present hard evidence to support its charge. Of course, we cannot know how the Bush administration will proceed, if at all. The United States might not wish to have its claims against Cuba investigated by a third party because either they cannot be supported with evidence of sufficient weight to prove its case, or it does not want to reveal intelligence sources. But after the intelligence debacle concerning alleged Iraqi weapons of mass destruction, most of the world's nations probably would be unwilling to accept only the assurances of the U.S. government that Cuba is involved in activities that violate the BWC without seeing hard evidence proving this contention.

What about the Bush administration's accusation that Cuba is exporting biotechnological know-how and equipment to nations of concern, such as Iran, that provides them with the means to acquire biological weapons? This is truly a difficult issue to resolve because of the dual-use characteristics of biotechnology know-how and equipment. In his May 10, 2002, speech, Castro freely admitted to Cuba having in place or being in the process of negotiating technology transfer agreements with 14 countries, including Iran (four transfers and four products) (Castro, 2002). He does not reveal what these transfers and products are, but if fermenters, separators, dryers, and other industrial equipment were being sold to Iran, they could be used for either civilian or military purposes. Thus, the question is, are the Cubans responsible for how the Iranians use the equipment sold to them? For instance, were the Iranians at some time in the future to use this equipment to acquire biological weapons in violation of the BWC, would Cuba be held wholly or partially responsible for this development?

On the face of it, as long as Cuba in good faith sells equipment that it believes will be used for civilian purposes, it is allowed to do so without international opprobrium. After all, in a similar situation of the past, the industries in Austria, Sweden, France, United Kingdom, Germany, and the United States, assumedly with the permission of their home governments, in the late 1980s sold the Saddam Husayn government many cultures of pathogens and biotechnological equipment required by Iraqi industries to manufacture vaccines, antibiotics, and other biological products. Were these countries responsible for the Iraqi government having diverted some of these cultures and equipment to manufacture BW agents used in Iraqi biological weapons? Probably not, since no legal challenge has been mounted to claim damages for this commerce.

The issue then is, does the U.S. government possess evidence indicating that Cuba is selling Iran, and other nations, equipment that it knows will be used for illicit purposes? If so, it could accuse Cuba of violating Article 3 of the BWC and seek recourse under international law as described above. But for now, no such evidence has been published. Until the U.S. government chooses to release any information it may have that implicates Cuba as a proliferant country, Cuba has the right to sell its biotechnology products to anyone it wants to.

REFERENCES

Alibek, K. and Handelman, S., *Biohazard: The Chilling True Story of the Largest Covert Biological Weapons Program in the World—Told From Inside by the Man Who Ran It*, Random House, New York, 1999.

Baker, G., Ed., *Cuban Biotechnology: A First Hand Report*, Center for Defense Information, Washington, DC, 2003.

Bolton, J.R., *Beyond the Axis of Evil: Additional Threats from Weapons of Mass Destruction*, Heritage Lectures 743, the Heritage Foundation, May 6, 2002a, http://www.mtholyoke.edu/acad/intrel/bush/bolton.htm, accessed on 11/30/03.

Bolton, J.R., *The U.S. position on the Biological Weapons Convention: Combating the BW threat*, Speech presented on August 27, 2002b, Tokyo American Center, Tokyo, Japan, http://www.state.gov/t/us/rm/13090.htm.

British Broadcasting Corporation, Fidel Castro Calls on USA to 'Act Calmly' Following 11 September Attacks, BBC Monitoring Latin America, September 14, 2001.

Castro, F., There Will be Weapons Much More... (II), *Granma Internacional (Internet version)*, May 14, (2002), http://www.granma.cu/ingles/mayo02-3/20respue2-i.html.

Center for Free Cuba website, 1991–2001, 2003, http://www.cubacenter.org/media/news_articles/um_report2.php3.

Cereijo, M., *Cuba: The Threat*, Guaracabuya, 1999, http://www.globalsecurity.org/wmd/library/news/cuba/oagmc024.htm.

Colwell, R.R., *Report on travel to Havana, Cuba*, June 28–July 4, 1997, Unpublished report to the American Society for Microbiology, 1998.

Council on Foreign Relations website, *Terrorism: Q&A*, 2003, http://www.terrorismanswers.com/sponsors/cuba_print.html.

Cubavision, Castro Refutes US John Bolton's Statements on Biological Warfare, FBIS LAP20020511000019, May 10, 2002.

Daily Trust, Africa News, April 30, 2002.

Defense Intelligence Agency, in coordination with the Central Intelligence Agency, Department of State Bureau of Intelligence, National Security Agency, and United States Southern Command Joint Intelligence Center 1998, *The Cuban Threat to U.S. National Security*, http://www.globalsecurity.org/wmd/library/news/cuba/cubarpt.htm, accessed on 11/30/03.

de la Fuente, J., *Nat. Biotechnol.*, **10**, 905–907 (2001).

Ford, C.W., Jr., Statement before the Senate Subcommittee on Western Hemisphere, Peace Corps, and Narcotic Affairs, Senate Committee on Foreign Relations on June 5, 2002, http://www.globalsecurity.org/wmd/library/news/cuba/cuba-020605-usia01.htm, accessed on 11/30/03.

Kaiser, J., *Science*, **282**, 1626–1628 (1998).

Lage, A., *Biotecnologia Aplicada*, **17**(1), 55–61 (2000).

Landau, A.K. and Smith, W.S., Cuba on the Terrorist List: In Defense of the Nation or Domestic Political Calculation? *International Policy Report*, Center for International Policy, November 2002.

Limonta, M., "Biotechnology and the Third World: Development Strategies in Cuba," in B.R. Bloom and A. Cerami, Eds., *Biomedical Science and the Third World: Under the Volcano*, Annals of the New York Academy of Sciences, Vol. 569, New York, 1989, pp. 325–334.

Marx, G., Cuba Accused of Harboring Basques; Spanish Group Not Exporting Terror, Havana Maintains, *Chicago Tribune*, September 29, 2002.

Ministry of Foreign Affairs of Cuba, *Declaration by the Ministry of Foreign Affairs: Cuba Has Nothing to Hide, and Nothing to Be Ashamed Of*, May 2, 2003, http://www.iacenter.org/cuba_may303.htm.

Nature, How Castro's Enthusiasm for Biotech Spurred Vaccine Development, 1999, **398**, 6726, http://www.nature.cm/nature/journal/v398/n6726supp/box/398022a0_bx1.html, accessed on 10/25/04.

Olson, A., Cuba Rejects Its Inclusion in U.S. List of Countries that Sponsor Terrorism, *Associated Press Worldstream*, May 8, 2003.

Powell, C.L., Press briefing on board plane en route Gander, Newfoundland, May 13, 2002, http://www.globalsecurity.org/wmd/library/news/cuba/cuba-020514-usia01.htm, accessed on 11/30/03.

Prensa Latina, 15 January, BBC Monitoring, January 18, 2002.

Ramkisson, H., *Science and Technology in Cuba Today*, CARISCIENCE, 1999, http://www.cariscience.org/aboutcs.htm, accessed on 12/2/03.

Reid-Henry, S., *Genet. Eng. News*, **22**(3), 13,15 (2002).

Satz, S., *Genet. Eng. News*, **20**(12), 1, 34 (2000).

Smith, W.S., *Los Angeles Times*, June 16, (2002).

SpecialOperations.Com website, *SpecialOperations.Com Guide to Terrorist Organizations*, 2001, http://www.specialoperations.com/Terrorism/SOCGuide/N_Z.htm.

Tamayo, J.O., *Miami Herald*, 23 June, 1999.

US Department of State, *Cuba's Renewed Support of Violence in Latin America*, Bureau of Public Affairs, Special Report No. 90, December 14, 1981, http://cuban-exile.com/doc_201-225/doc0224.htm.

US Department of State, Overview of State-Sponsored Terrorism, *Patterns of Global Terrorism: 1999Report*, 2000, http://www.state.gov/www/global/terrorism/1999report/sponsor.html, accessed on 11/30/03.

US Department of State, Overview of State-Sponsored Terrorism, *Patterns of Global Terrorism: 2001Report*, 2002, http://www.state.gov/s/ct/rls/pgtrpt/2001/html/10249.htm, accessed on 11/30/03.

Zilinskas, R.A., *Crit. Rev. Microbiol.*, **25**(3), 173–227 (1999).

D

DARK HARVEST

Alex Yabroff

On October 10, 1981, a package containing what looked like soil was found at Porton Down Chemical Defense Establishment in Southern England. Upon further investigation, the soil was determined to contain a concentration of approximately 10 *Bacillus anthracis* spores per gram (Monterey WMD-Terrorism Database, 2003). Four days later, on October 14, a second similar package containing soil was discovered at a Conservative Party meeting. Upon testing, however, no *B. anthracis* spores were identified in the package.

In a message to the local press, an unknown group called "Dark Harvest" claimed responsibility for the incidents. With its stated objective of returning the "seeds of death" to their source, the group elaborated that it had removed some three hundred pounds of soil from Gruinard Island, where the British government had performed field testing with *B. anthracis* during World War II and where viable spores persisted until a dedicated cleanup operation was conducted in the 1980s (Carus, 1998). Dark Harvest's solution to the problem of the contaminated island: The British government should bury the contaminated soil beneath concrete.

Dark Harvest is accurately described as a single-issue organization. Its actions did not involve the intentional harming of others but rather were carried out in order to send a political message, in the same way that the environmental activist group Greenpeace's March 2002 placement of toxic waste in front the U.S. Embassy in Manila protested American dumping, for example (Monterey WMD-Terrorism Database, 2003).

The group has not been heard from since.

REFERENCES

Carus, S., *Working Paper: Bioterrorism and Biocrimes. The Illicit use of Biological Agents Since 1900*, Center For Counterproliferation Research, Washington, DC, August 1998.

Monterey WMD-Terrorism Database, Center for Nonproliferation Studies, 1998–2003.

Encyclopedia of Bioterrorism Defense, Edited by Richard F. Pilch and Raymond A. Zilinskas
ISBN 0-471-46717-0 Copyright © 2005 Wiley-Liss

DARK WINTER

Michael Mair

BACKGROUND

On June 22–23, 2001, the Johns Hopkins Center for Civilian Biodefense Strategies in conjunction with the Center for Strategic and International Studies (CSIS), the Analytic Services (ANSER), Institute for Homeland Security, and the Oklahoma National Memorial Institute for the Prevention of Terrorism (MIPT), conducted a senior-level bioterrorism exercise entitled "Dark Winter," which simulated a smallpox attack against the United States. Dark Winter was designed to increase awareness of the threat posed by biological weapons among influential decision makers; examine the unique challenges that a well-coordinated, clandestine attack with a contagious disease would present to the National Security Council (NSC) members; and provoke more interest in and support for bioterrorism prevention and preparedness efforts among leaders in government.

EXERCISE PARTICIPANTS

The 12 main participants of Dark Winter portrayed the members of the NSC. Each is an accomplished individual who has served in high-level government and/or military positions (Table 1). In addition, the Honorable Frank Keating, at that time the governor of Oklahoma, portrayed himself. Five senior journalists, who were currently working for major networks or news organizations, also observed the simulated NSC deliberations and participated in a mock press conference as a part of the exercise. Finally, approximately 50 individuals with current or former policy or operational responsibilities related to biological weapons preparedness observed the exercise.

EXERCISE DESIGN

Dark Winter was a tabletop exercise in which high-level decision makers participated in mock NSC meetings where they were presented with a fictional smallpox attack and asked to react to it by establishing response strategies and making policy decisions. As the NSC, the participants' task was to advise and assist the president of the United States in coordinating the nation's response to the attack. Decisions made were incorporated into

Table 1. Roles of Key Participants in the Dark Winter Exercise

Role	Participant
President of the United States	The Honorable Sam Nunn
National Security Advisor	The Honorable David Gergen
Director of the Central Intelligence Agency	The Honorable R. James Woolsey
Secretary of Defense	The Honorable John White
Chairman, Joint Chiefs of Staff	General John Tilelli (U.S.A., Ret.)
Secretary of Health and Human Services	The Honorable Margaret Hamburg
Secretary of State	The Honorable Frank Wisner
Attorney General	The Honorable George Terwilliger
Director, Federal Emergency Management Agency	Mr. Jerome Hauer
Director, Federal Bureau of Investigation	The Honorable William Sessions
White House Communications Director	Mr. Paul Hanley
Governor of Oklahoma	The Honorable Frank Keating
Press Secretary to Governor Frank Keating (Oklahoma)	Mr. Dan Mahoney
Correspondent, *NBC News*	Mr. Jim Miklaszewski
Pentagon Producer, *CBS News*	Ms. Mary Walsh
Reporter, *British Broadcasting Corporation*	Ms. Sian Edwards
Reporter, *New York Times*	Ms. Judith Miller
Reporter, freelance	Mr. Lester Reingold

the evolving exercise to the extent possible so that key decisions affected the exercise's evolution and outcomes.

In an effort to maintain scientific accuracy, the scenario was designed after an analysis of available scientific and historical data from past smallpox outbreaks (O'Toole et al., 2002). On the basis of this analysis, two key assumptions were made that significantly affected the epidemic portrayed in Dark Winter: (1) the number of persons infected in the initial attack ($n = 3000$); and (2) the transmission rate or number of persons subsequently infected by each smallpox case prior

Encyclopedia of Bioterrorism Defense,   Edited by Richard F. Pilch and Raymond A. Zilinskas
ISBN 0-471-46717-0   Copyright © 2005 Wiley-Liss

to effective public health intervention (1:10). While these assumptions were based on the understanding of the smallpox virus at the time, they were not intended to be definitive mathematical predictors or models. In addition, the authors posited that the quantity of undiluted smallpox vaccine available during the scenario equaled the amount stockpiled by the Centers for Disease Control and Prevention (CDC) at the time of the exercise (approximately 15.4 million doses).

Dark Winter took place over the course of a day and a half. The scenario was divided into three segments and simulated a time span of nearly two weeks. Each segment portrayed an NSC meeting, and each meeting was set about a week apart in the story (December 9, 15, and 22, 2002). The meetings were structured like actual NSC meetings. Exercise participants began each segment with a review of all relevant information and received further information to help inform and shape their debates during the meetings through a variety of sources: deputies and/or special assistants (played by exercise controllers) conducted informational briefings and provided policy options; individual NSC members received memos providing updated information on issues or events that would normally fall within the purview of that specific individual's position or agency (e.g., the Director of Central Intelligence was provided with updated intelligence data over the course of the meetings); and NSC deliberations were interrupted with breaking television news coverage of the epidemic as events unfolded.

THE SCENARIO

The following is a brief description of the scenario—more in-depth information can be found in the Further Reading/Web Resources section below.

Background/Context

- A suspected lieutenant of Usama Bin Ladin has recently been arrested in Russia in a sting operation attempting to purchase plutonium and biological pathogens that had been weaponized by the former Soviet Union.
- UN sanctions against Iraq have ended and the United States suspects Iraq of reconstituting its biological weapons program.
- Iraqi forces have moved into offensive positions along the Kuwaiti border in the past 48 h—in response, the president of the United States has ordered an additional aircraft carrier battle group to the Persian Gulf.

Segment 1—December 9, 2002

Situational Information Presented to NSC Members

- The NSC meeting is convened to address the developing situation in Iraq—but just as the meeting

is getting underway, members are informed that there is a smallpox outbreak occurring in the United States (Fig. 1):
  - Oklahoma – 20 confirmed cases and 14 suspect cases
  - Georgia – 9 suspect cases
  - Pennsylvania – 7 suspect cases
- The smallpox virus is communicable from person to person.
- There is no effective treatment for smallpox once symptoms present.
- The U.S. stockpile of smallpox vaccine is 15.4 million doses, but that translates to an estimated 12 million usable doses.

Critical Debate Issues

- With limited doses of vaccine available, what is the best strategy to contain the outbreak?
- Who are priority vaccine recipients (e.g., military or civilian, all states or only those affected, critical personnel and family members)?
- Should the president activate the National Guard?
- How should the developing situation in Iraq be handled?
- What should the president tell the public?
- Is the United States at war?

Decisions

- NSC members agree that the public should be fully informed as soon as possible in order to maximize public confidence and adherence to disease containment measures and reduce the likelihood that these measures would need to be forcibly imposed.
- Members decide to implement a ring vaccination strategy (i.e., vaccination and monitoring of contacts of smallpox cases and their contacts) in order to focus vaccination efforts on those at highest risk of contracting smallpox while preserving as much vaccine as possible. They also decide to set aside sufficient doses of smallpox vaccine for the Department of Defense to meet its immediate needs (approximately one million doses).
- Members also decide that the deployment of an additional aircraft carrier battle group to the Persian Gulf should proceed, but that any further deployments should be delayed pending further developments.

Segment 2—December 15, 2002 (6 Days into the Epidemic)

Situational Information Presented to NSC Members

- Smallpox cases:
  - Two thousand smallpox cases have been reported in 15 states (Fig. 2).

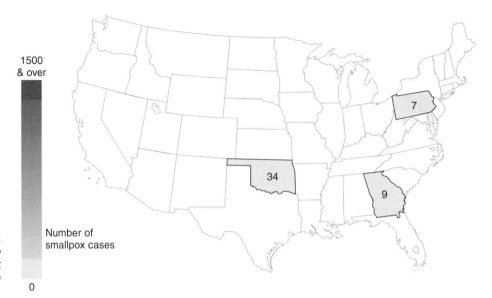

Figure 1. Map showing reported smallpox cases ($n = 50$) reported to the National Security Council during segment 1 (December 9, 2002) of the Dark Winter bioterrorism exercise.

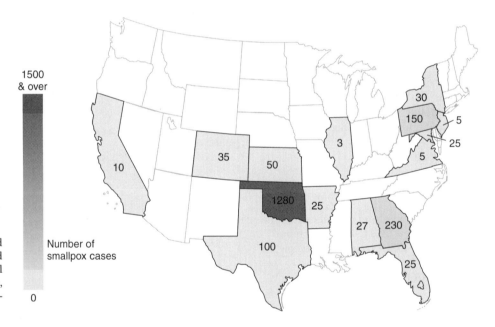

Figure 2. Map showing reported smallpox cases ($n = 2000$) reported to the National Security Council during segment 2 (December 15, 2002) of the Dark Winter bioterrorism exercise.

- Isolated cases have also been reported in Canada, Mexico, and the United Kingdom.
- Three hundred people have died.
- Only 1.25 million doses of smallpox vaccine remain.
- The epidemic has overwhelmed the health care systems in affected states and patient care is suffering.
- Many international borders are closed to U.S. trade and travelers.
- Food shortages have emerged in affected states due to travel problems and store closings.
- News agencies are reporting sporadic violence against minorities who appear to be of Arabic descent.
- The government response to the epidemic has received intense media coverage, scrutiny, and criticism.

- Additional smallpox vaccine will not be available for at least five weeks.

Critical Debate Issues

- How can homeland security and disease containment needs be met while also maintaining international commitments (e.g., should U.S. military deployments to the Persian Gulf continue)?
- What alternative epidemic controls (e.g., quarantine, travel restrictions) are possible and how should implementation of those measures be balanced against potential economic disruption and civil rights infringements?
- What assistance can the federal government offer states?
- What is the role of the National Guard?
- Is the United States at war?

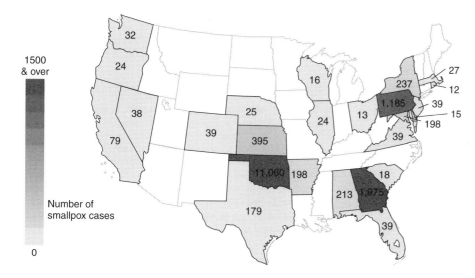

Figure 3. Map showing reported small-pox cases ($n = 16,000$) reported to the National Security Council during segment 3 (December 22, 2002) of the Dark Winter bioterrorism exercise.

Decisions

- NSC members decide to leave control of the National Guard as well as decisions on disease containment measures to state officials.
- Members decide to expedite a program to produce new smallpox vaccine despite unresolved liability issues.
- Members also decide to accept smallpox vaccine offered by Russia provided it passes safety evaluations.

Segment 3—December 22, 2002 (13 Days into the Epidemic)

Situational Information Presented to NSC Members

- Smallpox cases:
  - Sixteen thousand smallpox cases have been reported in 25 states (Fig. 3).
  - One thousand people have died.
  - Ten other countries are reporting cases of smallpox believed to be the result of international travel from the United States.
- It is unclear whether new smallpox cases are due to unidentified contacts of initial victims, contacts who were not vaccinated in time, ineffective vaccine, new smallpox attacks, or some combination thereof.
- There is no more smallpox vaccine and new vaccine will not be ready for at least four weeks.
- The national economy is suffering, food shortages are growing, and states are restricting nonessential travel.
- Canada and Mexico have closed their U.S. borders.
- While speculative, the predictions are extremely grim: Over the next 12 days, the total number of cases is expected to reach 30,000. Of those, approximately one-third is expected to die.
- There are no solid leads as to who is behind the attack.
- The exercise is ended when it is announced to NSC members that three major U.S. newspapers have

each received an anonymous letter calling for the United States to remove all of its military forces from Saudi Arabia and the Persian Gulf within one week or face additional biological attacks with smallpox, anthrax, and plague. The letters contain a genetic fingerprint matching the smallpox attack strain lending credibility to the threats.

Critical Debate Issues

- How can the epidemic be contained while awaiting new smallpox vaccine?
- How should the United States respond to the anonymous letters?
- What is the appropriate response should the United States discover who was behind the attack?

LESSONS LEARNED

The following lessons learned are based on an analysis of comments and decisions made by exercise participants during the exercise, subsequent Congressional testimony by exercise participants, and public interviews given by participants in the months following the exercise. The lessons learned reflect the analysis and conclusions of the Dark Winter contributors from the Johns Hopkins Center for Civilian Biodefense Strategies and do not necessarily reflect the views of the exercise participants or collaborating organizations. More in-depth discussion on these lessons learned can be found in the Further Reading/Web Resources section below.

1. Most leaders—even those with national security experience—are unfamiliar with the character of bioterrorist attacks, available policy options, and their consequences.
2. After a bioterrorist attack, the leaders' situational awareness and decisions depend on data and expertise from the medical and public health sectors.
3. The lack of sufficient vaccine or drugs to prevent the spread of disease severely limited management options.

4. To end a disease outbreak after a bioterrorist attack, decision makers require ongoing expert advice from senior public health and medical leaders.

5. The individual actions of U.S. citizens are critical to ending the spread of contagious disease; leaders must gain the trust and sustained cooperation of the American people.

6. Federal and state priorities may be unclear, differ, or conflict; authorities may be uncertain; and constitutional issues may arise.

Acknowledgments
This article has been adapted and certain sections reproduced with permission from an article previously published in *Clinical Infectious Diseases* entitled, "Shining Light on Dark Winter" (2002; 34: 972-83),© 2002 by the Infectious Diseases Society of America. All rights reserved.

REFERENCE

O'Toole, T., Mair, M., Inglesby, T. V., *Clin. Infect. Dis.*, **34**(7), 972–983 (2002).

FURTHER READING AND WEB RESOURCES

Information on Dark Winter from the Center for Biosecurity, University of Pittsburgh Medical Center, is available at http://www.upmcbiosecurity.org/pages/events/dark_winter/dark_winter.html (accessed on 1/5/04). Includes links to: (1) the full Dark Winter exercise script; (2) a Dark Winter briefing/slide presentation; and (3) *Shining Light on "Dark Winter"*—an in-depth write-up of Dark Winter published in *Clin. Infect. Dis.*, **34**(7), 972–983 (2002).

Information on Dark Winter from the ANSER Institute for Homeland Security is available at www.homelandsecurity.org/dark-winter/index.cfm, accessed on 1/5/04.

Information on Dark Winter from the Center for Strategic and International Studies is available at http://www.csis.org/press/ma_2001_0723.htm, accessed on 1/5/04.

DEFENSE THREAT REDUCTION AGENCY

GERALD L. EPSTEIN

BACKGROUND

The Defense Threat Reduction Agency (DTRA) is the U.S. Department of Defense's (DoD) focal point for reducing the present threat of weapons of mass destruction (WMD) and preparing for this threat in the future. It was formed in 1998 by consolidating several DoD agencies involved in understanding, deterring, preventing, and reducing potential WMD threats. Organizationally, the agency reports to the Assistant to the Secretary of Defense for Nuclear, Chemical, and Biological matters. As a combat support agency, however, it also reports to the Chairman of the Joint Chiefs of Staff, and it is prepared to provide direct support to U.S. military forces in wartime or emergency situations. Furthermore, DTRA will play a key role in supporting any response that DoD provides to civil authorities in response to WMD incidents. DTRA's fiscal year 2003 budget appropriation was $2.3 billion, out of a total DoD budget of $364 billion, and it is authorized up to 2141 staff (both military and civilian).

ACTIVITIES AND ORGANIZATION

DTRA's activities include anticipating future WMD threats, shaping the international environment to reduce current threats, deterring and defeating the use of WMD, and managing the consequences of a WMD event. The agency's mandate addresses nuclear, radiological, chemical, and biological weapons threats through six operational directorates:

- *Combat Support* provides expertise and capabilities to respond to the use of WMD. These capabilities support active military forces, including those of the newly established U.S. Northern Command, which as directed by the President or the Secretary of Defense has the mission of assisting civil authorities responding to WMD and other contingencies within the United States.
- *Technology Development* coordinates and oversees research and development activities to enhance the U.S. ability to control, reduce, and respond to WMD threats. These capabilities include both the means of destroying chemical and biological agents and production facilities, as well as the ability to assess WMD effects. (Research and development to mitigate the consequences of chemical and biological threats are the responsibility of the Chemical/Biological Defense Directorate, discussed below.)

- *On-Site Inspection* executes DTRA's responsibilities as the U.S. government focal point for implementing U.S. arms control inspection, escort, and monitoring activities. On-Site Inspection Directorate staff conduct treaty inspections abroad and escort foreign personnel who are inspecting and monitoring U.S. facilities.
- *Chemical/Biological Defense* supports research and technology development activities that allow U.S. military forces to survive and successfully complete their operational missions in environments that are contaminated with chemical or biological warfare agents. These R&D activities address both medical and nonmedical technologies, and they proceed through demonstration and rapid deployment of needed capabilities.
- *Cooperative Threat Reduction* works to help the countries of the former Soviet Union destroy or otherwise secure nuclear, chemical, and biological weapons and their associated infrastructure. These activities include establishing verifiable safeguards against the proliferation of such weapons, bolstering the security of former weapons facilities, and engaging former weapons researchers in legitimate, nonweapons-related activities.
- *Weapons Elimination*, DTRA's newest directorate, was formed to provide expertise needed to eliminate WMD in Iraq.

In addition to these operational directorates, DTRA has an Advanced Systems and Concepts Office that serves as an in-house "think tank," conducting studies and analyses related to anticipating, deterring, defeating, and responding to WMD events.

ROLE IN RESPONDING TO BIOTERRORISM

DTRA's responsibilities are to DoD, and they are not restricted to biological weapons. The agency does not have an explicit charter to counter bioterrorism against civilian targets, a mission that falls primarily to civilian agencies such as the federal Departments of Homeland Security and Health and Human Services, as well as state and local governments. Nevertheless, the capabilities, technologies, and systems that DTRA has developed to fulfill its military mission necessarily have applicability that extends beyond DoD. Since biological agents cannot distinguish between military and

Encyclopedia of Bioterrorism Defense, Edited by Richard F. Pilch and Raymond A. Zilinskas
ISBN 0-471-46717-0 Copyright © 2005 Wiley-Liss

nonmilitary targets, means to counter them typically have applicability both within and outside DoD. Furthermore, since military forces, logistics chains, and installations depend on national and not just military infrastructures, protecting DoD forces and preserving their mission capabilities under biological attack may require a more robust defense of the society at large. Finally, since U.S. military forces (especially those assigned to U.S. Northern Command) have an explicit mission to assist civilian authorities, those forces must be capable of providing such assistance, and DTRA's efforts to support those forces must therefore anticipate and meet the demands that might be put on them in that capacity.

Some DTRA activities that may therefore have direct applicability to bioterrorism defense include:

- *Training.* DTRA trains military first responders to manage the consequences of nuclear, chemical, and biological weapons.

- *Modeling.* DTRA develops and maintains computer modeling capabilities to predict and prepare for biological weapons incidents, among others. This expertise supports development of scenarios and exercises conducted by both military and civilian responders.

- *Analysis.* DTRA studies have examined issues such as "double counting" of civilian first responders who also serve in the National Guard or Reserve; systems approaches to counter smuggling of WMD into the United States; assessing a community's reaction to a bioterrorist attack; and tracing the origin of a biological agent used as a weapon.

- *"Reachback" and Operational Support.* DTRA maintains liaisons with military combatant commands, including U.S. Northern Command; maintains an operations center that can provide technical expertise on biological and other weapons on a 24-h basis; and can deploy Consequence Management Advisory Teams to the field that can then "reach back" to tap the agency's collective WMD expertise.

- *Device Detection / Defeat.* DTRA is developing technologies to detect and defeat terrorist devices.

- *Decontamination.* DTRA is developing models to accurately predict the persistence of biological agents on various surfaces, supporting efforts to develop decontamination approaches.

- *Threat Reduction.* DTRA's Cooperative Threat Reduction activities help reduce, eliminate, or otherwise secure biological weapons–related infrastructure in the former Soviet Union, lessening the contribution that such facilities and expertise might make to the proliferation of biological weapons.

- *Threat Anticipation.* In order to prepare for and counter future threats, DTRA works with the intelligence community and with scientific and technical leaders to identify emerging biological weapons threats and to understand the behavior of those who might employ them.

- *Biodefense Technologies.* DTRA is developing, demonstrating, and fielding medical and nonmedical technologies to help military forces counter biological threats; these technologies can also have applicability in the civil sector.

- *Vulnerability Assessment.* DTRA supports U.S. military forces and DoD agencies by conducting vulnerability assessments that help senior leaders protect critical assets. Assessment teams offer procedural and technical solutions to mitigate identified vulnerabilities. In addition, DTRA analyzes these assessments to identify trends, recommend priorities, and provide guidance on "lessons learned."

FUTURE OUTLOOK

DoD has created new organizations to execute its role in defending the U.S. homeland against terrorist attack and to interact with other federal, nonfederal, and nongovernmental institutions that share that mission. As noted, a new military combatant command, U.S. Northern Command, has been established to defend the U.S. homeland from external threats and, as directed by the President or the Secretary of Defense, to support civil authorities in responding to emergencies such as terrorist attack. The new position of Assistant Secretary of Defense for Homeland Defense has been created and assigned the responsibility of supervising DoD's overall homeland defense activities. At the same time, a new Department of Homeland Security has been created to centralize the civilian portion of the federal government's homeland security mission. As these new organizations mature, they will further develop their missions, operating procedures, and working relationships. Serving two of them directly and the third one indirectly, DTRA will continue to play a central role in the nation's efforts to reduce the threat of bioterrorism and weapons of mass destruction.

DEFENSE RESEARCH AND DEVELOPMENT CANADA - SUFFIELD

Lauren Harrison

OVERVIEW

The mission of Defence Research and Development Canada-Suffield (DRDC-Suffield) is to perform research and development in the Canadian defense sector, with a specific emphasis on military engineering and chemical and biological weapons. DRDC-Suffield is one of six research centers under Defence Research and Development Canada (DRDC), which is in turn a component of the Canadian Department of National Defence. The Director General of DRDC-Suffield, Dr. Robert Angus, reports directly to DRDC Canada.

DRDC-Suffield is located on the Suffield "Range" in Alberta, Canada, an area of approximately 3000 square kilometers and home to abundant wildlife, oil and gas reserves, and archaeological sites, as well as the Canadian Center for Mine Action Technologies, which hosts a program in military countermine research. DRDC-Suffield's chemical and biological research facilities include the following:

- *Counter Terrorism Technology Center (CTTC).* A recently established center that focuses on training military and first responders in emergency preparedness and response to a chemical or biological attack.
- *Chemical/Biological Forensic Reference Laboratory.* Established in 2003, this reference laboratory facility is part of the CTTC.
- *Canadian National Single Small-Scale Facility (CNSSSF).* A small-scale production facility for Schedule 1 chemicals under the Chemical Weapons Convention, manufactured for research, defense, and pharmaceutical functions.
- *Biological Aerosol Facilities.* Part of the Canadian Integrated Bio/Chemical Agent Detection System (CIBADS), established to create both field and chamber environments to test challenges to the aerosol process.
- *Biosafety Level 3 Facility*
- *Chemical Vapor Penetration Test Facility.* A facility housing testing and evaluation of protective clothing and equipment.
- *Inhalation Exposure Facility.* A facility where testing of the toxicity of inhaled aerosol particles is performed.

DRDC-Suffield also houses a number of other military and defense-related facilities, including the Range and Accuracy Site, Flash X-Ray Site, Non-Metallic Laboratory, Fuel-Air Explosive Test Site, the Weapons Test Center, Terrain Motion Simulator, Blast Tube and Undex Pond, and the Vehicle Concepts Group. In all, DRDC-Suffield houses close to 150 personnel, including 64 professionals and 48 technicians.

HISTORY

DRDC-Suffield was established in 1941 as a part of combined Allied efforts in the chemical and biological weapons offensive and defensive areas (Bryden, 1989). After World War II, DRDC-Suffield began to focus on problems in physics, including shock and blast studies, aerial targets, meteorology, vehicle mobility, and military engineering research. In addition to this research, DRDC-Suffield developed a program to research chemical and biological warfare agents and defense in order to provide specific defense equipment for the Canadian military.

RESEARCH AREAS

The DRDC-Suffield's research program has three main categories: military engineering, Technology Demonstration Project (TDP), Blast Force Protection, and chemical and biological weapons defense.

Military Engineering

DRDC-Suffield is involved in explosives and military engineering research, including obstacle clearing and minefield breaching, remotely piloted vehicles, weapons systems evaluations, verification technology, and explosive applications. Other research topics in this division include land mine detection, military robotics, hyper spectral imaging, robotic scanning for land mines, and related research.

TDP Blast Force Protection

This research program focuses on the development of protective equipment and strategies for dealing with explosives and blasts.

Chemical and Biological Weapons Defense

Since World War II, the DRDC-Suffield has developed a deep understanding of the toxicology of chemical agents and infectivity of biological agents, as well as the behavior of aerosols, liquids, and gases released in the atmosphere,

Encyclopedia of Bioterrorism Defense, Edited by Richard F. Pilch and Raymond A. Zilinskas
ISBN 0-471-46717-0 Copyright © 2005 Wiley-Liss

and applies the knowledge in related weapons applications and the development of appropriate safeguards against them. Major research components include Hazard Assessment, Detection and Identification, Physical Protection, Medical Countermeasures, and Verification Technology. Specifically, scientists engage in the following activities:

Programs

Demilitarization and Special Projects

DRDC-Suffield has developed the Chemical, Biological, and Radiological Detection Demilitarization Team (CBR Team) to provide risk assessment analysis and disposal of chemical, biological, and radiological waste. This team works primarily with Canada's Department of National Defence. The team focuses specifically on chemical warfare agents, physical protection systems, technical evaluation, neutralization of chemical and conventional weapons, agent sampling and identification, and the construction of material containment facilities.

Chemical and Biological Incident Management

The Chemical and Biological Incident Management program was created to provide both technical capability and advice to the military to protect against the use of chemical or biological weapons. This program conducts research in chemical and biological weapons defense, provides technical expertise and support to the government, works to develop defensive equipment, and works with private industry to market defensive equipment to the international community.

Chemical and Biological Agent Identification

DRDC-Suffield created the Chemical and Biological Agent Identification program in response to the military's need for a chemical, toxin, and biological identification and confirmation capability. This program works to develop new techniques to identify primarily toxin and biological agents, and secondly chemical agents.

DRDC-Suffield Electrospray Mass Spectrometry (ESI-MS) Database

This database contains the mass spectrometry profiles of over 50 chemical warfare (CW) agents.

Chemical and Biological Demilitarization and Site Remediation Projects

Project teams aid the Canadian Department of National Defence with the containment and removal of chemical and biological waste.

Chemical/Biological Training for the Canadian Forces

This program was developed to provide simulation of chemical and biological attacks for emergency response training.

Developments and Initiatives

Portable Aerosol Inhalers

DRDC scientists are working on a system to deliver therapeutic drugs to the lungs of people who have been exposed to biological agents, with the goal of adapting this therapy to the counterterrorism and military settings.

Automated Microchip Platform for Biochemical Analysis

A one-package biochemical analysis system is under development that integrates injection of the sample, mixing of sample with appropriate reagents, separation of target agent from contaminants, detection and identification of the target agent, and a waste elimination system.

Fluorescence Aerodynamic Particle Sizer (FLAPS)

This product was built by DRDC-Suffield to measure biological aerosols using flow cytometry techniques. Among the characteristics of this system is that it can distinguish between biological and nonbiological materials.

Canadian Integrated Biological/Chemical Agent Detection System (CIBADS)

Researchers have designed a biological and chemical agent detection system that detects both types of agents in real time. The system also has the unique capability to identify specific biological agents.

Mobile Atmospheric Sampling and Identification Facility (MASIF)

This comprehensive biological area defense system was designed for the military to detect a biological attack and then to collect and analyze aerosol samples. It was deployed in the 1991 Gulf War.

Canadian Aqueous System for Chemical-Biological Agent Decontamination (CASCAD)

Researchers have developed an aqueous decontaminant solution that is effective against both chemical and biological agents. Thus, the solution, which is a combination of a decontaminant concentrate and water or seawater, can be used to decontaminate material contaminated with several types of pathogens, as well as nitrogen mustard, Lewisite, and G and V type nerve agents.

RECENT HIGHLIGHTS

Given its prolific history of accomplishments, many examples of DRDC-Suffield success could be mentioned. A selection of its more recent activities includes the following:

- During the Gulf War, DRDC-Suffield provided defense equipment for the Canadian armed forces. This equipment included detectors, atropine, and a skin decontaminant.
- In April 2003, DRDC-Suffield hosted a multinational North Atlantic Treaty Organization (NATO) nuclear, biological, and chemical defense training exercise, which involved close to 70 participants from 13 countries. The purpose of the exercise was to train

participants on how to respond effectively to chemical and biological attacks, including limiting the effects of such an attack.

- During spring 2003, DRDC-Suffield signed an agreement with Nexia Biotechnologies, Inc., to jointly research a protein called Protexia, which may prove effective in mitigating the effects of nerve agents.

REFERENCE

Bryden, J., *Deadly Allies: Canada's Secret War 1937—1947*, The Canadian Publishers, Toronto, 1989.

WEB RESOURCES

Defence R&D Canada Corporate website, http://www.drdc-rddc.gc.ca/, accessed on 11/3/03.

Defence R&D Canada-Suffield website, http://www.suffield.drdc-rddc.gc.ca/AboutDRDC/index_e.html, accessed on 11/3/03.

NATO Conducts NBC Exercise at DRDC-Suffield, Canada News Wire, 5/2/03, available from http://www.newswire.ca/en/releases/archive/May2003/02/c7284.html, accessed on 11/11/03.

Nexia Biotechnologies, Inc. website, *Nexia collaborates with DRDC-Suffield on NEX-91 (Protexia) as a bioscavenger to counter chemical weapons*, 3/31/03, http://www.nexiabiotech.com/pdf/PR%202003-03-31%20DRDC%20English.pdf, accessed on 11/12/03.

DELIVERY METHODOLOGIES

Richard F. Pilch

INTRODUCTION

The delivery method of a biological agent is generally based on whether it is contagious, such as the smallpox virus or *Yersinia pestis*, or noncontagious, such as *Bacillus anthracis* or botulinum toxin.

CONTAGIOUS AGENTS

As a general rule, contagious agents require a comparatively low-tech delivery system that begins with the deliberate infection of a small group or individual, whether that group or individual is unaware of it or is knowingly infected as a so-called smallpox suicide bomber would presumably be. This group or individual then serves as a delivery device, spreading the disease by secondary transmission. Such an approach eliminates the need for mass production of the contagious agent (only a small amount is needed to initiate the chain of events potentially leading to an epidemic), specific formulation, or the design of an effective dissemination device. It should be noted, however, that the terrorist (or state) use of alternative delivery methods for contagious agents, for example dispersal via a simple aerosol device, cannot be ruled out and may in fact be more likely. Other possible means for spreading a contagious agent, such as the exploitation of zoonotic transmission or the contamination of illicit drugs, exist as well.

The threat of a smallpox suicide bomber has been the focal point of numerous media reports over the past year. A focused assessment is yet to be presented, however, to offer the public some perspective on the potential for success with this type of attack. Without vaccination, humans are susceptible to smallpox, indicating an underlying vulnerability of the United States and world population to this virus. But a terrorist would first have to want to carry out such an attack, and what is known about conventional suicide bombers does not translate well to the delivery of a biological agent in this way. For example, a key component of the suicide bombing tactic is the promise of a quick and honorable death, not offered in the case of a smallpox suicide bomber (Dolnik, 2003). Second, he or she would have to acquire the virus, an exceedingly difficult task. Even then, initiating an epidemic would be more complicated than simply injecting it into an operative, waiting for a rash to appear, and then sending that operative to public places. If this was to be done, it would be similar to variolation, the immunization technique employed against smallpox before Dr. Edward Jenner developed his breakthrough vaccine from cowpox in 1796 (this cowpox vaccine is the predecessor to the vaccinia vaccine used today). The process of variolation, which consisted of inoculating unexposed individuals—through incisions in their skin—with scabs or pus from mildly infected smallpox patients, began sometime before 1000 B.C., and effectively reduced the fatality rate of subsequent infections from 30 percent to approximately 1 percent (Tucker, 2001). Although a potential suicide bomber might develop fulminating smallpox from such an injection, he or she would be more likely to develop a mild infection with or without a rash that in most cases would not lead to the shedding of the virus and secondary spread (it should be noted, however, that if the virus were successfully injected intravenously rather than simply into the skin, these expectations would conceivably differ). Furthermore, he or she would probably be severely debilitated by the disease, and in all likelihood would be sufficiently immobilized to disrupt an attack. And the virus itself usually requires close contact (less than 2 m) for person-to-person spread and in fact is not very contagious, further hindering any possible transmission. The overall likelihood of success with this method of attack is therefore believed to be quite low.

NONCONTAGIOUS AGENTS

Noncontagious agents can be delivered using multiple methods, including injection, contaminated water, contaminated food or beverages, or the dissemination of a particulate aerosol. Each is discussed in turn below.

Injection

Delivery via injection was seen in the 1978 assassination of Bulgarian dissident Georgii Markov. A Bulgarian secret service agent used an umbrella with a concealed air gun system to fire a steel pellet into Markov's leg while he was waiting for a bus on Waterloo Bridge in London. The pellet was filled with the toxin ricin. Markov died three days later.

A fragmentary bomb with laced shrapnel has the same effect. For example, scientists of Japan's World War II–era biological warfare (BW) program "Unit 731" developed such bombs using *Clostridium perfringens*, the causative agent of gas gangrene. More recently, penetrating bone fragments from a suicide bomber in Israel infected a victim

Encyclopedia of Bioterrorism Defense,   Edited by Richard F. Pilch and Raymond A. Zilinskas
ISBN 0-471-46717-0   Copyright © 2005 Wiley-Liss

with hepatitis B (Braverman et al., 2002). Although it is highly unlikely that this transmission was deliberate, the event shows that explosive dissemination of an infectious agent in this way is in fact possible.

Water Contamination

Perhaps contrary to popular belief, a terrorist would have difficulty in effectively contaminating a water supply. In the simplest terms, a water supply can be divided into two systems, the pretreatment system and the posttreatment system. The pretreatment portion is almost always a closed system that carries water from its source (e.g., a reservoir, Lake Michigan, or the Potomac River), through multiple filters designed to remove particles as small as 0.03 μm in size, to a treatment plant. In the treatment plant, the water is chlorinated and often treated with ozone as well (Zilinskas and Carus, 2002). The posttreatment area involves the conduit from treatment to the consumer. Often, this involves storage in towers or reservoirs, which if unprotected could present a potential vulnerability (Zilinskas and Carus, 2002). Alternatively, a perpetrator could use a vacuum pump at a remote faucet or water fountain to force an agent back into the water supply (Hearings Before the Select Committee to Study Governmental Operations With Respect to Intelligence Activities of the United States Senate, 1976). Even if a terrorist targeted the posttreatment system, however, the large dilutional effect and residual chlorination would likely minimize the chance of success. A terrorist might also attempt to disable a water treatment mechanism, interrupting the flow of disinfectants and allowing nature to take over and contaminate the supply (Croddy, 2001).

Food or Beverage Contamination

A terrorist would have similar difficulty attacking a food or beverage supply. Food supply lines from "crop to consumer" have multiple nodes that a terrorist could potentially target. However, only if he or she targeted a node early in the supply line would a large-scale attack be possible. An insider threat inside a processing or distribution plant is of most concern in this regard. A terrorist operative covertly placed inside such a facility could conceivably mount a successful attack from within that, with the aid of a given company's own distribution system, would then reach a wide target population.

Airborne Dissemination

Except for a few very rare exceptions, aerosolized biological agents have to be inhaled to be effective, unlike classical chemical weapons such as mustard gas and nerve agents (e.g., sarin and VX) that can be absorbed through the skin (biological agents with skin effects include the trichothecene mycotoxins, along with a limited number of other toxins not generally considered in the context of BW). Thus, the aerosol dissemination of a BW agent almost always targets the human respiratory system, necessitating the distribution of proper-sized particles for their successful uptake into the host, where they can then initiate disease. Three general approaches exist for this type of delivery: point source, multiple point source, and line source delivery.

Point Source Delivery. Point source delivery traditionally employs a munition—for example, an artillery shell, bomb, or rocket (but possibly something as simple as a glass flask containing a biological agent that could be smashed to create a dispersive effect)—that delivers its payload from a stationary source. Impact or detonation causes a burster charge within the munition to explode and the payload to be released. The wind then directs the payload's spread over (or away from) a target population. Point source delivery is considered highly inefficient because approximately 95 to 99 percent of the agent is destroyed in the blast, and much of what survives is driven into the ground or broken down into very small particles that either disperse too widely or are inhaled and exhaled right back out again.

Instead, a terrorist might use a different approach. He or she might place the nozzle of a spraying device into the air intake duct of a building's air handling system and initiate flow, allowing fans within the system to circulate the biological agent throughout the building. Filtration devices inside these systems might offer some protection against this type of attack, however. In addition, if a terrorist were using *B. anthracis* spores, they might stick to the walls inside the system.

Aum Shinrikyo carried out several unsuccessful biological attacks using the point source approach (Center for Nonproliferation Studies, Undated). On two separate occasions, cult members used a sprayer system to release a wet anthrax formulation into a giant fan situated atop an eight-story building. Apparently, they used an avirulent strain of *B. anthracis* in these attempts and were therefore unsuccessful in causing casualties (Keim et al., 2001).

Multiple Point Source Delivery. Multiple point source dispersal is fairly self-explanatory. The classic example is what would be seen in a bombardment. A terrorist might use multiple dispersal devices coordinated by timing mechanisms. Aum employed this technique in an unsuccessful attempt to deliver botulinum toxin among a localized target population (Center for Nonproliferation Studies, Undated). Cult members positioned three briefcases equipped with small tanks, vents, and battery-powered fans in a Tokyo subway station, but upon activation, the released contents had no effect because an Aum member had sabotaged the operation by filling the tanks with water. Regardless, Aum was never able to acquire a toxigenic strain of *Clostridium botulinum*, such that even if the operator had loaded what he or she believed to be botulinum toxin into the tanks, the attack would nevertheless have been ineffective. It is widely held that this failed attack directly led to the group's decision to use sarin (and to deliver it in a relatively unsophisticated way) in the successful Tokyo subway attack, which took place only 5 days later.

Line Source Delivery. Line source distribution removes the static element of the dispersal system such that a moving delivery device releases a flow of agent over an extended period of time. Aum attempted this type of dispersal on multiple occasions as well, again unsuccessfully (Center for Nonproliferation Studies, Undated). As just one example, cult members drove a truck equipped with a custom-made spraying device around the Imperial Palace and Tokyo Tower, intent on distributing a wet anthrax solution. They again used a nonpathogenic strain for this, however, and in any event the nozzle on the truck had apparently clogged prior to the operation and was thus nonfunctional at the time of intended release.

The classic line source dispersal device is a crop duster, ideally flown crosswind upwind of a target so that the stream of released agent is carried by the wind over the target area (while the term "crop duster" is commonly used by the lay public, "aerial applicator" is the proper name for these aircraft in the agricultural industry; for the sake of simplicity, however, the lay term is used in this analysis). The goal with this approach, and generally with the use of any type of spray device, is to generate an aerosol cloud of the ideal particle size range of 1 to 5 μm in a high enough concentration to cover a broad area. The spraying mechanism of a crop duster, like that of other agriculture and painting equipment, consists of a hopper tank, a source of compressed air, one or multiple feeding tubes from the hopper tank, and nozzles for expulsion (Zilinskas and Carus, 2002). The compressed air propels material from the hopper tank through the tube or tubes and out the nozzles, which break up the dispersed agent unevenly to produce a wide range of particle sizes. Some of these are 1 to 5 μm, and are thus readily absorbed in the lungs. Most, however, are either too large and fall to the ground or get trapped by mucociliary defenses of the upper respiratory system or too small and float away or get breathed in and out. The average particle size produced by a crop duster is approximately 100 μm (a size that causes the particles to descend to the ground, as intended).

A terrorist could attach special nozzles with small orifices to deliver a more uniform size in the desired range. He or she would have to increase the system's pressure accordingly to adequately force the material through these smaller outputs, but as far as crop dusters are concerned, this does not present an insurmountable challenge: Most crop dusters are capable of delivering 40 pounds per square inch of pressure already, which is enough to overcome wall tension without alteration (Zilinskas and Carus, 2002). Nozzle adjustment, on the other hand, demands a great leap in terms of technical capability, and effectively eliminates the possibility of a "grab-and-go" scenario in which a crop duster is commandeered at an airfield and used immediately without modification.

Even if a terrorist were to successfully modify the aircraft, upon initiation of dispersal, a given biological preparation would clog the altered nozzles fairly quickly. This is particularly true if he or she were using a wet agent, as was the case in the Aum Shinrikyo attack described above. In addition, the propulsion of a given preparation through any type of sprayer creates a shearing effect that can kill 95 percent or more of the agent. Because crop duster hopper tanks hold from approximately 1100 to 3000 L of solution, however, the 5 percent that does survive might still be enough to have a devastating effect if a substantial amount of the total potential payload is released before the nozzles clog. A terrorist might also specially formulate a wet agent to improve survival, but this would demand a great leap in technical ability and still would only be expected to have a minimally appreciable effect.

Handling the plane itself is considered the final hurdle to conducting a crop duster attack. Loading the hopper tank is challenging; taking off requires considerable skill on the part of the pilot, and once airborne, the plane is very difficult to fly, especially with a full load at a low altitude.

WET VERSUS DRY AEROSOL

The challenges inherent in any type of wet aerosol delivery are significant but not insurmountable, as illustrated by the fact that both the former U.S. and former Soviet Union's BW programs were able to develop reliable methods for wet agent dispersal. United Nations Special Commission (UNSCOM) inspectors revealed that sprayers and holding tanks had been installed on a number of Iraqi military aircraft and land vehicles. Inspectors subsequently learned that in 1990, the Iraqis had modified a Mig-21 so that it could be remotely piloted, equipped it with a 2200 L belly tank from a Mirage F1 fighter plane, put in a spray mechanism, and field tested it with the anthrax simulant *Bacillus subtilis var. niger* (BG) in January 1991 (Zilinskas, 1997). The Iraqis also tested the system with water on three other occasions around that time, twice in December 1990 and once in January 1991 (Director of Central Intelligence, 2002). Although the results of the tests are unknown, the delivery system nevertheless represented a significant advance in Iraq's technical capability. Of course, the virtually limitless funding of dedicated state BW programs fueled these accomplishments, and therefore do not reflect the capabilities of most if not all terrorist organizations at this time.

A terrorist must first produce a dry agent before he or she can deliver it. This demands significant technical ability. However, if successfully produced, dry agents are superior to wet agents in terms of ease of dissemination and overall effect. Dry agents can also be stored for much longer periods of time in either case. With significant time, expertise, and money, a terrorist can take dry agents an additional step further by specially formulating them to prevent clumping due to electrostatic forces, as was the case in the anthrax letters. The anthrax letters proved that a terrorist can meet the technical demands of dry preparation, formulation, and aerosol dissemination, conceivably outside the construct of a state-level program. Thus, regardless of who or which group was responsible, these technical hurdles appear to be eroding.

CONCLUSION

Hurdles aside, the potential for the aerosol delivery of a biological agent is real. What effects might be expected from such a release? In this concluding segment, the aerosol dispersal of *B. anthracis* spores over Washington, DC, under optimal meteorological conditions is considered.

In a 1950 simulation, U.S. Army officials dispersed BG and monitored its spread to assess the potential impact of a comparable release of *B. anthracis* spores (Fothergill, 1958). The test employed off-the-shelf technology that has improved tremendously in the last half-century. Despite this limitation, a 2-mile dissemination line yielded a highly infectious area approximately 6 miles in length, with simulant traveling a maximum distance of 23 miles. In all, the release covered approximately 100 square miles, with an infectious area large enough to cover the entire metropolitan Washington, DC, area. The simulation lasted only 29 min.

A 1993 Office of Technology Assessment (OTA) study estimated that the release of 100 kg of *B. anthracis* spores upwind of the Washington, DC, area in such conditions could result in between 130,000 and 3 million deaths, a lethality matching or exceeding that of a hydrogen bomb (US Congress Office of Technology Assessment, 1993).

The repercussions of such an attack, if successful, would ultimately be profound. Thus, this hypothetical scenario serves as an effective illustration of the inverse relationship between probability and impact. This relationship is characteristic of the biological threats most commonly feared by security analysts and civilians alike today. If delivery of a low-probability but high-impact agent such as *B. anthracis* is successful, the worst in terms of civilian reaction and response can be expected. Biodefense efforts must therefore focus first on prevention and management of these low-probability high-impact threats, before limited resources are directed toward other perhaps more likely but far less devastating agents and delivery methods.

Acknowledgment

This entry draws extensively from Pilch, R., "The Bioterrorist Threat in the United States," in Howard, R.D. and Sawyer, R.L., eds., *Terrorism and Counterterrorism*, Revised ed. (New York: McGraw-Hill, 2003), with permission.

REFERENCES

Braverman, I., et al., *Israeli Med. Assoc. J.*, **4**, 528–529 July (2002).

Center for Nonproliferation Studies, *Chronology of Aum Shinrikyo's CBW Activities*, http://cns.miis.edu/pubs/reports/aum_chrn.htm, accessed on 12/15/03.

Croddy, E., *Chemical and Biological Warfare: A Comprehensive Survey for the Concerned Citizen*, Copernicus Books, New York, 2001, p. 81.

Director of Central Intelligence, *Iraq's Weapons of Mass Destruction Programs*, October 2002, available online at http://www.cia.gov/cia/publications/iraq_wmd/Iraq_Oct_2002.htm.

Dolnik, A., *Studies in Conflict and Terrorism*, March 2003.

Fothergill, L., *Armed Forces Chem. J.*, **12**(5), Sept/Oct (1958).

Hearings Before the Select Committee to Study Governmental Operations With Respect to Intelligence Activities of the United States Senate, *Unauthorized Storage of Toxic Agents*, U.S. Government Printing Office, Washington, DC, 1976, p. 113.

Keim, P., et al., *J. Clin. Microbiol.*, **39**(12), 4566–4567 December (2001).

Tucker, J., *Scourge: The Once and Future Threat of Smallpox*, Atlantic Monthly Press, New York, 2001.

US Congress, Office of Technology Assessment, *Proliferation of Weapons of Mass Destruction, OTA-ISC-559*, U.S. Government Printing Office, Washington, DC, 1993, pp. 53–55.

Zilinskas, R., *J. Am. Med. Assoc.*, **278**(5) August 6 (1997).

Zilinskas, R. and Carus, W., *Possible Terrorist Use of Modern Biotechnology Techniques*, Chemical and Biological Defense Information Analysis Center, April 2002 (For Official Use Only).

See also AEROSOL; ASSASSINATIONS; FOOD AND BEVERAGE SABOTAGE; and WATER SUPPLY: VULNERABILITY AND ATTACK SPECIFICS.

DEPARTMENT OF DEFENSE

Michelle Baker

INTRODUCTION

The Department of Defense (DoD) mission is to protect American interests both at home and abroad, a mandate that has remained unchanged since the department's inception. Following the attacks of September 11, 2001, and the subsequent creation of the Department of Homeland Security (DHS), DoD's responsibilities have shifted within this mandate to address the current priorities of the U.S. government as it prepares to meet the threat of terrorism and weapons of mass destruction (WMD).

BACKGROUND

Historically, the DoD has concentrated on research and development (R&D) that aim to support the soldier on the battlefield. But because the DoD is able to transfer much of its expertise concerning biological warfare (BW) prevention and response to the civilian sector through information sharing and exchange, and, furthermore, because the DoD is beginning to employ a portion of the department's resources and technology strictly for civilian use, it can now be stated that "[c]ivilian organizations may increasingly turn to the Department of Defense to leverage technology development efforts to support the needs of homeland security" (Johnson-Winegar, 2001a).

Because of its breadth, an all-encompassing discussion of the DoD's efforts regarding bioterrorism defense is beyond the scope of this work. What follows therefore is a review of certain aspects of the DoD's program that address the threat of bioterrorism.

JOINT SERVICE CHEMICAL AND BIOLOGICAL DEFENSE PROGRAM

Following Operation Desert Storm in 1991, the DoD became increasingly concerned about deficiencies in chemical and biological defense measures within the United States. In 1994, the Congress directed the DoD to coordinate its chemical and biological defense programs by creating the Joint Service Chemical and Biological Defense Program (CBDP), the objective of which is to prepare and equip soldiers to operate in a biologically or chemically contaminated environment, especially the battlefield. The "Defense Technology Objectives" detail priorities for R&D efforts to be undertaken by the CBDP in three operational areas are: contamination avoidance (detection, identification, warning, reporting, reconnaissance, and battle management), protection (individual, collective, and medical support), and consequence management (decontamination and restoration) (Johnson-Winegar, 2001b, 2002a).

In May 2001, the CBDP became an Acquisition Category 1D Program (Major Defense Acquisition Program), which increases its prominence within the DoD as the program now is overseen by senior department officials as a priority program (Johnson-Winegar, 2002b). In April 2003, the DoD announced the creation of the Joint Program Executive Office for Chemical and Biological Defense (JPEO-CBD) within the CBDP, designed

Encyclopedia of Bioterrorism Defense,   Edited by Richard F. Pilch and Raymond A. Zilinskas
ISBN 0-471-46717-0   Copyright © 2005 Wiley-Liss

to consolidate current chemical and biological defense programs by obtaining resources from existing Army, Navy, Air Force, and Marine programs. The office has programs covering all aspects of military and civil support team defense, and has become the single point of contact for all biological, chemical, nuclear, and radiological detection, vaccine acquisition, and medical diagnostic acquisition activities. Although the JPEO-CBD calls on expertise from all branches of service, the Army remains the "executive agent" for the CBDP (http://www.defenselink.mil/news/Apr2003/b04252003_bt277-03, accessed on 11/18/2003).

In FY2004, the CBDP will continue to support civilian programs designed to enhance consequence management following a WMD terrorism incident. Such funding will be utilized to purchase protection, detection, and training equipment; develop capabilities to detect, identify, and characterize biological agents in a civilian environment; establish communications capabilities providing access to federal, state, and local agencies; and evaluate the safety and effectiveness of such equipment (DoD-CBDP, 2003).

DEFENSE THREAT REDUCTION AGENCY AND DEFENSE ADVANCED RESEARCH PROJECTS AGENCY

Within the DoD, the Defense Threat Reduction Agency (DTRA) remains the focal point for technology design and acquisition intended to support U.S. efforts to meet the challenges of WMD. For more information on DTRA, please refer to its independent entry entitled "Defense Threat Reduction Agency" in this volume. Much of DTRA's current work centers on protecting U.S. targets against a chemical or biological attack, which includes applying results from research undertaken by the Defense Advanced Research Projects Agency (DARPA), the DoD's premier R&D agency. DARPA currently is expanding its biological research begun in the mid-1990s to address both threats disclosed by the Defense Intelligence Agency and "nonvalidated" threats identified by the Intelligence Community.

As part of its Biological Warfare Defense (BWD) program, DARPA has developed a framework for research efforts that includes preparedness, identification of biological agents, immediate consequence management, diagnosis and treatment of victims, and thorough decontamination. DARPA is also working to develop a natural immunity booster to help increase the effectiveness of vaccines for both the military and civilian population and plans to soon implement its Bio-ALIRT plan (Bio-Event Advanced Leading Indicator Recognition Technology), which detects a "covert biological attack" on the basis of "statistical, population-level analysis" derived through information on school and work absences, pharmaceutical purchases, and utilization of the poison-control center or other medical hotlines (Tether, 2003). In addition, DARPA is developing technology to protect people taking shelter inside buildings during a biological event by employing "smart" heating, ventilation, and air-conditioning (HVAC) systems capable of preventing the further spread of aerosol particles drawn in from the outside.

DARPA's work is not limited to research, however. For example, the Agency is involved in the design of the Unconventional Pathogen Countermeasures (UPC) program, which strives to increase the probability of effective response to a biological attack by expediting diagnosis and providing effective treatments against all types of pathogens (Tether, 2003). The foregoing are just a few examples of the Agency's pursuits.

U.S. ARMY MEDICAL RESEARCH INSTITUTE OF INFECTIOUS DISEASES

The U.S. Army Medical Research Institute of Infectious Diseases (USAMRIID) serves as the lead research laboratory for the U.S. Biological Defense Research Program (http://www.usamriid.army.mil, accessed on 11/16/2003). Scientists at USAMRIID research many types of biological threat agents for the purpose of developing vaccines, therapeutic drugs, and diagnostics for both laboratory and field use. In addition, USAMRIID designs and provides training programs and procedures for medical personnel responding to a biological event. USAMRIID houses several Biosafety Level 4 (BSL-4) laboratories that enable its scientists to conduct research on highly contagious and virulent pathogens. In the event of a domestic bioterrorism attack, USAMRIID has established procedures for collaborating with the Centers for Disease Control and Prevention (CDC) to assist the Federal Bureau of Investigation (FBI) in its investigations (General Accounting Office, 2001).

JOINT VACCINE ACQUISITION PROGRAM

As part of its medical countermeasures program to mitigate the effects of a biological attack, in 1997, the DoD established the Joint Vaccine Acquisition Program. This program is designed to ensure that the DoD complies with Food and Drug Administration (FDA) procedures when developing vaccines against select biological agents and to ensure that sufficient vaccines are produced to protect U.S. troops (Johnson-Winegar, 2001). By 2001, the DoD vaccine acquisition strategy was focused on the development of eight different vaccines: Anthrax Vaccine Adsorbed (AVA), smallpox vaccine, plague vaccine, tularemia vaccine, a multivalent botulinum vaccine, the next generation anthrax vaccine, ricin vaccine, and a multivalent Equine Encephalitis vaccine (Cohen, 2001). Following the attacks on September 11, DoD officials began collaborating with the Department of Health and Human Services (HHS) and DHS to guarantee the availability of enough vaccine not only to protect U.S. forces but also to guard U.S. civilians from the potential effects of such biological agents (Johnson-Winegar, 2001).

U.S. ARMY SOLDIER AND BIOLOGICAL CHEMICAL COMMAND

Many of the DoD's programs designed to integrate civilian and military resources are contained under the purview of the Homeland Defense Program at the U.S. Army Soldier and Biological Chemical Command (SBCCOM). Housed at Aberdeen Proving Ground, Maryland, the SBCCOM is designed to enhance military, federal, state, and local response to a domestic WMD terrorism incident. SBC-COM was designated lead DoD agency for WMD response under the Domestic Preparedness Initiative (1997) and charged with achieving greater interagency cooperation through training, exercises, expert assistance, and an improved response program. Since the inception of the Domestic Preparedness Program, SBCCOM has trained over 28,000 first responders in 105 cities throughout the country in effective response to WMD terrorism events (http://hld.sbccom.army.mil/ip/fs/hld_overview.htm, accessed on 11/19/2003).

The Domestic Preparedness Program was transferred from DoD to the Department of Justice in late 2000. On October 9, 2003, SBCCOM was reorganized into five new organizations: (1) Research, Development and Engineering Command; (2) Chemical Materials Agency; (3) Guardian Brigade; (4) PM Nuclear, Biological, and Chemical Defense under JPEO-CBD; and (5) Soldier Systems Care. These new units are evolving but are expected to soon be fully functional entities of the DoD (http://www.sbccom.army.mil/, accessed on 12/8/2003).

CHEMICAL BIOLOGICAL RAPID RESPONSE TEAM

In order to provide further DoD support to civil authorities, the Chemical Biological Rapid Response Team (CB-RRT) was designed to assist designated lead federal agencies in both preparedness procedures and consequence management. The CB-RRT is located within the SBCCOM operations center and is thus able to draw upon the expertise of many different areas within DoD (other divisions within the DoD are also equipped to handle emergency response, for example the U.S. Army Technical Escort Unit, which is discussed in detail elsewhere in this volume) (http://www2.sbccom.army.mil/cbrrt/fs_cbrrt.htm, accessed on 11/10/2003).

BIOLOGICAL COUNTERTERRORISM RESEARCH PROGRAM AND BIOLOGICAL DEFENSE HOMELAND SECURITY PROGRAM

By 2002, the Bush administration had recommended significant increases in funding for two key DoD programs: the Biological Counterterrorism Research Program and the Biological Defense Homeland Security Program. During FY2003, the Center for Biological Counterterrorism was established as part of the Biological Defense Homeland Security Program; it is led by a panel of scientists from the DoD, federal laboratories, academia, industry, and Intelligence Community. The Biological Defense Homeland Security Program is supporting R&D efforts to design a National Biological Defense System capable of coordinating all aspects of homeland defense against a biological attack. This program also established the Joint Service Installation Protection Project (JSIPP), a pilot project providing nine DoD installations with "contamination avoidance, protection, and decontamination equipment packages; emergency response capability for consequence management; integrated command and control network; and a comprehensive training and exercise plan" (Klein, 2002).

MILITARY IMPROVED RESPONSE PROGRAM AND BIOLOGICAL WEAPONS IMPROVED RESPONSE PROGRAM

The Military Improved Response Program (MIRP) is designed to identify the requirements of first responders in managing a chemical or biological event by addressing issues of personal protective equipment, mass casualty decontamination, and the development of a "generic response template" for an attack involving biological weapons (http://hld.sbccom.army.mil/ip/fs/mirp_fact_sheet.htm, accessed on 11/19/2003). This program also seeks to integrate military and civilian response, strengthening communication between the two entities (http://hld.sbccom.army.mil/about_us.htm, accessed on 11/19/2003).

The Biological Weapons Improved Response Program (BW IRP), much like MIRP, is designed to integrate multiagency response to a terrorist incident involving biological weapons by engaging the FBI, the Federal Emergency Management Agency (FEMA), HHS, and the military reserve and National Guard, among others (http://www2.sbccom.army.mil/hld/bwirp/bwirp_intro.htm, accessed on 11/19/2003).

IMPLICATIONS

The DoD is expected to supply significant expertise and technical support to federal, state, and local authorities in the event of a WMD terrorist attack on a civilian population. Allocation of funding for such activities has increased dramatically over the past five years: Although funding of the DHS and HHS exceeds the DoD by a large margin, the DoD's share is not insignificant—it has received over $100 million in 2003 (with another $78 million requested for FY2004) in order to "perform research and development related to chemical and biological threats" (OMB, 2003). The DoD will receive $172 million in emergency preparedness funds to combat terrorism involving WMD in 2004, along with $599 million for DoD-wide chemical and biological defense research, development, testing, and evaluation programs (OMB, 2003; DoD, 2003). Scientists involved in these programs are working to assist law enforcement, national security, and public health officials in understanding the weaponization, transportation, and dissemination of biological agents (White House, 2002).

As DoD continues to enhance current programs targeting biological defense, the responsibilities of many of its agencies will change. Once the DHS and the DoD are able to finalize program responsibilities and goals, the structure of civilian biological defense under the DoD should become more defined.

REFERENCES

Chemical and Biological Defense Program Annual Report to Congress, Vol. 1, US Department of Defense, April 2003.

Cohen, W., *Annual Report to the President and Congress 2001*, US Department of Defense, Government Printing Office, Washington, DC, 2002.

Department of Defense Budget, FY2004/2005, Office of the Under Secretary of Defense, Comptroller, US Department of Defense, February 2003.

Department of Defense Chemical and Biological Defense Program, Volume 1: Annual Report to Congress, US Department of Defense, April 2002.

General Accounting Office, *Bioterrorism: Federal Research and Preparedness Activities*, GAO-01-915, September 2001.

Klein, D., *Biological Terrorism: Department of Defense Research and Development*, Senate Committee on Armed Services Emerging Threats and Capabilities, US Senate, April 10, 2002, http://www.slu.edu/colleges/sph/csbei/bioterrorism/official/congress/testimony/KLEIN041802.htm.

Office of Management and Budget, *Report to Congress on Combating Terrorism*, September 2003.

Report on Activities and Programs for Countering Proliferation and NBC Terrorism: Executive Summary, Counterproliferation Program Review Committee, May 2002.

SBCCOM ONLINE, *Biological Weapons Improved Response Program (BW IRP): Introduction*, Homeland Defense, US Department of Defense, http://www2.sbccom.army.mil/hld/bwirp/bwirp_intro.htm.

SBCCOM ONLINE, *Chemical Biological Rapid Response Team (CB-RRT)*, US Department of Defense, http://www2.sbccom.army.mil/cbrrt/fs_cbrrt.htm.

SBCCOM ONLINE, *Homeland Defense: About Us*, US Department of Defense, http://hld.sbccom.army.mil/about_us.htm.

SBCCOM ONLINE, *Homeland Defense: Program Overview Fact Sheet*, US Department of Defense, http://hld.sbccom.army.mil/ip/fs/hld_overview.htm.

SBCCOM ONLINE, *Military Improved Response Program (MIRP)*, US Department of Defense, http://hld.sbccom.army.mil/ip/fs/mirp_fact_sheet.htm.

SBCCOM ONLINE, *SBCCOM Command Information: US Army Soldier and Biological Chemical Command (SBCCOM)*, US Department of Defense, http://www.sbccom.army.mil. (Note: This website has been restructured in December 2003).

Statement of Dr. Anna Johnson-Winegar, Deputy Assistant to the Secretary of Defense for Chemical and Biological Defense, 'Biological Terrorism,' Senate Committee on Government Affairs, US Senate, October 17, 2001a.

Statement of Dr. Anna Johnson-Winegar, Deputy Assistant to the Secretary of Defense for Chemical and Biological Defense, 'Biological Terrorism: Department of Defense Research and Development,' House Science Committee, Full Committee Hearing, 'Science of Bioterrorism: Is the Federal Government Prepared?' US House of Representatives, December 5, 2001b.

Statement of Dr. Anna Johnson-Winegar, Deputy Assistant to the Secretary of Defense for Chemical and Biological Defense, 'Biological Terrorism: Department of Defense Research and Development,' Senate Committee on Commerce, Science, and Transportation, Subcommittee on Science, Technology, and Space, 'Fighting Bioterrorism: Using America's Scientists and Entrepreneurs to Find Solutions,' US Senate, February 5, 2002a.

Statement of Dr. Anna Johnson-Winegar, Deputy Assistant to the Secretary of Defense for Chemical and Biological Defense, 'Chemical and Biological Equipment: Preparing for a Toxic Battlefield,' Subcommittee on National Security, Veterans Affairs, and International Relations, House Government Reform Committee," United States House of Representatives, October 1, 2002b.

Statement by Dr. Tony Tether, Director, Defense Advanced Research Projects Agency, Before the Subcommittee on Terrorism, Unconventional Threats and Capabilities, House Armed Service Committee, United States House of Representatives, United States House of Representatives, March 19, 2003, http://www.darpa.mil/body/NewsItems/pdf/31903.pdf.

The White House, *Fact Sheet: Bush Strategy to Defend Against Bioterrorism*, Washington File, International Information Programs, US Department of State, February 5, 2002; from *Defending Against Biological Terrorism*, Office of the Press Secretary, http://usinfo.state.gov.

WEB RESOURCES

Defense Program Implementation Plan Approved, News Release, US Department of Defense, April 24, 2003, http://www.defenselink.mil.

Defense Threat Reduction Agency, *Combating Terrorism Technology Program*, Fact Sheet, US Department of Defense, http://www.dtra.mil/news/fact/nw_combat_t_tech.html.

Defense Threat Reduction Agency, *Joint NBC Defense Board*, US Department of Defense, http://www.dtra.mil/cb/cb_joint.html.

First Joint Program Executive Office for Chemical and Biological Defense Formed, News Release, US Department of Defense, April 25, 2003, http://www.defenselink.mil.

Garamone, J., *Homeland Defense Chief Speaks of New Responsibilities*, American Forces Press Service, March 19, 2003, http://www.defenselink.mil.

Gilmore, G., *Radiological and Bioterror-Attack Exercise Starts May 12*, American Forces Press Service, May 5, 2003, http://www.defenselink.mil.

Joint Program Executive Office for Chemical Biological Defense website, US Department of Defense, http://www.jpeocbd.mil.

Kozaryn, L., *President Boosts Nation's 'Biodefense' Budget*, American Forces Press Service, February 5, 2002, http://www.defenselink.mil.

McGough, T., *CBIRF Marines Train at FDNY Academy*, Joint Task Force Civil Support, Northern Command, US Department of Defense, http://www.jtfcs.northcom.mil.

SBCCOM ONLINE, *US Army Technical Escort Unit: A Unique National Response Capability, America's Guardians- Escort with Pride!*, US Department of Defense, http://www2.sbccom.army.mil/teu/factsheet.htm.

Statement of Mr. Peter Verga, Special Assistant for Homeland Security, Armed Services Military Procurement Subcommittee, House Armed Services Committee, US House of Representatives, March 5, 2002, http://www.house.gov.

TOPOFF (Top Officials), Fact Sheet, US Department of State, July 24, 2003, http://www.state.gov.

'TOPOFF 2'- Week-Long National Combating Terrorism Exercise Begins May 12, 2003, US Department of Homeland Security, May 5, 2003, http://www.dhs.gov.

Technical Support Working Group, TSWG.gov. Chemical, Biological, Radiological, and Nuclear Countermeasures, http://www.tswg.gov/tswg/cbrnc/cbrnc_ma.htm.

See also DEFENSE THREAT REDUCTION AGENCY and FORT DETRICK AND USAMRIID.

DEPARTMENT OF HEALTH AND HUMAN SERVICES

Jennifer Runyon

INTRODUCTION

The mission of the U.S. Department of Health and Human Services (HHS) is to protect the health of all Americans and to provide essential human services, from disease and drug prevention to financial assistance and food and drug safety. An important component of this mission is providing grants; in fact, the HHS is the largest federal grant-making agency in the United States. The HHS has a range of responsibilities and programs related to the threat of bioterrorism, largely focusing on basic scientific research for biodefense, assisting states with preparedness and response activities, and strengthening the public health infrastructure. Specific programs are implemented by individual agencies within HHS with some crossover of goals and activities.

HHS SUBAGENCIES

The responsibilities of the individual agencies in supporting the HHS mission, and specifically its bioterrorism-related activities, are discussed below.

Centers for Disease Control and Prevention

The mission of the Centers for Disease Control and Prevention (CDC) is to promote health and quality of life by preventing and controlling disease, injury, and disability. In the biological weapons sphere, the CDC's main responsibilities lie in establishing and sustaining bioterrorism preparedness and safeguarding biological security. Since there is a substantial entry in this volume that addresses the CDC, here only an outline of its activities will be presented.

An important component of the CDC's preparedness activity is the Cooperative Agreement on Public Health and Bioterrorism Preparedness, which provides state and local public health agencies guidance and funding to strengthen preparedness planning, laboratory capability, surveillance and epidemiological capacity, the Health Alert Network (a CDC program intended to ensure communication capability on a local, state, and federal level), risk communication, information dissemination, education, and training. In this fifth budget year of the program, approximately $844 million will be distributed.

In response to the Public Health Security and Bioterrorism Preparedness and Response Act of 2002, the National Advisory Committee on Children and Terrorism

(NACCT) was created and housed within the CDC. Its main purpose is to advise the secretary of the HHS of any necessary changes that must be made to ensure that during a bioterrorism incident the public health system is as capable of caring for children as it is for adults.

The CDC regulates the possession and transfer of designated select agents, which are biological agents and toxins that have the potential to pose a severe threat to public health and national security. The Select Agent Program requires individuals and entities in possession of a select agent to register with the CDC and provide notification if transfer of the agent is desired. The Select Agent Program was started in response to the Anti-Terrorism and Effective Death Penalty Act of 1996, and was strengthened by the USA PATRIOT Act of 2001 and the 2002 Bioterrorism Preparedness Act.

Health Resources and Services Administration

The mission of the Health Resources and Services Administration (HRSA) is to improve and expand access to quality health care for all Americans. This includes enhancing hospital capacity and educating health care professionals in the event of any emergency, whether natural or deliberate. The National Hospital Bioterrorism Preparedness program is HRSA's primary program to counter the biological threat.

National Institutes of Health

The National Institutes of Health (NIH) is a research-based organization that acts as a steward of medical and behavioral research. It is composed of 27 institutes and centers, and of these the National Institute of Allergies and Infectious Diseases (NIAID) is the lead research and development (R&D) institute for biodefense. NIAID conducts and offers grants for basic research in the field of microbiology and complementary activities directed toward the understanding and management of global and emerging infectious diseases. Other institutes conduct and offer bioterrorism-related research opportunities, such as the National Institute on Alcohol Abuse and Alcoholism.

Office of the Assistant Secretary for Public Health Emergency Preparedness

The Office of the Assistant Secretary for Public Health Emergency Preparedness coordinates interagency activities between HHS, other federal departments, agencies,

Encyclopedia of Bioterrorism Defense,   Edited by Richard F. Pilch and Raymond A. Zilinskas
ISBN 0-471-46717-0   Copyright © 2005 Wiley-Liss

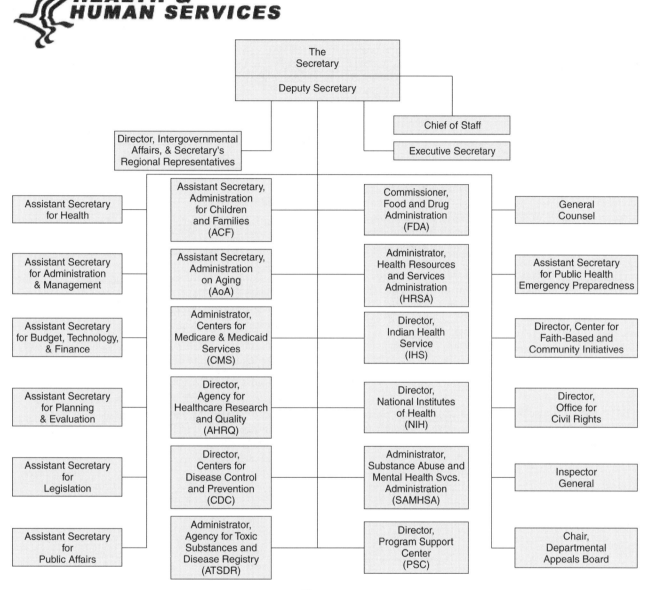

Figure 1.

and offices, and state and local officials responsible for bioterrorism preparedness activities. The assistant secretary is the principal advisor to the secretary of HHS on bioterrorism.

Substance Abuse and Mental Health Services Administration

The mission of the Substance Abuse and Mental Health Services Administration (SAMHSA) is to improve the quality and availability of prevention, treatment, and rehabilitative services in order to reduce illness, death, disability, and cost to society resulting from substance abuse and mental illnesses. SAMHSA's primary bioterrorism-related responsibility is to provide states with guidance and funding on how to incorporate mental health into preparedness activities.

Agency for Healthcare Research and Quality

The mission of the Agency for Healthcare Research and Quality (AHRQ) is to sponsor and conduct research that provides evidence-based information on health care outcomes, quality, cost, use, and access. Its main bioterrorism preparedness activity is to examine the interface between the clinical care delivery system and public health infrastructure.

HHS AFTER SEPTEMBER 11, 2001

The terrorist incidents of 2001 shifted the U.S. federal government's focus toward homeland security and bioterrorism defense. As a result, the HHS's budget allocation for bioterrorism preparedness and response has dramatically increased. In 2003, for example, money designated for enhancing bioterrorism preparedness at hospitals increased to 264 percent from the amount provided the previous year. This new focus also resulted in the creation of the Department of Homeland Security (DHS) from previously disparate domestic agencies and programs. Several programs that had been created and managed by HHS were subsequently moved to DHS, including the Strategic National Stockpile, the Metropolitan Medical Response System, and the National Disaster Medical System.

ORGANIZATIONAL TIES

The threat of bioterrorism requires that federal, state, and local governments both work together and collaborate with outside professional organizations, for example the National Association of County and City Health Officials (NACCHO), to establish an adequate level of preparedness and response. The shifting of certain programs to DHS, especially programs that require coordination with HHS programs, suggests close collaboration between these departments in the future. HHS has links to other federal departments as well, for example, food safety efforts performed in conjunction with the United States Department of Agriculture (USDA).

FUTURE

The HHS has seen an increase of funds for its mission since the September 11, 2001, terrorist attacks and the subsequent "anthrax letters," with an emphasis on activities designed to prepare the nation for a bioterrorism incident. This has resulted in an increase in interest and money in the public health system as a whole, with the intent of improving capacity to mitigate and manage both natural and deliberate disease outbreaks in the future. The HHS will continue to have the responsibility to address the threat of bioterrorism as part of its wider public health responsibilities, in the years to come.

FURTHER READING

Chemical and Biological Arms Control Institute, *Bioterrorism in the United States: Threat, Preparedness and Response*, Washington, DC, November 2000.

WEB RESOURCES

Centers for Disease Control and Prevention website, http://www.cdc.gov/.

Department of Health and Human Services website, http://www.hhs.gov/.

Health Resources and Services Administration website, http://www.hrsa.gov/.

National Institutes of Health website, http://www.nih.gov/.

Substance Abuse and Mental Health Services Administration website, http://www.samhsa.gov/.

See also CENTERS FOR DISEASE CONTROL AND PREVENTION'S BIOTERRORISM PREPAREDNESS PROGRAM; CONSEQUENCE MANAGEMENT; DEPARTMENT OF HOMELAND SECURITY; FOOD AND DRUG ADMINISTRATION; METROPOLITAN MEDICAL RESPONSE SYSTEM; and NATIONAL INSTITUTES OF HEALTH AND NATIONAL INSTITUTE OF ALLERGY AND INFECTIOUS DISEASES.

DEPARTMENT OF HOMELAND SECURITY

JENNIFER MITCHELL

INTRODUCTION

On June 6, 2002, President George W. Bush announced his plans to create the Department of Homeland Security (DHS). With the signing of the Homeland Security Act of 2002 on November 25, that plan became law (White House, 2002). Subsequent efforts marked the largest reorganization of the federal government in 50 years, involving the transfer of some 22 federal agencies into one single body (Bush, 2002).

BIODEFENSE EFFORTS AND RESPONSIBILITIES

Three directorates within DHS handle biological weapons–related issues: Border and Transportation Security (BTS), Emergency Preparedness and Response (EPR), and Science and Technology (S&T).

Border and Transportation Security

BTS is charged with inspecting plants and animals entering the United States, a task that includes searching passengers, luggage, and cargo for plants and animals that may carry disease agents (Department of Homeland Security, 2003). As roles become more clearly delineated within the DHS and other federal agencies, the United States Department of Agriculture's (USDA) Animal and Plant Health Inspection Service (APHIS) is expected to play a major role in BTS efforts by protecting plants and animals against deliberate attack (Acord).

Emergency Preparedness and Response

The EPR is responsible for preparedness and response efforts directed at meeting the challenges of biological attacks. This directorate coordinates the efforts of first responders and local, state, and federal governments to ensure an efficient and effective response and recovery strategy via its lead agency, the Federal Emergency Management Agency (FEMA) (Department of Homeland Security, 2003).

Science and Technology

Science and Technology is responsible for researching select biological agents in order to produce new and improved vaccines, diagnostics, and therapies. S&T also assists in the promotion of research and technology to develop biological and chemical detection equipment from the production stage to deployment. Various laboratories, including the Lawrence Livermore National Laboratory, Los Alamos National Laboratory, Sandia National Laboratory, and Plum Island Animal Disease Center (responsible for protecting animal husbandry in the United States), are charged with research and technology to prevent, detect, deter, and mitigate the threat and use of weapons of mass destruction (WMD) (Department of Homeland Security, 2003). Battelle Memorial Institute's Chemical and Biological Information Analysis Center (CBIAC) will provide the DHS with services related to chemical and biological defense, including access to databases, journals, and training programs (CBIAC, 2003).

RECENT DEVELOPMENTS

Despite the short history of the DHS, some recent initiatives are worth highlighting in the context of bioterrorism defense.

Project Bioshield

Most notable is Project Bioshield, an initiative intended to create bioterrorism countermeasures such as drugs, vaccines, and tests. Bioshield consists of three components: spending authority for the delivery of next-generation medical countermeasures, new National Institutes of Health (NIH) programs to accelerate the countermeasure research and development (R&D), and new Food and Drug Administration (FDA) emergency use authorization for promising medical countermeasures under development (White House, 2003). Legislation for Bioshield initially stalled in both the Senate and House, but on May 15, 2003, a draft legislation was approved, which authorized $5.6 billion over the next decade to be used to encourage commercial development and production of medical countermeasures against bioterrorism (Rovner, 2003). Regardless, concerns persist regarding the bill's mandatory funding mechanism and liability issues faced by manufacturers of experimental drugs not approved by the FDA, which could be used under the protocols of Bioshield (Rovner, 2003).

National Biodefence Analysis and Countermeasures Center

The DHS recently established the National Biodefense Analysis and Countermeasures Center (NBACC) at Fort Detrick, Maryland, to integrate national resources for

Encyclopedia of Bioterrorism Defense, Edited by Richard F. Pilch and Raymond A. Zilinskas
ISBN 0-471-46717-0 Copyright © 2005 Wiley-Liss

homeland security, drawing on resources from public health, law enforcement, and national security. NBACC's mission is to provide a unique, interdisciplinary capability to better defend against the full range of human, animal, and plant BSL-3 and BSL-4 biothreat agents. To this end, NBACC will encompass four primary centers including: (1) the Biothreat Assessment Support Center, (2) the Bioforensics Center, (3) the Bio-Countermeasures Test and Evaluation Center, and (4) the Biodefense Knowledge Center. The analytical capabilities of the NBACC will be functional in FY2004. For FY2004, the funding for DHS' science and technology program is set at $918 million, including $88 million that has been allocated to construct the headquarters for NBACC within Fort Detrick, Maryland. The current director of NBACC is Colonel Gerald W. Parker.

Bioterrorism Preparedness and Response Act of 2002

Another major development involving the DHS is the introduction of the Bioterrorism Preparedness and Response Act of 2002. This Act assists agencies in planning, coordinating, and reporting so that they are better prepared to respond to biological events. In addition, it promotes the development of biological weapons countermeasures, improves state, local, and hospital response to outbreaks (whether natural or deliberate), places controls on biological agents and toxins, and implements measures to protect the safety and security of the food, drinking water, and drug supply (U.S. Congress, 2002).

Biodefense Improvement and Treatment for America Act

The DHS has also introduced the Biodefense Improvement and Treatment for America Act, which consists of three components: biodefense, smallpox vaccine compensation, and vaccine affordability. If implemented, the Act will fund vaccine development, purchasing, and stockpiling; compensate first responders injured as a result of smallpox vaccination; and in theory improve immunization rates among health care personnel nationwide (Office of Senator Judd Gregg, 2003).

Operation Safe Commerce Grant

On June 24, 2003, the DHS's Transportation Security Agency commenced Operation Safe Commerce Grant. The pilot program's $28 million budget is directed toward surveillance of containers docking at ports in Los Angeles, Seattle, and New Jersey, from point of origin to point of destination, in an effort to guard against chemical, biological, radiological, and nuclear weapons (CBRN) smuggling (Secretary Ridge Announces the Awarding of $28 Million for Operation Safe Commerce, 2003).

FUTURE OUTLOOK

Thus far, the DHS has offered a mechanism for cooperation between 22 different federal agencies, state and local entities, and educational universities and institutions. Yet the future of the DHS and its ability to protect the United States remains to be seen. Although a multitude of well-designed initiatives have been put forward, funding problems persist, hindering effective implementation of these initiatives. Already its FY2004 budget is expected to be lower than that of the previous year, and it is unclear whether this trend will continue in the coming years. With sufficient time and funding, however, the DHS holds the promise of consolidating U.S. homeland defense in an efficient and effective manner.

REFERENCES

Chemical and Biological Defense Information Analysis Center, Homeland Security Services, 2003, www.cbiac.apgea.army.mil/hls.html.

Department of Homeland Security, *Prepositioned Equipment Program and Terrorism Early Warning Capabilities*, 2003.

Office of Senator Judd Gregg, *Senator Gregg Introduces Bio-Defense and Smallpox Compensation Legislation Today*, March 11, 2003.

Rovner, J., *House Committee Approves 'Bioshield' Funding Proposal*, GovExec.com, May 15, 2003.

Secretary Ridge Announces the Awarding of $28 Million for Operation Safe Commerce, Transportation Security Administration, June 24, 2003, http://www.tsa.gov.

White House, *President Details Project BioShield*, February 3, 2003.

U.S. Congress, H.R. 3448, *Public Health Security and Bioterrorism Preparedness and Response Act of 2002*, January 23, 2002.

FURTHER READING

Penrose, A., Statement before the Select Committee on Homeland Security, Subcommittee on Cyber Security, Science and Research and Development, US House of Representatives, October 30, 2003.

WEB RESOURCES

Clarke, D., *Up in the Morning and Out to School: Universities After Terror War Dollars*, UC-San Diego, CA, December 23, 2002, www.physicalsciences.ucsd.edu.

Department of Homeland Security, *DHS Organization: Department Components*, 2003, www.dhs.gov.

Department of Homeland Security Organizational Chart, http://www.senate.gov/~budget/republican/analysis/2002/homeland_security.pdf.

Department of Homeland Security's FY2004 Budget, http://www.whitehouse.gov/omb/budget/fy2004/homeland.html.

Department of Homeland Security website, http://www.dhs.gov.

FEMA, *R&R: Introduction and Background*, 2003, www.fema.gov.

Lawrence Berkeley Laboratory, *A First Line of Defense Against Disease Organisms*, May 7, 2002, www.lbl.gov.

National Strategy for Homeland Security, http://www.whitehouse.gov/homeland/book/index.html.

New, W., *Homeland Security Funds May Shrink in 2004, Key Aide Says*, GovExec.com, April 28, 2003.

Office for Domestic Preparedness, *About ODP*, 2002, www.ojp. usdoj.gov.

Sarkar, D., *Homeland to Streamline Filing*, GovExec.com, May 2, 2003.

U.S. Department of Justice, *Domestic Preparedness Equipment Technical Assistance Program (DPETAP)*, www.ojp.usdoj.gov, accessed on May 2003.

See also HOMELAND DEFENSE and UNITED STATES LEGISLATION AND PRESIDENTIAL DIRECTIVES.

DEPARTMENT OF JUSTICE

Don Prosnitz

INTRODUCTION

The mission of the Department of Justice (DOJ) is "... to enforce the law and defend the interests of the United States according to the law; [and] to ensure public safety against threats foreign and domestic...." In accordance with this mission, the DOJ has a responsibility to prevent and respond to terrorist attacks against U.S. citizens at home and abroad. This article briefly reviews the history of the DOJ and summarizes its role in countering biological, chemical, and radiological terrorism from prevention to prosecution.

HISTORY

The Judiciary Act of 1789 created the office of the Attorney General. The same act created the Supreme Court, circuit and district courts, the United States District Attorneys (now US Attorneys) and the United States Marshals. In 1879, the sole role of the Attorney General was to prosecute suits in the Supreme Court and give advice and counsel to the President and federal department heads. In 1861, the Attorney General was given control of the US Attorneys and the US Marshals. On June 22, 1870, nearly 100 years after the appointment of the first Attorney General, an Act of Congress formally established the Department of Justice as the government's official legal department.

Responsibility for federal prisons (now the Bureau of Prisons) was added to DOJ by the Three Prisons Act of 1891, but it took another 17 years before DOJ obtained its own investigatory agency. It was then that, Attorney General Charles Bonaparte created the forerunner of the Federal Bureau of Investigation (FBI) in response to a law preventing him from using secret service agents for investigations. This group officially became the Bureau of Investigation in 1909. The DOJ continued to grow throughout the twentieth century, acquiring the Immigration and Naturalization Service (INS, including the Border Patrol) in 1940 and the Drug Enforcement Administration (DEA) in 1973.

In recent years, DOJ has assumed a major role in training and equipping state, local, and tribal law enforcement organizations, including preparing them for terrorist attacks utilizing weapons of mass destruction (WMD), whether chemical, biological, radiological, or nuclear. In 1996, the FBI responded to the growing threat posed by WMD by creating the Hazardous Material Response Unit (HMRU) to gather and process evidence at scenes involving chemical, biological, and/or radiological materials. Responsibility for analysis of the material was moved to the FBI's newly created Chemical Biological Sciences Unit (CBSU) in April of 2002.

In 2000, the Nunn-Lugar-Domenici WMD training program was transferred from the Department of Defense to the DOJ, and the Office of Domestic Preparedness (ODP) was created. ODP's mission was to enhance the capability of first responders to handle WMD incidents. ODP owned facilities for training (such as the live chemical agent training school at the Center for Domestic Preparedness) and equipping first responders. Within the FBI, the WMD Countermeasures Unit (WMDCU) assumed the responsibility for training and exercises. At the same time, DOJ's National Institute of Justice began research on, and testing of, equipment for first responders. ODP was transferred to the Department of Homeland Security (DHS) in 2003.

Today, the Attorney General is recognized as the chief law enforcement officer of the federal government. In the last decade alone, the DOJ budget has nearly tripled. As of fiscal year 2003, the DOJ was composed of 140,000 employees with a budget of nearly $30 billion (these numbers do not reflect the 2003 transfer of some functions

Figure 1. The FBI is responsible for the forensic investigation of WMD crimes, such as the anthrax letter, pictured here, sent to Senator Leahy.

Encyclopedia of Bioterrorism Defense,   Edited by Richard F. Pilch and Raymond A. Zilinskas
ISBN 0-471-46717-0   Copyright © 2005 Wiley-Liss

to the DHS and the transfer of the Bureau of Alcohol, Tobacco, and Firearms to DOJ).

DOJ'S ANTITERRORIST ROLE TODAY

Terrorism is a federal criminal offense. As the chief law enforcement agency within the federal government, the DOJ has a major responsibility in combating and responding to terrorist acts.

The DOJ's role in combating terrorism and the terrorist use of WMD is split into three broad categories:

1. Prevention
2. Investigation
3. Criminal prosecutions and general legal advice.

Prevention activities are concentrated on detecting and accessing indications of terrorist planning and future attacks. It is primarily an intelligence function, although it does include tasks related to reducing vulnerabilities of sensitive and high-value targets. Investigations commence when there is a specific threat. Investigative activities also include the DOJ's role in responding to a WMD attack. Criminal prosecution and legal advice reflects DOJ's 214-year role as the federal government's legal arm. Subsequent sections will go into each of these three functions in more detail.

Prevention

The consequences of a successful attack using a WMD are potentially so catastrophic that the Department's primary antiterrorist objective is to "prevent, disrupt and defeat terrorist operations before they occur" (US Department of Justice FY 2001–2006 Strategic Plan). Prevention requires intelligence on terrorist plans and intentions. Domestic intelligence is specifically a law enforcement function assigned to the Attorney General.

The Department utilizes its law enforcement components (FBI, United States Marshal's Service, Drug Enforcement Agency, and the Bureau of Alcohol, Tobacco, and Firearms) to gather information on threats to the United States and on terrorist groups, plans, and intentions. Information-sharing among state and local law enforcement organizations and the federal agencies, such as the Secret Service and Customs and Border Protection, is facilitated by the FBI's Joint Terrorism Task Forces (JTTFs). JTTFs are set up in each of the FBI's 56 major field offices and in several smaller offices. In addition, each of the 96 U.S. attorneys has created an antiterrorism task force (ATTF) and appointed an antiterrorism coordinator to serve as a focus for antiterrorism activities in their respective districts. A National JTTF (NJTTF) coordinates and disseminates information from the local JTTFs. Information from foreign law enforcement organizations is provided through the FBI Legal Attaché (Legat) program. Legats are located in embassies throughout the world.

Effective intelligence analysis requires merging and sharing of intelligence information collected by multiple agencies. Recent legislation, court decisions, and DOJ policy have set new, less-restrictive guidelines for collecting and sharing intelligence information among agencies. Sections 203 and 504 of the USA Patriot Act of 2001 now permit the previously prohibited sharing and merging of foreign intelligence information and grand jury law enforcement information with "Federal law enforcement, intelligence, protective, immigration, national defense, or national security officials." Furthermore, under revised Attorney General Guidelines, the FBI may now conduct "online research even when not linked to an individual criminal investigation and may conduct preliminary inquiries whether to launch investigation of groups involved in terrorism (i.e., terrorism enterprise investigation)" (Fact Sheet Attorney General Guidelines: Detecting and Preventing Terrorist Attacks, 2002). Information-sharing is further facilitated by the newly created interagency Terrorist Threat Information Center.

If the information is deemed reliable enough to issue a warning to local law enforcement agencies (either broadly or by specific jurisdiction), the FBI issues threat warnings and general guidance through the National Law Enforcement Telecommunications System (NLET), JTTFs and other avenues.

The DOJ seeks to stop attacks by preventing terrorists from acquiring dangerous materials (i.e., hazardous chemicals, select biological agents, radiological sources) and barring them from being employed in sensitive occupations. The high risk individuals is charged with performing background checks on hazardous material drivers, and one of the responsibilities of the Foreign Terrorist Tracking Task Force (FTTTF, created in October 2001 by the Attorney General to prevent suspected foreign terrorists from entering the United States and tracking down those who do enter) is to perform background checks on foreign students seeking flight training. The Attorney General also has the responsibility to consult "criminal, immigration, national security and other electronic databases" and to conduct security risk assessments of individuals requesting access to select biological agents and toxins (Public Health Security and Bioterrorism Preparedness and Response Act of 2002). Most recently, Homeland Security Presidential Directive 6 *Integration and Use of Screening Information* (HSPD-6) directed the Attorney General to establish an organization to "consolidate the government's approach to terrorism screening" and to consolidate the government's watch lists and provide full-time operational support to law enforcement officers. The Terrorist Screening Center was formally created under FBI auspices to meet this directive in September 2003.

The DOJ is also called upon to make vulnerability assessments of various cyber and physical infrastructure, both public and private. [*Note*: As of October 2004, there are several proposed plans to reorganize federal counter terrorism intelligence activities.]

Investigation

Federal investigations regarding the use or potential use of WMD can be initiated in several ways. "The United States Attorney, as the chief federal law enforcement officer in his district, is authorized to request the appropriate federal investigative agency to investigate alleged or suspected violations of federal law... The grand jury may be used by the United States Attorney to investigate alleged or suspected violations of federal law" (*United States Attorney's Manual*, Section 9-2.010). WMD investigations include small-scale domestic use of biological weapons (e.g., the contamination of salad bars with *Salmonella typhimurium* by the Rajneeshees in The Dalles, Oregon), extortion, hoaxes (e.g., the mailing of "anthrax hoax letters" to abortion clinics), or the mishandling of select agents or toxins.

The FBI can also initiate a National Security Investigation (NSI). The Attorney General guidelines define three levels of NSI: (1) threat assessment, (2) preliminary investigation, and (3) full investigation. The category of threat assessment is new and designed to permit the FBI to be more proactive in preventing acts of terrorism. It enables the FBI to collect information concerning National Security Threats on "individuals, groups, and organizations of possible investigative interest" (Attorney General's Guidelines Regarding FBI National Security Investigations and Foreign Intelligence Collection, 2003). Each level of investigation requires a different level of approval and may utilize (generally) different categories of investigatory tools and techniques. The most intrusive techniques, such as performing physical searches and obtaining wiretaps, or installing trap and trace or pen register equipment, still require court approval. In terrorism cases involving foreign intelligence information, the DOJ may obtain the required approval by showing "probable cause" to the Foreign Intelligence Surveillance Court (FISA), a special court set up under the Foreign Intelligence Surveillance Act of 1978 (Bulzomi, 2003). Proceedings of this court are secret. The Office of Intelligence Policy and Review within DOJ is responsible for filing all requests to the FISA Court. Information obtained under a FISA court order may now, in some circumstances, be shared with other law enforcement officers to coordinate antiterrorist activities.

When a specific threat is received, or an incident involving WMD occurs, the FBI's Weapons of Mass Destruction Operations Unit (WMDOU) may be asked to evaluate the credibility of the threat or provide technical assistance to state or local authorities. WMDOU may, in turn, access its own experts, other agencies, and possibly the Behavioral Science Unit of the FBI for support in making an accurate assessment of the threat.

The response to a major terrorist attack that utilizes WMD requires coordination from a large number of federal, state, and local government agencies. A series of Presidential Decision Directives (PDDs), Homeland Security Presidential Directives (HSPDs), and interagency plans specify the response roles and responsibilities of each of the federal agencies. These include *US Policy on Counter Terrorism (PDD-39), Protection Against Unconventional Threats to the Homeland and Americans Overseas (PDD-62,) Management of Domestic Incidents (HSPD-5), National Preparedness (HSPD-8), the Federal Response Plan (FRP), U.S. Government Interagency Domestic Terrorism Concept of Operations Plan (CONPLAN), National Oil and Hazardous Substances Pollution Contingency Plan, Federal Radiological Emergency Response Plan, Mass Migration Response Plan, and Initial National Response Plan*. As of this writing (December 2003), most of these plans are under revision in order to reflect the responsibilities of the DHS.

The Federal Response Plan (FRP), first issued in 1992, describes how the federal government will provide assistance to state and local governments under the Robert T. Stafford Disaster Relief and Emergency Assistance Act. The terrorism annex to the FRP—first issued in 1997—is in response to the policies and agency guidelines laid down by PDD-39 (1995). PDD-62, issued in 1998, further builds on U.S. policy for responding to acts of terrorism. These documents provided the original framework for the federal government's response to a terrorist attack within the United States.

Fundamental to the government's response strategy is the concept of lead federal agencies (LFA). Under PDD-39 and the original FRP, the DOJ was responsible for crisis management, defined as "measures to identify, acquire, and plan the use of resources needed to anticipate, prevent, and resolve a threat or an act of terrorism." This responsibility was delegated to the FBI. The Federal Emergency Management Agency (FEMA, now incorporated into DHS) was assigned the role of LFA for consequence management. Consequence management was defined as "measures to protect health and safety, restore essential government services, and provide emergency relief to governments, businesses, and individuals affected by the consequences of terrorism." Consistent with the fact that a domestic terrorist WMD attack is a criminal act, PDD-39 also directed that DOJ be the overall LFA until the Attorney General transferred lead responsibility to FEMA and consequence management became the principal federal role.

The CONPLAN, issued in February 2001, provided details as to how a terrorist incident involving WMD would be managed at the incident site. Under the CONPLAN, the FBI designates an on-scene commander (OSC) "to manage and coordinate the response." The local FBI field office and Special Agent in Charge (SAC) are directed to set up a Joint Operations Office (JOC) and Joint Information Center (JIC). The FBI can then request through the Attorney General and the National Security Council the deployment of a Domestic Emergency Support Team (DEST). The DEST is composed of subject matter experts who can provide advice and assistance from their home agencies.

HSPD-5, issued in February 2003, eliminated the distinction between crisis and consequence management and designated the Secretary of Homeland Security as the "principal Federal official for domestic incident management... The Secretary is responsible for coordinating Federal Operations within the United States to prepare for, respond to, and recover from terrorist attacks." HSPD-5 also reaffirms that the Attorney General has "lead responsibility for criminal investigations of terrorist acts or terrorist threats... where such acts are within Federal criminal jurisdiction... as well as for related intelligence collections." HSPD-5 directs the Attorney General and the

Secretary of Homeland Security to establish relationships and mechanisms for cooperation. Finally, HSPD-5 directs the Secretary of DHS to develop a National Incident Management System (NIMS) and to prepare a new National Response Plan (NRP).

Although HSPD-5 eliminated the distinction between crisis management and consequence management, many of the structures set up under the FRP and CONPLAN remain. The Initial NRP (issued September 30, 2003) specifically states that authorities of federal officials defined in the FRP remain unchanged and support to law enforcement through groups such as the DEST will also remain unaltered. New structures are also created. The initial NRP sets up a National Homeland Security Operations Center as the national level "hub" to coordinate communications and information relating to a terrorist incident. The NRP also creates an Interagency Incident Management Group (IIMG) to facilitate incident management at the national level. The DHS may appoint a Principal Federal Officer (PFO) to represent the Secretary at the incident site and work with the OSC and FBI Special Agent in Charge. A new NRP plan will clarify many of these roles, but certainly the DOJ will retain lead status for any law enforcement response or action.

Investigation of a WMD incident requires advanced forensics. In the event of a WMD incident, the event location becomes a crime scene and the FBI's Hazardous Material Response unit may be called upon to retrieve evidence, including samples of biological, chemical, or radiological evidence. The FBI Laboratory Division's CBSU conducts its own forensics on these samples and may call upon other agencies (such as the Centers for Disease Control and Prevention, Department of Energy, and National Laboratories) or even private entities to help with advanced analysis. All analysis is performed consistent with accreditation standards and the federal rules of evidence. Standard forensics techniques (such as fingerprinting) can become especially challenging when the material is potentially contaminated with a biological agent. The material may need to be made safe without altering its forensic value.

Criminal Prosecution and General Legal Advice

The DOJ and US Attorneys are ultimately responsible for prosecuting domestic terrorists and those who would use WMD for criminal intent. The counterterrorism section of the Criminal Division of the DOJ supports federal attorneys prosecuting individuals accused of using WMD (under 18 U.S.C. 175 (biological weapons), 229 (chemical weapons), 831 (nuclear weapons), 2332a (all WMD)). In addition, the DOJ provides general legal advice to the president and other executive agencies regarding potential violations of federal law regarding WMD. This may include everything from advice on quarantines to suggestions for new legislation to aid in the prevention of terrorist acts and the prosecution of terrorists.

CONCLUSION

Local, state, and federal law enforcement all have major responsibilities in the realm of bioterrorism defense. With the creation of the DHS, the specific roles and responsibilities of the DOJ in preventing and responding to WMD terrorist attacks are under reversion, but unquestionably the DOJ, as the chief federal law enforcement agency, will continue in a lead role for many years to come.

AUSPICES STATEMENT

This work was performed under the auspices of the U.S. Department of Energy by University of California, Lawrence Livermore National Laboratory under Contract W-7405-Eng-48.

REFERENCES

Attorney General's Guidelines Regarding FBI National Security Investigations and Foreign Intelligence Collection (October 31, 2003) as redacted and released under FOIA and Attorney General's Guidelines on General Crimes, Racketeering and Terrorism Investigation, May 30, 2002.

Bulzomi, M.J., *c.f. Foreign Intelligence Surveillance Act—Before and After the USA Patriot Act*, FBI Law Enforcement Bulletin, June 2003.

Fact Sheet Attorney General Guidelines: Detecting and Preventing Terrorist Attacks, 5/30/02.

Public Health Security and Bioterrorism Preparedness and Response Act of 2002.

United States Attorney's Manual (2003), Section 9-2.010. (Available at www.usdoj.gov).

US Department of Justice, *FY 2001–2006 Strategic Plan* (2002).

FURTHER READING

Smith, J.S.E., *The Department of Justice, Its History and Functions*, W.H. Lowdermilk & Company, Washington, DC, 1904.

See also CRISIS MANAGEMENT; CONSEQUENCE MANAGEMENT; FEDERAL BUREAU OF INVESTIGATION; HOMELAND DEFENSE; INTELLIGENCE COLLECTION AND ANALYSIS; and UNITED STATES LEGISLATION AND PRESIDENTIAL DIRECTIVES.

DEPARTMENT OF STATE

ERIN HARBAUGH

INTRODUCTION

The Department of State (DOS) is the principal U.S. foreign affairs agency dedicated to advancing American interests and objectives abroad (Fig. 1). The DOS maintains 250 posts worldwide to support diplomatic relations with some 180 countries and a variety of international organizations. Its domestically posted Civil Service works in tandem with the Foreign Service, serving stateside assignment to provide logistical support to posts, conduct negotiations on international agreements, formulate and analyze reports from overseas, issue travel warnings and passports, provide assistance to American citizens traveling or residing abroad, and communicate with Congress on foreign policy initiatives.

RESPONSIBILITIES REGARDING BIOLOGICAL WEAPONS AND BIOTERRORISM

The DOS addresses a range of biological weapons issues through a variety of bureaus and offices. Its primary responsibilities with respect to these issues are concentrated in the areas of nonproliferation and arms control, counterterrorism, the promotion of peaceful uses of science and technology, and the education of other government agencies and the American public. Responsibilities of the primary DOS bureaus and offices involved in addressing the threat of bioterrorism, their ties to other DOS agencies and offices, and ties to other organizations are outlined below.

Bureau of Arms Control

The Bureau of Arms Control (AC) is responsible for supporting negotiations, implementation of existing agreements, making policy recommendations, and advising the Secretary of State on matters pertaining to biological weapons. The Bureau oversees the implementation of the Biological and Toxin Weapons Convention (BWC), Chemical Weapons Convention (CWC), as well as U.S. commitments to the Australia Group.

Bureau of International Organizational Affairs

The mission of the Bureau of International Organizational Affairs (IO) is to represent the interests and policies of the United States in multilateral organizations, namely the United Nations (UN). IO works in conjunction with other DOS offices and bureaus on issues related to treaty implementation and compliance, nonproliferation, counterterrorism, and the promotion of peaceful uses of science and technology.

Bureau of Intelligence and Research

The Bureau of Intelligence and Research (INR) utilizes all-source intelligence as a foundation for providing analysis of international events for DOS policymakers. INR also ensures that intelligence supports national security and foreign policy directives in the areas of treaty compliance and verification, as well as counterterrorism.

Bureau of Nonproliferation

The mission of the Bureau of Nonproliferation (NP) is to provide leadership in interagency activities to combat the spread of weapons of mass destruction (WMD) by leading negotiations for agreements on biological weapons and promoting transparency and controls over dual-use goods and technologies.

Bureau of Verification and Compliance

Efforts of the Bureau of Verification and Compliance (VC) are directed toward ensuring that the United States is in compliance with its international commitments related to biological and chemical weapons through the development of verification policy. The Bureau also seeks to ensure the compliance of other nations in fulfilling their respective commitments. VC cochairs a variety of interagency working groups, including the Interagency Verification and Compliance Analysis Working Group (VCAWG), which specifically addresses the BWC.

Regional Bureaus and Overseas Missions

Regional bureaus and individual missions work to assess the threat posed by the use of biological weapons in their respective geographical areas, and to alert American citizens abroad to such threats through warden messages and travel advisories. Overseas posts also serve to disseminate important security updates provided by the Department of Homeland Security (DHS) regarding

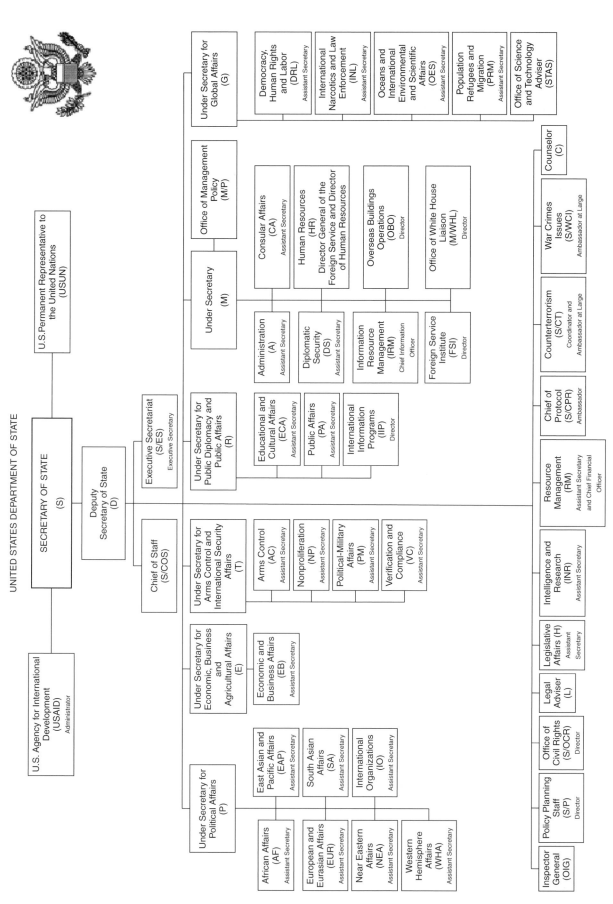

UNITED STATES DEPARTMENT OF STATE

Figure 1. Department of state organizational chart. Department of state website: http://www.state.gov/documents/organization/8792.pdf.

elevated threat levels. The Monterey WMD-Terrorism Database shows that at least fifteen anthrax hoaxes have targeted U.S. consulates and embassies worldwide since September 2001 (Monterey WMD-Terrorism Database, 2003), highlighting the susceptibility of U.S. posts abroad to such attacks (hoax or otherwise) in the future and demonstrating the significant role that regional bureaus and overseas missions must therefore play in addressing this threat.

Since 1998, the DOS has implemented a biological weapons countermeasure program to ensure that posts are prepared for a biological attack. The program provides a three-day supply of ciprofloxacin for each employee at a given facility, and accounts for the treatment of any American citizen present at the time of attack. In addition, many individual posts provide web resources and contact information for local health authorities to be used by American citizens abroad in the event of a biological attack.

NEW DEVELOPMENTS

In September 2003, the DOS announced the development of the BioIndustry Initiative (BII), a new nonproliferation program that seeks to counter the threat of bioterrorism through the targeted conversion of biological weapons research and production facilities in the former Soviet Union. A $1.7 million contract under the BII pairs the Boston-based Center for Integration of Medicine and Innovative Technology (CIMIT) with Moscow's International Science and Technology Center (ISTC) to redirect former Soviet facilities toward peaceful and ultimately self-sustainable commercial biotechnology applications. Other BII projects include the development

of an accelerated vaccine program to combat multidrug-resistant strains of *Mycobacterium tuberculosis* and other emerging and reemerging infectious diseases, development of toxicology testing services, and funding of applied research and academic programs.

THE FUTURE

For almost three decades, the DOS has played a critical role in developing and participating in nonproliferation efforts, including such regimes as the BWC and the Australia Group, to address the biological weapons threat. Now, DOS faces new challenges, not only regarding the implementation of these agreements and verification of compliance with them but also regarding threat reduction, biosecurity, and related issues, challenges expected to persist in the future as both the American and international biotechnology sectors continue to grow. Ultimately, the DOS will persevere in addressing these issues as part of its overall effort to curtail the proliferation of biological weapons, and the threat of bioterrorism, in the years to come.

REFERENCES

Monterey Institute's Center for Nonproliferation Studies website, http://cns.miis.edu (2003).

FURTHER READING

Center for the Integration of Medicine and Innovative Technology, http://www.cimit.org.

Department of State websites, http://www.state.gov, and http://travel.state.gov.

DETECTION OF BIOLOGICAL AGENTS

Rocco Casagrande

Margaret E. Kosal

INTRODUCTION

Although commonly associated with each other under the term "weapons of mass destruction" (WMD), biological agents differ markedly from chemical and nuclear weapons in several respects. The use of a biological weapon, in marked contrast to the use of nuclear and most chemical weapons (which will cause human casualties in minutes or hours), may not immediately be apparent. Also, unlike a nuclear or chemical weapon, a biological weapon produces deleterious effects that are delayed for several days or weeks, a factor that complicates detection of a biological attack. Although this delayed effect of biological agents presents complications, it also provides a window of opportunity to limit or eliminate the impact of an attack that is not afforded in an attack with a nuclear or chemical weapon, in which casualties are produced too quickly to help civilians initially exposed to the agent or blast.

The technology deployed in several major cities today aims to detect a biological attack several hours after it has occurred. This type of detection would enable the treatment of victims of an attack before symptoms appear, a time window in which antibiotic treatment against disease caused by the bacterial biological agents, such as *Yersinia pestis* and *Bacillus anthracis*, is most effective. Because victims of an attack can be more effectively treated the earlier the exposure to an agent is detected, this type of detection scheme is often termed *detect-to-treat*. Once an attack is detected, vaccines can be administered to prevent the spread of a contagious disease, reducing its impact and preventing secondary infections among critical health care workers. Also, the more quickly an attack is recognized, the more rapidly the Strategic National Stockpile of therapies and equipment can be mobilized and delivered to the affected area. Even if vaccines and antibiotics would be largely ineffective in treating or preventing the spread of some diseases several hours after victims are exposed (as would be the case with an attack using botulinum toxin), early warning enables the public health system to divert less critical patients elsewhere, recruit nurses and doctors into the area from unaffected areas, and obtain supportive equipment (such as respirators) prior to the appearance of critical shortages of beds, personnel, and equipment.

Although currently not feasible in a civilian setting, the detection of a biological attack may be completed as the attack is occurring ("real-time" detection), enabling additional measures to be taken in addition to those mentioned above. This type of detection scheme is either called *detect-to-warn*, because it may enable the evacuation of areas that will soon be affected (for example, those downwind of an aerosol release), or *detect-to-protect*, because those who cannot evacuate may either shelter in place or don protective clothing to prevent an exposure.

In the absence of such warning, worst-case scenarios could be realized. The first victims to show symptoms of the intentionally spread disease may be misdiagnosed as having influenza (since the symptoms of diseases caused by many biological agents of concern overlap with those of influenza) and receive the wrong type of treatment. The public health system may be overwhelmed by the sudden appearance of hundreds or thousands of critically ill patients. Worse still, many people may develop disease that could have been prevented by the timely delivery of antibiotics or vaccines, multiplying the impact of the attack by orders of magnitude.

But the detection of a biological attack is no easy task. The environment is filled with particles that overtly are similar to biological agents. Inorganic particulates, such as those produced by automobile exhaust, may be the same size as particles in a biological aerosol. And nonpathogenic microbes (such as mold spores) and organic excreta (such as sloughed-off skin cells) share many of the same characteristics of the microbes that comprise a biological aerosol.

Below, methods for the detection of biological agents in the environment are discussed, with particular attention paid to these and other potential sources of false positives.

NONDISCRIMINATORY METHODS

It is commonly held that the most effective method to reach tens of thousands of potential victims would be to distribute biological agents via an aerosol. If the agent in the aerosol cloud is disseminated in particle sizes between 1 and 5 μm, it will float for miles downwind suspended in the air. Persons located in the aerosol cloud's path may inhale sufficient number of pathogens to become infected

Encyclopedia of Bioterrorism Defense,   Edited by Richard F. Pilch and Raymond A. Zilinskas

ISBN 0-471-46717-0   Copyright © 2005 Wiley-Liss

and diseased. If this cloud were generated over a densely populated urban area, a high-quality biological aerosol could infect tens to hundreds of thousands of people; no other means of biological agent dissemination is likely to infect such a large number at one time (for more information, please refer to the entries entitled "Food and Beverage Sabotage" and "Water Supply: Vulnerability and Attack Specifics").

Because of this fact, the initial criterion for monitoring and surveillance of potential bioterrorist agents is often the discovery of biological aerosols. At the rudimentary level, detectors simply aim to detect the presence of an approaching cloud. If the detection of aerosol clouds can be accomplished from a distance, the devices that do so are called *standoff detectors*. Because there are many naturally occurring clouds, some of these detectors can be modified to offer a more refined assessment of the contents of the cloud, such as water droplets, inert inorganic material, dead biotic particulates, and live biological agents.

Cloud Recognition

One technique for the identification of approaching clouds that is familiar from weather reporting is the use of Doppler radar. Using reflected radio waves, the shape, size, directionality, and speed of a cloud can be monitored (Ember, 2002). The amount of time for the waves to return and the change in the waves' energy upon return to a receiver provides information about the target cloud's characteristics. For example, shape can offer clues to differentiate naturally occurring clouds from those that are ovoid or cigar shaped, which are indicative of aerosol release from a point source.

Another standoff tool for cloud detection and recognition, LIDAR, is based on the same physical principles as radar, except, instead of bouncing radio waves off a target, higher energy light waves are used. An acronym for "Light Detection And Ranging," LIDAR is occasionally attributed to "Laser Identification and Ranging." Using lasers that generate light waves in the infrared, the ultraviolet, and the visible portion of the electromagnetic spectrum supplies more detailed information about cloud characteristics as described below (Weibring et al., 2003).

Several versions of LIDAR-based detectors are currently fielded (Lee, 2002). The U.S. Army's Long Range Biological Standoff Detection System (LR-BSDS) uses LIDAR-based technology to detect aerosol clouds from long distances. The Short Range Biological Standoff Detection System (SR-BSDS) combines infrared LIDAR with ultraviolet light reflectance (UV). The addition of the UV capability provides enhanced discrimination capabilities because biologicals can be distinguished from nonbiological material on the basis of the excitation by UV light of fluorescent compounds naturally present in all living cells, such as the amino acid tryptophan, the coenzyme nicotinamide adenine dinucleotide (NADH), the cellular energy storage molecule adenosine triphosphate (ATP), and the vitamin riboflavin (Lognoli et al., 2002). It should be noted that since all living cells contain these molecules, clouds of mold spores, pollen, and certain agricultural fertilizers based on decaying organic matter can appear similar to a cloud of biological agents.

One of the simplest methods to detect whether a biological cloud is overhead is to draw the surrounding air into a device that counts particulates. The particle count is constantly tracked by these devices, and a sharp, sudden increase in the number of particles in the air would indicate that an aerosol has passed overhead. In contrast to other types of cloud detectors, this type of device has no standoff capability, as the cloud must pass through where the device is emplaced.

Aerosol Particle Sizers

As discussed above, biological agents disseminated as aerosols should have a mean particle size of between 1 and 5 μm. Aerosol particle sizers (APS) take advantage of these size characteristics for detecting bioterrorist agents. A strongly uniform particle distribution in the characteristic size range may be indicative of the presence of a biological agent.

In APS systems, particles are drawn through an orifice into a steady high-speed airflow. The velocity of the carrier air remains constant throughout, but the introduced particles accelerate at rates proportional to their size. Particle speed is measured by the time taken to impact a collector or pass through a laser light beam (Liu et al., 1995). While most particle sizers are fairly large and heavy systems, handheld analyzers are commercially available.

Flow cytometry is a sophisticated particle sizing technique in which particles are accelerated in a moving stream of fluid. Laser light scattering provides additional information with respect to the size, number, and, when combined with fluorescent dye molecules, chemistry of a sample. Combining this technique with UV or fluorescence-based detection yields information about both the size distribution of particles in a sample and the biological content of the material (Ho, 2002).

Cloud recognition and APS detectors rely on the agent cloud passing through the point where the detector is emplaced. Detectors of this type, in which the cloud or sample to be analyzed must be brought to them, are therefore called *point* detectors. Note, however, that some standoff detectors take advantage of the fact that particles scatter light differently on the basis of size. These detectors operate similarly to LIDAR systems but can determine the size of the particles that constitute the suspect cloud by the amount of light scattering that each particle produces.

All of the nondiscriminatory detectors work quickly enough to detect-to-warn. Potentially, people could be told to shelter in place, don protective equipment, or evacuate the path of an oncoming cloud detected by standoff detectors or by point detectors placed upwind.

DISCRIMINATORY METHODS

The methods described above make no attempt to determine the identity of the constituents of the suspect cloud. Dilute clouds of biological agents can share several characteristics with clouds of mold spores, pollen, and skin cells. Crowded urban environments are choked with skin cells, hair, and dust that kick up into clouds every time a vehicle passes.

Given the constant presence of these *confounders*, a detector in an urban setting must be able to discriminate between harmless biological substances in the air and true biological agents. To do this, engineers have turned to methods that are capable of identifying the contents of a sample. These methods take advantage of one of two facts of all living cells: (1) all microbes possess unique molecules on their surfaces that can be bound by components of the immune system, or (2) these unique molecules are encoded by unique stretches of DNA contained inside every cell.

The techniques described below are incorporated into a myriad of detection devices in various ways (because there exist many different devices that use variants of the same scheme, individual systems will rarely be discussed). In today's civilian biodefense setting, these techniques are usually not performed autonomously by machines. Currently, laboratory technicians use the techniques discussed below on samples sent to them from air samplers placed in strategic locations in cities. Whoever or whatever is performing the analysis, the methods discussed below can be extremely powerful and sensitive techniques for the detection of a biological attack.

Immunoassays

Immunoassay-based detectors take advantage of the specificity of the body's natural immune system. The immune system produces highly specific proteins, called *antibodies*, in response to the presence of foreign substances (called *antigens*). In theory, antibodies will bind only to a particular part of an antigen (the part of the antigen to which the antibody binds is called the *epitope*). By injecting model animals with an antigen in the laboratory, scientists can produce antibodies that will tightly bind to that antigen, and, ideally, only to that antigen. This specific binding is the foundation of immunological detectors.

Disposable handheld assay (HHA) test kits or tickets for detecting biological warfare (BW) agents have been available since the early 1990s. These tickets resemble and function similarly to home pregnancy tests. A sample is dabbed onto the absorbent surface of the ticket, which contains antibodies specific for a particular biological agent. As the sample migrates into the ticket, it encounters the antibodies and if the target antigen is present in the sample, the two will bind. This process is called *immunochromatography*. Sometimes these antibodies are tagged with an enzyme that causes a visible or fluorescent dye to be deposited onto a ticket. The detection limit with fluorescence-based tags is on the order of a thousand cells per milliliter (Iqbal et al., 2000; Lisi et al., 1982). Colloidal gold particles that generate a red indicator color without the need for a fluorescent light source have been used by the U.S. military and commercialized for general public, but these devices are less sensitive than other methods. Such immunochromatography tests ideally respond to only a particular pathogen. Tickets usually are single use, that is, they can only be used once before being discarded.

Many other immunoassays rely on two types of antibody for each target particle (Fig. 1). One antibody is tethered to a surface on which the sample is deposited. If the target particle is in the sample, it will bind to the antibody on the surface. The rest of the sample can be washed away, or the surface can be moved to another part of the device to purify the target particle by removing interfering material in the sample (Park et al., 2000). The second antibody is then added is tethered to the first surface-*bound* antibody, forming a "sandwich." The second antibody is modified with a component (such as a fluorescent molecule or an enzyme that creates a dye) to enable the detection of the second antibody's binding. If the target particle is present, the first antibody retains the target particle and the second antibody can bind, forming the sandwich and producing a signal wherever the first antibody is deposited on the surface. If the target particle is absent, nothing will bind to the first antibody; the second antibody then has nothing to bind to and is washed away, producing no signal.

The Biological Detector (BD) portion of the U.S. military's Biological Integrated Detection System (BIDS) includes immunoassay-based sensors as part of its suite of detectors (more information is available at http://www.sbccom.apgea.army.mil/products/bids.htm). Deploying immunoassay tests for 10 pathogens of terrorism concern, including *B. anthracis, Y. pestis*, botulinum toxin A, and staphylococcal enterotoxin B (SEB), the portable system requires substantial power, reagents, and warm-up time, and is portable (135 lbs) only as far as the generator-carrying vehicle on which it is mounted can travel.

Antibodies are sometimes used in conjunction with devices that can detect minute changes in their mass. Surface Acoustical Wave (SAW) systems are based on

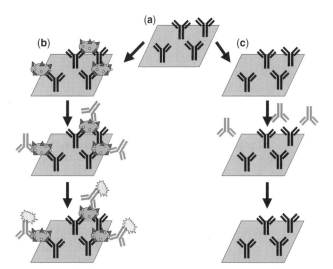

Figure 1. This cartoon depicts the use of a two-antibody sandwich to enable detection of a biological agent. (*a*) A surface is modified with the first antibody (Y-shaped objects). (*b*) If the target microbe is present, the surface-bound antibody binds to an antigen present on the target's surface (triangles). When the second antibody is added, it binds to a different antigen (dots) on the target and forms a sandwich. The second antibody is modified to yield a signal (explosion). (*c*) If the target microbe is not present, the second antibody cannot be localized to the surface and is washed away. No signal is produced.

piezoelectric materials (those that produce an electrical current when subjected to pressure or mechanical stress) coated with antibodies. If the antibodies were to bind to a target particle, this event would cause a change in the mass of the piezoelectric sensing crystals that in turn would change the frequency at which the crystal vibrates under an electric current (Uttenthaler et al., 2001). This change in frequency signals the presence of the biological agent to which the antibody decorating the piezoelectric material is designed to bind (Grate et al., 1993). Sensitivities on the order of one hundred thousand to a million microbes have been reported.

In a recent report, MIT Lincoln labs used living cells from the immune system to sense the presence of specific pathogens. The researchers used B-cells, which display receptor molecules on their surface that are very similar to antibodies, as the core element to the sensor. The B-cells were genetically engineered to emit light when the antigen bound to the receptor, and warned of the presence of small amounts of pathogen in as little as 5 min (Rider et al., 2003).

DNA-based Assays

DNA is a large molecule made up of two strands of repeating sugar molecules. Each sugar molecule is modified with one of four chemical components, called bases: adenine (A), cytosine (C), guanine (G), and thymine (T). In DNA, one chain is intertwined with the other, forming a double helix with the bases of each strand binding to each other as the axis of the helix. Base pairing is complementary, meaning that each type of base on the axis binds to only one other type (A binds only to T, and G binds only to C). The two strands of DNA will only bind to each other and form a double helix if the complementary base pairs can align across from one another. Since every animal and plant species contains DNA with unique base sequences, the analyst can tell if DNA from an organism of interest is present in a sample by testing it using a small piece of DNA, called a probe, that is complementary only to DNA unique to that organism.

It is a challenging process to select probe DNA. If a DNA sequence targeted by a probe is shared by only a small number of strains of a particular pathogen species, the strains that do not share this sequence will be missed by the probe (even though they might still be pathogenic). Similarly, if the chosen sequence is widely shared among strains of a species, the probe has the potential to respond to not only the target pathogen but also to vaccine strains or closely related species. A detection system based on such a probe would be likely to give false positive signals, that is, to give alarm although the target pathogen is not present. To try to avoid this problem, probes can be constructed to carry specific stretches of DNA that endow a pathogen with virulence, thus drastically reducing the probability that harmless microorganisms will be falsely identified as pathogens (Slezak et al., 2003).

DNA-based detectors can attempt to detect DNA from target pathogens either directly or through amplification (explained below). The two methods of DNA detection are often used together, with target DNA amplified prior to direct detection. Direct DNA-detection methods often

work very similarly to the antibody sandwich described above in that they also require two probes. The first probe is tethered to a surface, trapping target DNA from a sample. The second probe, modified to enable detection of the second binding event, will bind to the other side of the target DNA (Fig. 2). This method is used in DNA microarrays, or "gene chips," currently being investigated for biological agent detection. In the microarray, probe DNA from dozens to hundreds of genes from many target organisms can be displayed in discrete spots on a given surface (Wilson et al., 2002). If DNA from any of the displayed genes is present in a sample, it will bind to that particular spot; addition of a third probe sequence (labeled to enable fluorescent detection) will determine to which spots DNA in the sample is binding, indicating the presence of that particular biological agent (Cheng et al., 1996; Du et al, 2003). Alternatively, probe DNA can be used to decorate a piezoelectric material; a change in the vibration frequency indicates the presence of target DNA sequences.

DNA-based detection is typically combined with an amplification technique such as the polymerase chain reaction (PCR) in order to generate larger quantities of genetic material and increase the sensitivity of the detection (Jones et al., 2001; Belgrader et al., 1998). In PCR, a target stretch of double-stranded DNA is dissociated into two single strands and then bound by a short probe; once this binding occurs, an enzyme can elongate the probe by attaching complementary bases along the rest of the target strand, forming a copy of the strand that is complementary to the target. This cycle then repeats. Each cycle of strand dissociation, probe binding, and enzymatic elongation doubles the amount of target DNA. Note that since a given probe is designed to be complementary to sequences unique to

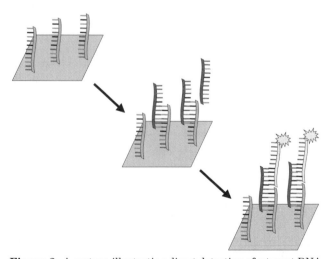

Figure 2. A cartoon illustrating direct detection of a target DNA sequence using two probes. A surface is modified with a single stranded DNA probe that has a sequence complementary to a portion of target DNA. If the target DNA is present, it will bind to the complementary sequence on the probe and become linked to the surface via the probe. A second probe is added that contains sequences complementary to another portion of the target DNA. This second probe is modified to enable visualization.

a target pathogen, it will only enable the amplification of the DNA of the target pathogen despite the possible presence of other DNA molecules in the sample. Therefore, if amplification occurs (amplification can be detected by making amplified DNA stretches from components that fluoresce, for example) the target pathogen is present. Theoretically, repeated cycles of amplification can enable the magnification and identification of a sample that contains one target sequence of DNA, an ideal level of sensitivity. In practice, usually between 10 and 100 molecules are required at least.

When first developed, each PCR cycle took about a half hour to complete, requiring several hours for the entire amplification process. In the last decade, significant technical breakthroughs in the machines that perform the cycling have reportedly reduced analysis times to less than 10 min, exclusive of sample preparation and optimization time.

PCR and other DNA amplification techniques, while extremely powerful, are not without drawbacks: They are often labor intensive (DNA must be extracted from the inside of the target organism, reagents must be added to the sample, etc.), offer marginal portability (typically the required equipment weighs more than 50 lbs.), and are demanding on power resources. Further, some organisms, such as *B. anthracis,* are notoriously resistant to disruption, complicating efforts to extract its DNA (Belgrader et al., 1999). In addition, the sample must contain DNA for PCR and other DNA amplification techniques to work; therefore, these techniques cannot be used for the identification of toxins such as botulinum toxin and ricin.

Recall that DNA-based assays and immunoassays work because they use reagents (antibodies or DNA probes) specific for each pathogen. This specificity has a price: There are many more pathogens usable by terrorists than are currently practical to test for, because each pathogen requires its own reagents. If a terrorist uses a pathogen that is unexpected, it is unlikely that the assay will be performed with reagents specific for the pathogen used, and its presence will be missed by the test.

Mass Spectrometry

The ability to characterize potential BW agents has been further enhanced by the use of mass spectrometry (MS) instruments. The main approach is to generate "holistic" whole-organism fingerprints (Beverly et al., 2000; Fuerstenau et al., 2001; Morgan et al., 2001).

In mass spectrometry, a sample is fragmented into charged pieces. The pieces may be proteins, peptides, characteristic fragments of cell walls, or other small molecules, all with different masses depending on their chemical composition. These masses are then measured in comparison to the other particles in the mass spectrometer. When a complex particle such as a bacterium enters the mass spectrometer, it is separated into hundreds of thousands of pieces, creating a complex pattern of molecular weight readings. An automated process analyzes the pattern and compares it to known MS patterns of already characterized bacteria, viruses, and proteins that are stored in a library. If a match is found, then the pathogen

in the sample is identified. Each pathogen species generates a characteristic pattern because each contains a unique combination of molecules, molecules which individually could be shared across species. In microseconds, MS detectors can simultaneously volatilize, take apart, and detect proteins or other fragments of a microorganism. Even spore-forming bacteria can be detected and identified using MS techniques (Snyder et al., 1999; Goodacre et al., 2000; Vaidyanathan et al, 2001). Unlike DNA-based assays and immunoassays, no pathogen-specific reagents are required, so MS can be used to identify multiple pathogens and provide clues to the identity of microorganisms that might not be expected in advance.

If two or more different microbes enter the mass spectrometer at the same time, the patterns may overlap, confounding identification of either microbe. Therefore, real-world applications of MS require the separation of mixed samples. When an MS detector is linked directly to a separation scheme, it can be used to detect, identify, and differentiate diverse biological agents (bacteria, toxins, and viruses). Automated separation devices are usually seamlessly integrated directly into the MS instrument.

A secondary approach to the detection of biological agents using MS is to examine the characteristic patterns associated with enzymatic metabolites produced by growing bacteria. During normal cellular processes, small volatile by-products are emitted by bacteria and fungi. The unique combination of these volatile biomarkers is not only specific to each bacterial species but can also be used to discriminate between strains of bacteria within a species. For example, identification among *Yersinia* species has been demonstrated, and biomarkers from *B. anthracis* and a nearest-neighbor species, *B. cereus,* have been identified and differentiated (Jantzen and Lassen, 1980; Leclercq et al., 1996; Fox et al., 1993).

Viruses and toxins do not emit volatile biomarkers (although a toxin's parent organism does), so utility is limited to bacterial and fungal species. Furthermore, genetically identical bacteria may yield slightly different biomarker patterns depending on the environment (organisms often change their component parts depending on growth conditions). Although MS currently is excellent at identifying biomarkers from a single species of bacteria grown under standardized conditions, a terrorist is unlikely to propagate microorganisms under the same local conditions used for reference strains. Most current MS detector systems are nevertheless bulky complex instruments (and therefore usually remote from initial response sites), expensive, and require trained laboratory personnel—characteristics that have thus far been challenging for field deployment.

Hybrid Systems

As noted, each previously described technique has its own advantages and limitations. Thus, there have been several attempts to use immunoassays, DNA-based methods, and MS in various combinations to capture the advantages of each technique employed while eliminating most limitations. For example, the Lawrence Livermore National Laboratory has developed the Autonomous Pathogen Detection System, which uses antibodies to collect and sort targets into another reaction chamber where PCR occurs, greatly increasing sensitivity (more information is available at http://greengenes.llnl.gov/bbrp/html/langloisabst.html). Furthermore, the device performs a double check for every possible alarm in that the target particle must bind to a specific antibody and be amplified via specific DNA probes.

In another publication, researchers report that they have developed a method to achieve the sensitivity of PCR-based techniques in the detection of toxins, which have no DNA. The researchers use an antibody sandwich-based approach; however, the second antibody, instead of being linked to a tag to allow visualization, is linked to a "bar-code" DNA molecule. Following formation of a sandwich of antibody-antigen-antibody, PCR is performed to amplify and detect the presence of the bar-code DNA (Nam et al., 2003).

DNA-based methods have been combined with MS to create a system called TIGER, which can identify bacteria regardless of their growing conditions (Stix, 2002). DNA probes are used that bind to a DNA sequence coding for the basic machinery of a cell, in other words, a sequence common to all life forms (every organism from man to oak trees to *E. coli* shares this DNA sequence). Although the probes bind to DNA common to almost all life, they direct amplification of a DNA region between the binding sites that is highly variable; in fact, almost every species has its own unique sequence. The amplified DNA derived from the unique region between the common regions is fed into a mass spectrometer to derive a unique mass/charge pattern for each piece of DNA. These patterns are examined to calculate the DNA sequence that produced the pattern, and these sequences are then compared against the thousands of sequences in data banks to determine the DNA's origin. This system has no reagents specific for a particular organism, so it can identify unexpected organisms in the sample (unlike conventional PCR). Additionally, since an organism does not change the sequence of its DNA in response to environmental factors such as growth conditions, this system avoids the potential confounders described with respect to conventional MS analysis above. However, unlike conventional MS analysis, this system does not work on toxins (which do not have DNA) and must be greatly modified to handle viruses (which do not have a copy of the common DNA sequence).

COMBINING DISCRIMINATORY AND NONDISCRIMINATORY METHODS

Currently, detectors that can discriminate pathogens from harmless particles in the air require expensive reagents or complicated sample preparation. In contrast, nondiscriminatory devices, although unable to distinguish between harmful particles and harmless ones, are generally cheap to operate and can take the punishment of constant usage. To gain the advantages of both types of detection schemes, engineers may use a nondiscriminatory device to trigger analysis by a discriminatory device. The cheap and robust nondiscriminatory detector can constantly scan the air for unusual clouds. When

it detects something suspicious, the nondiscriminatory device directs the discriminatory device to test the suspicious sample further. In this manner, costly reagents can be saved, and fragile or slow devices can only be used when absolutely needed.

BEYOND CHECKING THE AIR

In addition to an airborne biological attack, which can potentially expose thousands of civilians to a biological agent, mass casualties can be caused by the contamination of foods. Discriminatory detection techniques are suited for testing food because these methods can distinguish the biological material that comprises the food from pathogens present on the food sample (Hall, 2002). However, food is a complex material that requires extensive sample preparation to prevent the clogging of delicate instruments or the overwhelming of PCR-based tests with DNA derived from meat or vegetables in the sample. Furthermore, in commercial plants, food is processed in extremely large volumes and shipped quickly; methods must therefore be developed to test large amounts of material in extremely short periods of time.

In addition to checking the environment for the presence of a biological attack, these discriminatory detection methods can be used on patient samples in a hospital to determine whether a patient's symptoms are due to exposure to a BW agent or a naturally occurring pathogen such as the influenza virus (recall that the presenting symptoms are often very similar). If the attack is not detected through environmental sampling prior to a victim arriving in the emergency room, devices of this type can be used on patient samples and may provide the first warning that an attack has occurred and that hundreds or thousands more persons may be incubating the same disease agents. Alternatively, detectors can be used as diagnostic tools to distinguish the "worried well," people who are unexposed to a biological release but demand treatment because they think they have been exposed, from the actual victims of a biological attack, allowing treatment to get to those who need it most.

NONSAMPLING METHODS

All biological agent detection schemes described above are limited by the fact that they can only detect the presence of agents in the samples that they are analyzing. Detectors that analyze collected air will be useless if a biological attack is perpetrated by contaminating food or water, and detectors scanning the air in Seattle will not provide early warning for those living in Tacoma. Today, detectors are too expensive or unreliable to be deployed like smoke detectors, protecting people wherever they live or work. Therefore, other methods, which do not rely on the testing of samples, must be used to enable the early detection of a biological attack regardless of where or how it occurs.

The methods described below rely on the fact that a biological attack will not affect all of its victims with the same speed. In any disease exposure, some people will fall ill more quickly (such as the elderly, the immunocompromised, or those that received a larger infectious dose) and some will fall ill more slowly. If a system can identify that a biological attack has occurred by recognizing illness patterns in the first victims of an attack, the public health community can gain several days of advanced warning before the majority of people fall ill.

Syndromic Surveillance Systems

Some computer systems examine symptoms databases of patients presenting at emergency rooms with nonspecific illness, such as fever, labored breathing, or headache. These systems search the databases for patterns not normally associated with natural disease (such as numerous patients with flu-like symptoms outside of flu season or an unusual number of cases of pneumonia in young adults) to determine if a biological attack has taken place (Buehler, 2003). Because these systems search for telltale patterns of syndromes (groups of symptoms that are indicative of a disease), they are often called *syndromic surveillance systems* (for similarly obvious reasons, these systems are alternatively known as *prediagnostic* or *health indicator surveillance systems*).

There are several variants of this type of syndromic surveillance system deployed today—for example, NEDDS, ENCOMPASS, RODS, RSVP, LEADS, PILGRIM, and ESSENCE II—some of which use other sources of information in addition to that gathered on hospital patients. Some have been deployed during "high-risk" events over the past few years, such as political conventions and the Superbowl. Many of these systems are scouring data from hospitals in major metropolitan areas at this very moment (Lombardo et al., 2003).

As noted above, syndromic surveillance systems are not limited in their data pool to hospital records. For example, they can also monitor pharmacy sales of over-the-counter products such as aspirin or Imodium to identify population-wide symptoms not captured by hospital visits. Tracking over-the-counter medication can reveal much about the health of the general population because each medication can be correlated with a specific symptom or syndrome (No author, 2003). For example, a run on antidiarrheals may indicate an outbreak of Norwalk virus, whereas a buying spree of Theraflu may indicate that a flu epidemic is occurring. Another approach is to monitor the flow of calls to a toll-free health advisory system; this is the approach taken by PILGRIM.

Syndromic surveillance systems can be either *active* or *passive*. Passive systems patch into existing methods that hospitals use to track data on patients in their admitting process; there is no additional time burden on health care personnel. Active systems require a user to record information he or she otherwise would not be reporting, or to report it in a way that he or she otherwise would not do. For example, an active system may require that a nurse record a checklist of symptoms for each patient using a personal digital assistant (PDA) in addition to filling out a patient history on a hospital computer. Often, the data collected by active systems is more specific to a biological attack or unusual illness but requires additional resources that passive systems do not.

Sentinel Population Surveillance Systems

Sentinel population surveillance systems do not rely on the symptoms of individuals but focus instead on their behavior. These systems analyze data from large employers or schools to determine if greater numbers of people are calling in sick than would be expected. Additionally, some of these systems record highway tollbooth volume, because gravely ill people tend not to commute, in an attempt to identify illnesses before people report to emergency rooms (No author, 2003).

Another sentinel population consists of patients hospitalized with an infectious disease. The MIT Lincoln Laboratory has built a prototype system that uses data on the places these patients have been prior to their hospitalization to determine if a biological attack has taken place. If several patients suffering from a similar disease all visited the same location at nearly the same time, a release of a pathogen may have occurred at that location. Importantly, this system will not only identify that an attack occurred but also the likely time and place of the attack, indicating which populations may need the most urgent treatment (Hoffeld, 2003, personal communication).

Another sentinel population consists of wild animals and livestock. Certain biological agents are much more adept at infecting and killing animals than humans (*B. anthracis* is a prime example). Because of this characteristic, sick livestock or dead birds may be an early indicator of a biological attack, one that would appear while humans are still incubating the disease. Systems that track and analyze patterns of animal deaths may be able to give early warning that an attack took place.

Although presented here as two distinct types of systems, syndromic surveillance systems are often combined with sentinel population surveillance systems (such as in ESSENCE II) to create a more complete picture of the state of public health in the population.

THE FUTURE

Each system described above has its own advantages and disadvantages. A truly robust defense will likely only be achieved through the integration of multiple systems (nondiscriminatory methods to trigger discriminatory methods and supplemented by nonsampling surveillance systems to catch outbreaks that were caused by biological releases outside of areas where samples were taken). Obviously, the mass deployment of complicated devices throughout all major population centers will be costly.

The dream of the biological defense community is to have a "biological smoke detector," a cheap, reliable device that can be put in every home and every office. However, to be deployed in a home, a biological detector would have to have virtually no false alarms. Conventional smoke detectors have false alarms all the time, alerting people to smoke that is not related to a dangerous fire, such as the smoke made by oil in a frying pan. In this case, a person can immediately determine that a false alarm has occurred because he or she can see that there is no raging fire in the vicinity and act appropriately to the false alarm. Since biological agents in the air are invisible and odorless, a home owner will not be able to determine if an activated biological detector is giving a false alarm or not. Therefore, the goal of a biological smoke detector may have to be postponed until a detector with a false alarm rate bordering on zero has been developed.

Considering that a large-scale biological attack has never happened, it is uncertain whether all the expense of full-scale detector deployment will be worthwhile. What if another biological attack does not occur? The current stance of the United States is to act on the side of caution, because the widespread deployment of detection systems could save tens of thousands of lives if the worst-case scenario is realized. Further, these systems can be used beyond defense to boost public health and overall environmental monitoring. Disease surveillance systems and sentinel population surveillance systems can give the public health community greater information about the patterns and behavior of natural outbreaks of disease. For example, when a syndromic surveillance system was used during the presidential inauguration, a small flu outbreak was detected by the system and brought to the attention of the public health community (DARPA Press Release, 2001).

Discriminatory methods are currently being investigated for their ability to provide early warning of natural outbreaks of agricultural disease. Theoretically, a detection system assaying air collected from a room in which a cow incubating foot-and-mouth disease is held would be able to detect the virus several days before overt symptoms appear. Possibly, detectors such as these could be placed in every major feedlot, grain storage node, and auction barn to determine if diseased animals are present before the disease spreads beyond the facility.

In the further future, one can imagine devices based on the defense systems of today that are used to break the cycle of infectious disease. Commuters may go through a pathogen detector before getting on a subway, going into their office building, or entering into school, for example, and if any pathogen is detected on the breath of the commuter then he or she is sent home, possibly preventing all those he would have contacted from becoming infected. That future, however, remains a long way away.

REFERENCES

Belgrader, P., Benett, W., Hadley, D., Long, G., Mariella, R., Milanovich, F., Nasarabadi, S., Nelson, W., Richards, J., and Stratton, P., Clin. Chem., 44, 2191–2194 (1998).

Belgrader, P., Hansford, D., Kovacs, G., Venkateswaran, K., Mariella, R., Milanovich, F., Nasarabadi, S., Okuzumi, M., Pourahmadi, F., and Northrup, M.A., Anal. Chem., 71, 4332–4336 (1999).

Beverly, M.B., Voorhees, K.J., Hadfield, T.L., and Cody, R.B., Anal. Chem., 72, 2428–2432 (2000).

Buehler, J.W., Berkelman, R.L., Hartley, D.M., and Peters, C.J., Emerg. Infect. Dis. [serial online] 9(10), 1197–1204, (2003). Available from http://www.cdc.gov/ncidod/EID/vol9no10/03-0231.htm.

Cheng, J., Frotina, P., Surrey, S., Kricka, L.J., and Wilding, P., Mol. Diagn., 1, 183–200 (1996).

DARPA Press Release, DARPA Epidemiology Software Used During Presidential Inauguration, March 9, 2001. Available at http://www.darpa.mil/body/NewsItems/wordfiles/ENCOM-PASS_release.doc.

Du, H., Miller, B.L., and Krauss, T.D., J. Am. Chem. Soc., 125, 4012–4013 (2003).

Ember, L., Chem. Eng. News, 80, 23–42 (2002).

Fox, A., Black, G.E., Fox, K., and Rostovtseva, S., J. Clin. Microbiol., 31, 887–894 (1993).

Fuerstenau, S.D., Benner, W.H., Thomas, J.J., Brugidou, C., Bothner, B., and Siuzdak, G., Angew. Chem., 40, 542–544 (2001).

Goodacre, R., Shann, B., Gilbert, R.J., Timmins, É.M., McGovern, A.C., Alsberg, B.K., Kell, D.B., and Logan, N.A., Anal. Chem., 72, 119–127 (2000).

Grate, J.W., Martin, S.J., and White, R.M., Anal. Chem., 65, 940A (1993).

Hall, R.H., Microbes Infect., 4, 425–432 (2002).

Ho, J., Anal. Chim. Acta, 457, 125–148 (2002).

Hoffeld, R., personal communication with author (2003).

Iqbal, S.S., Mayo, M.W., Bruno, J.G., Bronk, B.V., Batt, C.A., and Chambers, J.P., Biosens. Bioelectron., 15, 549–578 (2000).

Jantzen, E. and Lassen, J., Int. J. Syst. Bacteriol., 30, 421–428 (1980).

Jones, M., Alland, D., Marras, M., El-Hajj, H., Taylor, M.T., and McMillan, W., Clin. Chem., 47, 1917–1918 (2001).

Leclercq, A., Wauters, G., Decallonne, J., El Lioui, M., and Vivegnis, J., Med. Microbiol. Lett., 5, 182–194 (1996).

Lee, K.J., Youngsikpark, B.A., Nunes, R., Pershin, S., Voliak, K., Appl. Opt., 41, 401–406 (2002).

Lisi, P.J., Huang, C.W., and Hoffman, R.A., Clin. Chim. Acta, 120, 171–179 (1982).

Liu, B.Y.H., Yoo, S.-H., and Chase, S., J. Inst. Environ. Sci., 38, 31–37 (1995).

Lognoli, D., Lamenti, G., Pantani, L., Tirelli, D., Tiano, P., and Tomaselli, L., Appl. Opt., 41, 1780–1787 (2002).

Lombardo, J., Burkom, H., Elbert, E., Magruder, S., Lewis, S.H., Loschen, W., Sari, J., Sniegoski, C., Wojcik, R., and Pavlin, J., J. Urban Health, 80, i32–i42 (2003).

McBride, M.T., Gammon, S., Pitesky, M., O'Brien, T.W., Smith, T., Aldrich, J., Langlois, R.G., Colston, B., and Venkateswaran, K.S., Anal. Chem., 75, 1924–1930 (2003).

Morgan, C.H., Mowry, C., Manginell, R.P., Frye-Mason, G.C., Kottenstette, R.J., and Lewis, P., Proc. SPIE—Int. Soc. Opt. Eng. (Adv. Environ. Chem. Sens. Technol.), 4205, 199–206 (2001).

Nam, J., Thaxton, C.S., and Mirkin, C.A., Science, 301, 1884–1886 (2003).

No author, "BioTerrorism–Syndromic Surveillance," Proceedings of the 2002 National Syndromic Surveillance Conference, New York, September 23–24, 2002; J. Urban Health, 80, i1–i140 (2003).

Park, M.K., Briles, D.E., and Nahm, M.H., Clin. Diagn. Lab. Immunol., 7, 486–489 (2000).

Rider, T.H., Petrovick, M.S., Nargi, F.E., Harper, J.D., Schwoebel, E.D., Mathews, R.H., Blanchard, D.J., Bortolin, L.T., Young, A.M., Chen, J., and Hollis, M.A., Science, 203, 213–215 (2003).

Slezak, T., Kuczmarski, T., Ott, L., and Torres, C., Briefs Bioinf., 4, 133–149 (2003).

Snyder, A.P., Maswadeh, W.M., Parsons, J.A., Tripathi, A., Meuzelaar, H.L.C., Dworzanski, J.P., and Kim, M.-G., Field Anal. Chem. Technol., 3, 315–326 (1999).

Stix, G. (2002) Sci. Am., November, 18–20.

Uttenthaler, E., Schraml, M., Mandel, J., and Drost, S., Biosens. Bioelectron., 16, 735–743 (2001).

Vaidyanathan, S., Rowland, J.J., Kell, D.B., and Goodacre, R., *Anal. Chem.*, **73**, 4134–4144 (2001).

Wilson, W.J., Strout, C.L., DeSantis, T.Z., Stilwell, J.L., Carrano, A.V., and Andersen, G.L., *Appl. Opt.*, **42**, 3583–3594 (2002).

FURTHER READING

Casagrande, R., *Sci. Am.*, October, 81–87 (2002).

Cunningham, A.J., *Introduction to Bioanalytical Sensors*, Wiley-Interscience, NY, 1998.

Ivnitski, D., Abdel-Hamid, I., Atanasov, P., and Wilkins, E., *Biosens. Bioelectron.*, **14**, 599–624 (1999).

Morse, S.S., "Detecting Biological Warfare Agents," in R.A. Zilinskas, Ed., *Biological Warfare: Modern Offense and Defense*, Lynne Rienner Publishers, Boulder, Colo., 2000, pp. 85–103.

Smithson, A.E. and Levy, L.-A., *Ataxi: The Chemical and Biological Terrorism Threat and the US Response*, The Henry L. Stimson Center, Washington, D.C., 1999, p. 185.

WEB RESOURCES

The National Institute of Justice's Law Enforcement Standards and Testing Program, *An Introduction to Biological Agent Detection Equipment for Emergency First Responders*, NIJ Guide 101-00, December 2001, http://www.ncjrs.org/pdffiles1/nij/190747.pdf.

DIANE THOMPSON: A CASE STUDY

MONTEREY WMD-TERRORISM DATABASE STAFF

OVERVIEW

In 1996, Diane Thompson, a laboratory technician at St. Paul Medical Center in Dallas, Texas, enticed a number of fellow hospital workers to consume pastries contaminated with the incapacitating agent *Shigella dysenteriae* type 2, leading to 12 illnesses, including four hospitalizations. After the incident, she was arrested and indicted on charges of tampering with a food product; she currently is serving a 20-year prison sentence.

THE INCIDENT

On October 29, 1996, Thompson sent an anonymous email inviting her colleagues to help themselves to a batch of pastries, including doughnuts and blueberry muffins, which she had placed in the hospital break room. Twelve employees consumed the pastries between 7:15 A.M. and 1:30 P.M. that day, and by 4:00 A.M. the next morning, all had begun suffering from symptoms of gastroenteritis. Five employees eventually sought medical care for their symptoms, and four were ultimately hospitalized (*Houston Chronicle*, 1996; Carus, 1999).

Upon investigation of the incident, officials found that Thompson had contaminated the pastries with a strain of *S. dysenteriae* stored in laboratory at which she worked and to which she had ready access (Kolavic, 1997). Authorities also discovered that this was not her first such offense: In 1995, Thompson had given food tainted with an unknown pathogen to John P. Richey, her boyfriend at the time. He had experienced symptoms similar to those of the October 29 victims, including diarrhea and fever, and had been hospitalized. However, Thompson had managed to fabricate his laboratory reports to prevent proper determination of the nature of his illness. Thompson also allegedly had contaminated a syringe and used it to take a sample of Richey's blood, and committed other hostile

acts against him as well, including slashing his tires and pouring sugar in the gasoline tank of his car. During the investigation of the St. Paul Medical Center case, Richey informed authorities that Thompson had admitted to contaminating him (*Houston Chronicle*, 1996; Carus, 1999).

Authorities arrested Thompson on November 8, 1996, and indicted her for tampering with both a food product and a government document. She was charged with aggravated assault and found guilty of five felony assault charges. Although her felony charges carried a maximum penalty of life in prison, Thompson was sentenced to only a 20-year term (*Houston Chronicle*, 1998).

IMPLICATIONS

This case underscores the importance of implementing dedicated biosecurity measures at culture collections and laboratories throughout the United States and world. Particularly, as this case demonstrates, insider threats are of great concern, and thus measures to ensure personnel reliability must be integral components of any biosecurity initiative for it to be effective.

REFERENCES

Carus, W.S., *Bioterrorism and Biocrimes: The Illicit Use of Biological Agents in the 20th Century*, Working Paper, Center for Counterproliferation Research, National Defense University, July 1999 revision.

Houston Chronicle, Bacteria Put in Pastries, November 12, 1996.

Houston Chronicle, Ex-lab Worker Tainted Food, September 12, 1998.

Kolavic, S.A. et al., *J. Am. Med. Assoc.*, **278**, 397, August 6, 1997, http://jama.ama-assn.org/cgi/content/short/278/5/396, accessed on 10/3/03.

Encyclopedia of Bioterrorism Defense,   Edited by Richard F. Pilch and Raymond A. Zilinskas
ISBN 0-471-46717-0   Copyright © 2005 Wiley-Liss

DIRECTOR OF CENTRAL INTELLIGENCE COUNTERTERRORIST CENTER

STEPHEN MARRIN

INTRODUCTION

The Director of Central Intelligence's (DCI's) Counterterrorist Center (CTC), located at Central Intelligence Agency (CIA) Headquarters in Langley, Virginia, provides a focal point for the United States foreign intelligence community's efforts to prevent, deter, disrupt, and destroy the bioterrorist threat. CTC was created in 1986 as an intelligence community body under the authority of the DCI in order to allow personnel from many different intelligence and security agencies to work together "to collect intelligence on, and minimize the capabilities of, international terrorist groups and state sponsors" (http://www.odci.gov/terrorism/ctc.html, accessed 12/13/03). The CTC focuses its attention on the structure and operations of foreign terrorist groups, including those that might conceivably develop or use biological weapons against the United States or its interests. While the CTC is primarily an operational center focused on the collection of information on and disruption of terrorist activities, it also contains intelligence analysts who "produce in-depth analyses of the groups and states responsible for international terrorism" and provide this analysis to national-level policymakers, intelligence collectors, law enforcement professionals, and military responders so that they can use the powers at their disposal to prevent attacks or mitigate the effects if prevention efforts should fail (http://www.odci.gov/terrorism/ctc.html, accessed 12/13/03).

CTC'S ROLE IN COORDINATING COUNTERTERRORISM EFFORTS

The CTC contributes to bioterrorism defense by coordinating the U.S. intelligence community's efforts to collect and analyze intelligence on terrorist groups and state sponsors of terrorism. In 1997, then-acting DCI George Tenet pointed out that CTC "includes personnel from CIA as well as 11 other departments and agencies (including) intelligence agencies, such as DIA and NSA, law enforcement, such as the FBI and Secret Service, and policymaking agencies such as the Department of State" (Tenet, 1997).

Yet, despite CTC's designation as an intelligence community entity operating under the auspices of the DCI, it is primarily a CIA-dominated center. In 1996, a House Permanent Select Committee on Intelligence staff study

observed that CTC was intended to be a shared intelligence community resource with substantial representation of staff from elsewhere in the intelligence community, but that this did not occur (Permanent Select Committee on Intelligence, 1996). Instead, as the report noted, because of its presence at CIA, CTC has "a distinct 'CIA' identity. [It is] predominantly staffed by CIA employees, and [is] dependent upon the CIA for administrative support and funding" (Permanent Select Committee on Intelligence, 1996). In the end, CTC is less of an intelligence community partnership than a CIA entity "into which a Community partnership is inserted." In 2002, the CTC's non-CIA intelligence community personnel consisted of only 52 employees (Tenet, 2002) out of approximately 1000 total staff (Calabresi and Ratnesar, 2002).

Nonetheless, the CTC facilitates intragovernmental communication, cooperation and coordination by bridging organizational seams between collection, analysis, and policymaking institutions. Traditionally, each intelligence and security bureaucracy transmits information vertically through its own institution in a process known as *stovepiping* before sharing the information laterally with other agencies in the government with similar missions. As a result of these communication stovepipes, the sharing of crucial information at the working level between intelligence collectors, analysts, and security or law enforcement personnel regarding the terrorist threat may be significantly delayed or perhaps not take place at all. The CTC bypasses these institutional stovepipes by providing a venue for personnel from different agencies to work together and share all the information available in different institutions to address counterterrorism issues.

According to a CIA report, "the exchange of personnel among agencies has been key to the flow of critical information between agencies and has strengthened the overall US Government counterterrorism capability. Information is also exchanged between agencies via several telecommunications systems that have dramatically decreased the lag time experienced between the receipt of terrorist threat information and the dissemination of that information to Federal, state, and local intelligence/law enforcement agencies" (Director of Central Intelligence Annual Report for the United States Intelligence Community, 1999). As a result, the CTC is "the paragon of interagency cooperation," which "is everything the rest of the intelligence community is not: coordinated, dynamic and designed for

Encyclopedia of Bioterrorism Defense, Edited by Richard F. Pilch and Raymond A. Zilinskas
ISBN 0-471-46717-0 Copyright © 2005 Wiley-Liss

the post-cold war threat," according to a *Time* magazine article (Calabresi and Ratnesar, 2002).

The CTC may be an effective mechanism for sharing bioterrorism threat information within the U.S. intelligence community and with other domestic security agencies, but it is not effective as a way to integrate foreign intelligence with bioterrorism-related intelligence collected domestically by law enforcement and other domestic security agencies. The 2001 terrorist attacks highlighted the need to integrate foreign and domestic intelligence to prevent terrorist attacks, but the CTC's ability to handle information collected by domestic security agencies, such as the FBI, is limited by legal prohibitions: Because of abuses uncovered in the 1970s, the U.S. intelligence community operates under carefully defined legal restrictions on its ability to collect or analyze information on U.S. persons. The CTC's FBI personnel—which in 2002 consisted of only 14 employees on rotation (Tenet, 2002)—primarily use the foreign intelligence provided by the CTC to track down terrorists both domestically and overseas, but in return provide little information to the U.S. intelligence community. Because of these limitations, in May 2003, the Terrorist Threat Integration Center (TTIC) was created to supplement the U.S. government's ability to share terrorist threat-related information between its foreign and domestic intelligence and security agencies. In addition, in 2004, the CTC will likely be moved out of CIA headquarters and be collocated with the TTIC and the FBI's Counterterrorism Division to centralize the government's intelligence on terrorists, according to Associate Director of Central Intelligence for Homeland Security Winston Wiley (Wiley, 2003).

CTC'S CONTRIBUTIONS TO BIOTERRORISM DEFENSE

The CTC contributes to bioterrorism defense by providing decision makers involved in bioterrorism defense policy-making and implementation with information that enables them to use their military, economic, and political power with greater effectiveness and precision. According to Mark Kauppi, the program manager of the Intelligence Community's Counterterrorism Training Program, "counterterrorism intelligence analysis aims to improve our understanding of terrorist activities (what they do), their motivation (why they do what they do), and organizational associations (how they are organized to carry out their activities). The goal in terms of intelligence products... is to, at a minimum: improve threat awareness of consumers, facilitate the disruption or destruction of terrorist organizations and their activities, [and] provide timely warning and accurate forecasting" (Kauppi, 2002). The CTC meets this goal by providing intelligence that "assesses the capabilities and intentions of key terrorist groups worldwide: their organization, infrastructure, leadership, support and financial networks, weapons acquisitions, capabilities, and operational intent to attack US facilities and personnel" (Director of Central Intelligence Annual Report for the United States Intelligence Community, 1999).

The CTC's analytical component, the Office of Terrorism Analysis (OTA), "monitors and assesses... emerging trends in terrorism with the parallel objectives of informing policymakers and supporting the intelligence, law enforcement, and military communities," according to the CIA's website (http://www.odci.gov/cia/di/org_chart_section.html, accessed 12/13/03). The website goes on to note that "OTA analysts travel throughout the world to gain firsthand knowledge of issues and situations," enabling them to:

- "Track terrorists and nonstate actors and serve as the primary conduit for domestic customers who need terrorism-related intelligence and analysis in support of the homeland security mission."
- "Analyze worldwide terrorist threat warning information and patterns to provide warnings that can prevent terrorist attacks."
- "Monitor and assess terrorism issues that cross regional boundaries, including worldwide trends and patterns, emerging and nontraditional terrorist groups, changing or new types of terrorist threats or operations methods, and collusion between terrorist groups and facilities."
- "Produce intelligence to identify, disrupt, and prevent international financial transactions that support terrorist networks and operations."

In addition, CTC analysts produce bioterrorism warning intelligence so that military and law enforcement agencies can act to prevent attacks or, if prevention efforts fail, so that the effects of an attack can be mitigated. According to Mark Kauppi, provision of warning intelligence is "the number one job of the counterterrorism analyst" (Kauppi, 2002). He goes on to describe three levels of warning: *tactical warning* that indicates an attack may come within hours or days, *operational warning* that indicates threats within several weeks or months, and *strategic warning* that warns of threats from six months to several years out. The CTC fulfills its warning function by "immediately disseminat[ing] information that warns of an impending terrorist operation to those who can counter the threat," according to the DCI's website (http://www.odci.gov/terrorism/ctc.html, accessed 12/13/03). While "strategic level warning in terms of terrorist trends has generally been quite good," according to Kauppi, policymakers and commanders primarily "want... tactical level warning: Who is going to hit what target, when, where, how, and why? ... Consumers want timely threat warnings that allow terrorist operations to be deterred, preempted, or disrupted" (Kauppi, 2002).

USE OF GENERATED INTELLIGENCE

Military, law enforcement, and homeland security authorities use the CTC's intelligence both offensively and defensively to mitigate the bioterrorism threat:

Intelligence for offensive purposes includes information that will assist operators to "exploit vulnerabilities within terrorist groups, weaken terrorist groups' infrastructures so that they will be unable to carry out plans, work closely with friendly foreign security and intelligence services around the world... and pursue major terrorists overseas and help the FBI render them to justice" (http://www.odci.gov/terrorism/ctc.html, accessed 12/13/03). According to Kauppi (2002), "analysts also provide an important service by carefully constructing detailed analyses of the structure of terrorist organizations that can assist operators and foreign security services in the dismantling of terrorist networks." He observes that neutralizing terrorists before they implement their plans avoids the difficulties of "predict[ing] when, where, and how terrorists will strike."

Intelligence can also have defensive purposes such as force protection, civil defense, or other programs that support homeland security. CTC analysts provide military consumers with intelligence related to the security of military facilities and personnel because "US military personnel overseas have long been prime targets of terrorists. The CTC contributes to protecting US military forces overseas by: determining the modus operandi of terrorist groups that may operate in the vicinity of US military units overseas, ensuring that military components overseas that need the information about terrorist groups receive it quickly and in usable form, and maintaining direct contact with military intelligence units at all the major commands" (http://www.odci.gov/terrorism/ctc.html, accessed 12/13/03). CTC also provides this kind of threat information to domestic civilian leaders so that they can implement defensive programs such as inoculations to mitigate the effects of a bioterrorist attack.

AUTHOR'S NOTE

Portions of this paper were taken from a July 2002 background paper entitled "Homeland Security and the Analysis of Foreign Intelligence" written for the Markle Foundation Task Force on National Security in the Information Age.

This entry was reviewed and approved by the CIA Publication Review Board. The review does not indicate an endorsement of the views or the accuracy of information herein.

REFERENCES

Calabresi, M. and Ratnesar, R., Can We Stop the Next Attack? *Time*, March 3, 2002.

CIA website: DCI Counterterrorist Center, http://www.odci.gov/terrorism/ctc.html, accessed on 12/13/03.

CIA website: Directorate of Intelligence: Organizational Chart, http://www.odci.gov/cia/di/org_chart_section.html, accessed on 12/13/03.

Director of Central Intelligence Annual Report for the United States Intelligence Community, May 1999, http://www.odci.gov/cia/reports/Ann_Rpt_1998/report.html.

Kauppi, M., *Defense Intell. J.*, **11-1**, 39–53 (2002).

Permanent Select Committee on Intelligence, "IC21: Intelligence Community in the 21st Century Staff Study," *House of Representatives. One Hundred Fourth Congress*, Chapter 12: Intelligence Centers, US Government Printing Office, Washington, DC, 1996, http://www.access.gpo.gov/congress/house/intel/ic21/ic21012.html.

Tenet, G., Testimony: Creation of Department of Homeland Security, United States, Cong. Senate Committee on Governmental Affairs, Hearing, June 27, 2002. Accessed transcript via Lexis Nexis.

Tenet, G., Testimony for a special hearing on counterterrorism, US Senate Committee on Appropriations, May 13, 1997, http://www.fas.org/irp/congress/1997_hr/sh105-383.htm.

Wiley, W.P., Testimony Before the Senate Governmental Affairs Committee, February 26, 2003, http://www.odci.gov/cia/public_affairs/speeches/2003/wiley_speech_02262003.html.

See also CENTRAL INTELLIGENCE AGENCY and INTELLIGENCE COLLECTION AND ANALYSIS.

DoD POLICIES ON FORCE HEALTH PROTECTION: MEDICAL DEFENSE AGAINST BIOLOGICAL WARFARE AGENTS (DoD DIRECTIVES, DoD INSTRUCTIONS)

Frank J. Lebeda

Opinions, interpretations, conclusions, and recommendations are those of the author and are not necessarily endorsed by the U.S. Army.

INTRODUCTION

The concept of force health protection (FHP) defines the medical defense approaches to maintain the health of the men and women who serve in the U.S. Armed Forces, to minimize or eliminate casualties, and to provide superior casualty care. Medical defense against biological warfare (BW) agents, whether encountered on a battlefield or as a result of bioterrorism, forms an important part of this strategy. Summarized here are directives, instructions, and other policy statements published by the U.S. Department of Defense (DoD) that apply to the medical aspects of FHP against BW threats (Tables 1 and 2). These statements focus on medical countermeasures, medical surveillance, health risk assessment, and preparedness training.

DOD ISSUANCES FOR FHP AGAINST BW AGENTS

Medical policies in the U.S. military to prevent disease can be traced back to 1777, when General George Washington ordered that his troops be variolated (the method of conferring active immunity that was employed prior to the advent of vaccination) against smallpox. More than 200 years later, this disease is again a focal point of FHP efforts, as policies and instructions were given in 2002 to begin vaccinating military personnel against this potential BW threat. This section will review some of the more recent DoD medical policy statements and some of the federal laws mentioned within these issuances that deal with these health threats.

The DoD defines BW agents as microorganisms or biologically derived poisons (i.e., toxins) that are developed into weapons and intentionally used to cause human disease or fatalities (DoD Directive 6205. 3, 1993). Driven most recently by the health concerns after Operation Desert Storm in 1991, new health care policies and plans are continuously being developed and implemented to enhance the protection of U.S. military personnel against diseases and environmental hazards, including BW agents (Mazzuchi et al., 2002). In this effort, the DoD and other agencies were directed by (Presidential Review Directive, National Science and Technology Council-5 (PRD/NSTC-5), 1998) to review their existing policies and develop plans to help establish new FHP programs. Recommendations were made to alter policies and doctrines concerning threat analyses, medical countermeasures (vaccines and therapeutic drugs), medical surveillance (record keeping, detection, and epidemiology), health risk assessments (occupational and environmental), and preparedness training.

DoD policies can be promulgated as DoD directives (DoDDs) that establish or guide activity of the Army, Navy, Air Force, Marines, National Guard, Reserve components, and other DoD organizational units. These broad policy statements reflect the requirements set forth by legislation, the President, or the Secretary of Defense (SECDEF). Senior military and civilian DoD officials issue memoranda that can also become policy statements. Within these directives, missions are defined and responsibilities are assigned, while policies, programs, and organizations are established or described. Individuals affected by DoD's health care policies, in addition to military personnel, can include essential civilians in the DoD and other federal departments when assigned as part of the U.S. Armed Forces, as well as beneficiaries of the military health system (MHS).

General guidelines for implementing these policies, outlining procedures, and assigning more detailed responsibilities are found in DoD instructions. Detailed technical instructions can come in the form of memos from the senior leadership, such as the 2002 memo from the Assistant Secretary of Defense for Health Affairs (ASD(HA)) that established policy for reinstating the administration of smallpox vaccine. Most of these issuances are currently

Encyclopedia of Bioterrorism Defense,   Edited by Richard F. Pilch and Raymond A. Zilinskas
ISBN 0-471-46717-0   Copyright © 2005 Wiley-Liss

Table 1. DoD Directives and Instructions Related to FHP Against BW Agents[a]

DoDD/DoDI	Number[b]	Title	Publication Date	Web Address[c]
Directive	5136.1	Assistant Secretary of Defense for Health Affairs	December 2, 1992	http://www.dtic.mil/whs/directives/corres/html/51361.htm
Directive	5160.5	Responsibilities for Research, Development, and Acquisition of Chemical Weapons and Chemical and Biological Defense	May 1985	http://www.dtic.mil/whs/directives/corres/html/51605.htm
Directive	6025.3	Clinical Quality Management Program in the Military Health Services	July 20, 1995	
Directive	6200.2	Use of investigational new drugs for force health protection	August 1, 2000	http://www.dtic.mil/whs/directives/corres/html/62002.htm
Directive	6205.2	Immunization requirements	1989	http://usmilitary.about.com/library/milinfo/dodreg/bldodreg6205-2i.htm?terms=6205.2
Directive	6205.3	DoD immunization program for biological weapons defense	1993	http://www.dtic.mil/whs/directives/corres/html/62053.htm
Directive	6490.2	Joint Medical Surveillance	August 30, 1997	http://www.dtic.mil/whs/directives/corres/html/64902.htm
Instruction	1322.24	Military Medical Readiness Skill Training	December 20, 1995	http://www.dtic.mil/whs/directives/corres/html/132224.htm
Instruction	6490.3	Implementation and Application of Joint Medical Surveillance for Deployments	August 7, 1997	http://www.dtic.mil/whs/directives/corres/ins1.html

[a]Tables 1 and 2 were modified from (Mazzuchi et al., 2002).
[b]DoDD and DoDI numbered series of categories relevant to BW defense.
2000—International and Foreign Affairs.
3000—Plans and Operations, Research and Development, Intelligence, and Computer Language.
5000—Acquisition and Administrative Management, Organizational Charters, Security, and Public and Legislative Affairs.
6000—Health.
(from http://www.fas.org/irp/doddir/dod/index.html)
[c]All websites were accessible in March 2003.

available on the worldwide web (see Tables 1 and 2 for more information).

As described in DoDD 5136.1 (1992), the ASD(HA) assists the SECDEF in DoD-related health policy issues and oversees the MHS, including all aspects concerning the delivery of health care services. The fundamental missions of the MHS are to support military readiness and peacetime health care, both of which include FHP (Bailey, 1999). FHP in turn involves three strategic programs: surveillance and casualty prevention programs designed to protect against endemic diseases and environmental hazards (including BW agents), and clinical programs that provide casualty care and management (Joint Publication 4-02, 2001).

Public Law 105-85 established the requirements of maintaining records of all health care services, including vaccinations, in a centralized location; conducting and recording pre- and postdeployment medical examinations; and recording changes in a service member's medical condition during deployment. FHP policy by the Joint Chiefs of Staff also takes into consideration host countries, multinational forces, and civilians who provide essential support (Joint Publication 3–11, 2000).

Policymakers are supplied with feedback regarding FHP plans and their implementations from DoD-generated reports, analyses by other government agencies,

and by nongovernment groups. Implementations of medical policy are summarized in congressionally mandated reports. For example, annual DoD reports on the Chemical and Biological Defense Program, in accordance with P.L. 103–160 Sec. 1523, are presented to Congress describing the overall organization of research and its progress toward protecting the health of military service members.

MEDICAL COUNTERMEASURES AGAINST BW AGENTS

The more recent DoD biodefense health policies began in 1976 with DoDD 5160.5, which assigned responsibility for chemical and biological defense research, development, testing, and evaluation. Revised in 1985 to update procedures for budgeting, programming, and operations, this directive also assigns responsibilities for DoD research, development, and the acquisition of countermeasures against chemical and biological agents. An example of a resulting implementation from DoDD 5160.5 is the Joint Medical Biological Defense Research Program. This program integrates DoD-supported internal and extramural basic research efforts leading to the development of vaccines, therapeutic drugs, and diagnostic tools to protect against and to deter the use of BW agents (Lebeda, 1997; DoD, 2002).

Table 2. DoD-issued Policy Statements Related to FHP Against BW Agents

DoD Issuances & Memoranda	Number	Title	Publication Date	Web Address[a]
ASD(HA) Memorandum	HA Policy 01–017	Updated policy for pre- and post-deployment health assessments and blood samples	October 25, 2001	http://chppm-www.apgea.army.mil/deployment/#DODandArmyPolicies
ASD(HA) Memorandum	None available	Clinical policy for the DoD smallpox vaccination program	November 26, 2002	http://www.smallpox.army.mil/resource/policies.asp
Joint Publication	3-11	Joint Doctrine for Operations in Nuclear, Biological, and Chemical Environments	July 11, 2000	http://www.dtic.mil/doctrine/jpoperations-seriespubs.htm
Joint Publication	4-02	Doctrine for Health Service Support in Joint Operations	July 30, 2001	http://www.dtic.mil/doctrine/jplogistics-seriespubs.htm
Joint Publication	5.00–2	Joint Task Force Planning Guidance and Procedures	April 13, 1999	http://www.dtic.mil/doctrine/jpplanning-seriespubs.htm
CJCS Memorandum	MCM-251-98	Deployment Health Surveillance and Readiness	December 4, 1998	Not currently available
CJCS Memorandum	MCM-0006-02	Updated Procedures for Deployment Health Surveillance and Readiness	February 1, 2002	http://www.dtic.mil/doctrine/other_directives.htm
Joint Instruction	AFJI 48-110; AR40-562; BUMEDINST 6230.15; CG COMDTINST M6230.4E	Immunizations and Chemoprophylaxis	November 1, 1995	http://afpubs.hq.af.mil/pubfiles
Army Regulation	525-13	Antiterrorism Force Protection: Security of Personnel, Information, and Critical Resources from Asymmetric Attacks	September 10, 1998	Not currently available
Army Medical Command Regulation	MEDCOM 525-4	Emergency Preparedness	December 11, 2000	http://chppm-www.apgea.army.mil/smallpox/

[a]All websites were accessible in March 2003.

Vaccination policies for all members of the armed forces, civilian employees of the DoD, and eligible beneficiaries of the MHS are stated in DoDD 6205.2 (1989). Programs are outlined for diseases that can be prevented by vaccination. Specified civilian personnel, for example at risk DoD laboratory workers who may become exposed to pathogenic organisms, may also need to use vaccines other than those that are routinely administered.

The main DoD policies that are specific for vaccinating the armed forces against known or potential BW agents are summarized in DoDD 6205.3 (1993). The policies involve "research, development, testing, acquisition, and stockpiling" of vaccines to be used for biological defense. These vaccines are (or will be) either licensed or classified as Investigational New Drugs (INDs) by the Food and Drug Administration (FDA). DoDD 6200.2 (2000) provides further policy and guidance for the use of INDs for FHP. This document identifies at-risk armed forces personnel and DoD-related civilians who need to be protected from nuclear, biological, and chemical (NBC) threats and endemic diseases.

DoDD 6200.2 also incorporates the mandates of Title 10 USC 1107, Executive Order 13139 (1999), and the FDA interim final rule 21 CFR 50.23. The SECDEF, under 10.1107, must notify service personnel in writing that the specified drug they are to receive is an IND or a drug whose use, under the circumstances, is not approved by the FDA. Notification must also be given for the rationale for administering the drug, possible adverse reactions produced by the drug, and any other information required by the Secretary of the Department of Health and Human Services. Title 10 USC 1107 also allows the president to waive the prior consent requirement for administering INDs to service personnel. The policy in E.O. 13139 outlines procedures for informed consent requirements and waiver provisions. The FDA's interim rule 21 CFR 50.23 permits the waiver of prior consent if personnel are already involved in a military operation in which exposures to NBC threats or other environmental hazards are imminent and likely to be lethal or cause serious illness. Just before Operation Desert Shield, this rule was invoked to allow the use of two INDs, pyridostigmine bromide and botulinum toxoid vaccine, to counter chemical and biological threats, respectively.

The joint instruction published in 1995 for vaccinations and chemoprophylaxis outlines implementation requirements under the Armed Forces Immunizations Program for protecting individuals from endemic diseases and potential BW threats. Also provided are requirements for tracking vaccinations for an individual's health record.

These vaccinations are intended for service personnel and selected civilians (federal employees and MHS-eligible family members).

SURVEILLANCE: RECORD KEEPING, DETECTION, AND EPIDEMIOLOGY

Medical (health) surveillance programs are designed to warn of and to reduce the incidence of illness. A memorandum by the Chairman of the Joint Chiefs of Staff (CJCS) (MCM-251-98, 1998) initially outlined a FHP plan that featured health surveillance as a key component. For joint operations, DoD Instruction (DoDI) 6490.3 (1997), by defining health surveillance requirements, implements policy, outlines procedures, and assigns responsibilities set forth in DoDI 6490.2 (1997). Medical surveillance is mandated for all members of the military services and reserve components before, during, and after military deployments. The main goals are to identify potential threats to health, which may include exposure to BW agents; to collect health data from multiple locations in the deployment area; and to communicate quickly the relevant data for storage, integration, and further assessment. The significant health risks are to be identified by the CJCS in coordination with the ASD(HA). In this Instruction, the Secretary of the Army is responsible for the operation and maintenance by the U.S. Army Center for Health Promotion and Preventive Medicine of a DoD Serum Repository that stores specimens for exclusive use in the diagnosis, prevention, and treatment of health problems associated with such deployments. Some of these FHP procedures are also referred to in joint doctrine (Joint Publication 5.00-2, 1999), while recent policies regarding health assessments before and after deployments are found in a memorandum by the ASD(HA) (HA Policy 01–017, 2001). The Defense Medical Surveillance System is responsible for maintaining these assessment data, which include both physical and psychological evaluations.

Instructions for assessing health readiness and conducting health surveillance in support of joint and unified command deployments were delivered in a memorandum from the CJCS (MCM-251-98, 1998). A more recent memorandum (MCM-0006-02, 2002) instructs that standardized health readiness assessments and health surveillances be conducted. This memo also places responsibility on the combatant command to decide which specific medical countermeasures, such as vaccines, antibiotics, or other drugs, are needed (Embrey, 2002). Epidemiology data are required to be analyzed and archived to report and track adverse reactions and illnesses that may occur after required vaccinations (DoDI 6205.2, 1986). Although each military service currently has its own vaccination-tracking program, the Preventive Health Care Application (PHCA) that is being developed will become the single standard system to help deliver and track clinical preventive services.

OCCUPATIONAL AND ENVIRONMENTAL HEALTH RISK ASSESSMENTS

Relevant to BW agents, guidance for joint medical surveillance on nuclear, biological, or chemical battle or operational environments contaminated by other factors, such as toxic industrial chemicals, was provided in 1997 by DoDD 6490.2 and DoDI 6490.3. Both documents emphasize the need for accurate communication about potential environmental and health risks to service members. "Environmental risk assessments" predict the frequency of disease occurrence, based on environmental exposures, in a population, while predictions from "health hazard assessments" are based on occupational exposures. The CJCS memo (MCM-0006-002; 2002) outlines the process of occupational and environmental health (OEH) risk assessment in such environments. A part of this assessment is an Environmental Baseline Survey (EBS) at the site of deployment, which identifies occupational and environmental health and safety hazards and estimates their severity. Technical guidance in conducting this survey is in Field Manual FM 3–100.4 (2000). As part of the FHP program, the EBS documents these hazards. Commanders are instructed to use these surveillance measures to determine appropriate responses and communicate these risks with their forces. Commanders can then make informed choices and weigh mission requirements during operational planning against occupational and environmental health hazards.

MEDICAL TRAINING: PREPARING FOR A BW EVENT

The DoD has developed programs that provide and monitor military medical skills training in accordance with DoDI 1322.24 (1995). Knowledge of BW threats and medical consequence responses to BW exposure are provided by the U.S. Army Medical Department. Courses have been developed in the joint medical management of biological casualties. Three reference texts that accompany some of these courses are cited in the Further Reading and Web Resources section below. The U.S. Army Medical Command, in response to MEDCOM Regulation 525-4, is preparing a pamphlet on emergency preparedness that provides guidance on responses to emergencies including those involving the use of NBC weapons. Health-care providers within the military must also be certified when assigned to military operations (DoDD 6025.3, 1995). The certification process includes training in BW defense measures.

CONCLUSION

These directives, instructions, and other policy statements are designed to enhance FHP and the preparedness of the U.S. Armed Forces. It is expected that some of these health-related policies developed by the DoD after 1990 for military personnel will have civilian applications in the post-September 11 era. It is also reasonable to expect that some of the implementations resulting from these DoD policies will help in the continuing struggle by the civilian public health sector to protect against current and emerging infectious diseases.

REFERENCES

Bailey, S., *Military Health System Overview Statement*, Submitted to the Personnel Subcommittee, Committee on Armed

Services, US Senate, First Session, 106th Congress, March 11, 1999.

Code of Federal Regulations (21 CFR 50.23), Title 21 Food and Drugs, Part 50 Protection of Human Subjects, Subpart B Informed Consent of Human Subjects, Sec. 50.23 Exception from general requirements, Vol. 1, Revised as of April 1, 2002, pp. 288–290.

Code of Federal Regulations (21 CFR 312.21) Title 21 Food and Drugs, Part 312 Investigational New Drug Application, Sec. 312.21 Phases of an Investigation, Vol. 5, pp. 61–62, Revised as of April 1, 2002.

Department of Defense, Chemical and Biological Defense Program, Annual Report to Congress and Performance Plan, July 2001.

Department of Defense, Chemical and Biological Defense Program, Vol. 1, Annual Report to Congress, April 2002.

Embrey, E., Deputy Assistant Secretary of Defense for Force Health Protection and Readiness, Department of Defense, Statement to the House Committee on Veterans Affairs Subcommittee on Health, February 27, 2002.

Lebeda, F.J., *Mil. Med.*, **162**, 156–161 (1997).

Mazzuchi, J.F., Trump, D.H., Riddle, J., Hyanms, K.C., and Balough, B., *Mil. Med.*, **167**, 179–185 (2002).

Presidential Review Directive, National Science and Technology Council-5, A National Obligation: Improving the Health of Our Military, Veterans, and Their Families, Executive Office of the President, Office of Science and Technology Policy, August 1998.

Public Law 103–160, National Defense Authorization Act for Fiscal Year 1994, Title 50 USC, Sec. 1522, Conduct of chemical and biological defense program, Sec. 1523, Annual report on chemical and biological warfare defense.

Public Law 104–201, National Defense Authorization Act for Fiscal Year 1997, Title XIV: Defense Against Weapons of Mass Destruction (WMD), Subtitle A: Domestic Preparedness, September 23, 1996.

Public Law 105-85, National Defense Authorization Act for Fiscal Year 1998.

Report 106–945, Conference Report to Accompany H.R. 4205, Enactment of Provisions of H.R. 5408, The Floyd D. Spence National Defense Authorization Act For Fiscal Year 2001.

FURTHER READING AND WEB RESOURCES

Eitzen, E., Pavlin, J., Cieslak, T., Christopher, G., and Culpepper, R., Eds., *Medical Management of Biological Casualties Handbook*, 4th ed., US Army Medical Research Institute of Infectious Diseases, Fort Detrick, MD, February 2001, http://www.usamriid.army.mil/education/instruct.html.

Franz, D., *Defense Against Toxin Weapons*, US Army Medical Research and Materiel Command, Fort Detrick, MD, 1997, http://www.usamriid.army.mil/education/toxdefbook.doc.

Headquarters Department of the Army, *Environmental Considerations In Military Operations*, (Field Manual 3–100.4, Marine Corps Reference Publication, MCRP4-11B), Headquarters United States Marine Corps, Washington, DC, June 15, 2000, http://www.adtdl.army.mil/cgi-bin/atdl.dll/fm/3-100.4/toc.htm.

Sidell, F.R., Takafuji, E.T., and Franz, D.R., "Medical Aspects of Chemical and Biological Warfare," in R. Zajtchuk, Ed., *Textbook of Military Medicine*, Office of The Surgeon General, TMM Publications; Borden Institute, Walter Reed Army Medical Center, Washington, DC, 1997, http://www.nbc-med.org/SiteContent/HomePage/WhatsNew/MedAspects/contents.html.

Treatment of Biological Warfare Agent Casualties (Field Manual No. 8–284 NAVMED P-5042, Air Force Manual (Interservice) No. 44–156, Marine Corps MCRP 4–11.1C). Headquarters Departments of The Army, The Navy, and The Air Force and The Commandant, Marine Corps, Washington, DC, July 17, 2000 (incorporates Change 1, 8 July 2002), http://www.adtdl.army.mil/cgi-bin/atdl.dll/fm/8-284/fm8-284.htm.

See also DEPARTMENT OF DEFENSE; DETECTION OF BIOLOGICAL AGENTS; RISK ASSESSMENT IN BIOTERRORISM; and SYNDROMIC SURVEILLANCE.

DUAL-USE EQUIPMENT AND TECHNOLOGY

ERICA MILLER

INTRODUCTION

Biological dual-use equipment and technology can be broadly defined as items and information that have both legitimate, civilian applications and potential illicit or weapons applications. "Equipment" refers to the physical tools and machinery that make research, development, and production in the life sciences possible. "Technology" refers to the processes, procedures, techniques, and information employed in order to properly use the above equipment. Although not covered here, a number of biological materials—bacteria, viruses, and toxins—are also dual-use, having legitimate as well as weapons applications.

BACKGROUND

Because much of the equipment and technology used in the life sciences can be considered dual-use, it is almost impossible to differentiate between that used for legitimate commercial, or academic biological, research and that used for biological weapons research and development. This also complicates the distinction between defensive biological activities, such as the creation of vaccines and detection systems to prevent or identify biological attacks, which are legal under the Biological and Toxin Weapons Convention (BWC), and offensive biological weapons programs, which are prohibited.

In the life sciences, there is no clear definition of exactly what equipment and technology have weapons applications. For the purposes of governance, however, dual-use equipment is typically characterized as meeting a certain level of specification, making it of greater utility in a weapons program. Capacity, speed, flow rate, material composition, and capability/function, among others, are examples of these specifications. Pieces of equipment that can be considered dual-use include centrifugal separators, fermenters, freeze-drying equipment, aerosol inhalation chambers (though these have comparatively limited, legitimate application), certain kinds of containment and protective equipment (e.g., biological safety cabinets, protective suits and hoods), and tangential flow filtration equipment. There are numerous techniques and technology platforms that can be considered dual-use, including bioinformatics, high-throughput screening techniques, toxicity screening, and even gene-splicing techniques. Differentiation between legitimate and illicit uses of dual-use equipment and technology remains difficult because each of these items and technologies has many peaceful uses in the fields of medicine,

pharmaceuticals, fermentation, agriculture, and most recently defense.

The "dual-use" classification is not unique to the biological sciences but is a common feature of nuclear and chemical research and development as well. The commercial and weapons applications of biology, however, share more common characteristics than their chemical or nuclear counterparts, making it considerably more complex to determine the purpose of a biological research and development program. In fact, any competent life scientist with access to the requisite biological materials, equipment, and technology could, if they so desired, successfully engage in work with biological weapons applications on some level.

Relative to nuclear and chemical dual-use items, biological dual-use equipment and technologies are attractive to a would-be proliferator because they are:

- Less expensive
- Easier to conceal in small facilities with few signatures
- Easier to acquire without attracting undue attention because of their multiple civilian uses.

SIGNIFICANCE

The rise to prominence of the biotechnology industry in the early 1970s has promoted the lawful dissemination of biological dual-use equipment and technology around the globe for use in legitimate, commercial ventures. It has also caused a dramatic increase in the number of students receiving advanced degrees in the life sciences. This industry, and those who work in it, has thus far remained centralized in Western nations, but less industrialized states are beginning to develop their own indigenous industries as well. While these trends have many positive implications, they also complicate the tasks of tracking, controlling, and preventing the spread of these items and technologies to potential proliferators, though it must be emphasized that simply because many governments and individuals around the world have access to the equipment and technology necessary to produce biological weapons, that does not necessarily mean that this capability will be exploited for these purposes.

Governance of biological dual-use equipment and technology is a necessity, given the risk of malicious application. Such governance is challenging, however, because these items are essentially ubiquitous and because misuse hinges largely upon intent, a characteristic difficult to both detect and quantify. There are, however, several governance mechanisms currently in use to manage the diffusion and misuse of equipment and technology. *Export control*, for example, by regimes such as the Australia Group (AG), is the primary system currently in place to monitor and manage the spread of dual-use equipment and technology. The AG, an informal, international mechanism composed of the world's most industrialized countries and leading exporters of biological and chemical dual-use technologies, performs this function by controlling the export of certain dual-use biological materials, equipment, and technology to destinations and end users of concern. A second governance mechanism in place is *domestic regulation*, for example the required registration and prohibition of misuse of facilities, pathogen collections, and personnel working with certain agents, equipment, and technologies. Despite the utility of export controls and domestic regulation, these mechanisms are currently confined to states that pose little proliferation risk and are less adept at preventing misuse by nonstate actors. Additionally, there remain a variety of ways in which a potential weaponeer could evade these mechanisms—such as acquisition on the black market or through illicit transshipment—to acquire the equipment and know-how to create a biological weapon.

As a final point, it should be noted that one governance mechanism that has not been widely employed in the biological arena is that of industry and facility inspections. The lack of mandate for this mechanism is due, in part to the absence of verification mechanisms within the BWC and to intellectual property concerns among those in industry. The dual-use dilemma also complicates the success of industry and facility inspections in identifying misuse.

FUTURE OUTLOOK

As the biotechnology revolution continues to progress in an increasingly globalized environment, more countries and nonstate actors will have access to the equipment and information necessary to produce peaceful products and, by nature, biological weapons as well. Changes in biological threat perception may compel more states to engage in research, development, and production of products relating to homeland security and biodefense, which will require an increasing number of individuals and facilities worldwide to undertake work with dangerous pathogens. In addition, the continued progress of the biotechnology revolution is likely to simplify and reduce the costs associated with employing freely available dual-use equipment and technology toward weapons applications.

Because of this forecast, many challenges exist for the future, including successfully applying governance mechanisms to nonstate actors and determining optimal methods for monitoring the worldwide dissemination of dual-use equipment and technology. Above all, however, it is vital to ensure that any management efforts undertaken do not hinder scientific freedom and the positive developments that will continue to emerge from advances in biotechnology.

REFERENCES

Zelicoff, A., "The Dual-use Nature of Biotechnology: Some Examples from Medical Therapeutics," in K. C. Bailey, Ed., *Director's Series on Proliferation 4*, Lawrence Livermore National Laboratory, 1994, pp. 79–84.

FURTHER READING

Office of Technology Assessment (OTA), "Technical Aspects of Biological Weapon Proliferation," *Technologies Underlying Weapons of Mass Destruction*, OTA-BP-ISC-115, U.S. Government Printing Office, Washington, DC, December 1993, pp. 71–117, http://www.wws.princeton.edu/~ota/disk1/1993/9344_n.html.

WEB RESOURCES

Australia Group, *List of Dual-use Biological Equipment for Export Control*, 2000, http://www.australiagroup.net/control_list/bio_equip.htm.

See also BIOSECURITY: PROTECTING HIGH CONSEQUENCE PATHOGENS AND TOXINS AGAINST THEFT AND DIVERSION; BIOTECHNOLOGY AND BIOTERRORISM; and INTERNATIONAL REGULATIONS AND AGREEMENTS PERTAINING TO BIOTERRORISM

DUGWAY PROVING GROUND

MICHAEL GLASS

The value of biological warfare will be a debatable question until it has been clearly proven or disproven by experience. The wide assumption is that any method which appears to offer advantages to a nation at war will be vigorously employed by that nation. There is but one logical course to pursue, namely, to study the possibilities of such warfare from every angle, make every preparation for reducing its effectiveness, and thereby reduce the likelihood of its use.

WBC Committee on Biological Warfare, February 1942 (established by the National Academy of Sciences at the request of Secretary of War Stimson)

HISTORY

Following the entry into World War II by the United States, its government realized a need for increased military capability in many areas, to include expanded knowledge in chemical and biological warfare (CBW). On February 6, 1942, President Franklin D. Roosevelt withdrew an initial 126,720 acres of Utah land from the public domain for use by the War Department. Six days later, Dugway Proving Ground was established. On March 1, the proving ground was officially activated, and weapons testing were under way by the following summer.

Dugway Proving Ground was authorized to fill the need for testing weapons and defenses in relation to chemical and biological weapons. Important projects early on included the testing of chemical weapons and novel dissemination methods. During this time, Dugway also performed pioneering munitions work on mortars and incendiary bombs.

Over the years, the proving ground underwent various name changes and periods of deactivation and reactivation. In 1943, the Army established biological warfare (BW) testing facilities at Dugway. The size of the installation was increased in 1945 when part of the Wendover Bombing Range was transferred to the proving ground. After the war, the proving ground combined with the Deseret Chemical Depot to form a single command called the *Dugway Deseret Command,* later renamed the *Western Chemical Center*. The installation was then placed on a standby status until 1950. In that year, the center resumed active status and acquired an additional 279,000 acres of land for exclusive use to support the mission. Work continued through the 1950s, with new responsibilities added as defense weaponry evolved.

In 1954, Dugway Proving Ground was confirmed as a permanent installation. The Fort Douglas–based Deseret Test Center and Dugway Proving Ground combined in 1968 and became known as the Deseret Test Center. In 1969, Dugway's mission changed by executive order as the U.S. offensive BW program was abolished. In 1973, the present Dugway Proving Ground became part of the U.S. Army Test & Evaluation Command (TECOM), headquartered at Aberdeen Proving Ground, Maryland. Another mission change occurred in 1990, when the United States discontinued its production of binary chemical weapons and hence inactivated its active offensive chemical warfare (CW) program. In 1999, TECOM became the Developmental Test Command aligned under the Army Test & Evaluation Command in Alexandria, Virginia.

Dugway Proving Ground now emcompasses approximately 798,855 acres, and in addition to chemical and biological defensive testing, conducts environmental characterization and remediation technology testing, battlefield smokes and obscurants and munitions testing, and reliability and survivability testing of all types of military equipment in a chemical or biological environment.

MISSION AND CAPABILITIES

The Dugway Proving Ground mission statement reads as follows:

To serve America's soldiers, citizens, and allies by operating our nation's premier defense proving ground to provide quality testing, data and information. To recognize, trust, empower, and develop a work force team dedicated to mission accomplishment. To anticipate and exceed customer expectations using recognized quality standards and processes to advance technology and support all aspects of chemical and biological defense, meteorology, smoke, obscurants, illumination, and munitions testing. To conduct all operations consistent with the highest standards of Army Values and environmental stewardship.

The Dugway Proving Ground has been recognized as the Department of Defense's (DoD) primary chemical and biological defense testing center under the Reliance Program. As a Major Range and Test Facility Base (MRTFB), the proving ground is uniquely capable of accomplishing a mission that not only has grown consistently over the past 60 years but also has transitioned its posture from what was once both offensive and defensive to what is now purely defensive in nature. The proving ground has developed an infrastructure to support the remoteness of the base while continuing to support the critical mission area. The full service community includes permanent housing, visitor accommodations, schools, recreational facilities (swimming pools, golf courses, and a bowling alley) and other amenities, all of which support the mission area, the West Desert Test Center. The West Desert Test Center, a subordinate command to Dugway Proving Ground, contains all of the facilities and staff required for chemical and biological defense testing and training.

There are four primary commodity areas that are utilized for testing and training at Dugway Proving Ground:

- Life Sciences (BW Defense Test and Training)
- Chemical Test (CW Defense Test and Training)
- Meteorology (Weather prediction, modeling and simulation)
- Munitions (Munitions testing and lot acceptance).

In addition to laboratories and chambers, the proving ground contains a number of outdoor test sites and grids. Outdoor testing using simulated biological and chemical agents supports the on-site laboratory work by placing existent technologies and procedures in a realistic operational environment for evaluation. Many of the proving ground test facilities are located in the Ditto Test Area, approximately 12 miles from the installation's main gate. The biological test facility and test grids are situated farther west, in the proving ground west desert area.

FACILITIES

Lothar Salomon Life Sciences Test Facility

The Lothar Salomon Life Sciences Test Facility (LSTF) and Baker area is a laboratory complex that houses the BW defense testing and training capability. Opened in late 1997, this facility contains more than 32,000 square feet of laboratory space certified up to biosafety level 3 (BSL-3). The new LSTF replaced the Baker Test Facility, a 1950s-era series of buildings now utilized for training.

The primary work undertaken by the LSTF group focuses on testing of BW detectors, although recent events have cause an increased focus on decontamination systems to sterilize equipment and facilities following the intentional release of a BW agent. Scientists and technicians perform all testing of biological toxins and pathogens rated BSL-3 and below (i.e., requiring handling in a BSL-3 facility or less) inside sealed containment chambers using appropriate safety hoods. In an effort to minimize the risk to personnel and the environment, many tests use simulants—killed agents or the least virulent strains of pathogens—in order to accurately assess the performance of an actual live threat agent. Because of obvious environmental and health reasons, only simulants are used in outdoor field tests.

In order to achieve the mission of the LSTF, the following equipment is utilized:

- Class II biosafety cabinets, used in baseline liquid agent testing to determine the threshold concentration levels for detection by biological detectors and decontamination technologies.
- Environmental chambers, where tests on biological detectors and decontamination technologies are performed using liquid biological simulants. Testers can control temperature and relative humidity, replicating potential environmental conditions while determining the limitations of the detector.
- An aerosol simulant exposure chamber for biological simulant aerosol generation and analysis. This 13 by 12 by 11.5-foot stainless steel chamber has an air lock door and pass-ports to maintain the environment. Technicians can alter temperature from 23 to 104 degrees F, relative humidity from ambient to 100 percent, and aerosolized simulant concentrations as desired. Detector and decontamination work is completed in this chamber, as well as training of personnel in a simulated BW environment.
- A containment aerosol chamber, an environmentally controlled chamber used in challenging biological samplers and detectors and testing decontamination technologies with aerosolized pathogens.

Because of recent terrorist attacks using biological agents, LSTF has been utilized in support of the federal response. Scientists helped determine the conditions for the safe decontamination of the Senate Hart Office Building. In addition, LSTF scientific expertise was used to understand the nature of the agent in the mailings. Recently, the LSTF's workload has increased so dramatically that the facility can no longer house all of the resources necessary for its completion. Thus, an annex has been approved that will substantially increase all levels of the laboratory in order to support the testing and training workload.

Combined Chemical Test Facility

The Combined Chemical Test Facility (CCTF) is a complex made up of more than 35,000 square feet of laboratory and administrative workspace. The 27 laboratories house state-of-the-art analytical as well as test equipment. In addition, the laboratories are equipped with 52 hoods for work with various forms of chemical agent. This facility supports work for testing and training using CW detection equipment, protection equipment and decontamination technologies.

The CCTF also houses a series of stainless steel chambers in which chemical agent and chemical simulants can be used. These environmentally controlled chambers allow for testing of large-scale decontamination, remote sensing, and individual and collective protection. In addition, the chambers are utilized to verify procedures for work in a chemically or biologically contaminated environment.

HOMELAND DEFENSE AND TRAINING

Over the years, the proving ground has been a site for training of soldiers in a high desert environment. This is a result of the large area that was dedicated for the original outdoor testing mission. In support of this testing, the foundation was laid for an advanced chemical and biological training program in a realistic environment. Environmental documentation, risk analysis, and resident subject matter experts in chemical and biological warfare defense provided the framework from which the program was built.

In 1999, the proving ground was identified within the military to provide CW and BW defense training at an advanced level for Special Mission units within the DoD required to have hands-on experience with agents and simulants to test equipment and tactics (e.g., the National Guard Civil Support Teams). By 2001, Dugway Proving Ground was recognized by the Department of Justice as a site to assist in the training of the 11 million–person emergency first responder community, a relationship that has allowed the development of a course designed primarily for hazardous material technicians, the Advanced Chemical and Biological Incident Response Course. These activities make up the core of the proving ground homeland defense initiatives.

Special Mission Units Testing

One of the proving ground homeland defense initiatives is to provide training and exercise programs for National Guard Weapons of Mass Destruction Civil Support Teams (NG WMD-CST). Assets within the test center coordinated by the Special Programs Division plan, prepare, and initiate challenging full-scale field exercises, complete with detailed laboratory training programs that enable response organizations to validate their tactics, techniques, and procedures during a chemical or biological weapons incident. The National Guard WMD-CST first took advantage of the opportunity to utilize Dugway's training resources in June of 1999. Three of the most challenging programs have been the Field Training Exercises, the Biological Sampling & Detection Course, and the Chemical/Biological Survey Course.

Field Training Exercises. Using controlled chemical and biological simulants of actual agents, the proving ground is able to realistically replicate terrorist incidents for field training exercises. As part of the training, CST units respond to scenarios created by role-players acting as civilian responders, victims, and terrorists. The exercises also include challenges in sampling and detection of live and dilute agents in the confines of the proving ground chemical and biological test laboratories. Throughout the training, units are evaluated by subject matter experts from the West Desert Test Center, who provide instantaneous feedback on the unit's planning, preparation, and execution of the mission's essential tasks. These 5, 10, or 14-day exercises are patterned after the U.S. Army Combat Training Center's "discovery learning" model.

Biological Sampling & Detection Course and the Chemical/Biological Survey Course. These four-day courses of instruction were developed on the basis of WMD-CST feedback and observed field training needs. Each course is designed to concentrate on both the operational and technical aspects of chemical/biological agent detection, sampling, and analysis. Conducted in laboratories and chambers at the LSTF and CCTF complexes, the courses offer students the opportunity to work with live agent under a safe and controlled environment. Both courses give students opportunities to test their skills and build their confidence in the actual detection and sampling of biological and chemical agents.

First Responder Training

The Advanced Chemical and Biological Incident Response Course was designed to offer the emergency first responder community the opportunity to operate specific procedures in a realistic chemically and biologically contaminated environment. Specifically, the course was designed for those responders who would find themselves in the "red zone" of a chemical or biological incident. Therefore, required procedures and equipment are emphasized to hone the skills of survey, detection, sampling, and decontamination. Oversight of this program has now been transitioned to the Department of Homeland Security. Several other courses are currently under development to address other needs within the emergency first responder community and assist in a robust homeland defense capability.

THE FUTURE

Since the terrorist attacks of 2001, the proving ground involvement in homeland defense and homeland security has increased dramatically. With the increased workload come the challenges of working with new entities (primarily the domestic first responder community at the state and local level) and addressing and planning for infrastructure to meet the needs of the country. The DoD chemical and biological community has recognized the situation and has begun to plan for the impact of future programs on the infrastructure of Dugway Proving Ground. In addition, the proving ground is positioning itself to be able to address these new mission areas with the "graduate level" training of special mission units with the DoD, other government agencies as well as state and local emergency first responders. This challenge is being addressed through coordinated efforts at the Office of the Assistant Secretary of Defense for Homeland Defense to maximize DoD assets in order to meet homeland defense requirements and assist in homeland security where needed. Dugway Proving Ground is uniquely situated and capable of supporting these efforts, and therefore plays a critical role in homeland defense.

E

EDGEWOOD CHEMICAL BIOLOGICAL CENTER (FORMERLY EDGEWOOD ARSENAL), ABERDEEN PROVING GROUND

Charlotte Savidge

INTRODUCTION

Located on the Aberdeen Proving Ground in Harford County, Maryland, Edgewood Chemical Biological Center (ECBC) is the United States's primary center for the research and development of chemical and biological defenses for the Army, Navy, Air Force, and Marines (http://www.ecbc.army.mil/about/index.htm, accessed 12/13/03). ECBC's mission is to "protect the warfighter and U.S. interests through the application of science, technology and engineering in chemical and biological defense." Since 1996, the mission has included protecting the U.S. homeland as well (Hinte, 2002).

BACKGROUND

The ECBC grew out of Aberdeen Proving Ground's Edgewood Arsenal, which was established in November 1917 to help the U.S. Army defend against chemical weapons employed by Germany in World War I, as well as to develop and produce chemical weapons for use by the U.S. Army. During the war, Edgewood Arsenal served as a chemical weapons research, development, and testing facility (http://www.apg.army.mil/aberdeen_proving_ground.htm, accessed 12/13/03), and successfully produced the chemical agents phosgene, chloropicrin, and mustard (http://www.apg.army.mil/garrison/safety-environ/restor/history.html, accessed 12/13/03). After World War I, Edgewood Arsenal, under the umbrella of the Chemical Warfare Services (CWS), continued to develop retaliatory and defensive chemical warfare capabilities, including protective clothing. During the Korean and Vietnam wars, the Edgewood Arsenal produced incendiaries, flamethrowers, smoke, chemical mortars, riot control agents, and protective devices. In 1969, after President Richard Nixon's reaffirmation of the United States's "no first use" policy for chemical warfare (CW) and dismantlement of its offensive biological weapons program brought an end to open-air testing of such weapons, Edgewood Arsenal discontinued their production indefinitely.

In 1972, after the United States signed the Biological and Toxic Weapons Convention (BWC), the army consolidated its defensive biological warfare (BW) research at Edgewood. Edgewood's notable achievements in biological weapons defense include standardization in 1957 of the M17, the U.S. Army's first biological agent sampling kit; development in 1960 of DS2 decontaminating agent, effective against all known toxic chemical and most biological agents; research in the 1970s on the Biological Detection and Warning System, for field detection of biological agents; and standardization in 1996 of the M31 Biological Integrated Detection System (BIDS) to detect biological agents on the battlefield (http://www.ecbc.army.mil/about/history.htm, accessed 12/13/03). Beginning in 1996, the ECBC also led the implementation of Congress's Domestic Preparedness Program, training more than 28,000 first responders in 105 communities in weapons of mass destruction (WMD) management (http://www.ecbc.army.mil/ip/brochures/ecbc_brochure.pdf, accessed 12/13/03).

In 2000, ECBC completed the Biological Attack Warning System (BAWS) Advanced Technology Demonstration Program and the M31A1 Biological Integrated Detection System (BIDS) Pre-Planned Product Improvement (P3I) acquisition program. BAWS uses a network of small, lightweight, remote battery/vehicle powered sensors, the positions of which are recorded by the Global Positioning System and linked to a central base station, the size of a briefcase, to provide early warning of a BW attack. BIDS P3I uses sensor technologies and data fusion methodology to help the Army detect and identify twice as many BW agents as its predecessors (http://www.ecbc.army.mil/ip/annual_report/2000_annual_report.pdf, accessed 12/13/03).

CURRENT RESPONSIBILITIES AND EFFORTS

Since September 11, 2001, the ECBC has been involved in chemical and biological defense for both the military and the federal government. In the biological arena, the ECBC's possesses capabilities in general biology, biotechnology, and biosafety. Current research and development efforts focus on integrated biological point detection, fully automated biological identification, and water and food contaminant detection (http://www.ecbc.army.mil/ps/svcs_detect_monitor.htm, accessed 12/13/03). ECBC also maintains a Critical Reagent Repository, which stores and validates all immunological and DNA-based reagents for the Department of Defense. In addition, ECBC continues to offer first responder training to enhance biological, chemical, and nuclear preparedness.

Encyclopedia of Bioterrorism Defense, Edited by Richard F. Pilch and Raymond A. Zilinskas
ISBN 0-471-46717-0 Copyright © 2005 Wiley-Liss

New Developments

In 2002, the ECBC designed, developed, and fielded mobile laboratories that first responders, the military, and federal agencies can use for integrated and standardized field sampling and analysis in the event of biological, chemical, or radiological terrorism (http://www.ecbc.army.mil/ip/annual_report/fy02_annual_rpt.pdf, accessed 12/13/03).

Currently, the ECBC is developing the Joint Biological Standoff Detection System (JBSDS), which will help detect and discriminate biological aerosols from a distance (www.jpeocbd.osd.mil/ca-jbsds.htm, accessed 07/05/04), and the Joint Biological Point Detection System (JBPDS), which uses BIDS and BAWS technologies to detect and identify airborne BW agents at very low levels and to communicate warnings and threat information (http://www.battelle.org/navy/chembio/jbpds.stm, accessed 12/13/03).

Organizational Ties

The ECBC's U.S. customers and partners include the Navy, the Air Force, the Office for Domestic Preparedness, the Program Manager for Chemical Demilitarization, the U.S. Army Engineering and Support Center at Huntsville, the Department of Justice, the Defense Threat Reduction Agency, the Integrated Material Management Center at Redstone Arsenal, the Federal Bureau of Investigation (FBI), the Central Intelligence Agency (CIA), the National Security Agency (NSA), the U.S. Army Soldier and Biological Chemical Command, the National Institute for Occupational Safety and Health, the National Institute of Standards and Technology, the Environmental Protection Agency, the Food and Drug Administration, Letterkenny Army Depot, Johns Hopkins University, and Virginia Polytechnic Institute and State University (Virginia Tech).

At the international level, the ECBC provides training, advice, and planning related to chemical and biological weapons to the United Nations Monitoring and Verification Inspections Center (UNMOVIC) and supports the Office of the Assistant Secretary of Defense for Strategy and Threat Reduction's Cooperative Defense Initiative with Jordan, Egypt, Saudi Arabia, Bahrain, Kuwait, the United Arab Emirates, Qatar, and Oman.

THE FUTURE

Over the next decade, ECBC will continue to develop technologies to aid in the early detection of BW agents. Specific projects include the Joint Service Wide Area Detector, the Wide Spectrum CB Detector, and the Joint Surface Contamination Detector. ECBC also aims to develop a Joint Decon Visualization Detector. To support its chemical defense activities, ECBC is establishing an Advanced Chemistry Laboratory, which is scheduled to open in 2005.

REFERENCES

Hinte, J. Ed., *Celebrating 85 Years of CB Solutions: Edgewood Biological Chemical Center*, p. 4, http://www.edgewood.army.mil/ip/brochures/85_year_brochure.pdf 2002.

FURTHER READING

Aberdeen Proving Ground website, http://www.apg.army.mil/aberdeen_proving_ground.htm, accessed on 12/13/03.

APG's History, APG website, http://www.apg.army.mil/garrison/safety-environ/restor/history.html, accessed on 12/13/03.

Battelle, Navy, Joint Biological Point Detection System, http://www.battelle.org/navy/chembio/jbpds.stm, accessed on 12/13/03.

ECBC Brochure, http://www.ecbc.army.mil/ip/brochures/ecbc_brochure.pdf, accessed on 12/13/03.

ECBC Detection/Monitoring, http://www.ecbc.army.mil/ps/svcs_detect_monitor.htm, accessed on 12/13/03.

ECBC History, http://www.ecbc.army.mil/about/history.htm, accessed on 12/13/03.

ECBC 2000 Annual Report, http://www.ecbc.army.mil/ip/annual_report/2000_annual_report.pdf, accessed on 12/13/03.

ECBC website, http://www.ecbc.army.mil/about/index.htm, accessed on 12/13/03.

Joint Program Executive Office Chemical and Biological Defense website, www.jpeocbd.osd.mil/ca-jbsds.htm, accessed on 7/5/04.

EDUCATION FOR BIODEFENSE

FRED WEHLING

INTRODUCTION

Education, the development of knowledge, skills, and attitudes for broad self-directed application, differs both conceptually and operationally from training, defined as the preparation in detail of facts, skills, and responses to be applied in specific situations. The most successful educational programs take the approach endorsed by the (United Nations, 2002): "Teach students how to think, not what to think." While a great deal of attention has been given to the training of first responders, clinicians, and law enforcement to cope with bioterrorist attacks, education for biodefense remains relatively undeveloped. Nevertheless, significant efforts have been made since 2001 to educate both professionals and the general public on the potential threats from and range of possible responses to bioterrorism.

PROFESSIONAL EDUCATION

The attacks of autumn 2001, which utilized letters containing anthrax bacteria spores, demonstrated the need for the education of physicians and other medical professionals on the diagnosis and treatment of conditions resulting from bioterrorism (Lane and Fauci, 2001). Since that time, the American Medical Association, the Centers for Disease Control and Prevention (CDC), and other organizations have made major efforts to increase awareness of the potential use of pathogens and toxins by terrorists and the need for health care providers to prepare for mass-casualty bioterrorism. Online learning and distance education are being utilized to reach medical audiences and provide health care workers with the background knowledge necessary for effective and creative responses to bioterrorist threats. While these efforts have concentrated on working professionals, medical and public health schools have also begun to develop curricula for bioterrorism and biodefense. In 2002, with CDC support, medical colleges in several states began development of Bioterrorism Medical Education Consortia to develop courses and materials for medical and allied health students, as well as physicians, nurses, laboratory technicians, and other health providers (UNMC, 2003).

At the same time, the establishment of the U.S. Department of National Homeland Security (now the Department of Homeland Security (DHS)) has stimulated the development of homeland security as a professional field. Before the events of September 2001 forced terrorism onto the national consciousness, very few

universities, notably Georgetown University, the Center for Nonproliferation Studies at the Monterey Institute of International Studies and St. Andrew's University in Scotland, offered advanced degrees concentrating on the subject. After 2001, many institutions of higher education were quick to offer seminars and specialized courses on terrorism, and homeland security began to emerge as an academic discipline. The National Academic Consortium for Homeland Security (NACHS), headquartered at Ohio State University, was created to promote research, advanced study, and policy analysis in this field (NACHS, 2003). Louisiana State University, the University of Nevada, Las Vegas (UNLV), the University of New Mexico (UNM), and Texas A&M University, to name only a few, initiated the development of graduate degree programs in homeland security, each with a specialized emphasis. UNLV and UNM, for example, are most likely to focus on analysis of and response to nuclear and radiological incidents, while the Homeland Security Leadership Development (HSLD) program at the Naval Postgraduate School concentrates on executive education for federal, state, and local government officials (HSLD, 2003). It remains to be seen if these new academic departments of homeland security will endeavor to become centers of excellence in biodefense education, or if they will rely on schools of medicine and public health to fill that role.

PUBLIC EDUCATION

The need for providing the general public with accurate information on bioterrorism was quickly recognized as an important means of limiting the shock and fear that most terrorist incidents intend to cause. However, public reaction to the events of 2001 also demonstrated that education faces an uphill struggle to counter the images of bioterrorism promulgated by inaccurate reporting, urban myths (including exaggeration of the dangers of vaccination against anthrax), and media sensationalism. Several nongovernmental organizations, including the Sloan Foundation, the Carnegie Endowment for International Peace, and the Nuclear Threat Initiative (NTI), were quick to respond to the anthrax incidents and hoaxes of 2001 by producing documentary films, K-12 curriculum materials, and online learning programs designed to educate citizens on the real risks of bioterrorism and the steps that could be taken against the threat (NTI, 2003).

Governments have also launched public educational programs on bioterrorism, but the results so far have been

Encyclopedia of Bioterrorism Defense,   Edited by Richard F. Pilch and Raymond A. Zilinskas
ISBN 0-471-46717-0   Copyright © 2005 Wiley-Liss

mixed. The initial efforts of the DHS to educate citizens on mitigating terrorist attacks, while well-intentioned, were criticized as inadequate and potentially confusing, possibly because they were actually efforts at public *training* on how to use duct tape and plastic sheeting to seal homes against contamination rather than public *education* on the nature and range of responses to biological warfare threats. More recent DHS publications and the revised Ready.gov website include basic information on pathogens, vaccinations, and other topics to help citizens become better informed about potential biological warfare threats, as well as better able to respond with self-protective measures.

Because the vast majority of citizens depend on public schools for education, integration of education for biodefense into public school curricula promises to be an effective means of mitigating both the medical and psychological consequences of bioterrorism. While a return to the early 1950s "duck and cover" era of civil defense education in response to the threat of nuclear attack is probably not desirable, age-appropriate materials can be used to teach elementary and middle students about pathogens, vaccines, diagnosis and treatment, first response, and protective measures. Courses in integrated science, biology, and health can provide the most effective modes of delivery at these levels. Secondary schools, community colleges, and universities can continue these topics and add specialized courses on terrorist organizations and motivations, the history of bioterrorism and biological warfare, and biological arms control and disarmament.

CONCLUSION

Overall, while the development of educational programs in bioterrorism defense is still in its early stages, education for biodefense has great potential for reducing the impact of biological attacks, increasing the effectiveness of first response and consequence management, and encouraging students and professionals to pursue careers in homeland security and bioterrorism defense.

REFERENCES

HSLD, Homeland Security Leadership Development program website, http://www.hsld.org/public/home.cfm, accessed on 4/22/03.

Lane, H.C. and Fauci, A.S., *J. Am. Med. Assoc.*, **286**, 2595–2597 (2001).

NACHS, National Academic Consortium for Homeland Security website, http://www.acs.ohio-state.edu/homelandsecurity/NACHS, accessed on 4/10/03.

NTI, Nuclear Threat Initiative online tutorials, prepared by the Center for Nonproliferation Studies, http://www.nti.org/h_learnmore/h_index.html, accessed on 4/22/03.

United Nations, *United Nations Study on Disarmament and Non-Proliferation Education–Report of the Secretary General*, UN General Assembly First Committee, 57th Session, October 9, 2002.

UNMC, *UNMC's Readiness and Response*, University of Nebraska Medical Center website, http://www.unmc.edu/bioterrorism, accessed on 4/10/03.

FURTHER READING

Bartlett, J.G. et al. Eds., *PDR Guide to Biological and Chemical Warfare Response*, Medical Economics, New York, 2002.

Croddy, E., Perez-Amendariz, C., and Hart, J., *Chemical and Biological Warfare: A Comprehensive Survey for the Concerned Citizen*, Copernicus Books, New York, 2001.

Null, G. and Feast, J., *Germs, Biological Warfare, Vaccinations: What You Need to Know*, Seven Stories Press, New York, 2003.

WEB RESOURCES

Center for Nonproliferation Studies (CNS), Educational Resources Guide, http://cnsdl.miis.edu/cnserd/.

Nuclear Threat Initiative (NTI) Teacher's Toolkit, http://www.nti.org/h_learnmore/h5_teachtoolkit.html.

United States Department of Homeland Security citizen readiness web page, http://www.ready.gov.

ENVIRONMENTAL PROTECTION AGENCY

ODELIA FUNKE

INTRODUCTION

The Environmental Protection Agency (EPA) has an important but limited role in bioterrorism defense. A number of federal agencies, including EPA, initiated counterterrorism preparedness planning in response to legislation and several Presidential Decision Directives put in place in the mid-to-late 1990s. Like other federal agencies, however, EPA did not focus on comprehensive planning for terrorist threats until after the destruction of the World Trade Center in September 2001. After the attack, EPA and other agencies greatly accelerated efforts to define their respective responsibilities for monitoring, prevention, and emergency response within the context of their missions, expertise, and legislative authorities. Numerous gaps and jurisdictional issues were identified in the following months, some requiring legislative resolution. While assessing these issues, agencies were also responding to new legislative requirements from Congress, including the creation of a new Department of Homeland Security (DHS), whose broad authorities, responsibilities, and priorities were still evolving.

EPA has clear authority and expertise in several areas important to bioterrorism defense, related primarily to its responsibilities for clean drinking water, emergency response to and cleanup of environmental hazards, and environmental monitoring. Bioterrorism is not a distinct category for EPA programs. EPA has experience with biological materials in connection with water contamination and site remediation, but the majority of EPA programs deal with chemicals rather than biological agents. The Agency treats chemicals, radioactive materials, and biologics as environmental pollutants to be prevented, controlled, monitored, and cleaned up.

In the first 12 to 18 months after the World Trade Center disaster, EPA undertook several steps to further assess its role in preventing and responding to terrorist attacks, including how to fill current gaps. EPA developed a new Homeland Security Center to lead an expanded research and development (R&D) effort; created a new office, the Office of Homeland Security, to coordinate Agency planning; and developed a Homeland Security Plan. EPA undertook program initiatives, particularly for drinking water, to further safeguard human health and environmental resources. The Agency has also participated in an ongoing coordination effort at the federal level, in conjunction with the Office (and then the Department) of Homeland Security, to sort out roles and responsibilities.

Under existing programs, EPA has played an important role in a wide variety of activities to assess, clean up, and decontaminate environmental sites, whether caused by spills or explosions or longer-term pollution. Its national response capabilities include expert teams, equipment, and laboratory support. Because of this expertise, EPA was one of many agencies involved in the emergency response efforts in New York City following the collapse of the World Trade Center towers. In a subsequent assessment of that response effort, EPA noted numerous areas requiring additional planning, coordination, and capacity building. For example, there were difficulties associated with having over a dozen federal, state, and local entities involved, and laboratory capacity was seriously overtaxed from the massive needs of the cleanup, a problem further compounded by demands from multiple entities. EPA learned valuable lessons about the cleanup of deliberately dispersed biological materials when it led the federal effort to decontaminate the Senate Hart Building after *Bacillus anthracis* spores were spread through the mail. In addition to the complex challenges of coordinating Congress, agencies, and local officials regarding this undertaking, EPA faced significant new technical challenges in planning and executing the decontamination.

Assigning responsibility for laboratory capacity was one of the major early issues that emerged for federal government resolution. The Centers for Disease Control and Prevention (CDC), which has experience with biohazards and with highly toxic materials, defines its role as responsibility for human, not environmental, contamination. CDC samples human tissue and fluids and runs the national surveillance system for human illness, but is not involved in environmental sampling. The military also has experience with biological and highly contagious agents but has not readily embraced a wider responsibility for providing national laboratory capacity in the event of a bioterrorist attack. Some have suggested that this should be an EPA responsibility. EPA, however, has traditionally not dealt with highly toxic or contagious biological agents, nor has it been funded to support this kind of work. The Agency's laboratory network and personnel have not been equipped to analyze bioterrorism materials. In 2003, EPA did not have a single laboratory designed or equipped to handle materials requiring the highest level of containment (e.g., the smallpox virus), and had very limited capacity for highly infectious materials such as *B. anthracis*. Issues for laboratory capacity include defining the requirements for analyzing weaponized biological and chemical agents,

Encyclopedia of Bioterrorism Defense,   Edited by Richard F. Pilch and Raymond A. Zilinskas
ISBN 0-471-46717-0   Copyright © 2005 Wiley-Liss

as well as assessing the number and location of resources potentially needed.

RESPONSIBILITIES OF THE EPA

EPA has defined its responsibilities in terms of three general categories: R&D, emergency response and cleanup, and water infrastructure.

Research and Development

EPA's new Center for Homeland Security leads the R&D effort for bioterrorism. It has laid out plans for several priority areas related to bioterrorism defense: infrastructure issues, including how to build facilities to better protect them from attack, developing methods for containing attacks in buildings and water systems, and methods for decontamination and cleanup of contained agents; risk assessment, developing methods to rapidly assess risks to humans in case of attack and to aid appropriate response (e.g., should the water system be shut down? should people evacuate? shelter in place? etc.); and monitoring, developing ways to detect biological agents in the air, on surfaces, and in water. The focus of the Center is research methods, not operations.

Under guidance from the DHS, EPA has some additional technical responsibilities. It coordinates closely with the Departments of Defense and Energy, as well as the CDC. EPA has the lead for a monitoring effort known as Biowatch, an initiative to conduct ambient monitoring for potential biocontaminants in a number of cities. These monitors are checked daily, and CDC laboratories quickly process the results. EPA is also working on an Indoor Air Task Force to better understand the issues and plan for enhanced protection, monitoring, and cleanup.

To plan its research activities, EPA has undertaken a broad-scale inventory of laboratory capacity (expertise, geographical coverage, equipment, etc.). EPA's laboratory capacity includes both government and contract laboratories in its network, but these are oriented toward environmental analysis of pollutants commonly found in the environment, not of weaponized chemicals or biologics. The inventory covers both its internal capacity and capabilities that can be accessed through the contract laboratory network. The completed inventory will include as much private sector information as possible, gathered through a public inquiry of who is interested in conducting the kind of laboratory work needed. EPA wants to include information about the capacity of other federal agencies as well. Planners puzzle over how to provide incentives for private sector laboratories (currently an important part of the network) to take on the additional capabilities, with associated burdens and costs, not knowing whether it will be used. Since laboratory capacity is an essential part of national preparedness, EPA must participate in federal planning for this need. EPA has recognized that however the issue of laboratory capacity is resolved, coordination with the Department of Health and Human Services (HHS), particularly with CDC, is essential. Bioterrorism defense has given greater urgency to data-sharing efforts between EPA and HHS.

Response and Cleanup

EPA has clear authority for cleanup/remediation of hazardous materials, authority that has been outlined in several laws beginning in 1968, prior to EPA's creation. Through a series of legislative actions over several years, EPA gained responsibilities for responding to oil discharges and spills along with hazardous substances releases, and for conducting emergency responses. The laws defining EPA's emergency response responsibilities are the Clean Water Act, Oil Pollution Act, Superfund (also known as CERCLA, the Comprehensive Environmental Response, Compensation, & Liability Act), Superfund Amendments & Reauthorization Act SARA, and Emergency Planning and Community Right-to-Know Act (EPCRA, or SARA Title III). These statutes require EPA to prepare for and respond to environmental releases and threats of release of oil, hazardous substances, pollutants, and contaminants that might present an imminent and substantial danger to public health or welfare, or to the environment.

EPA's National Contingency Plan contains EPA's plan for responding to the release of hazardous substances. EPA has experience (and clear responsibility) for environmental contaminants, not human biological agents used as weapons. EPA's planning after the attack of 2001 has turned greater attention to how EPA might respond to weapons of mass destruction (WMD), including biological as well as chemical and radioactive materials. On the basis of its experience in dealing with apartment assessment and cleanup issues in Manhattan, and with planning and executing the cleanup of anthrax spores at the Hart Building, EPA must also consider how to set standards and methods for decontamination and cleanup: How clean is "safe" regarding hazardous/biological agents in the environment, or in buildings? EPA must identify agents that would be efficacious in cleaning up an unknown variety of potential contaminants. And it must assess how to safeguard the surrounding people, environment, and buildings during decontamination.

A major lesson learned in the World Trade Center cleanup, and a focus of EPA bioterrorism response planning, is the need to build a cross-agency and cross-levels-of-government infrastructure for response. EPA's counterterrorism role has four aspects: coordination with key federal responders is essential; it is important to help state and local responders plan for emergencies, and to coordinate federal efforts with those entities; EPA will train first responders; and EPA will provide resources in case of a terrorist attack. On the basis of the World Trade Center experience, EPA has begun planning a more sophisticated information network, one that will bring together a wide range of geographical, environmental and health data, to feed its operations center and aid in emergency response analysis. Planning associated with EPA's role in bioterrorism monitoring and response again raises the issue of adequate laboratory capacity to meet evaluation needs associated with this new threat. EPA would be called upon to assess and/or clean up environmental contamination in case of attack. Further, EPA expertise would be appropriate for assessing options related to the disposal of contaminated biological materials so as to minimize hazards to human health and the environment.

Water Infrastructure

Heightened fears of terrorist attack have caused great concern about protecting drinking water resources and systems (e.g., source water, treatment plants, and distribution systems). Cyber attacks on control systems are of concern, in addition to physical attacks. EPA has long-term relationships with other government agencies, utilities, and associations to promote safe water, including both wastewater and drinking water. These same partners are key collaborators on security issues regarding how to use the best scientific information and technologies; supporting assessments of vulnerability to attack; taking action to improve security; and ensuring rapid, effective response if an event occurs.

EPA has undertaken significant actions for bioterrorism defense in its drinking water program. Soon after the 2001 attack, Congress provided EPA with approximately $90 million in supplemental funding for drinking water protection and passed the Public Health Security and Bioterrorism Preparedness and Response Act of 2002 ("Bioterrorism Act"), Title IV of which amends the Safe Drinking Water Act. A Water Protection Task Force was formed, creating a forum for identification of issues and coordination among the utilities, states, and EPA. EPA quickly sent a notice to utilities suggesting ways they might secure themselves. Funding has supported the creation of: new mechanisms for secure and rapid information-sharing; research to develop methods, tools, and training; and research to improve knowledge of both threats and possible countermeasures. Research was funded to develop a risk assessment methodology and model emergency response guidelines and operation plans. Research for "hardening" systems and modeling biochemical characteristics is ongoing. EPA initiated a grant program for large community water systems (serving over 100,000 people), and developed a strategy for medium and small systems, giving states money to coordinate activity and oversee development of tools for these systems. EPA also funded a third-party entity to run an information-sharing and analysis center, which allows sharing among systems and between the federal government and systems, creating a complex set of information-sharing mechanisms.

The Bioterrorism Act created three new sections for drinking water oversight. These new requirements address vulnerability assessments for drinking water systems, emergency response plans, and research needs. Vulnerability assessment provisions (as the name implies) require the larger drinking water systems to assess weak aspects of their infrastructure. Systems must submit the assessments to EPA, then develop emergency response plans taking these problems into account. Medium and small systems are encouraged to assess and plan as well. However, many feel that pointing out vulnerabilities can create targets, increasing the likelihood that someone will take advantage of an identified weakness. Because of the fear of information misuse (and opposition to giving EPA this information), Congress created a new information

protection regime for EPA. The law includes strict controls on the EPA's handling of vulnerability assessments. EPA personnel must be designated to handle the assessments, and they face strict criminal penalties for disclosure to nondesignated persons. Also under the new law, EPA and CDC are required to collaborate in developing methods and means for detecting biohazards in the drinking water. This effort will require answering difficult questions about what kinds of agents to look for, how to sample, and whom to notify if contamination is found. These organizations are also working on how to build awareness about these complex issues. CDC's focus is primarily on human health and contagion; EPA's focus is on evaluating the kinds of attacks that might occur, how to conduct environmental sampling, and whom to notify. Beyond the studies and assessments, the Agency wants to see actual improvements to water systems; implementation poses many challenges to water systems, including how to fund it. Federal requirements and funding are focused more on assessment/planning phases than on how to fund and implement improvements.

CONCLUSION

As the EPA seeks to define its role more clearly in coordination with other federal agencies, and under the lead of DHS, the Agency is working to:

- Fulfill its new statutory obligations for building and protecting the drinking water infrastructure,
- Develop methods for conducting rapid risk assessments,
- Complete an inventory/assessment of its laboratory capacity and capabilities,
- Expand laboratory capacity/capabilities, in coordination with other agencies,
- Enhance and expand EPA's information infrastructure to aid in gathering and analyzing information for emergency response situations, and for programmatic analysis and evaluation,
- Expand knowledge and capabilities for prevention,
- Expand capabilities and staffing for decontamination teams.

Some of these efforts require coordination of roles and responsibilities across multiple federal agencies, and perhaps with states and nongovernmental entities as well. Some planning and initiatives began prior to 2001 and were accelerated in the months after the World Trade Center attack; other initiatives, such as a more sophisticated network of national decontamination teams, cannot be undertaken without significant increases in funding. Serious budget constraints are likely to continue for the foreseeable future. The issues posed by homeland security needs are technically, legally, and politically complex. EPA and other agencies will undoubtedly be working to resolve definitional and authority issues, in addition to capacity problems, for years to come.

WEB RESOURCES

Biowatch, between May 30 and June 6, 2003. http://www.epa.gov/region03/ebytes/ebytes02_14_03.html.

Chemical Emergency Preparedness and Prevention, http://yosemite.epa.gov/oswer/ceppoweb.nsf/content/ct-epro.htm.

EPA's Homeland Security Research Center, http://www.epa.gov/ordnhsrc/index.htm.

EPA's Role in Counter-Terrorism Activities Fact Sheet, http://yosemite.epa.gov/oswer/ceppoweb.nsf/vwResourcesBy Filename/ct-fctsh.pdf/$File/ct-fctsh.pdf.

EPA's Strategic Plan for Homeland Security, www.epa.gov/epahome/downloads/epa_homeland_security_strategic_plan.pdf.

Homeland Security and the Indoor Environment, http://www.epa.gov/iaq/ohs.html.

Laws defining EPA's Emergency Response Program, www.epa.gov/oerr/oilspill/lawsregs.htm.

National Strategy for Homeland Security, www.whitehouse.gov/homeland/book/nat_strat_hls.pdf.

Presidential Decision Directives (PDD) 39, 62, 63, http://fas.org/irp/offdocs/direct.htm.

Public Health Security and Bioterrorism Preparedness and Response Act of 2002, www.epa.gov/safewater/security/security_act.pdf.

Water Infrastructure Security, www.epa.gov/safewater/security.

See also DEPARTMENT OF HOMELAND SECURITY; UNITED STATES LEGISLATION AND PRESIDENTIAL DIRECTIVES; and WATER SUPPLY: VULNERABILITY AND ATTACK SPECIFICS

EPIDEMIOLOGY IN BIOTERRORISM (INFECTIOUS DISEASE EPIDEMIOLOGY, OUTBREAK INVESTIGATION, EPIDEMIOLOGICAL RESPONSE)

Tyler Zerwekh

Stephen Waring

INTRODUCTION

The public health response to bioterrorism entails many distinct functions, including surveillance and epidemiology, laboratory diagnosis, medical response, and interagency communication between involved stakeholders. This entry focuses on surveillance and epidemiology, which together aim to decrease morbidity and mortality through the early detection and rapid diagnosis of disease in a community.

SURVEILLANCE

Surveillance is defined by the Centers for Disease Control and Prevention (CDC) as the ongoing systematic collection, analysis, and interpretation of health data essential to the planning, implementation, and evaluation of public health practice, and is closely integrated with the timely dissemination of these data to those who need to know. Surveillance can be characterized as either active or passive, depending on the methodology employed to enumerate cases. *Active* surveillance results from direct contact with the reporting agent or agency, and is used for monitoring endemic levels or recognizing epidemic levels of infectious disease. *Passive* surveillance involves notification and/or reporting of infectious diseases to authoritative figures, such as a local or state health department, generally by phone, mail, or both. Although passive surveillance is reasonably accurate for rare diseases, a lack of judicious reporting may lead to missed cases of some of the more common diseases. However, the word "passive" should be seen as a characterization of a technique and not an indicator of lower importance. It is important to note that active surveillance produces more complete information, but requires extensive time, resources, and money compared to passive surveillance methods. Both passive and active surveillance data are used in the analysis of trends. Once an unusual trend in surveillance data is identified, proper intervention methods can be implemented.

The major goals of surveillance related to bioterrorism are early detection of an event, enhanced disease tracking of that event, and effective intervention to minimize the event. Surveillance activities are the foundation of a successful public health response to bioterrorism, leading to better detection, evaluation, and implementation (CDC, 2001). All CDC-defined critical bioterrorism agents are classified as notifiable diseases, and are required to be reported to state health authorities.

EPIDEMIOLOGY

Epidemiology is defined as the study of the frequency, distribution, and determinants of disease in populations, with the ultimate goal of prevention and control. It is used to study past and future trends in health illness, and can be applied to describe the overall health of a particular community. One component of this is outbreak epidemiology, which is employed to control outbreaks of infectious diseases in the community. Infectious disease outbreak possibilities include spontaneous outbreaks of a known endemic disease, a spontaneous outbreak of an emerging or reemerging disease, a laboratory accident, or an intentional attack with a biological agent (Jernigan et al., 2002).

Epidemiologic investigation of disease outbreaks follows a defined process. First, laboratory and/or clinical evidence confirm that a disease outbreak has occurred. Next, a case definition is created for "confirmed" and "suspected" patients to assist in determining the attack rate. Case definitions are often subclassified into laboratory and clinical definitions, which determine the patient's confirmed or suspected status. Then, the incidence and prevalence of the outbreak is compared to background rates of the same disease to determine if the rate deviates from rates of previous years (Nelson et al., 2000).

However, the annual incidence of diseases caused by biological agents with terrorism potential is very low in the United States. Therefore, the detection of even one case of illness caused by a select agent as defined by the CDC should be considered a sentinel event warranting further investigation. Once the sentinel event has been identified for what it is, it may be characterized in terms of person, place, and time with respect to the potential source of the outbreak. Table 1 shows the cumulative number of cases of disease reported in the United States from 1992 to 1999 that were caused by potential bioterrorism agents (Pavlin, 1999).

The next step in the outbreak investigation process is the environmental investigation and laboratory analysis. These findings establish control measures to be implemented to prevent additional cases during an outbreak.

The last step of investigation involves conducting additional studies on the event and communicating the

Encyclopedia of Bioterrorism Defense,   Edited by Richard F. Pilch and Raymond A. Zilinskas
ISBN 0-471-46717-0   Copyright © 2005 Wiley-Liss

Table 1. Reported Cases of Conditions Caused by Critical Biologic Agents, by Geographic Region of Residence, National Notifiable Disease Surveillance System, United States, 1992–1999

Geographic Region[a]	Botulism		Brucellosis No. (%)	Cholera No. (%)	Plague No. (%)	Tularemia No. (%)
	Foodborne No. (%)	Other No. (%)				
New England	1 (0.5)	1 (0.7)	9 (1.1)	7 (3.1)	0 (0.0)	11 (1.2)
Middle Atlantic	9 (4.0)	4 (2.7)	20 (2.5)	16 (7.2)	0 (0.0)	17 (1.9)
East North Central	4 (1.8)	1 (0.7)	82 (10.1)	12 (5.4)	0 (0.0)	46 (5.2)
West North Central	2 (0.9)	1 (0.7)	37 (4.6)	3 (1.4)	0 (0.0)	296 (33.5)
South Atlantic	14 (6.3)	4 (2.7)	116 (14.3)	14 (6.3)	0 (0.0)	32 (3.6)
East South Central	10 (4.5)	2 (1.4)	20 (2.5)	0 (0.0)	0 (0.0)	24 (2.7)
West South Central	28 (12.6)	0 (0.0)	224 (27.6)	21 (9.4)	1 (1.3)	283 (32.0)
Mountain	21 (9.4)	5 (3.4)	65 (8.0)	28 (12.6)	66 (85.7)	114 (12.9)
Pacific	134 (60.1)	130 (87.8)	240 (29.5)	122 (54.7)	10 (13.0)	62 (7.0)
Total	223 (100.0)	148 (100.0)	813 (100.0)	223 (100.0)	77 (100.0)	885 (100.0)

[a]New England includes Maine, New Hampshire, Vermont, Massachusetts, Rhode Island, and Connecticut.
Middle Atlantic includes New York, New York City, New Jersey, and Pennsylvania.
East North Central includes Ohio, Indiana, Illinois, Michigan, and Wisconsin.
West North Central includes Minnesota, Iowa, Missouri, North Dakota, South Dakota, Nebraska, and Kansas.
South Atlantic includes Delaware, Maryland, District of Columbia, Virginia, West Virginia, North Carolina, South Carolina, Georgia, and Florida.
East South Central includes Kentucky, Tennessee, Alabama, and Mississippi.
West South Central includes Arkansas, Louisiana, Oklahoma, and Texas.
Mountain includes Montana, Idaho, Wyoming, Colorado, New Mexico, Arizona, Utah, and Nevada.
Pacific includes Washington, Oregon, California, Alaska, and Hawaii.
Source: Table courtesy of CDC

Table 2. Critical Bioterrorism Agents and Their Epidemiological Properties

Agent	Biological Agents			
	Incubation	Lethality	Persistence	Dissemination
Bacteria				
Anthrax	1–5 days	3–5 days fatal	Very stable	Aerosol
Cholera	12 h–6 days	Low with treatment High without treatment	Unstable Stable in saltwater	Aerosol Sabotage of water
Plague	1–3 days	1–6 days fatal	Extremely stable	Aerosol
Tularemia	1–10 days	2 weeks moderate	Very stable	Aerosol
Q fever	14–26 days	Weeks?	Stable	Aerosol Sabotage
Viruses				
Smallpox	10–12 days	High	Very stable	Aerosol
Venezuelan equine encephalitis	1–6 days	Low	Unstable	Aerosol Vectors
Ebola	4–6 days	7–16 days fatal	Unstable	Aerosol Direct contact
Biological Toxins				
Botulinum toxins	Hours to days	High without treatment	Stable	Aerosol Sabotage
Staphylococcal enterotoxin B	1–6 days	Low	Stable	Aerosol Sabotage
Ricin	Hours to days	10–12 days fatal	Stable	Aerosol Sabotage
Tricothecene mycotoxins (T2)	2–4 h	Moderate	Extremely stable	Aerosol Sabotage

findings to the public. The "anthrax letters" of 2001 and the subsequent epidemiological response highlighted the importance of this last step, and demonstrated the need for an efficient epidemiological response overall (Chang et al., 2003).

The epidemiologic curve will also play a major factor in the epidemiologic recognition of a bioterrorism event. This curve uses data acquired on cases over a period of time to elucidate patterns of disease occurrence and distinguish between a natural outbreak and an intentional event. The curve can also indicate the time of initial exposure, which allows for measurement of incubation periods (period from time of exposure to onset of disease). Once an incubation period is identified, further hypotheses can be formulated regarding the cause of the disease and its potential for person-to-person

Table 3. Listing of Epidemiological Clues that Could Signal a Biologic Event

1. Large numbers of ill persons with a similar clinical presentation, disease, or syndrome.
2. An increase in unexplained diseases or deaths.
3. Unusual illness in a population.
4. Higher morbidity and mortality in association with a common disease or syndrome or failure of such patients to respond to regular therapy.
5. Single case of disease caused by an uncommon agent, such as smallpox, Machupo hemorrhagic fever, pulmonary anthrax, glanders.
6. Several unusual or unexplained diseases coexisting in the same patient without any other explanation.
7. Disease with an unusual geographic, temporal, or seasonal distribution, that is, influenza in the summer, or Ebola hemorrhagic fever in United States.
8. Similar disease among persons who attended the same public event or gathering.
9. Illness that is unusual or atypical for a given population or age group.
10. Unusual or atypical disease presentation.
11. Unusual, atypical, unidentifiable, or antiquated strain of an agent.
12. Unusual antibiotic resistance pattern.
13. Endemic disease with a sudden, unexplained increase in incidence.
14. Atypical disease transmission through aerosols, food, or water, which suggests deliberate sabotage.
15. Many ill persons who seek treatment at about the same time.

transmission. Table 2 displays the critical bioterrorism agents and a summary of their epidemiological information.

A typical epidemiologic curve for an intentional bioterrorism event will demonstrate a very steep initial spike, with many cases presenting early. This initial spike represents persons exposed at the time of agent's release. The epidemic curve for communicable bioterrorism agents (e.g., the smallpox virus) would then have secondary and tertiary spikes, smaller than that of the initial outbreak, indicating transmission of disease from person to person.

Ultimately, epidemiology will provide important information to assist public health professionals and medical responders in identifying the scope of the attack, the source of exposure, the epidemic progress, and the relative success of intervention methods for any bioterrorism incident (O'Toole, 2001).

ADDITIONAL CLUES

In addition to a clear differential diagnosis and the formation of an epidemiologic curve, there are other epidemiological indicators of a bioterrorism attack having occurred. A complete list of such clues is shown in Table 3. It is possible that none of the listed clues will occur during a given bioterrorism event. However, the presence of one or more indicators on the list should tip off medical personnel and first responders to the possibility of such an event, offering information that may ultimately decrease morbidity and mortality.

THE FUTURE

The future of epidemiology in the field of bioterrorism-preparedness is focused on increasing syndromic surveillance activities, including passive and active surveillance. It is critical for local health departments and related organizations to become more proactive in reporting notifiable diseases to their respective state health department. The quicker data can be gathered regarding outbreaks, the earlier the epidemiologists can identify any unusual change in disease patterns and respond appropriately to those patterns by determining the source of disease and preventing further exposure to it. This requires a collaborative and expeditious effort from the local, state, and federal agencies involved in the response.

Additionally, education and training are at the forefront of strengthening this response. Training and retraining of epidemiologists focusing on epidemiological tools and principles, as well as the clinical aspects of potential bioterrorism agents (including the incubation period, presentation, symptomatology, diagnosis, and treatment of each agent), is essential for the improvement of the public health response to bioterrorism (Jernigan et al., 2002).

CONCLUSION

The epidemiological response plays a pivotal role in public health, both related and unrelated to bioterrorism. Epidemiologists must facilitate rapid determination that an attack has occurred, perform surveillance for additional case identification and tracking, and, finally, prevent spread of the disease through the implementation of effective intervention methods. This component of public health response, when executed correctly and expeditiously, can significantly reduce morbidity and mortality in exposed populations.

REFERENCES

Chang, M., Glynn, M.K., and Groseclose, S.L., *Emerg. Infect. Dis.*, **9**(5), 556–564 (2003).

Jernigan, D.B., Raghunathan, P.L., Bell, B.P. et al, *Emerg. Infect. Dis.*, **8**(10), 1019–1028 (2002).

Nelson, K.E., Williams, C.M., Graham, N.M.H, Master, C.F., *Infectious Disease Epidemiology: Theory and Practice*, Aspen Publishers, MD, 2000.

O'Toole, T., *J. Urban Health*, **78**(2), 396–402 (2001).

Pavlin, J.A., *Emerg. Infect. Dis.*, **5**(4), 528–530 (1999).

United States Department of Health and Human Services, *The Public Health Response to Biological and Chemical Terrorism: Interim Planning Guidance for State Public Health Laboratories*, Published July 2001, Available from CDC website, http://www.bt.cdc.gov/Documents/Planning/PlanningGuidance .PDF.

WEB RESOURCES

Bioterrorism Readiness Plan for Healthcare Facilities and Bioterrorism Resources, www.apic.org/bioterror.

CDC Emergency Preparedness and Response: Surveillance, 2001, http://www.bt.cdc.gov/episurv/index.asp

EQUINE ENCEPHALITIS, VENEZUELEN, AND RELATED ALPHAVIRUSES (SYNONYMS: NU (DESIGNATION IN PRE-1969 U.S. MILITARY PROGRAMS), VEE, VENEZUELAN EQUINE ENCEPHALOMYELITIS, VENEZUELAN EQUINE FEVER)

STEPHEN S. MORSE

INTRODUCTION

A number of RNA viruses are classified in the Alphavirus genus. These viruses can be found throughout the world. Eastern equine encephalomyelitis (EEE), Western equine encephalomyelitis (WEE), and Venezuelan equine encephalitis (VEE) viruses are among the alphaviruses native to the Americas, while Old World alphaviruses include Sindbis, Semliki Forest, O'nyong-nyong, and several others (ICTV, 2000). All alphaviruses are naturally transmitted by mosquitoes. However, under certain circumstances, VEE can also be transmitted by aerosol, a characteristic that in fact led to its development as a candidate biological weapon during the 1950s and 1960s. VEE has caused a large number of natural outbreaks in both horses and humans. EEE, WEE, and VEE cause similar clinical disease, and affect equines (horses, donkeys, burros, and mules), humans, and sometimes other mammalian species. Usual symptoms in humans include fever, myalgia (muscle aches), headache, and malaise; in more severe cases, encephalitis (inflammation of the central nervous system) may also occur. While mortality in equines can be high, VEE disease in humans tends to be highly incapacitating but with relatively low mortality.

This entry focuses on VEE virus as a representative alphavirus, with elucidation of EEE and WEE when necessary.

BACKGROUND

VEE virus was first isolated in 1938. In natural infections, it causes a severely incapacitating disease in humans, with relatively low mortality (generally <1 percent). The virus was successfully developed as a candidate weapon by the United States during the 1950s and 1960s (USAMRIID, 2001), and VEE virus was listed as one of the agents successfully weaponized by the United States as of 1969 (when the offensive program was terminated and stockpiles were destroyed) (USAMRIID, 2001). It is assumed that the former (pre-1969) U.S. program intended to develop VEE virus largely as an incapacitating agent for potential use against hostile troops. It could be delivered by aerosol, either wet or as a dry powder (USAMRIID, 2001). In addition to studies in animals, VEE virus was tested on human volunteers in "Operation Whitecoat" (Project CD-22) (Mole and Mole, 1998; Regis, 1999). It is believed that several other countries' biological warfare (BW) programs also have explored VEE virus (USAMRIID, 2001).

CHARACTERIZATION

VEE virus (actually a viral complex, a group of closely related viruses) is classified in the genus *Alphavirus* of the family Togaviridae (ICTV, 2000). Alphaviruses are small (approximately 70 nm in diameter) spherical enveloped viruses, containing a single strand RNA genome (ICTV, 2000). The term "enveloped" means that the viral particle is surrounded by a lipid envelope, derived from modified host cell membrane, which incorporates the surface proteins of the virus. Other relatives of VEE include Eastern and Western equine encephalomyelitis viruses (EEE and WEE), also found in the Americas. All of these viruses are naturally transmitted by various mosquito species, and can infect both horses and humans through the bite of an infected mosquito. EEE and WEE also infect birds, which are the natural reservoirs of these viruses; the natural reservoirs for VEE appear to be various rodents. With all the equine encephalitis viruses (EEE, VEE, and WEE), disease in horses can be severe, with fever, lethargy, and encephalitis, and a high mortality rate (Acha and Szyfres, 2003). In humans, VEE, EEE, and WEE cause similar clinical signs, although WEE often causes milder disease, and EEE may cause more severe disease, especially in children (Acha and Szyfres, 2003).

STRAINS AND SUBTYPES

Viruses of the VEE complex can be found throughout the Americas from Florida and Texas to Peru. Historically, different subtypes were originally recognized by sensitive serologic tests; gene sequencing has been used more recently (Rico-Hesse, 2000). Six major subtypes (I-VI) are known; subtype I itself has six major identified variants, of which I-A/B and I-C are historically the most significant. Viruses in the complex differ in virulence, and in their likelihood of causing large outbreaks. Some strains (classically subtype I, variants A/B or C), termed "epizootic," have historically caused most of the large outbreaks. Others, termed "enzootic" strains, are at times associated with sporadic human disease, but generally not with severe outbreaks. However, it has been suggested that new epizootic strains may evolve from enzootic strains (Rico-Hesse, 2000; Weaver et al., 1996).

The infection is still widely encountered as a natural disease, especially in horses, in Central and South America (Acha and Szyfres, 2003; Rico-Hesse, 2000). One major natural VEE outbreak, caused by a subtype I-A/B virus, began in 1969, swept through Ecuador to Central America, reached Guatemala and Mexico, and from there spread to Texas in June 1971 before being controlled (Acha and Szyfres, 2003). A more recent outbreak, caused by a subtype I-C virus, began in April 1995 in northwestern Venezuela, and spread to Colombia in September 1995. Over 14,000 suspected human cases were reported in Colombia, with 26 deaths (Acha and Szyfres, 2003), although some accounts have suggested higher numbers of cases (Weaver et al., 1996).

TRANSMISSION AND CLINICAL COURSE

All of the equine encephalitides are naturally transmitted by mosquitoes. However, unlike the other equine encephalitides, VEE can also be transmitted by aerosol under certain circumstances, a fact exploited by former biological weapons programs, including the former U.S. (pre-1969) offensive program (USAMRIID, 2001). Various past studies suggest the average human aerosol infectious dose to be about 10 to 100 pfu (plaque forming units, a measure of infectious viral particles as assayed in cell culture) (USAMRIID, 2001). After an incubation period of 2 to 6 days, VEE infection in humans usually begins as a flu-like illness, with rapid onset. Typical signs and symptoms include fever, chills, and muscle aches; headache and photophobia are common. Nausea, vomiting, and diarrhea are also often reported. The acute disease may typically last 24 to 72 h. During acute infection, there may be enough viruses in the blood to successfully infect mosquitoes, and therefore, precautions to prevent mosquitoes from having access to these patients are advised (especially in tropical areas). After the acute disease, full recovery may take up to two weeks. Encephalitis and serious neurological complications are relatively uncommon in healthy adults infected during natural outbreaks, but have been reported to be higher in children (up to 4 percent in one series), and then are associated with a more prolonged recovery and a higher mortality rate. However, it has been suggested that in a bioterrorism setting, the different route of exposure (inhalation) and potentially higher doses may increase the chances of exposed individuals developing encephalitis and more severe disease (USAMRIID, 2001).

DIAGNOSIS AND DETECTION

Several diagnostic tests, for both medical and environmental testing, are available. Virus isolation (growth on cell culture) is the traditional gold standard, but requires special expertise and takes several days to yield results. Nucleic acid detection, such as PCR (polymerase chain reaction, modified with a reverse transcriptase step to detect the viral RNA), is now becoming standard. Some automated PCR systems can now provide results in one hour or less. Both assays are sensitive; cell culture is theoretically capable of detecting one plaque forming unit and the PCR method approximately 10 to 100 viral copies. Virus isolation and specific nucleic acid tests can be used to identify any of the alphaviruses. Immunoassays (such as the handheld assays originally developed by the U.S. military) are also available for VEE, to detect viral antigen in the environment or in medical specimens, but are considered much less sensitive than viral isolation or PCR. Past infections in people or animals can be identified by serologic tests, such as Enzyme-Linked Immunosorbent Assay (ELISA), to detect the presence of antibodies in the blood.

PROPHYLAXIS AND TREATMENT

In the 1960s, the U.S. military developed a live attenuated VEE vaccine, known as TC-83. It has been used successfully to control equine outbreaks (Acha and Szyfres, 2003). More than 6000 humans have received the vaccine on an investigational basis (Acha and Szyfres, 2003). Most developed detectable antibody titers, though about 20 percent did not (Acha and Szyfres, 2003; USAMRIID, 2001). About a quarter also developed fever, muscle aches, and other symptoms, resembling a mild case of VEE (Acha and Szyfres, 2003; USAMRIID, 2001). C-84, an inactivated version of this vaccine, has also been developed, and can be used to boost the immunity in individuals who have previously received this vaccine (Acha and Szyfres, 2003; USAMRIID, 2001). There are also vaccines for EEE and WEE, based on inactivated viruses. In the last few years, a version of the VEE live vaccine has also been developed commercially ("AlphaVax") as a vaccine platform for other antigens, including HIV, Ebola, Marburg, and Lassa.

DECONTAMINATION AND INFECTION CONTROL

As enveloped viruses, VEE (and the other togaviruses) are relatively easily inactivated in the environment by standard decontamination or disinfection methods, such as sodium hypochlorite, alcohols, and other solvents, detergents, or heat (e.g., 80°C for 30 min).

In the laboratory, BSL-3 containment precautions are recommended for VEE. There have been 150 reported VEE laboratory infections, with one death, up to mid-1999 (U.S. Department of Health and Human Services, 1999).

BIOLOGICAL WEAPONS CONSIDERATIONS

U.S. stockpiles of VEE were destroyed when its offensive BW program was terminated in 1969 (USAMRIID, 2001).

Many other countries, including Russia, have conducted scientific research with VEE. It is likely that other countries have the capability to produce and weaponize this agent, and the status of VEE as a BW or bioterrorism agent is therefore unknown. VEE, together with the other alphaviruses such as EEE and WEE, is listed by the U.S. Centers for Disease Control and Prevention (CDC) as a Category B bioterrorist agent. This category is defined by CDC as follows: "second highest priority agents include those that: are moderately easy to disseminate; result in moderate morbidity rates and low mortality rates; and require specific enhancements of CDC's diagnostic capacity and enhanced disease surveillance."

THE FUTURE

Natural outbreaks with the New World alphaviruses, especially with VEE, occur periodically, and they are likely to continue in the future. They remain a concern for both animal and human health in the Americas. Past outbreaks have generally been successfully controlled by immunization of equines (and, as needed, also humans) in the affected area, controls on transport of equines, and mosquito control. The Pan American Health Organization (PAHO) monitors the disease in the western hemisphere. There is concern that resources for effective disease surveillance and control may be eroding in many places, especially in developing countries (Acha and Szyfres, 2003). The future of VEE as a BW or bioterrorism agent is less easily predicted, being dependent on the interests and capabilities of potential users. Although the alphaviruses (such as VEE) are not highly lethal to humans, they are relatively easy to obtain and can potentially impact both human and agricultural targets. As with most infections, good disease surveillance and control measures (such as immunization) remain the cornerstones of defense.

REFERENCES

Acha, P.N. and Szyfres, B., "Venezuelan Equine Encephalitis," in *Zoonoses and Communicable Diseases Common to Man and Animals*, Vol. 2, 3rd ed., Pan American Health Organization, Washington, DC, 333–345, 2003.

ICTV (International Committee on the Taxonomy of Viruses), in M.H.V. van Regenmortel, C.M. Fauquet, D. Bishop, E. Carsten, M. Estes, S. Lemon, J. Maniloff, M.A. Mayo, D. McGeoch, C. Pringle, and R. Wickner Eds., *Virus Taxonomy: Seventh Report of the International Committee on the Taxonomy of Viruses*, Academic Press, San Diego, CA, 2000.

Mole, R.L. and Mole, D.M., *For God and Country: Operation Whitecoat: 1954-1973*, Teach Services, Brushton, NY, 1998.

Regis, E., *The Biology of Doom*, Henry Holt, New York, 1999.

Rico-Hesse, R., *Vet. Clin. North Am.: Equine Pract.*, **16**(3), 553–563 (2000).

USAMRIID, "Venezuelan Equine Encephalitis," in M. Kortpeter, G. Christopher, T. Cieslak, R. Culpepper, R. Darling, J. Pavlin, J. Rowe, K. McKee Jr., and E. Eitzen Jr., Eds., *Medical Management of Biological Casualties Handbook*, 4th ed., US Army Medical Research Institute of Infectious Diseases, Fort Detrick, MD, 2001.

U.S. Department of Health and Human Services, Centers for Disease Control and Prevention and National Institutes of Health, *Biosafety in Microbiological and Biomedical Laboratories (BMBL)*, 4th ed., US Government Printing Office, Washington, DC, 1999.

Weaver, S.C., Salas, R., Rico-Hesse, R., Ludwig, G.V., Oberste, M.S., Boshell, J., and Tesh, R.B., VEE Study Group, *Lancet*, **348**, 436–440 (1996).

WEB RESOURCES

In addition to the references above, additional information can be found at the following websites:

CDC, www.cdc.gov (background information on VEE and other agents), accessed on 10/25/04.

ICTVdB—The Universal Virus Database, version 3, http://www.ncbi.nlm.nih.gov/ICTVdb (listings of viruses, and their characteristics and classification), accessed on 10/25/04.

ProMED-mail, www.promedmail.org (reports of infectious disease outbreaks worldwide), accessed on 10/25/04.

USAMRIID, www.usamriid.army.mil (the USAMRIID handbook cited under "References" can be downloaded from this site): http://www.usamriid.army.mil/education/bluebook.htm, accessed on 10/25/04.

ETHNIC WEAPONS

Raymond A. Zilinskas

W. Seth Carus

INTRODUCTION

Security experts have discussed the issue of so-called ethnic weapons for some time (Larson, 1970; Hammerschlag, 1974). An ethnic weapon can be defined as a weapon that will cause high morbidity and/or mortality rates amongst a targeted human population, which can be specified on the basis of ethnic or racial characteristics, while causing little or no damage to other populations. In this short section, the feasibility that scientists working for terrorists might successfully research and develop ethnic biological weapons is discussed.

BACKGROUND

It is necessary at the outset to make a note about terminology. The term "ethnic" in this context is probably incorrect. An ethnic group is one that is categorized by common national, cultural, and linguistic characteristics. It is doubtful whether genetic markers exist for these kinds of characteristics. Rather than "ethnic," a more apt term would be "racial." A racial group is one that possesses common traits that are transmissible from generation to generation and sufficient to characterize it as a distinct human population. Some, perhaps most, of these traits are genetically determined and therefore may have genetic markers. Conceivably, members of one race may be targeted for a biological attack based on specific genetic markers they possess. However, since the term "ethnic" weapon has crept into popular parlance, it is used here.

The issue of ethnic weapons reemerged as a subject of intense interest in the late 1990s. This was spurred by the successes of the Human Genome Project (HGP), which by that time had begun to generate data that conceivably could be applied to the development of biological weapons against specific human populations. As the HGP progressed, it generated a working draft of the information contained in the human genome (International Human Genome Sequencing Consortium, 2001; Venter et al., 2001), information that was easily accessible to anyone possessing a computer connected to the Internet. Finally in 2003, two scientific teams working separately published more or less complete maps of the human genome (National Human Genome Research Institute, 2004). One of the implications of this accomplishment is that scientists might now be able to utilize information generated by the HGP to identify genetic markers specific to certain populations, and to perform research for the purpose of developing pathogens or antigens that will preferentially harm individuals possessing these markers.

Also of importance to the consideration of ethnic weapons is the fact that a host of smaller projects are being undertaken in parallel to the HGP, the goals of which are to map the genomes of viruses, bacteria, fungi, insects, and worms. At the time of this writing (January 2004), 169 bacterial genomes have been fully sequenced (and an additional 425 bacterial genomes have been partially sequenced) (Integrated Genomics Inc., 2004), as have the genomes of 1230 viruses (National Center for Biotechnology Information, 2004). It is reasonable to assume that the maps of thousands of pathogen genomes will be published in the next five years.

Findings from pathogen genome research certainly will be useful to scientists involved in projects to improve human and animal health. For example, such research has revealed previously unknown genes that affect immunity. As has been explained: "Whole-genome data provides insight into all the features of microorganisms, including access to virtually every single antigen that may provoke an immune response" (Halim, 2000).

But these findings may be misused as well. From the information generated so far by whole-genome research, it is already possible to identify certain genetic characteristics of microorganisms that cause them to be pathogenic. The possibility, then, is that an ill-willed scientist might link data concerning the susceptibility of certain racial groups to infection with data from pathogen genome research to develop pathogens that preferably will cause disease in a targeted population.

DISCUSSION

Two major problems present formidable barriers to an ethnic weapon being realized. First, it would require a substantial effort by a team of highly qualified scientists to successfully undertake any project to research, develop, and produce an ethnic weapon. For starters, a team assembled for this purpose would require scientists possessing expertise in bacteriology, bioinformatics, biophysics, gene therapy, human genetics, molecular biology, and virology. As the project progressed, there would likely be a need for additional expertise to produce an actual weapon.

Second, scientists currently cannot genetically differentiate one human population (e.g., Australian aborigines) from another (e.g., Maya Indians) because the genetic differences between the two are so small that present-day biotechnology techniques cannot detect them. The tiny difference between any two persons is estimated to rest in only 2 to 10 million nucleotide bases, which is less than 0.4 percent of the approximately 3 billion nucleotide bases that constitute the human genome. Further, current indications are that *intragroup* genetic variability is greater than genetic variability between groups (Krebs and Daniel, 2000).

While these two barriers are substantial, given the rapical rate of biotechnological developments, they might be breached by tomorrow's scientists. The HGP and its offshoot and successor programs can eventually be expected to generate sufficient data to enable investigators to determine genetic differences between human populations. This knowledge will be exceedingly useful to scientists attempting to devise better ways to prevent, diagnose, and treat disease. However, the same data might also be utilized by those who harbor ill will toward a specific population group to attempt design biological or toxin agents that would preferably harm members of that group. An example of how host genetics may influence the outcome of disease can be found in a study demonstrating that African-American men are at a substantially greater risk of contracting disseminating coccidiomycosis (a very serious systemic fungal disease) than Caucasian men (Louie et al., 1999). This finding appears to indicate the existence of genetic differences between races that could be utilized by future weapons scientists for the purpose of designing pathogens able to preferentially attack the racial population of their choice. Of course, until the biochemical pathways are fully mapped, it remains possible that such correlations result from environmental factors or interactions between genetics and the environment.

Note must be made of the qualifier "preferably" in the preceding paragraph, because no ethnic weapon is likely to be truly specific. The reason is that the heterogeneity of racial and ethnic groups is so large as to render almost any conceivable ethnic weapon nonspecific. For example, it might be possible to design a weapon that would preferably affect Maya Indians, but it inevitably would strike down other racial groups as well, albeit at a lower frequency.

CONCLUSION

Support for the hypothesis that genetic differences between populations do exist and are being identified can be found in a report of the British Medical Association, which deals in part with ethnic weapons (British Medical Association, 1999). One of its conclusions is that ethnic weapons are not a practical possibility at this time, but that it would be complacent to assume that ethnic weapons would never be developed in the future. In the final analysis, while the conclusion made by the British Medical Association appears reasonable, the possibility of an ethnic weapon being developed in the next five years is vanishingly small. Most probably, the scientific knowledge and capabilities necessary to even begin the development of a highly specific ethnic weapon will not be available for at least 10 years. Even if done, it would appear that such a weapon would not be sufficiently specific to offer any meaningful advantage over the usual, more easy to develop, nonspecific biological weapons.

REFERENCES

British Medical Association, *Biotechnology, Weapons and Humanity*, Harwood Academic Publishers, New York, 1999.

Halim, N.S., *Scientist*, **14**(8), 10 (2000).

Hammerschlag, R., "Ethnic Weapons," *Chemical Weapons and U.S. Public Policy: A Report of the Chemistry and Public Affairs*, American Chemical Society, Washington, DC, 1974, pp. 19–24.

Integrated Genomics Inc., GOLD: Genomes OnLine Database HomePage, http://wit.integratedgenomics.com/GOLD/, accessed 01/08/04.

International Human Genome Sequencing Consortium, *Nature*, **409**, 860–921 (2001).

Krebs, M.A. and Daniel, D., "The Age of Biology and the Responsible Ancestor," in R.A. Zilinskas and P.J. Balint, Eds., *The Human Genome Project and Minority Communities: Ethical, Social, and Political Dilemmas*, Praeger, Westport, CN, 2000, pp. 1–10.

Larson, C.A., *Milit. Rev.*, **50**, 3–11 (1970).

Louie, L., Susanna, N., Rana, H., Royce, J., Duc, V., Benson, S.W., Ronald, T., William, K., *Emerg. Infect. Dis.*, **5**(5), 672–680 (1999).

National Center for Biotechnology Information, *Viral Genomes*, http://www.ncbi.nlm.nih.gov/genomes/VIRUSES/viruses.html, accessed on 01/08/04.

National Human Genome Research Institute, *International Consortium Completes Human Genome Project*, http://www.genome.gov/11006929, accessed on 01/08/04.

Venter, J.C., Adams, M.D., Myers, E.W., Li, P.W., Mural, R.J., Sutton, G.G., Smith, H.O., Yandell, M., Evans, C.A., Holt, R.A., Gocayne, J.D., Amanatides, P., Ballew, R.M., Huson, D.H., Wortman, J.R., Zhang, Q., Kodira, C.D., Zheng, X.Q.H., Chen, L., Skupski, M., Subramanian, G., Thomas, P.D., Zhang, J.H., Miklos, G.L.G., *Science*, **291**, 1304–1351 (2001).

F

FATALITY MANAGEMENT

Dennis McGowan

INTRODUCTION

In most communities in the United States, the sudden and unexpected death of a previously healthy person is considered a matter for investigation by a coroner or medical examiner (ME). Hospital deaths from diagnosed, identifiable illness are often excluded from the reporting process since the clinical community can classify the manner of death as natural and attribute the cause of death to a specific pathologic process. Release of a biological agent can produce deaths that are initially underappreciated and unreported, and may slip through the system until a disease pattern emerges, suggesting that a deliberate attack has occurred. Unlike a structural collapse, terrorist bombing, or natural disaster, the release of a biological agent makes no noise, attracts no immediate attention, and has the potential to cause extensive injury before being discovered. This delay significantly complicates approaches to mass fatality management, placing an added burden on what is sure to be an overtaxed response capability during such an event.

INFECTION CONTROL

In the event of a disease outbreak, whether natural or deliberate, infection control is essential. The staffs of hospitals and the family members of those ill at home may be at significant risk of infection from contact with patients afflicted by infectious diseases. Whether the disease is transmitted through air or by physical contact with the victim, the risk of infection may persist after the death of the victim, mandating awareness on the part of both health care providers and emergency responders of the hazards of postmortem handling. Workers at morgues and funeral establishments who subsequently handle these remains are subject to exposure to pathogens as well. Most workers in emergency services and the health care community routinely wear surgical or examination gloves when touching a patient, but families are not typically aware of the need for such protection and may leave themselves exposed. However, a full suite of protection consisting of disposable gloves, gowns, masks, and eye protection is not always employed during initial patient contact until suspicions of an infectious disease are raised.

IDENTIFICATION OF AN OUTBREAK

During a disease outbreak, it is possible that local clinical laboratories will not have provided a pathogen-specific diagnosis at the time of the first deaths. Hospital personnel may recognize, however, that the death of young, otherwise healthy patients is unusual, and may report the situation to the local ME's office as a precaution. Regardless, since no one medical facility will likely be treating an overwhelming number of cases during the early stages of a biological event, the index of suspicion among first responders and physicians may remain relatively low. As Emergency Medical Services (EMS) personnel log a growing number of calls regarding cases of severe illness requiring transport to the hospital, however, and as the ME's office receives reports of similar deaths from independent sources, the indicators of a bona fide outbreak begin to accumulate.

EARLY STEPS IN MANAGEMENT

Once the increasing caseload is identified and verified, it will be reported to the local public health authority. For safety, the ME will begin to isolate the remains of the victims, placing them in a first body bag that is in turn fitted inside a second body bag. Both are then decontaminated with household bleach. EMS supervisors will advise all responders to take appropriate precautions to protect themselves against possible infection. The donning of gloves and masks now becomes mandatory. It is early in the response process, but awareness of the seriousness of the situation is growing among first responders.

The public health office will begin notifying all area hospitals of the potentially alarming syndrome, and ask for immediate reporting by every health care facility of patients whose symptoms fit the emerging pattern. While the public health department begins real-time surveillance of the event, its laboratory is assigned the task of screening specimens from hospitals and the ME's office. As a Level A member of the national Laboratory Response Network (LRN), the public health department laboratory conducts rapid "rule out" testing that identifies a potentially dangerous pathogen. If it appears to pose a serious threat, and if the laboratory does not have the capability to identify the agent specifically, specimens are forwarded to a higher level, more sophisticated Level B laboratory in the network, for specific identification. If necessary, samples may be forwarded to one of the two

Encyclopedia of Bioterrorism Defense, Edited by Richard F. Pilch and Raymond A. Zilinskas
ISBN 0-471-46717-0 Copyright © 2005 Wiley-Liss

Level C laboratories currently existing in the United States. As each successive laboratory completes testing at its level of capability, more information is gained as to whether the pathogen is a serious public health threat. This "rapid rule out or forward" protocol is designed to produce a pathogen-specific identification in the shortest possible time.

While the laboratories work to identify the agent, the county health commissioner contacts the state health department to determine the extent of illness statewide. If evidence indicates that surrounding states may also be affected, the commissioner notifies the U.S. Public Health Service and the Centers for Disease Control and Prevention (CDC) in Atlanta. The CDC's laboratory sends for specimens from infected patients and enters the testing process.

In addition to the growing number of disease victims, the ME's office must continue to investigate standard cases, such as homicides, suicides, accidents, and other deaths that fall under the statutory purview of the office. In order to avoid cross-contamination and unnecessary risk to the staff, the suspect cases may be isolated and exams conducted in a small autopsy room apart from the main suite. Personnel use standard personal protective equipment supplemented with powered respirators and protective hoods for improved particle barricading and filtration efficiency.

ADAPTING THE RESPONSE

Once laboratory specimens yield a pathogen-specific diagnosis, a second threshold of awareness is crossed. As required, the Federal Bureau of Investigation (FBI) is notified of the outbreak and its possible cause, and a federal investigation into the source of the outbreak is initiated. Patient and family member interviews are conducted by both medical and law enforcement personnel seeking epidemiological clues to the origin of the outbreak.

As the number of deaths increases, the ME's office may seek prosecutorial concurrence to limit its examinations, for example, by performing autopsies only on young, previously healthy individuals whose deaths can be attributed directly to the disease. These cases serve as a representative selection of decedents, such that their autopsy reports can be used at a later date by the criminal justice system to effect prosecution if the incident is determined to be a deliberate act and perpetrators are apprehended. Meanwhile, hospital isolation needs continue to grow, and limited morgue space approaches full capacity.

Treatment for the newly identified disease, if available, is provided to confirmed cases, and prophylaxis, if it exists, is distributed to first responders, hospital personnel, and others at heightened risk of infection. In the private sector as well, physicians begin administering medications to ill patients and their contacts. Media coverage of the event has by this point likely induced a heightened concern among the public, however, requiring already stressed private and hospital physicians to sift through a perhaps overwhelming number of "worried well" (uninfected persons who believe themselves to been exposed) as they seek out and identify infected or exposed individuals.

If the disease agent is contagious, it is conceivable that at this point public health officials will seek to implement quarantine measures above and beyond those of individual isolation. Since laws permitting public health authorities to invoke orders of quarantine vary widely from state to state, and are virtually untested, it is uncertain how effective they will be in limiting the reach of an outbreak.

If the number of fatalities continues to increase, the capacity of morgues may be exceeded. In order to manage an event were this to occur, the ME may have to detail teams of personnel to remove decedents, perhaps moving them to refrigerated trailers brought to the facility. In the face of having to accommodate a high number of corpses, the ME may focus his or her activities only on verifying the identity of each victim, shifting the bulk of the scientific process from autopsy to quick external examination, and drawing blood samples to confirm the presence of disease. Even if local assets are supplemented by assistance from nearby counties and the state, a point might be reached after which federal assistance is required. In this case, the governor will request large-scale aid, including the dispatch of a federal team of disaster mortuary specialists to assist the ME with the task of identifying and processing bodies.

Depending on the agent causing the outbreak, public health authorities will have to determine whether remains can be released to funeral homes for embalming and conventional burial or be disposed off by mandatory cremation or mass burial. Laws that allow public health authorities to mandate cremation or mass burial, like those permitting involuntary quarantine, vary widely from state to state and are largely untested.

As victim and family interviews continue, a profile develops of the illness that allows epidemiologists to track the causative pathogen to a possible point of origin. Samples are collected at this location, and portable testing equipment is transported to the site. If the disease agent is identified at the suspect site, the third and final threshold, the realization that an attack has occurred, is reached.

CONCLUSION

Over the past two decades, American experience with mass fatalities resulting from natural disasters, transportation accidents, and conventional acts of terrorism has prepared the public safety and health care communities to deal with such events. However, a bioterrorism attack differs from these disasters by virtue of both its covert origin and its potential for spread during the necessary management activities. Adequate preparation is imperative in order to prevent potential complications of fatality management during a bioterrorism event, necessitating that proper training, tools, and support be provided to those who will be asked to carry out this task. If this is done, they will be able to perform their duties with skill, dedication, and dignity.

See also Consequence Management and Laboratory Response to Bioterrorism.

FEDERAL BUREAU OF INVESTIGATION

RANDALL MURCH

INTRODUCTION

The Federal Bureau of Investigation (FBI) is a principal investigative and law enforcement agency of the United States Department of Justice (USDOJ). The FBI is also a member of the U.S. Intelligence Community because of its assigned missions and responsibilities in counterintelligence and counterterrorism, and has become more prominent in this role in recent years in view of attacks on the U.S. homeland and interests abroad, and the rise of terrorism as a threat to U.S. national security.

BACKGROUND

The FBI was founded in 1908 but only became known as the Bureau of Investigation in 1935. In 1924, J. Edgar Hoover was appointed director of the young agency and its small number of investigators, a number that has substantially grown over the near century since its inception. Today, the FBI is staffed with approximately 11,000 Special Agents responsible for investigations and related activities as well as the supervision and management of key operational and operational support functions of the FBI. Its nearly 16,000 support employees provide assistance to the Special Agents and work in a breadth of administrative, management, financial, logistics, language, identification, intelligence and investigative analysis, forensic science, technical, information systems, and legal functions required by the FBI and its customer base. The FBI has over 300 investigative categories ranging from violent crime to organized crime, environmental crime to espionage, civil rights to critical infrastructure protection, and crimes on Native American lands to counterterrorism, and further supports as one of its principal missions the USDOJ and US Attorney's Offices in the prosecution of individuals, organizations, and groups for violations of federal law.

Though headquartered in Washington, DC, the FBI's investigations are principally conducted from its 56 field offices and their 400 resident agencies covering the United States and its territories. These field offices and their resident agencies work closely with other federal agencies as well as state and local police agencies within their assigned geographical jurisdictions. In addition, the FBI has personnel stationed in approximately 45 countries around the globe in order to facilitate investigations that involve foreign countries and to establish cooperation with foreign police agencies and internal security services. The FBI Laboratory was created in 1932 and is today one of the largest and most comprehensive forensic laboratories in the world, with strong ties to the national and international forensic science and crime laboratory communities. The FBI also possesses strong capabilities in technical surveillance (Federal Bureau of Investigation, 2003; Federation of American Scientists, 2003).

PROGRAMS AND RESOURCES

Since 1996, Presidential Decision Directives 39, 62, and 63, and a number of new U.S. laws have assigned lead or key responsibilities of the U.S. counterterrorism and critical infrastructure protection programs to the FBI. This shift in responsibility initially coincided with several major terrorist events in which the United States was a target, a desire to increase security for international sporting events and national political conventions, and a groundswell of concern over the use of infectious diseases and chemical agents as terrorist weapons. (Alibek, 1999; Monterey Institute 2003; Preston 2002) As a result, several new programs and organizational structures were created to enhance counterterrorism in national operational and technical divisions within the FBI, offering substantial resources to support the heightened demands on investigations and specialized response activities. Following the events of September 11, 2001, and the "anthrax attacks" in Florida, Washington, DC, and New York City, the FBI corporately shifted its focus, programs, and resources to make the prevention and investigation of and response to global terrorism its top priority. New investigative resources, an increased number of joint terrorism task forces, and new intelligence analysis capabilities will buttress and extend the FBI's contributions to national efforts against global terrorism as these efforts move forward.

FBI Laboratory

Since 1996, the FBI laboratory has been a leader in the development of a national, interagency forensic capability for the investigation of and response to terrorist or criminal events involving the threatened or actual use of weapons of mass destruction (WMD), particularly biological and chemical agents (Murch, 2001). A number

Encyclopedia of Bioterrorism Defense,   Edited by Richard F. Pilch and Raymond A. Zilinskas
ISBN 0-471-46717-0   Copyright © 2005 Wiley-Liss

Figure 1. Eurblern of the FBI's Hazardous Materials Response Unit.

Figure 2. Training Exercise: Investigation of a makeshift laboratory by FBI personnel.

Figure 3. Personnel decontamination following an FBI HMRU training exercise.

of observations and assessments by the laboratory led to this status, including:

- The FBI laboratory assessed that there was insufficient scientific expertise on and understanding of the forensic investigation of biological and chemical terrorism focusing on (1) the attribution of such acts and (2) the steps leading up to them.

- The laboratory further observed that there was no nexus in the U.S. government to integrate forensic science and crime scene investigation with government experts in closely related disciplines and organizations.

- The laboratory anticipated (correctly as it turns out) that the investigation and attribution of biological terrorism events would rely heavily on forensic evidence because of the ease of unobtrusively or clandestinely acquiring, producing, and releasing biological weapons (and chemical weapons as well).

- The laboratory recognized that its established outreach with the national and international crime laboratory communities, U.S. national laboratories, and academic institutions could be redirected to aid in establishing a new national program and associated opportunities.

As a result of these findings, the FBI laboratory reasoned that it should leverage and extend its considerable array of capabilities in forensic science and apply it aggressively toward WMD terrorism.

Hazardous Materials Response Unit

Just before the 1996 Olympics in Atlanta, Georgia, the Hazardous Materials Response Unit (HMRU) Fig. 1 was created and resourced as a new program in the FBI laboratory. Concomitantly, the FBI initiated its leadership position in special events management. The new HMRU quickly set about to organize, for the first time, available U.S. government technical WMD response and public health resources into a cohesive WMD response community for the Olympics as a facet of the overall planning effort. As part of this community building, HMRU joined together with the Centers for Disease Control and Prevention (CDC), the Naval Medical Research Institute (NMRI), and the U.S. army's Engineering Research, Development and Education Command (ERDEC, now Soldier Biological and Chemical Command, SBCCOM) to successfully establish the first combined field laboratory for WMD biological and chemical analysis on the grounds of the CDC. At that time, HMRU also established a firm partnership with the U.S. Army Medical Research Institute for Infectious Diseases (USAMRIID), which continues to this day, and other Department of Defense (DoD) components as well. The Olympics further provided HMRU and its partners the opportunity to begin to develop business models for effective joint response to and investigation of WMD events, with particular emphasis on those involving biological and chemical weapons.

From 1996 to present day, the HMRU has expanded its community building to include the U.S. Department of Agriculture (agricultural bioterrorism), the U.S. Intelligence Community, the Department of Energy National Laboratories, the National Institutes of Health (NIH), the Federal Emergency Management Administration (FEMA), the Environmental Protection Agency (EPA), the intergovernmental counterterrorism technology consortium known as the Technical Support Working Group (TSWG), the National Academy of Sciences (NAS), and such professional organizations as the Association of Public Health Laboratories. Since 1996, HMRU and its federal partners have responded to dozens of suspected and actual bioterrorism, criminal, and serious environmental crime events.

At its inception, HMRU was comprised of three part-time forensic experts from the FBI laboratory and a summer intern, and used equipment borrowed from the DoD. Today, HMRU is based at Quantico, Virginia, and is staffed by more than 50 Agent and Support personnel possessing investigative, scientific, forensic, and hazardous materials response and training backgrounds Fig. 2. The continued growth of this unit is expected. HMRU has extensive capabilities to deploy quickly to the field, with its own transportation, communications, field laboratories and analytical instrumentation, hazardous materials safety and decontamination capability, and crime scene search equipment. HMRU also works and trains closely with the FBI's Bomb Data Center (its mission: training for and rendering safe improvised explosive devices) to be able to successfully deal with improvised dispersal devices. In addition, HMRU also works and trains with other specialized FBI components, and has extensive connections to the national public safety community as well Fig. 3.

Over the past five years, HMRU has equipped and trained Evidence Response Teams (ERTs, crime scene teams) in FBI field offices for very rapid assessment of and response to possible biological and chemical events. After September 11 and the sending of the anthrax letters, HMRU evolved this program into fully trained and equipped Hazardous Materials Response Teams (HMRTs) that now exist separately from the ERTs. By the end of September 2004, 27 FBI field offices will have HMRTs. Parallel to HMRU's formation of the HMRTs, the FBI's Counterterrorism Division, located at FBI Headquarters, has established and trained weapons of mass destruction coordinators in all 56 field offices, who maintain close contact with law enforcement, fire and hazmat, public health, and environmental agencies within their field office's jurisdictions.

HMRU has extensive and ongoing relationships with other pertinent federal, state, local, and professional organizations. The FBI laboratory, through HMRU and the FBI Washington Field Office, has drawn heavily on some of the nation's leading scientists in key fields to help with, consult on, and review the science applied to the investigation of the anthrax attacks of October 2001.

RECENT DEVELOPMENTS

In 2002, the FBI laboratory created a second unit, the Chem-Bio Sciences Unit (CBSU), to lead research, development, test and evaluation, and validation programs for the innovation and discovery of new methods and technologies for the forensic investigation of biological and chemical terrorism. A former chief of the HMRU currently heads CBSU.

Also, in 2002, the FBI laboratory moved most of its work force and facilities from FBI Headquarters in Washington, DC, to a new facility on the ground of the FBI Academy, Quantico, Virginia. This new facility provides for expanded laboratory, office, training, logistics, and staging spaces for HMRU and CBSU. It also provides for better integration with forensic units and experts of the FBI laboratory that have been designated to assist with WMD incidents.

In addition, in 2002, the FBI led the creation of the Scientific Working Group for Microbial Genomics and Forensics (SWGMGF). The FBI laboratory's Senior Scientist for Biology currently chairs this group. The SWGMGF formalizes a "community of interest" from a long-standing informal network of research and applied scientists from both government stakeholder and academic programs who have experience in closely related fields and an interest in advancing and validating the science of microbial forensics. The national science policy community is also represented on SWGMGF. Members meet or communicate regularly to share information and to coordinate, assess, and develop protocols, policies, standards, programs, and strategies for the new scientific discipline of microbial forensics (Budowle et al., 2003).

The SWGMGF, as well as HMRU and CBSU, are closely linked with a new initiative of the U.S. Department of Homeland Security that is being called the National Biodefense Analysis and Countermeasures Center (NBACC). This important, multidisciplinary program seeks to build, staff, and equip a national biodefense center at Ft. Detrick, Maryland, (near USAMRIID), which includes a Bioforensics Center. The FBI has been a key stakeholder and active participant in NBACC from its inception.

CONCLUSION

Microbial forensics itself has its origins in U.S. government biological weapons counterproliferation programs, but has been brought to the forefront in recent years by circumstances and events, as well as with the creation and evolution of the HMRU, CBSU, and now SWGMGF. The accurate and valid forensic identification and characterization of biological weapons evidence is an important capability for U.S. leadership and policymakers for source attribution, association, or exclusion purposes, not unlike the use of forensic science in the criminal justice system (Murch, 2003). As it is with other U.S. government agencies, the prevention, early detection and warning, interdiction, disruption, defeat, mitigation, investigation and attribution of biological and all forms of WMD terrorism, and its associated participants, materials, transactions, and acts, are top priorities for the FBI.

REFERENCES

Alibek, K. and Handelman, S., *Biohazard*, Random House, New York, 1999.

Budowle, B., Schutzer, S.E., Einseln, A., Kelley, L.C., Walsh, A.C., Smith, J.A.L., Marrone, B.L., Robertson, J., and Campos, J., *Science*, **301**, 1852–1853 (2003).

Federal Bureau of Investigation, 2003, Internet Home Page, www.fbi.gov.

Federation of American Scientists, 2003, www.fas.org/irp/agency/doj/fbi/fbi_hist.

Monterey Institute of International Studies, 2003, Center for Nonproliferation Studies, http://cns.miis.edu/research/cbw.

Murch, R.S., 'Forensic Perspective on Bioterrorism and the Proliferation of Bioweapons," in S.P. Layne, T.J. Beugelsdijk, C. Kumar, and N. Patel, Eds., *Firepower In the Lab*, Joseph Henry Press, Washington, DC, 2001, pp. 203–213.

Murch, R.S., *Biosecur. Bioterror.*, **1**(2), 117–122 (2003).

Preston, R., *The Demon in the Freezer*, Random House, New York, 2002.

See also DEPARTMENT OF JUSTICE and INTELLIGENCE COLLECTION AND ANALYSIS.

FOOD AND BEVERAGE SABOTAGE

Jeremy Sobel

INTRODUCTION

The sabotage of food or beverages by contamination, with the intention of assassinating individuals, incapacitating armies, or demoralizing populations, has been practiced since antiquity. Today, the vast and complex food supply systems of nations continue to be vulnerable to deliberate contamination (Sobel et al., 2002a; Khan et al., 2001). Contamination of food or beverages with biological or chemical agents may serve the objectives of terrorists who seek to create panic, threaten civil order, or cause economic losses. Sabotage of crops or livestock may result in similar consequences.

Over 76 million cases of foodborne illnesses are estimated to occur in the United States yearly (Mead et al., 1999). Over 1000 outbreaks of foodborne disease are reported to the Centers for Disease Control and Prevention (CDC) each year, and these represent but a fraction of actual events. The public health system and food safety regulatory apparatus have evolved over the past century to address such outbreaks; specific epidemiological, laboratory, and legal approaches have been developed to detect, investigate, and control these events (Sobel et al., 2002b). The same personnel that handle naturally occurring foodborne disease in the course of their routine duties would almost certainly be the first to respond to an act of bioterrorism involving food (Sobel et al., 2002a).

This entry focuses on the public health and human illness aspects of sabotage of foods and beverages, not including water (for a discussion of this, please see the entry entitled "Water Supply: Vulnerability and Attack Specifics"). The reader is referred to other sources on the topic of food security, which entails baseline protection of the food supply from deliberate contamination (WHO, 2002; Lee et al., 2003).

VULNERABILITY OF THE FOOD SUPPLY

International and governmental authorities have long recognized the threat of terrorism to the food supply (WHO, 1970, 2002; US General Accounting Office, 1999; Food and Drug Administration, 2003), and indeed biological contamination of foods by terrorists and criminals has occurred in the United States in recent decades (Török et al., 1997; Kolavic et al., 1997; Phills et al., 1972). The modern food supply comprises thousands of classes of foods, both domestically produced and imported. Ever-more centralized production and processing and wide distribution of products has resulted in unintentional foodborne disease outbreaks that increasingly occur over large, dispersed geographic areas, a situation that may delay recognition of an outbreak and complicate identification of the contaminated food (Hedberg et al., 1994; Sobel et al., 2001). Deliberate contamination of foods could produce a similar situation.

The potential consequences of an attack on the food supply can be inferred from examples of unintentional foodborne disease outbreaks. In 1985, over 170,000 persons were infected with *Salmonella typhimurium* resistant to nine antimicrobial agents from contaminated pasteurized milk from a dairy plant in Illinois (Ryan et al., 1987). In 1994, about 224,000 persons in the United States were infected with *Salmonella enteritidis* from contaminated ice cream (Hennesy et al., 1996). And in 1996, over 7000 children in Sakai City, Japan, were infected with *Escherichia coli* O157:H7 from contaminated radish sprouts served in school lunches, an outbreak that resulted in broad-reaching psychological trauma, including suicide (Mermin and Griffin, 1999). However, as the mailings of envelopes containing *Bacillus anthracis* spores in the United States have demonstrated, even limited dissemination of biological agents using simple means and causing relatively few illnesses can produce considerable public anxiety and pose a significant challenge to the public health system (CDC, 2001), and even contamination with no cases of illness can result in severe economic loss (Grigg and Modeland, 1989).

Full-proof protection of the food supply is impossible. Prevention falls under the rubric of food security and entails physical protection of the food supply along the "farm-to-table" continuum, including all stages of production, processing, transport, storage, and retail (WHO, 2002; Lee et al., 2003). This challenge rests

Encyclopedia of Bioterrorism Defense,   Edited by Richard F. Pilch and Raymond A. Zilinskas
ISBN 0-471-46717-0 Copyright © 2005 Wiley-Liss

principally with food safety regulatory agencies, industry, and law enforcement. Approaches include identifying high-risk foods and critical control points at which contamination could be carried out in the complex web of production and commerce, and then executing appropriate control measures. Should an attack occur, preparedness entails maximizing capacity to detect and investigate the consequent outbreak with the objective of identifying the contaminated food and removing it from circulation, advising the public, and apprehending the perpetrators (Sobel et al., 2002a; Khan et al., 2001).

POTENTIAL THREAT AGENTS

The list of pathogens, toxins, and chemicals that could cause disease by ingestion is extensive. Further, laboratory-based diagnosis and surveillance systems are geared toward the identification of agents that cause disease in natural settings. Given these points, attention must be focused on those pathogens and toxins considered to be of most concern from a public health and national security standpoint.

The CDC's strategic plan for Bioterrorism Preparedness and Response includes a list of critical biological agents for public health preparedness (CDC, 2000). The highest priority category of agents, Category A, includes one naturally occurring foodborne toxin, *Clostridium botulinum* neurotoxin, which produces a flaccid paralysis that can result in death from respiratory arrest if untreated (Shapiro et al., 1998; CDC, 2000; Hatheway, 1990), and *B. anthracis*, which uncommonly produces a high-mortality gastrointestinal illness in the developing world (Sirisanthana and Brown, 2002). The full potential of damage for malicious contamination of food with *B. anthracis* (Erickson and Kornacki, 2003), as well as other Category A agents *Yersinia pestis* (Butler et al., 1982) and *Francisella tularensis* (Reintjes et al., 2002; Tarnvik and Berglund, 2003), is not known.

The category of second-most critical biological agents for public health preparedness, Category B, consists of organisms and toxins that are moderately easy to disseminate, cause moderate morbidity and low mortality, and require specific enhancement of diagnostic and surveillance capacities. This category includes several foodborne pathogens, many of which are listed in Table 1. With proper therapy, these organisms are rarely lethal. Beyond this list are a variety of foodborne pathogens that could potentially be used, including viral and parasitic agents such as hepatitis A virus and *Cryptosporidium parvum*.

Naturally occurring biological toxins and synthetic chemicals can produce illness by ingestion, including aflatoxins and trichothecene mycotoxins, saxitoxin, tetrodotoxin (Lee et al., 2003), and the Category B agents ricin and staphylococcal enterotoxin B (Franz and Jaax, 1997). Assorted chemical agents, many of which are available in the form of pesticides, cleaning compounds, or industrial solvents, could be used to contaminate foods or beverages as well. The CDC list includes blood agents such as cyanide; heavy metals including arsenic, lead, and mercury; and corrosive industrial chemicals (CDC, 2000).

DETECTION OF AN ATTACK

Unless announced by the perpetrator, an attack will most likely be recognized by epidemiologic investigation of an outbreak. The potential for hoaxes is well recognized, and outlandish claims might accompany a small-scale contamination. As with any foodborne outbreak, early recognition and investigation is vital if the food vehicle has wide distribution, and prevention of additional cases may depend on identifying and recalling the yet-unconsumed food product. Additionally, prompt suspicion of the terrorist nature of the event will help direct the criminal investigation and bring into play the full array of federal resources available to counter bioterrorist attacks (U.S. General Accounting Office, 1999). In a suspected bioterrorism event, the Federal Bureau of Investigation (FBI) will assume overall leadership of the response (Institute of Medicine, 1999).

Outbreaks may be reported by astute clinicians cognizant of a cluster of patients with similar symptoms. Reporting such clusters immediately to public health authorities remains the fastest mode of detection, and training clinicians to rapidly report suspicious syndromes and disease clusters is a cornerstone of preparedness for both biological terrorism and natural epidemics. Where cases are geographically dispersed, laboratory-based surveillance systems may detect increases in illnesses. For foodborne diseases, the Public Health Laboratory Information System (PHLIS) electronically collects data on foodborne enteric pathogens, many of them on CDC's biological agents list (Bean et al., 1992). Computerized algorithms such as the *Salmonella* Outbreak Detection Algorithm (SODA) analyze disease trends for increases in the incidence of specific serotypes compared to historical baselines (Hutwagner et al., 1997; Mahon et al., 1997). The national molecular subtyping network PulseNet performs pulsed-field gel electrophoresis "fingerprinting" on isolates of select foodborne bacterial pathogens from patients, foods, and farm animals, and has detected many common-source outbreaks that occurred over widespread geographic areas without the focal increase in case counts required by less-sensitive systems (Stephenson, 1997; Swaminathan et al., 2001; Sivapalasingam et al., 2000). In recent years, syndromic surveillance systems have been developed in several metropolitan areas, which electronically monitor in near-real-time the rates of specific syndromes characterized, for example, by diarrhea, flu-like illnesses, pneumonia, or neurological symptoms based on input from emergency medical services calls, emergency room admission or discharge diagnoses, and other health care data (Greenko et al., 2003; Pavlin, 2003). A unique surveillance system exists for botulism. A clinician suspecting a case must contact the state public health department in order to obtain the specific therapy, botulinum antitoxin, which is available in the United States only from CDC (Shapiro et al., 1998; CDC, 1998).

Table 1. Some Potential Foodborne Biological Terrorist Agents and Select Characteristics

Agent	Availability	Minimum Infectious Dose, Secondary Transmission	Clinical Syndrome	Case-fatality	Other Characteristics of Microbe or Illness
Botulinum toxin	Organism ubiquitous in environment; cultures require anaerobic conditions	$Ld_{50} = 0.001$ µg/kg (Hatheway, 1990)	Descending paralysis, respiratory compromise	5% (treated) (Shapiro et al., 1998)	95% of patients require hospitalization 60% of patients require intubation
Salmonella serotypes (Excluding *S. typhi*)	Clinical and research labs, culture collections, poultry, environmental sources	10^3 organisms (Blaser and Newman, 1982) limited 2° transmission	Acute diarrheal illness, 1–3% chronic sequelae	>1% (Benenson, 1995)	Organism hardy, prolonged survival in the environment
Salmonella typhi	Clinical and research labs	10^5 organisms (Blaser and Newman, 1982) Secondary transmission possible	Acute febrile illness, protracted recovery, 10% relapse, 1% intestinal rupture (WHO 1970).	10% untreated, 1% treated (Benenson, 1995)	Clinical syndrome unfamiliar in United States Long incubation period (1–3 weeks) Produces asymptomatic carrier rate in 3% of cases
Shigella spp.	Clinical and research labs	10^2 organisms (Dupont et al., 1989) Secondary transmission possible	Acute diarrhea, often bloody	For most common species in United States, <1% (Benenson, 1995)	
Shigella dysenteriae Type 1	Clinical and research labs	10–100 organisms (Dupont et al., 1989) Secondary transmission possible	Dysentery, seizures	Up to 20% (treated) (Benenson, 1995)	Causes dysentery, toxic megacolon, hemolytic-uremic syndrome, convulsions in children
E. coli O157:H7	Clinical and research labs, bovine sources, farms	>50 organisms (Mead and Griffin, 1998; Tilden et al., 1996) Secondary transmission possible	Acute bloody diarrhea, 5% HUS (hemolytic uremic syndrome), longer-term complications	1% (Mead and Griffin, 1998)	Long-term sequelae: hypertension, stroke, renal insufficiency/failure, neurologic complications (Mead and Griffin, 1998; Griffin et al., 1994).
Vibrio cholerae	Clinical and research labs	10^8 organisms (Tauxe, 1992) Secondary transmission possible	Acute life-threatening dehydrating diarrhea	Up to 50% untreated 1% treated (Bennish, 1994).	Historically, causes massive waterborne epidemics in areas with poor sanitation

RECOGNITION OF A FOODBORNE DISEASE EVENT AS A TERRORIST ATTACK OR A CRIMINAL ACT

Unusual relationships between person, time, and place of an outbreak, or unusual or implausible combinations of pathogens and food vehicles, are epidemiologic clues to a deliberate, covert act of contamination (Treadwell et al., 2003). Such features may be absent in an event of deliberate contamination, however, and also may occur naturally in unintentional outbreaks. Thus, epidemiologic features alone cannot prove a terroristic act; rather, they inform investigators and may prompt consultation with law enforcement agencies that may then confirm or refute the possibility of malicious contamination.

The adequacy of response will depend on public health officials' capacity to respond to *all* foodborne disease outbreaks. Hence, a cornerstone of preparedness is improving the public health infrastructure for detecting and responding to *unintentional* outbreaks: ensuring robust surveillance, improving laboratory diagnostic capacity for patient and food product samples, increasing trained staff for rapid epidemiologic investigations, and enhancing effective communications. Preparedness for such a situation additionally requires the capacity to respond to extraordinary demands on emergency services and medical resources.

DIAGNOSIS

A key factor in rapid diagnosis of the etiologic agent during the investigation of unexplained foodborne disease is ordering the appropriate diagnostic laboratory test. This requires that clinicians be familiar with the likely agents and their clinical presentations, that they not be hampered in ordering tests by cost concerns, and that they know how to contact public health sector consultants when needed.

Most foodborne pathogens on CDC's Strategic Plan for Bioterrorism Preparedness and Response are detectable by routine culture practices in state public health laboratories. Botulism is diagnosed in some state and municipal laboratories and at the CDC, because identification of toxins requires testing of appropriate samples in specialized laboratories. The CDC has developed the national Laboratory Response Network (LRN) for bioterrorism, specializing in diagnosis of biological agents that includes public health, military, veterinary, and commercial laboratories, to provide standardized protocols for diagnosis and reagents; make initial, rapid diagnoses; and then refer specimens to appropriate specialty laboratories at CDC and elsewhere. The LRN provides surge capacity to handle increased numbers of samples anticipated in a bioterrorism event; for example, over 120,000 samples were collected in the course of the anthrax mailings investigations of 2001.

RESPONSE

In the United States, county, municipal, and, in some cases, state health departments are typically the first to be informed of outbreaks and the first to respond to and investigate them. Generally, the state public health laboratory and a few municipal laboratories play a primary role in diagnosing the etiology of a given outbreak. Outbreaks with cases distributed over a wide geographic area without clustering, however, may be recognized first by CDC through national laboratory-based surveillance systems, in which case CDC may play a coordinating role for a multistate investigation.

The objectives of the epidemiologic investigation of an outbreak of foodborne disease would not greatly change if intentional contamination were suspected. Identification of the etiologic agent, vehicle of transmission, and manner of contamination remain the most important aspects of an investigation, followed by timely implementation of control measures, including removal of the contaminated food from circulation and the proper treatment of exposed persons (Sobel et al., 2002b). The familiar components of the investigation include formulation of case definitions; case finding; pooling and evaluation of data on potential exposures in different geographic locations; rapid development of standardized instruments and execution of case-control studies to identify specific food vehicles; collection of laboratory samples; transport and processing; collating information from tracebacks; coordination with law enforcement, food safety regulatory agencies, and agencies involved in emergency medical response; and standardization of treatment and prophylaxis recommendations. The CDC and the federal food regulatory agencies—the Food and Drug Administration (FDA) and United States Department of Agriculture (USDA)—routinely collaborate on tracebacks of contaminated foods implicated in many of the approximately 1000 foodborne disease outbreaks reported annually in the United States, a norm that would be followed in a bioterrorism event.

A sophisticated bioterrorist attack on the food supply could produce many casualties. In the United States, the medical components of the response to such an event are part of overall bioterrorism response preparedness and have been described elsewhere (e.g., see the entries entitled "CRISIS MANAGEMENT" and "CONSEQUENCE MANAGEMENT") (Khan et al., 2000). Adequate stocks of antimicrobial drugs, antitoxins, and other medications, as well as ventilators and other medical equipment, are maintained in nationwide stockpiles in order to ensure rapid delivery if needed. However, a bioterrorism attack targeting a food that is distributed over a vast geographic area could challenge this delivery system owing to its widespread nature. The effectiveness of the medical response will therefore depend on timely epidemiologic surveillance data collected by public health investigators to direct medical resources to the casualties and their caretakers in good time.

COMMUNICATIONS

Swift communication between health care providers, public health officials at various levels, and government agencies is an absolute requirement for a rapid, effective response to a bioterrorist attack on the food supply. Communication patterns similar to those used in the coordination of multistate outbreak investigations will likely be effective for incidents of intentional contamination of food (Sobel et al., 2002b).

Clinicians, clinical laboratory staff, and coroners who identify suspected cases or clusters of illness must have lists of appropriate local contacts in order to notify the public health sector of their findings. Local health departments should notify state public health departments even as they begin their investigation locally. There are standing modalities used routinely to inform public health officials at the state and federal level of ongoing outbreaks, and to coordinate multistate investigations. In the case of an intentional contamination of food, these communication systems would function as they do in regular outbreaks. Depending on the food affected, the FDA's or USDA's regulatory authorities would be engaged rapidly during a bioterrorist event linked to food. Communication between public health officials and food industries would be coordinated with the appropriate regulatory agency, which would be positioned to request a recall of contaminated food from the market.

Communication to the public is also a critical consideration. Intense media coverage of a bioterrorist event is to be expected, and skill and experience are therefore required to transmit accurate information through the media about the nature and extent of the event, the suspected or implicated foods, and measures to take to prevent exposure or consequences of exposure. Ultimately, the accuracy, timeliness, and consistency of the information provided may in part determine the success of control measures.

CONCLUSIONS

Sabotage of food by terrorists and criminals has occurred in the United States, perhaps the best-known case being the deliberate contamination of salad bars with *S. typhimurium* in 1984 by the Rajneeshee cult in order to test the possibility of swaying a local election in Oregon by sickening voters. But such cases are remarkably rare. Regardless, a multiplicity of suitable biological and chemical agents exists, and the vast contemporary food supply is vulnerable. Prevention therefore requires enhancement of food security. Since an outbreak caused by food sabotage would most likely be detected and handled by the existing public health system, minimization of casualties requires a robust standing public health infrastructure capable of detecting, investigating, and controlling all foodborne disease outbreaks, intentional and unintentional.

REFERENCES

Bean, M.H., Martin, S.M., and Bradford, H., *Am. J. Public Health*, **82**, 1273–1276 (1992).

Benenson, A.S, Ed., *Control of Communicable Diseases Manual*, 16th ed., American Public Health Association, Washington, DC, 1995.

Bennish, M.L., "Cholera: Pathophysiology, Clinical Features, and Treatment," in I.K. Wachsmuth, P.A. Blake, and O. Olsvik, Eds., *Vibrio cholerae and Cholera, Molecular to Global Perspectives*, ASM Press, Washington, DC, 1994, pp. 229–256.

Blaser, M.J. and Newman, L.S., *Rev. Infect. Dis.*, **4**, 1096 (1982).

Butler, T., Fu, Y.S., Furman, L., Almeida, C., and Almeida, A., *Infect. Immun.*, **36**, 1160–1167 (1982).

Centers for Disease Control and Prevention, *Botulism in the United States, 1899–1996. Handbook for Epidemiologists, Clinicians, and Laboratory Workers*, Centers for Disease Control and Prevention, Atlanta, GA, 1998.

Centers for Disease Control and Prevention, *MMWR*, **49**, 1–14 (2000).

Centers for Disease Control and Prevention, *MMWR*, **50**, 941–948 (2001).

Dupont, H.L., Levine, M.M., Hornick, R.B. et al., *J. Infect. Dis.*, **159**, 1126, (1989).

Erickson, M.C. and Kornacki, J.L., *J Food Prot.*, **66**, 691–699 (2003).

Franz, D.R. and Jaax, N.K., "Ricin Toxin," in F.R. Sidell, E.T. Takafuji, and D.R. Franz, Eds., *Medical Aspects of Chemical and Biological Warfare*, Borden Institute, Walter Reed Army Medical Center, Washington, DC, 1997, pp. 631–642.

Greenko, J., Mostashari, F., Fine, A., and Layton, M., *J. Urban Health*, **80**(2 Suppl. 1), i50–i56 (2003).

Griffin, P.M., Bell, B.P., and Cieslak, P.R. et al., "Large Outbreak of *Escherichia coli* O157:H7 Infections in the Western United States: the Big Picture," in M.A. Karmali and A.G. Golglio, Eds., *Recent Advances in Verocytotoxin-Producing Escherichia coli Infections*, Elsevier Science B.V., New York, 1994.

Grigg, B. and Modeland, V., *FDA Consumer*, **July-August**, 7–11 (1989).

Hatheway, C.L., *Clin. Microbiol. Rev.*, **3**, 66–98 (1990).

Hedberg, C.W., MacDonald, K.L., and Osterholm, M.T., *Clin. Infect. Dis.*, **18**, 671–682 (1994).

Hennesy, T.W., Hedberg, C.W., Slutsker, L. et al., *N. Engl. J. Med.*, **334**, 1281–1286 (1996).

Hutwagner, L.C., Maloney, E.K., Bean, N.H., Slutsker, L, and Martin, S.M., *Emerg. Infect. Dis.*, **3**, 395–400 (1997).

Institute of Medicine, "Committee on R&D Needs for Improving Civilian Medical Response to Chemical and Biological Terrorism Incidents," *Chemical and Biological Terrorism. Research and Development to Improve Civilian Medical Response*, National Academy Press, Washington, DC, 1999.

Khan, A.S., Morse, S., and Lillibridge, S., *Lancet*, **356**, 1179–1182 (2000).

Khan, A.S., Swerdlow, D.L., and Juranek, D.D., *Public Health Rep.*, **116**, 3–14 (2001).

Kolavic, S.A., Kimura, A., Simons, S.L., Slutsker, L., Barth, S., and Haley, C., *J. Am. Med. Assoc.*, **278**, 396–398 (1997).

Lee, R.V., Harbison, R.D., and Draughon, F.A., *Food Prot. Trends*, **23**, 664–674 (2003).

Mahon, B., Ponka, A., Hall, W. et al., *J. Infect. Dis.*, **175**, 876–882 (1997).

Mead, P.S. and Griffin, P.M., *Lancet*, **352**, 1207–1212 (1998).

Mead, P.S., Slutsker, L., Dietz, V., McCaig, L.F., Bresee, J.S., Shapiro, C. et al., *J. Emerg. Infect. Dis.*, **5**, 607–625 (1999).

Mermin, J.H., Griffin, P.M., *Am. J. Epidemiol.*, **150**, 797–803 (1999).

Pavlin, J.A., *J. Urban Health*, **80**(2 Suppl. 1), i107–i114 (2003).

Phills, J.A., Harrold, A.J., Whiteman, G.V., and Perelmutter, L., *N. Engl. J. Med.*, **286**, 965–970 (1972).

Reintjes, R., Dedushaj, I., Gjini, A., Jorgensen, T.R., Cotter, B., Kieftucht, A., D'Ancona, F., Dennis, D.T., Kosoy, M.A.,

Mulliqui-Osmani, G., Grunow, R., Kalaveshi, A, Gashi, L., and Humolli, I., *Emerg. Infect. Dis.*, **8**, 69–73 (2002).

Ryan, C.A., Nickels, M.K., Hargrett-Bean, N.T. et al., *J. Am. Med. Assoc.*, **258**, 3269–3274 (1987).

Shapiro, R., Hatheway, C., and Swerldlow, D., *Ann. Intern. Med.*, **129**, 221–228 (1998).

Sirisanthana, T., Brown, A.E., *J. Emerg. Infect. Dis.*, **8**, 649–651 (2002).

Sivapalasingam, S., Kimura, A., Ying, M. et al., "A Multi-State Outbreak of *Salmonella* Newport Infections Linked to Mango Consumption, November-December 1999," Latebreaker Abstract, *49th Annual Epidemic Intelligence Service (EIS) Conference*, Centers for Disease Control and Prevention, Atlanta, GA, 2000. Schedule Addendum, Latebreaker Abstracts.

Sobel, J., Khan, A.S., and Swerdlow, D.S., *Lancet*, **359**, 874–880 (2002a).

Sobel, J., Griffin, P.M., Slutsker, L., Swerdlow, D.L., and Tauxe, R.V., *Public Health Rep.*, **117**, 8–19 (2002b).

Sobel, J., Swerdlow, D.L., and Parsonnet, J., "Is There Anything Safe to Eat?," in J.S. Remington and M.N. Schwartz, Eds., *Current Clinical Topics in Infectious Diseases*, Vol. 21, Blackwell Scientific Publications, Boston, MA, 2001.

Stephenson, J., *J. Am. Med. Assoc.*, **277**, 1337–1340 (1997).

Swaminathan, B., Barrett, T.J., Hunter, S.B., and Tauxe, R.V., *J. Emerg. Infect. Dis.*, **7**, 382–389 (2001).

Tarnvik, A. and Berglund, L., *Eur. Respir. J.*, **21**, 361–373 (2003).

Tauxe, R.V., *J. Am. Med. Assoc.*, **267**, 1388–1390 (1992).

Tilden, J., Young, W., McNamara, A.M. et al., *Am. J. Public Health*, **86**, 1142–1145 (1996).

Török, T., Tauxe, R.V., Wise, R.P. et al., *J. Am. Med. Assoc.*, **278**, 389–395 (1997).

Treadwell, T.A., Koo, D., Kuker, K., and Khan, A.S., *Public Health Rep.*, **118**, 92–98 (2003).

US Food and Drug Administration, *Risk Assessment for Food Terrorism and Other Food Safety Concerns*, October 13, 2003 www.cfsan.fda.gov/~dms/rabtact.html.

U. S. General Accounting Office, *Food Safety: Agencies Should Further Test.Plans for Responding to Deliberate Contamination*, GAO/RCED-00-3, October 27, 1999.

WHO, *Health Aspects of Chemical and Biological Weapons*, Report of a WHO group of consultants, Annex 5, Sabotage of Water Supplies, World Health Organization, Geneva, 1970, 113–120.

WHO, *Terrorist Threats to Food, Guidelines for Establishing and Strengthening Prevention and Response Systems*, World Health Organization, Geneva, 2002.

See also DETECTION OF BIOLOGICAL AGENTS; FOOD AND DRUG ADMINISTRATION; FOOD AND WATERBORNE PATHOGENS; and WATER SUPPLY: VULNERABILITY AND ATTACK SPECIFICS.

FOOD AND DRUG ADMINISTRATION

MICHELLE BAKER

INTRODUCTION

The U.S. Food and Drug Administration (FDA) within the Department of Health and Human Services (HHS) is accountable for the safety and security of human and veterinary drugs, biological products, much of the U.S. food supply, devices that emit radiation, cosmetics, and medical devices. The eight offices of the FDA, encompassing centers for drug research, veterinary medicine, and food safety, reflect the various responsibilities that fall under its jurisdiction. Since the terrorist attacks of September 11, 2001, the FDA has become increasingly vigilant in its monitoring of products, both produced domestically and imported into the United States, a commitment that has also served to improve interagency cooperation and coordination vital to adequate bioterrorism preparedness and response.

BACKGROUND

As a direct result of the September 11 attacks, the Public Health Security and Bioterrorism Preparedness and Response Act (the Bioterrorism Act) was signed into law in June 2002. This Act, designed to identify and correct weaknesses within several federal agencies as to their bioterrorism preparedness capabilities, has provided the FDA with direction regarding its mandate in the prevention of terrorism involving the food supply. Title III of the Bioterrorism Act specifically designates the FDA as being responsible for the safety and integrity of the nation's food and drug supply, and recognizes FDA's oversight of 80 percent of the entire U.S. food supply (http://www.cfsan.fda.gov/~dms/fssrep.html, accessed on 10/3/2003). The FDA must provide for the security of both finished food products within United States and the facilities involved in domestic and imported food manufacture and processing (http://www.cfsan.fda.gov/~dms/secltr.html, accessed on 10/3/2003). While this is not a drastic change in policy from the pre-September 11 era, the appearance of new threats and challenges to the U.S. food supply has forced a reassessment of the administration's current practices and funding dedicated to such efforts.

To coordinate a response to the regulations of the Bioterrorism Act, the FDA has adopted a stepwise approach consisting of several overarching strategies: awareness, prevention, protection, response, and recovery. *Awareness* examines the relationship between the FDA and other federal, state, local, and tribal agencies, with the goal of enhancing the dissemination of potentially useful information concerning the U.S. food supply among these parties. *Prevention* focuses on the need to raise awareness of specific threats against the food supply. *Protection* seeks to impede terrorist access to the food supply. And *Response* and *Recovery* are required following a terrorist action to affect a rapid response and coordinated recovery in order to recertify the safety and security of the food supply as quickly and effectively as possible (http://www.cfsan.fda.gov/~dms/fssrep.html, accessed on 10/3/2003).

FOOD SECURITY

In a July 2003 progress report to the secretary of HHS, the FDA commissioner presented the following 10-point program implemented as part of the Administration's Homeland Security plan and designed to provide specific benchmarks in the development of a food security strategy:

1. To facilitate the implementation of any new programs, the FDA formed an expanded investigative and scientific team in the months following September 11, 2001. New hires were quickly placed into food safety field activities, and the number of workers stationed at ports around the United States was increased. In addition, the FDA has both hired new scientists and retrained staff scientists to participate in investigations to detect biological, chemical, or radiological agents in domestic or imported food (http://www.cfsan.fda.gov/~dms/fssrep.html, accessed on 10/3/2003).

2. To further protect the security of the U.S. food supply, the FDA has developed an Import Strategic Plan (IPS) to monitor the events that constitute the life cycle of imported products. The FDA has doubled its presence at U.S. borders, drastically increasing food import examinations since 2001. In addition to this enhanced domestic capacity, the FDA is seeking to increase the reliability of foreign inspections before goods enter the United States (http://www.cfsan.fda.gov/~dms/fssrep.html, accessed on 10/3/2003).

3. The FDA has published specific regulations corresponding with the requirements of the Bioterrorism Act. In October 2003, HHS Secretary Tommy

Thompson announced two regulations regarding food security. The first provision requires that beginning on December 12, 2003, food importers must supply the FDA with prior notice of both human and animal food imports. Advanced notice of a given shipment must be provided to the FDA no less than two hours before the notifiable product arrives by land transport, four hours before arrival by air or rail, and eight hours before arrival by water. The second regulation requires that domestic or foreign facilities involved in any stage of the production or manufacturing of food entering the United States must be registered with the FDA. Exceptions exist for facilities handling meat or poultry covered by the U.S. Department of Agriculture (USDA), as well as restaurants or other retail food establishments (http://www.hss.gov/news/press/2003pres/20031009.html, accessed on 10/10/2003). This regulation meant that almost 400,000 domestic and foreign facilities were required to register with the FDA by December 12, 2003 (http://www.cbsnews.com/stories/2003/10/09/attack/main577392.shtml?cmp=EM8705, accessed on 10/10/2003).

4. The FDA is assisting both food production and manufacturing facilities and restaurants and food importers in the implementation of regulations providing protection against bioterrorism. FDA guidance to businesses unaffected by new regulations encourages these establishments to perform background checks on restaurant and supermarket employees, safeguard water supplies, and maintain a level of surveillance over any salad bar or exposed food. In March 2003, the FDA published the final versions of these guidance documents. In addition, the FDA in conjunction with St. Joseph's University in Pennsylvania is currently developing a training curriculum on the security and protection of food, which will soon be distributed to industry representatives (http://www.cfsan.fda.gov/~dms/fssrep.html, accessed on 10/3/2003).

5. The FDA has employed Operational Risk Management (ORM) to create a "vulnerability assessment for foods." The assessment is designed to identify areas most in need of public health resources while also recognizing areas in which increased safety measures are necessary to strengthen the security of the food industry. In conjunction with this vulnerability assessment, FDA provides Congress with information concerning the greatest challenges and security threats facing the food industry (http://www.cfsan.fda.gov/~dms/fssrep.html, accessed on 10/3/2003).

6. The FDA participated in Operation Liberty Shield in March 2003, which was designed to increase the agency's ability to create a comprehensive security plan during periods of national elevated alert status and to analyze the effectiveness of current FDA measures to counter bioterrorism (http://www.cfsan.fda.gov/~dms/fssrep.html, accessed on 10/3/2003).

7. The FDA's Office of Crisis Management (OCM) was established in order to coordinate the various FDA centers and other federal, state, and local agencies responsible for response to a crisis involving FDA-regulated products. Interaction between agencies, facilitated by OCM, has been integral to emergency response exercises conducted over the past two years. Cooperative measures were further strengthened by the implementation of an Inter Agency Agreement (IAG) between the FDA and the U.S. Army. The two entities are currently working toward a capability to jointly deploy mobile laboratories capable of detecting food adulteration (http://www.cfsan.fda.gov/~dms/fssrep.html, accessed on 10/3/2003).

8. Laboratories both within and affiliated with the FDA have begun to develop increasingly sophisticated methods for the detection of biological and chemical agents in food. These methods are being developed in conjunction with the Department of Defense (DoD), with emphasis being placed on the detection of agents most likely to be employed in a terrorist attack. In addition to the interactions between FDA and DoD laboratories, the FDA has partnered with the USDA, Centers for Disease Control and Prevention (CDC), Environmental Protection Agency (EPA), Department of Energy (DOE), and various states to develop a nationwide Food Emergency Response Network (FERN). The laboratories within this network seek to analyze food samples following biological, chemical, or radiological terrorism event. As of June 2003, the FERN network encompassed 63 laboratories in 27 states and 5 federal agencies (http://www.cfsan.fda.gov/~dms/fssrep.html, accessed on 10/3/2003).

9. The FDA continues to strengthen its research infrastructure, which presently is composed of intra-agency laboratories, collaborative laboratories (e.g., the National Center for Food Safety and Technology), and research contractors at outside agencies. Projects undertaken by these laboratories have become more narrowly focused as the FDA has determined its top priorities under the bioterrorism initiative (http://www.cfsan.fda.gov/~dms/fssrep.html, accessed on 10/3/2003).

10. As demonstrated by the preceding activities, the FDA often relies on interagency support to achieve its mandate as directed by the Bioterrorism Act. This requires that the FDA engage in regular communications with most other federal agencies, including the Department of Homeland Security and Customs and Border Control. Emergency preparedness exercises involving the FDA also necessitate collaboration with agencies such as the USDA, DoD, EPA, Central Intelligence Agency (CIA), and Federal Bureau of Investigation (FBI). In addition to these relationships,

the FDA has reinforced its food security efforts with both Canada and Mexico through existing partnerships and newly formed working groups (http://www.cfsan.fda.gov/~dms/fssrep.html, accessed on 10/3/2003).

TERRORISM COUNTERMEASURES

On May 30, 2002, the FDA announced a new regulation, the aim of which is to expedite approval of drugs and other products developed to reduce or mitigate the effects of biological, chemical, or radiological terrorism agents. Such drugs would be ready for approval by the FDA before undergoing the usual multiphase clinical trial requirements, such that while preclinical animal testing would be performed to establish baseline efficacy and toxicity, human testing would not be required (http://usinfo.state.gov/topical/pol/terror/02053105.htm, accessed on 10/6/2003). While some analysts fear that these new drugs will not undergo the same scrutiny of other drugs on the market, others stress that this regulation relieves the concern of potentially unethical human testing involving lethal agent, which is requisite in the standardized drug approval process. Efforts are largely focused on drugs perceived as critical for development: antiviral agents against the smallpox virus, a safer smallpox vaccine, antitoxins to prevent or mitigate the effects of *Bacillus anthracis*, antimicrobials to treat anthrax or plague, and therapies for viral hemorrhagic fever (http://www.fda.gov/oc/bioterrorism/role.html, accessed on 2003). These efforts are being carried out in conjunction with FDA research to provide the U.S. military and first responders with appropriate medicines and equipment to effectively counter a bioterrorist attack.

CURRENT STATUS

The FDA has hired over 800 new employees directly involved with food safety and security, 655 of whom serve in the field. By mid-2003, the FDA had conducted some 62,000 food inspections, compared to the 12,000 inspections performed during all of 2001, and had staffed 90 ports of entry into the United States, up from 40 in 2001 (http://www.hhs.gov/news/press/2003pres/20030204.html, accessed on 10/6/2003). Of the $116.3 million in the FDA 2004 fiscal year budget, $20.5 million has been allocated toward activities specifically mandated under the Bioterrorism Act: $5 million to improve laboratory preparedness at the federal, state, and local level; $5 million to improve food monitoring and inspections at the state level; and $10.5 million to implement the federal registry and the prior notice of imported food shipments requirement that began on December 12, 2003. With the FDA having such a vital role in protecting the United States against a bioterrorist attack, its continued funding on this scale or larger seems assured for many years to come.

REFERENCES

U.S. Department of Health and Human Services, Food and Drug Administration, Center for Food Safety and Applied Nutrition, *Summary of Title III, Subtitle A of the Public Health Security and Bioterrorism Preparedness and Response Act of 2002*, July 17, 2002, http://www.cfsan.fda.gov/~dms/sec-ltr.html, accessed on 3/10/03.

U.S. Department of Health and Human Services, Food and Drug Administration, Center for Food Safety and Applied Nutrition, *Progress Report to Secretary Tommy G. Thompson: Ensuring the Safety and Security of the Nation's Food Supply*, July 23, 2003, http://www.cfsan.fda.gov/~dms/fssrep.html.

U.S. Department of Health and Human Services, *HHS Creates Food Security Research Program, Increases Import Exams More than Five Times to Protect Nation's Food Supply: Progress Report Details Ongoing Efforts to Enhance the Nation's Food Security*, July 23, 2003, http://www.hss.gov/news/press/2003pres/20030723.html.

U.S. Department of Health and Human Services, *HHS Issues New Rules to Enhance Security of the U.S. Food Supply*, October 9, 2003, http://www.hss.gov/news/press/2003pres/20031009.html.

U.S. Department of Health and Human Services, Food and Drug Administration, *FDA's Counterterrorism Role*, http://www.fda.gov/oc/bioterrorism/role.html.

U.S. Department of State, International Information Programs, *U.S. Regulators Clear Path for Speedier Anti-Bioterror Drug Approval*, May 30, 2002, http://usinfo.state.gov/topical/pol/terror/02053105.htm. Washington Post, FDA Urges Safeguards for U.S. Food Supply, 4 January 9, (2002).

WEB RESOURCES

Kerry, B., *U.S. Response: House Passes Bioterrorism Response Bill*, Global Security Newswire, May 23, 2003, http://www.nti.org/d_newswire/issues/2002/5/23/10p.html.

National Food Processors Association, *NFPA Encourages FDA to Develop 'Effective and Efficient Regulations' to Implement Bioterrorism Act*, July 7, 2003, http://www.nfpa-food.org/News_Release/NFPAPressRelease070803.htm.

National Food Processors Association, *Compliance with FDA Bioterrorism Regulations 'Will be Top Priority for Industry,' Says NFPA*, October 9, 2003, http://www.nfpa-food.org/News_Release/NFPANewsRelease100903.htm.

U.S. Department of Health and Human Services, Food and Drug Administration, *Counterterrorism*, http://www.fda.gov/oc/mcclellan/counterterrorism.html.

U.S. Department of Health and Human Services, Food and Drug Administration, *The Food and Drug Administration's Strategic Action Plan, Protecting and Advancing America's Health: Responding to New Challenges and Opportunities*, August 2003, http://www.fda.gov/oc/mcclellan/strategic.html.

U.S. Department of Health and Human Services, Food and Drug Administration, *Prior Notice of Imported Food Shipments Proposal*, January 29, 2003, http://www.fda.gov/bbs/topics/NEWS/2003/NEW00866.html.

U.S. Department of Health and Human Services, *President's Budget Includes Vital Food Supply Protections, Fiscal Year 2004 Proposal Maintains Historic Commitments to Cities, States, and Hospitals*, February 4, 2003, http://www.hss.gov/news/press/2003pres/20030204.html.

See also DEPARTMENT OF HEALTH AND HUMAN SERVICES and PHARMACEUTICAL INDUSTRY.

FOOD AND WATERBORNE PATHOGENS

Zygmunt F. Dembek

Donald L. Noah

INTRODUCTION

Food and waterborne pathogens contribute significantly to the burden of disease in the United States. The U.S. Department of Agriculture Economic Research Service estimates that household medical costs and productivity losses for diseases caused by five leading foodborne pathogens is as high as $6.7 billion per year (Buzby et al., 1996). Many foodborne pathogens, whether bacterial, viral, parasites, or toxins, could cause disease if purposefully introduced into water or food sources. Such pathogens, characterized by low-infective dose, high virulence, widespread availability, and stability in food products or potable water, include *Campylobacter jejuni*, various *Salmonella* and *Shigella* species, enterohemorrhagic *Escherichia coli*, *Cryptosporidium parvum*, and *Vibrio cholerae*. These microorganisms are currently listed by the Centers for Disease Control and Prevention (CDC) as Category B agents, meaning that they are of the second highest priority in terms of risk posed to national security. Others included in the CDC listing of biological threat agents that could potentially cause food or waterborne disease include *Bacillus anthracis*, *Clostridium botulinum*, *Brucella* species, and staphylococcal enterotoxin B (SEB). Agents not listed by the CDC, for example mycotoxins, may cause such diseases as well.

AGENT REVIEW

The major food and waterborne pathogens are briefly reviewed below.

Campylobacter jejuni

Campylobacter jejuni has been identified as the most commonly reported bacterial cause of foodborne infection in the United States. Chronic sequelae associated with *C. jejuni* infections include Guillain–Barré syndrome and arthritis. In the United States, infants have the highest age-specific isolation rate for this pathogen. This is attributed to infants being more susceptible to infection on first exposure and their parents being more inclined to seek medical treatment for their youngest offspring, which leads to an increased case-identification rate (Altekruse et al., 1999). Infected individuals who recover are immune to the pathogen, which explains why children younger than 2 years have a significantly higher incidence rate than older persons.

Reservoirs of *C. jejuni* include wild fowl and rodents. The intestines of poultry are easily colonized with *C. jejuni*, and the organism is a commensal inhabitant of the intestinal tract of cattle (Altekruse et al., 1999). For humans, poultry is the primary source of the pathogen. The infective dose of this pathogen is between 100 and 1000 cells. After exposure, there is a 3- to 5-day incubation period before onset of illness and a 1-week recovery from campylobacteriosis.

Salmonella Species

The human gastrointestinal tract is the natural habitat of thousands of *Salmonella* serotypes. A number of these serotypes can cause gastroenteritis, manifested by diarrhea, abdominal pain, vomiting, fever, chills, headache, and dehydration. Other disease presentations include enteric fever, septicemia, and localized infections. The most highly pathogenic *Salmonella* species for humans is *S. typhi*. This bacterium is the causative agent of typhoid fever (about 2.5 percent of salmonellosis in the United States), the symptoms of which include septicemia, high fever, headache, and gastrointestinal illness. Poultry is a principal reservoir of Salmonellae. Water, shellfish, raw salads, and milk also are commonly implicated as vehicles for this pathogen.

Salmonella typhimurium was the pathogen used in 1984 by the religious cult the Rajneeshees to contaminate a number of restaurant salad bars in The Dalles, Oregon, with the ultimate goal of swaying a local election. This was the first contemporary bioterrorist attack in the United States, and resulted in 751 illnesses but no deaths (Török et al., 1997). An outbreak of salmonellosis of natural origin involving over 5770 cases occurred in northern Illinois in 1985, ultimately determined to have been caused by the accidental comingling of raw milk and the pasteurized product (CDC, 1985). Similarly, a 1985 outbreak affecting over 16,000 persons with antimicrobial-resistant salmonellosis in Minnesota was thought to have been caused by cross-contamination of raw milk and a pasteurized milk product that was sold to the public (Ryan et al., 1997).

Shigella dysenteriae

Shigella dysenteriae is implicated as causing about 25,000 cases of illness each year in the United States. The only significant reservoir for this pathogen is humans, with infants and young children most susceptible, partially due to toiletry behaviors and child care practices (CDC, 1986). Four serogroups (A through D) cause some 80 percent of shigellosis cases in the United States. Although not particularly hardy in the environment, *Shigella* is highly infectious and can be very persistent in a close-community environment (CDC, 1983, 1994a). Its

Encyclopedia of Bioterrorism Defense,   Edited by Richard F. Pilch and Raymond A. Zilinskas

ISBN 0-471-46717-0   Copyright © 2005 Wiley-Liss

infectious dose is only between 10 and 100 organisms; thus, the contamination of food, water, or milk with small number of the organism can cause substantial outbreaks. It is readily transferred person-to-person by close contact and fomites, and can be transmitted by insect vectors (primarily flies) as well (Islam et al., 2001).

After exposure, there is a 1- to 3-day incubation period before symptoms of illness appear, such as diarrhea, abdominal pain, vomiting, fever, and chills. Sick persons shed *Shigella* organisms for 3 to 5 weeks after symptoms cease, contributing to greater probability of person-to-person spread than is the case with other enteric pathogens such as *Salmonella* species and *V. cholerae*.

An example of *S. dysenteriae* having been deliberately dispersed occurred in Texas in 1996, when a hospital employee deliberately contaminated pastries with this pathogen and gave them to coworkers (Kolavic et al., 1997). As a result, 16 persons became ill, though none died. The perpetrator was caught and received a 20-year prison sentence (Anonymous, 1998).

Escherichia coli

E. coli O157:H7, which produces two Shiga toxins, has emerged as a major cause of serious pediatric illness throughout the world. The illness it causes can exhibit bloody diarrhea and hemolytic uremic syndrome (HUS), which is indicated by the presence of microangiopathic hemolytic anemia, acute renal failure, and thrombocytopenia (Karmali et al., 1985). Children younger than 5 years are at greatest risk for developing HUS when infected with *E. coli* O157:H7.

The major source of *E. coli* O157:H7 is beef cattle. During July 2002, the Colorado Department of Public Health and Environment (CDPHE) identified an outbreak of *E. coli* O157:H7, which linked 28 illnesses in Colorado and six other states to the consumption of contaminated ground beef products. Of the seven patients who were hospitalized, five developed HUS (CDC, 2002).

Cryptosporidium parvum

Cryptosporidium parvum, an obligate intracellular protozoan, forms oocysts that may be carried in food and drinking water. Persons involved in recreational activities in contaminated waters may become infected as a result of the inadvertent ingestion of oocysts (CDC, 1998, 1997, 1994b; Mac Kenzie et al., 1994). The infectious dose is very low: presumably, a single organism can cause infection. Symptoms of intestinal cryptosporidiosis can range from no symptoms to severe watery diarrhea. Pulmonary and tracheal cryptosporidiosis in humans, which generally results from person-to-person spread via the aerosolization of oocysts shed during acute infection, is associated with coughing and frequently a low-grade fever (Casemore et al., 1994). These symptoms are often accompanied by severe intestinal distress. The pathogenic mechanism of *C. parvum* remains unclear.

Vibrio cholerae

Vibrio cholerae is the causative agent of cholera, which is an acute diarrheal illness that can present as either a mild illness or as a severe disease characterized by profuse watery diarrhea, vomiting, and leg cramps. In the absence of treatment, the severe form of cholera can result in a rapid loss of body fluids leading to dehydration, shock, and death. Fecal discharge by infected individuals can lead to the pathogen being transmitted to other humans by contaminated food or water, an effect that can be mimicked in a bioterrorism scenario (CDC, 2003).

Poor hygiene, especially inadequately treated water and wastewater, greatly facilitates the spread of cholera through populations. A cholera outbreak in Hamburg, Germany, in 1892 that affected about 17,000 people, of whom 8605 died, reportedly was caused by drinking water having been contaminated by bacteriologists who were studying the pathogen at that time (Strong, 1942). A century later, in 1992, 75 passengers aboard a flight from Lima, Peru to Los Angeles, California, contracted cholera from eating in-flight a cold seafood salad; 10 were hospitalized and one died (Eberhart-Phillips et al., 1996). This outbreak demonstrates the potential for airline-associated spread of cholera from epidemic areas to other parts of the world.

Japan's World War II–era biological warfare (BW) program, led by Unit 731, developed and used *V. cholerae* as a biological weapon, causing widespread disease among the Chinese population (Gold, 1996).

Bacillus anthracis

Bacillus anthracis, reviewed extensively elsewhere in this volume, is the causative agent of two forms of foodborne or waterborne anthrax: oropharyngeal and gastrointestinal. The organism is of greatest bioterrorism concern through aerosol dispersal, and is not normally thought of as having bioterrorism potential as a foodborne bacterial contaminant, mainly because the infective dose is quite high (Inglesby et al., 2002). Nevertheless, it should not be ignored as a possible foodborne or waterborne threat agent since *B. anthracis* spores are easy to disperse in food and beverages and are exceedingly hardy, being able to resist the destructive effects of heating and other treatments. In addition, since early diagnosis of gastrointestinal anthrax is difficult, it may lead to this form having a higher mortality than the other forms of anthrax.

Clostridium botulinum and Botulinum toxin

Clostridium botulinum produces botulinum toxin, which is the causative agent of botulism, of which there are three types: classic, wound, and infant botulism. Botulinum toxin is the most potent natural toxin known to medicine, with a human lethal dose of about 1.0 μg/kg (Arnon et al., 2001).

There are seven antigenic types of botulinum toxin, denoted by the letters A through G. Most human disease is caused by types A, B, and E. Botulinum toxins A and B are often associated with home canning. Food tainted with toxin appears normal by visual examination; thus, the cook is often the first victim due to sampling the food during preparation. After ingesting the toxin, there is a 12- to 36-h incubation period, followed by vision blurring, swallowing and speech difficulties, and descending paralysis (Arnon et al., 2001).

Prior to the availability of modern respiratory intervention equipment and treatments, the botulism mortality rate exceeded 80 percent. Currently, botulism has a mortality rate of approximately 10 percent (Arnon et al., 2001). Successful treatment utilizes aggressive trivalent (A, B, E) antitoxin therapy as well as ventilatory support. In an outbreak investigation, early case diagnosis is very important for patient survival. The toxin can be found in a sick person's food, feces, and serum samples, all of which may be tested in a standard mouse model for the presence of botulinum toxin (Arnon et al., 2001). There is little evidence of victims acquiring immunity, even after severe infection.

Brucella Species

Brucella species are discussed in a separate entry.

Staphylococcal Enterotoxin B

Staphylococcal enterotoxin B (SEB) is discussed in a separate entry.

Mycotoxins

As the name suggests, mycotoxins are metabolites produced by various fungus species that can be found throughout the world as contaminants of stored cereal grains (Li et al., 1999; Jemmali et al., 1978). These toxins include aflatoxin B1, ochratoxin, and trichothecenes (a group of mycotoxins that include deoxynivalenol (DON), nivalenol (NIV) and, most notoriously, T-2 toxin (Schollenberger et al., 1999). The fact that low concentrations of these toxins are found naturally in commercially available cereal-based foods, including bread and related products, noodles, breakfast cereals, baby and infant foods, and rice, indicates that a ready substrate for their growth is available everywhere (Hsia et al., 1988). Since this is the case, the deliberate contamination of these foodstuffs is possible. Large disease outbreaks caused by naturally occurring tricothecenes have occurred in the past. For example, in 1944, the population of Orenburg, Russia, ate moldy grain and bread that later was found to be contaminated with T-2 toxin (Akhmeteli, 1977). As a result, a large number of that population became afflicted with an illness called alimentary toxic aleukia, and at least 10 percent of victims died. Similarly, an outbreak affecting 130,000 people in Anhui province, China, in 1991 was attributed to the consumption of moldy wheat and barley (Li et al., 1999). Mycotoxins from *Fusaria* spp., including DON and NIV, have been detected in corn samples in Linxian, China, and epidemiological studies have demonstrated a positive correlation between the ingestions of these toxins and the incidence of esophageal cancer among the exposed population (Hsia et al., 1988; Luo et al., 1990).

REFERENCES

Akhmeteli, M.A., *Ann. Nutr. Aliment.*, **31**, 957–975 (1977).

Altekruse, S.F., Stern, N.J., Fields, P.I. and Swerdlow, D.L., *Emerg. Infect. Dis.*, **5**, 28–35 (1999).

Anonymous, Hospital employee sentenced to 20 years for poisoning co-workers, *Abilene Reporter-News*, September 12, 1998, http://www.reporternews.com/texas/poison/0912.html.

Arnon, S.S., Schechter, R., Inglesby, T.V., et al., *J. Am. Med. Assoc.*, **285**, 1059–1070 (2001).

Buzby, J.C., Roberts, T., Lin, C.T.J., and MacDonald, J.M., *Bacterial Foodborne Disease: Medical Costs and Productivity Losses*, Agricultural Economics Report No. 741, USDA, 1996.

Casemore, D.P., Garder, C.A., and O'Mahony, C., *Folia Parasitol.*, **41**(1), 17–21 (1994).

CDC, *MMWR*, **32**, 250–252 (1983).

CDC, *MMWR*, **34**, 200 (1985).

CDC, *MMWR*, **35**, 753–755 (1986).

CDC, *MMWR*, **43**, 657 (1994a).

CDC, *MMWR*, **43**, 561–563 (1994b)

CDC, *MMWR*, **46**, 4–8 (1997).

CDC, *Morb. Mortal. Wkly. Rep.*, **47**, 856–860 (1998).

CDC, *Morb. Mortal. Wkly. Rev.*, **51**, 637–639 (2002).

CDC, *Morb. Mortal. Wkly. Rev.*, **52**, 1093–1095 (2003).

Eberhart-Phillips, J., Besser, R.E., Tormey, M.P., Koo, D., Feikin, D., Araneta, M.R., Wells, J., Kilman, L., Rutherford, G.W., Griffin, P.M., Baron, R. and Mascola, L., *Epidemiol. Infect.*, **116**, 9–13 (1996).

Gold, H., "Creating Pathology," *Unit 731 Testimony*, Chapter 3, Yenbooks, Singapore, 1996, pp. 67–85.

Hsia, C.C., Wu, J.L., Lu, X.Q. and Li, Y.S., *Cancer Detect. Prev.*, **13**, 79–86 (1988).

Inglesby, T.V., O'Toole, T., Henderson, D.A., Bartlett, J.G., Ascher, M.S., Eitzen, E., Friedlander, A.M., Gerberding, J., Hauer, J., Hughes, J., McDade, J., Osterholm, M.T., Parker, G., Perl, T.M., Russell, P.K. and Tonat, K., Working Group on Civilian Biodefense, *J. Am. Med. Assoc.*, **287**, 2236–2252 (2002).

Islam, M.S., Hossain, M.A., Khan, S.I., Khan, M.N., Sack, R.B., Alber, M.J., Huq, A. and Colwell, R.R., *J. Health Popul. Nutr.*, **19**, 177–182 (2001).

Jemmali, M., Ueno, Y., Ishii, K., Frayssinet, C., and Etienne, M., *Experientia*, **34**, 1333–1334 (1978).

Karmali, M.A., Petric, M., Lim, C., Fleming, P.C., Arbus, G.S. and Lior, H., *J. Infect. Dis.*, **5**, 775–782 (1985).

Kolavic, S.A., Kimura, A., Simons, S.L., Slutsker, L., Barth, S. and Haley, C.E., *J. Am. Med. Assoc.*, **278**, 396–398 (1997).

Li, F.Q., Luo, X.Y., and Yoshizawa, T., *Nat. Toxins*, **7**, 93–97 (1999).

Luo, Y., Yoshizawa, T., and Katayama, T., *Appl. Environ. Microbiol.*, **56**, 3723–3726 (1990).

MacKenzie, W.R., Hoxie, N.J., Proctor, M.E., Gradus, M.S., Blair, K.A., Peterson, D.E., Kazmierczak, J.J., Addis, D.G., Fox, K.R., Rose, J.B. and Davis, J.P., *N. Engl. J. Med.*, **331**, 161–167 (1994).

Ryan, C.A., Nickels, M.K., Hargrett-Brean, N.T., Potter, M.E., Endo, T., Mayer, L., Langkop, C.W., Gibson, C., McDonald, R.C. and Kenney, R.T., *J. Am. Med. Assoc.*, **258**, 3269–3274 (1987).

Schollenberger, M., Suchy, S., Jara, H.T., Drochner, W. and Muller, H.M., *Mycopathologia*, **147**, 49–57 (1999).

Strong, R.P., "Cholera," *Stitt's Diagnosis, Prevention and Treatment of Tropical Diseases*, Chapter XVII, The Blakiston Company, Philadelphia, PA, 1942, pp. 590–650.

Török, T.J., Tauxe, R.V., Wise, R.P., Livengood, J.R., Sokolow, R., Mauvais, S., Birkness, K.A., Skeels, M.R., Horan, J.M. and Foster, L.R., *J. Am. Med. Assoc.*, **278**, 389–395 (1997).

FURTHER READING

Cary, J.W., Linz, J.E., and Bhatnagar, D., *Microbial Foodborne Diseases: Mechanisms of Pathogenesis and Toxin Synthesis*, CRC Press, Boca Raton, FL, 1999.

Cliver, D.O. and Riemann, H.P., *Foodborne Diseases*, 2nd ed., Academic Press, New York, 2002.

Hunter, P.R., *Waterborne Disease*, John Wiley & Son, Hoboken, NY, 1997.

See also Food and Beverage Sabotage and Water Supply: Vulnerability and Attack Specifics.

FORT DETRICK AND USAMRIID

Norman M. Covert

INTRODUCTION

Fort Detrick (previously Camp Detrick) is a U.S. Army military reservation, encompassing 1212 acres within the city limits of Frederick, Maryland. It was established on April 10, 1943, as Camp Detrick, the top-secret biological warfare (BW) research complex, at a former municipal airfield 45 miles northwest of Washington, DC, named for Major Frederick L. Detrick, M.D. (1889–1931), World War I surgeon with the 28th Aero Squadron, 3rd Pursuit Group, in France and later surgeon of the 104th Aero Squadron, Maryland National Guard. Major Detrick died in June 1931, prior to his unit's first encampment at the airfield (August 1931), which would later bear his name. Memorialization as Detrick Field was unrelated to the biomedical research and development mission, although it has been termed in historical documents as apropos.

President Franklin D. Roosevelt authorized George W. Merck, head of the War Research Service, to appoint Dr. Ira Baldwin of the University of Wisconsin as the first director of science in January 1943. The nation's foremost biomedical research and development team was assembled in a concerted effort to rapidly increase the nation's knowledge and capabilities in BW. President Richard M. Nixon declared an end to offensive BW research in the United States by executive order in November 1969.

CURRENT STATUS

Fort Detrick is a multimission U.S. Army installation, serving units of the Army, Navy, Air Force, and U.S. Marine Corps, as well as other governmental agencies. It is the premier biomedical research and development reservation of the Department of Defense. Fort Detrick and military biomedical facilities are managed through the United States Army Medical Research and Materiel Command (USAMRMC) in consultation with the Surgeon General of the U.S. Army and Assistant Secretary of Defense (Health Affairs). The installation contains a separate research campus supporting the National Institutes of Health's (NIH) National Cancer Institutes (NCI) and the National Institute of Allergies and Infectious Diseases (NIAID). The United States Department of Agriculture (USDA) maintains its Foreign Disease and Weed Sciences Location at Fort Detrick as well. The organization was an outgrowth of the former Crops Division of the defunct U.S. Army Biological Laboratories.

USAMRIID

The primary military research facility on the installation is the U.S. Army Medical Research Institute of Infectious Diseases (USAMRIID), recognized as the nation's leading biocontainment facility. The physical plant was built in 1969. Today, USAMRIID research suites support animal research protocols requiring up to biosafety level 4 (BSL-4), the highest barrier level. Research in its containment suites requires all scientists and support personnel to be protected by available vaccines, toxoids, or other means of prophylaxis. An elaborate pass box system utilizing decontaminating showers and ultraviolet lights for protection allows personnel to enter suites and change into sterile positive airflow protective suits. Air handling and filtration systems in the negative airflow atmosphere prevent contamination of the environment inside or outside the suites.

All solid waste from the suites is processed through an autoclave (a simple steam pressure cooker), from which it is taken to a second, larger autoclave. It is processed again before being taken from the building for disposal in the medical incinerator at temperatures up to 1800°F. A further means of safety in the suites requires scientific personnel to work with organisms or exposed animals inside safety cabinets, each having self-contained negative air flow systems. All liquid waste in the suites is pretreated before disposal in a special contaminated sewer system. This separate sewer system was developed during World War II, with the current system being installed in 1952. Major renovations were completed about 1996. All liquid waste from the laboratories is carried underground to a processing station, where it traverses a system of heat exchangers, exposing it to more than 250°F for about 20 min. After it is allowed to cool, the sterilized liquid is disposed in the sanitary sewer system. Fort Detrick has operated its own water and sewer treatment facilities since World War II. The installation uses more than 2.5 million gallons of water each day.

Aside from its array of sophisticated research suites, USAMRIID is world renowned for its broad expertise in viral and bacterial agents that have been identified as threats to military personnel as well as civilians of the United States and its allies. Its research mission includes developing defensive strategies and prophylactic and therapeutic products to overcome naturally occurring diseases in a theater of war and diseases artificially

Encyclopedia of Bioterrorism Defense, Edited by Richard F. Pilch and Raymond A. Zilinskas
ISBN 0-471-46717-0 Copyright © 2005 Wiley-Liss

introduced into the battlefield by overt or covert delivery systems.

USAMRIID is a lineal descendant of the former U.S. Army Medical Unit (1954–1969) and elements at Camp Detrick of the Walter Reed Army Hospital Medical Clinic. The latter units conducted defensive research studies in addition to providing medical support services for military members, all laboratory and technical support personnel, and members of the "White Coat" human volunteer study group composed primarily of soldiers affiliated with the Seventh-day Adventist church and classified as conscientious objectors.

Other Missions

In addition to its Department of Defense biomedical support mission, Fort Detrick hosts management and support units for military medical supply and procurement functions of the U.S. Army, U.S. Navy, and U.S. Air Force. It also manages and provides Presidential and U.S. military satellite communications functions at the primary gateway stations maintained by elements of the 1110th U.S. Army Signal Battalion and the Direct Communications Link (Hotline to Moscow).

HISTORY

Fort Detrick has remained as the center of U.S. BW defense preparations since World War II. It was authorized as the U.S. Biological Warfare Laboratories in April 1943, charged to develop defensive mechanisms to protect the United States, its military forces, and those of the allied nations from biological attack. Then-Camp Detrick's secondary mission was the development of offensive biological weapons, giving the United States the capability of responding "in kind" should an enemy use biological weapons against the United States or its allies.

The U.S. BW research program was founded on a realistic assessment of enemy potentialities as the nation rapidly mobilized for war after the bombing of Pearl Harbor, Hawaii, in December 1941. Secretary of War Henry Stimson sought the counsel of members of the National Academy of Sciences and the National Research Council, which included Mr. Merck and prominent U.S. scientists. The distinguished advisors were mobilized under the banner of the War Bureau of Consultants (WBC), which concluded in February 1942 that BW represented a bona fide threat to the United States. The committee's conclusions ring as true in 2003 as they did in 1942:

> The value of biological warfare will be a debatable question until it has been clearly proven or disproven by experience. The wide assumption is that any method which appears to offer advantages to a nation at war will be vigorously employed by that nation. There is but one logical course to pursue, namely, to study the possibilities of such warfare from every angle, make every preparation for reducing its effectiveness, and thereby reduce the likelihood of its use.

The War Research Service (WRS) immediately directed that a thorough review be made of "all known pathogenic agents," according to Mr. Merck in his post-World War II report (January 1946) to Secretary of War Robert P. Patterson. In that first year of mobilization in 1942, the scientists recommended a nationwide alert to protect all utility systems, agriculture, livestock, and the movement of such foodstuffs to market. There was no question of threat from possible biological or chemical contamination by enemy agents or fifth column operatives. Mr. Merck pointed out in his later report that the committee had been convinced that the use of pathogens as weapons of sabotage, terror, or war was in fact in the realm of possibility, and certainly was an area of vulnerability that required a national commitment.

Secretary Stimson's initial interest in biological weapons clearly stemmed from concerns about a BW attack by Germany, which had used biological and chemical weapons in World War I. German agents used *Burkholderia mallei* in 1915 to contaminate a shipment of mules headed for France from the New York Port of Debarkation, for example, and special German chemical warfare units in Belgium and France operated fogging devices to spread such agents as chlorine and phosgene against Allied soldiers. These episodes clearly indicated that Germany would have little reticence in selecting first-use of weapons of mass destruction (WMD) in global warfare.

Intelligence reports after the surrender of Nazi Germany indicate that it had lost interest in biological weapons as part of its military arsenal. This was influenced by the entry of the United States as an active combatant in November 1942, opening the Western Front in North Africa. Mounting losses on the Eastern Front proved to be another, as the Russian Army held and then counterattacked the bogged-down German forces. No less of an influence was the intensive implementation of the Nazi's so-called Final Solution, the extermination of some 8 million Jews and other political opponents in what became known as "The Holocaust," which demanded significant resources from other areas of operation. And German intelligence had apparently learned of the existence of the U.S., British, and Canadian BW programs, despite extensive efforts to maintain secrecy at Camp Detrick. Fears thus never materialized that German "Buzz Bombs" might be filled with such pathogens as *Bacillus anthracis*.

Ironically, the concern over Nazi Germany led the Allies to overlook the threat from the Empire of Japan, which conducted BW research from the mid-1920s until 1944. Lieutenant General Shiro Ishii directed the expansive Japanese program from its inception, obscuring its work under the guise of a water purification unit (Unit 731) concentrated at Harbin and Pingfan, Manchuria. Allied leadership was unaware in 1941 of Japanese aerial attacks, which dropped clay Uji bombs filled with fleas carrying *Yersinia pestis* on Chinese villages, and other BW atrocities in China and Manchuria. The reports were obscured amid the political and military quagmire that characterized the war in China. The strained relationship between Chinese leader Generalissimo Chiang Kai-shek and Lt. Gen. Joseph W. Stillwell, head of the China-Burma-India Theater of Operations, was such that BW use by the Japanese was seen as having little overall impact on military operations. The reports became lost in the so-called fog of war. The Japanese BW threat finally reached high levels of concern among Allied principles in the last months of the war. U.S. military agents in Manchuria discovered the extensive BW research facilities at Harbin, which were linked to the Japanese network of military facilities, including numerous prisoner of war camps. The Russian Army captured the site in late August 1945.

A comprehensive report on Japanese BW research and development was compiled and verified in late 1947 from hundreds of interviews conducted by Camp Detrick personnel and intelligence agents of the Supreme Commander Allied Powers (SCAP) in Japan. Lt. Gen. Ishii and the staff of Unit 731 escaped prosecution for war crimes. The decision not to prosecute them in the War Crimes Tribunal in Japan was made jointly by General of the Army Douglas MacArthur and his SCAP staff, the State Department, and the War Department, in the name of national security. A State Department cable cited the need to prevent the Union of Soviet Socialist Republics (USSR) from learning the extent of the U.S., British, and Canadian BW research programs. However, intelligence documents now reveal that the Allies wanted to end the work of the War Crimes Tribunal. The documents present a basis for charges of malfeasance on the part of the General MacArthur and senior U.S. government officials at the State Department and War Department.

Camp and Fort Detrick research ultimately covered microbiology, chemistry and physics, plant sciences and entomology, medical and veterinary sciences, engineering, mathematics, and statistics. Nearly 2000 scientific articles were published by Fort Detrick researchers in public peer reviewed journals, and remain definitive scientific discussions on diseases of significance. A similar number of articles containing classified material and specific information reports were written by Fort Detrick scientists, but have remained mostly unavailable to the larger scientific community. However, many of these were declassified in 1977, making them available to a large portion of the world's research centers.

THE WAR ON TERROR

Fort Detrick's mission continually crosses paths with its past secretive and often storied activities in the offensive biological weapons era. A storm of media coverage engulfed Fort Detrick in the fall of 2001, when it served as both savior and sacrificial lamb as the United States suffered the realities of BW. Now called "bioterrorism," and striking as the nation still reeled from the events of September 11, 2001, the media flurry began with the mailing of letters containing B. anthracis spores from Trenton, New Jersey, to a Florida tabloid newspaper (The Sun); the New York Post; ABC, CBS, and NBC network news offices in Manhattan; and the Washington, DC, offices of Senators Tom Daschle (D-SD) and Frank Leahy (D-Vt). Civilian health departments, including the

U.S. Centers for Disease Control and Prevention (CDC), were quickly overwhelmed, but Fort Detrick's expertise remained untapped until the DoD was finally called upon to process samples taken during the ensuing investigation and cleanup operation.

The search for the perpetrator(s) of the mail attacks floundered from the beginning. While several promising leads appeared in Trenton, no bona fide suspects were identified until the Federal Bureau of Investigation (FBI) profile finally pinpointed a former Fort Detrick virologist, whose exaggerated credentials caused investigators and the public to remain focused on him for more than two years. The FBI ultimately conceded that it possessed no solid evidence to charge the former scientist, but by that time his career had suffered an irreparable blow. At the time of the editing of this entry (March 2005), the FBI continues to investigate numerous foreign and domestic leads, but has obtained little information of substance to identify, arrest, or prosecute those responsible for the crime. Regardless of the outcome, the events following the fall 2001 anthrax letters demonstrate that the expertise of Fort Detrick scientists can be viewed by both government officials and the public as a liability rather than an asset in times of crisis, risking the underutilization of precious resources during such times.

In its mission to support the Department of Homeland Security and the federal investigation into the letters, the Army continues to track the strain of B. anthracis used (identified as a variant of the Ames strain), and has found success in its effort to develop an improved anthrax vaccine. Its research into numerous other diseases having military significance and potential use to terrorist organizations persists as well. Today, the national effort to develop and stockpile vaccines, toxoids, and other prophylactic and therapeutic measures for the treatment and prevention of diseases continues to grow, as do efforts to upgrade the capabilities of emergency services personnel in local communities, all based to some extent on the knowledge and experienced gained by scientists and medical personnel at Fort Detrick and USAMRIID.

FURTHER READING

Covert, N.M., *Cutting Edge: A History of Fort Detrick, Maryland*, Public Affairs Office, Headquarters, U.S. Army Garrison, Fort Detrick, MD, 1997.

Harris, Sheldon, H., *Factories of Death, Japanese Biological Warfare 1932–45 and the American Cover-up*, Routledge, London, 1994.

Mole, Robert L. and Mole, Dale M., *For God and Country, Operation Whitecoat: 1954–1973*, Seventh-day Adventist Church, Teach Services Inc., Brushton, NY, 1998.

Regis, Edward, *The Biology of Doom: The History of America's Secret Germ Warfare*, Henry Holt and Company, Inc., New York, 1999.

See also DEPARTMENT OF DEFENSE.

G

GLANDERS (*BURKHOLDERIA MALLEI*)

BRIDGET K. CARR

EDWARD EITZEN

Opinions, interpretations, conclusions, and recommendations are those of the authors and are not necessarily endorsed by the U.S. Army.

SCIENTIFIC NAME

The highly infectious zoonotic disease known as glanders is caused by a nonmotile, non-spore-forming, aerobic gram-negative bacillus recently reclassified (1993) as *Burkholderia mallei* (Yabuuchi et al., 1992). The first account of the disease in humans was published in 1821 (Robins, 1906), though its recognition as a syndrome in the medical community was much earlier. Loeffler and Schutz first identified the causative organism in 1882 (Loeffler, 1886; Bartlett, 1998). Former taxonomic genera include *Pseudomonas* (members of this genus are motile), *Loefflerella, Pfeifferella, Malleomyces, Actinobacillus, Corynebacterium, Mycobacterium,* and *Bacillus*. Glanders is less commonly known by other names, including equinia, malleus, droes, and farcy. Farcy is a term originating from the cutaneous manifestation in horses that was once believed to be a separate condition wherein nodular abscesses ("farcy buds") ulcerate, regional cutaneous lymphatic vessels become thickened and indurated ("farcy pipes"), and ooze a glanders-typical yellow-green gelatinous pus ("farcy oil") (Bartlett, 1998). Pure "farcy" without ulceration of the mucous membranes is rare, if not just a temporary stage of glanders, as is the opposite condition (Robins, 1906). *Burkholderia mallei* is closely related to *Burkholderia pseudomallei*, the causative agent of melioidosis. The genetic homology between the two approaches 70 percent (Henning, 1956; Steele, 1979), causing some scientists to consider them biotypes or isotypes.

DISEASE

Glanders is a highly contagious and often fatal disease of solipeds including horses, mules and donkeys, and is characterized by ulcerating granulomatous lesions of the skin and mucous membranes. Aristotle first described the disease in horses in 330 B.C., and gave it the name malleus, meaning hammer or mallet. It was associated with clustered horses, particularly army horses and mules. Before control and eradication of the disease, its association with domesticated equids was so commonplace that the combination of horses and their glanders appeared in early literature, including brief mentions in Shakespeare's *The Taming of the Shrew* and in Dumas' *The Three Musketeers*. Humans, goats, dogs, cats, rabbits, and carnivorous predators living in close proximity to infected equids or carcasses have been naturally infected (Loeffler, 1886; Kovalov, 1971). Camels have also been infected, and are associated with human disease (Kovalov, 1971). Glanders was eradicated by 1960 in most countries, but is still found in parts of Africa and the Middle East.

Heavy losses of horses and the infrequent but deadly transmission to humans in the late nineteenth century led several countries to consider glanders control and eradication programs. Early programs in some countries involved destroying only clinically ill equids, with compensation, and meticulous disinfection of the premises of such cases. Despite these tactics, glanders would reemerge in new or remaining animals in stables and barns that once housed glandered equids, and countrywide cases would continue to rise. The notion of a carrier-state began to be accepted.

By 1890, the mallein test was discovered (see below). Control and eradication programs would soon incorporate the mallein testing of all contact equids, followed by quarantine and recommendation for slaughter of all test-positive animals. This failed in some locations at first because of lack of enforcement and incentive to owners for killing non–clinically ill animals. Some horse owners would deliberately hide contact animals to avoid testing, or would sell these and asymptomatic test-positive animals to unsuspecting individuals to salvage economic loss (Rutherford, 1906). Inexpensive steam transportation aided spread of the disease by the ready availability of shipment of glanders carrier equids to uninfected regions and countries. The United States was blamed for the exporting of glandered horses to Cuba in 1872 (Robins, 1906) and for the great increase of glanders cases in Canada near the turn of the twentieth century, where tens of thousands of U.S. horses were shipped annually (Rutherford, 1906; Robins, 1906).

Once control programs offered indemnity to test-positive and contact animals, and popular belief accepted the existence of a carrier state, eradicating glanders progressed more rapidly. Eliminating glanders in livestock also effectively eradicated human glanders in those countries with successful eradication programs.

Glanders remains a reportable livestock and human disease in many countries, and eradication programs are still ongoing in some countries attempting to free themselves of the disease. In over 5000,000 equids tested in Turkey between 2000 and 2001, for example, less than 2 percent tested positive and were destroyed. Only one of these, a mule, showed clinical signs of infection. Glanders is also internationally reportable to the 164-member Office International Des Epizooties (OIE) in accordance with the International Animal Health Code (OIE, 2003). Between 1996 and 2002, glanders in livestock has been reported in Bolivia, Belarus, Brazil, Eritrea, Ethiopia, Iran, Latvia, Mongolia, Myanmar, Pakistan, and Turkey. During the same time frame, glanders in humans has been reported in Cameroon, Curacao, Sri Lanka, Turkey, and the United States (laboratory-acquired) (OIE, 2002).

Humans exposed to infected horses have contracted glanders in occupational, hobby, and lifestyle settings.

Because *B. mallei* is easily aerosolized in the laboratory and very few organisms are required to cause disease, laboratory workers studying the organism have also been infected accidentally (Howe and Miller, 1947; CDC, May 2000). Infection by ingesting contaminated food and water has occurred; however, it does not appear to be a significant route of entry for infections in humans (Loeffler, 1886; Kovalov, 1971; Jennings, 1963). Human-to-human transmission is known. Glanders has drawn interest as a possible biological warfare (BW) agent in the biological weapons programs of several countries.

Disease progression and pathology in humans and horses are similar, though the clinical presentation of any two cases, horse or human, may vary significantly, even in cases related by direct transmission (Loeffler, 1886; Robins, 1906; Rutherford, 1906; Howe and Miller, 1947). Generalized symptoms may include fever, myalgia, headache, fatigue, diarrhea, and weight loss. Glanders is typically progressive, characterized by nodular and ulcerative lesions of the skin, mucous membranes, and various organs, depending upon the route of infection and the path(s) of lymphatic and hematogenous spread. Thick nasal exudate, particularly with respiratory involvement, may also be present in humans and particularly equids.

Acute and chronic infections have been described, though distinguishing the two is not at all definitive. Further, these descriptions originated long before a viable treatment was available, and before most countries had eradicated the disease. Robins states in his fine 1906 review of 156 cases of chronic glanders in humans that chronic disease may be interrupted with acute symptoms, and that a disease with an acute onset may run a chronic course. For the purpose of his review, Robins regarded a chronic case as one with duration of at least 6 weeks, adding "... one has to recognize that the distinction between acute and chronic glanders, as between acute and chronic forms of any infectious disease, is a matter of convenience." Most historical literature attempting to distinguish the two in humans and equids classifies a more fulminant and rapidly fatal clinical course (within 2–4 weeks) as an acute form of glanders, which is more often the case with untreated acute pneumonic and frank septicemic infection, whether primary or recurrent (Batts-Osborne et al., 2001; Neubauer et al., 1997; Howe and Miller, 1947). Chronic infections are most common in horses, where they comprise the majority of cases (Kovalov, 1971). An acute disease course is more common in donkeys and humans.

Glanders manifestations can be quite variable. At least six forms of infection have been described, though they are far from exclusive. They include localized, nasal (equids), pulmonary, septicemic, disseminated, and the aforementioned chronic infections. The most important distinction is whether or not the infection is localized, which is unusual except early in the infectious process. The various forms are largely explained by route of infection, regional lymphatic inflammation and drainage, and loci of dissemination and embolism via hematogenous or lymphatic spread. With disease progression and chronicity, all forms may be recognized.

Localized infections are just that, and are characterized by pus-forming nodules and abscesses that ulcerate and drain for long periods of time. Lymphangitis or regional lymphadenopathy may develop. Increased mucus production from affected ocular, nasal, and respiratory mucosa is often present. Localized infections typically disseminate, leading to pulmonary, septicemic, or disseminated infection.

The nasal form of glanders classically described in equids is essentially a localized infection of the nasal cavity characterized by yellowish-green unilateral or bilateral nasal discharge, with or without nodules or ulcers on the nasal mucosa. Regional lymphadenopathy and lymphangitis most often accompany nasal signs. Laryngeal, tracheal, and lower respiratory tract pathology is often present however, even if microscopically, supporting the concept of delineating various forms of glanders as somewhat historical.

A pulmonary infection typically produces pneumonia, pulmonary abscess, pleuritis, and pleural effusion, with associated signs and symptoms such as cough, dyspnea, chest pain, and mucopurulent sputum. Nasal exudate and cervical lymphadenopathy may also be present if the upper respiratory tract is involved. Nonspecific signs and symptoms often accompany respiratory infections, such as fatigue, fever, chills, headache, myalgias, and gastrointestinal signs. Pulmonary infection is typical of most laboratory-acquired infections due to aerosolization and inhalation of the organism.

Septicemic glanders may produce a myriad of signs consistent with a highly pathogenic bacterial septicemia. Without aggressive treatment, *B. mallei* septicemia runs an acute course and may lead to death in 7 to 10 days. Donkeys are particularly susceptible to *B. mallei* septicemia, as this form or syndrome manifests in a great majority of those naturally and experimentally infected.

Cutaneous or mucosal infections may spread, leading to disseminated infections. Cutaneous manifestations include multiple papular or pustular lesions that may erupt anywhere on the body. Dissemination to internal organs produces abscesses in virtually any organ but most commonly in the spleen, liver, and lungs. Disseminated infections are associated with septic shock and high mortality, though they may also produce a more chronic, indolent course of infection.

Before the discovery of antibiotics, chronic infection in humans occurred less often than rapidly progressive and acutely fatal infections (Robins, 1906). Chronic infection in horses was quite common, and was less so in mules. Such infections may be intermittently latent. Chronic infection can be expected in surviving untreated or failed-treatment human patients, as spontaneous cure—if it occurs—is exceedingly rare. Abscess of any organ or tissue, including the skin, mucous membranes, spleen, liver, kidney, brain,

bone, and lungs, may occur chronically. Weight loss and emaciation may increase with chronicity, with or without other clinical signs. Recurrences and remissions over several years may occur in humans and particularly in horses. The longest recorded human infection is 15 years (Robins, 1906). Horses may become long-term asymptomatic carriers and as such are hazardous sources of infection.

HISTORY OF GLANDERS AS A BIOLOGICAL WEAPON

Glanders was one of the first agents of BW to be used in the twentieth century. During World War I, Germany employed an ambitious biological sabotage campaign in several countries on both western and eastern fronts, including the United States, Russia, Romania, France, and Mesopotamia. Cattle, horses, mules, and other livestock destined to be shipped from the United States to her allies were also targeted and inoculated with *B. mallei* cultures (Wheelis, 1998). One saboteur was an American-educated surgeon named Anton Dilger, who was a member of the German Army in 1914 but was sent home to live with his parents in Virginia after a nervous breakdown (the United States was still neutral at this time). He brought with him strains of *Bacillus anthracis* and *B. mallei*, and with his brother Carl set up a laboratory to grow the organisms in a private home in Chevy Chase, Maryland. The organisms were delivered to another German saboteur in Baltimore, who inoculated horses awaiting shipment to the allies in Europe. German agents infected 4500 mules in Mesopotamia with *B. mallei*, and a German agent was arrested in Russia in 1916 with similar intentions. French cavalry horses were targeted with glanders as well (Smart, 1997). Attempts were also made to contaminate animal feeds in the United States. The Central Powers successfully infected large numbers of mules and horses on the eastern front in Russia, which had the effect of impairing artillery movement as well as troop and supply convoys. Human cases increased in Russia during and after the war, in concert with animal infections.

The Japanese offensive BW program Unit 731 used *B. mallei* to deliberately infect horses, civilians, and Prisoners of war (POWs) in occupied Manchuria leading into and during World War II. The "research" involved inhumane experimentation and field trials using a variety of different organisms. During experiments with glanders in 1937, there were two deaths at Ping Fan involving Japanese laboratory workers accidentally exposed to the organism (Regis, 1999).

In response to perceived BW threats from Japan and Germany, in 1942, the United States began work on BW agents at Camp Detrick (now Fort Detrick), Maryland, under the leadership of George Merck and Ira Baldwin (Regis, 1999). The scientists at Detrick studied glanders (known as agent LA) as a possible BW agent in the 1943–1945 time frame, but did not weaponize it. Between November 1944 and September 1953, there were seven laboratory-acquired human infections from *Malleomyces mallei* (the taxonomic name of glanders at that time) in the U.S. BW program employees at Camp Detrick. The first six of these were reported in a case series by Howe and Miller, which remains the largest human case series in the U.S. medical literature (Howe and Miller, 1947). The seventh case was not previously published; all seven original case files were reviewed for this chapter. Another laboratory-acquired systemic case of glanders occurred in March of 2000 in the course of U.S. defensive research on *B. mallei* (CDC, May 2000).

In 1972, the United States signed the Convention on the Prohibition of the Deployment, Production, and Stockpiling of Bacteriological (Biological) and Toxin Weapons and on Their Destruction (BWC), which banned development, production, stockpiling, acquisition and retention of biological agents, toxins, and the weapons to deliver them (Smart, 1997). All offensive BW work at Fort Detrick had stopped by this time, and any remaining biological weapons were destroyed by 1973. However, research aimed at the defense against bacterial agents and toxins that may be used in warfare, including *B. mallei*, continues to be conducted in the United States.

A report by the Monterey Institute of International Studies suggests that Russia may have the capability to use *B. mallei* as a BW agent. There are no known historical or current attempts for acquisition and use by terrorists (Monterey Institute of International Studies, 2002).

IMPORTANT ASPECTS OF WEAPONIZATION

Although glanders has been given comparatively little attention as a BW or bioterrorism threat relative to other agents, it could be of significant concern if weaponized. In 1947, Rosebury and Kabat believed *B. mallei* to be a possible BW agent because of its high infectivity, the high degree of incapacitation of those infected, and the availability of the agent (Rosebury and Kabat, 1947). *Burkholderia mallei* is quite infectious by the respiratory route, as demonstrated by the history of laboratory-acquired infections, but is not contagious (i.e., transmissible from person to person) via aerosol. Obtaining access to the organism should not be difficult for a determined bioterrorist, either from an infected animal or from a laboratory or commercial culture collection. As its clinical symptoms are protean and nonspecific, and because glanders is relatively unknown in the west, postattack diagnosis and treatment may be delayed, even in regions or countries with the most advanced medical facilities. Delayed diagnosis and therapy could lead to significant morbidity and mortality. Since equids and some other animals are susceptible, glanders could cause a combination of human and animal disease,

and possibly further spread from animals to humans. Treating glanders in humans may be complicated by the relative paucity of knowledge and experience regarding effective treatment. However, glanders is curable, and premedication (prior to the presentation of symptoms) may be an option to protect individuals if exposure is suspected, though genetic engineering could produce a virulent organism with atypical antibiotic resistance. For all these reasons, if glanders were to be cultivated, concentrated, and delivered as a wet or dry bacterial aerosol, significant casualties could result (Christopher et al., 1998).

PATHOGENESIS, CLINICAL COURSE, AND PROGNOSIS IN HUMANS

Even during its peak near the turn of the twentieth century, human glanders was uncommon—but very well documented. What is known about the clinical course of glanders is based on reports of hundreds of cases published well before the antibiotics era, and a small series of cases that occurred in the United States since the discovery of sulfonamides. The earlier reports describe a nearly always fatal disease of short (a few days to weeks) to long (months to years) duration, usually acquired from close contact with infected equids. The most recent cases were laboratory-acquired, and all patients survived. The following represents a compilation of information that spans 175 years.

Burkholderia mallei most often enters the body through abrasions or openings in the skin, particularly where occupationally exposed on the hands and forearms, face, and neck. An abrasion is not always present, however, at least grossly. Normal, intact skin is said to resist penetration of the organism; however, in several human infections, the affected insisted there was no wound or penetration during the likely exposure interval. Thus, a patient history in which there is no recollection of exposure to horses or of abrasion should not preclude glanders as a differential diagnosis. Organisms may also enter through oral, nasal, and ocular mucous membranes, as well as via inhalation. The latter has occurred in several laboratory-acquired infections; however, at least one laboratory-acquired case most likely occurred through cutaneous exposure. When they are present, the most characteristic feature of the disease is glanders nodes—small papular to egg-sized abscesses—which are very slow to heal if they open.

The incubation period is variable, ranging from less than a day to several weeks. Cutaneous and mucous membrane exposure generally leads to symptoms in 3 to 5 days, though without direct inoculation of the organism the duration may be longer (Robins, 1906). Inhalational exposure may incur a slightly longer range of about 7 to 21 days (Howe and Miller, 1947; Robins, 1906). In the absence of a wound, the first symptoms are likely constitutional. Once absorbed, the organism generally travels through lymph channels, first to regional lymph nodes, often causing irritation (lymphangitis, lymphadenitis) en route. If unchecked, organisms may enter the bloodstream and thus be carried anywhere throughout the body. Without proper treatment, the course of disease may range from one that is acute and rapidly fatal to a very slow, protracted course with alternating remissions and exacerbations.

Constitutional signs and symptoms typically occur early in the course of disease and some may persist through treatment. These may be exceedingly severe, leaving the patient in a state of extreme prostration. The more common findings include fever, low-grade fever only in the afternoon or evening, chills with or without rigors, severe headache, malaise, generalized myalgias (particularly of the limbs, joints, neck, back and flank), dizziness, nausea, vomiting, diarrhea, tachypnea, diaphoresis (including "night sweats"), altered mental status, and fatigue. Other nonspecific findings may be tender lymph nodes, sore throat, chest pain, blurred vision, splenomegaly, abdominal pain, photophobia, and marked lacrimation. Any or many of these signs may be present. Clinical courses are discussed in greater detail below as associated with route of entry or disease spread.

Infection of the respiratory tract may be anticipated after aerosol exposure, or secondarily as a consequence of disseminated infection. Pulmonary abscess and pleuritis are common sequelae. Symptoms may take up to 2 to 3 weeks to develop, and include tender cervical lymph nodes, fatigue, lymphangitis, sore throat, pleuritic chest pain, cough, fever (often greater than 102°F), chills, tachypnea, dyspnea, and mucopurulent discharge. Nonspecific signs are also usually present, such as night sweats, rigors, myalgia, severe headache, tachycardia, nausea, weight loss, dizziness, and mucosal eruptions. Some of the latter symptoms may indicate disseminated infection. Imaging studies may show diffuse or localized infiltration depending upon the stage of infection. Miliary to necrotizing nodules, or a localized (lobar to bilateral) bronchopneumonia, are other potential radiographic signs. Forming abscesses may be well circumscribed and circular, later becoming cavitated with evidence of central necrosis. Pleural irritation may also be visible on imaging studies. Acute bronchopulmonic or pneumonic disease if untreated tends to have a rapid onset of symptoms and was once said to be almost uniformly fatal within 10 to 30 days (Howe and Miller, 1947).

A review of eight laboratory-acquired infections occurring at Camp Detrick shows that the most common recorded symptoms (experienced by at least four) include, in order of most common occurrence, afternoon to evening low-grade fever, malaise, fatigue, headache, myalgias including backache, lymphadenopathy, and chest pain. These eight cases include six laboratory-acquired infections described by Howe and Miller in 1945, a previously unpublished case which occurred in 1953, and the most

recent case which occurred in 2000. Inhalation is suspected as the route of exposure for the first seven, whereas percutaneous exposure is believed to have led to the eighth (year 2000) case. According to case records of the first seven patients, the most common initial symptoms reported by the patient included fever, myalgia, headache, backache, and fatigue. One complained of a sore throat. Upon physical examination, cervical lymphadenopathy was present in three of the seven patients. For six of the seven presumed exposed by inhalation of infectious aerosols, radiographic signs were consistent with bacterial abscess or pneumonitis, and the primary site of infection was believed to be pulmonary. For the patient with clear chest radiography, signs and symptoms included fever, night sweats, malaise, headache, backache, drowsiness, dizziness, blurred vision, excess lacrimation, photophobia, nausea and vomiting, and enlarged spleen. Though inhalation was believed to be the route of entry, no obvious signs of respiratory tract infection were recorded even when signs suggesting disseminated infection (enlarged spleen) were present. Gastro-intestinal signs and symptoms were also present in 2 others exposed by inhalation; these included diarrhea, indigestion, flatulence and belching.

Septicemic glanders results from the seeding of *B. mallei* into the bloodstream, whether as a primary event, secondary to a local or pulmonary infection, or as a relapse to chronic or latent infection. Septicemia may be passing and lead to protracted disseminated infection, or may be fulminant and rapidly fatal. The thromboembolic process of glanders was well described by the early 1900s as follows (Robins, 1906; Loeffler, 1886). *Burkholderia mallei* causes damage and subsequent death of the endothelial cells lining the vessels. As they detach, the endothelial lining is predisposed to thrombosis. The thrombi serve as an excellent culture medium, and seed the bloodstream with bacteria. The embolic process may be realized by the patient as sharp stinging pain in the receiving part or tissue of the body. Robins describes one quite protracted chronic infection in which the patient was always aware of pain before multiple impending dissemination sites. Bacteremia is transient; however, the more acute or sudden the onset of a septicemic course the more likely *B. mallei* may be isolated from the blood. Bacteremia is also more likely shortly before and during the appearance of multiple eruptions and pustules, if they occur.

Historically, mortality rates have been reported to be 95 percent without treatment and up to 50 percent with treatment. A more recent analysis estimates the mortality rate for localized disease to be 20 percent when treated, and the overall mortality rate to be 40 percent (Batts-Osborne et al., 2001). Since the near-eradication of glanders and the development of worthy antibiotics, even these may be high estimates. Successful treatment (cure) was achieved in 100 percent of the eight U.S. laboratory-acquired cases, despite three of the eight cases (37 percent) experiencing a delay in effective treatment of 2 months. It is noteworthy that a brief period of "apparent recovery" is a common clinical feature, which can easily lead to delayed treatment and complications. Four of the eight patients were successfully treated with sulfadiazine for at least 20 days. The first two who received delayed treatment still recovered with only 10 days of sulfadiazine, though recovery was protracted. The most recent case (patient eight) had disseminated disease, which included abscesses of the spleen and liver, and was gravely ill, requiring ventilatory assistance before improving on a prolonged course of several antibiotics. These recent cases imply that prognoses range from good with localized infection and prompt treatment to guarded with septicemic infection.

TRANSMISSION AND EPIDEMIOLOGY

Naturally acquired cases of glanders are sporadic and rare; most countries have eradicated the disease. It is still infrequently reported in Northern Africa, South America, Europe, and Asia (OIE, 2002). Sporadic cases are also seen in South America. Because of serologic cross-reactivity with *B. pseudomallei*, the distribution (and prevalence) of *B. mallei* by serologic means alone cannot be determined. Though human outbreaks have been reported in Austria and Turkey, a human epidemic has not been reported (Neubauer et al., 1997).

In nature, the horse is the reservoir of *B. mallei*, and may also serve as the amplifying host. Donkeys are considered to be more prone to develop acute forms of glanders, and horses are more prone to develop chronic forms and latent disease. The mule, a crossbred animal resulting from the mating of horse and donkey, is susceptible to both acute and chronic forms and to latent infections (Henning, 1956; Steele, 1979; Verma, 1981). Humans are an accidental host. *Burkholderia mallei* is not found in water, soil, or plants.

Zoonotic transmission of *B. mallei* from equid to human is uncommon even with close and frequent contact to those infected. This may be explained by low concentrations of organisms from infection sites and a difference between humans and equids regarding susceptibility to virulent strains. During World War II, cases of human glanders were rare despite a 30 percent prevalence of infection among horses in China (Kortepeter, 2001). In Mongolia, between 5 and 25 percent of tested animals were reactive, yet no human cases were reported. With successful transmission, however, humans are quite susceptible to infection.

Exposures are typically occupational, hobby, or lifestyle-associated. Naturally infected humans have included veterinarians and veterinary students, farriers, flayers (hide workers), transport workers, soldiers, slaughterhouse personnel, farmers, horse-fanciers and caretakers, and stable hands. Subclinical or unapparent infections in horses and mules have posed a hidden risk to humans. Laboratory workers have also been sporadically infected. In contrast to zoonotic transmission, culture aerosols are highly infectious to laboratory workers. In the Howe and Miller case series, the six infected workers represented 46 percent of the personnel actually working in the laboratories during the year of occurrence.

Usually, transmission of the pathogen is by direct invasion of the nasal, oral, and conjunctival mucous membranes, by inhalation into the lungs, and by invasion of abraded or lacerated skin. Areas of the body most often exposed include the arms, hands, and face. Considering the affinity for warm and moist conditions (Loeffler, 1886), *B. mallei* may survive longest in stable bedding, manure, feed and water troughs (particularly if heated), wastewater, and in enclosed equine transporters. Transmission via handling contaminated fomites such as grooming tools, hoof trimming equipment, harnesses, tack, feeding and husbandry equipment, bedding, and veterinary equipment has occurred. Such equipment stored away from any contact with equids for at least 3 months, even without disinfection, is not likely to serve as a source of infection.

DETECTION AND DIAGNOSIS

Definitive diagnosis of glanders is dependent upon isolation and positive identification of the causative pathogen. Physical findings that support the differential diagnosis of glanders will most likely be linked to the potential route of infection. With pulmonary involvement (most likely originating from *aerosol* exposure), suspect clinical signs and symptoms include oropharyngeal injection, headache, chest pain, fever, rigors, night sweats, fatigue, cough, nasal discharge, and diagnostic imaging studies that support either localized or lobar pneumonia or bronchopneumonia, miliary nodules, lobar infiltrative pneumonia, and consolidation (early) or cavitating (later) pulmonary lesions. Neurologic signs may also be present, with or without obvious pulmonary signs. With cutaneous involvement and regional lymphadenopathy (most likely originating from *percutaneous* exposure to infected equids or contaminated fomites), clinical signs and symptoms include lymphadenopathy with or without ulceration, and single or multiple cutaneous eruptions that may be slow healing, particularly along lymphatic vessel pathways. For presentation at autopsy, suspect findings include disseminated nodular and ulcerative disease, particularly involving the spleen, lungs, and liver. Cultures of nodules in septicemic cases will usually establish the presence of *B. mallei*. These presentations support glanders as a differential diagnosis and prompt further testing to rule out *B. mallei* infection.

A clinical picture consistent with glanders should prompt immediate notification of local animal health authorities to explore potential cases of glanders in livestock, particularly equids. The converse is also true; glanders as a potential differential diagnosis in livestock warrants immediate notification of local regulatory animal and public health authorities. Outbreaks of cutaneous ulcerative disease in sheep, goats, and swine accompanying suspect human cases would be more consistent with a *B. pseudomallei* (melioidosis) outbreak than *B. mallei*. Because of the rarity of natural glanders infection, bioterrorism should also be immediately suspected, particularly in regions from which glanders has been eradicated. Human glanders without animal exposure or more than one human case is presumptive evidence of a biological attack. With this suspicion, regional public health authorities can initiate an appropriate emergency public health response for disease prevention, environmental decontamination, epidemiologic investigation, and criminal investigation (CDC, 2000; PHLS, 2003).

Because *B. mallei* has a high potential for aerosol or droplet production and laboratory-acquired infection, biosafety level three personnel and primary containment precautions are indicated for activities attempting to rule out *B. mallei* infection. Aseptically collected exudates from abscesses, cutaneous and mucous membrane lesions, sputum, and blood, as well as aspirates from preerupting nodules and abscesses, are excellent culture sources. Blood cultures are often not productive. Bacteremia is more likely during febrile peaks (and acute disease), thus sampling during such peaks may enhance chances for a productive culture. Among the eight U.S. laboratory-acquired infections, blood cultures were attempted at least once within several weeks of initial presentation. In at least the first seven, special media were used to enhance growth of *B. mallei*. All were negative. In the eighth case, a positive blood culture was obtained 2 months after initial presentation during an acute septicemic relapse in which the patient was in a guarded condition (Srinivasan et al., 2001).

Organism Identification

Several relatively new molecular method diagnostic capabilities exist today to reliably confirm specific identification of *B. mallei* within several hours. These include polymerase chain reaction (PCR)-based assays and DNA gene sequencing (Bauernfeind, 1998; Ramisse et al., 2003, Gee et al., 2003). The latter methods, as phenotypic testing and 16S ribosomal RNA gene-sequence analysis, identified *B. mallei* from other *Burkholderia* species in the U.S. laboratory-acquired infection in 2000 (Srinivasan et al., 2001). A PCR procedure was also developed to discriminate *B. mallei* and *B. pseudomallei*, and is based on differences detected in ribosomal DNA sequences (Bauernfeind et al., 1998).

PCR-based techniques and DNA gene sequencing are increasingly used in clinical settings and public health laboratories for bacterial identification (Patel, 2001). Automation of sequencing and improved efficiencies of reagents has also reduced the cost per test as well as the time required for identification. Further, because killed bacteria or their templates may be used, these techniques also have the great advantage of reducing the risk of

exposure and infection to laboratory personnel compared to conventional methods (Bauernfeind et al., 1998). These methods are not yet widely available for *B. mallei* identification; however, the current interest in biodefense research is prompting a continued increased capability based on recent publications (Bauernfeind et al., 1998; Gee et al., 2003; Ramisse et al., 2003; Sacchi et al., 2002; Sirisinha, 2000).

Pulsed-field gel electrophoresis (PGE) and ribotyping have been used to identify strains of *B. pseudomallei* in outbreaks (Inglis et al., 1999). These methods have also been used to differentiate pathogenic *B. pseudomallei* strains from less virulent strains (Pitt et al., 2000). PGE and ribotyping may be as useful for identification and virulence testing of *B. mallei*, though these methods may be more labor intensive and time consuming than gene sequencing. Gas-liquid chromatography of cellular fatty acids was used to help identify the organism in the laboratory-acquired infection in the United States in the year 2000 as a *Burkholderia* genus.

Imaging Studies

Radiographic imaging is useful to monitor pulmonary infection. Early radiographic signs are typically infiltrative or support early abscess formation. Segmental or lobar infiltrates are common. With time, pulmonary abscesses tend to undergo central degeneration and necrosis that radiographically resembles cavitation. These may be single or multiple. Unilateral or bilateral bronchopneumonia may be seen, as may a smattering of miliary nodules. Because of the potential for disseminated disease, computed tomography imaging (CT Scan) is quite useful for monitoring deep tissues and visceral organs.

TREATMENT AND PREVENTION

Because human cases of glanders are rare, limited information exists regarding effective antibiotic treatment. *B. mallei* infection responds to antibiotic therapy; however, recovery may be slow after a delayed diagnosis or with disseminated disease. Sulfonamides were used successfully in the first six U.S. laboratory-acquired infections. The seventh was successfully treated with the tetracycline compound aureomycin. Disseminated glanders in a stable hand who had only indirect contact with horses was also treated successfully with aureomycin in Austria in 1951 (Domma, 1953). Doxycycline was used with success in the laboratory-acquired infection in 2000 (Srinivasan et al., 2001). The sulfonamide sulfadiazine has been found to be effective in other animals as well. In vitro susceptibility tests have found *B. mallei* to be usually sensitive to sulfonamides, aminoglycosides, ciprofloxacin, novobiocin, several tetracyclines, imipenem, and ceftazidime (Kenny et al., 1999; Batmanov, 1993; Al-lzzi and Al-bassan, 1990; Russell et al., 2000).

Susceptibility testing of the *B. mallei* isolate in the laboratory-acquired infection case in 2000 demonstrated sensitivity to the latter three (Srinivasan et al., 2001). The patient was treated with imipenem and doxycycline for 2 weeks, and improved rapidly. A 6-month course of doxycycline and azithromycin followed, though retrospective susceptibility testing found that the organism was resistant to azithromycin. Diagnostic imaging of the patient's splenic and hepatic abscesses through the 6-month course showed their near-complete resolution. Susceptibility to streptomycin and chloramphenicol in vitro has been inconsistent, with some reporting sensitivity and others reporting resistance (Kovalov, 1971; Kenny et al., 1999; Batmanov, 1993). *Burkholderia mallei* is considered to be resistant to penicillin, bacitracin, and chloromycetin (CDC May 2000; Kovalov, 1971; Neubauer et al., 1997).

Recommendations for antibiotic therapy depend on the infection site and severity. Localized disease should be treated with at least a 2-month, and preferably a 6-month, course of antibiotics based on sensitivity. Without susceptibility test results, oral tetracycline at 40 mg/kg per day in three divided doses, trimethoprim-sulfamethoxazole at 4 and 20 mg/kg per day, respectively, in two divided doses, or amoxicillin and clavulanate at 60 mg/kg per day in three divided doses may be used. Combined therapy for at least the first month should be considered for patients with any mild systemic symptoms. For visceral disease without pulmonary involvement, prolonged treatment up to a year is recommended. For severe disease, such as with septicemia or pulmonary involvement, parenteral combined therapy of ceftazidime at 120 mg/kg per day in three divided doses plus trimethoprim-sulfamethoxazole at 8 and 40 mg/kg per day in four divided doses for 2 weeks is recommended, with oral therapy afterward for at least 6 months. Abscesses may be surgically drained, depending upon their location (Batts-Osborne et al., 2001).

Because of the intractable nature of glanders, long-term follow-up and possibly prolonged, tailored therapy is indicated for infections that are slow to clear. Patients should be followed at regular intervals for at least 5 years after recovery. Diagnostic imaging is useful to follow the reduction and resurgence of abscesses, serology may help to monitor the clearing of antibody, and inflammatory markers may also suggest resurgence of a latent infection. Patients should be advised of the lifelong risk of relapse, and to alert their health care providers of their previous history, particularly if they develop a febrile illness. This becomes even more important when potentially dealing with a genetically engineered strain of *B. mallei*.

There is no immunoprophylaxis for glanders. Antibiotic therapy may offer some protection, however, in the face of a *B. mallei* strategic attack. Prophylaxis with doxycycline and ciprofloxacin given before and coincident with intraperitoneal inoculation in rodents caused the minimum lethal dose to rise several thousandfold but did not completely protect against infection (Russell et al., 2000). Limitations to this approach are similar to other biological agents in that the organism may be engineered to be resistant to the anticipated antibiotic regimen.

INFECTION CONTROL AND DECONTAMINATION

The greatest risk for glanders exposure to humans outside of a biological attack is infected equids, particularly the asymptomatic horse. When glanders infection is considered as a differential diagnosis in countries with ongoing or completed eradication programs, local and state public health and veterinary authorities should be contacted immediately. Where human infection has occurred, potential exposure to infected equids should be investigated by a team approach involving patient care personnel, public health officials, and local veterinarians. A horse suspected as a possible human exposure source should be tested, and if positive, humanely destroyed in accordance with the local regulatory animal health authority. Facilities and transporters traced back to positive equine cases should be quarantined and disinfected in accordance with the local animal health authority. Stall bedding, feed, and manure in the vicinity of infected livestock should be burned.

In the event of deliberate release of *B. mallei*, emergency response personnel entering a potentially heavily contaminated area should wear protective gear, which includes a mask with a biological filter. Decontamination procedures for the patient include the removal and containment of outer clothing. Such clothing should be regarded as contaminated or "high risk" and handled in accordance with local protocol for the same. All waste should be managed in accordance with biosafety level three containment protocols. Patient showers are indicated, preferably in a facility for which decontamination and containment can be managed. The risk of acquiring infection from contaminated persons and their clothing is thought to be low (PHLS, 2003). Prophylactic treatment with ciprofloxacin or doxycycline may help to prevent infection in those potentially exposed, including emergency responders.

Environmental contamination will decline over time as a result of sunlight exposure and drying. Monitoring highly contaminated areas is indicated, however, and the advice of FAD (Foreign Animal Disease) experts should be sought. *Burkholderia mallei* can be destroyed by heating to at least 55°C for 10 min, and by ultraviolet irradiation. It is susceptible to several disinfectants, including 1 percent sodium hypochlorite, at least 5 percent calcium hypochlorite, 70 percent ethanol, 2 percent glutaraldehyde, iodine, benzalkonium chloride, at least 1 percent potassium permanganate, at least 3 percent solution of alkali, and 3 percent sulfur-carbolic solution. Phenolic and mercuric chloride disinfectants are not recommended (Kovalov, 1971; Gilbert, 1998).

Because human-to-human transmission has occurred nosocomially and with close personal contact, standard precautions are recommended. These include use of disposable gloves, face shields, surgical masks, and when appropriate, surgical gowns to protect mucous membranes and skin. Personnel, microbiological, and containment procedures for biosafety level three are advised in the laboratory. Appropriate barriers to direct skin contact with the organisms are mandatory at all times (CDC and NIH, 1999; CDC, 1996). Family contacts should be advised of blood and body fluid precautions for patients recovering at home. Barriers protecting mucous membranes, cuts and sores, and potential skin abrasions from genital, oral, nasal, and other body fluids are recommended.

Import restrictions for equids are also in place in many countries. Veterinary health authorities may require testing within a few weeks of shipment and again at the place of disembarkation, as well as documentation of the animal's location in the exporting country for the 6 months before shipment (OIE, 2003). Restrictions vary by country and glanders-free status under the International Animal Health Code. The most current information regarding import and export should be sought from the regional animal health authority.

CURRENT AND FUTURE IMPLICATIONS

Glanders is currently a Category B disease of concern for bioterrorism as classified by the CDC. This is because the agent is believed to be moderately easy to disseminate, dissemination would result in moderate morbidity and low mortality, and enhancements to current diagnostic capabilities and disease surveillance would be required to diagnose the disease rapidly and accurately.

With *B. mallei* as a contender for use as a BW or terrorism agent, raising the clinical index of suspicion for glanders in humans is in order. The rarity of recent human cases may make glanders a difficult diagnosis even in regions with exceptional medical facilities. As is the case with many rare diseases, final diagnosis and appropriate treatment is often delayed, with sometimes disastrous results. Without a higher index of suspicion, diagnostic laboratories may not conduct tests appropriate to detect *B. mallei*, as happened in the eight U.S. laboratory-acquired infection in 2000 (Srinivasan et al., 2001).

Further studies are needed to fully assess the usefulness of 16S rRNA sequencing as a tool in epidemiologic investigations. Further studies are needed to assess the potential of using the subtle variations in the 16S rRNA gene sequence as a subtyping method for virulence and toxin production as well.

The genetic homology between *B. mallei* and *B. pseudomallei* may cause confusion with melioidosis, especially in endemic areas for *B. pseudomallei*, presenting an additional challenge and inviting further research. The capability to distinguish virulent strains from nonvirulent, naturally occurring strains would also be useful. Finally, more research on antibiotic susceptibilities

to *B. mallei* is warranted. Specifically considering an aerosol threat from a virulent strain, studies to distinguish the effectiveness of therapeutic agents for treating septicemic and pulmonary infections are indicated. The potential for prophylactic treatment regimens should also be investigated.

Aerosol dissemination of *B. mallei* would likely cause disease in humans, equids, goats, and possibly cats in the vicinity. Unintentional infection may first manifest in equids or humans. Thus, teaming with animal health officials in the event of a suspected outbreak will expedite identification and control of an event. Though a formal surveillance system for glanders does not exist in the United States, local and state veterinary and public health authorities would be among the first to recognize a potential outbreak regardless of intent. These agencies would then work in concert with the United States Department of Agriculture (USDA), the CDC, and the Department of Health and Human Services (DHHS) to control and eradicate the disease.

REFERENCES

Alibasoglu, M., Yesilders, T., Calislar, T., Inal, T., and Calsikan, U., *Berl Munch Tierarztl Wochenschr*, **99**, 57–63 (1986).

Al-lzzi, S.A. and Al-bassan, L.S., *Comp. Immunol. Microbiol. Infect. Dis.*, **13**, 5–8 (1990).

Allen, H., *J. R. Army Vet. Corps*, **1**, 241–245 (1929).

Bartlett, J.G., "Glanders," in S.L. Gorbach, J.G. Bartlett, N.R. Blacklow, Eds., *Infectious Diseases*, 2nd ed., WB Saunders Company, Philadelphia, PA, 1998, pp. 1578–1580.

Batmanov, V.P., *Antibiot. Khimioter.*, **38**, 18–22 (1993).

Batts-Osborne, D., Rega, P.P., Hall, A.H., McGovern, T.W., *Emed. J.*, 1–12 (2001). Available at http://www.emedicine.com/emerg/topic884.htm, accessed July 2003.

Bauernfeind, A., Roller, C., Meyer, D., Jungwirth, R., Schneider, I., *J. Clin. Microbiol.*, **36**, 2737–2741 (1998).

Blood, D.C. and Radostits, O.M., "Diseases Caused by Bacteria," in *Veterinary Medicine*, 7th ed., Balliere Tindall, London, 1989, pp. 733–735.

CDC, *Am. J. Infect. Control*, **24**, 24–52 (1996).

CDC, *MMWR*, **24**, 532–535 (2000).

CDC, *MMWR*, **49**, 1–26 (2000).

CDC, NIH, *Biosafety in Microbiological and Biomedical Laboratories (BMBL)*, 4th ed., U.S. Department of Health and Human Services, Washington, DC, 1999. Available at http://www.cdc.gov/od/ohs/biosfty/bmbl/bmbl-1.htm, accessed July 2003.

Christopher, G.W., Cieslak, T.J., Pavlin, J.A., Eitzen, E.M., Jr., *J. Am. Med. Assoc.*, **5**, 412–417 (1998).

Domma, K., *Wiener Tierarztliche Monatsschrift*, **40**, 426–432 (1953).

Gee, J.E., Sacchi, C.T., Glass, M.B., De, B.K., Weyant, R.S., Levett, P.N., Whitney, A.M., Hoffmaster, A.R., and Popovic, T., *J. Clin. Microbiol.*, **41**, 4647–4654 (2003).

Gilardi, G.L., "Pseudomonas," in E.H. Lenette, A. Bulawo, W.J. Hausler, and H.J. Shadomy, Eds., *Manual of Clinical Microbiology*, 4th ed., American Society for Microbiology, Washington, DC, 1985, pp. 350–372.

Gilbert, R.O., Ed., "Glanders," *Foreign Animal Diseases*, Pat Campbell & Associates, and Carter Printing Company, Richmond, VA, 1998. United States Animal Health Association. Available at http://www.vet.uga.edu/vpp/gray_book/FAD/gla.htm, accessed July 2003.

Henning, M.W., *Animal diseases in South Africa*, Central News Agency, Johannesburg, South Africa, 1956, pp. 159–181.

Howe, C. and Miller, R.R., *Ann. Intern. Med.*, **26**, 93–115 (1947).

Inglis, T.J., Garrow, S.C., Adams, C., Henderson, M., Mayo, M., and Currie, B.J., *Epidemiol. Infect.*, **123**, 437–443 (1999).

Jana, A.M., Gupta, A.K., Pandya, G., Verma, R.D., and Rao, K.M., *Indian Vet. J.*, **59**, 5–9 (1982).

Jennings, W.E., "Glanders," in T.G. Hull, Ed., *Diseases Transmitted from Animals to Man*, 5th ed., Charles C. Thomas Publisher, Springfield, IL, 1963, pp. 264–292.

Katz, J.B.C., Hennager, S.G., Nicholson, J.M., Fisher, T.A., and Byers, P.E., *J. Vet. Diagn. Invest.*, **11**, 292–294 (1999).

Kenny, D.J., Russell, P., Rogers, D., Eley, S.M., and Titball, R.W., *Antimicrob. Agents Chemother.*, **43**, 2773–2775 (1999).

Kortepeter, M., Ed., *USAMRIID's Medical Management of Biological Casualties Handbook*, 4th ed., U.S. Army Medical Research Institute of Infectious Diseases, Fort Detrick, MD, 2001, pp. 44–52.

Kovalov, G.K., *Zhurnal Midrobiologii Epidemiologii I Immunobiologii*, **1**, 63–70 (1971).

Krieg, N.R. and Holt, J.G., Eds., "Gram-negative Aerobic Rods and Cocci," in *Bergey's Manual of Systematic Bacteriology*, Vol. 1, Williams & Wilkins, Baltimore, MD and London, 1984, pp. 174–175.

Loeffler, F.A., *Trans. Imp. Health Off.*, **1**, 141–198 (1886).

Mendelson, R.W., *U. S. Armed Forces Med. J.*, **7**, 781–784 (1950).

Miller, W., Pannel, L., Cravitz, L., Tanner, A., and Ingalls, M., *J. Bacteriol.*, **55**, 115–126 (1948).

Minett, F.C., "Glanders (and Melioidosis)," in A.W. Stableforth and I.A. Galloway, Eds., *Infectious Diseases of Animals*, Vol. 1, *Diseases Due to Bacteria*, Butterworths Scientific Publications, London, 1959, pp. 296–309.

Mohammad, T.J., Sawa, M.I., and Yousif, Y.A., *Indian J. Vet. Med.*, **9**, 1517 (1989).

Monterey Institute of International Studies, "Chemical and Biological Weapons: Possession and Programs Past and Present," *The Chemical & Biological Weapons Resource Page*, Center for Nonproliferation Studies, Monterey, CA, 2002. Available at http://cns.miis.edu/research/cbw/possess.htm. Accessed August 2003.

Neubauer, H., Meyer, H., and Finke, E.J., *Revue Internationale Des Services De Sante Des Forces Armees*, **70**, 258–265 (1997).

OIE, *Terrestrial Animal Health Code—2003*. Available at http://www.oie.int/eng/normes/mcode/A_00086.htm, accessed September 2003.

OIE, *Data on Animal Diseases*, 2002. Available at http://www.oie.int/eng/maladies/en_mal.htm, accessed October 2003.

Patel, J.B., *Mol. Diagn.*, **6**, 313–321 (2001).

PHLS 2003: Glanders and Melioidosis Interim Public Health Laboratory Service (PHLS), *Guidelines for Action in the Event of a Deliberate Release*. Health Protection Agency (HPA) for England and Wales. May 29, 2003. Available at http://www.phls.org.uk/topics_az/deliberate_release/menu.htm, accessed July 2003.

Pitt T.L., Trakulsomboon, S., and Dance, D.A., *Acta. Trop.*, **74**, 181–185 (2000).

Popov, S.F., Mel'nikov, B.I., Lagutin, M.P., and Kurilov, V., *Mikrobiol. Zh.*, **53**: 90–92 (1991).

Ramisse, V., Balandreau, J., Thibault, F., Vidal, D., Vergnaud, G., and Normand, P., *Int. J. Syst. Evol. Microbiol.*, **53**, 739–746 (2003).

Regis, E., *The Biology of Doom*, Henry Holt and Company, New York, 1999.

Robins, G.D., *Stud. R. Victoria Hosp. Montreal*, **1**, 1–98 (1906).

Rosebury, T. and Kabat, E.A., *J. Immunol.*, **56**, 7–96 (1947).

Russell, P., Eley, S.M., Ellis, J., Green, M., Bell, D.L., Kenny, D.J., and Titball, R.W., *J. Antimicrob. Chemother.*, **45**, 813–818 (2000).

Rutherford, J.G., *Special Report on Glanders* (from the Veterinary Director-General and Live Stock Commissioner), Canada Department of Agriculture, Health of Animals Branch, September, 1906, pp. 1–24.

Sacchi, C.T., Whitney, A.M., Mayer, L.W., Morey, R., Steigerwalt, A., Boras, A., Weyant, R.S., and Popovic, T., *Emerg. Infect. Dis.*, **8**, 1117–1123 (2002).

Sirisinha, S., Anuntagool, N., Dharakul, T., Ekpo, P., Wongratanacheewin, S., Naigowit, P., Petchclai, B., Thamlikitkul, V., and Suputtamongkol, Y., *Acta. Trop.*, **74**, 235–245 (2000).

Smart, J.K., "History of Chemical and Biological Warfare: an American Perspective," in R. Zajtchuk and R.F. Bellarmy, Eds., *Textbook of Military Medicine*, Borden Institute and Office of the Army Surgeon General, Walter Reed Army Medical Center, Washington, DC, 1997, pp. 16, 64.

Srinivasan, A., Kraus, C.N., DeShazer, D., Becker, P.M., Dick, J.D., Spacek, L., Bartlet, J.G., Byrne, W.R., and Thomas, D.L., *N. Engl. J. Med.*, **4**, 256–258 (2001).

Steele, J.H., "Glanders," in J.H. Steele, Ed., *CRC Handbook Series in Zoonoses, Section A: Bacterial, Rickettsial and Mycotic Diseases*, Vol. 1, CRC Press, Boca Raton, FL, 1979, pp. 339–362.

Verma, R.D., *Indian Vet. J.*, **58**, 177–183 (1981).

Verma, R.D., Sharma, J.K., Venkateswaran, K.S., and Batra, H.V., *Vet. Microbiol.*, **25**, 77–85 (1990).

Vesley, D. and Hartman, H.M., *Am. J. Public Health*, **78**, 1213–1215 (1988).

Weyant, R.S., Moss, C.W., Weaver, R.E., Hollis, D.G., Jordan, J.G., Cook, E.C., and Daneshvar, M.I., "*Pseudomonas pseudomallei*," in W.R. Hensyl, Ed., *Identification of Unusual Pathogenic Gram-Negative Aerobic and Facultatively Anaerobic Bacteria*, 2nd ed., Williams & Wilkins, Baltimore, MD, 1995, pp. 486–487.

Wheelis, M., *Nature*, **395**, 213 (1998).

Wilson, G.S. and Miles, A.A., "Glanders and Foie Melioidosis," in E. Arnold, Ed., *Topley and Wilson's Principles of Bacteriology and Immunity*, London, England, 1964, pp. 1714–1717.

Xie, X., Xu, F., Xu, B., Duan, X., and Gong R, *A New Selective Medium for Isolation of Glanders Bacilli*, Collected Papers of Veterinary Research, Control Institute of Veterinary Biologics, Ministry of Agriculture, Peking, China, Vol. 6, 1980, 83–90.

Yabuuchi, E., Kosako, Y., Oyaizu, H., Yano, I., Hotta, H., Hashimoto, Y., Ezaki, T., and Arakawa, M., *Microbiol. Immunol.*, **12**, 1251–1275 (1992).

WEB RESOURCES

Department for Environment Food and Rural Affairs (DEFRA), United Kingdom, 2003 Glanders and Farcy. Available at http://www.defra.gov.uk/animalh/diseases/notifiable/disease/glanders.ht, accessed July 2003.

OIE, *Manual of Standards for Diagnostic Tests and Vaccines 2000*, 2000. Available at, http://www.oie.int/eng/normes/mmanual/A_00076.htm, accessed September 2003.

H

HAMAS (HARAKAT AL-MUQAWWAMA AL-ISLAMIYYA, ISLAMIC RESISTANCE MOVEMENT)

ANJALI BHATTACHARJEE

OVERVIEW

Emerging shortly after the 1987 outbreak of the first Palestinian *intifada* ("uprising"), HAMAS has positioned itself as one of the leading resistance movements in the disputed territories of Gaza and West Bank. On the basis of strong religious and social foundations, HAMAS serves as a strong internal opposition to the Palestinian Liberation Organization (PLO), which has now assumed a more bureaucratized, quasi-governmental form in the Palestinian Authority (PA). HAMAS leaders claim that the PA has become too corrupt and does not represent the true needs of the Palestinian populace. Consequently, members of HAMAS reject the secular nationalist and socialist ideology propagated by older PA leaders, and in its place promote a radical Islamist doctrine in an effort to bring about social and political change. In addition to its political agenda of liberating Palestine from Israeli occupation, it has sought to mobilize popular support and Islamize Palestinian society by setting up several educational, medical, and charitable facilities in the West Bank and Gaza strip. Currently HAMAS's membership is in the tens of thousands, and although there is no official membership roster, members and sympathizers are known to be prevalent throughout the disputed territories, Arab nations, and other parts of the world.

BACKGROUND

The roots of HAMAS can be traced back to the Muslim Brotherhood (Ikhwan al-Muslimin) movement, which originated in Egypt in the late 1920s and thence spread across the Arab world since it appealed to young Muslims seeking an organization and ideology that might aid their struggles for political independence. The Muslim Brotherhood had a branch in the Gaza Strip called the *Islamic Center*, which aggressively promoted a return to basic Islamic values. HAMAS first turned to violence during the so-called knives war of 1989, when members of the group conducted a stabbing campaign against Israeli nationals. As a result, over 400 HAMAS members were deported to southern Lebanon, where they were eventually trained in explosives and warfare tactics by the local Shi'ite group Hizb'allah. The military wing of HAMAS, the Izz al-Din al-Qasam Brigades, has been primarily responsible for the group's attacks against Israel.

SELECTED CHRONOLOGY

HAMAS has claimed responsibility for numerous attacks in the Israeli territory. A selection of the major incidents involving nonconventional warfare agents is provided below:

- On 21 August 1995, a handwritten statement from Abu Ahmad (aka "Muhammad Salah"), head of the military wing of HAMAS, indicated that his organization had attempted to recruit members in the United States with expertise in chemical weapons. Approximately 27 individuals were allegedly recruited by the group to manufacture weapons on the basis of their fields of expertise. The operation was called the "Palestine Organization/Enlistment," and was initiated by Musa Abu Marzuq, the leader of the Muslim Brotherhood in the United States. Abu Ahmad was instructed by Marzuq to collect the names of Palestinians residing in the United States who had expertise in "chemical materials, toxins, physics, military education, and knowledge of computers." HAMAS provided the materials used in the manufacturing of these weapons, such as "remote-control activation, agricultural pesticides, and basic chemical materials."

- According to statements made by Steve Emerson, director of the documentary "Jihad in America," in 1998, Israeli officials uncovered plans to poison Israeli food and water supplies with chemical and biological agents. This discovery was supposedly made when members of HAMAS were arrested in Hebron, in the course of which their plans to poison Israeli food and water were allegedly found in computer records.

- A joint investigation into HAMAS activities conducted by the Israeli security forces and the Palestinian National Authority in 1999 uncovered evidence that HAMAS was planning a chemical attack against Israel. The investigation resulted in the arrest of HAMAS activist Alif 'Ayin, who later confessed that he had received charts, formulas, tables, and "detailed" instructions on the methods used in the manufacture of chemical weapons from sources abroad while living in Israel. The purpose of the attack was believed to be the disruption or prevention of Israeli elections.

- On 19 June 1999, the newspaper *Yidyout Ahranout* reported that members of HAMAS planned to

Encyclopedia of Bioterrorism Defense, Edited by Richard F. Pilch and Raymond A. Zilinskas
ISBN 0-471-46717-0 Copyright © 2005 Wiley-Liss

poison water supplies in Israel. Ibrahim Ghawsha, a spokesman for HAMAS, denied the allegations and claimed that the confession sourced in the report had been the result of "severe torture" at the hands of Israeli authorities.

- In its "Washington Whisper" section of the 6 March 2000 issue of *U.S. News and World Report*, Paul Bedard reported that HAMAS may have used unspecified chemical weapons at some point in its history.

- On 29 November 2000, Emergency Net News cited the *Al-Jadida* newspaper to the effect that HAMAS was in possession of "a dangerous chemical weapon." The report stated that Ibrahim ibn Awda, the alleged bombmaker for the Izz al-Din al-Qasam Brigades who had been killed during an explosion in Nablus on 23 November 2000, had produced this chemical weapon at some point. To date, it remains unclear whether Ibn Awda had in fact ever worked on chemical weapons or who, if anyone, in HAMAS possessed such a weapon.

- On 1 December 2001, a bomb, later confirmed to have contained chemical agents, was detonated by a HAMAS suicide bomber at a pedestrian mall in Jerusalem. A spokesperson claimed that nails contained within the bomb had been dipped in rat poison or some other form of pesticide, but that the contaminant had burned off during the explosion. At the same time, Izz al-Din al-Qasim proclaimed on HAMAS's website that the group now possessed a new weapon that would "create a situation of fear in the Zionist security services."

- On 10 December 2001, Israeli authorities thwarted a suicide bomber who reportedly intended to detonate a "crude chemical weapon" in Haifa.

- On 27 March 2002, a HAMAS suicide bomber detonated his explosives in front of the Park Hotel in Netanya, Israel. Some months later, on 5 June 2002, Israeli Defense Forces' (IDF) Major General Aharon Zeevi claimed that the perpetrator had planned for the bomb to release cyanide gas (hydrogen cyanide) upon detonation. Although the bomb did kill 29 Israelis, it apparently failed to release the gas because of technical problems. Sources claimed that Al Naja University in Nablus and another laboratory on the West Bank had carried out the cyanide research. Major General Zeevi also claimed that the work had been performed with the help of Hizb'allah, Iran, and Syria.

- On 16 June 2002, Israeli Television Channel 2 stated that HAMAS was planning to use chemical weapons in the near future. The report stated that HAMAS had in the past employed chemical weapons in its suicide attacks, but that the chemicals used had been routinely consumed in the blast (see, e.g., the 10 December 2002 entry above) and that the group was now working with stronger chemicals to overcome this problem. According to the report, these weapons, once developed, would be used to perpetrate a mass casualty attack in Israel.

- On 1 August 2002, HAMAS militant 'Abbas al-Sayyid was arrested in Tulkarim. The ensuing investigation revealed that he had intended to disperse a chemical agent, believed by Israeli officials to be cyanide, in a suicide attack. Technical difficulties, however, had postponed the attack.

ANALYSIS

From the brief chronology outlined above, it remains unclear whether HAMAS will pursue the acquisition of unconventional warfare agents, particularly biological warfare (BW) agents. Thus, in order to predict whether or not HAMAS might undertake a BW option in the future, the group's goals and motives must be understood. HAMAS's declared goal is to claim Israeli territory back from the "occupying forces" for the Palestinian people. In a more realistic sense, several talks have been underway for years between the Israeli and Palestinian authorities to negotiate a peaceful settlement of the disputed territory. HAMAS differentiates itself from other Palestinian opposition groups—excepting Islamic Jihad—by advocating fundamentalist Islamic agendas and by rejecting any secular ideologies. It promotes a host of social services in the West Bank and Gaza strip in order to demonstrate its committed support for the Palestinian populace.

In order to achieve its political goals, HAMAS has conducted numerous terrorist attacks that have brought international attention to its cause. Shootings, knifings, assaults, and bombings are all among the different tactics undertaken by HAMAS members. Such terrorist attacks serve two ends: First, they cause significant fatalities, thereby ensuring that their enemies, the Israelis, feel the same pain as their Palestinian "victims"; second, they attract attention to the perceived plight of the Palestinian people. Bombings, especially suicide bombings, have already achieved these results in a dramatic way, both by generating the largest number of fatalities of various types of attack in the West Bank and by instilling fear and panic among Israelis.

From a strictly rational perspective, it would appear that using a biological weapon would not serve HAMAS's interests at present. Although HAMAS members have allegedly experimented with chemicals, the chemicals are thought to have only been delivered in conjunction with conventional bombs. In these cases, the poison is said to have been consumed in the blasts and thus rendered useless, as noted above. There have been no confirmed reports that HAMAS members have attempted to weaponize any BW agent. Furthermore, two important factors deter the group from doing so, namely, (1) territorial proximity to its intended target(s) and (2) the potential for Israeli retaliation. Palestinians and Israelis do not have a large spatial barrier separating their communities. Therefore, a contagious biological agent delivered among the Israeli community would place Palestinians at significant risk, and even a noncontagious agent might have a boomerang effect on the Palestinian population if delivered ineffectively or during inappropriate meteorological conditions. Moreover, it appears highly doubtful that HAMAS has the necessary resources to accurately develop a biological weapon, and even if such a weapon were to be successfully developed and used against Israel, the retaliation from Israeli authorities would almost certainly be swift and deadly, making the investment of time, money, and resources in this type of unknown and untested weapon wasteful for the group, particularly when cheaper, proven methods are readily available.

There is a flipside to this argument, however. Though rationally it does not make sense for HAMAS to pursue a biological weapons capability, not every individual within a given group is a rational thinker. It is quite plausible that a HAMAS member or splinter faction might seek to cause greater panic or suffering, or simply to demonstrate the group's scientific prowess, by using such an agent, especially considering that several HAMAS members are known to have strong scientific backgrounds. Ultimately, only time will tell whether members of HAMAS, whether working alone or in conjunction with the group's leadership, choose to follow this path in the future.

FURTHER READING

Dolnik, A. and Bhattacharjee, A., *Terror. Political Violence*, **14**(3), 109–128.

Islamic Resistance Movement, *Biographies of Palestinian HAMAS Leaders*, March 3, 1998, Internet, available from FBIS, document ID FTS19980307000607, accessed on 7/1/99.

Mishal, S. and Sela, A., *The Palestinian HAMAS: Vision, Violence and Coexistence*, Columbia University, New York, 2000.

Monterey WMD-Terrorism Database, Center for Nonproliferation Studies, 1998–2003.

Preparations for CB Attack Viewed, The Jerusalem Report in English, March 29, 1999, Internet, available from FBIS, document ID FTS19990320000602, accessed on 6/28/99.

United States District Court for the Southern District of New York,, *The Extradition of Mousa Mohammed Abu Marzuq*, May 8, 1996.

WEB RESOURCES

Federal Information Systems Corporation, Prepared Statement of Steven Emerson Before the Senate Judiciary Committee Subcommittee on Terrorism, Technology, and Government Information Subject-Foreign Terrorists in America: Five Years After the World Trade Center Bombing, February 24, 1998, Internet, available from http://www.geocities.com/CollegePark/6453/emerson.html, accessed on 8/13/03.

HEMORRHAGIC FEVER VIRUSES

Stephen A. Berger

INTRODUCTION

Ninety-three distinct human diseases are caused by viruses. Twelve of these are characterized by high fever, severe multisystem illness, and hemorrhage. As a group, these "hemorrhagic fevers" share many epidemiological and clinical features. Few are treatable or preventable through vaccination. All are caused by RNA-containing viruses that occur naturally in mammals. Propagation, maintenance, and identification can only be performed by highly specialized laboratories, operating under maximum standards of biosafety. Several of these diseases are highly contagious and would have a devastating effect if used as agents of bioterrorism. The short incubation period and overt illness associated with bleeding from multiple body sites would add to their psychosocial impact if introduced into a civilian or military community.

SCIENTIFIC NAMES

Hemorrhagic fever viruses considered potential agents of bioterrorism are found in four genera:

1. *Filoviridae.* Ebola virus, Marburg virus
2. *Arenaviridae.* Junin virus (Argentine Hemorrhagic Fever, AHF), Machupo virus (Bolivian Hemorrhagic Fever, BHF), Sabia virus (Brazilian Hemorrhagic Fever), Lassa fever virus, Guanarito virus (Venezuelan Hemorrhagic Fever, VHF)
3. *Bunyaviridae.* Crimean-Congo Hemorrhagic Fever (CCHF) virus, hantaviruses, Rift Valley Fever (RVF) virus
4. *Flaviviridae.* Omsk Hemorrhagic Fever (OHF) virus, Yellow Fever (YF) virus

Three additional viral diseases (dengue, Colorado tick fever, and Kyasanur Forest disease) may occasionally present with a hemorrhagic fever syndrome, but are not considered to have bioterror potential.

PATHOGENS

Filoviridae

General interest in the hemorrhagic fever viruses began in 1995 with an epidemic of Ebola in Kikwit, Zaire (Fig. 1). In fact, outbreaks of Ebola had been described as early as 1976, but the advent of cable television and potential for spread beyond Africa were lacking at the time, and thus the world hardly took notice. To date, a total of 13 outbreaks have been reported, involving six countries and approximately 1600 victims. Potential for natural disease exists in 15 countries, all in sub-Saharan Africa. In 1989, Ebola virus (Reston strain) was identified in monkeys (*Cynomolgus macaques*) imported from the Philippines to the United States; at least four human contacts seroconverted without clinical illness. Seven years later, two imported monkeys died of Ebola in Texas.

Marburg disease was first described in 1967 in Marburg, Germany, and the former Yugoslavia among laboratory workers handling vervet monkeys originating from Uganda. The disease did not reappear again until 1975, when Australians hitchhiking in Zimbabwe were infected, with additional secondary infections among health care workers while the index cases were receiving treatment in South Africa. The next cases were described in Kenya in 1980 and again in 1987. A tourist returning to Sweden from Kenya in 1990 was also thought to have acquired the disease. A case of laboratory infection with Marburg virus in Russia was described in the literature in 1994. The world's first extensive outbreak of Marburg fever occurred in the Democratic Republic of Congo (DRC) from 1998 to 1999, with additional cases in the area during 2000.

Arenaviridae

Four of the hemorrhagic fevers endemic to South America are caused by arenaviruses. Each is acquired from rodents and is limited to a single country: Junin (Argentina), Machupo (Bolivia), Sabia (Brazil), and Guanarito (Venezuela). Lassa fever is reported in 15 countries but is thought to occur in all of West Africa, from Nigeria to Senegal. According to some estimates, as many as 500,000 cases may occur yearly. Disease rates peak during January to April ("the dry season"), and all age groups are susceptible to Lassa infection.

Encyclopedia of Bioterrorism Defense, Edited by Richard F. Pilch and Raymond A. Zilinskas
ISBN 0-471-46717-0 Copyright © 2005 Wiley-Liss

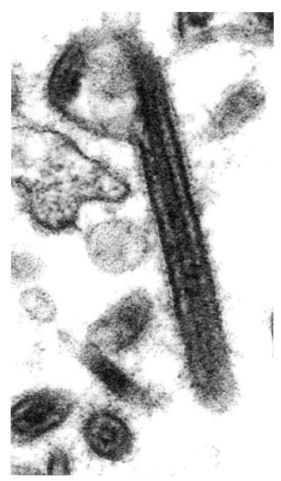

Figure 1. Colorized transmission electron micrograph of the Ebola Virus, Public Health Image Library (PHIL), id # 1835; Credit: CDC/C. Goldsmith.

Bunyaviridae

CCHF is a more widely distributed, albeit lesser-known disease that was initially described in the Crimea in 1944, and later equated with an illness that occurred in the Congo in 1956. Fifty-one countries are considered as potentially endemic for the disease. Widely scattered cases have been confirmed in Europe and Asia, most commonly among adult males working with livestock. Evidence of the virus has been found among ticks in Africa, Asia, the Middle East, and Eastern Europe. Similarly, hantavirus infections have been reported in 90 countries in Europe and Asia, with an estimated 200,000 patients requiring hospitalization every year. The hantavirus genus includes a number of distinct viruses (Hantaan, Puumala, Dobrava, Seoul, etc.), which vary in geographic distribution and clinical severity. (A less common but more widely publicized group of hantaviruses associated with lung infection in the Americas is beyond the scope of this review.)

RVF is endemic or potentially endemic in 34 countries. An outbreak in Egypt during 1977 to 1978 involved 18,000 human cases and 598 fatalities. An outbreak in Kenya, Somalia, and Tanzania during 1997 to 1998 involved an estimated 89,000 humans. The virus was first encountered in West Africa in 1974 among mosquitoes in Senegal, and human disease in the area was first reported in 1987 during an epidemic in southern Mauritania. Rift Valley fever was first reported outside of the African continent (Saudi Arabia and Yemen) in 2000. Outbreaks usually occur first in animals, often heralded by outbreaks of unexplained abortion amongst livestock. Human infection is seen mainly among farmers and others at occupational risk. Sheep are more susceptible than cattle; goats are least susceptible. Infected bats have been identified in Guinea.

Flaviviridae

OHF was first reported in the Omsk district, Russia, during 1943 to 1944, and since then has been encountered only in Russia and Romania. Disease incidence peaks from April to June and September to November and, in Russia, is limited to western and northwestern Siberia. Cases during the winter are associated with hunting activities. Most patients are rural residents, hunters, and agricultural workers.

Yellow fever occurs in both South America and Africa, and currently affects between 100 and 500 patients per year. Sporadic outbreaks are largely limited to jungle environments ("sylvatic yellow fever"), with only rare instances of spread into cities ("urban yellow fever").

USE AS A WEAPON

History

One of the milestones in modern bioterrorism was the introduction of sarin gas into an urban subway system by the Aum Shinrikyo cult in 1995. It was subsequently claimed that members of the cult had traveled to Zaire (present day Democratic Republic of Congo) in a failed attempt to acquire samples of Ebola virus. This is the only know instance of attempted acquisition of a hemorrhagic fever virus for illicit purposes, and as of January 2004, there have been no allegations of development or stockpiling of these agents by hostile governments or criminal groups.

Former biological warfare (BW) programs of both the United States and the former Soviet Union are known to have investigated hemorrhagic fever viruses for offensive application. The agent of choice by the United States in this regard was yellow-fever virus, which was considered but not successfully weaponized before the offensive program was terminated in 1969. The former Soviet Union was successful in weaponizing Marburg virus, and is known to have researched Ebola virus, Bolivian and Argentinean hemorrhagic fever viruses, and Lassa fever virus as well.

Important Considerations

The obvious potential of these viruses as agents of bioterrorism is outweighed by difficulties in safety, propagation, storage, and delivery. From the point of view of persons who might seek to weaponize hemorrhagic fever viruses, there are few vaccines to protect workers and hardly any effective treatments if a worker were to contract hemorrhagic fever. All of these agents can only be grown in tissue culture, eggs or animals, in facilities having rigorous standards of biosafety. Most would not survive in the environment and cannot be delivered in aerosol, water, or food. Indeed, many are only transmissible through the bites of arthropods. Although this would obviate dissemination in the setting of a mass attack, some agents (CCHF, RVF) could theoretically be introduced into cattle populations via infected ticks or mosquitoes.

TRANSMISSION

In nature, most of these diseases are maintained and transmitted among mammals, including:

- *Rodents.* Calomys species (AHF, BHF), unknown species (Brazilian hemorrhagic fever), field mouse and vole (hantaviruses), rat (hantaviruses), Mastomys species (Lassa), muskrat (OHF), cane mouse (VHF)
- *Hares.* CCHF
- *Cattle, Sheep, or Goats.* CCHF, RFV
- *Primates.* YF, Marburg, possibly Ebola
- *Bats.* Possibly Marburg and Ebola

Aerosols of animal excreta represent the principal vehicle for infection in AHF, Brazilian hemorrhagic fever, CCHF, hantaviruses, Lassa fever, and VHF. In addition, BHF and Lassa fever are naturally acquired from contaminated food. As such, the potential for delivery of these agents through aerosol or food may enhance their threat potential. Four diseases in this group are generally transmitted through the bites of mosquitoes (RVF and YF) or ticks (CCHF, OHF), and at least five have been acquired through contact with infected patients (BHF, CCHF, Ebola, Lassa fever, Marburg disease). In many cases, people have acquired Ebola through washing and preparing the bodies of relatives who had died of the disease. It is prudent to assume that all hemorrhagic fever viruses might be acquired through contact with patients, patient specimens, and laboratory culture material.

Person-to-person transmission occurs with Lassa fever, especially in the hospital environment, by direct contact with blood (including inoculation with contaminated needles), pharyngeal secretions, or urine, or alternatively by sexual contact. Person-to-person spread may occur during the acute phase of fever when the virus is present in the throat. The virus may be excreted in the urine of patients for three to nine weeks from the onset of illness. The virus can be transmitted via semen for up to three months.

Nosocomial outbreaks of Lassa fever have been described in Nigeria, Liberia, and Sierra Leone. Nosocomial transmission has also been documented for AHF, BHF, and Ebola. Laboratory infection has occurred while handling Marburg virus and Brazilian hemorrhagic fever virus.

CLINICAL COURSE

As a group, the hemorrhagic fevers are all characterized by fever, headache, arthralgia, and muscle or back pain. Most are also associated with abdominal pain, nausea, vomiting, and a rash. Neurological signs ranging from confusion to seizures and coma may be prominent. Additional features may include photophobia, lymphadenopathy, sore throat, a macular or petechial rash, or the appearance of a faucal enanthem. The hallmark of all of these diseases is hemorrhage, which may involve the oral or nasal mucosa, the lungs, or the gastrointestinal tract. As illness progresses, renal, hepatic, cardiac, or pulmonary failure may develop.

All of these diseases are associated with severe morbidity and high case/fatality ratios. Reported death rates by disease are as follows: AHF 30 to 40 percent; BHF 18 percent; Brazilian hemorrhagic fever 67 percent (only three cases have been reported); CCHF 30 percent; Ebola 50 to 88 percent; hantavirus infections 0.1 percent (Puumala virus) to 10 percent (Hantaan virus); Marburg 60 percent; OHF 0.5 to 2.5 percent; RVF 0.5 to 2 percent; VHF 19 to 33 percent; and YF 10 to 60 percent.

Permanent sensorineural deafness is common following Lassa fever, and blindness may occur following Rift Valley fever. In most cases, however, survivors of hemorrhagic fever are surprisingly free of long-term or permanent sequelae.

DETECTION AND DIAGNOSIS

Specific diagnosis can be accomplished through isolation of virus from patient specimens, serologic testing, detection of viral RNA or antigens, and occasionally through electron microscopy. Whole blood or serum for viral isolation should be placed on dry ice or otherwise suitably frozen for shipment to designated laboratories. Serum for antibody analysis can be stored in an ordinary refrigerator, or even at normal ambient temperatures if processed immediately. The importance of safety and containment in sample collection, packaging, and transport cannot be overemphasized.

As a rule, serum, whole blood, and biopsy material from involved tissues will be useful in the diagnosis of any of these diseases. Additional specimens that have been suggested include liver (AHF, BHF, Brazilian hemorrhagic fever, Ebola, Lassa fever, Marburg, VHF, YF), spleen (AHF, BHF, Brazilian hemorrhagic fever, Ebola, Lassa, Marburg, VHF), cerebrospinal fluid (CCHF, RVF), and throat washings (Lassa fever).

Laboratories working at biosafety level 3 (BSL-3) are appropriate for the diagnosis of hantavirus, RFV, and YF. All of the other diseases discussed in this session require BSL-4 facilities.

TREATMENT AND PROPHYLAXIS

The approach to any patient with suspected or confirmed hemorrhagic fever must include strict isolation, with particular attention to blood, laboratory specimens, and other secretions. As such, persons involved in transport and analysis of specimens must also adhere to proper precautions. When dealing with diseases transmitted by mosquitoes or ticks, appropriate measures will also include netting, insecticides, and so on.

Therapy consists largely of monitoring and correction of vital organ function (renal, cardiac, and pulmonary) in addition to replacement of blood loss. A single drug, intravenous ribavirin, has been used successfully in the treatment of AHF, BHF, Brazilian hemorrhagic fever, hantavirus infections, and Lassa fever. An initial dose of 2.0 g (30 mg/kg in children) is followed by 1 g (15 mg/kg) every 6 h for four days and then 0.5 g (7.5 mg/kg) every 8 h for 6 days. Specific immune globulin has been used in some cases of AHF and BHF.

Vaccines are available for the prevention of AHF, hantavirus infections, and YF. The AHF vaccine is given intramuscularly as a single 1.0-ml dose of attenuated virus. The hantavirus vaccine is an inactivated preparation given as 0.5 ml at 0 and 30 days. The YF vaccine is an attenuated virus given as 0.5 ml subcutaneously. The reader should become familiar with potential toxicities and contraindications before receiving any of these vaccinations.

FURTHER READING

Anonymous, *Morb. Mortal. Wkly. Rep.*, **37**(Suppl. 3), 1–16 (1998).

Borio, L., Inglesby, T., Peters, C.J., Schmaljohn, A.L., Hughes, J.M., Jahrling, P.B., Ksiazek, T., Johnson, K.M., Meyerhoff, A., O'Toole, T., Ascher, M.S., Bartlett, J., Breman, J.G., Eitzen, E.M., Jr., Hamburg, M., Hauer, J., Henderson, D.A., Johnson, R.T., Kwik, G., Layton, M., Lillibridge, S., Nabel, G.J., Osterholm, M.T., Perl, T.M., Russell, P., Tonat, K., *J. Am. Med. Assoc.*, **287**(18), 2391–2404 (2002).

Buchmeier, M.J. and Bowen, C.J., "Arenaviridae: The Viruses and their Replication," in D.M. Knipe and P.M. Howley, Eds., *Fields Virology*, Vol. 2, 4th ed., Lippincott, Williams & Wilkins, Philadelphia, PA, 2001, pp. 1635–1668.

Colebunders, R. and Borchert, M., *J. Infect.*, **40**, 16–20 (2000).

Feldmann, H., Slenczcka, W., and Klenk, H.D., *Arch. Virol.*, **11**, 77–100 (1996).

Gear, J.H., *Rev. Infect. Dis.*, **11**(Suppl. 4), S777–S782 (1989).

WEB RESOURCES

US Centers for Disease Control and Prevention, *Biosafety Levels*, http://www.cdc.gov/od/ohs/biosfty/bnmbl4/bmbl4s7f.htm.

See also AUM SHINRIKYO AND THE ALEPH and CDC CATEGORY C AGENTS.

HIZBALLAH (HEZBOLLAH, HIZB' ALLAH, AND HIZBULLAH, ISLAMIC JIHAD, REVOLUTIONARY JUSTICE ORGANIZATION, ORGANIZATION OF THE OPPRESSED ON EARTH, AND ISLAMIC JIHAD FOR THE LIBERATION OF PALESTINE)

SKYLER JOHN CRANMER

OVERVIEW

Hizballah means "Party of God." As such, Hizballah's ideology is based on a fundamentalist interpretation of Islam. The group adheres to a Manichean notion of the world as being divided between oppressors (*mustakbirun)* and oppressed (*mustad'ifin)*, and sees the relationship between these sides as a battle between good and evil that must be fought and won by the oppressed at any cost. Hizballah seeks to create an Islamic state ruled by Islamic law, believing that Islamic law is the one true law and is, by its divine nature, universal (making any state or group adhering to any other law, evil). Further, the group sees all Islamic peoples as part of the Party of God, such that the plight of any Islamic people is the plight of all Islamic people. This has provided the ideological basis for Hizballah's intervention in the Palestinian conflict: the group is "coming to the rescue" of its compatriots. Hizballah regards the United States as the "Great Satan," and all other western countries as simply "evil" (Hajjar, 2002).

Hizballah has an extensive governmental structure that consists of political, military/terrorist, economic, and humanitarian branches. The highest governing body is the *Majlis al-Shura* (Consultative Council), which is led by Secretary General Hassan Nasrallah. Hizballah is headquartered in Beirut, Lebanon, with areas of operation that include heavy involvement in the Biqa' Valley, Harmil, the southern suburbs of Beirut, and Southern Lebanon. Cells are also believed to have been established in Europe, Africa, South America, North America, and Asia. Hizballah's exact strength is unknown, but it is estimated to have a few hundred active members and several thousand supporters (U.S. Department of State, 2002).

Hizballah's primary state affiliation is with Iran. Because Hizballah is both ideologically based upon and partially founded by the Islamic establishment in Iran, Iran carries much influence with the group. If Hizballah's leadership is unable to resolve an issue, the matter is referred to its supreme leader, Iran's Ayatollah Khamene'i. Hizballah also receives financial aid, training, weapons and explosives, and political and diplomatic support from Iran.

The Lebanese government regards Hizballah as important to its society and affords the group extensive protection and privileges. Hizballah also controls a number of seats in the Lebanese Parliament. The group has also received diplomatic, political, and logistical support from Syria.

Hizballah may have ties with other terrorist organizations as well. A 2001 U.S. Department of State report accuses Hizballah of providing training and weaponry to HAMAS, the Palestinian Islamic Jihad (PIJ), and the Palestinian Liberation Organization (PLO) in the West Bank and Gaza (Hajjar, 2002).

Hizballah remains an active terrorist group that poses an immediate threat to Western entities in the Middle East (Israel in particular) and has global reach.

HISTORICAL BACKGROUND

Hizballah was officially founded in 1982 in response to the Israeli invasion of Lebanon, but the group's history can be traced to an earlier date. In 1975, Musa al-Sadr created a Shi'ite militia named Afwaj al-Muqawamah al-Lubnaniya (AMAL: Legions of the Lebanese Resistance). Following the Israeli invasion of 1982, radicals within AMAL objected to what they believed to be accommodation of Israel by the AMAL leadership. The radicals left AMAL and, aided by Syria's willingness to host the training of some 1000 members of the Iranian *Pasdaran* (Revolutionary Guard) under their command in the Biqa' valley, founded Hizballah. Hizballah then began a long campaign of partisan and terrorist attacks against Israeli targets, both military and civilian, in the occupied territory of Southern Lebanon.

Hizballah began its political activities in 1989, and has remained heavily involved in Lebanese politics and the Lebanese government ever since (Ranstorp, 1997). The group gained credibility with its constituent population after the Israeli withdrawal from Southern Lebanon in May 2000, for which it claimed responsibility.

SELECTED CHRONOLOGY

While a complete listing of Hizballah attacks is beyond the scope of this writing, it is useful to note several major attacks as well as the range of attack methodology employed by the group. Hizballah has been responsible for a number of high-profile attacks, including the suicide truck bombings of the U.S. Embassy in Beirut in April

1983 and the U.S. Marine barracks in Beirut in October 1983, the 1985 hijacking of TWA Flight 847, and perhaps also the bombing of the Israeli Embassy in Argentina in 1992, and the 1994 bombing of the Israeli cultural center in Buenos Aires. In the course of its history, Hizballah has carried out suicide truck bombings, car bombings, hijackings, hostage taking operations, grenade attacks, shootings, and missile/rocket attacks (U.S. Department of State, 2002).

BIOLOGICAL WEAPONS CAPABILITY

To date, there is no information indicating that Hizballah either has been or currently is in possession of biological weapons of any kind. However, Israel has accused Hizballah of possessing chemical weapons, an allegation that Hizballah has forcibly denied (Dakrouib, 2003).

MOTIVATION TO PURSUE BIOLOGICAL WEAPONS

The sole biological weapons alarm that has been raised with regard to Hizballah was put forth by former director of the Central Intelligence Agency (CIA), R. James Woolsey. In an interview with the Center for Security Policy on 17 March 1998, Woolsey commented that "trying to have a verification regime that would, on a routine basis, go into pharmaceutical facilities and look at them would really only penalize the people who are... behaving themselves and staying within the law.... You're not going to find what Hezbollah is doing with biological weapons that way" (Woolsey, 1998). This rather elusive and ominous comment has led to much speculation, but no corroborating evidence from any party, United States or otherwise, has since been brought to bear.

IMPLICATIONS

Hizballah is an organization of at least some concern regarding future prospects for biological proliferation. Its fundamentalist mindset feeds into an apparent belief that Islamic dominance in the Middle East region,

if not the world, must be established by any means necessary. Hizballah has also shown its willingness and resourcefulness in employing a wide variety of means to attack its targets. Further, Hizballah is closely involved with Iran, which is believed by many analysts to be developing biological weapons on at least a rudimentary level (U.S. Arms Control and Disarmament Agency, 1996) and that therefore could potentially supply Hizballah with such weapons (although no such transfer has yet been alleged or confirmed). The fact that biological weapons are inherently difficult to control, however, and that Hizballah's constituency lives in close proximity to its most likely target in Israel, both militate against its biological weapons use in the foreseeable future. Further, Hizballah has been dedicating a great deal of attention to legitimizing itself as a political force. Thus, although it has yet to renounce terrorism and seek international recognition, Hizballah's political success may eventually preclude it from direct involvement in acts of terrorism of any kind (Ackerman and Snyder, 2002).

REFERENCES

Ackerman, G. and Snyder, L., Would They if They Could? *Bull. Atomic Scientists*, May/June, 2002.

Dakrouib, H., *Hezbollah Threatens Israel over Flights in Lebanon, Denies Having Chemical Weapons*, Associated Press World Politics, February 2, 2003.

Hajjar, S., *Hizballah: Terrorism, National Liberation, or Menace?*, Strategic Studies Institute - U.S. Army War College, August 2002, 8–17.

Hizballah, Program, *Al-Safir*, February 16, 1985.

Ranstorp, M., *Hizb' Allah in Lebanon: The Politics of the Western Hostage Crisis*, Saint Martins Press, New York, 1997, pp. 33–39.

U.S. Arms Control and Disarmament Agency, *Adherence to and Compliance with Arms Control Agreements*, U.S. Government Printing Office, Washington, DC, 1996, p. 68.

U.S. Department of State, Patterns of Global Terrorism: 2001 Report, U.S. State Department, March 2002, p. 95.

Woolesy, R.J., *Filmed Interview Conducted with the Center for Security Policy on 17 March*, 1998.

HOMELAND DEFENSE (DISEASE PREVENTION, EPIDEMIC CONTROL)

AMY E. SMITHSON

INTRODUCTION

Just as terrorists have become disturbingly adept at making and inflicting harm with bombs, no one can dismiss the possibility that terrorists could master the technical intricacies associated with releasing disease agents in a manner that would cause massive casualties or significant losses in plant and animal life. Therefore, concerted efforts are afoot to ready frontline U.S. emergency personnel to contend with an intentionally inflicted disease disaster. These efforts center around public health and health care facilities, particularly on the strengthening of plans and capabilities to prevent, prepare for, mitigate, respond to, and recover from a major disease outbreak in the human population. Programs have also been launched to bolster readiness for agroterrorist attacks. This entry will review such efforts at the local, state, and federal levels.

DISCUSSION

Although terrorist attacks are likely to continue to involve bombs and guns far more frequently than germs, emergency personnel still need to become better prepared to deal with disease catastrophes because disease erupts naturally (Institute of Medicine, 1992; World Health Organization, 2000). Controlling the spread of a contagious disease will be very difficult in an era when people tend to live in megacities and can travel between continents in just hours. Traditional methods of control (e.g., quarantine) will be difficult to implement before a contagious disease has already spread, but policymakers are being encouraged to institute legislative reforms that would aid decision making during this type of crisis (Model State Emergency Powers Act, 2001). Among the other challenges confronting public health and medical personnel are how to identify a disease outbreak in sufficient time to take lifesaving intervention, to provide swift health care screening and prophylaxis to large numbers of people, and to give more sophisticated medical care to those who contract the disease (Institute of Medicine, 1999; Smithson and Levy, 2000).

Research and Laboratory Involvement

One of the most important federal bioterrorism readiness efforts underway is accelerated research to discover, develop, and test new vaccines and antibiotics. The National Institutes of Health (NIH), select Department of Defense (DoD) agencies, the Food and Drug Administration (FDA), university researchers, and the U.S. biotechnology and pharmaceutical industries are key players in these drug development efforts. The Centers for Disease Control and Prevention (CDC) is overseeing improvements to communications capabilities in public health departments as well as to state and regional laboratories so that more of these laboratories can identify exotic diseases. The CDC also established and maintains the National Strategic Stockpile (SNS) of drugs and medical equipment that would be deployed in the event of a serious disease outbreak.

Training and Readiness

In 1996, Senators Sam Nunn and Richard Lugar championed domestic preparedness programs to train and equip local responders for all types of terrorist attacks. At that time, the Department of Health and Human Services (HHS) also inaugurated the establishment of Metropolitan Medical Response System teams in larger U.S. cities, an endeavor that involved planning and training for terrorist-induced epidemics (Institute of Medicine, 2002). After the 2001 anthrax letter attacks, the HHS dispersed additional funds to help states and local jurisdictions with bioterrorism readiness. This preparedness funding has given rise to several innovative concepts to heighten bioterrorism readiness and response capabilities. Some U.S. cities are working toward full implementation of these concepts, which are organizationally complex and have yet to expand nationwide.

Surveillance

Traditional disease surveillance capabilities, which rest largely with laboratory analysis and clinicians, can be augmented by improving laboratory capabilities and providing physicians with better diagnostic tools. Taking another approach, New York City pioneered syndrome surveillance, which harnesses available data and compares it with historic data, tracking indicators of flu-like illness since the initial symptoms of exposure to many biowarfare agents closely resemble the symptoms of influenza. An upsurge in syndromic cases could tip officials off to a disease outbreak long before physician or laboratory diagnosis occurs, which can in turn trigger more intensive laboratory testing and, if appropriate, epidemiological surveys to pin down the identity of the mystery illness. This early warning can allow officials to activate response capabilities at the earliest possible moment.

The components of syndrome surveillance include sentinel nursing homes, watchdog public hospitals that report daily on the number of certain types of hospital admissions and discharges, and sentry laboratories that

Encyclopedia of Bioterrorism Defense,   Edited by Richard F. Pilch and Raymond A. Zilinskas
ISBN 0-471-46717-0   Copyright © 2005 Wiley-Liss

report on the number of samples submitted for certain tests. The sale of selected over-the-counter medications that consumers purchase for home treatment of flu-like symptoms is monitored. Daily patterns of 911 calls are tallied, focusing on seven call types correlated with flu-like illness. Unexplained deaths due to possible infectious disease among individuals from ages 2 to 49 are reviewed daily. Other indicators that can be monitored include levels of school absenteeism, sick calls into employee health clinics for public safety and public transportation personnel, and outpatient calls to hotlines used by health maintenance organizations to direct people to medical care. Finally, sentinel veterinarians can also be incorporated to keep tabs on epizootic diseases.

Treatment

Once the cause of a disease outbreak is identified, local officials will know which vaccines or antibiotics to administer but may not have a sufficient supply to treat an eighth, a quarter, a half, or all of the citizens in a large U.S. city. To cut costs, hospitals usually keep on hand just enough medicine to last two or at most three days. Local officials would probably request the dispatch of the Strategic National Stockpile, which should arrive within 12 hours of deployment, accompanied by a small staff to help with distribution. In the interim, local officials would endeavor to provide medical treatment to symptomatic patients as well as prophylaxis to emergency responders, and, if possible, their families. To do so, local authorities may tap into a local cache of key emergency medications. Local officials are using different strategies to cut the costs of maintaining an emergency supply of medicines. For example, manufacturers do not guarantee the efficacy of a drug past its expiration date, but testing has shown that several medications are still 90 percent viable 10 years after their expiration dates; thus, emergency stockpiles can be tested for persisting efficacy. In addition to shelf-life extension testing, the local drug cache can be extended indefinitely by housing the medications at a local hospital that agrees to use and then restock antibiotics and other drugs prior to their expiration dates.

Patient and Hospital Management

Another daunting problem facing emergency response personnel is that an epidemic could rapidly overwhelm hospital capacity, bringing health care systems to the verge of collapse. To prevent local health care systems from buckling under the strain, hospitals throughout a region need to engage in advance detailed planning, agreeing how burdens would be shared between hospitals during a disease disaster. For example, hospitals would need to agree which ones would convert over to care for symptomatic patients, allowing the hospitals with advanced trauma, burn, cardiac, and other critical capacities to remain open to contend with the routine medical problems that would continue to occur during the outbreak. Comorbidity must be addressed in this context as well. Regionally, hospitals would need to ascertain how to obtain important supplies aside from medications (e.g., linens, intravenous fluids, food) that would enable them to operate at above-average capacity for the duration of the disaster. The regional hospital plan would also need to include steps to improve and routinely test command, control, and communication systems among hospitals.

To divert the tidal wave of frightened, possibly sick people away from hospitals, a regional hospital plan needs to address surge capacity at the hospitals as well as medical outposts away from them. Cities are attempting to create an overflow capacity adjacent to or nearby hospitals, in buildings or temporary field care centers. Some cities have standing arrangements to take over large indoor arenas and stadiums in an emergency. Health care can also be taken to the citizens in their neighborhoods by setting up treatment centers in familiar locations, such as fire stations, schools, local health clinics, and the heating/cooling centers used during temperature extremes. Mobile field care centers would also be deployed. In these outposts, cross-disciplinary teams of emergency medical technicians, paramedics, nurses, physicians, and mental health workers would conduct medical exams, dispense antibiotics and vaccines, and provide counseling as appropriate. Symptomatic patients would be sent to regular or field hospitals. This strategy would be suitable for noncommunicable diseases, but problematic if a contagious disease were in play. For those circumstances, where it is key to keep person-to-person contact to a minimum, one city is planning to commandeer fast-food restaurants and administer drive-through prophylaxis.

Since severe personnel shortages are expected, especially among nurses, a regional hospital plan would also address how to tap into local personnel reservoirs. After calling on temporary service companies, officials are considering assigning emergency medical technicians to hospitals for routine medical care, freeing up regular staff to work with critical cases and in the intensive care unit. Emergency medical technicians from prisons and coal mines may also be drafted for hospital service. Cities would probably call for federal help with manpower at the same time that they request the strategic pharmaceutical stockpile. Federal officials contend that vanguard assets, such as Disaster Medical Assistance Teams, could be there within 12 to 24 hours. However, if disease is erupting in several areas simultaneously, local officials need to have contingency plans to provide health care for an extended period of time without substantial federal aid.

Other key aspects of a regional hospital plan include providing adequate security for the hospitals and their key supplies; contending with the management of large numbers of casualties that may require special handling because of religious or cultural affiliations or that may be so contaminated that special handling would be ineffectual; and orchestrating an aggressive, cohesive public information campaign to quell panic so that citizens will have confidence that their health care needs will be met. The terms of regional hospital burden sharing and the other features of a regional hospital plan need to be set far in advance of a health care crisis. At present, very few U.S. metropolitan area hospital planning committees have begun to broach the level of detail and collaboration required, much less put the finishing touches on such plans.

Agroterrorism Issues

While most individuals think first about the human toll that a bioterrorist attack could cause, one must not forget that diseases that damage and kill plants and livestock have also been turned into weapons. The former USSR, for example, employed roughly 10,000 scientists and technicians in its anticrop and antilivestock bioweapons programs (Alibek and Handelman, 1999). Agroterrorist attacks have resulted in the loss of few lives, but it is widely recognized that the release of disease agents such as foot and mouth disease virus, rinderpest virus, wheat fungus, and potato virus could have significant economic repercussions (Pate and Cameron, 2001; Brown, 1999). Since roughly 13 percent of the U.S. gross domestic product and about 17 percent of U.S. employment is related to agricultural activities and because the ability to feed the U.S. population has self-evident national security implications, policymakers have considered a number of options to enhance regular agricultural disease surveillance and response capacities (Horn, 2000).

Just as physicians, nurses, and laboratory technicians are on the forefront of detecting and identifying an outbreak of human disease, so are farmers and veterinarians the vanguard of any efforts to monitor the health of livestock and crops. A plethora of diseases can affect animals and plants, making proper initial training and continuing education of these key caretakers essential to disease surveillance. In addition, recommendations to harden the nation's agriculture sector against terrorism include increasing: (1) the ability of field personnel to screen imported seeds and livestock feed; (2) the number and capabilities of laboratories to identify relevant diseases rapidly; (3) safeguards at key agricultural chokepoints, such as feedlots; and (4) the response capabilities of special teams that would deploy quickly to curtail, eradicate, and clean up following a disease outbreak should one occur (Parker, 2002). Several branches of the U.S. Department of Agriculture, in conjunction with the Department of Homeland Security (DHS) at the federal level and the food and agriculture departments at the state level, are leading these efforts.

REFERENCES

Alibek, K. and Handelman, S., *Biohazard*, Random House, New York, 1999.

Brown, C., "Economic Considerations of Agricultural Diseases," in T.W. Frazier and D.C. Richardson, Eds., *Food and Agricultural Security: Guarding against Natural Threats and Terrorist Attacks Affecting Health, National Food Supplies, and Agricultural Economies*, New York Academy of Sciences, New York, 1999, pp. 92–95.

Horn, F., "Agricultural Terrorism," in *Hype or Reality?: The "New Terrorism" and Mass Casualty Attacks*, B. Roberts, Ed., Chemical and Biological Arms Control Institute, Alexandria, VA, 2000, pp. 109–115.

Institute of Medicine, *Emerging Infections: Microbial Threats to Health in the United States*, National Research Council, National Academy Press, Washington, DC, 1992,

Institute of Medicine, *Chemical and Biological Terrorism: Research and Development to Improve Civilian Medical Response*, Committee on R&D Needs for Improving Civilian Medical Response to Chemical and Biological Terrorism Incidents, National Academy Press, Washington, DC, 1999.

Institute of Medicine, *Preparing for Terrorism: Tools for Evaluating the Metropolitan Medical Response System Program*, Committee on Evaluation of the Metropolitan Medical Response System Program, National Academy Press, Washington, DC, 2002, pp. 23–24, 52–74.

Model State Emergency Powers Act, Center for Law and the Public's Health at Georgetown and Johns Hopkins Universities, With the National Governors Association, National Conference of State Legislatures, Association of State and Territorial Health Officials, National Association of City and County Health Officers, and National Association of Attorneys General, Georgetown University Law Center, Washington, DC, 2001.

Parker, H.S., *Agricultural Bioterrorism: A Federal Strategy to Meet the Threat*, McNair Paper 65, National Defense University, Washington, DC, 2002.

Pate, J. and Cameron, G., *Covert Biological Weapons Attacks against Agricultural Targets: Assessing the Impact against US Agriculture. Executive Session on Domestic Preparedness*, Harvard University, Boston, MA, August 2001, http://bcsia.ksg.harvard.edu/BCSIA_content/documents/Covert_Biological_Weapons_Attacks_Against_Agricultural_Targets.pdf.

Smithson, A.E. and Levy, L.A., *Ataxia: The Chemical and Biological Terrorism Threat and the US Response*, Stimson Center, Washington, DC, October 2000, pp. 244–271.

World Health Organization, *Overcoming Antimicrobial Resistance: World Health Report on Infectious Diseases 2000*, World Health Organization, Geneva, 2000.

WEB RESOURCES

Centers for Disease Control and Prevention, Office of Public Health Emergency Preparedness & Response, www.bt.cdc.gov/.

Johns Hopkins Center for Civilian Biodefense, www.hopkinsbiodefense.org/.

Monterey Institute, Center for Nonproliferation Studies, www.cns.miis.edu/.

See also AGRICULTURAL BIOTERRORISM; CONSEQUENCE MANAGEMENT; CRISIS MANAGEMENT; DETECTION OF BIOLOGICAL AGENTS; METROPOLITAN MEDICAL RESPONSE SYSTEM; PUBLIC HEALTH PREPAREDNESS IN THE UNITED STATES; SCIENTISTS, SOCIETIES, AND BIOTERRORISM DEFENSE IN THE UNITED STATES; and SYNDROMIC SURVEILLANCE.

HUMAN IMMUNODEFICIENCY VIRUS

Monterey WMD-Terrorism Database Staff

INTRODUCTION

Shortly after Human Immunodeficiency Virus (HIV), the causative agent of Acquired Immunodeficiency Syndrome (AIDS), first appeared in the 1980s, a Soviet Union–supported disinformation campaign alleged that the virus had been created by Army scientists experimenting with recombinant DNA technology at Fort Detrick, Maryland, former home of the United States's offensive biological weapons program. Epidemiologists were quick to demonstrate that this argument could not be supported by scientific evidence, however, and the rumor was ultimately attributed to propaganda and discarded. Almost two decades later, however, the threat of HIV as an apparent weapon has reemerged, this time in the context of bioterrorism.

HIV possesses a number of characteristics that make it poorly suited as a biological warfare (BW) agent. Its prolonged incubation results in a delayed effect are measured not in days but years from the time of initial exposure. Even when effects are felt, they progress slowly, and slower still with treatment. And treatment regimens have improved at such a rapid pace that the lethality of the disease has been significantly reduced over the past decade. But potential terrorist weapons differ in quality from traditional warfare agents because they are meant to achieve different ends. Thus, from a terrorism perspective, HIV possesses multiple attributes that make it particularly appealing. First, it continues to be perceived by the general public as a devastating and deadly virus, thus increasing its ability to generate panic. Second, it is transmissible from person to person, such that an initial pool of infection can propagate itself, causing long-term psychological and economic damage. And perhaps most importantly, the ease with which it may be delivered makes the virus particularly intimidating on an individual level: While widespread contamination with HIV is difficult to achieve, a simple needlestick is enough to cause infection in an unprotected person.

Despite these attributes, HIV has received relatively little attention in the biodefense arena. Precedents do exist for its illicit use, however, suggesting that this threat may warrant a greater degree of concern. This entry reviews four cases of deliberate contamination with HIV and one case of threatened HIV use in order to gain some insight into the virus's potential application in criminality and in terrorism.

CASE ONE: GRAHAM FARLOW

In July 1990, Geoffrey Pearce, an Australian prison guard, was attacked and injected with HIV-positive blood by Graham Farlow, an inmate at Sydney's high-security Long Bay prison. Pearce tested positive for HIV four weeks later. Farlow died of AIDS before he could stand trial.

At the time of the incident, Farlow was HIV-positive but asymptomatic. He allegedly attacked Pearce after Pearce opened a security gate for him. Pearce felt a stab and saw that Farlow had dropped a needle. Testing proved that the blood inside was Farlow's and was indeed HIV-positive. Farlow was diagnosed with full-blown AIDS nine months after the incident, and died two months later. Despite being treated withzidovudine (AZT), Pearce contracted the virus and died in 1997, seven years after the attack.

CASE TWO: BRIAN T. STEWART

On February 6, 1992, Brian Stewart injected his 11-month old son with HIV-infected blood at St. Joseph's Medical Center-West in Lake Saint Louis, Missouri. He did not want to pay child support and apparently believed that his son would die quickly upon contamination. Stewart was charged with first-degree assault and sentenced to life in prison.

Stewart was a medical technician who drew blood for patients at Barnes Hospital in St. Louis. He and Jennifer Jackson, the boy's mother, were married in July 1993 (they divorced in 1996). Their son was born 13 months after they first met. Jackson disclosed that Stewart had refused to pay child support and even claimed that the boy had not been fathered by him.

At the time of the incident, the 11-month old boy was being treated at St. Joseph's Medical Center-West in Lake Saint Louis for a respiratory condition. Witnesses claim to have seen Stewart walk into his son's room carrying a lab coat and then emerge from the room a few minutes later. For years afterward the boy was chronically ill, and in May 1996, doctors finally diagnosed him with AIDS. An investigation was opened, and after two years a case

Encyclopedia of Bioterrorism Defense,   Edited by Richard F. Pilch and Raymond A. Zilinskas
ISBN 0-471-46717-0   Copyright © 2005 Wiley-Liss

was brought against Stewart. Although the case was built on circumstantial evidence, Stuart was found guilty as described. In the end, it was believed that Stewart had used a syringe filled with infected blood drawn from an HIV-positive patient at Barnes Hospital, his place of work, to infect his son.

CASE THREE: "IWAN E"

In June 1992, "Iwan E," a Dutch man, injected his former girlfriend, "Gina O," with HIV-infected blood. Doctors diagnosed her with HIV three months later. Authorities stated that "Iwan E" had been angry at "Gina O" for breaking up with him, and had proceeded toward her with blood that he had drawn from an HIV-positive friend. "Iwan E," on the other hand, claimed that "Gina O" had threatened to stab him and that he had only acted in self-defense. In May 1994, "Iwan E" was convicted and sentenced to 10 years in jail.

CASE FOUR: RICHARD SCHMIDT

On October 23, 1994, in Lafayette, Louisiana, Dr. Richard Schmidt injected Janice Trahan, his ex-girlfriend, with HIV-contaminated blood. Schmidt was apparently angry with Trahan for breaking off their 10-year affair. Trahan tested positive for HIV six months later, and in February 1999, Schmidt was convicted of attempted murder and sentenced to 50 years in prison.

Schmidt, a gastroenterologist, and his nurse Trahan, both married, began their affair in the early 1980s. Although he had performed numerous abortions on her, she did give birth to one child during the course of their relationship. Schmidt threatened several times to hurt Trahan if she moved on to someone else. Trahan eventually left her husband, but Schmidt refused to end his own marriage, and thus in 1994, Trahan ended the affair.

Schmidt went to Trahan's house on August 4, 1994, to give her a "Vitamin B-12" shot against her will. She tested HIV positive six months later, and accused Schmidt of having deliberately infected her. Schmidt denied the charge, but upon further testing, investigators found that Trahan was infected with the same strain of HIV as one of Schmidt's patients. Schmidt's defense attorney argued that the genetic evidence, based on an evolutionary analysis known as phylogenetics, did not hold up in a court of law. The jury disagreed, however, and convicted Schmidt of attempted second-degree murder. Unsuccessful appeals to the Louisiana Supreme Court and the U.S. Supreme Court followed, and Schmidt is now serving out his 50-year sentence.

THREATENED USE: THE JUSTICE DEPARTMENT

The animal rights group, the "Justice Department," began a campaign of violence in October 1993. Initially, efforts involved incendiary bombings against businesses and individuals involved in animal testing, hunting sports, the fur industry, and breeding. Victims were located in various cities in the United Kingdom (Postlewaite, 1993).

On January 1, 1994, it was reported that the Justice Department had mailed postal tubes containing firebombs to 13 different targets, ranging from researchers and animal breeders to drug stores carrying products known to be tested on animals. Included in at least some of these packages were one or more clean syringes. Also included was a note from the group stating that it was in the possession of HIV-infected blood and that next time the blood would be used in the mail bombs. Two of the bombs did detonate, one injuring four veterinary laboratory employees (Combe and Muir, 1993) and another, disguised in a home perm product at "Boots and the Chemist" (a U.K. drug store chain), causing some damage to store property (Verity, 1993; Press Association Newsfile, 1993a, b).

The security at all 1100 Boots and the Chemist stores was increased as a result of these incidents. Consumers who had purchased any hair-coloring kits or perm kits from the stores on either December 27 or 28 were advised to remove the product from their homes and place it outside before calling the authorities (Postlewaite, 1993; Press Association Newsfile, 1993a, b).

Police announced that they were searching for two suspects known to have been in one of the stores on December 28. The suspects were described as females in their twenties. Police stated that they could not confirm that the two women were responsible for the bombings, but that they wanted to find the women "to eliminate them from our inquiry" (Press Association Newsfile, 1994).

In a statement to the press, Robin Webb, spokesman for the Animal Liberation Front (ALF), claimed that he had received a phone call on December 26, 1993, from someone claiming to represent the Justice Department warning that firebombs had been placed in five or six stores owned by the Boots chain (Verity, 1993; Press Association Newsfile, 1993a, b). Webb further stated that "[t]he ALF takes every precaution not to harm life. The Justice Department argues that if the animals could fight back there would be a lot of dead animal abusers already. We do not know who they are" (*The Observer*, 1994).

The perpetrator(s) were never identified.

(Case Note: An animal rights group calling itself the Justice Department sent letters containing razor blades coated with rat poison to medical researchers in Canada on October 25, 1999; it is not known whether this is the same Justice Department as the group in the United Kingdom.)

REFERENCES

Bellos, A., *Boots Attack Blamed an Animal Group, The Daily Telegraph*, December 29, 1993.

Carus, W.S., *Bioterrorism and Biocrimes: The Illicit Use of Biological Agents in the 20th Century*, Working Paper, Center for Counterproliferation Research, National Defense University, July 1999 revision.

Combe, V. and Muir, Hugh, *Third Firebomb at Boots Store, The Daily Telegraph*, December 30, 1993.

Grmek, M., *History of AIDS: Emergence and Origin of a Modern Pandemic*, Princeton University Press, Princeton, NJ, 1990.

Postlewaite, J., *Animal Activists Put Store on Firebomb Alert, Evening Standard* (London), December 29, 1993.

Press Association Newsfile, *Store Firebombs Blames on Animal Rights Faction*, December 28, 1993a.

Press Association Newsfile, *Third Boots Firebomb Found*, December 29, 1993b.

Press Association Newsfile, *Police Seek Women in Firebomb Inquiry*, January 4, 1994.

The Observer, Animal Rights Bombers: We've Only Just Started, January 2, 1994.

Verity, E., *Bomb Fanatics Launch a Third Strike at Boots, Daily Mail* (London), December 30, 1993.

WEB RESOURCES

Centers for Disease Control and Prevention website, http://www.cdc.gov/hiv/dhap.htm.

Dalk, K., *Guilty Sequence: In Court, Scientists Map a Murder Weapon*, Genome News Network, *1/24/03*, http://gnn.tigr.org/articles/01_03/hiv.shtml, accessed on 10/14/03.

MEDLINEplus Medical Encyclopedia, http://www.nlm.nih.gov/medlineplus/ency/article/000594.htm

Monterey WMD-Terrorism Database, Center for Nonproliferation Studies, 1998–2003, http://cns.miis.edu

Reuters Online, *Prison Guard Says He Got AIDS Virus after Attack*, 9/1/90, http://www.aegis.com/news/re/1990/RE900901.html, accessed on 10/9/03.

Robinson, B., *A Father's Hate*, ABCnews.com, 11/19/02, http://abcnews.go.com/sections/us/DailyNews/HIV_injected021119.html, accessed on 10/10/03.

I

INFLUENZA (FLU)

Yannick Pouliot

INTRODUCTION

Influenza refers to the disease associated with the highly contagious influenza viruses. These viruses cause acute infection of the upper and lower respiratory tract (nose, throat, bronchial tubes, and lungs) in many avian and mammalian species, including man. Furthermore, influenza viruses can be zoonotic (i.e., transmittable from animal to man). The existence of a natural pool of emerging strains of influenza in the form of irreducible animal reservoirs guarantees yearly epidemics.

Along with pneumonia (often associated with influenza infection), influenza is the sixth leading cause of death in the United States, with an infection rate ranging from 100 to 150 million people and claiming between 10,000 and 40,000 lives annually. Seven percent of all physician visits involve flu-related treatment (Health Care Financing Administration, 1999).

Influenza primarily affects the elderly and the very young. Although the virus initially infects epithelial cells of the upper respiratory tract, its effects are realized systemically. The main symptoms of influenza include fever, coughing, headaches, myalgia, and swings in body temperature. Morbidity and mortality result from cardiopulmonary and respiratory complications due to opportunistic infections (principally pneumonia, but also bronchitis and other respiratory ailments), rather than from influenza itself. In the overall population, the disease has low mortality. While immunocompromised individuals, the elderly, and juveniles are typically at highest risk, for unknown reasons certain viral strains are associated with unusually high mortality in different populations, such as healthy young adults (see below).

The ability of influenza viruses to infect economically important animal species such as chickens and hogs is also a significant consideration (Poultry Disease Bulletin Committee, 1971), illustrated by the loss of 10 million chickens in the United States in 1983 and over 1 million chickens in Hong Kong in 1997 due to epidemics of particularly lethal strains of influenza.

BACKGROUND

Influenza epidemics have been recorded throughout history. They occur regularly each winter, whereas worldwide epidemics (pandemics) appear irregularly.

The morbidity and mortality of both epidemics and pandemics is highly variable. Comprehensive immunity is not possible because of novel viral subtypes resulting from genetic mutations that alter the hemagglutinin and neuraminidase surface antigens, such that recombinant viruses become able to evade host immunity, causing reinfection.

Among epidemics, the 1918 influenza pandemic ("Spanish flu") holds the sad record for the single most devastating human infective event ever, even outclassing plague and smallpox (Taubenberger et al., 1997). It caused at least 20 million deaths worldwide, with about 500,000 deaths occurring in the United States; 28 percent of the U.S. population was infected (Frost, 1920). Unfortunately, the technology of the time did not permit the unambiguous isolation and identification of the responsible viral subtype. While the subtype was later reconstituted through the recovery of viral genomes from its victims, the molecular properties responsible for its high lethality remain unclear (Gibbs et al., 2001).

THE MICROBIOLOGY OF INFLUENZA VIRUSES

Classification

The three species of influenza viruses form the *Orthomyxoviridae* family. The virion is a spherical or filamentous enveloped particle, 80 to 120 nm in diameter (see Fig. 1).

The helically symmetric nucleocapsid is composed of a nucleoprotein and a multipartite genome of eight single-stranded antisense RNA segments, each carrying a distinct gene. The envelope carries the hemagglutinin attachment protein and the neuraminidase enzyme.

Three influenza strains exist (A, B, and C) on the basis of antigenic classification of the nucleoprotein, chosen because of its antigenic stability. Conversely, the hemagglutinin and neuraminidase antigens are highly variable and exhibit continuous antigenic drift. Thus, a large number of subtypes exist (see, e.g., the Influenza Database of Los Alamos National Laboratory, http://www.flu.lanl.gov/ Macken et al., 2001). Antibodies directed against these two surface antigens are responsible for host immunity, although antibodies against hemagglutinin have the primary responsibility in this regard. Types A and B cause yearly regional epidemics. The most severe worldwide epidemics are always due to Type A.

Encyclopedia of Bioterrorism Defense, Edited by Richard F. Pilch and Raymond A. Zilinskas
ISBN 0-471-46717-0 Copyright © 2005 Wiley-Liss

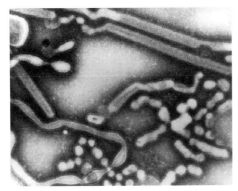

Figure 1. Transmission electron micrograph of influenza strain A, early passage. Photograph by Dr. Erskine Palmer, courtesy of the U.S. Centers for Disease Control and Prevention. (Original: http://phil.cdc.gov/Phil/detail.asp?id=279. See http://phil.cdc.gov/Phil/publicdomain.asp for copyright availability (this image is in the public domain)).

Pathogenesis

The virus is transmitted in aerosols of respiratory secretions. Following infection of mucosal epithelial cells of the upper respiratory tract, the virus causes cellular destruction and inflammation. Mortality is usually low (below 0.1 percent), but can rise as high as 2.5 percent, as was the case of the 1918 pandemic (Marks and Beatty, 1976; Rosenau and Last, 1980).

Monitoring

Since 1948, the World Health Organization (WHO) has maintained a worldwide influenza surveillance program intended to detect the emergence of new subtypes with the intent of designing targeted vaccines.

Diagnosis

The diagnosis of influenza is typically circumstantial, as many infections respiratory diseases share the symptoms of influenza (most notably the common cold). Diagnosticians usually rely on the presence of a known influenza epidemic in the area when making this determination. Definitive diagnosis depends on detecting the virus proteins directly using an enzyme-linked immunosorbent assay (ELISA) or indirectly from a rise in host antibody titer.

Treatment

Prevention via immunization is both safe and very effective if the correct strain and subtype are targeted by the vaccine. For this reason, vaccines are usually targeted toward three strains/subtypes anticipated by the WHO to become the dominant forms of influenza in the coming epidemic seasons. However, such correct prognostication is not 100 percent reliable and has resulted in vaccines that were ineffective against what became the dominant strain/subtype (see Fig. 2) (Cox, 2001).

Treatment approaches are limited. If detected early, newer drugs such Tamiflu (oseltamivir phosphate; Roche Laboratories and Gilead Sciences) and Relenza (zanamivir; GlaxoWelcome) are effective against influenza. They dramatically decrease both the duration of the infection and the seriousness of the symptoms. However, both drugs are effective only if given within two days of the inception of infection. This severely limits their usefulness given the present lack of accurate and cost-effective diagnostic tests capable of rapid detection of the virus. Older drugs such as Flumadine (rimantadine hydrochloride) and Symmetrel (amantadine hydrochloride) are also reasonably effective at diminishing the impact of infection, but because of their mechanism of action, cause greater side effects than Tamiflu and Relenza.

POTENTIAL WEAPONS APPLICATION

Influenza virus has never been part of a national biological warfare (BW) arsenal, presumably because its

Figure 2. Accuracy of predominant epidemic influenza strains vs. vaccine strains. Legend: For each yearly flu season, the degree of antigenic match between the predominant epidemic influenza strain(s) and the strains selected for vaccine preparation for that season is depicted. Here, "1" indicates no antigenic match and "4" indicates near-perfect antigenic match. In some seasons, more than one strain became epidemic (e.g., 1992–1993). Since vaccine composition is determined roughly one year in advance, three strains are selected for vaccination to increase the likelihood of correct prediction, as well as to account for the possibility of more than one strain predominating. Still, because of the necessary time lag between strain selection and vaccine delivery, prediction accuracy varies. A strain can remain undetected by the time the vaccine composition is determined, and yet become epidemic subsequently. This was the case for the A/Sydney/5/1997 strain, which was not detected until the summer of 1997, too late for inclusion into the vaccine mixture. *Source*: World Health Organization, as reported in Cox, 2001.

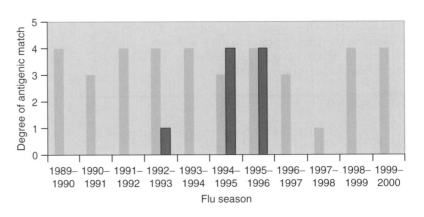

normally low mortality and relative lability hinder its weaponization. However, influenza virus was considered by Heinrich Himmler as a possible weapon against the looming D-day invasion by the Allies (Geissler, 1999). It was also considered as a potential BW agent by Japan's Imperial Army during the Second World War (Williams and Wallace, 1989). And, according to an unsubstantiated report, influenza viruses were deliberately released into the open environment off the coast of southern California by the U.S. Navy in 1945 in order to gain an understanding of the susceptibility of the United States to BW attacks (Williams and Wallace, 1989).

Influenza viruses are not specifically listed by the U.S. Centers for Disease Control and Prevention as a biological weapon or potential bioterrorism agent. However, they are listed as Priority Area pathogen (Category C) by the National Institute of Allergy and Infectious Diseases's (NIAID) Biodefense Research Program (http://www.niaid.nih.gov/biodefense/bandc_priority.htm).

Weaponization and Terrorism Potential

There is no evidence that influenza has ever been weaponized or considered in a weapons context by terrorist organizations. This is presumably due to the danger to the aggressor posed by the highly infectious nature of the virus and the difficulty of immunizing national populations covertly.

Conversely, the potential use of influenza for terrorist applications must be considered because of its ease of development, high medical and economic impact, and limited countermeasures. Of particular concern is the potential of newer techniques such as reverse genetics, which permit the synthesis of virions completely in vitro (Neumann et al., 1999), thus obviating the need to obtain an isolate of the virus, a major stumbling block in the analysis of the 1918 strain (Hatta et al., 2001). The widespread availability of such techniques may further increase the likelihood of appearance of synthetic strains of influenza created for nefarious purposes.

FUTURE OUTLOOK

Predicting Influenza Strain A Evolution

Methods based on mathematical models of influenza evolution are being developed that have the potential to help predict dominant strains/subtypes of influenza (Bush et al., 1999), and may also have applicability in helping to determine whether an epidemic is likely to be of natural or synthetic origin. However, the ultimate success of such methods will depend, in part, on the existence of comprehensive data sets of influenza virus gene sequences beyond current levels, thus strengthening the argument for more robust surveillance (see below).

Global Laboratory for the Surveillance of Influenza

Because of both the theoretical impossibility of conferring immunity against all possible subtypes of influenza viruses and the inevitability of naturally occurring epidemics, the WHO maintains a global influenza surveillance program. However, the intent of this program has focused on supporting vaccine design rather than real-time monitoring capable of supporting the containment of epidemics. For this and other reasons, the U.S. Institute of Medicine and the National Academy of Engineering have recently proposed a plan to enhance the surveillance of influenza by leveraging genomics-era technologies to create a global surveillance laboratory and associated institutions (Layne et al., 2001). Such a laboratory would have the potential of very quickly detecting and identifying new strains, thus enabling appropriate responses with stockpiled drugs and/or immunization, if a vaccine is available.

Microarray Monitoring

An important advance in the detection and typing of influenza has been the application of DNA microarrays. These miniature detection systems are capable of holding vast amount of genetic data, enabling the very rapid and precise identification of pathogens and their molecular properties (Li et al., 2001). Microarrays are likely to play an increasing role in influenza monitoring in the future.

REFERENCES

Bush, R.M., Bender, C.A., Subbarao, K., Cox, N.J., and Fitch, W.M., *Science*, **286**, 1921–1925 (1999).

Cox, N., "Expanding the Worldwide Influenza Surveillance System and Improving the Selection of Strains for Vaccines," in S.P. Layne, T.J. Beugelsdijk, and C.K.N. Patel, Eds., *Firepower in the Lab. Automation in the Fight Against Infectious Diseases and Bioterrorism*, Joseph Henry Press, Washington DC, 2001, pp. 47–54.

Frost, W.H., *Public Health Rep.*, **35**, 584 (1920).

Geissler, E., "Biological Warfare Activities in Germany, 1923-45," in E. Geissler and J.E. van Courtland Moon, Eds., *Biological and Toxin Weapons: Research, Development and Use from the Middle Ages to 1945*, SIPRI Chemical and Biological Warfare Studies, No. 18, Oxford University Press, Oxford, 1999, pp. 91–126.

Gibbs, M.J., Armstrong, J.S., and Gibbs, A.J., *Science*, **293**, 842–1845 (2001).

Hatta, M., Gao, P., Halfmann, P., and Kawaoka, Y., *Science*, **293**, 1840–1842 (2001).

Health Care Financing Administration, Influenza Fact Sheet for Health Care Providers, 1999.

Layne, S.P., Beugelsdijk, T.J., Patel, C.K.N., Taubenberger, J.K., Cox, N.J., Gust, I.D., Hay, A.J., Tashiro, M., and Lavanchy, D., *Science*, **293**, 1729 (2001).

Li, J., Chen, S., Li, J., Chen, S., Evans, D.H., *J. Clin. Microbiol.*, **39**, 696–704 (2001).

Macken, C., Lu, H., Goodman, J., and Boykin, L., "The Value of a Database in Surveillance and Vaccine Selection," in A.D.M.E. Osterhaus, N. Cox, and A.W. Hampson, Eds., *Options for the Control of Influenza IV*, Elsevier Science, Amsterdam, 2001, pp. 103–106.

Marks, G. and Beatty, W.K., *Epidemics*, Scribner, New York, 1976.

Neumann, G., Watanabe, T., Ito, H., Watanabe, S., Goto, H., Gao, P., Hughes, M., Perez, D.R., Donis, R., Hoffmann. E., Hobom, G., and Kawaoka, Y., *Proc. Natl. Acad. Sci.*, **96**, 9345–9350 (1999).

Poultry Disease Bulletin Committee, *A Manual of Poultry Diseases*, Texas A & M University, College Station, TX, 1971.

Rosenau, M.J. and Last, J.M., *Maxcy-Rosenau Preventive Medicine and Public Health*, Appleton-Century-Crofts, New York, 1980.

Taubenberger, J.K., Reid, A.H., Krafft, A.E., Bijwaard, K.E., and Fanning, T.G., *Science*, **275**, 1793–1796 (1997).

Williams, P. and Wallace, D., *Unit 731: Japan's Secret Biological Warfare in World War II*, Hodder and Stoughton, London, 1989.

WEB RESOURCES

Influenza Tutorial from Tulane University, http://www.tulane.edu/~dmsander/WWW/335/Orthomyxoviruses.html, accessed on 5/28/03.

US Centers for Disease Control and Prevention's Influenza website, http://www.cdc.gov/ncidod/diseases/flu/overview.htm.

http://www.bt.cdc.gov/agent/agentlist.asp, accessed on 8/13/03.

http://www.flu.lanl.gov/, accessed on 8/13/03.

http://www.niaid.nih.gov/biodefense/bandc_priority.htm, accessed on 8/13/03.

INTELLIGENCE COLLECTION AND ANALYSIS

STEPHEN MARRIN

INTRODUCTION

The U.S. intelligence community contributes to bioterrorism defense by covertly collecting information related to the bioterrorist threat overseas, integrating the disparate streams of intelligence via all-source analysis, and providing the finished intelligence analysis to national security decision makers involved in creating and implementing bioterrorism defense policies. The purpose of intelligence is to provide information that assists decision making, and in the case of national security to provide decision makers with information that enables them to wield their power—whether military, economic, or political—with greater precision and effectiveness.

To support bioterrorism defense policies, intelligence agencies provide national security decision makers with information on the biological weapons threat to the United States and its interests in general, and the threat posed by terrorist groups in particular, so that they can use military or law enforcement powers to preempt, disrupt, or destroy the threat before an attack occurs. In addition, intelligence agencies provide warning of possible bioterrorist attacks—optimally, with actionable intelligence that provides time, place, or agent—so that homeland security efforts can be focused on the most effective ways to mitigate the effects of the attack if prevention efforts fail.

INTELLIGENCE COMMUNITY OVERVIEW

The U.S. intelligence community is a complicated amalgam of all or part of multiple agencies and departments that covertly collect and analyze information on other countries and transnational groups in order to supplement the overt collection of information by State Department foreign service officers and Defense Department military attaches. The U.S. government institutionalized its peacetime covert intelligence collection and analysis capabilities with the passage of the National Security Act of 1947, which created both the Central Intelligence Agency (CIA) and the National Security Council (NSC) and charged the CIA with "coordinating the nation's intelligence activities and correlating, evaluating, and disseminating intelligence which affects national security" (*CIA Factbook on Intelligence*, 2002).

Over the past half-century, the intelligence community has expanded greatly with the creation of numerous intelligence agencies other than the CIA, and as a result of this growth, intelligence has become "a vital element in every substantial international activity of the U.S. government" (http://www.intelligence.gov/2-customers.shtml, accessed 12/13/03). The workings of the intelligence community are coordinated by the Director of Central Intelligence (DCI), who according to a CIA publication is "the primary adviser to the President and the National Security Council on national foreign intelligence matters" and "is simultaneously Director of the CIA and the leader of the Intelligence Community" (*CIA Factbook on Intelligence*, 2002).

The operations of the intelligence community can best be understood through the intelligence cycle, which is defined as "the process of developing raw information into finished intelligence for policy makers to use in decision-making and action" (*CIA Factbook on Intelligence*, 2002). The intelligence cycle portrays the different steps in intelligence production, and can be understood as governmental learning in action. In sequence, the cycle's five steps are:

- *Planning and Direction.* According to a CIA publication, "the intelligence process begins when policy makers or military commanders express a need for information to help them accomplish their missions... These needs are then used to guide collection strategies and the production of appropriate intelligence products" (CIA, Undated). In other words, the entire intelligence mission is predicated on the policy makers' stated needs for information.

- *Collection.* Collection is defined as "the gathering of raw data from which finished intelligence is produced" (CIA, Undated). *Open source intelligence* (OSINT), which consists of unclassified, publicly accessible information, provides a stable context in which to place interpretation of classified information. Within this unclassified context, different kinds of intelligence provide complementary insights into the operations of a potential adversary. *Imagery intelligence* (IMINT) collected via airborne or space platforms can assist in assessing military capabilities, while *signals intelligence* (SIGINT)—consisting of intercepted and translated communications—and *human intelligence* (HUMINT) provide insight into adversary intentions.

Encyclopedia of Bioterrorism Defense,   Edited by Richard F. Pilch and Raymond A. Zilinskas
ISBN 0-471-46717-0   Copyright © 2005 Wiley-Liss

- *Processing.* Once the intelligence information is collected, it is then processed before general dissemination to the rest of the intelligence community. Processing entails "converting the vast amount of information collected to a form usable by analysts ... through a variety of methods including decryption, language translations, and data reduction" (*CIA Factbook on Intelligence*, 2002). This processed information is usually referred to as "raw" intelligence to distinguish it from all-source intelligence analysis (below).

- *All Source Analysis and Production.* The purpose of intelligence analysis "is to minimize the uncertainty with which U.S. officials must grapple in making decisions about American national security and foreign policies" (http://www.odci.gov/cia/di/ intel_analysis_section.html, accessed 12/13/03). To accomplish this, all source analysts integrate information acquired by the government (both overtly and covertly) with "open source" information in order to produce finished intelligence that meets the information needs of their consumers. The kind of information sources most relied upon and the format of the product depend on the needs of the consumer.

- *Dissemination.* Finished intelligence is provided primarily through written products and briefings to policy makers, who "then make decisions based on the information, and these decisions may lead to the levying of more requirements, thus triggering the Intelligence Cycle" (*CIA Factbook on Intelligence*, 2002).

Each of the organizations that constitute the intelligence community performs one or more steps in the intelligence cycle. These include:

- *Central Intelligence Agency (CIA).* CIA's Directorate of Operations case officers collect HUMINT by recruiting foreign nationals, and CIA's Directorate of Intelligence analysts produce departmentally objective all-source intelligence analysis covering the range of intelligence disciplines—political, military, and economic. CIA provides this finished intelligence to the president, vice president, NSC, and other consumers throughout the executive branch to support their foreign policy and national security information needs.

- *Defense Intelligence Agency (DIA)* DIA supports the intelligence needs of the military by specializing in the collection and analysis of military intelligence. It collects HUMINT, both overtly through its Defense Attaches, who are based overseas in embassies, and covertly through its Defense HUMINT Service. DIA also centralizes the intelligence community's focus on measurement and signature intelligence, or MASINT, which includes the use of nuclear, optical, radio frequency, acoustics, seismic, and materials sciences to locate, identify, or describe certain kinds of targets (CIA, Undated).

- *National Reconnaissance Office (NRO).* The NRO researches, develops, acquires, and operates the nation's intelligence satellites. While the NRO collects IMINT, it does not process it.

- *National Geospatial-Intelligence Agency (NGA).* NGA, formerly the National Imagery and Mapping Agency, processes and interprets IMINT acquired from the NRO and other imagery collectors and provides geospatial intelligence in all its forms" to national level policy makers and the military. (http://www.intelligence.gov/1- members_nima.shtml, accessed 7/6/04).

- *National Security Agency (NSA).* NSA provides the U.S. government with the bulk of its SIGINT by intercepting foreign communications, breaking their encryption codes, and translating the resulting "take."

- *Military Service Units.* The intelligence components of the Army, Navy, Air Force, Marines, and Coast Guard operate select collection platforms and provide intelligence analysis tailored to the needs of the individual services. For example, the Air Force collects IMINT and the Office of Naval Intelligence operates ocean surveillance systems. Each service's intelligence component provides intelligence to its own operators, specializing in the capabilities of opposing force weaponry and the effectiveness of their tactics.

- *Departmental Intelligence Units.* Intelligence units within the State, Treasury, Energy, and Homeland Security Departments create intelligence products tailored to the informational requirements of their respective departments. For example, the State Department's Bureau of Intelligence and Research focuses on foreign policy in general and contributes to decisions regarding visa denials and the sharing of intelligence with foreign governments. In addition, the Department of Homeland Security's (DHS) Information Analysis and Infrastructure Protection Directorate "map(s) the vulnerabilities of the nation's critical infrastructure against a comprehensive analysis of intelligence and public source information" (http://www.intelligence.gov/1- members_dhs.shtml, accessed 12/13/03). Each departmental unit also disseminates products on the basis of their respective specialized expertise to the rest of the intelligence community.

- *Federal Bureau of Investigation(FBI).* As the principal investigative arm of the Department of Justice (DOJ), the FBI investigates violations of federal criminal statutes and as a result is not a member of the intelligence community in its entirety, according to the DCI's website (http://www.intelligence.gov/1- members_fbi.shtml, accessed 12/13/03). The website goes on to note that the part of the FBI that is a component of the intelligence community is its National Security Division, which "plays an important role in protecting the US from foreign intelligence activities." The FBI's domestic counterterrorism program, however, is not considered part of the foreign intelligence community.

INTELLIGENCE COLLECTION AND ANALYSIS

The intelligence community supports bioterrorism defense policies by providing national security decision makers with specific information on both the biological weapons capabilities of terrorist groups and states and the intentions of those who might employ biological weapons to target the United States or its interests. In the aftermath of the September 11, 2001, terrorist attacks, the intelligence community underwent a number of changes associated with its expanded counterterrorism mission, and while future changes may be forthcoming, the three components that currently provide focused attention on the bioterrorist threat are:

- *The DCI Center for Weapons Intelligence, Nonproliferation, and Arms Control (WINPAC).* Formerly the DCI's Non-Proliferation Center, WINPAC is the intelligence community's focal point for collection and analysis of information on weapons of mass destruction (WMD), and "provides US policy makers and warfighters [with] intelligence support to protect the US and its interests from all foreign weapons threats" (http://www.odci.gov/cia/di/org_chart_section.html, accessed 12/13/03).

- *The DCI Counterterrorist Center (CTC).* The CTC is the intelligence community's focal point for collection and analysis on the structure and operations of foreign terrorist groups, including those that might conceivably develop or use biological weapons against the United States or its interests (http://www.odci.gov/terrorism/ctc.html, accessed 12/13/03).

- *The Terrorist Threat Integration Center (TTIC).* Created in May 2003, the TTIC "enables full integration of US Government terrorist threat-related information and analysis" by combining elements of CTC, the FBI's Counterterrorism Division, the DHS, the Defense Department, and other U.S. government agencies to "close any gaps separating the analysis of foreign-sourced and domestic-sourced terrorist threat-related information," according to Associate Director of Central Intelligence for Homeland Security Winston Wiley (Wiley, 2003). TTIC produces a range of intelligence products for policy makers, including the daily Threat Matrix. Mr. Wiley has also stated that once TTIC is operating at full capacity, "it will be collocated with CTC and FBI's Counterterrorism Division at a yet-to-be-acquired site" to centralize the government's intelligence on terrorists (Wiley, 2003).

To address the biological capabilities of terrorist groups, the intelligence community collects and analyzes information on terrorist abilities to acquire and produce biological agents indigenously, as well as information on what delivery mechanisms could be used to distribute these agents. In addition, intelligence agencies also keep tabs on the status of biological agents and delivery devices developed by states, assessing their command and control in order to ascertain the likelihood that terrorists might successfully purchase, steal, or otherwise acquire these assets.

Assessments of intentions are also performed in order to provide decision makers with a more accurate and precise measure of the bioterrorist threat. Intelligence regarding military capabilities has historically been much easier to acquire than intelligence regarding intentions because military capabilities can be measured in ways that intentions cannot. Traditionally, IMINT has provided the best source of information for assessing military capabilities because weaponry and its supporting infrastructure, including production facilities, can be found, counted, and monitored for activity. IMINT is less useful in tracking biological weapons capabilities, however, because biological weapons production facilities can be smaller than those used to produce other WMD and are in most cases dual-use in nature. As John Lauder, then-Special Assistant to the DCI for Nonproliferation, pointed out in 1999, biological weapons "overlap with legitimate research and commercial biotechnology," such that "even supposedly 'legitimate' facilities can readily conduct clandestine [biological weapons] research and can convert rapidly to agent production" (http://usembassy-australia.state.gov/hyper/WF990304/epf410.htm, accessed 12/13/03). As a result, it is easier for both states and terrorist groups to hide biological weapons activities and, if they are detected, to deceive intelligence agencies into believing existing programs are for commercial or medical research purposes only. Bioterrorism intelligence analysis production must therefore rely heavily on HUMINT and SIGINT in the end, because these forms of intelligence provide the best indicators of a given adversary's intentions.

AUTHOR'S NOTE

Sections of this entry were adapted from a part of a July 2002 background paper entitled "Homeland Security and the Analysis of Foreign Intelligence," written for the Markle Foundation Task Force on National Security in the Information Age. The entry was reviewed and approved by the CIA Publication Review Board. The review does not indicate an endorsement of the views or the accuracy of information herein.

REFERENCES

Central Intelligence Agency, *A Consumer's Guide to Intelligence*, Undated.

CIA Factbook on Intelligence, 2002, http://www.odci.gov/cia/publications/facttell/textonly.htm.

Wiley, W.P., Testimony Before the Senate Governmental Affairs Committee, February 26, 2003, http://www.odci.gov/cia/public_affairs/speeches/2003/wiley_speech_02262003.html.

WEB RESOURCES

CIA website, DCI Counterterrorist Center, http://www.odci.gov/terrorism/ctc.html, accessed on 12/13/03.

CIA website, Directorate of Intelligence, Frequently Asked Questions, http://www.odci.gov/cia/di/intel_analysis_section.html, accessed on 12/13/03.

CIA website, Directorate of Intelligence, Organizational Chart, http://www.odci.gov/cia/di/org_chart_section.html, accessed on 12/13/03.

CIA website, United States Intelligence Community—What We Do: The Consumers of Intelligence, http://www.intelligence.gov/2-customers.shtml, accessed on 5/9/03.

CIA website, United States Intelligence Community—Who We Are: Federal Bureau of Investigation, National Security Division, http://www.intelligence.gov/1-members_fbi.shtml, accessed on 12/13/03.

CIA website, United States Intelligence Community—Who We Are: The Department of Homeland Security (DHS), Information Analysis and Infrastructure Protection Directorate, http://www.intelligence.gov/1-members_dhs.shtml, accessed on 12/13/03.

CIA website, United States Intelligence Community—Who We Are: National Geospatial-Intelligence Agency, http://www.intelligence.gov/1-members_nima.shtml, accessed on 7/6/04.

State Department website, *Excerpts: Lauder Says Biological Weapons Threat 'Is Growing'*, March 4, 1999, http://usembassyaustralia.state.gov/hyper/WF990304/epf410.htm.

See also CENTRAL INTELLIGENCE AGENCY and DIRECTOR OF CENTRAL INTELLIGENCE COUNTERTERRORIST CENTER.

INTERNATIONAL COOPERATION AND BIOTERRORISM PREPAREDNESS

ROGER ROFFEY

BACKGROUND

The risk of terrorists using biological or chemical agents might well have increased after the attacks in the United States on September 11, 2001, and the subsequent delivery of *Bacillus anthracis* spores through the mail. This perceived increase in risk could be due to a number of reasons, for example:

- The attacks have demonstrated that terrorists can and will go as far as causing mass deaths.
- A moral barrier has been passed with the successful delivery of a live biological agent to kill.
- The media impact and attention given to such terrorist acts have been monumental.
- The most recent targets were not selected to increase political pressure for change and to convince a specific population of the terrorists' cause, as has been the case in the past; rather, other factors have become more important.

Because terrorism has shifted to a more global, network-based threat over the past many years, active and enhanced international cooperation at all levels is now an absolute necessity in order to effectively deal with the phenomenon. Efforts to this end by multiple international organizations will therefore be discussed so as to gain an accurate understanding of the current state of international preparedness.

THE EUROPEAN UNION

The European Union (EU) rapidly responded to the terrorist attacks of September 11, 2001, and (1) an action plan for the fight against terrorism, (Council of the European Union, 2004) and a programme to enhance cooperation on preparedness and response to chemical and biological threats, which includes increased information sharing, arrangements for safe transportation of samples, inventory of laboratories and response teams, the generation of a list of nuclear, biological and chemical (NBC) experts available 24 hours a day, the creation of vaccine and antibiotic stockpiles (including plans for the stockpiling of smallpox vaccine) and enhanced cooperation on research and procurement of vaccines and antibiotics (Commission of the European Communities, 2001; Gouvras, 2002); and (2) an EU program covering health, research, agriculture, energy, and environment, for improving cooperation in preventing and limiting the consequences of chemical, biological, radiological, and nuclear terrorist threats, which was adopted in December 2002 (Council of the European Union, 2002). The latter includes a Civil Protection Mechanism (set up in 2001) and a Monitoring and Information Center (MIC). These efforts build upon a Communicable Disease Network, which was set up in 1998, and an Early Warning and Response System (EWRS), to which has been added since June 2002 the rapid alert system called RAS-BICHAT which are connected to other disease reporting networks and provides support to the World Health Organization (WHO) disease surveillance system (Petersen and Catchpole, 2001). In addition, the EU Joint Research Center and EUROPOL have studied the threat of, and vulnerabilities to, NBC terrorism. The EU has also 2003 adopted a strategy against the proliferation of WMD (Council of the European Union, 2003).

THE NORTH ATLANTIC TREATY ORGANIZATION

In 2000, the North Atlantic Treaty Organization (NATO) established a weapons of mass destruction (WMD) center in Brussels to improve coordination on force protection and to harmonize counterterrorism and counterproliferation capabilities. The NATO countries have recently launched several WMD initiatives that focus largely on defense against biological warfare (BW). The Alliance is also stepping up its cooperation with the EAPC (Euro Atlantic Partnership Council) countries through the Partnership Action Plan against Terrorism (under which Sweden and Finland recently launched an initiative). There is also a WMD-protection initiative in NATO including deployable laboratories, event response teams, a disease surveillance system that has been further developed into a CBRN battalion.

Encyclopedia of Bioterrorism Defense,   Edited by Richard F. Pilch and Raymond A. Zilinskas
ISBN 0-471-46717-0   Copyright © 2005 Wiley-Liss

THE WORLD HEALTH ORGANIZATION

WHO, the United Nations's specialized agency for international health, has taken steps to strengthen public health preparedness and response efforts regarding the deliberate use of biological and chemical agents. Actions are implemented in the framework of the Global Health Security with a "network of networks" approach based on the Global Outbreak Alert and Response Network, which is a technical collaboration of institutes and networks enabling rapid identification, confirmation, and response to outbreaks. It includes WHO Regional Offices and Collaborating Centers/Laboratories, Military Lab Networks, and the Global Public Health Intelligence Network (GPHIN) (WHO, 2002 a, b, c). As a result of external funding, WHO can now deploy rapid-response medical teams and equipment within 24 hours of a major outbreak as well (*Insurance Today*, 2003). For political reasons, however, the WHO cannot be expected to play a central role in BW nonproliferation activities.

THE UNITED NATIONS

In addition to the activities of the WHO, the United Nations (UN) is involved in antiterrorism efforts on a number of fronts. Multilateral negotiations within the UN framework have resulted in 12 international antiterrorism conventions, but progress remains slow due to a dispute over the definition of the term terrorism. The UN Secretary General's Disarmament Advisory Board has stated that urgent cooperative measures for emergency response in the biological weapons field include improving surveillance, improving systems for the early detection of and rapid response to the outbreak of diseases, and creating a global vaccine bank. The UN Security Council adopted a resolution 1540 (26 April 2004) to keep weapons of mass destruction from terrorists and requires states to adopt laws to prevent non-state actors from manufacturing, acquiring or trafficking in nuclear, biological or chemical weapons, the materials to make them, and missiles and other systems to deliver them. The Organization for the Prohibition of Chemical Weapons (OPCW) in The Hague plays a supportive role and has made an inventory of capacities available in the fight against terrorism (UN, 2002).

Negotiations to achieve a verification protocol to the Biological and Toxin Weapons Convention (BWC), ongoing since 1994, collapsed at the Fifth Review Conference in July 2001, after the United States declined to continue negotiations (Roffey, 2001). The conference had to be discontinued to prevent total failure, such that main problems were set aside to enable a continued multilateral process (Tucker, 2002; Boyd, 2002; UK, 2002). It was agreed that annual meetings of one week in duration, plus two weeks of expert meetings, would be held each year commencing in 2003 until the Sixth Review Conference in 2006 to discuss (BWC, 2002):

- The adoption of necessary national measures to implement the prohibition set forth in the Convention, including the enactment of penal legislation.

- The delineation of national mechanisms to establish and maintain the security and oversight of pathogenic microorganisms and toxins.

- The enhancement of international capabilities for responding to, investigating, and mitigating the effects of cases of alleged use of biological or toxin weapons or suspicious outbreaks of disease.

- The strengthening and broadening of national and international institutional efforts and existing mechanisms for the surveillance, detection, diagnosis, and combating of infectious disease affecting humans, animals, and plants.

- The content, promulgation, and adoption of codes of conduct for scientists.

Suggestions have been made to strengthen the international biological arms control regime outside the BWC. Thus, a number of proposals have been presented to strengthen the legal framework, covering storage, handling, and transfer of pathogenic microorganisms; imposing liability on anyone who publishes recipes for producing biological agents; and making possession of biological weapons a crime under international law (Meselson and Robinson, 2002). A further option is to initiate negotiations on a multilateral convention banning biological and chemical terrorism.

NONGOVERNMENTAL ORGANIZATIONS

Nongovernmental organizations (NGOs) have urged governments to prepare a response capability to the humanitarian disaster that a WMD event could cause, have stressed the need for education and training in order to achieve an adequate level of preparation (Prescott et al., 2002), and have recommended the establishment of independent rapid response teams for civilian emergency relief. This last need surfaced during the Iraq crisis of 2003, resulting in the formation of the Swiss/European WMD Task Force composed of Polish and Swiss experts in the field (Woodall, 2003). A new global activity is the Biological Weapons Proliferation Prevention (BWPP), initiated by a group of eight NGOs to strengthen the norm against BW by tracking government actions, including compliance with the BWC (Bioweapons Prevention Project, 2002). An ever-increasing amount of information is available on the Internet to inform policymakers and the public alike, and outbreak reports are increasingly placed on the web as well, accelerating and enhancing distribution and preventive measures by such means as the Program for Monitoring Emerging Diseases (ProMED-mail) (Woodall, 2001).

DISCUSSION

Intelligence sharing is an area where cooperation has thus far been limited. With the global threat now firmly established, however, this must change. Methods have to be improved for information gathering as well as for developing a common risk assessment and evaluation protocol similar to that used for chemical industry accidents. In this connection, training, simulations, and various other exercises are good areas for cooperation between countries and agencies. The Internet is also being used to provide information and assistance to both authorities and the general public to prepare for a bioterrorism attack (see, e.g., the U.S. Department of Homeland Security website).

There are a number of official agency websites (national, regional, and international) listing current outbreaks, some of which distribute restricted lists to subscribers. Because of the overwhelming amount of information available, mechanisms have further been implemented to sift through reports and identify those of regional and global significance. For example, the Canadian government has initiated the GPHIN, which subscribes to nearly one thousand sources worldwide; so far, however, it is only available to Canada and the WHO (Woodall, 2001).

Traditional epidemiological surveillance systems are not effective enough to provide early warning in the event of a bioterrorist attack. There is a need for more rapid and reliable communication systems, for both risk assessment and public health information, which can convey data from multiple web-based sources during an emergency in order to integrate consequence management and remedial services on a local, regional, national or transnational level (Kun and Bray, 2002). In times of crisis, clinicians must be able to communicate rapidly with each other, with biomedical researchers, and with public health experts, and regular updates must be provided.

A number of different systems for rapid alert and epidemiological surveillance are being developed and tested in the United States and EU, some of which allow for surveillance of unusual clinical syndromes or unexpected patterns of occurrence of uncommon syndromes. Many of these systems, both military and civilian, are still experimental, however. In the United States, the Epidemiology and Laboratory Capacity (ELC) program and the National Laboratory Response Network (LRN) have been initiated to develop a national strategy to strengthen the public health infrastructure for infectious disease surveillance and response. Similar initiatives are needed on a global scale. To overcome this, it has been proposed that an independent system, the Global Incident Analysis and Alerting System (GIAAS), be created to provide immediate information and analysis to those responsible for readiness and response to major biological, chemical, or radiological incidents, whether natural or deliberate. Twelve regional alerting centers, each responsible for a specific geographical area, would be established, and these would in turn be connected to regional and local centers, international centers for disease research, and internationally recognized laboratories providing a prime laboratory focal point for each region (Price et al., 2002).

In the EU, there is now a need to rapidly implement a biological, chemical, radiological, and nuclear protection program and further improve surveillance, central EU coordination, and the network of qualified laboratories (including BSL-4 labs). Planning must be improved for emergency distribution of medical supplies, especially antibiotics and vaccines, and cross-border cooperation in

the training of personnel, as well as exercises involving threats or use of BW agents, must be ongoing. Thus, much needs to be done before effective EU coordination becomes a reality, and to this end the EU and G7 countries agreed in 2002 on a number of initiatives to strengthen preparedness and response, including a multination exercise in June 2003 and the establishment of a "Richter-like" scale for attacks in order to facilitate communication between authorities.

Scientific research and development plays a crucial role in the fight against bioterrorism, as exemplified by the 10 year, $6 billion initiative of the United States named Project Bioshield (Fox, 2003). International cooperation with such initiatives is vital and should be promoted. Other examples of scientific cooperation on a large scale can be found in the area of disease research. For example, over a million individual personal computers (PCs) were at one point connected to a virtual supercomputer to enable the rapid screening of millions of chemical compounds for medical protection against smallpox (New York Times, 2003). Without question, there is a need to find a realistic balance between the conflicting needs of scientific openness and national security. Science can be misused for terror, but scientific cooperation and openness is critical for scientific progress. Scientists have initiated such discussions, and hopefully will take the lead in finding a common ground (Zilinskas and Tucker, 2002).

Enhanced efforts are needed concerning biosecurity (i.e., how to constrain malignant application of biology and biotechnology), without damaging the generation of essential knowledge. Biosecurity measures must be applied worldwide; otherwise, weak points can be identified and exploited by terrorists (Barletta et al., 2002; Kwik et al., 2003). It is, however, vitally important to keep biotechnology work and transfer as transparent as possible, from intangible technologies and data collection and processing to knowledge, technology and skills, as they are central to R&D in this area. Increased cooperation is required for implementation of the enhanced export control measures on biological agents, equipment, and technology set forth by the Australia Group to counter the risk of terrorist acquisition of these items, but such measures must be balanced so as not to hamper developments in biotechnology sciences and businesses (Brugger, 2002).

Threat reduction initiatives and cooperative efforts such as the U.S. Department of Defense Cooperative Threat Reduction Program, the EU Joint Action from 1999 on disarmament, and the G8 Global Partnership (Boese, 2002) play vital roles in limiting proliferation of knowledge and material from, for example, WMD programs of the former Soviet Union. The idea of cooperative threat reduction aimed at counterproliferation is now necessary in other regions as well, such as post-war Iraq. A new NATO-Russia Council has been established, where Russia sits as an equal partner, to forge joint policy on a list of nine issues topped by counterterrorism and actions to stop the spread of WMD. There is also a formal agreement between the United States and Russia in the area of bioterrorism, an area of cooperation that should be pursued among the EU, Russia, and the newly independent states in the near future.

REFERENCES

Barletta, M., Sands, A. and Tucker, J. *Bull. Atomic Sci.*, **58**(3), 57–62 (2002).

Bioweapons Prevention Project, New civil society watchdog monitors biological weapons, Press release, Geneva, November 11, 2002.

Boese, W., G-8 leaders agree to fund Threat Reduction Programs, *Arms Control Today*, July/August, 2002, p. 20.

Boyd, K., BWC Review Conference Meets, Avoids Verification Issues, *Arms Control Today*, December 2002, p. 21.

Brugger, S., Australia Group concludes new chem-bio control measures, *Arms Control Today*, July/August, 2002, p. 21.

BWC, *Fifth Review Conference of the States Parties to the Convention on the Prohibition of the Development*, Production and Stockpiling of Bacteriological (Biological) and Toxin Weapons and on their Destruction, Final Document, Geneva, November 19—December 7, 2001 and November 11—22, 2002, BWC/CONF.V/17.

Commission of the European Communities, *Programme of cooperation on preparedness and response to biological and chemical agent attacks*, Health Security, Luxembourg, 17 December, 2001, Doc G/FS D(2001) GG.

Council of the European Union, *Adoption of the Program to Improve Cooperation in the European Union for Preventing and Limiting the Consequences of Chemical, Biological, Radiological or Nuclear Terrorist Threats*, Brussels, November 21, 2002, Doc. 14627/02.

Council of the European Union, *Fight Against the Proliferation of Weapons of Mass Destruction - Draft EU Strategy Against Proliferation of Weapons of Mass Destruction*, Brussels, 3 Dec. 2003, Doc. 15656/03.

Council of the European Union, *EU Plan of Action on Combating Terrorism*, Brussels, 15 June, 2004, Doc. 10586/04.

Fox, J.L., *Business Regul. News*, **21**(3), 216 (2003).

Gouvras, G., *IEEE Eng. Med. Biol.*, September/October, 112–115 (2002).

Insurance Today, WHO sets up a rapid response force as world faces bio-terror reality, January 23, 2003.

Kun, L.K. and Bray, D.A., *IEEE Eng. Med. Biol.*, September/October, 69–85 (2002).

Kwik, G., Fitzgerald, J., Inglesby, T.V. and O'Toole, *Biosecur. Bioterror.*, **1**(1), 1–9 (2003).

Meselson, M. and Robinson, J.P., "A Draft Convention to Prohibit Biological and Chemical Weapons under International Law," in R. Yepepes-Enriquez and L. Tabassi, Eds., *Treaty Enforcement And International Cooperation in Criminal Matters: With Special Reference to the Chemical Weapons Convention*, Asser Press, The Hague, T.M.C, 2002, pp. 457–469.

New York Times, Smallpox researchers seek help from millions of computer users, February 5, 2003.

Petersen, L.R. and Catchpole, M., *Br. Med. J.*, **323**, 818–819 (2001).

Prescott, G., Doull, L., Sondorp, E., Bower H. and Mozumder A., Hope for the best, prepare for the worst: How humanitarian organizations can organize to respond to weapons of mass destruction, Program for Evidence-based Humanitarian Aid, Merlin and London School of Hygiene and Tropical Medicine.

Price, R., Woodall, J. and Netesov, S., *ASA Newsletter*, 02–6, December 27, **93**, 3–6 (2002).

Roffey, R., Implications of the protocol to the BTWC for the control and verification of BW, *Proceedings of the 7th International Symposium on Protection against Chemical and Biological Agents*, Swedish Defence Research Agency, Stockholm, Umeå, Sweden, June 10–15, 2001.

Tucker, J., *Nonproliferation Rev.*, **9**(1), 112–121 Spring, (2002).

UK Government Green Paper, *Strengthening the Biological and Toxin Weapons Convention: Countering the Threat from Biological Weapons*, Presented to Parliament by the Secretary of State for Foreign and Commonwealth Affairs by Command of Her Majesty, Published by the Stationary Office Limited, London, April 2002.

United Nations, OPCW capabilities relevant to the global struggle against terrorism, Note by the Secretariat, Office of the Deputy Director-General, S/294/2002, February 6, 2002.

WHO, Public Health Response to Biological and Chemical Weapons, WHO Guidance, World Health Organization, Geneva, 2004.

WHO, *Wkly. Epidemiol. Rec.*, August 23, 2002, **34**(77), 281–288, August 23, (2002).

WHO Global Outbreak Alert & Response Network (2002), http://www.who.int/csr/outbreaknetwork.

Woodall, J.P., *Cad Saude Publica, Rio de Janeiro*, **17**(4), 147–154 (2001).

Woodall, J.P., Civilian relief after release of weapons of mass destruction, *ASA Newsletter*, No. 04, 1–9, February 28, 2003.

Zilinskas, R.A. and Tucker, J.B., *J. Homeland Secur.*, December (2002), http://www.homelandsecurity.org/journal/articles/tucker.html#endref1.

INTERNATIONAL REGULATIONS AND AGREEMENTS PERTAINING TO BIOTERRORISM

ANGELA WOODWARD

INTRODUCTION

This entry provides an overview of five current international regulations and agreements relevant to bioterrorism defense as well as two draft conventions under consideration by the academic and policy community. They are:

1. The 1925 Protocol for the Prohibition of the Use in War of Asphyxiating, Poisonous or Other Gases, and of Bacteriological Methods of Warfare (more commonly called the Geneva Protocol)
2. The 1972 Convention on the Prohibition of the Development, Production and Stockpiling of Bacteriological (Biological) and Toxin Weapons and on Their Destruction (more commonly called the Biological and Toxin Weapons Convention or BWC)
3. The 1993 Convention on the Prohibition of the Development, Production, Stockpiling and Use of Chemical Weapons and on their Destruction (more commonly called the Chemical Weapons Convention or CWC)
4. The 1976 Convention on the Prohibition of Military or any other Hostile Use of Environmental Modification Techniques (ENMOD)
5. The Australia Group
6. The draft Convention to Prohibit Biological and Chemical Weapons under International Criminal Law
7. The draft Model Convention on the Prohibition and Prevention of Biological Terrorism.

In addition, references are provided for the following:

1. United Nations Conventions on Terrorism
2. United Nations Counter-Terrorism Committee
3. United Nations Department for Disarmament Affairs (UNDDA)
4. United Nations Monitoring, Verification and Inspection Commission (UNMOVIC)
5. United Nations Office on Drugs and Crime (UNODC)
6. United Nations Special Commission (UNSCOM).

GENEVA PROTOCOL

The 1925 Protocol for the Prohibition of the Use in War of Asphyxiating, Poisonous or Other Gases, and of Bacteriological Methods of Warfare (Geneva Protocol) bans the use of biological and chemical weapons during war.

Background

The use of "poisons" as a means of warfare dates back millennia, with agreements to limit their use dating back to 1675. The large-scale use of toxic chemicals and chemical weapons during World War I caused horrendous suffering. Thus, states subsequently resolved to ban their use under international law.

The protocol was negotiated from May 4 to June 17, 1925, at a conference under the auspices of the League of Nations in Geneva, Switzerland. It was intended to be a supplemental agreement to a convention on the supervision of the international trade in arms and ammunition, which has never entered into force. The protocol was adopted on June 17, 1925, and entered into force on February 8, 1928. Both English and French are the protocol's authentic texts. The treaty's depositary is the government of France, which also hosted the treaty's only Conference of States Parties to date, from January 7 to 11, 1989.

Current Status

There are currently 133 States Parties and one signatory to the Geneva Protocol, which is considered part of customary international law and thereby binding on all states, even those that have not joined the treaty. Those states that have registered a reservation to the protocol—predominantly, the right to "retaliate in kind" to the use of biological or chemical weapons against them—have been urged to remove them by the General Assembly and successive Review Conferences of the BWC. This has altered its effect at customary international law to a no-first-use agreement. Regardless, the Geneva Protocol remains the cornerstone of international humanitarian law banning the use of chemical and biological weapons.

States Parties to the Geneva Protocol (133)

Afghanistan, Albania, Algeria, Angola, Antigua and Barbuda, Argentina, Australia, Austria, Bahrain, Bangladesh, Barbados, Belgium, Benin, Bhutan, Bolivia, Brazil, Bulgaria, Burkina Faso, Cambodia, Cameroon, Canada, Cape Verde, Central African Republic, Chile, China, Côte d'Ivoire, Cuba, Cyprus, Czech Republic, Democratic People's Republic of Korea, Denmark, Dominican Republic, Ecuador, Egypt, Equatorial Guinea, Estonia, Ethiopia, Fiji, Finland, France, Gambia, Germany, Ghana, Greece, Grenada, Guatemala, Guinea-Bissau, Holy See, Hungary, Iceland, India, Indonesia, Iran, Iraq, Ireland, Israel, Italy, Jamaica, Japan, Jordan, Kenya, Kuwait, Lao People's Democratic Republic, Latvia, Lebanon, Lesotho, Liberia, Libya, Liechtenstein, Lithuania, Luxembourg, Madagascar, Malawi, Malaysia, Maldives, Malta, Mauritius, Mexico, Monaco, Mongolia, Morocco, Nepal, Netherlands, New Zealand, Nicaragua, Niger, Nigeria, Norway, Pakistan, Panama, Papua New Guinea, Paraguay, Peru, Philippines, Poland, Portugal, Qatar, Republic of Korea, Romania, Russian Federation, Rwanda, Saint Kitts and Nevis, Saint Lucia, Saint Vincent and the Grenadines, Saudi Arabia, Senegal, Serbia and Montenegro, Sierra Leone, Slovakia, Solomon Islands, South Africa, Spain, Sri Lanka, Sudan, Swaziland, Sweden, Switzerland, Syria, Tanzania, Thailand, Togo, Tonga, Trinidad and Tobago, Tunisia, Turkey, Uganda, Ukraine, United Kingdom, United States, Uruguay, Venezuela, Vietnam, and Yemen.

(As of December 31, 2003)

States Signatories to the Geneva Protocol (1)

El Salvador

(As of December 31, 2003)

FURTHER READING

Roberts, A. and Guelff, R., *Documents on the Laws of War*, 3rd ed., Oxford University Press, Oxford, 2000.

WEB RESOURCES

Harvard Sussex Program on Chemical and Biological Warfare Armament and Arms Limitation, http://www.sussex.ac.uk/spru/hsp/index.html, accessed on 12/29/03.

International Committee of the Red Cross, www.icrc.org, accessed on 12/29/03.

International Committee of the Red Cross, Treaty database, www.icrc.org/ihl, accessed on 12/29/03.

BIOLOGICAL AND TOXIN WEAPONS CONVENTION

The BWC explicitly bans the development, acquisition, retention, production, and stockpiling of biological and toxin agents except for prophylactic, protective, or other peaceful purposes. The BWC further bans equipment or means of delivery designed to use such biological agents.

Background

The renunciation by the United States of offensive biological weapons research in 1969 and toxins in 1970 was a catalyst for negotiations on a treaty to supplement the no-use policy of the 1925 Geneva Protocol by specifically banning the development of biological weapons. The convention was negotiated in the UN Committee on Disarmament (now the Conference on Disarmament) from 1969 to 1971, with text adopted on December 16, 1971, by the UN General Assembly. It was opened for signature on April 10, 1972, and entered into force on March 26, 1975, once 22 States Parties had either ratified or acceded to it. It has a triple depositary system comprising the governments of the United Kingdom, United States, and the Russian Federation, with obligations of depository nations specified in Article 14.

The treaty is directed toward "microbial or other biological agents, or toxins, whatever their origin or method of production, or types and in quantities that have no justification for prophylactic, protective or other peaceful purposes," and of "weapons, equipment or means of delivery designed to use such agents or toxins for hostile purposes or in armed conflict" (Article 1). This is known as a "general purpose criterion" definition, whereby the development of agents for defensive purposes is permitted. However, the development of weaponized agents, even if under the auspices of defensive research, is widely considered to be prohibited under the treaty. States Parties are obliged to destroy or divert to peaceful purposes any existing "agents, toxins, weapons, equipment and means of delivery" within nine months of joining the treaty (Article 2), and must adopt appropriate national implementation measures, which is understood to include penal sanctions to prohibit and prevent banned activity occurring on its territory (Article 4). The treaty's verification provisions require consultation and cooperation, bilaterally and multilaterally, to deal with any concerns of noncompliance (Article 5), combined with politically binding compliance procedures such as annual Confidence-Building Measure declarations and declarations on compliance prepared for each Review Conference. These procedures have been used unsatisfactorily to date, and recourse to the UN Security Council under Article 6 in case of alleged violation of the BWC has not been used at all, despite credible, documented instances of noncompliance. Where a non-compliance concern constitutes a threat to, or breach of, international peace and security, the matter may additionally be brought to the attention of the UN Security Council under the UN Charter provisions. Negotiations for a verification protocol, which would have established a verification organization and regularized inspections, broke down in 2001.

Current Status

There currently are 151 States Parties and 16 signatories to the treaty. A series of five Expert Meetings and Meetings of States Parties to discuss key aspects of treaty implementation are being held under a "new process" agreed upon at the resumed session of the Fifth Review Conference in 2002 (the first session was held in 2001). The results of these meetings will be reported to the Sixth Review Conference, which is due to be held before December 2006. A change in the political climate, especially in the United States, is required before negotiations can resume on improving multilateral verification of the treaty. In the meantime, a global civil society coalition has formed to enhance transparency and monitor treaty implementation: the BioWeapons Prevention Project.

States Parties to the Biological Weapons Convention (151)

Afghanistan, Albania, Algeria, Antigua and Barbuda, Argentina, Armenia, Australia, Austria, Bahamas, Bahrain, Bangladesh, Barbados, Belarus, Belgium, Belize, Benin, Bhutan, Bolivia, Bosnia and Herzegovina, Botswana, Brazil, Brunei-Darussalam, Bulgaria, Burkina Faso, Cambodia, Canada, Cape Verde, Chile, China, Colombia, Congo, Costa Rica, Croatia, Cuba, Cyprus, Czech Republic, Democratic People's Republic of Korea, Democratic Republic of Congo, Denmark, Dominica, Dominican Republic, Ecuador, El Salvador, Equatorial Guinea, Estonia, Ethiopia, Fiji, Finland, France, Gambia, Georgia, Germany, Ghana, Greece, Grenada, Guatemala, Guinea-Bissau, Holy See, Honduras, Hungary, Iceland, India, Indonesia, Iran, Iraq, Ireland, Italy, Jamaica, Japan, Jordan, Kenya, Kuwait, Lao People's Democratic Republic, Latvia, Lebanon, Lesotho, Libya, Liechtenstein, Lithuania, Luxembourg, Malaysia, Maldives, Mali, Malta, Mauritius, Mexico, Monaco, Mongolia, Morocco, Netherlands, New Zealand, Nicaragua, Niger, Nigeria, Norway, Oman, Pakistan, Palau, Panama, Papua New Guinea, Paraguay, Peru, Philippines, Poland, Portugal, Qatar, Republic of Korea, Romania, Russian Federation, Rwanda, Saint Kitts and Nevis, Saint Lucia, Saint Vincent and the Grenadines, San Marino, Sao Tome and Principe, Saudi Arabia, Senegal, Serbia and Montenegro, Seychelles, Sierra Leone, Singapore, Slovenia, Slovakia, Solomon Islands, South Africa, Spain, Sri Lanka, Sudan, Suriname, Swaziland, Sweden, Switzerland, Thailand, The Former Yugoslav Republic of Macedonia, Timor Leste, Togo, Tonga, Tunisia, Turkey, Turkmenistan, Uganda, Ukraine, United Kingdom, United States, Uruguay, Uzbekistan, Vanuatu, Venezuela, Vietnam, Yemen, and Zimbabwe.

(As of November 10, 2003)

States Signatories to the Biological Weapons Convention (16)

Burundi, Central African Republic, Côte d'Ivoire, Egypt, Gabon, Guyana, Haiti, Liberia, Madagascar, Malawi, Myanmar, Nepal, Somalia, Syrian Arab Republic, United Arab Emirates, and the United Republic of Tanzania.

(As of November 10, 2003)

FURTHER READING

Sims, N., *The Diplomacy of Biological Disarmament: Vicissitudes of a Treaty in Force, 1975-85*, Macmillan Press, London, 1988.

WEB RESOURCES

Biological and Toxin Weapons Convention website, www.opbw.org, accessed on 12/29/03.

Biological Weapons Convention: Collection of National Implementation Legislation, www.vertic.org, accessed on 12/29/03.

BioWeapons Prevention Project (BWPP), www.bwpp.org, accessed on 12/29/03.

Bradford Project on Strengthening the Biological and Toxin Weapons Convention, www.brad.ac.uk/acad/sbtwc, accessed on 12/29/03.

Harvard Sussex Program on Chemical and Biological Warfare Armament and Arms Limitation, http://www.sussex.ac.uk/spru/hsp/index.html, accessed on 12/29/03.

International Committee of the Red Cross (ICRC), Advisory Service on International Humanitarian Law, *1972 Convention on the Prohibition of Bacteriological Weapons and their Destruction, Fact Sheet*, ICRC, Geneva, Switzerland, January 2003, www.icrc.org, accessed on 12/29/03.

International Committee of the Red Cross, Treaty database, www.icrc.org/ihl, accessed on 12/29/03.

List of States Parties to the Convention on *The Prohibition of the Development, Production and Stockpiling of Bacteriological (Biological) and Toxin Weapons and on Their Destruction*, November 10–14, 2003, Geneva, Switzerland, BWC/MSP.2003/INF.2, www.opbw.org, accessed on 12/29/03.

CHEMICAL WEAPONS CONVENTION

The CWC bans the development, production, stockpiling and use of chemical weapons and obliges states parties to destroy existing chemical weapon stockpiles. This treaty is relevant in the context of bioterrorism defense because toxins, which are defined as chemicals of biological origin, fall under its purview. In addition, the CWC serves as a model for future treaties in that it provides for verification of compliance on the part of its members.

Background

During negotiations for the BWC in the early 1970s, states committed themselves to further negotiate an agreement banning the use and production of chemical weapons in the future. After some 20 years of planning, the convention was finally negotiated in the Conference on Disarmament in Geneva, Switzerland, and was opened for signature on January 13, 1993, in Paris, France. It entered into force on April 29, 1997, once 65 States Parties had ratified or acceded to it. The UN Secretary General is the treaty's depositary, with obligations specified in Article 23. While traditionally viewed as an arms control agreement, the CWC is also part of international humanitarian law governing the conduct of hostilities, whereby the right of parties to an armed conflict to choose methods or means of warfare is not unlimited.

The CWC complements the Geneva Protocol by, in addition to reinforcing the Geneva Protocol's ban of chemical weapons use, explicitly prohibiting the development, production, acquisition, stockpiling, retention or transfer of these weapons. States Parties also may not assist in preparations for the use of chemical weapons or any other activity prohibited by the treaty. Other treaty obligations include requirements to destroy existing chemical weapons—including those abandoned on other state's territory—along with any associated production facilities; adopt national implementation measures sufficient to give effect to the treaty; and establish a National Authority to facilitate and coordinate treaty implementation activities and to liaise with the Organization for the Prohibition of Chemical Weapons (OPCW), the treaty's verification organization.

States Parties' compliance is assessed according to the verification provisions, detailed in the treaty's verification annex. States Parties must provide a baseline and subsequent annual declarations of compliance and accept routine inspections of industrial facilities as well as any challenge inspections (to determine the facts in a case of suspected noncompliance) and ad hoc inspections (to investigate specific allegations of chemical weapons use) that may be authorized. A comprehensive definition of chemical weapons is contained in Article 2. Certain restricted chemicals may be used for peaceful purposes, with these activities also subject to the treaty's strict verification procedures.

Current Status

As of December 31, 2003, there are 158 States Parties and 22 signatories to the CWC. However, many states suspected of developing chemical weapons remain outside the treaty. The rate of adoption of effective national implementation measures to enforce the treaty remains slow, although the OPCW is facilitating necessary assistance to states. The OPCW continues to work toward universalizing the convention by promoting and encouraging ratification by states signatories and accession by other states.

States Parties to the Chemical Weapons Convention (158)

Afghanistan, Albania, Algeria, Andorra, Argentina, Armenia, Australia, Austria, Azerbaijan, Bahrain, Bangladesh, Belarus, Belgium, Belize, Benin, Bolivia, Bosnia and Herzegovina, Botswana, Brazil, Brunei Darussalam, Bulgaria, Burkina Faso, Burundi, Cameroon, Canada, Cape Verde, Chile, China, Colombia, Cook Islands, Costa Rica, Côte d'Ivoire, Croatia, Cuba, Cyprus, Czech Republic, Denmark, Dominica, Ecuador, El Salvador, Equatorial Guinea, Eritrea, Estonia, Ethiopia, Fiji, Finland, France, Gabon, Gambia, Georgia, Germany, Ghana, Greece, Guatemala, Guinea, Guyana, Holy See, Hungary, Iceland, India, Indonesia, Iran, Ireland, Italy, Jamaica, Japan, Jordan, Kazakhstan, Kenya, Kiribati, Kuwait, Kyrgyzstan, Lao People's Democratic Republic, Latvia, Lesotho, Liechtenstein, Lithuania, Luxembourg, Malawi, Malaysia, Maldives, Mali, Malta, Mauritania, Mauritius, Mexico, Micronesia (Federated States of), Monaco, Mongolia, Morocco, Mozambique, Namibia, Nauru, Nepal, Netherlands, New Zealand, Nicaragua, Niger, Nigeria, Norway, Oman, Pakistan, Palau, Panama, Papua New Guinea, Paraguay, Peru, Philippines, Poland, Portugal, Qatar, Republic of Korea, Republic of Moldova, Romania, Russian Federation, Saint Lucia, Saint Vincent and the Grenadines, Samoa, San Marino, Sao Tome and Principe, Saudi Arabia, Senegal, Serbia and Montenegro, Seychelles, Singapore, Slovakia, Slovenia, South Africa, Spain, Sri Lanka, Sudan, Suriname, Swaziland, Sweden, Switzerland, Tajikistan, Thailand, The former Yugoslav Republic of Macedonia, Timor Leste, Togo, Tonga, Trinidad and Tobago, Tunisia, Turkey, Turkmenistan, Uganda, Ukraine, United Arab Emirates, United Kingdom, United Republic of Tanzania, United States, Uruguay, Uzbekistan, Venezuela, Vietnam, Yemen, Zambia, and Zimbabwe.

(As of December 31, 2003)

States signatories to the Chemical Weapons Convention (22)

Bahamas, Bhutan, Cambodia, Central African Republic, Chad, Comoros, Congo, Democratic Republic of Congo, Djibouti, Dominican Republic, Grenada, Guinea-Bissau, Haiti, Honduras, Israel, Liberia, Madagascar, Marshall Islands, Myanmar, Rwanda, Saint Kitts and Nevis, and Sierra Leone.

(As of December 31, 2003)

WEB RESOURCES

Harvard-Sussex Program on Chemical and Biological Weapons Armament and Arms Limitation, www.fas.harvard.edu/~hsp and http://www.sussex.ac.uk/spru/hsp/index.html, accessed on 12/29/03.

International Committee of the Red Cross (ICRC), Advisory Service on International Humanitarian Law, *1993 Chemical Weapons Convention, Fact Sheet*, ICRC, Geneva, Switzerland, January, 2003, www.icrc.org, accessed on 12/29/03.

International Committee of the Red Cross, Treaty database, www.icrc.org/ihl, accessed on 12/29/03.

Organization for the Prohibition of Chemical Weapons (OPCW), www.opcw.org, accessed on 12/29/03.

ENVIRONMENTAL MODIFICATION CONVENTION

The ENMOD prohibits the hostile use of the environment as a means of warfare. This includes the deliberate manipulation of natural processes of the environment, such as deliberately caused floods, hurricanes, and earthquakes.

Background

The use of biological and chemical agents to effect environmental change for military advantage during the 1960s, such as the use of the 'Agent Orange' defoliant herbicide in Vietnam, and the fear that new weapons for disrupting environmental processes may be developed during the 1970s, led to calls for a ban on environmental modification weapons to be specifically laid down in treaties. The treaty complements the 1997 Additional Protocol I to the 1949 Geneva Convention, prohibiting damage to the environment during armed conflict, which was negotiated at the same time.

The treaty was negotiated at the Conference of the Disarmament Commission and was adopted by the UN General Assembly on December 10, 1976. It was opened for signature on May 18, 1977, in Geneva, Switzerland, and entered into force on October 5, 1978, after 20 states had ratified or acceded to it. The UN Secretary General is the treaty's depositary. The treaty bans the "military or any other hostile use of environmental modification techniques having widespread, long-lasting or severe effects as a means of destruction, damage or injury to any other state party" (Article 1). States Parties have agreed to nonbinding understandings to define the extent, duration, and severity criteria for prohibited activities. Research and development of environmental modification techniques, along with their use for peaceful purposes, are not prohibited under the treaty. Suspected noncompliance may be brought to the attention of the UN Security Council under Article 5, or under the UN Charter provisions where this activity constitutes a threat to, or breach of, international peace and security.

Current Status

As of December 31, 2003, there are 69 States Parties and 17 signatories to the convention. Two Review Conferences have been held according to the provisions in Article 8, the first in 1984 and the second in 1992. No complaints of noncompliance have been brought to the UN Security Council to date. This fact, along with the fact that no further Review Conferences have been held since 1992 when they may be held as frequently as five-yearly, indicates the decreasing relevance of this treaty to States Parties. The availability of other treaties with comprehensive procedures for resolving allegations of environmental damage, combined with the precedent of the Security Council establishing the UN Compensation Commission, a subsidiary organization to resolve complaints and assess compensation for environmental damage caused by Iraq during the Gulf War, have further undermined ENMOD's relevance.

States Parties to the Environmental Modification Convention (69)

Afghanistan, Algeria, Antigua and Barbuda, Argentina, Armenia, Australia, Austria, Bangladesh, Belarus, Belgium, Benin, Brazil, Bulgaria, Canada, Cape Verde, Chile, Costa Rica, Cuba, Cyprus, Czech Republic, Democratic People's Republic of Korea, Denmark, Dominica, Egypt, Finland, Germany, Ghana, Greece, Guatemala, Hungary, India, Ireland, Italy, Japan, Kuwait, Laos People's Democratic Republic, Lithuania, Malawi, Mauritius, Mongolia, Netherlands, New Zealand, Niger, Norway, Pakistan, Panama, Papua New Guinea, Poland, Republic of Korea, Romania, Russian Federation, Saint Lucia, Saint Vincent and Grenadines, Sao Tome and Principe, Slovakia, Solomon Islands, Spain, Sri Lanka, Sweden, Switzerland, Tajikistan, Tunisia, Ukraine, United Kingdom, United States, Uruguay, Uzbekistan, Vietnam, and Yemen.

(As of December 31, 2003)

States Signatories to the Environmental Modification Convention (17)

Bolivia, Democratic Republic of Congo, Ethiopia, Holy See, Iceland, Iran, Iraq, Lebanon, Liberia, Luxembourg, Morocco, Nicaragua, Portugal, Sierra Leone, Syria, Turkey, and Uganda

(As of December 31, 2003)

FURTHER READING

Roberts, A. and Guelff, R., *Documents on the Laws of War*, 3rd ed., Oxford University Press, Oxford, 2000.

WEB RESOURCES

Genomics Gateway, www.brad.ac.uk/acad/sbtwc/gateway, accessed on 12/29/03.

International Committee of the Red Cross (ICRC), Advisory Service on International Humanitarian Law, *1976 Convention on the Prohibition of Military or any Hostile Use of Environmental Modification Techniques, Fact Sheet*, ICRC, Geneva, Switzerland, January 2003, www.icrc.org, accessed on 12/29/03.

International Committee of the Red Cross, Treaty database, www.icrc.org/ihl, accessed on 12/29/03.

Sunshine Project, www.sunshine-project.org, accessed on 12/29/03.

AUSTRALIA GROUP

The Australia Group is an informal arrangement between 33 participant states and one intergovernmental organization (the European Commission) to harmonize national export licensing in order to prevent the diversion of specified chemicals, biological agents, and dual-use chemical and biological manufacturing facilities, and equipment for illegitimate purposes.

Background

The imposition of export controls on dual-use chemical agents and equipment became more widespread following verified instances of chemical weapon use, especially by Iraq and Iran during the first Gulf War (1983–1988). Yet, the absence of uniformity in the scope or application of states' national export criteria and licensing arrangements weakened their efficacy, as those determined to acquire these materials circumvented national export controls. This led Australia, in April 1985, to propose a meeting of those states that had introduced export licensing in order to determine the feasibility of harmonizing their national measures and to facilitate cooperation between them on export control issues. In 1990, as the potential for their diversion became more apparent, the group added criteria for dual-use materials relevant to biological weapons programs.

Australia Group members have agreed to require licenses for the export of specified chemical weapon precursors and dual-use chemical manufacturing facilities and equipment related technology; plant pathogens; animal pathogens; biological agents, including toxins, with human effects; and dual-use biological equipment. The group regularly reviews its "Common Control Lists," recording which agents and equipment shall be subject to participants' national export controls, and agrees on amended lists at its annual meetings held in Paris, France. It supports export licensing and monitoring intended to deter the diversion and proliferation of restricted agents and equipment while not unduly inhibiting the legitimate trade in these materials: Licenses are refused only where there is a specific concern that diversion may occur. Participants agree to adhere to a set of "Guidelines for transfers of sensitive chemical or biological items" in considering export licenses. States participating in the group are States Parties to both the BWC and CWC, and view these export controls as part of their national measures to implement the obligation under these treaties not to facilitate the development, production, or acquisition of chemical or biological weapons. That the group does not promote technology transfer, cooperation, or exchange for development or peaceful purposes as provided for in these conventions has led to criticism from some states outside the group.

Current Status

In addition to the current members, other states and relevant intergovernmental organizations may be accepted to join the group if they declare adherence to its principles and guidelines. The rate of accession has slowed, however, with the latest member, Bulgaria, joining in 2001. Other nations that once were part of the Soviet Union are considering joining, such as Kazakhstan and Ukraine.

Australia Group Participants (34)

Argentina, Australia, Austria, Belgium, Bulgaria, Canada, Czech Republic, Republic of Cyprus, Denmark, European Commission, Finland, France, Germany, Greece, Hungary, Iceland, Ireland, Italy, Japan, Republic of Korea, Luxembourg, Netherlands, New Zealand, Norway, Poland, Portugal, Romania, Slovak Republic, Spain, Sweden, Switzerland, Republic of Turkey, United Kingdom, and United States.

(November 30, 2003)

WEB RESOURCES

Arms Control Association, www.armscontrol.org, accessed on 12/29/03.

Nuclear Threat Initiative, www.nti.org, accessed on 12/29/03.

The Australia Group, www.australiagroup.net, accessed on 12/29/03.

DRAFT CONVENTION TO PROHIBIT BIOLOGICAL AND CHEMICAL WEAPONS UNDER INTERNATIONAL CRIMINAL LAW

The Draft Convention to Prohibit Biological and Chemical Weapons under International Criminal Law is a model agreement defining specific activities involving biological or chemical weapons as international crimes.

Background

The Harvard-Sussex Program on Chemical and Biological Warfare Armament and Arms Limitation devised the concept for the agreement in 1996 and developed an initial draft during workshops held in 1997 and 1998, with participants including international experts on biological weapons control and international law. The draft treaty was further revised in November 2001. It is intended to fill the gaps in the prohibition and prevention of the development, production, stockpiling, and use of chemical and biological weapons under existing international law.

Current treaties banning chemical and biological weapons are primarily directed at state activities and only oblige States Parties to adopt penal sanctions for offenses committed on their territory, with the CWC additionally requiring the extension of penal sanctions to states parties' citizens outside of their territory. This proposed treaty would ensure that individuals committing specified crimes involving biological and chemical weapons would be subject to penal sanctions in the jurisdiction of States Parties, even if the offense is committed outside its territory or by a foreign national. By setting an international standard for the prosecution and investigation of such offenses, the treaty would strengthen and harmonize national implementation measures for chemical and biological weapons–related offenses, which are currently implemented in a disparate fashion by states.

Current Status

The treaty remains in the draft stage. It must be endorsed or adopted by the UN General Assembly or a diplomatic conference of states before it can be opened for signatures. The draft treaty may be amended or renegotiated during this process. Following the Netherlands presentation of the draft treaty to the Public International Law Working Group (COJUR) of the Council of the European Union in January 2002, delegations agreed to submit proposals on the draft treaty to their governments for consideration, with a view to its adoption.

REFERENCES

"A draft convention to prohibit biological and chemical weapons under international criminal law" (editorial), *CBW Conventions Bulletin*, **42**, 1–2 (1998).

"International criminal law and sanctions to reinforce the BWC" (editorial), *CBW Conventions Bulletin*, **54**, 1–2 (2001).

WEB RESOURCES

Harvard Sussex Program on Chemical and Biological Warfare Armament and Arms Limitation, http://www.sussex.ac.uk/spru/hsp/index.html, accessed on 12/29/03.

DRAFT MODEL CONVENTION ON THE PROHIBITION AND PREVENTION OF BIOLOGICAL TERRORISM

The Draft Model Convention on the Prohibition and Prevention of Biological Terrorism (Draft Model Convention) is a model agreement defining activities that constitute biological terrorism as international crimes.

Background

The Draft Model Convention was proposed to stimulate the academic and policy debate on how biological terrorism may be more effectively dealt with under international law and, therefore, the national law of States Parties. It lays out offenses relating to biological terrorism and defines them as international crimes, and requires States Parties to ensure that they are transformed into offenses and penalties at the national level. Ultimately, the convention comprises deterrence, prevention, detection, and interdiction elements by criminalizing the hostile use of biological agents. It requires a licensing system for legitimate biological activities involving listed dangerous pathogens, and establishes a mechanism to promote biosafety and biosecurity standards for listed pathogens. It also aims to strengthen information gathering and analysis on illicit activity.

Current Status

The Draft Model Convention remains an academic document for furthering the discussion on prohibiting and preventing biological terrorism. There has been no attempt to have it endorsed or adopted by states with a view to it being opened for signature.

WEB RESOURCES

Draft Model Convention on the Prohibition and Prevention of Biological Terrorism: A textual summary, Center for Nonproliferation Studies, http://cns.miis.edu/research/cbw/biosec/pdfs/textsum.pdf, accessed on 12/29/03.

Draft Model Convention on the Prohibition and Prevention of Biological Terrorism, cns.miis.edu/research/cbw/biosec/pdfs/dcon.pdf, accessed on 12/29/03.

Iraq Weapons Inspections Database, www.vertic.org/onlinedatabase/unmovic/dsp_unmovicBackground.cfm, accessed on 12/29/03.

SIPRI Fact Sheet, Iraq, the UNSCOM Experience, http://editors.sipri.se/pubs/Factsheet/unscom.html, accessed on 12/29/03.

United Nations, United Nations Treaty Collection, Treaty Event, Multilateral Treaties on Terrorism, http://untreaty.un.org/English/tersumen.htm, accessed on 12/29/03.

United Nations, United Nations Treaty Collection, Conventions on Terrorism, http://untreaty.un.org/English/Terrorism.asp, accessed on 12/29/03.

United Nations Counter-Terrorism Committee, www.un.org/Docs/sc/committees/1373/, accessed on 12/29/03.

United Nations Department for Disarmament Affairs, http://disarmament.un.org, accessed on 12/29/03.

United Nations Monitoring, Verification and Inspection Commission, www.unmovic.org, accessed on 12/29/03.

United Nations Office on Drugs and Crime, www.unodc.org/unodc/index.html, accessed on 12/29/03.

United Nations Special Commission, www.un.org/Depts/unscom, accessed on 12/29/03.

IRAN

Benjamin D. Heath

INTRODUCTION

On January 29, 2002, U.S. President George W. Bush declared that Iran, along with Iraq and North Korea, comprised an "Axis of Evil" that threatened the global interests of the United States (Fig. 1) (Bush, 2002). According to President Bush, Iran was deserving of this designation because of its weapons of mass destruction (WMD) programs as well as its long-standing support for terrorist groups. Some analysts fear that Iran might combine its interests in WMD and terrorism and provide its terrorist proxies with biological weapons for use against Iranian enemies such as the United States. To gauge the reality of this fear, it is necessary to separate the rhetoric and unsubstantiated allegations against Iran from those facts that can be supported by solid evidence.

BRIEF HISTORY OF BIOLOGY IN IRAN

Iran has traditionally maintained one of the most advanced biotechnology programs in the developing world. Iran's foray into modern biology commenced in the 1920s, when two of its premier research centers, the Pasteur Institute and the Razi Institute for Serums and Vaccines, were established. Over the next 50 years, both institutes have conducted quality work in the fields of vaccine production and disease research. In the 1970s, the Razi Institute was recognized by the World Health Organization (WHO) as one of just four institutes worldwide that was capable of conducting advanced work on poliomyelitis.

Figure 1. Map of Iran. *Source*: www.iran.embassy.gov.au/general/p-iran-profile.htm.

After the Islamic Revolution, the national budget for biotechnology in Iran was reduced to a bare minimum. As a result, research stagnated and there was a corresponding shortfall in equipment and qualified scientists. These problems continue to plague Iran's biotechnology sector to this day. According to one leading scientist, Iran's 46 biological research institutions are currently staffed at just 4.3 percent capacity (Islamic Republic of Iran Broadcasting Channel Four, 2000/2001). Funding shortfalls are one reason why the sector does not attract students to enter the biosciences. Other reasons why biotechnology languishes are the antiquated patent, legal, and customs systems that effectively discourage innovative new work in the field. Despite these setbacks, the Iranian biotechnology sector has managed to retain a handful of quality research institutions that continue to perform sophisticated research, primarily in the fields of human and animal vaccine production as well as agricultural biotechnology. Also, Iranian institutes have sought to expand their international ties to help attract new sources of funding and scientists. Many of the leading Iranian institutes have forged relationships with the WHO, as well as sister institutes in countries such as Canada, Cuba, France, Greece, and India.

IRAN AND BIOLOGICAL WEAPONS

The official Iranian position on biological weapons is one of complete abhorrence. Iran is the only country in the Middle East that is a party to every relevant international arms control agreement. In terms of the Biological and Toxin Weapons Convention (BWC), Iran has hosted mock inspections for a possible future inspections mechanism at its Razi Institute and has played a constructive role in seeking to strengthen the Convention.

Despite these factors, many in the United States remain convinced that Iran maintains a secret biological warfare (BW) program in contradiction of its signed international agreements. However, information to substantiate these allegations is negligible, and most nongovernmental experts in the field readily admitting that little is known about the putative program. Judging from the fact that intelligence reports regarding Iran's alleged BW program consist entirely of vague estimates on capabilities that are repeated nearly verbatim year in and year out, the same dearth of reliable information also seems to plague

Encyclopedia of Bioterrorism Defense, Edited by Richard F. Pilch and Raymond A. Zilinskas
ISBN 0-471-46717-0 Copyright © 2005 Wiley-Liss

U.S. intelligence organizations. The case against Iran is as follows.

Analyst Anthony Cordesman and the U.S. Department of Defense believe Iran initiated a secret offensive BW program sometime in the early 1980s (Cordesman, 1997; US Department of Defense, 1996). Alternatively, Jane's Information Group and the Russian Foreign Intelligence Service believe that Iran did not begin offensively directed work on biological agents until the late 1980s or early 1990s (*Jane's Foreign Report*, 1993; *Jane's Strategic Weapons*, 1995). While neither of the two Jane reports does call the program a biological weapons program, Russia calls it a "military-applied biological program," suggesting that Iran might use biological agents for defensive purposes; to help develop gas masks, decontamination techniques, and other defensive equipment. According to a variety of sources, the offensive BW program has concentrated on mycotoxins, such as the T2 mycotoxin, and may have worked with *Bacillus anthracis* (anthrax), botulinum toxin, and ricin (Eisenstadt, 1995; Cordesman, 1997). These sources suggest that research on these agents is embedded within Iran's biotechnology industry as well as its many universities (U.S. Department of Defense, 1996). Other industries, many in the engineering fields, are said to focus on developing advanced biological production equipment such as fermenters (Samsami, 1999). These sources believe that Iran is recruiting bioweapons scientists from the former Soviet Union to work on developing an extensive BW arsenal, and that it now is in the latter stages of research and development (Miller and Broad, 1998). Finally, these sources suggest that Iran is able to produce small quantities of biological weapons agents, that it may be able to weaponize some agents, and that it is attempting to develop long-range delivery systems for these weaponized agents (DeSutter, 1997).

For the most part, these allegations cannot be corroborated in the open-source environment. To begin with, there is nothing in the public domain that corroborates the allegation that Iran began a BW program in the early 1980s. There are no defector accounts, nor are there documented accidents such as the 1979 anthrax incident in Sverdlovsk or the 1971 smallpox outbreak in Aralsk. The two Jane reports and Russian estimates, like the suggestions that Iran may have focused on the weaponization of T2 mycotoxin, appear to be drawn from attempted Iranian purchases in 1988 and 1989 of T2 mycotoxin-producing *Fusarium* subspecies (spp.) from research institutions in Canada and the Netherlands. While it is true that Iranian institutes actually did attempt to make these purchases, no open source evidence links Iran's interest in *Fusarium* to a BW program. A more plausible explanation is that Iran sought samples of such fungi to apply to its agricultural biotechnology programs. Indeed, according to information released by a number of Iranian research institutions, advanced work is now being conducted on *Fusarium* spp. to study its potential as a biopesticide against broomrape (a damaging weed), as is research into genetically modifying Iranian agricultural products to make them resistant to *Fusarium*. Similar research is conducted without problems by universities throughout North America in an

effort to limit the damage to wheat crops caused by these fungi (Wells, 2002; Tranberg, 2003).

In terms of *B. anthracis*, botulinum toxin, and ricin (the other BW agents typically associated with Iran), no open source evidence exists that links these agents to an offensive BW program. One report emerged in the early 1990s claiming that Iran had imported 120 tons of castor beans and suggested that Iran was planning on using them to produce ricin (Eisenstadt, 1998). However, this report was based on a statement by an anonymous Central Intelligence Agency (CIA) official and therefore could not be independently corroborated. While this CIA official feared that Iran might use castor beans to produce ricin, no subsequent reports have emerged substantiating this allegation. If the castor bean purchase actually took take place (which is also in doubt because no additional reports have ever emerged on this front as well), using them to produce vast quantities of ricin may not have been their intended purpose. After all, most analysts consider ricin only to be useful for individual assassination, not as a mass casualty weapon. A more likely explanation is that Iran bought the castor beans for the reasons most other countries seek them, above all to produce castor oil for use in industry.

As for Iran's alleged ties with former Soviet BW scientists, evidence does exist that Iranian research institutes like the Pasteur Institute have employed a small number of former Soviet scientists. However, no evidence exists in the open sources to link those scientists to Iran's alleged BW program. To the contrary, the research by the very few former Soviet biologists who work in Iran, particularly those accused in the Western media of having ties to an Iranian BW program, appears to be conducted in the open and to involve only legitimate and benign projects.

Unlike the unsubstantiated allegations that make up most of the charges against Iran, there is one area of confirmed research being conducted in Iran that could have significant BW applications. According to the Iranian Research Organization for Science and Technology (IROST), researchers there have succeeded in producing *Bacillus thuringiensis* on a pilot plant scale. IROST technicians have also succeeded in creating effective dissemination equipment to evenly spray *B. thuringiensis* over a target at the rate of 1/7 kg/ha. This project was originally funded by the United Nations Development Program (UNDP) as well as the United Nations Education, Scientific, and Cultural Organization (UNESCO) to use *B. thuringiensis* as a biopesticide to combat malaria vectors such as mosquitoes. IROST claims it has succeeded in developing a product capable of controlling 100 percent of mosquitoes in target sites for 15 to 25 days after application (www.irost.org, accessed 8/13/03). While this is indeed a worthy and important project that could have a significant impact on reducing malaria in Iran and elsewhere, the techniques used for producing and disseminating *B. thuringiensis* are nearly identical to those used for producing and disseminating *B. anthracis*. Although no open source evidence exists that suggests Iran has utilized these techniques for *B. anthracis*, the capability certainly is there.

IRAN AND TERRORISM

Shortly following the overthrow of the Shah in 1979, Iran embarked on a campaign to export the Islamic Revolution to other countries in the Middle East. Through this policy, Iran began funding militant Shiite groups such as Lebanon's Hizballah as well as violent groups in Iraq, Kuwait, Bahrain, and other Gulf countries. In countries such as Kuwait and Bahrain, these groups attempted to covertly rally the majority Shiite populations in order to foment revolutions against their Sunni governments. In Lebanon, Hizballah openly waged war against the Israeli forces occupying the southern third of the country, occasionally employing terrorist methods such as hostage taking and suicide bombing to achieve its goals. These tactics, particularly after the 1983 car bombing of the U.S. Marine barracks in Beirut, caused the United States to label Hizballah a terrorist organization. When Iran continued to fund and arm Hizballah and other radical Shiite groups throughout the Middle East, the United States labeled it as a state sponsor of terrorism. However, it bears mentioning that aside from the United States and Israel, few other entities list Hizballah as a terrorist organization. For instance, while the European Union (EU) lists Hizballah's senior intelligence officer as a terrorist, it does not include Hizballah on its official list of terrorist groups (http://www.deltur.cec.eu.int/terroristlist03052002.pdf, accessed on 8/13/03). Therefore, although Iran continues to fund Hizballah, the EU and other countries are likely to be less inclined to label Iran as a state sponsor of terrorism than, for instance, the United States.

Nevertheless, fears over whether Iran would help arm groups such as Hizballah with biological weapons appear to have first emerged in the late 1980s after the United States began accusing Iran of using chemical weapons on the battlefield. The first known instance in which this topic was formally addressed in the U.S. government came during testimony before Congress by then-CIA Director William Webster. Webster informed the Senate that there was no reason to believe Iran would provide WMD to terrorist groups (Hearing of the Full Senate Governmental Affairs Committee and the Permanent Subcommittee on Investigations, 1989). Similar statements have since been made by the Defense Intelligence Agency (DIA) and CIA in 1996 and the State Department in 1997. Since 1997, no direct statement has been made concerning Iran and the transfer of biological weapons to terrorist organizations. However, the Department of State has, in its 2002 *Patterns of Global Terrorism*, declared that no terrorist group appears to have acquired biological weapons. This, by default, suggests that Iran has, at least to the knowledge of the U.S. government, not transferred any biological agents to terrorist groups.

CONCLUSION

As has been described, the public record on Iran and biological weapons has long been rife with allegations unsubstantiated by evidence. While many American citizens appear to be willing to accept official U.S. government statements condemning Iran as a biological weapons proliferator, this does not seem to be the case elsewhere around the world. Skepticism appears to be widespread because of the United States's record of acting on classified evidence that later could not be substantiated, such as the bombing of the Al-Shifa pharmaceutical factory in Sudan and now the post-war effort in Iraq. Because no indisputable evidence of Iranian BW efforts has yet been offered in the open source environment, many around the world argue that there are no grounds to believe such research is being conducted. Regardless, if for argument's sake the existence of such research is conceded, it is still unlikely that Iran or any other country would transfer biological weapons to terrorist groups because such transfer eliminates control over where a given weapon is stored, how it is deployed, and against whom it is used. Ultimately, because serious questions remain concerning whether Iran actually possesses an offensive BW program, because no evidence exists that Iran has ever transferred biological weapons to terrorist groups in the past, and because Iran is unlikely to transfer such weapons to a terrorist group in the future, the threat Iran poses in the realm of bioterrorism remains negligible.

REFERENCES

Cordesman, A.H., *Iranian Chemical and Biological Weapons*, CSIS Middle East Dynamic Net Assessment, July 30, 1997.

DeSutter, P., National Defense University Strategic Forum, *Deterring Iranian NBC Use*, Number 110, April 1997.

Eisenstadt, M., *Iran's Military Capabilities and Intentions: An Assessment*, Transmitted by Federal News Service, November 9, 1995.

Eisenstadt, M., *The Deterrence Series: Chemical and Biological Weapons and Deterrence, Case Study 4: Iran*, Chemical and Biological Arms Control Institute, 1998, p. 2.

Hearing of the Full Senate Governmental Affairs Committee and the Permanent Subcommittee on Investigations, Capitol Hill Hearing, Federal News Service, February 9, 1989.

Islamic Republic of Iran Broadcasting Channel Four, *Biotechnology: A National Priority*, Nov. 2000—Feb. 2001, quoted in "Iran: Developments in Science and Technology," FBIS Report IAF20010410000067, April 4, 2001.

Jane's Foreign Report, Iran's Mustard and Nerve Gas, May 20, 1993, volume/issue: 000/2256.

Jane's Strategic Weapons—De-Militarisation Markets, Section 6—Middle East & North Africa, 6.4 Iran, January 1, 1995, p. 66.

Miller, J. and Broad, W.J., *The Germ Warriors: A Special Report: Iranians, Bioweapons in Mind, Lure Needy Ex-Soviet Scientists*, New York Times, December 8, 1998, p. A1.

President Bush, George.W., State of the Union Address, Office of the Press Secretary, January 29, 2002, http://www.whitehouse.gov/news/releases/2002/01/20020129-11.html.

Samsami, S., *Clerical Regime's Quest for Biological Weapons & Germ Arsenal*, National Council of Resistance of Iran, January 26, 1999.

Tranberg, J., Market preference can't bend science, *StarPhoenix* (Saskatoon), March 13, 2003, p. A17.

US Department of Defense, *Proliferation: Threat and Response*, April 1996, http://www.defenselink.mil/pubs/prolif/me_na.html, accessed on 8/13/03.

Wells, J., *Can. Bus. Curr. Aff.*, Jan/Feb, 74–80 (2002).

WEB RESOURCES

http://www.irost.org/, accessed on 8/13/03.

Iran Opposition Says Teheran in Deadly Weapons Drive, CNN, February 4, 1999, www.cnn.com, accessed on 8/13/03.

Office of the Secretary of Defense, Iran: Objectives, Strategies and Resources, *Proliferation: Threat and Response*, (US Department of Defense, Washington, DC, 1996), p. 16, http://www.defenselink.mil/pubs/prolif/, accessed on 8/13/03.

IRAQ

Robyn W. Klein
Andy Oppenheimer

INTRODUCTION

This entry consists of two sections. The first considers Iraq's history of terrorist support and examines the current status of terrorism in postwar Iraq. The second reviews Iraq's former offensive biological weapons program in order to ascertain what capability might have been made available to supported terrorist groups had Saddam Husayn so wished. To date, no evidence demonstrating that this type of transfer actually took place during Saddam Husayn's regime has been published.

HISTORY OF SUPPORT OF TERRORISM

In the months leading up to the United States's invasion of Iraq in 2003, top Bush administration officials often cited Saddam Husayn's ties to dangerous terrorist groups as a principal justification for ousting the dictator and his regime. Such suspected links between the Husayn government and international terrorism were not a new development. In fact, Iraq had been included on the U.S. Department of State's list of nations that sponsor terrorism for much of Saddam Husayn's more than three decades of de facto supremacy.

Even so, postwar debate continues amid allegations that, in an effort to grow support for invading Iraq, the Bush administration overstated the true extent of the Saddam Husayn regime's association with particular terrorist groups. Mindful of this ongoing controversy, what is certain about the former Iraqi government's support for terrorism is that, at minimum, Saddam Husayn provided sanctuary in Iraq as well as material and logistical support to various terrorists and terrorist organizations. In addition, the regime engaged its own state security agents in terrorist-like missions to eliminate opponents around the world, including a thwarted 1993 plot to assassinate former U.S. President George Bush and the Emir of Kuwait.

The most common thread in Saddam Husayn's support for terrorism was his backing for groups that were pitted against his perceived foes, especially Israel and Iran. This pattern, whereby Husayn fostered destabilizing efforts aimed toward his projected enemies, was less a reflection of any ideological accord with terrorist groups and more the manifestation of a despotic ruler seeking to foil his rivals and increase his own foothold on power. Those terrorist groups that Saddam Husayn openly supported, as well as those groups with more ambiguous ties to the regime, include the following:

Mujahedin-E-Khalq Organization

The Mujahedin-e-Khalq Organization (MEK) was formed in the 1960s by well-educated Iranians determined to overthrow the shah's regime and purge Western influence from Iranian political affairs. During the 1970s, the group carried out numerous terrorist attacks inside Iran, including assaults against both Iranian leadership targets and U.S. personnel working in the country. MEK dissidents were expelled from Iran after the 1979 Islamic revolution, and eventually found a convenient and hospitable new home base for their operations in neighboring Iraq.

The Iraqi leadership provided financial, material, and logistical support to MEK forces in return for their assistance, when called upon, to suppress adversaries of the Husayn regime within Iraq, such as the Kurdish populations in the north. This arrangement allowed the MEK to use Iraq as a launching pad for continued attacks in Iran and at Iranian embassies around the world throughout the 1990s.

The U.S. Department of State alleged in 2000 that, among the various manifestations of regime support for the group, profits from the United Nations mandated Oil-for-Food program were used for the construction of a 6.2-sq km complex for the MEK in Fallujah, Iraq. Further, MEK forces in Iraq prior to the U.S.-led invasion were estimated to number at least 5000 fighters and perhaps many more. As the American forces pushed into Iraq in the early spring of 2003, however, MEK representatives agreed to a still effective ceasefire with the coalition.

Palestinian Terror Organizations

The Husayn regime openly provided financial support to the families of Palestinian militants involved in attacks against Israel. In particular, the Arab Liberation Front (ALF) received funding and instructions from the Husayn regime, and operated as a conduit through which the regime paid the families of suicide bombers responsible for strikes on Israeli targets. Saddam Husayn also

Encyclopedia of Bioterrorism Defense,   Edited by Richard F. Pilch and Raymond A. Zilinskas
ISBN 0-471-46717-0   Copyright © 2005 Wiley-Liss

provided directed financial support and backing to other Palestinian terrorist groups, especially after violence between Palestinians and Israelis escalated in 2000. According to an April 2003 U.S. Department of State report, these groups included the Palestinian Liberation Organization (PLO), the Popular Front for the Liberation of Palestine (PFLP), HAMAS, the Palestine Islamic Jihad organization, and the May 15 Organization.

In addition, the Husayn regime long offered Iraq as a haven for Palestinian terrorists and prominent terrorist group leaders wanted by Israel or other nations for criminal prosecution. International fugitive Abu Nidal of the Fatah-Revolutionary Council and the Abu Nidal Organization (ANO) and Abu Abbas of the Palestine Liberation Front (PLF) are two of the most well-known terror organizers who lived securely within Iraq's borders for years.

Some intelligence reports also indicate that Palestinian militants received forged documents and logistical support from the Iraqi regime, along with military hardware and explosives training at Iraqi bases (Laub, 2003). It appears, however, that there was an ebb and flow to this type of operational assistance over the years, with financial backing and the granting of a haven being the more consistent types of support that the regime provided to Palestinian terrorists.

Ansar al-Islam

Kurdish Islamic extremists in Iraq formed the group Jund al-Islam in September 2001 and immediately declared a holy war on Kurdish secular ruling parties in northern Iraq. Soon after, the members of Jund al-Islam joined forces with other small Islamic extremist units from Iraq and Iran to form a unified group with the name Ansar al-Islam, meaning "Soldiers of God." Prior to the U.S.-led invasion of Iraq in 2003, Ansar al-Islam's membership likely grew into the several hundred with the group's operations based in Iraq's northern "no-fly" zone, which had been created by the international community to keep Kurdish populations safe from repression by the Iraqi regime.

Some analysts brand Ansar al-Islam to be a cell of al-Qa'ida, contending that al-Qa'ida provided large sums of cash to the group in 2001 in order to support their efforts to wage holy war. Several reports also charge that Islamic terrorists from the Middle East and Europe were trained at Ansar al-Islam's camp in northern Iraq and that the site provided a haven to followers of Usama Bin Ladin as they fled Afghanistan after the U.S. invasion there beginning in 2001. At least one Ansar al-Islam leadership figure, however, has denied any such connection with al-Qa'ida (Hosenball, 2003).

The relationship between Saddam Husayn's regime and Ansar al-Islam remains ambiguous and controversial. One line of reasoning maintains that any connection between the two would have been unlikely given Ansar al-Islam's goal of establishing a Muslim theocracy in Iraq, which certainly would be in opposition to the secular Husayn regime. Another argument is that a weak union did indeed exist whereby the Iraqi government provided Ansar al-Islam with some clandestine financial support and training assistance, in part motivated by the regime's desire to destabilize its Kurdish adversaries in the northern protected region of the country. Bolstering this notion, U.S. Secretary of State William Powell alleged to the UN Security Council in February 2003 that, "Baghdad has an agent in the most senior levels of the radical organization Ansar al-Islam that controls this corner of Iraq" (Powell, 2003).

In March 2003, during its war to oust the Husayn regime, U.S. forces destroyed Ansar al-Islam's headquarters in northeastern Iraq. The group's facility was a large complex composed of many buildings and underground tunnels. U.S. officials alleged for some time thereafter that Ansar had operated a "poison" and perhaps chemical weapons factory at the site. Americans investigating the facility also noted that "foreigners" (individuals other than Iraqis and Iranians) were among the dead.

Al-Qa'ida

Uncovering a strong link between Saddam Husayn's regime and the al-Qa'ida terrorist network remains a difficult and contentious task. In February 2003, U.S. Secretary of State Colin Powell spoke before the UN Security Council and described a suspected relationship that at least included the regime's creation of a "permissive environment" for al-Qa'ida movement in Iraq.

Powell's presentation included a case against Abu Mus'ab al-Zarqawi, who was depicted as a vital link between al-Qa'ida and the Iraqi regime. Zarqawi is an alleged associate of Usama Bin Ladin with expert knowledge of chemical weapons. According to U.S. intelligence reports, Zarqawi not only spent time training with Ansar al-Islam in northern Iraq, but also was permitted safe passage into Husayn's tightly controlled capital of Baghdad in order to receive medical attention for injuries suffered during U.S. attacks on Afghanistan in 2001. Reports suggest Zarqawi was joined in Baghdad at that time by other al-Qa'ida members—none of whom, many U.S. officials argued, would have been able to operate freely without the support of the authoritarian Husayn regime. Other less substantiated allegations include a claim that from Baghdad Zarqawi orchestrated the murder of a U.S. diplomat in Amman and was involved in a foiled plot to attack targets in London with ricin that was possibly manufactured at the Ansar al-Islam camp in northern Iraq (Tyler, 2003; Van Natta et al., 2003).

In June 2003, however, the chairman of the monitoring group appointed by the UN Security Council to track al-Qa'ida reported that his team had not yet discovered any genuine link between the ousted Husayn regime and al-Qa'ida (Chandler, 2003).

Kurdistan Workers' Party (PKK)

The Kurdish Workers Party (PKK) is a Marxist–Leninist separatist group that formed in 1973 and waged a 15-year guerilla war against Turkey beginning in the mid-1980s. The PKK's goal was to win an independent democratic

Kurdish homeland in the southeastern territories of Turkey. The group took advantage of the northern "no-fly" zone created after the 1990 Gulf War by basing thousands of armed PKK militants in northern Iraq and conducting vicious attacks across the Turkish border. The U.S. Department of State, the Turkish president, and other sources stated in the early 1990s that Saddam Husayn's regime provided at least modest financial assistance to PKK forces, perhaps in addition to the provision of some military training, supplies, and a haven in Iraqi territories.

CURRENT TERRORIST RELATIONS IN POSTWAR IRAQ

Iraq is currently a nation in transition, with a U.S.-led occupation force providing government services alongside a burgeoning post-Husayn civil and military administration. Terrorist activity occurs daily against United States and coalition forces, as well as against UN officials, aid workers, foreign contractors, Iraqi police, and Iraqi civilian reformers working with the coalition forces. Large-scale terrorist attacks have included the August 2003 bombing of the UN headquarters in Baghdad that killed at least 23 and the September 2003 mosque bombing in Najaf that killed dozens of Iraqi civilians together with Ayatollah Mohammed Baqir al-Hakim, the hugely popular leader of the largest Shiite political group in Iraq.

Postwar terrorist activity appears to emanate from a mix of sources, most likely consisting of Islamic militants from international terror groups such as al-Qa'ida, former Husayn regime and Baath Party loyalists, and possibly intranational groups that are vying for power or are vehemently opposed to the occupation. Foreign fighters—including Yemenis, Saudis, Jordanians, Lebanese, Algerians, and Syrians—are allegedly moving in and out of the country through Iraq's porous borders with relative ease, which in turn seems to encourage recurring waves of violence and terror attacks.

U.S. military officials also have noted the reemergence of terrorist activity conducted by Ansar al-Islam adherents who survived the coalition's attack on the group's headquarters in northern Iraq during the heavy combat stage of the war. In addition, a little-known extremist group named Al-Jami'at al-Salafiyya al-Jihadyya has claimed responsibility for several attacks since the end of the war in 2003, while previously unknown groups such as the Armed Vanguards of the Second Army of Mohammed and the Armed Islamic Movement for al-Qa'ida also declare responsibility for current terrorist activity in Iraq.

The level to which any of these groups may be coordinating terrorist activity together remains unclear. The likelihood that current terrorist activity stems from the command and control of international terrorist organizations is currently ill defined as well. The capture of Saddam Husayn by U.S. forces in December 2003, however, leaves only remnants of his regime's leadership yet to be taken into custody. Any remaining activity from former regime loyalists is, therefore, apt to be decentralized while declining in the future.

IRAQ'S OFFENSIVE BIOLOGICAL WEAPONS PROGRAM

Until the U.S.-led war to effect regime change in Iraq began in spring 2003, it was believed that Iraq had retained substantial growth media and biological warfare (BW) agents from pre-1991 stocks and that the regime was capable of resuming biological weapons production on short notice (in weeks) from existing facilities.

Following the occupation of Iraq, searches by a series of teams for Iraq's weapons of mass destruction (WMD) proved fruitless. Trailers initially believed to be mobile biological laboratories were located, but their real purpose now is unclear. It appears likely that in fact they were used to produce hydrogen gas for balloons. An interim report in September 2003 from the Iraq Survey Group (ISG) stated that although no biological weapons had been found, and although there was no evidence of active chemical or nuclear weapons programs, some evidence had been accumulated suggesting that clandestine biological weapons research was conducted after 1991 (Kay, 2003). However, actual findings were in effect limited to a single vial of live *Clostridium botulinum* Okra B stored in a scientist's home, which could be used to produce botulinum toxin. The ISG stated that deliberate dispersal and destruction of material and documentation related to weapons programs began preconflict and ran trans-to-post conflict.

History

In the decade before the first Gulf War in 1991, Iraq invested more resources into nuclear, biological, and chemical weapons, and missile programs than any other developing country. These weapons programs were integrally linked with Saddam Husayn's personal and strategic ambitions: to remain in power, to threaten Israel, and to dominate the Persian Gulf and the Arab world.

Though Iraq had for many years possessed the requisite medical, veterinary, and university facilities for a BW program, offensive efforts appear to have begun in earnest around 1985. Research and development (R&D) was initiated at a laboratory located on the Salman Pak peninsula. Bacterial strains were purchased from culture collections overseas, including the American Type Culture Collection (ATCC) in the United States, as were fermenters and other bioprocessing equipment. Initially research was concentrated on the characterization of *Bacillus anthracis* and *C. botulinum* to establish growth and sporulation conditions as well as storage parameters. With the shift from R&D to production in 1988, efforts became centered at the al-Hakam facility 80 miles from Baghdad, which attained large fermenters from abroad and began production of these two agents along with aflatoxin. In addition, 10 L of the toxin ricin were produced in 1988 at Salman Pak, which after testing proved unsuccessful for weapons purposes. The foot and mouth disease vaccine facility at Daura was partially converted to biological weapons agent production as well (Pike, 1998).

By 1990, Iraq had filled a total of 25 al-Husayn warheads and 166 400-pound aerial bombs with *B. anthracis*, botulinum toxin, or aflatoxin. During the Gulf War, the 25 al-Husayns were deployed to four locations, where they remained throughout the war. Ultimately, no biological weapons were used during the Gulf War.

In the 1990s, Iraq reluctantly acknowledged production of approximately 20,000 L of botulinum toxin solution, 8425 L of slurry containing *B. anthracis* spores, and 2200 L of aflatoxin. Baghdad was also found to have researched the weapons potential of such agents as ricin, *Clostridium perfringens*, camelpox virus, rotavirus, *Brucella melitensis*, enterovirus 70, trichothecene mycotoxins, and the plant pathogen (anticrop agent) wheat cover smut.

Current Status

While the many aspects of Iraq's WMD program remain to be clarified, the findings of successive inspections in the 1990s, 2002, and 2003, as well as postwar interviews of captured Iraqi scientists and government officials, offer some insight into the state's technical capability to produce biological weapons.

United Nations Special Commission. Following the 1991 Gulf War, inspectors of the United Nations Special Commission (UNSCOM) operated in Iraq until November 1998. Initially, Iraq categorically denied that it possessed an offensive BW program. Only in 1995 was the truth revealed, yet Iraq continued to make every effort to obscure its past, obstruct the dismantlement of its existing assets, and retain capabilities for the future. UNSCOM destroyed some BW facilities, including the al-Hakam facility, and much of the growth media required for biological agent production, but reported that as much as 17 tons of media remained unaccounted for. In addition, because no production documents were provided to support Iraq's statements on quantities of *B. anthracis* spore solution, botulinum toxin, *C. perfringens*, and aflatoxin, the percentage of each bulk agent that was ultimately destroyed remains unclear. After UNSCOM inspectors were withdrawn, it became much more difficult for outsiders to assess Iraqi capabilities (see Table 1).

United Nations Monitoring, Verification, and Inspection Commission. Inspectors reentered Iraq in December 2002 under the terms of UN Security Council Resolution 1441. The United Nations Monitoring, Verification, and Inspection Commission (UNMOVIC), headed by Dr. Hans Blix, found that although there were many discrepancies in the Iraqi documentation of materials not accounted for when the original inspectors left in 1998 (these discrepancies were never resolved), more than 100 biological samples collected at different sites failed to reveal any evidence of biological weapons R&D or production. In addition, examinations by hundreds of facilities by UNMOVIC inspectors had failed to find any sign of biological weapons–related activity.

Iraq Survey Group. Shortly after the U.S. invasion of Iraq, the ISG was established and given the mandate to find Iraqi WMD and related facilities. The ISG claims that, post-1998, Iraq began new research on *Brucella* spp. and Crimean-Congo hemorrhagic fever (CCHF) virus, and that work on ricin and aflatoxin continued. Further, the

Table 1. Iraq's Suspected BW Status Prior to the 2002 UNMOVIC Inspections and the 2003 War

- Stockpile of BW munitions included over 150 R-400 aerial bombs, and 25 or more special chemical/biological Al-Husayn ballistic missile warheads.
- Biological weapon sprayers for Mirage F-1 aircraft.
- Mobile production facility with capacity to produce "dry" biological agents (i.e., with long shelf life and optimized for dissemination).
- Seventeen metric tons of BW growth media not accounted for.
- Smallpox virus possession possible; tested camelpox prior to Gulf War.
- Technical expertise and equipment to resume production of *Bacillus anthracis* spores (anthrax), botulinum toxin, aflatoxin, and *Clostridium perfringens* (gas gangrene).
- BW munitions for missile and aircraft delivery in 1990–1991 Gulf War. Possible loading of al-Husayn ballistic missile warheads and R-400 aerial bombs with *B. anthracis*.
- Research on BW dissemination using unmanned aerial vehicles.
- Violation of obligations under UNSC Resolution 687, which mandates destruction of Iraq's biological weapon capabilities.
- Ratification of the BTWC on April 18, 1991, as required by the Gulf War ceasefire agreement.

Source: Center for Nonproliferation Studies at the Monterey Institute of International Studies. http://cns.miis.edu/research/wmdme/iraq.htm#fn3

ISG "unraveled a clandestine network of laboratories and facilities within the security service apparatus," but has yet to determine the extent to which this network was tied to large-scale military efforts or biological weapons (Kay, 2003).

Postwar Captures. Two of the leading Iraqi scientists believed to be behind Iraq's BW program, Dr. Rihab Taha (dubbed "Dr. Germ" by western media) and Dr. Huda Salih Mahdi Ammash (dubbed "Mrs. Anthrax"), are currently in U.S. custody. Taha studied plant toxins at the University of East Anglia, United Kingdom, during 1981 to 1984, and departed after earning a doctorate. She was the director of the BW research laboratory at Salman Pak, but was then promoted to be the director of the al-Hakam facility and its approximately 150-person staff. Several Iraqi defectors and UN inspectors reportedly believe that Ammash was deeply instrumental in rebuilding aspects of Iraq's BW production capability during the mid-1990s (Global Security Newswire, 2003).

LESSONS FROM IRAQ'S BIOLOGICAL WEAPONS PROGRAM

The Iraqi program offers important lessons on the establishment of a biological weapons capability:

- At least initially, agents were acquired largely from western culture collections, underscoring the need for implementation of stringent transfer and recipient validation requirements by these facilities.
- Education and training in the production of biological weapons was apparently gained in the west in a number of cases, including in the case of

Dr. Taha. In addition, according to Iraqi scientists and IIS sources, the Iraqi Intelligence Service (IIS) played a prominent role in sponsoring students for overseas graduate studies in the biological sciences, providing an important avenue for furthering BW-applicable research.

- Illicit BW-related activities proved difficult to track because of the lack of signatures inherent to dual-use development and production facilities.
- Technical hurdles to the production and delivery of biological weapons appear to have been overcome in many respects, suggesting that with proper time, training, and funding, a dedicated state-level program can have some success in establishing a biological weapons capability.

REFERENCES

Chandler, M., Chairman, *Report of the Monitoring Group Established Pursuant to Security Resolution 1363*, S/2003/669, June 26, 2003, http://ods-dds-ny.un.org/doc/UNDOC/GEN/N03/398/55/PDF/N0339855.pdf?OpenElement/.

Global Security Newswire, *Iraq: United States Denies IAEA Access to Tuwaitha Nuclear Complex*, Nuclear Threat Initiative (Online), May 6, 2003, http://www.nti.org/d_newswire/issues/2003/5/6/2s.html.

Hosenball, M., A Radical Goes Free, *Newsweek*, January 27, 2003.

Kay, D., Statement on the Interim Progress Report on the Activities of the Iraq Survey Group (ISG) before the House Permanent Select Committee on Intelligence, the House Committee on Appropriations, Subcommittee on Defense, and the Senate Select Committee On Intelligence, October 2, 2003, http://www.fas.org/irp/cia/product/dkay100203.html.

Laub, K., Iraq Aids Palestinian Terrorist Groups, but Israeli Intelligence Finds No Major Ties to Al-Qaida, *Associated Press*, February 1, 2003.

Pike, J. Biological Weapons Program: History, *Federation of American Scientists*, November 3, 1998 (Updated version), http://www.fas.org/nuke/guide/iraq/bw/program.htm.

Powell, C.L., Secretary Powell at the UN: Iraq's Failure to Disarm, New York City, February 5, 2003, http://www.state.gov/p/nea/disarm/.

Tyler, P.E., Intelligence Break Led US to Tie Envoy Killing to Iraq Qaeda Cell, *New York Times*, February 6, 2003.

Van Natta, D., Jr., Johnston, David, and Johnston, A. Terror Lieutenant With a Deadly Past, *New York Times*, February 10, 2003.

WEB RESOURCES

Bush, George W., Remarks by the President on Iraq, *Cincinnati Union Terminal*, October 8, 2002, http://www.c-span.org/.

Center for Nonproliferation Studies, Foreign Suppliers to Iraq's Biological Weapons Program, 2003, http://cns.miis.edu/research/ wmdme/flow/iraq/.

Center for Nonproliferation Studies, Weapons of Mass Destruction in the Middle East: Iraq: Biological, 2003, http://cns.miis.edu/research/wmdme/iraq.htm#fnB3/.

Center for Nonproliferation Studies, Iraq Special Collection: Biological Weapons Issues, http://cns.miis.edu/research/iraq/jn98bio.htm, accessed on 3/30/04.

International Institute for Strategic Studies, Iraq's Weapons of Mass Destruction: A Net Assessment, *IISS Strategic Dossier*, September 2002, http://www.iiss.org/news-more.php?itemID = 88/.

Iraq's Weapons of Mass Destruction: the Assessment of the British Government, The Stationery Office Limited, ID 114567 9/2002 776073, accessed on 9/24/02.

Johannson, E., Ed., Iraq: The UNSCOM Experience, *SIPRI Fact Sheet*, October 1998, http://editors.sipri.se/pubs/Factsheet/unscom.html.

United Nations Monitoring, Verification, and Inspection Commission (UNMOVIC), Documents, Twelfth Quarterly Report, UN Document S/2003/232, (2003), February 28, 2003, http://www.unmovic.org/.

U.S. Department of State, *Support for the Mujahedin-e-Khalq (MEK)*, Updated March 24, 2000, http://www.state.gov/.

U.S. Department of State, *Patterns of Global Terrorism 2000*, http://www.state.gov/s/ct/rls/pgtrpt/2000/2419.htm (posted on April 22, 2001).

U.S. Department of State, *Patterns of Global Terrorism 2003*, http://www.state.gov/s/ct/rls/pgtrpt/2003/ (revision posted on June 22, 2004).

ISLAM, SHI'A AND SUNNA

SAMMY SALAMA

KATHLEEN THOMPSON

INTRODUCTION

Fourteen centuries ago, an important schism occurred within the Islamic *umma* (nation of believers). From this rift, two main factions emerged: the *Sunna* and the *Shi'a*. Since then, myriad events in the Middle East have shaped the worldview and attitudes of the minority *Shi'a* population. Knowledge of this early Islamic history is vital to understanding the dynamics of present-day Iran and Iraq, both predominantly *Shi'a* nations. In addition to generating the concepts of *Sunna* and *Shi'a*, the schism sparked yet another idea that would have a tremendous impact on the course of Middle East history: *shahada,* or martyrdom. Gaining fame in the late 1980s, this concept of *shahada* is now used by various *Shi'i* and *Sunni* militant groups to justify suicide bombings as a legitimate method of warfare. To best understand the motives behind self-martyrdom, one needs to return to the roots of the *Shi'i-Sunni* conflict and explore how early Islamic history continues to impact the attitudes and religious interpretations of militant Islamist groups that rely upon suicide operations as integral parts of their military operations.

THE ORIGINS OF SHI'I ISLAM

Shi'i Islam grew out of a rift that developed within the Islamic community after the death of the Prophet Muhammad in 632 A.D. The majority of Muhammad's followers believed that his successor should be selected from his tribe, the Quraysh, by elders close to the Prophet. The elders chose Abu Bakr, the father-in-law of the Prophet and the first adult male convert to Islam, to lead the Islamic community (Munson, 1988). The successor, or *Khalifa* (Caliph), would rule the community according to Islamic law, or *Shari'a*, which had been established under Muhammad. Those who believed in this method of succession became known as *Sunni*; from the Arabic word *Sunna*, meaning custom or tradition. Thus, the term *Sunni* has the connotation of being a follower of the example of the Prophet (Esposito, 2002). Today, *Sunnis* comprise approximately 90 percent of the Muslim population (*Congressional Quarterly,* 2000).

ALI (THE FIRST IMAM)

Unlike their *Sunni* counterparts, a minority of Muslims believed that Muhammad's eldest male relative, *Ali ibn Abi Talib*, should succeed him (Esposito, 2002). The cousin and son-in-law of the Prophet, Ali was viewed by this faction, or *shi'a*, to be the true leader of the community (known as the *Imam*) (Munson, 1988). In addition to being a blood relative of the Prophet, Ali had repeatedly proven himself to be devoted to the nascent religion. In fact, a number of Ali's supporters regarded him

Encyclopedia of Bioterrorism Defense, Edited by Richard F. Pilch and Raymond A. Zilinskas
ISBN 0-471-46717-0 Copyright © 2005 Wiley-Liss

as the Prophet's intended heir (Cleveland, 2000). Thus, *Shi'i* Islam grew out of the belief that only descendents of Ali (and thus of the Prophet) could be leaders of the Islamic community (*Congressional Quarterly*, 2000). Initially, members of this faction were called *Shi'a Ali* (partisans of Ali); later their name was shortened to *Shi'a*.

THE EARLY CALIPHATE

The *Sunni* majority would select four caliphs in close succession in the three decades following Muhammad's death. The first caliph, Abu Bakr, was followed first by 'Umar ibn al-Khattab and then by 'Uthman ibn 'Affan. Finally, in 656 A.D., the *Sunni* elders selected Ali as the fourth caliph. Ali would rule only four-and-a-half years, until 661. These first four caliphs are known as the "rightly guided" Caliphs, or *al-khulafa' al-rashidun*. This era, during which Islam took root across the Arabian Peninsula, is viewed by many *Sunnis* as a continuation of the "golden age" of the Prophet Muhammad. For most *Shi'a*, however, the true height of the Islam occurred only during Muhammad's rule in Medina (622–632) and Ali's brief reign as caliph (656–661) (Munson, 1988).

JIHAD AND THE ADVENT OF ISLAMIC FUNDAMENTALISM

Ali's term as caliph was marred by controversy. His legitimacy as successor to the Prophet was challenged by Mu'awiya ibn Abi Sufyan, a member of the powerful Ummayad tribe. Mu'awiya, the governor of modern Syria, accused Ali of supporting the assassination of his cousin, the late caliph 'Uthman (*Congressional Quarterly*, 2000).

In 657 A.D., the two men's armies met at the battle of Siffin. Neither side could claim victory, although according to Shi'i tradition, Mu'awiya's troops faced certain defeat. The battle ended before Ali was able to secure his victory, however, when Mu'awiya's troops carried copies of the *Qu'ran* onto the field of battle, calling for arbitration that would "Let God decide." Ali consented, but the arbitration, like the battle, was inconclusive. Consequently, Ali and Mu'awiya remained locked in a struggle for leadership (Cleveland, 2000).

In the wake of the arbitration, a *Shi'i* faction known as the *Kharijites*, or *Khawarij* (from the Arabic *kharaja*, meaning to exit or secede), emerged (Esposito, 2002). The *Kharijites* had initially regarded Ali as the rightful successor to Muhammad, and considered Mu'awiya to be an infidel for his opposition to Ali's reign. After Ali consented to arbitration, the *Kharijites* became convinced that he, too, was guilty of *takfir*, or unbelief: By submitting to arbitration, Ali had flouted God's right to decide the victor of the battle at Siffin (Cleveland, 2000).

The *Kharijites* withdrew their support from Ali and formed their own religious community that preached strict adherence to the *Qu'ran* and the *Sunna* (examples of the Prophet). Central to the fundamentalist beliefs of the *Kharijites* was the concept of *jihad*. This concept existed long before the advent of the *Kharijites*. The *Qu'ran* places great emphasis on *jihad*, the Arabic term for "struggle."

According to the *Qu'ran*, followers of Islam must struggle to remain on the path of God and the Prophet. Over time, there have been various interpretations of what form this struggle should take. Often, *jihad* is believed to be a peaceful struggle, an effort to foster virtue and charity in the face of evil and adversity. Most interpretations of *jihad* focus on improvement within an individual or the Islamic community. Fundamentalist and militant interpretations take a more aggressive view, believing that *jihad* justifies force and violence in order to spread God's word to nonbelievers and to combat perceived injustices (Esposito, 2002).

The *Kharijites* placed a great emphasis on violent *jihad*, believing that force was often necessary to "command the good and forbid evil," as they construe the *Qur'an* to mandate. The *Kharijites* viewed the world in dualist terms as "us against them," with "them" being both non-Muslims and those Muslims who deviate from the *Kharijite* vision of Islam. The *Kharijites'* fundamentalist interpretation of the *Qu'ran* led them to equate the majority of other Muslims with infidels. In order to protect and purify Islam, the *Kharijites* believed that it was their duty to kill any unbelievers who did not repent. One of their more notable victims was Ali himself, who the *Kharijites* assassinated in 661 A.D., as punishment for his consent to arbitration with Mu-awiya. Ali's assassination took place in the southern Iraqi town of Najaf, and his shrine there remains among the most sacred sites in *Shi'i* Islam. The assassination itself was carried out by sleeper cells that infiltrated Ali's ranks, waiting patiently for the opportunity to strike.

Although the *Kharijites* comprised a small minority of Muslims that remained on the fringes of the Islamic community, the impact of fundamentalism and violent *jihad* has echoed throughout Islamic history (Esposito, 2002). In particular, it has taken center stage in the twentieth century with the resurgence of extremist *Sunni* Islamic fundamentalist groups (Cleveland, 2000). Present-day terrorist groups, such as al-Qa'ida and the Egyptian Jihad Group, have adopted strict interpretations of the *Qur'an* and embraced the practice of *takfir*, casting all Muslims who disagree with them as apostates. Like the *Kharijites*, these groups organize sleeper cell formations that assimilate within various communities and target opposing political leaders and civilian institutions that they regard as infidels or nonbelievers.

HASAN (THE SECOND SHI'I IMAM)

Following Ali's assassination, his eldest son, Hasan, became the second *Imam*. Hasan quickly ceded the caliphate to Mu'awiya, who launched the Ummayad dynasty that would rule the *Sunni* Muslim community from 661 until 750 A.D. Although Hasan did not act as caliph, *Shi'a* Muslims continued to regard him as their *Imam* until his death, sometime around 670 A.D. Shi'i tradition states that Hasan abdicated to Mu'awiya in order to prevent another conflict, but with the understanding that Hasan would one day be Mu'awiya's heir to the caliphate. Mu'awiya then allegedly went back on his word and had Hasan assassinated (Munson, 1988).

HUSAYN (THE THIRD SHI'I IMAM) AND MARTYRDOM

Shortly before his own death, Mu'awiya named his son, Yazid, as his successor. Yazid came to power in 680 A.D. Hasan's younger brother, the third *Imam* Husayn, refused to declare his loyalty to the new caliph and took haven in the Sacred Mosque of Mecca. Supporters of his late father, Ali, encouraged Husayn to organize a revolt against new caliph's regime in Kufa. Located in southern Iraq, Kufa had been the seat of Ali's rule (Munson, 1988). *Shi'i* tradition holds that Husayn accepted the call to rebellion in order to combat the growing hedonism of the Umayyads and to guide the Muslim people back to the path of Muhammad (Cleveland, 2000).

According to *Shi'i* accounts, Husayn left for Kufa knowing that he would not return. It is reported that before departing Mecca, he proclaimed that he would become a martyr in the struggle against the Umayyad "government of injustice and tyranny." In fact, Husayn met his end before reaching Kufa. The caliph's forces lashed out at the *Shi'i* leaders in Kufa, routing and killing Husayn and his small band of close supporters at Karbala in southern Iraq. Although his head was taken to Damascus as a sign of Yazid's victory, Husayn's body was entombed in Karbala and, like his father's tomb in Najaf, remains one of the most important holy sites in the *Shi'i* faith (Munson, 1988).

The Arabic term for martyr is *shahid*, meaning "witness" (just as the Greek word *martyr* literally means "witness") (Esposito, 2002). As a result of his death at Karbala, Husayn is revered in the *Shi'i* faith as *Sayyid al-Shuhada'*, Lord of the Martyrs (Munson, 1988). The death of Husayn illustrates the strong bond between the concepts of martyrdom and *jihad*. To sacrifice oneself in the struggle for Islam guarantees one entry into Paradise. Thus, the example set by Husayn has served as a source of faith and inspiration to the *Shi'i* community for several centuries. In the twentieth century in particular, his example has been heralded as a battle cry in numerous conflicts, from the Palestinian struggle against Israel to the Iran-Iraq War (Esposito, 2002).

MODERN APPLICATIONS OF SHAHADA (MARTYRDOM): ORIGINS OF SUICIDE BOMBINGS

On November 11, 1982, a new method of warfare in the eyes of the public was debuted: the human bomb. Seventeen-year-old Ahmad Qasir, a *Shi'i* from southern Lebanon, drove a white Mercedes Benz automobile filled with explosives into the Israeli military headquarters in the Lebanese city of Tyre, killing 141 people. This was the beginning of a new era, the inception of the deadly trend of suicide bombings (Jaber, 1997). Over the course of the next three years, there would be more than 30 similar attacks against Israeli targets. The *Shi'a* suicide bombers drove cars, carried suitcases, and in one instance even detonated a bomb while riding a donkey. The legitimacy for these instances of "self martyrdom" stems directly from the death of Husayn Bin Ali. The image of Husayn and his small band of devotees marching bravely toward certain death at the hands of Yazid and his vast army has

had a particularly profound impact on *Shi'a* resistance movements in the twentieth century (Jaber, 1997).

Shi'i theology glorifies the act of "self martyrdom" against overwhelming odds, claiming to provide the "martyr" with a direct path to heaven. What is more, the *shuhada'* (martyrs) see their actions as continuing Husayn's tradition of "self martyrdom". This Hizballah fighter echoes these sentiments:

> I, Salah Mohammed Ghandour, known by my alias *Malak* [angel], ask God to grant me success. I will be meeting the master of martyrdom Imam Husayn, this great Imam who taught all the free people how to avenge themselves on their oppressions. I shall, *inshaallah* [God willing], shortly after saying these words, be meeting my God with pride, dignity and having avenged my religion and all the martyrs who preceded me on this route. In a short while I shall avenge all the martyrs and oppressed of Jabal Amel, South Lebanon, as well as the children and sons of the Intifada in ravished Palestine (Jaber, 1997).

While religious fervor may have inspired the first suicide bombings, strategic acumen has perpetuated their use. Suicide bombs are notoriously cost-effective and enable groups to inflict extensive destruction upon the enemy while incurring relatively minimal damage themselves. Consequently, groups such as Hizballah consider suicide bombing to be the ultimate method of attack, religious ideology aside (Jaber, 1997). In the early 1980s, the growing trend of *Shi'a* suicide bombings in Lebanon took a heavy toll on the Israeli military, killing more than 650 soldiers in a three-year span. These bombings marked the beginning of the end for the Israeli occupation of Lebanon, and on June 10, 1985, Israeli forces withdrew from most of Lebanon, limiting their presence to a "Security Zone" in the South (from which they fully withdrew in the year 2000).

HOW THE SHI'I CONCEPT OF MARTYRDOM REACHED SUNNI PALESTINE

On December 16, 1992, Israeli Prime Minister Yitzhak Rabin sought to disrupt Hamas and Islamic Jihad's activities by deporting 415 Islamic militants to a deserted area in Southern Lebanon. The deportations came in response to Hamas's abduction and murder of Israeli border policeman Nissim Toledano, but immediately drew sharp international criticism and romanticized the Islamic militants in the eyes of their constituents in the West Bank and Gaza Strip. Popular support for Hamas and Islamic Jihad rose dramatically within the Palestinian community, (Bakogeorge, 1992) and the militants' desire for revenge against Israel only grew more potent during their time in exile (Fisk, 1993).

Stranded in "no man's land" just north of Israeli-occupied Southern Lebanon, the Palestinian deportees initially had little or no support system in Lebanon. The Lebanese government resented their presence, contesting Israel's use of its territory as a prison. The Palestinian exiles, however, soon received support from another powerful force within Lebanon. Hizballah, the perpetual thorn in Israel's side, provided the Palestinians with food and medicine.

The advent of the relationship between Hizballah and the Palestinian militants represents an important juncture in the Palestinian-Israeli conflict. It was during their exile in Southern Lebanon that the Palestinian Islamists learned the value of *shahada* from the Hizballah religious leaders. Until 1992, the Palestinians had relied upon secular nationalism as the justification for their resistance to Israeli occupation. From the birth of the Palestinian Liberation Organization (PLO) in 1964 until the Islamist exile in 1992, all Palestinian fighters were regarded as *fida-iyin*, or freedom fighters. Although the *fida-iyin* were willing to sacrifice their lives for the cause, their intent was not suicide, which Islam considers *haram* (forbidden).

With Lebanon continuing to voice its opposition to harboring the militants, Israel was forced to readmit the Islamic militants nine months after the initial deportation. The Palestinian *Sunni* Islamists returned to the West Bank and Gaza with Hizballah's *Shi'i* paradigm of *shahada*, giving rise to Palestinian suicide bombings (Fig. 1). Sacrificing one's life for the liberation of Palestine ceased to be solely a nationalistic obligation of the *fida-iyin*, and became a religious duty as well, rewarded with a direct path to heaven. In the years that followed, many young Palestinian men and women follow this path.

The first wave of Palestinian suicide bombings began in April 1994. The attacks came in response to the massacre of 29 Muslims in the Tomb of the Patriarchs in Hebron, carried out by Jewish settler Baruch Goldstein on February 25, 1994. Islamic Jihad and Hamas unleashed an unprecedented onslaught of attacks on Israeli military and civilians, killing more that 40 Israelis in five separate operations. Since that time, the suicide bomber has become the weapon of choice for Palestinian radicals. Over the course of the current *intifada*, more than 70 confirmed suicide bombings and 50 other attacks have been carried out, which have killed more than 800 Israeli civilians and soldiers and wounded thousands more.

OTHER SUNNI MILITANTS ADOPT THE PARADIGM OF SHAHADA

The hundreds of casualties inflicted by *Sunni* Palestinian suicide bombers in Israel have inspired other radical Islamist groups such as al-Qa'ida, the Egyptian Jihad Group, Ansar al-Islam, and others to adopt the Shi'i paradigm of martyrdom. Thus, suicide bombers have come to represent a legitimate method of warfare across the spectrum of Islamic militancy. On August 7, 1998, two suicide car bombs struck the American embassies in Tanzania and in Kenya, killing hundreds of civilians and injuring thousands more. These attacks marked the first stages in a pattern of al-Qa'ida suicide terror attacks, culminating in the deaths of some 3000 Americans on September 11, 2001. Similar attacks by al Qaida and associates have been carried out more recently in Riyadh, Casablanca, and Bali, and by the fall of 1993 (this should be by the fall of 2003), in postwar Iraq, suicide car bombings had become a weekly fixture, targeting such locations as the Jordanian embassy, the UN headquarters, the Turkish Embassy, Iraqi hotels, and the Shrine of Imam Ali in Najaf (Fig. 2).

IMAMS AND TWELVER SHI'ISM

Just as Ali and his two sons were the first three *imams*, their direct male heirs were regarded as *imams* also. According to the Shi'i faith, *imams* are not chosen by the people. Rather, they are ordained by God, a concept known as *nass*. Only one *imam* rules at a time. As God's representative on Earth, he carries out God's will, graced with virtue and infallibility (Bill and Robert, 1994).

The largest *Shi'i* sect, Twelver *Shi'ism*, is so named for its belief that 12 *imams* succeeded Muhammad through the line of Ali. The first 11 *imams* all died as martyrs (Munson, 1988). Tradition holds that the twelfth *imam*, Muhammad al-Muntazar, did not die but rather disappeared mysteriously around 878 A.D. (*Congressional Quarterly*, 2000) (other sources place this date at 874 A.D.). From the time of his disappearance in the 870s until

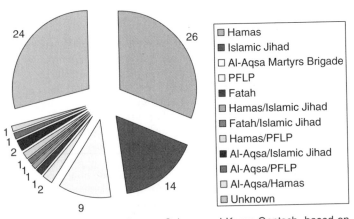

This chart was created by Sammy Salama and Karen Gentsch, based on "Suicide and other bombing attacks in Israel since the Declaration of Principles (September 1993)" entry on the Israeli Ministry of Foreign Affairs web page at http://www.mfa.gov.il/ as viewed on 9/29/03.

Figure 1. Suicide bombings in Israel, April 1994–August 2003.

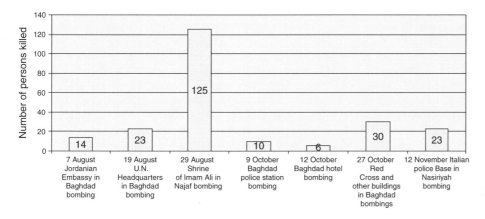

Figure 2. Major suicide bombings in postwar Iraq during 2003.

approximately 940, the twelfth *imam* existed in a state of lesser occultation (*al-ghayba al-sughra*). During this period, he remained hidden from all but four devotees who would relay his messages to his followers (Munson, 1988).

The twelfth *imam* moved into the state of greater occultation (*al-ghayba al-kubra*) in 940 A.D. In this state, he will remain completely hidden from all of humanity until the end time draws near and he reveals himself in order to bring justice to the world. Known as the Hidden *Imam* or *al-Mahdi* (the rightly guided one), the twelfth *imam* is regarded as the messiah of the *Shi'i* faith (Munson, 1988).

IJTIHAD AND SUNNI SCHOOLS OF THOUGHT

While in hiding, *al-Mahdi* is represented by individuals known as *mujtahidin* (Bill and Robert, 1994). During the twentieth century, the *mujtahidin* also became known as *ayatollahs*. From the Arabic phrase *ayat Allah*, literally meaning "the verse of God," ayatollahs are revered as the most learned scholars of *Shi'i* Islam. The term *mujtahid* itself comes from the Arabic word *ijtihad*, meaning "independent judgment" (Munson, 1988). An important concept in both Shi'i and Sunni Islam, *ijtihad* refers to the practice of using reason to apply the lessons of the *Qur'an* to matters not directly addressed in the holy text. Highly educated Islamic scholars (the *mujtahidin*) thus use their theological and scholarly background in *Qur'anic* studies to interpret God's will (Cleveland, 2000).

Like the question of succession following the Prophet Mohammed's death, the practice of *ijtihad* represents a source of tension between the *Sunna* and the *Shi'a*. In the *Sunni* faith, this practice of learned interpretation gave rise to a number of schools of religious and legal thought. Four of these schools, the *Hanafi*, the *Maliki*, the *Shafi'i*, and the *Hanbali* schools, remain in operation to this day. Despite variations in their respective interpretations of Islam, each *Sunni* school of legal thought recognizes the others as valid conceptions of *Sunni* Islam.

Although *ijtihad* has had a profound impact on early *Sunni* theology and modern *Sunni* legal thought, the *Sunna* believe that *ijtihad* must conform to strict limits. The *Qu'ran* must be narrowly interpreted in order to preserve its true message. Consequently, most *Sunni* scholars believe that the practice of *ijtihad* has outlived its usefulness (*Congressional Quarterly*, 2000). According

to the *Sunna*, the *Qur'an* is constant and cannot be reinterpreted.

The *Shi'a*, on the other hand, contend that the *Qur'an* must remain open for reinterpretation depending on the needs of the time. The *Shi'a* believe that as the world changes and progresses, the *Qur'an* must be reinterpreted to explain and govern changing realities.

EXTREMIST ISLAMIC SECTS AND IMPLICATIONS FOR WMD TERRORISM

In the wake of the September 11, 2001, attacks, militant Islamists have become a major focus in the war on terror. Islam in its purest form is a religion of peace (as its name implies, drawn from the Arabic *salam*, meaning peace). Nonetheless, deep-rooted divisions within Islam have given rise to numerous interpretations of the *Qu'ran* and the teachings of the Prophet. Al-Qa'ida, whose leaders adhere to a militant interpretation of *Sunni* Islam, has proven its willingness to carry out mass-casualty attacks against civilian targets. Will al-Qa'ida or a similar militant Islamist group make the leap from mass-casualty attacks to attacks involving weapons of mass destruction? And if so, what type of weapon will the group choose?

Another possible threat lies in the escalation of tactics of militant *Shi'i* groups. The *Shi'i* extremists who gave rise to the modern phenomenon of martyrdom may perceive weapons of mass destruction (WMD) as the key to the ultimate suicide mission. Biological weapons, with their tremendous potential for causing indiscriminate casualties, could also rank high on a would-be terrorist's list. On the other hand, the religious bent of *Shi'i* extremists, and more broadly Islamist extremists in general, may ultimately deter them from using weapons with such devastating consequences.

IRAN

The Safavid Empire and the Shi'i Conversion of Iran

The *Safavid* line traces its origins to either the Kurdish or Turkish tribes of Central Asia. In the late thirteenth century, a member of this line established a *Sunni* religious brotherhood in the northwest region of Iran, drawing heavily upon *Sufism* (Islamic mysticism). The popularity of the brotherhood spread quickly throughout the Caucasus. In 1494, the rule of the order passed to a child named Isma'il. Only seven years old at the time of his ascent to power, Isma'il launched a number of military campaigns that would ultimately transform the brotherhood into an empire. In 1501, the city of Tabriz, in what is today northwestern Iran, fell to the Safavid forces. Isma'il, barely a teen, immediately declared himself *shah* (king). By 1510, Isma'il had conquered most of what is modern Iran and expanded his sovereignty eastward across Afghanistan to the Oxus River (Cleveland, 2000).

The Safavid campaigns relied heavily on their Turkish followers, known as *Qizilbash* (meaning "Red-Headed Ones," referring to their brightly colored head gear). The Qizilbash expanded the Safavid Empire westward, across much of Anatolia and Iraq. In 1514, however, the Safavid army met the Ottomans at the battle of Chaldiran. Relying greatly on mounted archers, the Safavids were quickly defeated by the gunpowder artillery of the Ottomans. Although Isma'il was unable to penetrate the eastern borders of the Ottoman Empire, he succeeded in creating his own powerful empire based in Iran, an empire in which religion played a pivotal role.

The roots of *Shi'ism* in the Safavid brotherhood are unclear. The brotherhood had been born of *Sunni* Muslim beliefs in the late 1200s, yet by the time that Isma'il declared himself *shah* in 1501, he was a devout follower of Twelver *Shi'ism*. It is possible that the Safavid rulers who preceded him may have converted the brotherhood from *Sunni* the *Shi'i* Islam, but it is equally probable that Isma'il induced the conversion himself. As a child, he lived for a while with a *Shi'i* ruler in northwestern Iran, a contact that may well have catalyzed this conversion.

Whatever the roots of his belief might have been, Isma'il felt very strongly about the Twelver *Shi'i* faith. In fact, upon claiming the throne, *Shah* Isma'il decreed that Twelver *Shi'ism* would be the mandatory religion of the Safavid Empire. The self-proclaimed *shah* carried out his decree by force. *Sunni* establishments were disbanded, and all those living under Safavid rule were forced to convert for fear of execution. In order to foster the growth of *Shi'ism* in Iran, Isma'il recruited Islamic legal scholars and *ulama*, or Islamic theological scholars, from *Shi'i* dominated areas such as Lebanon. By importing well-educated *Shi'i* scholars, Isma'il established a religious base that eventually gave rise to an influential caste of Iranian *Shi'i ulama*.

Isma'il promoted a somewhat modified version of Twelver *Shi'ism* among his followers. The *shah* claimed to be a descendant of the seventh *Imam* and to have divine guidance. What is more, he claimed to represent *al-Mahdi*, the hidden twelfth *Imam* (Cleveland, 2000). The burgeoning Twelver *Shi'i ulama* of Iran supported such claims on behalf of Isma'il and subsequent Safavid *shahs*, most likely in order to secure their own position in the empire. The *shahs* and the *ulama* maintained a peaceful balance until Afghan armies began to attack the empire in 1722, bringing an end to Safavid rule.

The Afghan invaders shattered the power base of the *shah*, while the *Shi'i ulama* managed to maintain a significant hold on power. The *Sunni* Afghans to the east (in addition to *Sunni* Ottoman Turks to the west) were eventually defeated by Iranians under the command of Nadir Khan. In 1736, after driving back the invading forces, Nadir Khan declared himself the new *shah* and then attempted to integrate his *Shi'i* Empire into the Sunni faith. The backlash on behalf of the *Shi'i ulama* culminated in the *shah's* assassination in 1747. Following the death of Nadir Khan, the *Shi'i ulama* maintained a prominent position in Iran, but an irreparable rift had developed between the *ulama* and the crown (Munson, 1988).

Iran After the Safavids

The idea of centralized governance collapsed with the fall of the Safavid Empire in Iran. Despite his success against the Afghan armies, Nadir Khan was unable to fully consolidate his control of Iran. Following his assassination in 1747, a number of military and tribal leaders would emerge and rule over various factions in Iran. In 1794, Turkish tribal leader Fath Ali Shah launched the Qajar dynasty. The Qajars would rule Iran until the 1920s, but the government remained decentralized and its grip on power tenuous.

In 1901, an event occurred that would dramatically alter Iran's position in the world. The irresolute Qajar ruler, Muzzafir al-Din Shah, granted William D'Arcy of Britain the oil rights for the majority of Iran, excluding five provinces in the North. The Iranian economy had suffered greatly under Qajar rule, and the *shah* needed funds to fuel the lavish lifestyle of the court. Sixteen percent of any oil revenue would be given to the Iranian government as recompense for the oil rights, but despite these revenues, the Iranian economy continued to deteriorate. *Bazaris* (merchants), *ulama*, soldiers, and aristocrats alike developed a deep resentment of the monarchy and its concessions to the West. The discontent of these groups gave rise to the constitutional revolution in Iran.

In 1906, under pressure from protesters throughout Iran, Muzzafir al-Din Shah called together a constituent assembly (*al-Majlis*). The assembly resulted in the adoption of the Fundamental Law, which granted the elected legislature control over foreign affairs and contracts, and the Supplementary Fundamentals Laws, which among other things outlined citizens' rights and established Twelver *Shi'ism* as the official state religion (Cleveland, 2000).

The Role of Imperialism in Shaping Modern Iran

Soon after he signed the Fundamental Law, Muzzafir al-Din Shah died. He was succeeded by Muhammad Ali Shah, who was responsible for signing the Supplementary Fundamental Laws. Despite his initial cooperation with the constituent assembly, Muhammad Ali Shah's true desire was to strengthen the Qajar's hold on power. In the midst of this revolution and succession, Britain and Russia established their respective "spheres of influence" in Iran, with Britain claiming the southeast and Russia the north. Central Iran was to remain neutral.

Muhammad Ali Shah pointed to the British and Russian interference as a sign of the constitutionalists' inability to limit foreign influence in Iran. The royalists attacked the secular constitutionalists as atheists, thus forcing the *ulama* away from the constitutional movement. In 1908, civil war erupted between the royalists and the constitutionalists. While the constitution was ultimately preserved, the central government broke down, and the country once again deteriorated at the hands of tribal chieftains vying for power.

Russia withdrew its troops from Iran in 1917, though both Russia and Britain maintained substantial influence over Iran for the next several decades. In 1921, an Iranian military commander named Reza Khan deposed the Qajar ruling family with the help of *al-Majlis*. By 1925, he had secured for himself the title of *shah*, launching what would become the Pahlavi dynasty. Reza Shah infused his reign with secular ideology and heavily promoted Western behavior and culture. Although the *shah's* programs reinvigorated the beleaguered Iranian economy and education system, their secular nature alienated the majority of the *ulama*.

While the rift between the monarchy and the *ulama* widened yet again, ties between Iran and Germany strengthened. Although Reza Shah claimed to be neutral at the outset of World War II, Britain and the Soviet Union invaded Iran once again to ensure that the nation did not fall into German hands. Reza Shah's government and military collapsed, and he fled to South Africa. His son, Muhammad Reza, assumed the throne, but the British and the Soviets remained the true power brokers, and throughout the Second World War, Iran functioned as a logistical support base for the Allied Powers. It was at this time that the United States became intimately involved in Iranian affairs (Cleveland, 2000).

The Islamic Republic of Iran

In the aftermath of World War II, the Soviets attempted to maintain a military presence in part of Iran. Iran's new supporter, the United States, eagerly helped Muhammad Reza Shah's emaciated government engineer the Soviet withdrawal. In 1951, yet another constitutionalist coup was mounted against the monarchy, this time led by Muhammad Mosaddiq, the prime minister and leader of the National Front coalition. In August 1953, with the help of the U.S. Central Intelligence Agency, the *shah* succeeded in wresting control back from Mosaddiq. Upon reclaiming his throne, the *shah* initiated a system of close surveillance and harsh punishment for suspected members of the opposition. Under the reinstated monarchy, political freedom represented an unnecessary hazard.

Over the course of the next two decades, the government of Muhammad Reza Shah would eliminate many political freedoms in order to preserve its own hold on power. At various times throughout the 1960s and 1970s, a number of groups stepped forward to oppose the *shah's* repressive government. One such group was led by the Ayatollah Ruhollah Khomeini. Opposed to the corruption and Western influence that pervaded the monarchy, Khomeini called for a return to Islamic values in the state. Arrested in 1963, he was sent to exile in Turkey in 1964, then on to Iraq and finally to France in 1978.

Discord flourished inside Iran as the *shah* passed numerous pro-Western reforms. A low-security service, SAVAK, stepped up the repression of opposition elements. In 1975, the *shah* instituted a one-party system in Iran, securing the government's identity as a totalitarian regime (Cleveland, 2000). Throughout his time in exile, Khomeini continued to verbally attack the monarchy. The unrest came to a head in 1979, when supporters of Ayatollah Khomeini and other Iranian opposition groups succeeded in overthrowing the *shah's* brutal regime and established a *Shi'a* theocracy: the Islamic Republic of Iran.

IRAQ

The Development of Shi'i Islam in Iraq

In 1508, Safavid armies conquered most of what was a divided Iraq. Following the deaths of Ali and Husayn centuries before, the southern parts of Iraq had become rooted in the *Shi'i* faith, while *Sunni* Islam had flourished near Baghdad (the former capital of the *Sunni* Abbasid caliphate, Baghdad had remained a center of *Sunni* theological and legal study under the Ottoman Empire). The insult of the Safavid invasion was compounded by Shah Isma'il's demand that his new subjects convert to Twelver *Shi'ism*. As had been done in Iran, *Sunni* institutions throughout Iraq were destroyed. *Sunni* Ottoman forces refused to tolerate Iraq's forced conversion to *Shi'ism*, and in 1534, Suleiman the Magnificent defeated the Safavid forces. He reestablished Ottoman rule in Iraq, and with it the *Sunni* faith.

History would repeat itself in 1624, however, when the Safavids once again occupied Baghdad during the reign of Shah Abbas. In 1638, Ottoman Sultan Murat IV finally succeeded in defeating the Safavid occupation forces. The restoration of Ottoman rule again meant the restoration of *Sunni* institutions. The area would remain under Ottoman control for almost three centuries, until the fall of the Ottoman Empire during the First World War.

In the sixteenth and seventeenth centuries, the constant battle between *Sunna* and *Shi'a* for dominance in Iraq resulted in scores of deaths on both sides. The forced conversions carried out by the Safavids inflicted high numbers of *Sunni* casualties. Likewise, the *Shi'a* suffered heavy losses at the hands of Ottoman counterattacks. *Sunni* outrage at *Shi'i* rule in Baghdad, the former capital of the caliphate, was matched only by *Shi'i* outrage at Sunni rule in Najaf and Karbala. Consequently, the modern state of Iraq emerged as a society torn between two factions of Islam: a majority *Shi'i* population dominated by a powerful *Sunni* minority, both with long and bloody memories (Cleveland, 2000).

Shi'ism in Modern Iraq

Although they constitute a majority of the population, the *Shi'a* of Iraq have long been subjugated by the ruling power, be it the Ottomans, the British, or most recently the Ba'th Party. Following the U.S.-led invasion of Iraq, the *Shi'a* will play an integral part in the establishment of a new Iraqi government. At present, there are four predominant *Shi'i* factions in Iraq: the "Quietists," the al-Sadr faction, the Supreme Council for the Islamic Revolution in Iraq (SCIRI), and the Da'wa Party.

The 'Quietists'. The "Quietist" movement evolved under the leadership of Sayyid Abu al-Qasim al-Khu'i in the first few decades of Ba'thist rule. Following al-Khu'i's death in 1992, the Grand Ayatollah 'Ali al-Sistani assumed control of the movement. Quietism, a long-standing *Shi'i* tradition, discourages religious clerics from becoming directly involved in politics. According to the quietists, the clergy may monitor and debate politics, but it should refrain from becoming directly involved in the development of policy. Thus, the quietist view is a challenge to the ruling elite in Iran, where the Grand Ayatollah Ruhollah Khomeini launched Wilayat al-faqih, or the rule of the jurisprudent. In fact, Khomeini's establishment of clerical rule is considered anathema to a majority of leading clerics in both Iraq and Iran.

Under the rule of Saddam Husayn, quietism served as a survival mechanism for a number of Iraqi *Shi'a*. Their opposition to Wilayat al-faqih portrayed the quietists as posing a limited threat to Husayn's regime. Quietists, however, do not necessarily renounce politics completely. With the threat of Ba'thist rule no longer looming over their heads, the Iraqi *Shi'a* may turn to al-Sistani, one of the *maraji' al-taqlid* (sources of emulation), for political guidance. Having been subject to minority rule for so long, al-Sistani and the quietists may find that they must take a more active role in politics or lose their popular support (Fuller, 2003).

Thus far, the quietists have succeeded in reining in most Iraqi *Shi'a*, declaring that patience is vital to adopting a constitution and ending occupation (Berenson and Ian, 2003). The quietists themselves are growing more vocal, however. Although he refuses to openly participate in politics, al-Sistani exerts significant influence over the fledgling government in Baghdad, and has occasionally breached the divide between the religious and political spheres. In June 2003, al-Sistani issued a *fatwa*, or religious edict, demanding that the Iraqi constitution only be drafted by representatives chosen in a general election. L. Paul Bremer, the chief U.S. administrator in Iraq, was forced to accept al-Sistani's terms in order to stave off additional violence against occupation forces. Additionally, al-Sistani's silence on a number of political issues is often interpreted as tacit consent and considered vital for the survival of any new Iraqi government (MSNBC, 2003).

The al-Sadr Faction. The *al-Sadr faction* is led by Muqtada al-Sadr, whose father, the popular Sayyid Muhammad Sadiq al-Sadr, was assassinated by the Iraqi government in 1999. Following the overthrow of the Saddam Husayn, the *Shi'a* dominated area of Baghdad once known as 'Saddam City' was renamed 'Sadr City' in honor of Sayyid Muhammad Sadiq al-Sadr. Another member of the al-Sadr family, the extremely influential Ayatollah Muhammad Baqir al-Sadr, was also killed by the Iraqi government in 1980. Muqtada al-Sadr, young and inexperienced, has relied on his family name to attract followers. The core of the al-Sadr faction is the Jama'at al-Sadr al-Thani (Association of the Second al-Sadr). Muqtada al-Sadr has used this association as a

vehicle to propel himself to fame among the *Shi'a* of Iraq, primarily in and around Baghdad (Fuller, 2003).

Muqtada al-Sadr is known for fostering discord within the *Shi'a* community, often targeting the quietists. Following the U.S.-led invasion, al-Sadr declared that al-Sistani and two other prominent clerics, Baqir al-Hakim and Muhammad Ishaq al-Fayyad, should leave Iraq. Following the death of 'Abd al-Majid al-Khu'i in April 2003, rumors circulated that al-Sadr was involved in the murder, though no evidence of this has been brought to light (Fuller, 2003). Al-Sadr's suspected role in the death of al-Khu'i, a powerful, moderate cleric, has heightened tensions between his camp and that of Ayatollah al-Sistani (Finn and Avi, 2003).

Muqtada al-Sadr has targeted a number of leading clerics on the basis of their ethnicity. He is leading a "nativist" campaign to drive clerics of foreign descent out of Iraqi politics. Several prominent Iraqi clerics were actually born elsewhere, such as al-Sistani, who is of Iranian origin, and the nativist line touted by al-Sadr thus eschews Iranian interference in Iraqi *Shi'i* politics. Should al-Sadr's faction emerge as the dominate Shi'i force in the state, Iraq will pose a serious challenge to Iran's influence over Shi'i Muslims in the region (Fuller, 2003).

Supreme Council for the Islamic Revolution in Iraq. The Supreme Council for the Islamic Revolution in Iraq (SCIRI) was founded in Tehran in 1982, by then-exiled Ayatollah Muhammad Baqir al-Hakim. SCIRI has received considerable support from the Iranian government, and lingering ties to the regime could endanger SCIRI's popularity among Iraqis. While al-Hakim created a vast power base for SCIRI, including the 10,000-strong Badr Brigade that followed him from Iran to Iraq in spring 2003, he also developed a number of enemies, mostly resulting from his ties to Iran and his support for Iran during the 1980–1988 Iran-Iraq war (Fuller, 2003).

Though some believed that al-Hakim would be to Iraq what Khomeini had been to Iran, al-Hakim adopted a rather moderate stance upon returning after 20 years of exile. Al-Hakim denounced the possibility of a puppet government installed by the United States, but encouraged Iraqi *Shi'a* to work with the United States in order to develop a new Iraqi government (CNN, 2003). Also, he began to promote a separation of church and state in the future Iraqi government. Al-Hakim then turned his focus toward more spiritual issues. His brother, Abdul Aziz al-Hakim, assumed leadership of SCIRI's political objectives, joining the interim Governing Council (Mite, 2003).

Under the leadership of al-Hakim, SCIRI developed ties with the United States, despite U.S. reservations over SCIRI's relationship with Iran. SCIRI is also an important part of the Iraqi National Congress (INC), which has received substantial backing from the United States for several years (Fuller, 2003). But SCIRI suffered a serious blow when al-Hakim was killed by a car bomb on August 29, 2003, outside the sacred Imam Ali Mosque in Najaf. More than 120 Iraqi *Shi'i* civilians were killed in the blast, for which no one has claimed responsibility (Mite, 2003).

Following the assassination of al-Hakim, more than 400,000 *Shi'a* attended his funeral ceremonies in Najaf, a display of great support and solidarity that highlighted the immense popularity of al-Hakim and his SCIRI organization. Further, the assassination, in combination with bombings of the shrine of Ali, has driven many *Shi'i* groups in Iraq to form their own armed militias to defend themselves from threats. As a result, in recent weeks, there have been numerous bloody clashes among *Shi'i* militias and between armed *Shi'a* and coalition forces. If the United States and the Iraqi Governing Council are unable to find a way to persuade the Iraqi *Shi'a* to disarm their militias or to limit their involvement to domestic *Shi'i* matters, the systemic cycle of violence will only continue.

The Da'wa Party. Founded in 1957 by Sayyid Muhammad Baqir al-Sadr and his associates, the Da'wa (Call to Islam) Party predates the other major *Shi'i* movements in Iraq. The party staged multiple attacks on Saddam Husayn's regime in an effort to establish an Islamic state in Iraq, and due to its consequent suffering at the hands of the regime the Iraqi *Shi'i* populace developed a great reverence for the party.

Although it played an active role in the founding of SCIRI in 1982, the Da'wa leadership's support for Wilayat al-faqih led to a rift within the organization. Members opposed to the rule of the clergy and Iranian influence in the Da'wa movement broke with the party, creating two main factions of the Da'wa movement that persist today. One faction, Islamic Da'wa, remains affiliated with SCIRI and maintains a base of power in Iran (Fuller, 2003). The influence of extremist Lebanese Shi'i leaders is also evident in this branch of the Da'wa party (*Economist*, 2003). The other faction is based in London and espouses a more "nativist" approach (Fuller, 2003).

Da'wa's activities on the ground are currently being led by Shaykh Muhammad Nasiri (*Economist*, 2003). The true extent of the party's influence in post-Saddam Iraq is still unknown. Thus far, the Da'wa leadership has cooperated with occupation forces, but may well align itself more closely with the al-Sadr faction should popular opinion continue to turn against the U.S.-led rehabilitation (Fuller, 2003).

CONCLUSIONS

With the ongoing American occupation and nation-building effort in Iraq, and the parallel efforts of Iran's ambitious nuclear program, it appears self-evident that these two predominantly *Shi'i* nation states will be among the central focal points of the U.S. foreign policy in the Middle East and internationally for years to come. These two countries are also likely to receive much media attention and coverage. Thus, understanding the roots of the *Shi'a* breakaway and early Islamic history is necessary in order to comprehend the modern postures and attitudes of the predominantly *Shi'i* Iran and Iraq. Current attitudes and world views are influenced by the history of the

Shi'a faction, which from the days of Ali and Husayn to the recent assassination of Muhammad Bakr al-Hakim has endured a path of constant strife against the most numerous and sometimes hostile Muslim Sunni majority. Recognizing these attitudes and concerns, and opting to address them by reaching out to the long victimized *Shi'a* populations of Iraq and Iran, may just be the difference between success and failure both in rebuilding modern Iraq in the aftermath of the Ba'th removal and in hindering Iran's nuclear aspirations.

Acknowledgment
This article is drawn in part from the piece entitled "The Historic Roots of Current Terrorist Tactics and Methods" written by Sammy Salama and Kathleen Thompson, and published as research story of the week by the Center for Nonproliferation Studies on 11/21/03. The paper may be accessed at http://www.cns.miis.edu/pubs/week/031121.htm.

REFERENCES

Bakogeorge, P., Deportations Fuel Fundamentalism in Gaza, *Ottawa Citizen*, December 31, 1992, http://web.lexis-nexis.com.

Berenson, A. and Ian, F., Iraq's Shiite Factions Battle Over Mosques, *International Herald Tribune*, October 15, 2003.

Bill, J.A. and Robert, S., *Politics in the Middle East*, Harper Collins College Publishers, New York, 1994, p. 156.

Cleveland, W.L., *A History of the Modern Middle East*, Westview Press, Boulder, CO, 2000, pp. 15–16, 35, 53–57, 140–144, 182–188, 280–291.

CNN.com, Al-Hakim Had Called for Iraqi Unity, August 29, 2003, http://us.cnn.com/2003/WORLD/meast/08/29/sprj.irq.alhakim .obit.ap/

Congressional Quarterly, "The Middle East," 2000, pp. 201–208, 239.

Economist, Political Forces, From the Economist Intelligence Unit, Country ViewsWire, August 5, 2003, http://www.economist.com/countries/Iraq/profile.cfm?folder = Profile-Political%20Forces.

Esposito, J.L., *Unholy War*, Oxford University Press, Oxford, 2002, pp. 26–28, 37, 41, 69.

Finn, E. and Zevilman, A., A Guide to Iraq's Shiite Clerics, *Slate*, May 15, 2003, http://slate.msn.com/id/2082980/.

Fisk, R., Palestinians deportees still live in tents after being exiled by Israel, (From *The Independent*) *Vancouver Sun*, August 6, 1993, http://web.lexis-nexis.com.

Fuller, G.E., *Islamist Politics in Iraq after Saddam Husayn*, United State Institute of Peace, Washington, DC, August 2003, pp. 2–3, www.usip.org.

Jaber, H., *Hezbollah: Born with a Vengeance*, Columbia University Press, New York, 1997, p. 75.

Mite, V., Iraq: SCIRI Head Killed in Al-Najaf, RFE/RL, August 29, 2003. http://www.rferl.org/nca/features/2003/08/29082003153334.asp

MSNBC, U.S. Confident of Iraq Plan Approval, November 19, 2003, http://www.msnbc.com/news/992125.asp?0cv = CB10.

Munson, H. Jr., *Islam and Revolution in the Middle East*, Yale University Press, New Haven, 1988, pp. 10, 17, 22–23, 26–31.

ISLAMISM

Jeffrey M. Bale

DEFINITION

A radically anti-secular and anti-Western political current of contemporary Islamic thought with both revolutionary and revivalist characteristics.

BACKGROUND

The importance of Islamism in the context of biological warfare (BW) is that it has been openly embraced by various Muslim terrorist groups that have expressed some interest in acquiring and utilizing biological agents as weapons against their opponents. Among these groups are transnational networks such as al-Qa'ida (the Base or Foundation), the Tanzim al-Jihad (Jihad Group) in Egypt, al-Jihad al-Islami (Islamic Jihad) and al-Harakat al-Muqawwama al-Islamiyya (HAMAS: the Islamic Resistance Movement) in Palestine, the Groupe Islamique Armé (GIA: Armed Islamic Group) in Algeria, Hizballah (the Party of God) in Lebanon, Jemaah Islamiyah (the Islamic Community) in Malaysia and Indonesia, and certain terrorist organizations operating in disputed Kashmir.

Up until the 1950s, the term "Islamism" was synonymous with "Islam" in various Romance languages, and it is still sometimes misleadingly used in this older sense today. However, in the course of the 1970s and 1980s, the word was increasingly employed by academic specialists on Middle Eastern politics to refer solely to a newer type of radical political Islam, as opposed to Islam in general, and the etymologically related term "Islamists" was seen as having a close correspondence to the Arabic plural *islamiyyin* (literally "Islamic ones" but better understood as "true Muslims"), a designation often applied to Islamists, both by their enemies (sarcastically) and by themselves. Islamists also use other Arabic plurals to refer to themselves, such as *asliyyin* ("the authentic ones"), *mu'minin* ("the faithful"), and *mutadayyinin* ("the pious"), whereas their opponents are more likely to label them as *muta'sabin* ("fanatics" or "zealots") or *mutatarrifin* ("radicals" or "extremists").

Given this etymological confusion, it is not surprising that the precise meaning and significance of this particular politicized current of Islamic thought have often been misinterpreted. First, Islamism has at times been falsely equated or conflated with Muslim fundamentalism in general. Fundamentalism is perhaps best defined as any attempt by religious movements to return to what they regard as the pure, uncorrupted foundational elements of their own religious traditions, which in practice generally involves an ostensibly literal interpretation of sacred texts and a strict adherence to the tenets supposedly laid down by divinities or the authoritative religious figures claiming to speak for them. In the Islamic context, this generally means adhering as strictly as possible to Qur'anic injunctions and emulating the example set by the Prophet Muhammad during his life in seventh-century Arabia, as set down in the canonized accounts (*ahadith*) of his recorded statements and his customary behavior (*sunna*). However, Islamic fundamentalism has assumed a very wide variety of forms in different historical eras and geographical regions, ranging from pious but generally quietist movements such as the Tabligh-i Jamaat (Association for the Propagation of the Faith) movement in Pakistan to overtly activist movements such as the Muwahidin (Unitarians, as in Unity of God) movement in Saudi Arabia—which is generally referred to by outsiders as the Wahhabiyya movement after its founder Muhammad ibn 'Abd al-Wahhab (1703-1792)—and it need not embody any explicitly political agenda. Nevertheless, Islamism can be categorized as one particular subset of activist Islamic fundamentalism.

Second, the term Islamism has sometimes been used to refer to explicitly political movements of all types in the Muslim world, or at least those that claim to place Islam at the center of their political agendas. This too is problematic, since almost every political movement that has arisen in the Muslim world—with the exception of those rooted, however consciously or partially, in Western secular ideologies such as nationalism, communism, fascism, or liberalism—have claimed to be "Islamic." It therefore makes little sense to lump all such political movements into the Islamist category, irrespective of whether they are relatively liberal and democratic or culturally conservative and authoritarian. To characterize both the moderate Muhammad Iqbal and the radical Sayyid Qutb as Islamists would be to elide or gloss over crucial analytical distinctions of the sort that alone permit observers to draw meaningful distinctions between political thinkers. Similarly, no sensible person would place "political Christian" thinkers such as the proponents of "liberation theology," the liberal Catholic theologian Hans Küng, Christian Reconstructionists who seek to impose Biblical law, and the mystical fascist Corneliu Codreanu of the Rumanian Garda de Fier (Iron Guard) within the same taxonomic category. In short, Islamism is not synonymous with "political Islam"—it is only one of many doctrinal currents that fall within that broader category, along with Liberal Islam and Moderate Reformist Islam. Although today's Islamists often explicitly claim to represent and defend the interests of "the oppressed" (*al-mustad'afun*) and downtrodden, thereby assuming a superficially "progressive" or "egalitarian" coloration, Islamism actually occupies a position near the far right end of the "political Islam" spectrum.

Third, Islamism has often been confused with Muslim traditionalism, which has assumed and continues to assume a variety of forms. Indeed, many learned Muslim scholars who define themselves as traditionalists view

Islamism as a modern-day form of sectarian, violence-prone Kharijism—a reference to the Khawarij, or "those who go out," the radicals who first abandoned and later assassinated the fourth *khalif* 'Ali ibn Abi Talib (656–661), the Prophet Muhammad's cousin and son-in-law. They believe that today's Islamists have not only distorted and perverted the authentic values and tenets of traditional Islam, but also that they have foolishly adopted intolerant attitudes and carried out violent actions that can only produce a state of terrible disunity (*fitna*) within the Muslim community (*umma*), something that Muhammad himself considered to be a very great sin. Not surprisingly, representatives of the "official" religious establishments in many Muslim countries, which depend upon government patronage and are subject to state repression, also tend to characterize the Islamists in this pejorative fashion. Hence despite their claims to represent Islamic orthodoxy, Islamist ideologues are in reality often doctrinal innovators and practitioners of forms of exegetical interpretation (*ijtihad*) that others view as distinctly unorthodox if not idiosyncratic.

In fact, despite its uncompromising demands for a restoration of the supposedly "true" Islamic faith and the political unity of a renewed Muslim community cleansed of unbelievers and apostates, Islamism proper is a relatively recent political variant of Islamic thought. It was in many respects a product of the development of mass society in the early twentieth century, and was influenced, at least indirectly, by various European revolutionary doctrines from that era. Hence it should not be confused with earlier Islamic reform movements such as Wahhabism or Salafism, even though the evolving doctrines of these two movements subsequently became important currents within the Islamist ideological milieu. Nor should Islamism be mistaken for other political doctrines developed by Muslims in the late nineteenth and early twentieth centuries in an effort to resist European domination, such as pan-Islam (*al-wahda al-islamiyya*), Ottomanism (*Osmancilik*), or transnational ethno-cultural ideologies such as pan-Arabism (*al-wahda al-'arabiyya*), pan-Turkism (*Türkçülük*), and pan-Turanism (*Turancilik*), much less with Western-style nationalism (*al-qawmiyya*). Its first true exponent was arguably Hasan al-Banna' (1906–1949), the leader of the Egyptian Jamiàt al-Ikhwan al-Muslimin (Society of the Muslim Brothers, better known as the Muslim Brotherhood) that arose in Egypt at the end of the third decade of the twentieth century as a counterweight to both British control and a succession of national governments increasingly influenced by the West.

IDEOLOGICAL CHARACTERISTICS OF ISLAMISM

The principal ideological characteristics of Islamism in all of its forms are a radical rejection of Western secular values, an intransigent resistance to Western political, economic, social, and cultural influence over the Muslim world, an extreme hostility towards less committed and militant Muslims, and an affirmation of the importance of creating a truly Islamic state modeled on the strictest tenets of the *shari'a*, which was itself supposedly derived from the *Qur'an* and the exemplary behavior of Muhammad himself, his most loyal companions (both the original Meccan "émigrés" (*muhajirun*) and their Medinan "supporters" (*ansar*)), and their devout immediate successors, who are collectively known as the "virtuous forefathers" (*al-salaf al-salihin*) of the faith. Since this is an inherently uncompromising doctrine, it is misleading to speak of "moderate Islamism" and "radical Islamism," and even more ridiculous to talk about "democratic" Islamism—the true distinction is between Islamists who are willing to resort to gradualist and seemingly accomodationist political tactics such as participating in elections, however temporarily, and the "jihadists" who unceasingly advocate the waging of *jihad*, in the sense of an armed struggle against unbelievers (*kuffar*). (Nevertheless, some Islamists who consciously adopted, accomodationist tactics in an effort to increase their influence sometimes ended up being co-opted and largely neutralized by the state, as was the case of the Muslim Brothers in Jordan.) It is the jihadists, especially those with a global rather than a local or national political focus, who are most likely to attempt to acquire biological agents and employ them in acts of terrorism.

There are nonetheless several distinct ideological currents within Islamism. Apart from al-Banna', who did not completely eschew armed struggle but deployed his organization primarily to make a "long march through the institutions" by infiltrating the apparatus of the state, engaging in systematic proselytizing (da'wa), and providing education and other necessary social services to the destitute masses in the hopes of gradually wooing them away from loyalty to the corrupt Egyptian government and limiting their exposure to secular Western values, several other key Islamist thinkers should be identified. The South Asian Sayyid Abu al-A'la Mawdudi (1903–1979) portrayed the Islamic movement as a type of revolutionary party but nonetheless devoted most of his attention to the nature of the future Islamic state rather than actual methods of seizing power. The Egyptian Sayyid Qutb (1906–1966) insisted that modern Muslim societies had become so corrupt and spiritually bankrupt that they were as ignorant and barbarous (*jahili*) as the polytheistic tribal societies in pre-Islamic Arabia, which meant that they had to be destroyed and replaced by a true Islamic state and society. The Egyptian Muhammad 'Abd al-Salam al-Faraj (1952–1982) was even more radical, since he insisted that armed jihad was the sixth "Pillar of Islam" and branded all so-called Muslims who did not adopt his own uncompromising views as infidels, a process known as *takfir*, thereby marking them as legitimate targets for violence. A well-known Shi'ite Islamist, the Ayatollah Ruhollah Khomeini (1902–1989), developed the doctrine of "guardianship by religious scholars" (*vilayet-e faqih*) in order to justify the imposition of strict political rule by himself and other mullahs after the Iranian Revolution, though in theory they were only acting as regents of the Hidden Imam (*al-Imam al-Gha'ib*) until his long-awaited return. Another Iranian Islamist, 'Ali Shariati (1933–1977), attempted to incorporate certain left-leaning "anti-imperialist" elements into an explicitly Islamic political framework. 'Usama Bin Ladin and his Egyptian deputy Àyman al-Zawahiri further transformed the Islamist strategy of violence by insisting that the principal target of their *jihad* should be the "satanic" United States rather than the corrupt Muslim puppet governments under its sway, since if the head was cut off the limbs would necessarily wither and die. It is precisely because of its oft-proclaimed goal of overthrowing existing regimes and radically transforming contemporary society by means of the forcible restoration and imposition of the *shari'a* that one can speak of Islamism as a revolutionary and seemingly totalitarian political doctrine.

ISLAMISM AND WMD USE

Given their utopian goals and non-negotiable demands, as well as their burning desire to attack more militarily powerful Western societies in order to compel them to abandon their "imperialistic" efforts to "exploit" and control the Muslim world, it is hardly surprising that Islamist terrorist groups have displayed few if any qualms about carrying out mass-casualty attacks or that they have already engaged in periodic efforts to acquire weapons of mass destruction (WMD), including biological agents. Since these groups are unable to defeat the United States and its allies in a conventional war, they are necessarily forced to adopt an asymmetrical strategy to achieve their objectives. One key element of that strategy is suddenly smiting the "Great Satan" and its "crusader" allies on their own soil, whether by means of spectacular conventional terrorist attacks or unconventional attacks with chemical, biological, radiological, or nuclear (CBRN) weapons.

A few examples should suffice to highlight this potential danger. Not only have the leaders of al-Qa'ida already attempted to obtain nuclear devices and radioactive materials that can be used to manufacture an explosive radiological dispersal device (or "dirty bomb") and overseen the development of crude chemical weapons, they have also included recipes for making toxins such as botulinum toxin and ricin in the chapters of their training manuals dealing with assassination. They have apparently initiated efforts to produce standard BW agents as well, as U.S. troops confirmed when they discovered laboratories under construction in Afghanistan that seem to have been designed for the production, testing, and eventual weaponization of *Bacillus anthracis* and other biological agents.

Nor is al-Qa'ida unique in this respect. In recent months, the members of certain transnational Islamist terrorist networks, several of whom were Algerians linked to the Groupe Salafiste pour la Prédication et le Combat (GSPC: Salafist Group for Preaching and Fighting) and some of whom were reportedly trained by Chechen fighters in Dagestan or Georgia, were arrested in Britain (or other European countries) and found to be in possession of ricin toxin. Various toxins were also apparently being manufactured by a Kurdish Islamist group formerly operating in northern Iraq, Ansar al-Islam (the Supporters of Islam), and Pakistani scientists have reportedly collaborated with regional Islamists in their efforts to develop biological and chemical weapons. Various Islamist spokesmen have repeatedly threatened to employ biological weapons, and some of their followers appear to have perpetrated hoaxes by mailing threatening letters and powder to their enemies. At some point, these threats and hoaxes may actually give way to the actual deployment of biological agents.

CONCLUSION

Given their sectarian religious worldviews, Islamist terrorist groups appear to be far more willing—and therefore likely—to violate traditional moral and ethical taboos by employing mass-casualty terrorism and making use of WMD than their nationalist/separatist or secular ideological (Marxist, anarchist, neo-fascist) terrorist counterparts. After all, to the extent that violent extremist groups are absolutely convinced that they are doing God's bidding, virtually any action that they decide to undertake can be justified, no matter how heinous, since the "divine" ends are thought to justify the means. However, there is not necessarily any direct correlation between religious extremism and a particular terrorist group's decision to employ WMD. Many other factors are also undoubtedly involved, so the most that can be said is that under certain circumstances religious extremism can be a very important contributory factor in permitting a group to rationalize its development and use of BW agents. Whether Islamists turn out to be more willing to cross the WMD threshold than other types of religious militants remains to be seen, but the pronouncements of key Islamist leaders and the recent series of "toxic terror" arrests in Europe both suggest that this may well be the case.

REFERENCES

Abu-Rabi', I., *Intellectual Origins of Islamic Resurgence in the Modern Arab World*, SUNY, Albany, GA, 1996.

Dekmajian, R.H., *Islam in Revolution: Fundamentalism in the Arab World*, Syracuse University, Syracuse, NY, 1995.

Esposito, J.L., Ed., *Voices of Resurgent Islam*, Oxford University, New York, 1983.

Jansen, J.J.G., *The Neglected Duty: The Creed of Sadat's Assassins and Islamic Resurgence in the Middle East*, MacMillan, New York, 1986. [English translation of Muhammad 'Abd al-Faraj's *Al-Farida al-Gha'iba*]

Kepel, G., *Muslim Extremism in Egypt: The Prophet and the Pharaoh*, University of California, Berkeley, CA, 2003.

Khomeini, R., *Islam and Revolution 1: Writings and Declarations of Imam Khomeini*, KPI, London, 1985. [English translation]

Lamchichi, A., *L'islamisme politique*, Harmattan, Paris, 2001.

Maddy-Weitzman, B. and Efraim I., Eds., *Religious Radicalism in the Greater Middle East*, Frank Cass, London, 1997.

Mawdudi, Sayyid Abu al-A'la. *Jihad in Islam*. Lahore: Islamic Publications, 1976 [1939].

Mitchell, R.P., *The Society of the Muslim Brothers*, Oxford University, New York, 1993 [1969].

Qutb, S., *Milestones [Along the Path]*, International Islamic Federation of Student Organizations, Salamiyya, Kuwait, 1978. [English translation of *Ma'alim fi al-Tariq*]

Rahnema, A., Ed., *Pioneers of Islamic Revival*, Zed, London, 1994.

Rubin, B., Ed., *Revolutionaries and Reformers: Contemporary Islamist Movements in the Middle East*, SUNY, Albany, 2003.

Shariati, A., *Red Shi'ism*, Free Islamic Literatures, Houston, TX, 1980 [English translation].

Sivan, E., *Radical Islam: Medieval Theology and Modern Politics*, Yale University, New Haven, CT, 1990.

Wiktorowicz, Quintan, Ed., *Islamic Activism: A Social Movement Theory Approach*. Bloomington: Indiana University, 2004.

ISRAEL

ANDY OPPENHEIMER

INTRODUCTION

This entry discusses Israel's biological weapons–related efforts for both defensive (vaccines, antidotes, detection and decontamination equipment, and protective clothing) and offensive (deliverable biological agents) purposes.

Israel certainly has the capacity and resources to develop and produce biological weapons, and has long been suspected of maintaining such a program (Fig. 1). A 1993 report by the U.S. Congress Office of Technology Assessment stated that Israel is "generally reported as having an undeclared offensive biological warfare (BW) program," with extensive research conducted since the 1950s at the Israel Institute for Biological Research (IIBR) in Nes Ziona, some 10 km south of Tel Aviv. It is widely believed that the IIBR is Israel's main center for both offensive and defensive BW research. Whether Israel possesses an actual production capability above a pilot scale at this or any other facility is unknown. Israel is not a signatory to the Biological and Toxin Weapons Convention (BWC).

HISTORY

Israel's BW program was born as a result of Prime Minister David Ben-Gurion's fears for Israel's long-term security, destined to be eternally in jeopardy because of the geopolitical realities of the ongoing conflict with its Arab neighbors, following its establishment as a nation-state in 1949. Ben-Gurion's desire to find scientists who could "increase the capacity to kill masses or cure masses" led to the recruitment of both chemists and microbiologists, including Professor Ernst Bergmann, director of the Sieff Institute in Rehovot who apart from being Israel's main nuclear pioneer was influential in the development of chemical and biological weapons as well, and Aharon and Ephraim Katachalsky, both macromolecular chemists and members of the Jewish paramilitary organization Haganah.

Even before this, a group of students under the microbiologist Alexander Keynan had urged Ben-Gurion to establish a BW unit within the Israel Defence Force's Science Corps (HEMED). The new unit, HEMED BEIT, was decades later implicated in a plot to "poison" the reservoirs of German cities as an act of revenge for the Holocaust. Avner Cohen, in his major report for the Center for Nonproliferation Studies' *Nonproliferation*

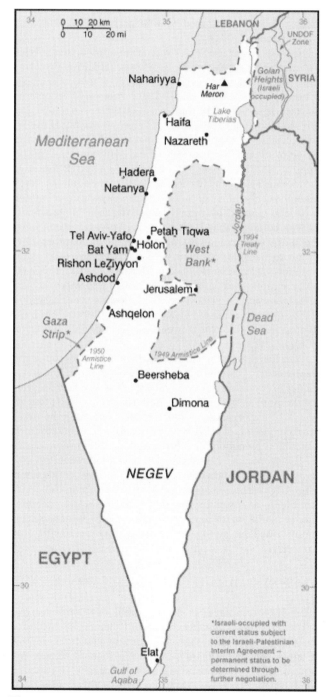

Figure 1. Map of Israel, *CIA World Factbook*, CIA website: http://www.cia.gov/cia/publications/factbook/geos/is.html.

Encyclopedia of Bioterrorism Defense,   Edited by Richard F. Pilch and Raymond A. Zilinskas
ISBN 0-471-46717-0   Copyright © 2005 Wiley-Liss

Review (2001), catalogs reported uses of biological weapons by Jewish forces (presumably HEMED BEIT) during the 1948 war in Palestine. Cohen notes allegations by Israeli historian Uri Milstein that "in many conquered Arab villages, the water supply was poisoned to prevent the inhabitants from coming back." A typhoid epidemic in Acre in 1948 was similarly attributed to deliberate contamination of its water. And four Israeli soldiers captured by Egyptian forces in 1948 were accused of contaminating wells in Gaza with the causative agents of dysentery and typhoid; they allegedly confessed to this activity and were executed (Cohen, 2001).

In the 1950s, HEMED was transformed into civilian research centers known as "machons." IIBR resulted from the merger of two machons, one of which was a continuation of HEMED BEIT. The initial IIBR laboratory, located in a fenced orange grove outside Nes Ziona, had previously housed HEMED BEIT, and HEMED BEIT's supervisor, Alexander Keynan, became the Institute's first director. The IIBR contained (and still contains) both a highly classified research center ("Machon 2") and an academic scientific institute. At the time of its establishment, chemical and biological weapons development was not banned by international regimes; nevertheless, it was believed that IIBR's broader research focus would be more effective in attracting top-level scientists hesitant to become involved with its program than solely weapons-related work. Former Mossad agent Victor Ostrovsky claims that lethal testing was performed on Arab prisoners at the IIBR in the following years, and in 1954, defense minister Pinchas Lavon allegedly proposed using tested biological weapons for special operations (Cohen, 2001).

The IIBR has since grown into an extensive, top-secret, high-security complex that does not appear on maps, and is not accessible even to members of the Knesset's (Israeli parliament's) foreign affairs and defense committees. The years of suppression of information regarding the arrest and conviction of IIBR scientist and deputy director Prof. Avraham Klingberg as a Soviet spy in 1993 are indicative of the secrecy surrounding IIBR. Even following the 1992 crash in Amsterdam of an El Al cargo aircraft transporting dimethylmethylphosphonate (DMMP, a dual-use chemical that can be used in the manufacture of the chemical nerve agent sarin) and destined for IIBR, Israeli officials maintained an air of secrecy regarding the facility and its activities.

CURRENT STATUS

The IIBR employs a staff of 300, ostensibly responsible for "investigating viral and bacterial pathogens to design new strategies for vaccine development." Its Center for the Diagnosis of Infectious Diseases produces "novel diagnostic assays... antigens, antibodies, and reagents." Other divisions are involved in synthesizing drugs, air pollution, meteorology, physical surface chemistry, and biosensors and detection—much of which can, in theory, be applied to both defensive and offensive activities.

There is no evidence indicating that Israel is engaged in the ongoing production or stockpiling of biological weapons. Analysts have noted that the high level of secrecy and lack of democratic oversight at IIBR may be suggestive of biological weapons research, but not indicative per se. Non-Israeli publications have made many claims about its chemical and biological weapons capabilities, but the government of Israel, as part of its traditional policy of deliberate ambiguity, has neither confirmed nor denied those reports. And the United States has only publicly acknowledged that Israel is "co-operating closely with the American military laboratories... for protection against biological weapons." As with Israel's nuclear ambiguity, the very existence of Nes Ziona may serve as a deterrent without its true function being declared (Oppenheimer and Eldridge, 2003).

AGENTS AND EXPERTISE

IIBR scientists initially researched the causative microorganisms and vectors of plague, typhus, and rabies. Today, scientists at the Institute are involved in the identification/isolation and production of antigens, antibodies, and conjugates for the diagnosis of viral, rickettsial, and leptospiral infectious diseases, along with extensive research on toxins such as Staphylococcal enterotoxin B. Another area of research is the remediation of soils and water using biological organisms to metabolize contaminants (Federation of American Scientists Military Analysis Network, Undated). Other academic institutes are engaged in biomedical research in Israel, such as the Department of Chemical Immunology and Cell Biology at the Weizmann Institute of Science in Rehovot, but there is no evidence that they are involved in anything that could be construed as offensive in nature (Oppenheimer and Eldridge, 2003).

Israel's relatively new biotechnology industry is an offshoot of the U.S. biotechnology industry, resulting largely from the establishment in Israel of subsidiaries of foreign pharmaceutical companies. Nes Ziona receives both data from university laboratories and U.S. capital to support research. From the mid-1980s, the IIBR has focused on genetic engineering techniques, human and animal diagnostics, and agricultural biofertilization.

In 1998, an article in the *Sunday Times* reported that research into a so-called ethnic bomb had been conducted at Nes Ziona (Mahnaimi and Colvin, 1998). This research was described during the South African Truth and Reconciliation Committee hearings, where it was claimed that the apartheid regime and its ally Israel had cooperated on such a project. According to the testimony, scientists had pinpointed a particular characteristic in the genetic profile of certain Arab communities, particularly in Iraq, and were trying to engineer BW agents to attack only those bearing the distinctive genes. The genetically modified agent could be spread by aerosol or direct contamination, for example by sabotaging a community's water supply. *Jane's Foreign Report* (1998) quoted unnamed South African sources as saying Israeli scientists had used some of the South African research in an attempt to develop an "ethnic bullet" against

Arabs, adding that the Israelis had discovered aspects of the Arab genetic makeup by researching on "Jews of Arab origin, especially Iraqis." A confidential Pentagon report the year before warned that biological agents could be genetically engineered to produce new lethal weapons, and U.S. Secretary of Defense William Cohen independently received reports of countries working to create "certain types of pathogens that would be ethnic-specific" (*Jane's Foreign Report*, 1998). The British Medical Association (BMA) became so concerned about the lethal potential of genetically based biological weapons that it opened an investigation into the matter, and in a 1999 report predicted that genetically engineered organisms might be developed to target a particular ethnic group within "five or 10 years" and further warned of their attractiveness to terrorists (British Medical Association, 1999).

BIODEFENSE

The Israeli biodefense program is closely linked to U.S. efforts in this realm, particularly since September 11, 2001, and the subsequent "anthrax letters." Israel possesses advanced expertise in personal protection and extensive capabilities in meeting the demand for operational military use of protective gear. The IIBR and various independent companies design and manufacture protective clothing and respirators (every Israeli citizen has a gas mask), detection and decontamination equipment, and filtration systems. The entire population was on high alert for a chemical or biological attack by Iraq in both the 1991 Gulf War and the 2003 Operation Iraqi Freedom, and drills in schools and other public buildings continue to be conducted on a regular basis. The many suicide bombings in Israel have so far not included biological agents (although one such bombing did transmit hepatitis B to a victim via contaminated bone fragments, it is unlikely that this was an intended result). However, HAMAS has laced bombs with rat poison in the past.

The Israeli Health Ministry has completed the production of enough doses of smallpox vaccine to immunize the entire population if necessary. In addition, some 100 health system personnel have been given a booster of smallpox vaccine in order to establish a source for vaccinia immune globulin (VIG). About half of the Israeli population has at one time or another received the vaccine, either as infants or when drafted into the Israel Defense Force.

REFERENCES

British Medical Association, *Biotechnology, Weapons and Humanity*, Harwood Academic Publishers, New York, 1999.

Cohen, A., Israel and Chemical/Biological Weapons: History, Deterrence and Arms Control, *Nonproliferation Review*, Fall-Winter 2001, Center for Nonproliferation Studies at the Monterey Institute of International Studies, http://cns.miis.edu/pubs/npr/vol08/83/83cohen.pdf.

Federation of American Scientists Military Analysis Network, *Army Science and Technology Master Plan (ASTMP), III. International Research Capabilities and Long-Term Opportunities, K. Biological Sciences*, 1997, http://www.fas.org/man/dod-101/army/docs/astmp/aE/E3K.htm.

Jane's Foreign Report, Genetic Warfare, October 29, 1998, vol issue: 000/2518.

Mahnaimi, U. and Colvin, M., Israel Planning 'Ethnic' Bomb as Saddam Caves In, *Sunday Times* (London), November 15, 1998.

Oppenheimer, A.R. and Eldridge, J., Israel: NBC Capabilities, *Jane's Nuclear, Biological, and Chemical Defence 2003*, Jane's Information Group.

WEB RESOURCES

Carnegie Analysis: Chemical and Biological Weapons in the Middle East, Carnegie Endowment for International Peace, April 2002, http://www.ceip.org/files/nonprolif/templates/article.asp?NewsID=2669.

Israel's Biological Weapons, Nuclear Threat Initiative, Center for Nonproliferation Studies at the Monterey Institute of International Studies, 2003, August 2002, http://www.nti.org/e_research/e1_israel_bwabstracts.html.

J

JOINT TASK FORCE CIVIL SUPPORT

MICHELLE BAKER

BACKGROUND

The Joint Task Force Civil Support (JTF-CS) is a newly created standing unit designed to coordinate disaster management response with the Department of Defense (DoD) and the Federal Emergency Management Agency (FEMA), which now is part of the Department of Homeland Security (DHS). Subordinate to the Joint Force Headquarters Homeland Security (JFHQ-HLS) and ultimately the U.S. Northern Command, JTF-CS is positioned to play an integral role in supporting civilian forces upon the order of the Secretary of Defense by providing logistical and technical support following a chemical, biological, radiological, nuclear, or high-yield explosive (CBRNE) attack. Authority for such a mission is granted under the policy guidance "Department of Defense Consequence Management Support to Domestic Incidents Involving Chemical, Biological, Radiological, Nuclear and High Yield Explosives (CBRNE-CM)," signed by U.S. Secretary of Defense Paul Wolfowitz in March 2001, which clearly designates the current mandate for JTF-CS (http://www.senate.gov/~armed_services/statemnt/2001/010501lawlor.pdf, accessed on 9/24/2003); however, the Robert T. Stafford Disaster Relief and Emergency Assistance Act, which allows the president to provide localities with federal assistance, offers flexibility regarding the deployment of JTF-CS and other agencies to disaster areas (http://www.jtfcs.northcom.mil, accessed on 9/22/2003). While JTF-CS has yet to be deployed in such a role (despite its activation for the first time following the attacks on September 11, 2001), it routinely conducts training exercises to prepare for CBRNE consequence response and management.

STRUCTURE

The creation of JTF-CS was determined to be a priority after the 1996 pipe bombing incident at the summer Olympics in Atlanta, Georgia, which demonstrated that the DoD and civilian authorities were not adequately prepared to undertake consequence management following a large-scale terrorist attack. Following an appeal by Defense Department officials to modify the structure of civilian and military response to such an incident, JTF-CS was created under the umbrella of the DoD in 1999, with full implementation of the program completed on April 1, 2000 (http://www.senate.gov/~armed_services/statemnt/2001/

010501lawlor.pdf, accessed on 9/24/2003). The DoD sought to set up an organization devoted entirely to the assistance of civil authorities, subordinate to civilian control, and trained for rapid response to a terrorist attack involving weapons of mass destruction (WMD), a structure designed to prevent the difficulty of dual responsibility for some military officers when assisting with disaster relief. As part of this mandate, the JTF-CS worked (and continues to work) closely with local and state officials in order to coordinate disaster preparedness efforts, enabling the military and other federal organizations to respond more quickly to potential emergencies (http://www.defenselink.mil/cgi-bin/dlprint.cgi? http://www.defenselink.mil/news/Aug 1999/n08171999_9908175.html, accessed on 9/22/2003). While JTF-CS was provided only a $4 million budget for the 2000 fiscal year, according to the former commander of the task force the demand for JTF-CS services and expertise in planning and preparation for a CBRNE event was far greater than expected (http://www.defenselink.mil/news/Jan2000/n01132000_20001132.html, accessed on 9/24/2003, and http://www.senate.gov/~armed_services/statemnt/2001/010501lawlor.pdf, accessed on 9/24/2003).

In response to the attacks on September 11, 2001, and following the creation of the DHS, the command structure of JTF-CS within the DoD shifted significantly. JTF-CS was placed directly under the JFHQ-HLS, which provides homeland defense and civil support capabilities to federal agencies responding to crisis situations (http://www.northcom.mil/index.cfm?fuseaction=news.factsheets&textonly=true&factsheet=4, accessed on 9/23/2003). The U.S. Northern Command, the lead agency presiding over JTF-CS and JFHQ-HLS, has consolidated control over missions designed to defend the United States and U.S. interests. Although JTF-CS was transferred to its current position under the Northern Command in 2002, the goals of the organization have changed little since its inception (http://www.northcom.mil/index.cfm?fuseaction=s.who_mission, accessed on 9/22/2003).

In order to fulfill its directive to provide a comprehensive and timely disaster management, JTF-CS has developed a consequence management doctrine, performs training and exercises simulating emergency situations, and identifies logistical requirements that must be met in order for the organization to complete its assignment. These activities involve coordination with many other agencies, including the FEMA, Department of Justice (DOJ), Department of Energy (DOE),

Encyclopedia of Bioterrorism Defense, Edited by Richard F. Pilch and Raymond A. Zilinskas
ISBN 0-471-46717-0 Copyright © 2005 Wiley-Liss

Centers for Disease Control and Prevention (CDC), and both state and local law enforcement agencies (http://www.jtfcs.northcom.mil, accessed on 9/22/2003). To facilitate this interagency cooperation, JTF-CS employs a "liaison directorate" responsible for coordination between JTF-CS and each of the 10 federal agencies responsible for the 12 Emergency Support Functions following a major domestic disaster. This structure also allows smaller civil support contingents to respond to a situation without the entire organizational presence (http://www.senate.gov/~armed_services/statemnt/2001/010501lawlor.pdf, accessed on 9/24/2003).

Currently, there are 160 military and civilian personnel under the command of JTF-CS, combining representatives from the Army, Navy, Marines, Air Force, Coast Guard, Reserves, National Guard, civil servants, and civilian contractors. If there were to be an actual CBRNE incident, however, several thousand additional military personnel conceivably could be placed under its control. JTF-CS is therefore the primary military support for the Lead Federal Agency (FEMA or another agency, depending on the circumstance) in any disaster (http://www.jtfcs.northcom.mil/pressreleases/pressrelease8.08.02.htm, accessed on 9/26/2003).

PREPAREDNESS EFFORTS

The JTF-CS trains consequence managers to provide local officials with appropriate medical expertise, transportation for ill or injured citizens, chemical and biological detection and decontamination capabilities, and logistical support following a CBRNE incident (http://www.jtfcs.northcom.mil/pressreleases/pressrelease8.08.02.htm, accessed on 9/26/2003). In August 2003, JTF-CS participated in a homeland security exercise (Exercise Determined Promise-03 (DP03)) in Clark County, Nevada, headed by the Northern Command (NORTHCOM), which focused on interagency coordination during a domestic disaster. The exercise simulated a biological attack involving the deliberate release of aerosolized *Yersinia pestis* on the Las Vegas strip. JTF-CS coordinated all aspects of DoD support to local, state, and federal agencies, involving some 950 military personnel in all (http://www.jfcom.mil/newslink/storyarchive/2003/pa-081803.htm, accessed on 9/22/2003, and http://www.jtfcs.northcom.mil/pressreleases/news082403jtfcsrealworld deployment.htm, accessed on 9/22/2003). In addition to designated simulations such as DP03, JTF-CS has participated in consequence management planning for domestic events, for example, offering support to the National Capitol Police during the President's 2001 State of the Union Address (http://www.senate.gov/~armed_services/statemnt/2001/010501lawlor.pdf, accessed on 9/24/2003).

THE FUTURE

Despite the popularity of the JTF-CS, particularly as civilian authorities acknowledge their inability to adequately respond to an attack involving chemical or biological weapons, some have questioned the role of the military in local consequence management following a domestic terrorist attack. Though providing training and support to local authorities, JTF-CS remains subordinate to civilian authorities in any disaster situation and therefore cannot directly participate in law enforcement activities. The ability of JTF-CS to mobilize troops to undertake roles traditionally held by civilians further conflicts with the original intent of The Posse Comitatus Act of 1878, which precludes military from assuming law enforcement duties unless authorized by either Congress or the Constitution (http://www.prospect.org/print-friendly/print/V12/19/dreyfuss-r.html, accessed on 9/26/2003). Such an expansion of powers is not without precedent, however, as Congress under the Reagan administration allowed for an increase in domestic military power if faced with the threat of nuclear terrorism, a decision that was later expanded to allow for greater military involvement in the event of a chemical or biological attack (Broad and Miller, 1999). Regardless, prior to 1999, the creation of JTF-CS was perceived by some as a means for strengthening military power in domestic situations beyond any legislative authorization.

Although JTF-CS has yet to demonstrate its full capabilities as the primary military support in a CBRNE situation, training exercises have displayed the value of coordination and cooperation between military and civilian agencies when responding to such an incident. The use of military forces in a domestic situation is a contentious issue for those concerned about the potential misuse of military power, but the benefits of military support to overwhelmed state and local law enforcement in a potential CBRNE situation are perceived as crucial to forming a rapid consequence management capability.

REFERENCES

Broad, W.J. and Miller, J., Pentagon Seeks Command for Emergencies in the US, *New York Times*, January 28, 1999, A21.

Dreyfuss, R., The Home Front, *American Prospect*, **12**(19), November 5, 2001.

Garamone, J., DoD Examines Joint Task Force Concept for Civil Support, American Forces Information Service, August 17, 1999, http://www.defenselink.mil/news/Aug1999/n08171999_9908175.html.

Garamone, J., Task Force Counters Terrorist WMD Threat, *American Forces Information Service*, January 13, 2000, http://www.defenselink.mil/news/Jan2000/n01132000_20001132.html.

Golden, R., *JTF-CS 'Fights a Bio-contagious Disease' During Viento Feo Exercise*, Joint Task Force Civil Support, http://www.jtfcs.northcom.mil/pressreleases/pressrelease8.08.02.htm.

Graham, B., Pentagon Plans Domestic Terrorism Team; Critics Fear Too Much Military Interference in Civilian Emergency Response, *Washington Post*, February 1, 1999, A02.

Joint Task Force Civil Support, 2003, http://www.jtfcs.northcom.mil.

Lawlor, B.M., *Statement of Major General Bruce M. Lawlor, USA, Commander, Joint Task Force Civil Support, US Joint Forces*

Command, Before the Senate Armed Services Committee on Status Update of JTF-CS, Senate Armed Services Committee, May 1, 2001, http://www.senate.gov/~armed_services/statemnt/2001/010501lawlor.pdf.

USJFCOM Public Affairs, *USJFCOM to Support Homeland Security Exercise*, United States Joint Forces Command, http://www.jfcom.mil/newslink/storyarchive/2003/pa081803.htm.

Walker, R., *JTF-CS Deploys to Clark Country for DP-03*, Joint Task Force Civil Support, http://www.jtfcs.northcom.mil/pressreleases/news082403jtfcsrealworlddeployment.htm.

WEB RESOURCES

U.S. Northern Command, *Fact Sheet*, http://www.northcom.mil.

U.S. Northern Command, *Mission*, http://www.northcom.mil.

See also CONSEQUENCE MANAGEMENT; DEPARTMENT OF DEFENSE; DEPARTMENT OF HOMELAND SECURITY; NORTHCOM (U.S. NORTHERN COMMAND); and WEAPONS OF MASS DESTRUCTION CIVIL SUPPORT TEAMS.

K

KOREA, DEMOCRATIC PEOPLE'S REPUBLIC OF

Daniel A. Pinkston

INTRODUCTION

After 35 years of Japanese colonial rule, Korea was divided into Soviet and American zones at the close of World War II in August 1945. The division was to be temporary, but two separate governments and states were established in 1948. The Democratic People's Republic of Korea (DPRK, or North Korea) tried to unify the peninsula by force when its forces invaded the Republic of Korea (ROK, or South Korea) on June 25, 1950. North Korea almost succeeded before the United Nations intervened. As the tide turned against DPRK forces, China intervened to support North Korea, and the war reached a stalemate before an armistice was signed in July 1953 (Fig. 1). During the Korean War, China and North Korea accused the United States of having employed biological weapons against its soldiers and civilians (International Scientific Commission, 1952); a charge that has been discredited to the satisfaction of most of the world's governments (Leitenberg, 1998). Nevertheless, the Chinese and North Korean governments resurrect this charge when they feel it is politically advantageous to do so.

Although China and the Soviet Union assisted the DPRK during the Korean War, the North Korean leadership was disappointed with the level of assistance, and later questioned the credibility of Chinese and Soviet alliance commitments. The DPRK recovered from the war more quickly than South Korea, but began to fall behind economically by the late 1970s. By the late 1980s, North Korea was falling behind in the conventional weapons balance as well, despite Pyongyang's superior numbers in terms of both military hardware and personnel (a numerical advantage that it maintains today). While South Korea has been steadily gaining qualitative advantages in its conventional forces while maintaining a robust bilateral security alliance with the United States, North Korea's security alliance with the former Soviet Union has expired, and the bilateral security treaty that China and North Korea signed in 1961 is probably no longer credible even though neither side has officially renounced it.

North Korea acceded to the Biological and Toxin Weapons Convention (BWC) on March 13, 1987, but the country nevertheless is suspected of having an active biological weapons program. According to South Korea's Ministry of National Defense, former North Korean leader Kim Il Sung issued a directive in the 1980s declaring that "poisonous gas and bacteria can be used effectively in war" (Republic of Korea Ministry of National Defense, 2000). Kim already in 1961 had made a declaration to pursue the development of chemical weapons (Republic of Korea Ministry of National Defense, 2000), and one source claims that Kim ordered the development of biological weapons in the early 1960s (Tom-sam, 1999).

BIOLOGICAL WEAPONS HISTORY

There are reports that North Korea acquired a turnkey plant for agar production from East Germany and that Pyongyang began producing *Bacillus anthracis*, *Vibrio cholerae*, and *Salmonella typhi* in the 1980s (Tom-sam, 1999; Son-ho, 1998). By 1992, the South Korea's Agency for National Security Planning was reporting that North Korea had produced 13 types of pathogens, including *B. anthracis, V. cholerae*, and *S. typhi*, and that Pyongyang had the capacity to produce 1000 tons of biological weapons agents per year (*Segye Ilbo*, 1992). However, one prominent North Korean defector has described a rumor circulating within the North Korean military to the effect that there had been an accidental release of or exposure to biological warfare (BW) agents in North Korea and that its military now is disinclined to pursue biological weapons (Interview, 2001).

CURRENT BIOLOGICAL WEAPONS STATUS

Little is known about the current status of the North Korean biological weapons program. North Korea certainly has the infrastructure to produce BW agents, but it is unclear whether any agents have been weaponized and, further, which of its systems (rockets, missiles, aircraft, or UAVs) are equipped to deliver such agents. There have been reports that North Korea might have obtained the smallpox virus from the former Soviet Union in the 1980s (Broad and Miller, 1999), and there is speculation that North Korea could have weaponized it. In May 2002, a high-level U.S. official said that if North Korea has not weaponized any BW agents, Pyongyang has the capability to do so within weeks (Bolton, 2002).

Encyclopedia of Bioterrorism Defense, Edited by Richard F. Pilch and Raymond A. Zilinskas
ISBN 0-471-46717-0 Copyright © 2005 Wiley-Liss

Figure 1. North Korea. Source: *CIA Factbook*.

HISTORY OF TERRORIST SUPPORT

There is no evidence that North Korea has sponsored or engaged in terrorist activities since 1987 (U.S. Department of State, 2003). However, North Korean agents have hijacked and bombed commercial aircraft in mid-flight, kidnapped individuals, and assassinated South Korean government officials. Some of the most notorious incidents include the infiltration of 31 commandos who nearly reached the South Korean presidential residence in an attempt to assassinate South Korean President Park Chung Hee in 1968; the attempted assassination of South Korean President Chun Du Hwan in Burma in 1983 that killed 17 South Korean officials; and the bombing of a Korean Airlines Boeing 707 over the Andaman Sea in 1987 that killed 115 people. North Korea has directed almost all of its past terrorist activities at South Korea, or, in some cases, at third targets with the intention of undermining international confidence in South Korea. Furthermore, North Korea has kidnapped a number of Japanese citizens to assist North Korean intelligence agencies in language training and other activities (see, e.g., Korean Central News Agency, 2002; Witter, 2002; Watts, 2002). And Pyongyang provided refuge for Japanese Red Army members who hijacked a Japan Airlines aircraft to North Korea in 1970.

According to the U.S. Department of State, North Korea has sold weapons to terrorists groups, but the types of weapons and the recipients are unclear. North Korea remains on the U.S. Department of State's list of states that sponsor international terrorism (U.S. Department of State, 2003). However, Pyongyang has signed six of the 12 international conventions and protocols on terrorism.

In October 2000, the United States and North Korea signed the "Joint U.S.-DPRK Statement on International Terrorism," whereby Pyongyang agreed that "terrorism should be opposed in all its forms, including terrorist acts involving chemical, biological, or nuclear devices or materials" (U.S. Department of State, 2000).

CURRENT TERRORIST RELATIONS

North Korea has not conducted any terrorist activities since 1987, but there is evidence that terrorist groups have acquired North Korean small arms, though it is unclear whether North Korea has supplied these weapons directly or if terrorist groups have acquired the weapons through intermediaries in the international arms market. In the past, North Korean agents have conducted operations that can be characterized as terrorism, most of which have, as noted, been targeted against South Korean and Japanese citizens. According to Japanese reports, in the summer of 2002, North Korea disbanded a unit responsible for spy ship operations in Japanese territorial waters. The unit had a staff of about 1500 and was probably responsible for illegal activities such as drug smuggling and the kidnappings of Japanese citizens (*Asahi Shimbun*, 2002; *Japan Economic Newswire*, 2002).

CONCLUSIONS

While North Korea has the capability to produce biological weapons, it is unclear whether North Korea has weaponized any biological agents, and the country's delivery capabilities for biological weapons are uncertain.

North Korea does not appear to have significant, if any, ties to international terrorist organizations, but the country's dismal economic performance over more than a decade suggests that there are economic incentives for Pyongyang to sell BW materials or technologies to other states or nonstate actors. Whether this has occurred is not known.

REFERENCES

Asahi Shimbun, Pyongyang Shuts Spy Ship Section, October 5, 2002, http://www.asahi.com/english.

Bermudez, J.S., *The Armed Forces of North Korea*, I.B. Taurus, London, 2001.

Broad, W.J. and Miller, J., Government Report Says 3 Nations Hide Stocks of Smallpox, *New York Times*, June 13, 1999.

Chang Jun, Ik., *Pukhan Haek-Missile Chŏnjaeng*, Somundang, Seoul, 1999.

International Scientific Commission, *Report of the International Scientific Commission for the Investigation of the Facts Concerning Bacterial Warfare in Korea and China*, Academia Sinica, Peking, 1952.

Interview with Lee Ch'ung Kuk by Daniel A. Pinkston, Seoul, April 7, 2001.

Japan Economic Newswire, N. Korea Dissolves Dept. in Charge of Abductions: Daily, October 3, 2002, in Lexis-Nexis, http://www.lexis-nexis.com.

John, R.B., U.S. Undersecretary of State for Arms Control and International Security, Beyond the Axis of Evil: Additional Threats from Weapons of Mass Destruction, Heritage Lectures, Number 743, May 6, 2002, p. 3.

Korean Central News Agency, DPRK-Japan Pyongyang Declaration Published, September 17, 2002, http://www.kcna .co.jp.

Leitenberg, M., *The Korean War Biological Warfare Allegations Resolved*, Occasional paper # 36, Center for Pacific Asia Studies, Stockholm University, Stockholm, 1998.

Nanato, D.K., *North Korea: Chronology of Provocations, 1950–2003*, CRS Report for Congress, Congressional Research Service, updated March 18, 2003.

Republic of Korea Ministry of National Defense, *Defense White Paper 2000*, Ministry of National Defense, Seoul, 2000.

Republic of Korea Ministry of Unification, *Pukhan'gaeyo 2000*, Ministry of Unification, Seoul, December 1999.

Segye Ilbo, Pukhan Saenghwahangmugi Shilch'onbaech'i/Angibu Kukkambogo [North Korea Deploys Bio-Chemical Weapons/ National Security Planning Agency Report for National Assembly Audit], October 24, 1992, in KINDS, http://www.kinds .or.kr.

Son-ho, Y., *Pukhan*, November 1998, pp. 38–51, in "North Korea's Technology for the Development of Weapons of Mass Destruction and Its Ability," Foreign Broadcast Information Service Document ID: SK2111131598.

Tom-sam, P., *Pukhan*, January 1999, pp. 62–71, in "How Far Has the DPRK's Development of Strategic Weapons Come?" Foreign Broadcast Information Service Document ID: FTS19990121001655.

U.S. Department of State, *Patterns of Global Terrorism: 1999*, April 2000, http://www.usis.usemb.se/terror/rpt1999/sponsor .html#NK.

U.S. Department of State, *Joint U.S.—DPRK Statement on International Terrorism*, Office of the Spokesman, October 6, 2000, http://secretary.state.gov/www/briefings/statements/2000/ ps001006.html.

U.S. Department of State, *Patterns of Global Terrorism—2002*, Office of the Coordinator for Counterterrorism, April 30, 2003, http://www.usis.usemb.se/terror/rpt2002/overview_of_state-sponsored_terrorism.html.

Watts, J., North Korea Apologises to Japan for Bizarre Tale of Kidnap and Intrigue: Eight out of 12 People Snatched in the 70s and 80s Are Dead, *The Guardian (London)*, September 18, 2002, p. 3, in Lexis-Nexis, http://www.lexis-nexis.com.

Witter, W., Pyongyang Admits Kidnapping Japanese, *Washington Times*, September 18 2002, p. A1, in Lexis-Nexis, http://www.lexis-nexis.com.

KURDISTAN

GAIL H. NELSON

OVERVIEW

Kurdistan is a region of 191,600 sq km, straddling the international boundaries of Turkey, Syria, Iraq, Iran, and the former Soviet Union. Kurdistan's notional boundaries represent the approximate limits of Kurdish settlements and enclaves (Fig. 1). It is populated by over 20 million ethnic Kurds, including 10 million in Turkey, five million in Iran, four million in Iraq, 500,000 in Syria, and 200,000 in the former Soviet Union. The majority of Kurds are Sunni Muslims and they are the fourth most numerous people in the Middle East. The relationship between the Kurds and their host nations has been one of continuous confrontation since antiquity. The 1920 Treaty of Sevres envisaged Kurdish autonomy and the possible creation of a Kurdish state, but the 1922 abolition of the Turkish Sultanate and the rise of nationalism under Kemal Ataturk led to the 1923 Treaty of Lausanne, in which there was no mention of an independent Kurdistan.

The triad of Iran, Iraq, and Turkey pursued anti-Kurdish policies throughout the post-World War II period, but their de facto alliance was disrupted by the 1958 Iraqi revolution. Baghdad initially supported Kurdish aspirations, but its good relations with the Kurds gradually floundered. The Iraqi army mounted a ferocious campaign against the Kurds in 1961–1962. Kurdish repression continued after the overthrow of the regime by the Ba'th Party in 1963. Since the Shah of Iran saw the Iraqi regime as a threat, Tehran offered support to the Iraqi Kurds. Iraq reciprocated with support to Kurdish separatists in Iran following the 1979 fall of the Shah and the anti-Kurdish onslaught of the Ayatollah Khomeini. There were also clashes in Turkey. Martial

Figure 1. Kurdistan.

law was declared in Anatolia in 1984. Kurdish separatists were clashing with the armed forces in Turkey, Iraq, and Iran by 1985. The most serious incident occurred in March 1988, when the Iraqi government responded to Kurdish advances by bombing the town of Halabja with chemical weapons that are believed to have included mustard agent and the nerve agents sarin, tabun, and VX. More than 5000 people died from the attack. It was the largest chemical attack that had ever been launched against a civilian population (Gutman and Rieff, 1999).

Violent clashes increased during the 1990s, particularly between the Kurds and the Turkish army. After the 1991 defeat of Iraq during Operation Desert Storm, the Kurds of northeastern Iraq pressed for autonomy and promptly were attacked by the Iraqi army. Many Kurds were killed and vast numbers fled into the mountains.

TERRORIST SUPPORT

Factions

Kurdistan Workers Party (KWP). The roots of the Partiya Karkaren Kurdistan (PKK: Kurdistan Worker's Party (KWP)) can be traced to Kurdish members of the 1960s leftist group Dev-Genç (Revolutionary Youth), which met in Ankara in 1974. Their leader was Abdullah Ocalan, alias "Apo" (father) in Kurdish, hence the name Apocu, that is Apo-ite, by which the group became known. The PKK was established on 27 November 1978, in the Lice District of Diyarbakir province. Attracting educated Kurds from poor backgrounds, it became notorious for its terrorist attacks not only against Turks but also against Kurdish groups considered tribalist, "feudal," or "bourgeois." The PKK advocated a fully independent, socialist Kurdistan that would include all the Kurdish-speaking areas in Turkey, Iraq, and Iran, although it dropped its insistence on socialism after the collapse of the East Bloc. It is now prepared to abandon terrorism in favor of regular political activity in Turkey if it is given legal status as a political party—a proposal that Ankara steadfastly refuses to grant.

The PKK is plagued by internal feuds that have resulted in factional murders. It is well organized in Europe despite having been outlawed in Germany and France. Its international political face is the Brussels-based National Liberation Front of Kurdistan. It has a strong propaganda

Encyclopedia of Bioterrorism Defense, Edited by Richard F. Pilch and Raymond A. Zilinskas
ISBN 0-471-46717-0 Copyright © 2005 Wiley-Liss

arm that includes the Kurdish television channel MED-TV, which is broadcast from London but managed from Brussels. PKK funds derive from drug trafficking, illegal migration, black marketing, and the Kurdish Diaspora, including the 500,000 Kurds living in Europe.

The PKK launched its initial terrorist campaign on 15 August 1984, first calling itself the Hazan Rizgariya Kurdistan (HRK: Kurdish Freedom Brigades), then the Kurdistan National Liberation Front (ERNK), and then the Peoples Liberation Army of Kurdistan (ARGK). It claims to maintain a Popular Army of 5000 to 10,000 soldiers. The PKK's modus operandi includes the targeting of local government officials, civic leaders, educators, police, and military patrols. It normally confines its operations to the countryside, but has also carried out several assassinations in Istanbul and Ankara. Moderate Kurds are also the targets of PKK vengeance. Together, the PKK's terrorist campaigns and Turkish counterinsurgency operations have claimed over 30,000 lives.

Kurdish Workers Party (PKK) Iraq. The PKK-Iraq is based in northern Iraq and is committed to bettering the conditions for the Kurds bordering Turkey. It suffered cross-border raids from Turkish troops after the Gulf War in 1991.

Kurdish Democratic Party of Iran (KDPI). The Kurdish Democratic Party of Iran (KDPI) was originally formed as an illegal organization after World War II out of the earlier Association for the Resurrection of Kurdistan, but it was practically liquidated when a Kurdish rebellion in Iran was crushed in 1966–1967. The KDPI has since operated as an insurgent movement seeking autonomy from Iranian rule, which has led to frequent rebellions and clashes with Iranian forces.

PKK Leadership

Abdullah Ocalan has been the leader of the PKK since its inception in 1978 (Fig. 2). The Turks demonize him as a threat to modern-day multiethnic Turkey, but he is viewed by the Kurds as the defender of Kurdish culture and political aspirations. Ocalan was born in Omerli, near the Syrian border, in 1948. He studied Political Science while attending university in Ankara. By 1973, he had organized a Maoist group to foment revolution, and founded the PKK in 1978 as an extreme left nationalist group with the aim of establishing an independent Kurdish state along Marxist lines. Ocalan fled Turkey in 1980 and took refuge in both Damascus and the Biqa' Valley, where he set up his PKK training camps. In 1998, under intense pressure from Turkey, Syria closed the camps and expelled Ocalan. Ocalan then sought political asylum in numerous countries, including Greece. He was finally captured in Kenya and spirited back to Turkey in 1999. He dropped his demand for Kurdish independence while in captivity, asserting that violent conflict can end only if Ankara grants Kurdish autonomy and cultural-linguistic freedoms. Ocalan was sentenced to death by a Turkish three-judge panel, but his sentence was commuted to life

Figure 2. Abdullah Ocalan, leader of the PKK.

in prison after Turkey's parliament abolished the death penalty in 2002.

THE PKK AND BIOLOGICAL WEAPONS

Intent

There is no indication that PKK leaders have adopted a policy for the use of weapons of mass destruction (WMD) to achieve their political ends. However, rogue elements within the movement may attempt to carry out unsanctioned acts of bioterrorism in an effort to generate fear and focus wider media attention on their cause, as exemplified in the cases described below.

Capability

Two cases involving the PKK and biological weapons have been reported:

- On 28 September 1997, a former PKK member claimed to have been assigned the task of building 12 small bombs containing rat poison, a half-liter sarin bomb, and a potassium cyanide bomb, which he said were meant to target Turkey's national monuments and tourist resorts. He also claimed to have handled sarin and botulinum toxin precursors. The man had reportedly broken away from the PKK after several terrorist attacks in which he took part failed. The PKK publicly discounted this report and proclaimed its opposition to the use of chemical and biological weapons.

- On 10 June 1998, three PKK members attempted to sell cobra venom, a black market product worth $2000 per vial, to Turkish undercover agents. Turkish police believe the venom was smuggled into Turkey from Azerbaijan as part of a PKK money-making scheme.

KURDISTAN TODAY

Kurdistan is an inherently problematic regional concept involving all nearby host nations in a perpetual conflict with their Kurdish minorities. Tensions are particularly high in Turkey because of the continued suppression of Kurdish culture and widespread resistance to granting political freedoms to the Kurdish minority. The Kurdish portions of Turkey will therefore remain a breeding ground for extremism and terrorism as long as the Kurds are denied basic human rights. PKK members are still provided with official and unofficial havens throughout the Middle East and Europe, but the PKK's military strength appears to be in decline even though some sources estimate that 5000 fighters can still be activated. Ankara is undertaking a major irrigation project in southeastern Anatolia in a desperate gamble to bolster that region's economy and defeat Kurdish separatism. Nonetheless, economic ventures will not be enough to pacify Kurdish aspirations.

IMPLICATIONS

The Kurdish quest for autonomy cannot be suppressed further without precipitating a protracted rebellion, and as a result, various host nations must recognize the legitimate rights of this powerful minority through liberal reforms. Meanwhile, the Kurds will remain a source of extremist activity and a breeding ground for regional instability. Despite the incidents described above, it is unlikely that the PKK will pursue a WMD capability to achieve its political objectives. The region of Kurdistan will, however, continue to be a perpetual flashpoint for crisis managers.

REFERENCE

Gutman, R. and Rieff, D., *Crimes of War: What the Public Should Know*, W.W. Norton and Company, New York, 1999.

FURTHER READING

Anderson, E., *An Atlas of World Political Flashpoints*, Pinter Publishers, London, 1993.

Crefeld, M.V., *The Encyclopedia of Revolutions and Revolutionaries*, Facts on File, New York, 1996.

Economist, An Ancient Tragedy: How 25m Kurds Pursue a Goal that Nobody Else Wants Them to Reach, February 18, 1999.

Economist, A New and Bitter Brew in the Middle East, October 8, 1998.

Economist, Can Turks and Kurds Settle? December 17, 1998.

Economist, So Where is Kurdistan? June 8, 2000.

Economist, The Battle-Lines in Turkey, January 11, 2001.

Economist, The Merits of Keeping Ocalan Alive, November 25, 1999.

Economist, Turkey and Its Kurds: A Turn for the Worse, January 31, 2002.

Economist, Turkey and Its Kurds: Getting Just A Little Better, October 24, 2002.

Economist, Turkey and Its Kurds: Back to Bloodshed? February 13, 2003.

Economist, Turkey on Trial: The Kurdish Separatist Leader is in the Dock—So is Turkey, June 3, 1999.

Economist, Turkey's Kurds: Down but far from Out, July 30, 1998.

Economist, *Turkey's Kurds: No Change?* April 18, 2002.

Economist, Eying Turkey Nervously, March 25, 2003.

Longman Current Affairs, *Border and Territorial Disputes*, Longman Publishers, United Kingdom, 1992.

Longman Current Affairs, *Revolutionary and Dissident Movements*, Longman Publishers, United Kingdom, 1991.

Longman International Reference, *World Directory of Minorities*, Longman Group UK Ltd., United Kingdom, 1990.

Shaikh, F., *Islam & Islamic Groups: A Worldwide Reference Guide*, Longman Publishers, 1992.

Usher, R., Nationalists without a Nation, CNN, November 20, 2000.

Witschi, B., Who is Abdullah Ocalan? CNN, November 20, 2000.

L

LABORATORY RESPONSE TO BIOTERRORISM (LABORATORY SURVEILLANCE, PUBLIC HEALTH LABORATORY CAPACITY)

TYLER ZERWEKH

ROBERT EMERY

BACKGROUND

The phrase "laboratory response to bioterrorism" is used to describe a single laboratory or network of laboratories equipped to quickly and efficiently identify biological agents (bacteria, viruses, or toxins) that might be used in a bioterrorism event. To respond to such an event, laboratories rely on rapid diagnostic techniques such as microscopy, agglutination, culturing, antibiotic susceptibility, and polymerase chain reaction (PCR) to identify and subsequently confirm the presence of a particular agent. The necessity for a systematic and cohesive laboratory response is critical for a successful response to an attack, and will serve to ultimately reduce the associated morbidity and mortality of an exposed population. Rapid laboratory recognition and confirmation of bioterrorism agents expedite the initiation of medical response, allowing timely clinical diagnosis and immediate initiation of prophylaxis to exposed persons.

By the late 1990s, little work had been done to develop a laboratory response to bioterrorism. Owing to an alarming increase in hoaxes, however, the need for a laboratory response component to bioterrorism became evident, and the Centers for Disease Control and Prevention's (CDC) National Laboratory Response Network (LRN) was created. When, for the first time in the new millennium, a legitimate bioterrorism attack occurred during autumn 2001 in the form of the "anthrax letters," the conceptual need for a rapid laboratory response became a reality (Heller et al., 2002).

Much has been achieved since 1999, including the creation of the LRN laboratory procedures manual for the various diagnostic techniques for testing potential bioterrorism agents. More importantly, with the enactment of the Public Health Security and Bioterrorism Preparedness and Response Act of 2002, additional measures have been directed toward further development and refinement of the laboratory response to bioterrorism.

TYPES OF BASIC LEVEL LABORATORIES

In developing a laboratory response to bioterrorism, a setting that can safely accommodate a variety of biological agents must be established. The CDC/NIH *Biosafety in Biomedical and Microbiological Laboratories* (BMBL) guidebook provides the foundation for safe handling of infectious agents in laboratories. The BMBL describes the facilities, equipment, and practices considered necessary to safely handle biological agents. Safety recommendations are classified according to biosafety levels (BSLs), ranging from Level 1 to Level 4. Biological agents are classified into one of the four BSL categories, designated in ascending order by the degree of protection that needs to be provided to personnel who work with these agents, as well as the community and environment.

BSL 1 laboratories are basic level laboratories with minimal specific recommendations for special safety precautions, and permit work with microorganisms that are nonpathogenic in normal, healthy adults.

BSL 2 laboratories are capable of handling moderate-risk organisms that pose percutaneous or mucous membrane exposure hazards. Special procedures are taken at this level to minimize aerosolization or splash risk. Recommendations include performing work with infectious agents inside a certified biosafety cabinet (BSC), classified from level I through III in increasing order of protection (with subclassifications of level II and III). A BSC utilizes laminar airflow and HEPA (High Efficiency Particulate Arresting) filtration technology to contain and filter out dangerous particles. These design features protect the laboratory work inside the cabinet from accidental contamination ("product protection") and prevent the release of infectious organisms into the surrounding environment ("personal protection").

BSL 3 laboratories place more emphasis on primary and secondary barriers to minimize infectious aerosol exposures, and further provide additional precautions against ingestion and autoinoculation of the laboratory worker. BSL 3 laboratories are of particular interest to the laboratory response to bioterrorism because this level of protection and practice aids in the prevention of the transmission of potentially serious or lethal diseases with aerosol transmission potential.

BSL 4 laboratories are the most complex BSLs because they implement all BSL recommendations from BSL 1 to 3 and further include precautions such as separated structures, showering facilities for laboratory exit, and full-body, air-supplied positive-pressure personnel suits. BSL 4 laboratories are designed for work with exotic agents that pose a high risk of life-threatening disease or generally have no vaccine or prophylactic treatment available.

Encyclopedia of Bioterrorism Defense, Edited by Richard F. Pilch and Raymond A. Zilinskas
ISBN 0-471-46717-0 Copyright © 2005 Wiley-Liss

The majority of laboratory response to bioterrorism work will be performed at the BSL 2 and BSL 3 levels. At the very least, all handling and testing will be performed in a BSL 2 facility. This means that all work performed after reception of the sample, including labeling, is to be performed inside a BSC. Response activities that may result in the production of aerosols or that involve the manipulation of large quantities of the biological agent in question should be referred to a BSL 3 laboratory.

LABORATORY RESPONSE NETWORK

Background

In 1997, the CDC created the LRN in an effort to provide an organized response system for the detection and diagnosis of suspected biological agents based on each constituent laboratory's testing capabilities and facility construction. Each of the public and private laboratories that constitute the LRN can perform analyses according to standard protocols developed by the CDC in collaboration with the Association of Public Health Laboratories and the American Society for Microbiology. Furthermore, they are categorized into four levels—A, B, C, or D—based on an independent classification created for response capabilities. The primary goal of this pyramidal classification system (Fig. 1) is to progressively increase diagnostic capability while increasing worker safety through each level. A lower-level laboratory is to refer potential bioterrorism samples for analysis to the next higher level if its diagnostic capabilities have been exhausted (Pavlin et al., 2002).

Types of LRN Laboratories

The primary goal of a Level A laboratory is to rule out hoaxes and non-threats by determining whether a biological agent is actually present. If a Level A laboratory cannot rule out or otherwise dismiss the potential agent as a hoax or non-credible threat, then the agent in question must be referred to the next highest laboratory, Level B. Work performed in Level A laboratories is considered BSL 2, and the laboratory must have a certified Class II BSC. During the anthrax mail attacks of 2001, an increased workload was placed on local and state public health laboratories charged with analyzing suspicious packages. This resulted in the creation of Level A laboratories by several public and private institutions and organizations to screen for suspicious packages within their institution or regions. The adoption of Level A laboratories at these institutions allowed immediate analysis of potential bioterrorism agents while alleviating worker apprehension and decreasing the amount of false-positives referred to higher-level laboratories during the time period. These Level A laboratories were able to successfully rule out hoaxes and non-threat packages and letters while minimizing workdays lost and relieving much of the burden on Level B and Level C laboratories.

Level B laboratories, sometimes referred to as core capacity laboratories, perform agent isolation and diagnostic testing, thus minimizing false negatives and providing the initial identification of the biological agent in question. Level B laboratories are most likely to be city health department clinical laboratories. These laboratories will receive the first clinical specimens, such as environmental swabs or patient samples, in order to make a preliminary identification. Level B laboratories must work at BSL 2 recommendations, but most Level B laboratories have the capability and facility infrastructure for BSL 3 practices and therefore operate on that level. Once identified, Level B laboratories refer a positive bioterrorism specimen to the next highest-level laboratory, Level C.

Level C laboratories are responsible for speciation of the suspected biological agent through PCR methodologies such as reverse transcriptase PCR (RT-PCR); performing follow-up molecular typing for agent comparison; and performing toxicity testing as appropriate. These laboratories

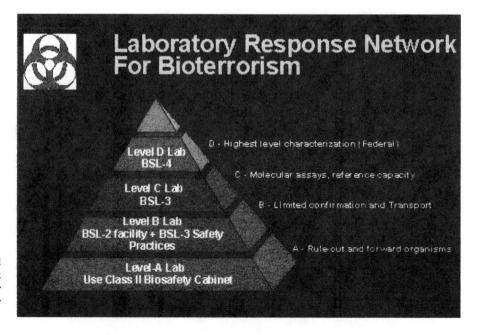

Figure 1. Centers for Disease Control National Laboratory Response Network for bioterrorism. *Source*: Centers for Disease Control, National Laboratory Response Network.

must operate at the BSL 3 level. Additional responsibilities include the evaluation of reagents and testing procedures in order to assist and enhance Level B laboratory response.

A Level D laboratory is the equivalent of the highest security level laboratories at present day United States Army Medical Research Institute of Infectious Diseases (USAMRIID) and the CDC. These laboratories can perform all tests utilized in Levels A, B, and C, and can further detect genetic recombinants of the suspected agent as well as perform additional investigational analyses such as antibiotic susceptibility, host and vector patterns, and genome storage. Additionally, Level D laboratories develop and refine rapid and valid analyses to disseminate to lower-level laboratories in order to enhance response. All Level D laboratories possess BSL 4 containment facilities that maximize worker and environmental protection from suspected biological agents, thus permitting handling and diagnosis of Level 4 organisms.

CATEGORIES OF BIOLOGICAL AGENTS

The CDC has identified 46 bacteria, viruses, and toxins as biological threat agents. These agents are classified into categories based on the likelihood of dissemination, the pathogenicity and availability of the organism, and the likelihood of the organism to cause paranoia or fear.

Category A agents have been defined as organisms that pose the greatest threat to national security if released. Category A agents can be readily disseminated, result in high mortality rates, have the potential for major public health impact, cause mass paranoia/hysteria, and may possess a person-to-person transmission risk. Special preparedness procedures must be implemented for response to Category A agents. Category A agents include *Bacillus anthracis* (anthrax), *Yersinia pestis* (plague), *Francisella tularensis* (tularemia), *Clostridium botulinum* neurotoxin (botulism), Variola major (smallpox), and hemorrhagic fever viruses, such as Ebola and Marburg.

Agents classified as Category B are of the second highest priority. These agents are moderately easy to disseminate, result in moderate morbidity and mortality, and require specific enhancements of the CDC's diagnostic capacity and surveillance for efficient and expedient laboratory response. Category B includes *Brucella* species (brucellosis), *Burkholderia mallei* and *B. pseudomallei* (glanders and melioidosis, respectively), *Coxiella burnetii* (Q fever), ricin, Staphylococcal enterotoxin B, *Rickettsia prowazekii* (typhus), alphaviruses such as Venezuelan equine encephalitis, eastern equine encephalitis, and western equine encephalitis, and food and waterborne pathogens such as *Cryptosporidium parvum* and *Salmonella* spp.

Category C agents include emerging pathogens that could potentially be manufactured for bioterrorism use because of availability of the agent, ease of production, and the potential for high morbidity and mortality rates. Examples of Category C agents are Nipah virus, hantaviruses, and multidrug-resistant *Mycobacterium tuberculosis*.

Currently, the CDC has developed Level A laboratory procedures for most Category A agents. Procedures for the analysis of Category A agents can be found on the CDC bioterrorism website at http://www.bt.cdc.gov, and include screening procedures for *B. anthracis, Y. pestis, F. tularensis,* and botulinum toxin, as well as *Brucella* spp. (Category B). Generally, Level A laboratories will perform basic staining analyses such as gram and spore staining, along with other basic microbiological tests such as motility, hemolysis, and broth growth. Level B laboratories will be responsible for further detection and identification through more complex and specific analyses, such as direct fluorescent antibody (DFA) microscopy, agglutination, and some PCR analyses. Additionally, Level B laboratories will begin to examine antibiotic susceptibilities and resistances at this level. Level C laboratories will be responsible for initial confirmatory analyses such as cell culturing, F1 antigen detection methods, mouse inoculations, and advanced agent-specific PCR analyses. Finally, Level D laboratories will be responsible for organism genotyping and all additional analyses not previously performed in laboratories A, B, and C, such as histopathology, plasmid typing, and refinement of Level A, B, and C laboratory response procedures and tests (Nulens and Voss, 2002). Table 1 summarizes LRN laboratory level capabilities for detection of suspected bioterrorism agents.

SURVEILLANCE AND EARLY DETECTION

The fall 2001 anthrax letters provided a wealth of information pertaining to the laboratory response to bioterrorism. Any laboratory that handles clinical or environmental specimens and wishes to serve as part of the LRN should possess the following attributes: knowledge of its own current BSL qualification; knowledge of protocols developed regarding chain of custody and sample preservation; knowledge of proper laboratory procedure, development, and implementation regarding the identification, preservation, and transfer of potential bioterrorism agents; knowledge of the location of the next highest-level reference laboratory, as identified in the LRN; and knowledge of the basic clinical and physical characteristics of potential bioterrorism agents (Heller et al., 2002).

The majority of current research involving laboratory response to bioterrorism has focused on increasing laboratory surveillance and improving field detection. Advances in laboratory surveillance include, for example, the development of email lists and integrated online technology for hospital reporting. One such online resource, the interactive email listserv newsletter ProMED (Program for Monitoring Emerging Diseases), has allowed for greater dispersal of information on worldwide emerging and reemerging infectious disease outbreaks, the laboratory capacity to identify causative agents, and epidemiologic patterns of infectivity. Specifically, ProMED enables public health laboratory professionals to disseminate new laboratory techniques and methodologies during an emerging disease outbreak investigation. This instant posting of information via email provides an avenue for immediate disbursement

Table 1. Laboratory Level Capabilities for Detection of Bioterrorism Agents

Laboratory	Level	Anthrax	Plague	Tularemia	Brucella	Smallpox	Botulism
A	Detection	BSL 2 Perform gram stain. If nonhemolytic and nonmotile, send to Level B lab.	BSL 2 Perform gram stain. If Yersinia species, send to Level B lab.	BSL 2 and 3 Perform gram stain. If growth on chocolate agar then Tularemia species. Send to Level B lab.	BSL 2 and 3 Perform gram stain. If rapidly urea positive in slant, send to Level B lab.	BSL 3 and 4 Human fibroblast cell culture with cytopathic effect, but negative for *Varicella*. Send to Level D lab.	BSL 2 Test not done in clinical labs.
B	Detection and identification	DFA Culture PCR Send to Level C lab	DFA Culture PCR Antibody susceptibility Send to Level C lab	DFA Culture Latex agglutination PCR Send to Level C lab	DFA Culture Speciation PCR test Antibody susceptibility Send to Level C lab	–	–
C	Detection and identification	Evaluate PCR	Mouse inoculation Evaluate PCR	Mouse inoculation Glycerol test	DFA Evaluate PCR	–	Mouse bioassay Subtype
D	Detection and identification	Develop PCR markers; serology, typing	Histopathology Plasmid typing	Differentiate vaccine strain and sample. Microagglutination Compare PCR	Develop new PCR tests and primers.	PCR test on dead agent. Evaluate genetic factors.	Development of rapid analysis to Level C and disseminate to Level B labs.

Source: Centers for Disease Control and Prevention. *Public Health Preparedness and Response for Bioterrorism*, (1999).

of critical information. Additionally, early recognition of a bioterrorism exposure can be posted to ProMED, enabling appropriate personnel to take necessary precautions, help educate health officials, and aid in answering technical questions. Following the 2001 anthrax mailings, ProMED was able to provide immediate notification of the possibility of disease, offer laboratory techniques to rapidly identify both the causative organism and effective prophylactic therapy, and even epidemiologically establish precedence for a potential bioterrorism incident.

Current research toward the development of field biological agent detectors has been initiated and developed in the military theater. The Defense Advanced Research Projects Agency (DARPA) has pursued newer and more efficient methods of performing diagnostic testing on suspected biological agents, and is now in the process of developing a detector that is handheld, can identify 20 different biological agents, costs less than $5000, and has a limited false negative rate compared to present detection devices. "Hand-held agent indicator rapid identification biological strips" have been developed that utilize antigen identification techniques from an environmental swab and can provide results in 15 to 30 min. Initially, these field detectors were met with some criticism, primarily due to the high frequency of false-positives; however, refinement of the product and increasing device specificity, combined with the heightened awareness that has come with the anthrax mailings of October 2001, has generated a surge of rapid analysis field detector purchases.

There is still a need for standardized training of multiple public health laboratories in order to further improve laboratory response capability. Formalized training must be offered so that laboratory employees from different geographical regions can coordinate and respond immediately, in their respective regions, in the event that their services are called into action, rather than having to await the arrival of an exotic disease specialist to perform the necessary battery of laboratory analyses. More research is also needed to further expedite the identification of pathogens. Research is being conducted to develop a "real-time" PCR analysis that investigates the suspected agent after each cycle of target strand replication rather than at the endpoint of the analysis. Initial results have proved fruitful.

CONCLUSION

Laboratory response plays a significant role in preparedness and response efforts in the realm of bioterrorism defense. Public health laboratories must perform a critical function in response to an attack by efficiently identifying the biological agent and rapidly performing confirmatory analyses and sustained active laboratory surveillance. Appropriately trained microbiologists implementing core analyses will be called upon to rule out a suspected agent. If the agent cannot be ruled out, timely referral to the next appropriate public health laboratory will be essential. Ultimately, the continued development, practice, and refinement of laboratory response procedures for the analysis of potential bioterrorism agents will aid first response efforts by decreasing the interval between agent release or patient exposure and the medical response to those exposed, thereby decreasing morbidity and mortality.

REFERENCES

Heller, M.B., Bunning, M.L., et al. *Emerg. Infect. Dis.*, **8**(10), 1096–1102 (2002).

Nulens, E., and Voss, A., *Clin. Microbiol. Infect.*, **8**(8), 455–466 (2002).

Pavlin, J.A., Gilchrist, M.J., Osweiler, G.D., and Woollen, N.E., *Emerg. Med. Clin. North Am.*, **20**(2), 331–350 (2002).

FURTHER READING AND WEB RESOURCES

http://www.bt.cdc.gov/labissues/index.asp - Centers for Disease Control Bioterrorism website. Accessed on 5/5/03.

http://www.cdc.gov/od/ohs/biosfty/bmbl4/bmbl4toc.htm - Biosafety in Microbiological and Biomedical Laboratories, United States Department of Health and Human Services Centers for Disease Control and Prevention, and National Institutes of Health, 4th ed., US Government Printing Office, Washington, DC, May 1999, accessed on 6/6/03.

http://www.phppo.cdc.gov/nltn/default.asp - National Laboratory Response Network website. Accessed on 5/20/03.

See also DETECTION OF BIOLOGICAL AGENTS and SYNDROMIC SURVEILLANCE.

LARRY WAYNE HARRIS

Lauren Harrison

INTRODUCTION

Larry Wayne Harris, a trained microbiological technician from Lancaster, Ohio, has been implicated in two cases of possession of biological materials that are of particular concern in a weapons context. In 1995, Harris was charged with misrepresenting himself while ordering three vials of freeze-dried *Yersinia pestis*, the causative agent of plague, from the American Type Culture Collection (ATCC) in Rockville, Maryland (Macy, 1998; Egan, 1998). Later, in 1998, Harris was arrested for suspected possession of *Bacillus anthracis*, the causative agent of anthrax; however, the substance in his possession was ultimately found to be a harmless veterinary vaccine strain of the bacteria (CNN.com, 1998). Harris is classified as a "lone actor" in that he is not believed to maintain ties with any terrorist groups or states, though it should be noted that Harris is known to have had ties with Aryan Nations and other neo-Nazi groups before reportedly renouncing his racist views in the mid-1990s (Stern, 2000).

BACKGROUND

According to Aryan Nations founder Richard Butler, Harris was a member of the organization in the early 1990s and climbed to the rank of lieutenant before leaving in 1995. He also belonged to the Christian Identity Church, which teaches white supremacy (Windrem, 1998), and claims to have been a member of the National Alliance, a neo-Nazi organization, as well (Ruth, 1998). Professionally, Harris received an Associate Degree in Biophysics from Ohio State University in 1995 and later became a certified member of the American Society for Microbiology (Stern, 2000; Windrem, 1998). His enrollment at Ohio State followed a stint as a mechanic in the military and what he claims were a number of postmilitary research projects related to national defense (Stern, 2000). He claims to have worked for the CIA from 1985 to 1990, a point that remains unsubstantiated to date (Windrem, 1998).

SEQUENCE OF EVENTS

1995

On May 5, 1995, Larry Wayne Harris ordered three vials of freeze-dried *Y. pestis* from the ATCC for $240,

using his employer's state certification (Vick, 1995). At the time, Harris presented himself as a researcher by using a fake letterhead with the name "Small Animal Microbiology Laboratory" and his home address printed on the top. When he called on May 10, 1995 to check on his order, the sales representative with whom he spoke became suspicious that he was not qualified to handle such materials. The representative conveyed this worry to the Centers for Disease Control and Prevention (CDC) in Atlanta, Georgia, and the CDC contacted Harris. According to CDC Office of Health and Biosafety Chief Richard Knudsen, "It [was] only by accident that [the CDC] learned this guy lived in a home and he was a little bit off his rocker" (Macy, 1998).

On May 12, 1995, Harris's home in Lancaster, Ohio, was raided. During the raid, he willingly led authorities to the three vials, still sealed in their original container and stored in the glove compartment of his 1989 Subaru. The raid also uncovered weapons and explosives, as well as a certificate declaring Harris a lieutenant in the Aryan Nations. Discovering Harris's neo-Nazi ties caused authorities to begin considering the incident a possible case of domestic terrorism, and only days later Harris was arrested for using false documents in order to acquire the pathogen (it was not illegal to possess human pathogens at the time).

Harris claimed that he had acquired the material to find a cure for plague, which he believed would soon be spread throughout the United States by Saddam Husayn's "super-germ-carrying rats." This was not the first record of such claims on Harris's part: In February 1993, he had telephoned the CIA, CDC, and FBI to inform them of an Iraqi plot to smuggle biological warfare (BW) agents into the United States in the "private parts" of young Iraqi women (Stern, 2000). When no one believed him, Harris purportedly felt compelled to spread information about the Iraqi threat to the public and to engage in the development of a novel plague vaccine.

Harris was convicted of fraud on April 22, 1997, (Stern, 2000) and sentenced to 200 hours of community service and 18 months probation (*Kansas City Star*, 1998).

1998

On February 18, 1998, Harris and William Leavitt, owner of two biology laboratories (one in Logandale, Nevada, and the other in Frankfurt, Germany), were arrested for possession of *B. anthracis*. Acting on a tip from an informant, the FBI officials arrested the two men outside the Green Valley Professional Office Building in Henderson, Nevada. The informant, Ronald G. Rockwell, a medical researcher with two felony criminal convictions for attempted extortion in 1981 and again in 1982, reportedly observed Harris carrying a white foam cooler into the building that, Harris informed him, contained the causative agent of anthrax. The informant also reported seeing 8 to 10 flight bags marked "biological" in the trunk of Harris's car.

Harris, Leavitt, and Rockwell were arrested by a Las Vegas SWAT unit, after which Harris and Leavitt were sent to the local hospital as a precaution. Four U.S. Army specialists and a technician from the hazardous materials unit of the Army's Dugway Proving Ground in Utah recovered the cooler and 40 Petri dishes from the building, then sealed Harris's and Leavitt's white Mercedes Benz in plastic and moved it to Nellis Air Force Base in Clark County, Nevada, for further analysis.

After testing of the confiscated materials at the United States Army Medical Research Institute of Infectious Diseases (USAMRIID) at Fort Detrick, Maryland, revealed that the *B. anthracis* recovered was in fact a harmless veterinary vaccine strain (Stern, 2000; Claiborne, 1998), charges against Harris and Leavitt of conspiracy and possession of a biological agent for use as a weapon were dropped on February 23, 1998. Nevertheless, Harris was successfully charged with violation of his probation in connection with the 1995 wire fraud conviction.

MOTIVATIONAL FACTORS

Harris has made numerous statements that call question to his true intentions. He has been quoted describing the potential release of *B. anthracis* both into a subway system by throwing a light bulb filled with spores onto one of the tracks (in essence, a description of a pre-1969 CIA field test using a simulant for *B. anthracis* that has been well documented in the public domain) and into the air by spraying it from a car (effectively describing one of the many failed attacks with *B. anthracis* attributed to the Aum Shinrikyo cult in the early 1990s) (Windrem, 1998). Harris has also been quoted as saying that he planned to attack the New York subway system with *Y. pestis*, and Rockwell informed authorities that Harris had claimed to possess enough *B. anthracis* to "wipe out" Las Vegas (Stern, 2000). Harris continues to be frequently cited on militia and Christian Identity web pages (Windrem, 1998).

CONCLUSION

While Larry Wayne Harris is no longer associated with specific white supremacist groups, his former associations in this regard underscore the potential for groups of this nature, like that of nearly all terrorist groups, to utilize the skills of trained scientists for illicit purposes should they so desire. Furthermore, Harris's actions

suggest that even individuals with relatively limited educational background in the field of biological weapons can have a reasonable amount of success in certain aspects of attaining a BW capability, in this case the acquisition of a viable pathogen. As a result, following the 1995 incident, new legislation was enacted by the U.S. Congress making the transfer of certain pathogens across state borders without CDC clearance a criminal offense. Though a positive step toward effective biosecurity, these regulations are limited in that they do not regulate the secondary transfer of biological agents, agents already in a person or group's possession, or agents isolated from nature (Stern, 2000).

REFERENCES

Egan, M. "Two Arrested in Nev. In Biological Agent Probe," *Reuters*, February 19, 1998.

Macy, R. FBI Arrests Two in Nevada for Possession of Biological Agent, *Las Vegas Review-Journal*; Internet, available from www.lvrj.com/lvrj_home/1998/Feb-19-Thu-1998/news/arrest.html accessed on 2/19/98.

CNN.com, "One Suspect in Anthrax Case Released from Custody," *CNN* (21 February 1998); Internet, available from cnn.com, accessed on 2/24/98.

Plague Bacteria Is Basis of Charge, *Washington Post* (17 May 1995): A14.

Ruth, R. Microbiologist Seen as "Yarn-spinner", *Columbus Dispatch* (24 March 1998); Internet, available from www.dispatch.com accessed on 4/3/98.

Kansas City Star, Scientist Who Ordered Germs Pleads Guilty, (23 April 1997); Internet, available from www.kcstar.com accessed on 6/1/98.

Stern, J. "Larry Wayne Harris," in Jonathan B. Tucker, ed., *Toxic Terror: Assessing Terrorist Use of Chemical and Biological Weapons*, MIT Press (2000), pp. 227–246.

Two Held in Anthrax Scare Cleared of Felony Charges, *Houston Chronicle* (24 February 1998): A2; Internet, available from www.chron.com accessed on 6/1/98.

Vick, K. Man Gets Hands on Bubonic Plague Germ, But That's No Crime, *Washington Post* (30 December 1995): D1, D4.

Windrem, R. "The Man Who Talks Too Much: No Stranger to Law Officials, Larry Wayne Harris Trips Again," MSNBC; Internet, available from www.msnbc.com accessed on 2/25/98.

FURTHER READING

Clairborne, W. Vials Seized in Las Vegas Are Tested for Anthrax, *Washington Post*, February 21, 1998, Internet, available from www.washingtonpost.com, accessed on 2/24/98.

Clairborne, W. Vials Seized by FBI in Las Vegas Are Found to Contain 'Harmless' Anthrax Vaccine, *Washington Post*, February 22, 1998, Internet, available from www.washingtonpost.com, accessed on 2/24/98.

Fleeman, M. "Men Held in Possible Anthrax Plot," *AP Online*, 20 February 1998.

"Harris: Biologist Turns Terrorist," *United Press International* (19 February 1998) 4:17 PM EST.

John O'Neill, "Hearings on Global Proliferation of Weapons of Mass Destruction," 31 October - 1 November 1995; cited in Anthony Fainberg, "Debating Policy Priorities and Implications," *Terrorism with Chemical and Biological*

Weapons, Brad Roberts ed. (CBACI: Alexandria, 1997) pp. 80–81.

Man Suspected of Buying Plague Bacteria, *St. Louis Post Dispatch*, May 17, 1995.

Ruth, R. Bubonic Plague Case Trial Set, *Columbia Dispatch*, March 12, 1997, p. B1.

Ruth, R. Deal May Be Dropped in Plague Case, *Columbia Dispatch*, April 2, 1996, p. D1.

Ruth, R. Suspect in Germ Case Won't Go to Jail, *Columbus Dispatch* (23 April 1997): 1A.

The RAND-St. Andrews Terrorism Chronology: Chemical/Biological Incidents 1968–1995; #19950500.

Tom, D.A. FBI Searches Researcher's Home Again, *Houston Chronicle* (23 February 1998): A2; Internet, available from www.chron.com, accessed on 6/1/98.

Vick, K. Plea Bargain Rejected in Bubonic Plague Case, *Washington Post*, April 3, 1996, p. A8.

Vick, K. Ohio Man Gets Probation in Bubonic Plague Case, *Washington Post*, April 23, 1997, p. B3.

LAWRENCE LIVERMORE NATIONAL LABORATORY

J. Patrick Fitch

INTRODUCTION

The Lawrence Livermore National Laboratory (LLNL) Chemical & Biological National Security Program (CBNP) provides science, technology, and integrated systems for chemical and biological security. Its approach is to develop and field systems that dramatically improve the nation's capabilities to prevent, prepare for, detect, and respond to terrorist use of chemical or biological weapons.

BACKGROUND

History underscores the importance of preparing to defend against terrorism. A well-documented example of a terrorist organization is the Aum Shinrikyo cult (Olson, 1999). After several flawed attempts at terrorism using botulinum toxin, *Bacillus anthracis* spores, and other biological agents, the cult released the chemical nerve agent sarin in Tokyo's subway in March 1995, injuring more than 1000 people. These terrorist acts served to catalyze and focus the early LLNL program on civilian counterterrorism. In 1995, LLNL began the CBNP using laboratory-directed research and development (R&D) investments and a focus on biodetection. The 1996 Nunn-Lugar-Domenici Defense Against Weapons of Mass Destruction Act initiated several U.S. nonproliferation and counterterrorism programs, including the Department of Energy (DOE) Chemical and Biological Nonproliferation Program (also known as the CBNP). In 2002, the Department of Homeland Security (DHS) was formed. The DOE CBNP and many of the LLNL CBNP activities have been transferred to the new Department as part of the Science and Technology Directorate's Biological and Chemical Countermeasures Program. The LLNL activities represented about half of the program that was transferred to DHS from DOE.

LLNL has a long history in national security, including nonproliferation of weapons of mass destruction (WMD). In biology, LLNL had a key role in starting and implementing the Human Genome Project and, more recently, the Microbial Genome Program. In addition, a medical device program at LLNL began in the early 1990s. The collective culture of the CBNP program at LLNL blends a national security mission focus, the competitive spirit of the human genome "race," and strong partnerships with industry and end users so that systems developed by CBNP are both useful and commercially

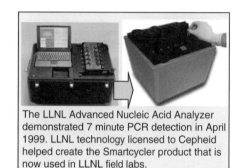

The LLNL Advanced Nucleic Acid Analyzer demonstrated 7 minute PCR detection in April 1999. LLNL technology licensed to Cepheid helped create the Smartcycler product that is now used in LLNL field labs.

Figure 1.

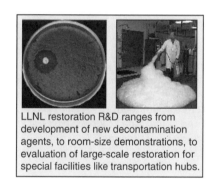

LLNL restoration R&D ranges from development of new decontamination agents, to room-size demonstrations, to evaluation of large-scale restoration for special facilities like transportation hubs.

Figure 2.

viable. Blending these three cultures supports a strategy of identifying new, high-impact systems and taking the systems from concept to operational proof of principle.

In addition to the CBNP-specific expertise, LLNL has more than 1000 scientists and engineers with relevant expertise in biology, chemistry, decontamination, instrumentation, microtechnologies, atmospheric modeling, and field experimentation. More than 100 LLNL scientists and engineers work full time on chemical and biological national security projects. The program also leverages special facilities that are part of the LLNL infrastructure, including the National Atmospheric Release Advisory Capability (NARAC), Forensic Science Center (FSC), Biosecurity and Nanoscience Support Laboratory (BSNL), Fermentation Laboratory, Pathogen Laboratories, Micro Technology Center (MTC), and Center for Accelerator Mass Spectrometry (CAMS).

Encyclopedia of Bioterrorism Defense,   Edited by Richard F. Pilch and Raymond A. Zilinskas
ISBN 0-471-46717-0   Copyright © 2005 Wiley-Liss

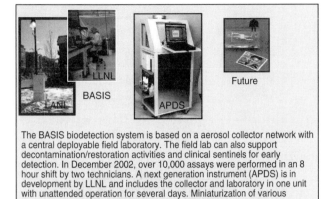

The BASIS biodetection system is based on a aerosol collector network with a central deployable field laboratory. The field lab can also support decontamination/restoration activities and clinical sentinels for early detection. In December 2002, over 10,000 assays were performed in an 8 hour shift by two technicians. A next generation instrument (APDS) is in development by LLNL and includes the collector and laboratory in one unit with unattended operation for several days. Miniaturization of various components improve performance and reduce costs.

Figure 3.

DEMONSTRATION PROGRAMS

Critical shortcomings remain in the nation's ability to prevent, prepare for, detect, and respond to chemical or biological terrorism. Thus, the LLNL CBNP emphasizes collaborative demonstration programs with end users as part of a spiral development strategy. These programs provide near-term capabilities, opportunities for dialogue among scientists and end users, focus for R&D investments, and paths for pilot-to-regional-to-national systems. This development strategy is strengthened by involvement in both operational systems and basic research in pathogen biology, material science, computer algorithms, and instrument development.

Current and emerging collaborative demonstration programs include the following:

BASIS

Biological Aerosol Sentry and Information System (BASIS) is an environmental monitoring system originally designed for the 2002 Salt Lake City Winter Olympics. BASIS is a joint project of LLNL and Los Alamos National Laboratory (LANL) with significant participation by law enforcement and public health organizations. BASIS was successfully deployed at the Olympics, and the LLNL team has been deployed to several other sites. In the past two years, several hundred thousand assays have been performed by LLNL in support of "counterbioterrorism" using a BASIS-like architecture with no false alarms. The BASIS team won an R&D100 award in 2003.

SENTINEL

Currently available environmental testing methods cannot identify low-level releases and can only detect known pathogens. Thus, laboratory testing of sentinel populations for early detection of bioterrorism (SENTINEL) has been developed as a strategy that complements BASIS-type environmental monitoring by performing surveillance of the population through direct, high-throughput testing of clinical samples of opportunity. LLNL has demonstrated 1000 samples (10,000 assays) performed in an 8-hour shift with only two technicians.

SENTINEL is an emerging program in collaboration with health providers to directly detect pathogens and toxins in samples. Preliminary data by LLNL and an industrial collaborator indicates that the strategy might also be applied to presymptomatic detection of host response biomarkers. LLNL is pursuing a basic science program known as Pathomics to establish the links between pathogenicity and host response (Fitch et al., 2002a). A SENTINEL strategy is useful in public health

Atmospheric modeling and experiments are being used to assess sensor placements and to provide feedback to first responders in urban environments. Photograph of August 7, 1998 tire fire in Tracy, CA with LLNL/NARAC-simulated particle positions in red (top) and city-wide simulations of a plume.

Figure 4.

applications (e.g., influenza) even in the absence of bioterrorism, and may also serve to address the genetically modified threat.

LINC

Local Integration of NARAC in Cities (LINC) provides responders in local government with modeling and prediction tools for decision support. Composite views of aerosol plume prediction with important local features, including schools and fire and police stations, are available to the users. Both local and reach-back capabilities are provided. The first pilot city for LINC evaluation, which began in September 2002, is Seattle, Washington. At present, there are five LINC cities.

Bio-Forensics

Bio-Forensics is a joint LLNL, Northern Arizona University, and LANL project to make specific forensic tools and data available to the broader community. The focus is on supporting law enforcement and the Centers for Disease Control and Prevention's (CDC) PulseNet Laboratories for foodborne illnesses. Round-robin comparisons of different methods and assays will precede the principle deliverables of (1) a database for a variety of end users focused on law enforcement, and (2) strain-sensitive markers and validated assays for *Salmonella* and *Escherichia coli* for CDC PulseNet.

PROTECT

PROTECT is an Argonne National Laboratory, Sandia National Laboratories (SNL), and LLNL program for protection of key facilities, focused primarily on chemical attacks. LLNL is supporting PROTECT through both NARAC and biodetection, with NARAC performing modeling and predictions for outside-facility and scenario studies, including operational support of deployed systems.

OPCW Laboratory Designation

The LLNL was selected by the U.S. Department of State as the second U.S.-designated laboratory for the Office for the Prohibition of Chemical Weapons (OPCW). The first U.S. OPCW-certified laboratory is at the U.S. Army Edgewood Chemical Biological Center (ECBC) in Edgewood, Maryland. The LLNL's OPCW designation was received in 2003, and at present the laboratory has implemented all the technical, safety, and procedural systems required and has been ISO-17025 approved. The ECBC group provided substantial assistance to LLNL in this process.

Restoration of Operations

Restoration of Operations is demonstrating strategies for decontamination and restoration of operations for major transportation facilities (Imbro et al., 2003). The activity builds on LLNL restoration planning for chemical warfare agent attacks on transportation facilities, decontamination reagents including L-Gel, sampling strategies to support decontamination, high-throughput sample processing, accelerated viability testing, and published studies to establish "How clean is clean enough?" LLNL and SNL jointly execute the project with a focus on airports and establishing templates for restoration strategies.

Model Cities

Model Cities is an LLNL, SNL, and LANL project to better understand geographic and other local or regional factors that influence prevention, preparation, and response. The goal is to create common "templates" that can be applied in many locations and in combination to create more comprehensive regional plans. The initial test beds for the concept have been the other demonstration projects (e.g., BASIS and LINC) and a tri-lab demonstration in Albuquerque, New Mexico, in December 2002. Integration and evaluation of multiple detection schemes for wide-area, special-facility, and epidemiological surveillance were accomplished.

SCIENCE AND TECHNOLOGY AREAS

Demonstration programs help focus LLNL's science and technology (S&T) investments. The S&T is managed in four areas: (1) decontamination and restoration, (2) modeling and prediction, (3) instrumentation, and (4) applied science. Some current S&T activities include the following:

Decontamination and Restoration

The goal of this focus area is to provide the S&T and systems approach to quickly restore civilian facilities to operation (Raber et al., 2002). Decontamination in a civilian setting requires fundamentally different technology from that for most military applications. Rapid and effective means of decontamination are needed for equipment, facilities, and large urban areas. LLNL developed and is licensing the L-Gel decontamination technology. Several new decontamination chemistries are in development and testing, including vaporous hydrogen peroxide. In addition, appropriate protocols for efficient restoration are being investigated (Raber et al., 2001).

Modeling and Prediction

The goal of this focus area is to develop predictive urban-environment modeling tools for local and other users for response, planning, and vulnerability assessments. Advances in computing algorithms and hardware now make it possible to model airflows over very complex terrain. LLNL is developing tools for modeling such flows in urban environments, including around buildings and in subways, to permit real-time prediction of agent dispersal during an actual event and to determine in advance how to best respond.

Instrumentation for Biological Detection

The goal of this focus area is to provide highly sensitive and accurate instruments for early warning, treatment triage, and detection of contaminated areas. LLNL is developing and integrating instruments with substantial increases in detection performance (McBride et al., 2003). LLNL technology for rapid DNA detection was licensed to industry and enabled a successful product that has been applied to counterterrorism. LLNL has developed prototypes for a new instrument that is capable of 100 simultaneous assays; detecting viruses, toxins, spores, and vegetative bacteria; and autonomous operation for several days at a time. Prototypes of this instrument have been tested with aerosolized live agents and were deployed in a limited capacity in 2002. Several next-generation instruments that are part of LLNL's R&D portfolio address major current shortcomings, including cost per assay, operational complexity, and real-time response.

Applied Science

The goal of this focus area is to provide biological and chemical support for detection and other countermeasures. The availability of DNA and RNA sequence information has enabled rapid development of biological signatures that are highly specific and sensitive (Fitch et al., 2002b). Techniques are also being developed for ligand signature discovery as well as subspecies-level bioforensics. LLNL has invented several high-throughput approaches for vetting signatures (both nucleic acid and ligand) including computational screening of potential signatures, automated strain-panel testing, complex environmental sample testing and, more recently, pathogen-associated function and host-associated response using genomic and proteomic tools (pathomics).

SUMMARY

The LLNL program is developing new systems to improve the ability to prepare for, detect, and respond to a terrorist event. Improvements come from a blend of new technologies and operational concepts that are vetted in experiments in the field with public health, law enforcement, and other collaborators. Several of the LLNL systems are already operational and deployed, specifically NARAC, BASIS, and the FSC. The next technology development focuses on cost reduction and shortening response times. As LLNL looks to the future, its most promising systems are dual-use in that they are valuable in the absence of terrorist events and invaluable during one.

Acknowledgments

This work was performed under the auspices of the U.S. Department of Energy by Lawrence Livermore National Laboratory under Contract W-7405-ENG-48. The work would not have been possible without the interest and support of our sponsors at DHS (J. Vitko, G. Parker, E. George, et al.), DARPA (A. Alving, R. Gibbs, S. Buchsbaum, et al.), and other organizations. We have numerous collaborations at other laboratories, especially the Sandia and Los Alamos national laboratories, and with industry, academia, and other government agencies, especially the CDC and DoD. Our team at LLNL is both talented and dedicated. We are organized by discipline with P. McCready, A. Burnham, B. Colston, and D. Imbro leading the biological, chemical, instrumentation, and systems and deployments areas, respectively. LLNL discipline organizations provide their best talent in computations, biology, physics, engineering, chemistry and environmental sciences and we acknowledge their generous support. V. Hambrick provides outstanding coordination and administrative support for the program, and we would be lost without her.

REFERENCES

Fitch, J.P., Chromy, B.A., Forde, C.E., Garcia, E., Gardner, S.N., Gu, P., Kuczmarksi, T.A., Melius, C., McCutchen-Maloney, S.L., Milanovich, F.M., Motin, V.L., Ott, L.L., Quong, A., Quong, J., Rocco, J.M., Slezak, T.R., Sokhansanj, B.A., Vitalis, E.A., Zemla, A.T., and McCready, P.M., Biosignatures of pathogen and host, *Proceedings of the IEEE Workshop on Genomic Signal Processing and Statistics (GENSIPS)*, Raleigh, NC, October 12–13, 2002a.

Fitch, J.P., Gardner, S.N., Kuczmarski, T.A., Kurtz, S., Myers, R., Ott, L.L., Slezak, T.R., Vitalis, E.A., Zemla, A.T., and McCready, P.M., *Proc. IEEE*, **90**(11), 1708–1721 (2002b).

Imbro, D.R. et al., Paper presented at the *Chemical and Biological National Security Program Summer Meeting*, Washington, DC, June 3–5, 2003.

McBride, M.T., Gammon, S., Pitesky, M., O'Brien, T.W., Smith, T., Aldrich, J., Langlois, R.G., Colston, B., and Venkateeswaran, K.S., *Anal. Chem.*, **75**(8), 1924–1930 (2003).

Olson, K.B., *Emerg. Infect. Dis.*, **5**(4), 513–516 (1999), [online www.cdc.gov/ncidod/EID/vol5no4/pdf/olson.pdf].

Raber, E., Hirabayashi, J.M., Mancieri, S.P., Jin, A.L., Folks, K.J., Carlsen, T.M., and Estacio, P., *Risk Anal.*, **22**(2), 195–202 (2002).

Raber, E., Jin, A., Noonan, K., McGuire, R., and Kirvel, R., *Int. J. Env. Health Res.*, **11**(2), 128–148 (2001).

WEB RESOURCES

Lawrence Livermore Chemical and Biological Countermeasures website, http://www.llnl.gov/hso/chembioprog.html.

Lawrence Livermore Factsheet on the *BASIS* program, http://greengenes.llnl.gov/bbrp/html/mccreadyabst.html.

LLNL Nonproliferation, Arms Control, and International Security website, http://www.llnl.gov/nai/who.html.

LIBERATION TIGERS OF TAMIL EELAM (TAMIL TIGERS)

Praveen Abhayaratne

OVERVIEW

The Liberation Tigers of Tamil Eelam (LTTE) is widely recognized as one of the world's most ferocious terrorist organizations. The group is especially notorious for its ability to seemingly strike targets at will and its advanced military and organizational adaptability. The LTTE has functioned as a guerrilla force, terrorist organization, conventional army, political body, and governing authority. It has also conducted the highest number of suicide attacks in the world, and has claimed the lives of two heads of state and countless others as well as fought and survived battles with both the Indian army from 1987 to 1989 and the Sri Lankan army since 1983. The LTTE portrays itself as the sole representative of the Tamil people in their struggle for Independence from the Sinhalese-dominated government in Sri Lanka. Its fierce commitment is apparent in its cadres, all of whom carry a cyanide capsule to consume in the event that they are captured.

Ideology

The LTTE is a nationalist and separatist group that is striving to establish an independent homeland for the Tamil ethnic minority in Sri Lanka.

Headquarters and Areas of Operation

The LTTE's present headquarters is located in Killinochchi, in the northeastern region of Sri Lanka. Prior to 1987, the LTTE headquarters were located in, and its main operations were launched from, the jungles of northeastern Sri Lanka and from Tamil Nadu, India (Swamy, 2002).

Strength

Approximately 10,000 to 15,000 LTTE "armed combatants" are located within Sri Lanka, Additionally, an unknown number of overseas sympathizers participate in fund-raising, weapons procurement, and propaganda activities (Swamy, 2002).

Group Ties

The LTTE is supported by a number of front organizations that conduct propaganda and handle the financial activities for the group, including:

- World Tamil Association (WTA)
- World Tamil Movement (WTM)
- Federation of Associations of Canadian Tamils (FACT)
- The Ellalan Force
- The Sangillan Force

The group also receives support from the Tamil diasporas in Australia, France, Norway, the United Kingdom, Canada, Germany, Switzerland, and the United States.

Current Status

As of June, 2004, the LTTE was engaged in peace negotiations with the Sri Lankan government. Both groups have observed a permanent cease-fire established in February 2002. The LTTE has been accused by the Sri Lankan government and the Sri Lanka Monitoring Mission of committing a number of cease-fire violations, which have strained the peace negotiations. The LTTE is known to have continued its weapons procurement efforts, its recruitment drives, and its training activities during the cease-fire (Davis, 2003).

BACKGROUND

There are contradicting accounts of the formation of the LTTE, but the official website of the organization provides an inaugural date of May 5, 1976, and credits its leader, Veluppillai Prabhakaran, with the group's establishment. The Tamil New Tigers, a splinter groups from an earlier militant Tamil youth movement, preceded the LTTE. Both the Tamil New Tigers and the LTTE were secret organizations with respect to membership, but the groups themselves were well known in the areas in which they operated. By the mid-1980s, the LTTE managed to consolidate itself as the primary Tamil group by overpowering competing Tamil nationalist groups, and by 1986 the LTTE was the dominant Tamil political party (Joshi, 1996). Today, the group proclaims to be the only Tamil organization that is continuing a real struggle for self-determination based on the Tamil people's mandate for a separate state (ElanWeb, 2004).

Under the guise of humanitarian functions, the LTTE's overt organizations support Tamil separatism by lobbying foreign governments and the United Nations. Large Tamil expatriate communities in North America (Canada), Europe (United Kingdom, Germany, Switzerland), and Asia (India, Singapore) are known to channel contributions to the LTTE through such overt organizations, as well as directly through local branches of the organization. The LTTE maintained its political headquarters in London until it was labeled as a terrorist group and subsequently closed in February 2001.

As an insurgent organization, the LTTE is structured as a conventional "liberation army" with a strict political and military hierarchy. Veluppillai Prabhakaran is the organization's supreme commander, and all group members are accountable to him. Prabhakaran heads the central governing committee, which oversees the two-tier military and political structure of the LTTE. Together, these two tiers include an amphibious group (the Sea Tigers), an airborne group (the Air Tigers), an elite fighting wing (the Charles Anthony Brigade), a suicide unit (the Black Tigers), a highly secretive intelligence wing, an international secretariat, and a political office headed by S.Thamilchelvam and Anton Balasingham. As a terrorist organization, the LTTE maintains a cellular structure to prevent infiltration (Raman, 2002). Upon the completion of their training, all LTTE cadres are handed a cyanide capsule, which is to be worn around their neck (Chalk, 1999).

Group Leadership

Veluppillai Prabhakaran was born on November 26, 1954, in Jaffna, to a conservative Tamil family. His father was strict and demanded absolute discipline from his children, a trait that his youngest son Veluppillai inherited. The family moved to Velvettiturai (VVT) in northeastern Sri Lanka, a coastal town that was notorious as a "smugglers paradise." The contacts Prabhakaran made during his time there proved useful later, when LTTE cadres needed to escape to South India or smuggle weapons and supplies. Prabhakaran was influenced by the Indian freedom struggle and its militant leaders, especially Subash Chandra Bose. Prabhakaran prepared himself from a young age for military struggle by routinely practicing his marksmanship, physical training, and self-inflicted torture. He was fascinated by revolvers and always carried one. He initially gained fame as the triggerman in the first major political assassination by the militant Tamils, which targeted the pro-government mayor of Jaffna. Prabhakaran has always been very suspicious of others and careful with his movements, and as a result has not been apprehended as of June 2004 (Swamy, 2002).

SELECTED CHRONOLOGY

The following is a chronological listing of important events in the group's history, with a particular focus on weapons of mass destruction (WMD) (see Monterey WMD-Terrorism Database, 2003; BBC: South Asia, 2003):

- *1983–1987.* At some point within this time period, the LTTE threatens to poison Sri Lanka's main

export crop (tea) with "leaf curl," a crop disease, in Assam, India.
- *18 June 1990.* The LTTE attacks a Sri Lankan Army camp in Eastern Kiran with canisters filled with "poison gas."
- *21 May 1991.* Former Indian premier Rajiv Gandhi is assassinated by LTTE suicide bomber.
- *1 May 1993.* Sri Lankan President Ranasinghe Premadasa is assassinated by an LTTE suicide bomber.
- *September 1994.* LTTE members threaten to poison tea designated for export.
- *23 May 1995.* Sri Lankan troops prepare for a chemical weapons attack when intelligence reports indicate that the LTTE are planning to attack the military's northern bases with toxic gas.
- *19 July 1995.* The LTTE allegedly uses poison gas in an attack on a police station of the Special Task Force in Tikkodi, Batticaloa district, killing one and injuring three policemen.
- *31 January 1996.* LTTE suicide bombers rammed a truck loaded with explosives into the Central Bank in Colombo, killing 91 people and injuring 1400.
- *21 December 1996.* LTTE members lace stamps destined to be used in Sri Lankan army camps with cyanide.
- *15 October 1997.* A truck packed with a large quantity of explosives detonated close to the World Trade Centre building in Colombo killing 18.
- *25 January 1998.* The LTTE staged a suicide bombing, devastating Sri Lanka's holiest Buddhist shrine, the Temple of the Tooth, killing 13 people.
- *18 December 1999.* At least 15 people are killed in two bomb blasts at election rallies in Colombo; President Chandrika Kumaratunga receives minor injuries in one of the blasts.
- *24 July 2001.* The LTTE stage a devastating suicide attack against the main air base and the only international airport in Sri Lanka, killing 12 people and destroying 13 aircraft.

ASSESSMENT OF WMD CAPABILITY

Agent Selection

In the past, the group has been associated with the use or potential use of arsenic, cyanide, "poison gas," yellow-fever virus, *Schistosoma* worms, and leaf curl.

Delivery

The following possible delivery methods have been suggested: agent-filled canisters, food, nails, knives, and stamps.

Summary

The LTTE has threatened to use varied chemical and biological agents, but to date has only used "poison gas" and cyanide (Monterey WMD-Terrorism Database, 2003). The group is renowned for its expertise in manufacturing land-mines, which have proven to be a devastating weapon against both Sri Lankan government forces and the Indian army. The LTTE has been given explosives and training in bomb making by the Indian intelligence agency, the Research and Analysis Wing (RAW). A few of its senior members have received training from the Popular Front for the Liberation of Palestine (PFLP) in Lebanon (Goshal, 2001). Its bombings of government infrastructure and military camps have demonstrated a sophisticated understanding of explosives technology.

There are reports indicating that the group attempted to equip land-mines with cyanide. Other known uses of WMD have been in unsophisticated forms, namely, infecting export crops and poisoning water supplies with (mostly unspecified) biological agents. The group has allegedly used "poison gas" in an attack on a police station, and has threatened to use unknown chemical agents against the security forces (Monterey WMD-Terrorism Database, 2003).

ASSESSMENT OF WMD INTENT

The LTTE is considered very capable of weaponizing low-end chemical and biological warfare agents given both the training, education, and background of its cadres and supporters and its extensive overseas arms procurement resources. The reported incidents involving these agents indicate that the LTTE resorted to such means out of necessity and after careful calculation. Their psychological impact on enemy forces, the lack of conventional explosives, the need to get more "bang for the buck," and the desire to sabotage the Sri Lankan economy, all seem to have been motivating factors for using and threatening to use chemical and biological weapons. However, taking into account the extent to which the group relies on the support of the international community for its cause, as well as its purported role as the representatives of the Tamil people, the LTTE could conceivably lose support and legitimacy if it resorted to the use of such weapons in the future. Presently, the LTTE focuses on proving its capability as an administrative governing organization for the purpose of securing both development funds and international recognition as a representative of the Tamil people. Given the organization's successful military operations and political flexibility, however, it is conceivable that it could acquire a biological weapons capability if it so desired.

CONCLUSION

Although there are a handful of documented cases of the LTTE's usage and threats to use biological and chemical weapons, the group's current form and aspirations as a legitimate governing authority make it unlikely that it will elect to pursue such weapons in the future. The principal existing threat posed by LTTE violence stems from the actions of its elite Black Tigers suicide squad, which has repeatedly demonstrated its ability to strike vulnerable targets at will (Perera, 2003).

REFERENCES

BBC: South Asia, *A Chronology of Key Events*, 2003, http://news.bbc.co.uk/1/hi/world/south_asia/1166237.stm.

Chalk, P., Liberation Tigers of Tamil Eelam's (LTTE) International Organization and Operations—A Preliminary Analysis, *Canadian Security Intelligence Publication*, Commentary No.77, 1999.

Davis, A., *Jane's Intelligence Review*, **15**(4) April 2003.

ElamWeb, A Struggle for Justice, http://www.eelamweb.com/publication/justice/, accessed on 11/21/04.

Goshal, B., *Asia Defense Journal*, December 2001.

Joshi, M., *Studies in Conflict and Terrorism*, **19**, 19–42 (1996).

Monterey WMD-Terrorism Database, Center for Nonproliferation Studies, 2003, http://cns.miis.edu.

Perera, A., *Inter Press Service*, July 16, 2003, http://www.ipsnews.net.

Raman, B., *The LTTE: The Metamorphis*, South Asia Analysis Group, Paper No. 448, 2002.

Swamy, M.R.N., *Tigers of Lanka: From Boys to Guerrillas*, 3rd ed., Konark Publishers, Delhi, 2002.

FURTHER READING

Gunaratna, R., LTTE Child Combatants, *Janes Intelligence Review*, 1998.

Eelam website: http://www.eelamweb.com/.

LIBYA

Michelle Karch

INTRODUCTION

After leading the Revolutionary Command Council that in 1969 overthrew the ruling monarchy, Mu'ammar Al-Qadhdhfi established a dictatorship in Libya that embraced a foreign policy of hostility toward the West and Israel, expansionist tendencies, and Nasser's Arab unity (Fig. 1). Soon afterward, Qadhdhfi turned to unconventional terrorist groups for assistance, due in part to an ineffective conventional force and a much needed defense against Islamist militants and dissident groups opposed to Qadhdhfi's regime. Qadhdhfi's extremist attitudes and inflammatory statements, such as "national liberation can only be achieved through armed struggle" (Jane's Sentinel Security Assessment, 2002), ultimately caused the relationship between Libya and terrorism to become embedded in the Western mindset. Libya is currently one of seven state sponsors of terrorism as designated by the U.S. Department of State, and is suspected of having an offensive biological warfare (BW) program in some capacity. In late December 2003, it was revealed that ongoing negotiations between Libya and American and British officials had resulted in complete disclosure of the state's unconventional weapons capabilities, which included dual-use biological research, development, and production facilities, but no biological weapons per se (Slevin and Frankel, 2003).

LIBYA AND TERRORISM

Following a military loss to Chad in 1978, Qadhdhfi turned to terrorist and revolutionary organizations in order to support his expansionist motives and quell internal conflict, an action that has since extended to organizations as varied as the Red Brigades in Italy, the Irish Republican Army, IRA, in England, and the Sandinistas in Nicaragua. Though Qadhdhfi debates ever having supported terrorism per—"I supported liberation—not terrorist movements" (Leymouth and Wally, 2003)—Libya has provided organizations like those above with training facilities, weapons, oil, asylum, and financial assistance.

High-profile terrorist acts that have been linked to Libya include:

- *1984 London.* A British cop is killed and 11 persons wounded when members of the Libyan People's Bureau fire on anti-Qadhdhfi demonstrators.
- *1985 Rome and Vienna.* Attacks are carried out against individuals in line at the airport counters of Israeli airlines, with 20 killed. Palestinian terrorist group Abu Nidal Organization (also known as Fatah Revolutionary Council), which has strong ties to Libya, is held responsible.
- *1986 West Berlin.* A nightclub is bombed, killing three.
- *1988 Lockerbie.* Pan Am Flight 103 is bombed, killing 270.
- *1989 Niger.* French airline UTA Flight 772 is bombed, killing 171.

Both the United Nations and individual countries have imposed sanctions on Libya for such affiliations, including initial sanctions in 1973; Libya's placement on the U.S. State Department's terrorism list in 1979; sanctions levied by the Reagan administration in 1986 in response to the airport attacks in Vienna and Rome Sanctions, an increase in 1992 in response to the Lockerbie bombing; and the Iran-Libya Sanctions Act (ILSA), passed by U.S. Congress in 1996, which imposed sanctions against any foreign investor or company that invested more than $40 million into Libya's oil sector (ILSA was extended for an additional five years on August 3, 2001). In addition, President Reagan ordered air strikes on Libya following the West Berlin incident of 1986.

Qadhdhfi's employment of more traditional means to influencing policy in the region, rather than influence through the support of terrorism, is pertinent to this discussion as well. Libya has regularly offered neighboring countries the conditional support of conventional forces as a means of influencing policy in the region. As he began to come out against terrorism in the late 1990s, however, Qadhdhfi shifted his regional alliances from the Middle East toward African countries: "Now we are Africa. We are no longer Libya" (Blanche, 2003). Later, in 2002, Libya withdrew from the Arab League because of its handling of the Iraqi and Palestinian crises.

In recent years, Qadhdhfi has made several attempts to reconcile with Western countries by disassociating Libya's ties with terrorist organizations. In 1999, Libya directly assisted in negotiating with the Pilipino terrorist group Abu Sayyaf to release hostages, and in the same year provided a total of $33 million in compensation to family members of the victims of the UTA Flight 772 bombing. In early 2003, a Libyan diplomat became President of the United Nations Commission on Human Rights (Blanche, 2003), and on August 15, 2003, Libya admitted responsibility for the 1988 Pan Am Flight 103 bombing, offering $2.7 billion dollars in compensation to surviving family members. The United

Encyclopedia of Bioterrorism Defense,   Edited by Richard F. Pilch and Raymond A. Zilinskas
ISBN 0-471-46717-0   Copyright © 2005 Wiley-Liss

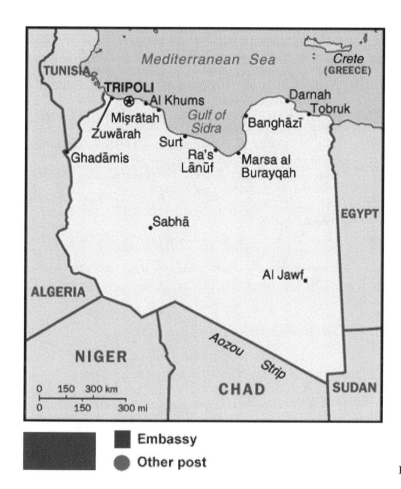

Embassy

Other post

Figure 1.

Nations responded to this action by suspending sanctions against Libya, and the United Nations Security Council (UNSC) will soon consider a resolution to lift sanctions completely despite the fact that the United States has indicated that its sanctions on Libya will remain in place (Schweid, 2003).

After September 11, Libya assisted in the war on terrorism by providing valuable intelligence to the United States, and Qaddafi has further called on Muslim organizations to provide aid to the United States, referring to the attacks as "horrifying" (www.terrrorismanswers.com/sponsors/libya.html, accessed 8/13/03). The United States has acknowledged Libya's assistance by placing the opposing Libyan Islamic Fighting Group, which has known ties with al-Qa'ida, on its official list of terrorist organizations. It is unclear whether these efforts are genuine offerings on the part of Libya; indeed, its cooperation may be due at least in part to the financial strain on Libya's economy as a result of decreased oil revenues in the late 1990s. It is theorized that Qadhdhfi may simply be reaching out in return for much needed foreign investment, assistance, and trade, and much skepticism has been expressed regarding the extent to which Libya's terrorist links have in fact been severed. Regardless, Libya remains party to all 12 international terrorism-related conventions and protocols.

LIBYA AND BIOLOGICAL WEAPONS

It is unclear when Libya began its pursuit of an offensive biological weapons capability, but the state is believed to presently be in the early stages of research and development on this front. Progress is thought to be extremely limited because of the lack of an indigenous biological scientific and technical capability and the presence of international sanctions that have served to hamper Libya's ability to receive foreign technical assistance and equipment. Which biological agents Libya is interested in as to a weapons capacity is unclear.

Libya signed the Geneva Protocol in 1971 and the Biological and Toxins Weapons Convention (BWC) in 1982, but subsequently did not submit data declarations in accordance with the United Nations for confidence-building measures (www.stimson.org/cbw, accessed 8/13/03). While Qadhdhfi adamantly denies the presence of an offensive BW program in Libya, it has been suggested that he in fact has sought scientific expertise and equipment in the realm of chemical and biological weapons from South Africa in 1995 and may have worked with Romania, Iraq, and other countries on cooperative BW efforts (www.cisis-scrs.gc.ca/eng/msicdocs/200005_e.html, accessed 8/13/03; www.cbw.sipri.se/cbw, accessed 8/13/03). Some analysts have further argued that Rabta, a dedicated chemical warfare (CW) facility in Libya, may also house an offensive biological research effort (Sinai, 1997).

In the final analysis, while Libya's efforts in the biological realm are largely unknown, Qadhdhfi's vow of transparency holds some promise that more will be learned in 2004.

REFERENCES

Biological Weapons Proliferation Concerns, Stimson Center, www.stimson.org/cbw, accessed on 8/13/03.

Blanche, E., Ghadaffi's Uncertain African Empire, *Jane's Terrorism & Security Monitor*, February 1, 2003.

Does Libya Sponsor Terrorism, Council on Foreign Relations, www.terrrorismanswers.com/sponsors/libya.html, accessed on 8/13/03.

Jane's Sentinel Security Assessment, Libya: External Affairs, June 26, 2002.

Leymouth, W., The Former Face of Evil, *Newsweek*, January 20, 2003.

Libya, the West and terrorism: a special JID report, *Jane's Intelligence Digest*, November 23, 2001.

Mark, C., *Libya*, CRS Issue Brief for Congress, December 21, 2001, Congressional Research Service, Library of Congress.

Perspectives—Biological Weapons Proliferation, www.cisis-scrs.gc.ca/eng/msicdocs/200005_e.html, accessed on 8/13/03.

Schweid, B., U.S. to Maintain Sanctions on Libya, *Philadelphia Inquirer*, August 17, 2003.

Sinai, J., Libya's Pursuit of Weapons of Mass Destruction, *Nonproliferation Review*, Spring-Summer 1997.

Slevin, P. and Frankel, G., Libya vows to give up banned weapons, *Washington Post*, December 20, 2003.

WEB RESOURCES

Libya: Biological Weapons Capability, www.cbw.sipri.se/cbw, accessed on 8/13/03.

Overview of State-Sponsored Terrorism, U.S. Department of State, www.state.gov, accessed on 8/13/03.

LOS ALAMOS NATIONAL LABORATORY

GARY RESNICK

BACKGROUND

The Los Alamos National Laboratory (LANL), from its origins as a secret Manhattan Project laboratory, has been providing science-based solutions to complex national security challenges for over half a century (Fig. 1). The laboratory has been managed since its establishment by the University of California and is currently part of the Department of Energy (DOE), National Nuclear Security Agency (NNSA). Its cutting edge multidisciplinary scientific and engineering research and development initiatives have resulted in advances in technology and knowledge ranging from the development of the first nuclear weapons to sequencing the human genome. Early research by the LANL into the biological hazards associated with production of nuclear materials stimulated the development of a robust biological sciences infrastructure. This infrastructure, along with supporting capabilities (e.g., computational and material sciences), began to focus on the biological threat facing the nation in the early 1980s. Current research and development activities are performed on behalf of the Department of Homeland Security (DHS) and collaborating federal, state, and local entities in the fight against bioterrorism.

Figure 1. Los Alamos National Laboratory.

RESPONSIBILITIES AND EFFORTS IN BIOLOGICAL THREAT REDUCTION

Begun in 1997, the NNSA Chemical and Biological National Security Program (CBNP) created a long-term, programmatic commitment to the development of countermeasures to protect the United States from biological threats. The CBNP program transitioned to the DHS in 2003 and became the nucleus of a comprehensive biological threat countermeasures program.

Because no one policy, operational, or technical approach holds the promise of being 100 percent efficacious in meeting the challenges of biological threats (both natural and deliberate), LANL pursues a full spectrum threat intervention strategy that involves the support of multiple federal agencies, including the NNSA, Department of Defense (DoD), Federal Bureau of Investigation (FBI), United States Department of Agriculture (USDA), and Environmental Protection Agency (EPA), in performing their respective missions in biological threat nonproliferation (e.g., enforcement and strengthening of the Biological and Toxin Weapons Convention, or BWC), counterproliferation, and counterterrorism research, development, and operational deployment.

Activities are focused on development of technology and knowledge in the areas of systems analysis, modeling and simulation, biological threat forensics and attribution, signatures and assays, biodetection, response and restoration, and operational demonstrations. The approach is science-based with strong emphasis on the enabling disciplines of predictive and material (nanotechnology) sciences. The LANL strategy leverages a broad and robust basic research program.

BASIC RESEARCH DRIVING INNOVATION

LANL maintains basic research capabilities in such diverse disciplines as genomics, proteomics, environmental microbiology, pathogen characterization, host/pathogen interactions, chemistry, physics, material science, computational science, and modeling and simulation, all of which have relevance to bioterrorism defense.

Encyclopedia of Bioterrorism Defense, Edited by Richard F. Pilch and Raymond A. Zilinskas
ISBN 0-471-46717-0   Copyright © 2005 Wiley-Liss

Biological Foundations

LANL's biological foundations area research strengths include cellular biology, molecular synthesis, structural biology, systems biology, protein purification and crystallography, immune response, DNA forensics, genomics, environmental microbiology, measurement science and diagnostics, and biophysics. Distinctive capabilities include the Structural Genomics Consortium (SGC), National Flow Cytometry and Sorting Research Resource (NFCSRR), the Center for Genome Studies (CGS) and Integrated Spectroscopy Resource (ISR).

- The *Mycobacterium tuberculosis* Structural Genomics Consortium focuses on high throughput protein production and structural determination.

- The National Flow Cytometry and Sorting Research Resource focuses on cytometry technologies. Capabilities include DNA fingerprinting, pathogen point detection, analysis of single nucleotide polymorphisms (SNP), DNA fragment sizing, and microsphere-based analysis of molecular assemblies.

- The Center for Genome Studies is a medium-capacity genome sequencing facility with the ability to produce nearly 10 million bases of raw sequence per week. Additional capabilities include cloning, sequencing, and data management issues associated with both mammalian and bacterial organisms.

- The Integrated Spectroscopy Resource is a 6000-sq ft laboratory and 2000-sq ft annex dedicated to protein structure and dynamics, hyperspectral imaging, and photodynamics.

Facilities are available for working under biosafety level 2 (BSL-2) containment with select agents in a secure environment. A new BSL-3 facility, near completion at the time of this writing, will enhance current research and strain archive capabilities. The facility is 3200 sq ft, of which approximately 624 sq ft are designed for work at the BSL-3 level.

Microbial sequencing and data analysis are closely linked to DOE missions in national security, energy, and environment. DOE founded the Joint Genome Institute (JGI) with participating members from three national laboratories—the LANL, Lawrence Berkeley National Laboratory (LBNL), and Lawrence Livermore National Laboratory (LLNL)—and established a production genomics facility in Walnut Creek, California. The JGI provides very high throughput, low-cost draft sequencing to its member laboratories in support of national needs. The LANL has developed a specialized role in the JGI that focuses on microbial finishing. Draft sequencing of a microbial genome at the JGI's production genomics facility results in about 90 to 95 percent genome coverage from random sequencing reads. The finishing at the LANL brings the sequence coverage up to 100 percent at a high level of accuracy (less than 1 in 10,000 base pair error rate). Recently, the LANL has finished several microbes that are closely related to *Bacillus anthracis*. The sequence data for biological threat agents and their near neighbors is fed into LANL's sequence analysis pipeline to develop unique sequence signatures for each agent, which can then be used for detection and bioforensics applications. Other genome targets include *Yersinia pestis*, *Francisella tularensis*, *Burkholderia mallei*, *Brucella* sp., and *Coxiella burnetii*.

A suite of proteomic capabilities at LANL support biodefense research. Pathogen gene chips produced in the LANL's microarray facility help to discover pathogen virulence factors and to elucidate host-pathogen interactions. A pathogen proteome can also be separated on two-dimensional gels and each resolved protein identified by mass spectrometry. In addition, the LANL is a world leader in phage display technology, producing libraries of single chain antibodies that can rapidly be selected for binding to desired protein ligands. Protein ligands identified by this approach can be used for detection as well as for basic research with the pathogen.

Environmental Background and Interaction

Environmental background and interaction capabilities are fundamental in understanding natural variability, which in turn is essential in monitoring and simulating aberrations or events. Additionally, the response to a threat agent may depend on natural forces such as air currents, soil conditions, ocean currents, or combinations thereof. For these reasons, complex systems in environmental background and interaction, from the cellular interaction level up to national transport model level, are directly relevant in CBTR (Chemical and Biological Threat Reduction). The LANL capabilities include geochemistry, geodynamics, surface and subsurface flow and transport, atmospheric and oceanic simulation, atmospheric instrumentation, characterization and analysis of plume detection, spatial analysis using GIS (geographic information system), and modeling of wildfires (Fig. 2).

Environmental background research capabilities overlap and complement other area capabilities. Discerning between naturally occurring and deliberately released bacteria, viruses, and toxins draws on biofoundation capabilities. Variations in the environmental background affect population dynamics, for example, the impact of a chemical or biological agent on the resident human population. In addition, chemical technologies capabilities are used to characterize compositions of soil, atmosphere, and oceans (see the below discussion of epidemiology and modeling and simulation for more information on environmental background models).

Figure 2. Sample plume analysis.

Environmental and agriculture resources are also at risk, making environmental background and modeling and simulation capabilities important in reducing anticrop, antianimal, and ecologic threats.

Epidemiology, Population Dynamics, and Medical Surveillance

The LANL's capabilities in epidemiology include medical surveillance, population statistics, medical data management, and modeling. Additional capabilities overlap and complement response, training, and deployment in treatment and management methodology capabilities.

Chemical Technologies

Chemical technologies capabilities include research in explosives, chemical destruction technologies, chemistry and phase behavior of energetic materials, and combustion chemistry. The chemical technologies capabilities overlap and complement sensor technology area capabilities in laser-induced breakdown spectroscopy, catalysis and separation science, surface science spectroscopy, and X-ray diffraction. Additional sensor technology complementary capabilities include direct solids analysis, automated volatile and semivolatile analysis, mechanical fabrication, chemical analysis automation, safe gas-sampling from high-pressure environments, headspace gas-sampling and analysis, physical chemistry, and applied spectroscopy (which further includes optical diagnostics of interfaces, novel electronic and optical materials, ultrafast time-resolved spectroscopy, and integrated ferroelectric devices).

The National Stable Isotope Resource (SIR) is a unique LANL resource that synthesizes compounds labeled with the stable isotopes ^{13}C, ^{15}N, ^{17,18}O, ^{33,34}S, and ^{77}Se, and distributes labeled compounds that are not available from commercial sources to accredited investigators for biomedical applications. Labeled materials are currently being used as analysis standards by the Centers for Disease Control and Prevention (CDC) Laboratory Response Network laboratories for analysis of chemical agent breakdown products.

Applied Physics Technologies

Applied physics technology capabilities include research in biophysics, sensors, imaging, spectroscopy, lasers, plasmas, advanced light sources, shock wave modeling, and advanced oxidation/reduction technologies for chemical and biological agent destruction. In terms of a biophysics focus, applied physics technology aims to investigate the relationships between structure, dynamics, and function of biological phenomena over a wide range of scales (from biomolecules to whole organisms); achieve a detailed interplay between high-resolution physical measurement and large-scale computational modeling and analysis of complex systems; and make extensive use of detection, imaging, and reconstruction techniques (X-ray crystallography, single-molecule electrophoresis, high-speed photon-counting optical imaging, and magnetic resonance imaging).

Single-Molecule Detection for Biochemical Analysis. The basis of this approach is to monitor a DNA or RNA sequence. These techniques have been developed with the desired properties of high sensitivity, specificity, homogeneous assay format, and high throughput.

Improved Microscopy for Bioanalysis. A novel approach for confocal microscopy (microscopy employing sequential illumination to reduce out-of-focus blur) has been developed that combines available illumination, detection, and data processing technologies to produce an imager with a number of advantages: reduced cost, faster imaging, improved efficiency and sensitivity, improved reliability, and much greater flexibility.

Low Temperature, Atmospheric-Pressure Plasma Jet. The atmospheric-pressure plasma jet (APPJ) provides a high flux gas stream of chemically active species (such as atomic oxygen or hydrogen) diluted in a flow of helium, which reacts with biological and/or chemical species present on a surface to either kill pathogens or convert chemical agents to less toxic or nontoxic reaction products. APPJ technology, a winner of the 1999 R&D100 Award, has been licensed to outside industry and can be commercially produced in large quantities for sensitive equipment decontamination.

Indoor, Pulsed Ultraviolet Fluorescence Imaging Detection, and Sterilization of Biological Agents. Enzymes that are present in pathogens will emit characteristic fluorescence emission when excited at certain ultraviolet (UV) wavelengths. Using a band-selective camera to analyze and image the fluorescence emission that follows multiple single-wavelength UV pulses, the identity, location, and viability of contamination can be mapped inside a facility. After mapping the contamination and identifying its location, pulsed UV emission may be used for sterilization and decontamination.

Materials Science

Materials development will lead the way to new methods of detecting, analyzing, responding to, and decontaminating chemical and biological threat agents. The LANL hosts capabilities that enable progress in the development of nanotechnologies, decontamination technologies, sensors, and fuel cells. Research capabilities include composite brittle materials, compatibility and polymers development, and nano-bio-micro-interfaces.

Materials science overlaps and complements chemical technologies at LANL, as well as spectroscopy efforts included in biofoundations and sensors technology.

- The Center for Integrated Nanotechnologies (CINT) is an interlaboratory center furthering research, collaboration, and capability access in nanoscale materials and structures. Specific LANL capabilities supporting CINT include the Los Alamos Neutron Science Center (LANSCE), National High Magnetic Field Laboratory (NHMFL), Electron Microscopy

Laboratory (EML), Ion Beam Materials Laboratory (IBML), Scanning Probe Microscopy Laboratory (SPML), and Quantum Research Institute at Los Alamos.

- LANSCE will have the world's most intense source of polarized cold neutrons, which, along with neutron reflectometry and other neutron spectroscopies, will be essential for the study of complexity in nanomaterials.
- The National Science Foundation (NSF) pulsed field facility represents a unique tunable nanoscale probe important for exploring nanostructured semiconductors, quantum systems, and complex materials.

Predictive Science

Convergence of advances in such disciplines as high-speed computing, modeling of complex systems, and high-throughput laboratory analysis holds the promise of generating new means for the performance of biological countermeasures research and operational activities, from acceleration of innovation through in silico biological research to modeling entire urban response systems. A future element of threat reduction will be the successful transformation of raw data to "actionable knowledge," a valuable resource used in decision making, and the LANL's computational abilities, which include high-performance computing, complex systems, Knowledge Discovery in Data Mining (KDD), and information management (databases), contribute greatly to this process.

Features illustrating LANL's computational science capabilities include the following:

- The laboratory recently completed construction of its Q computer, a 30 teraOp supercomputer, which at 30 trillion calculations per second is one of the world's two largest computers. Q is part of the Nicholas C. Metropolis Center for Modeling and Simulation.
- A LANL research team was recently highlighted through a 2002 R&D100 Award for GENetic Information Extraction tool (GENIE), an evolutionary computation software system for automatic feature extraction in digital imagery.
- KDD, or "information analysis and modeling" or "machine learning," is the nontrivial extraction of implicit, previously unknown, and potentially useful information from data. LANL's capabilities include a secure and collaborative information system, processing and analysis, high-level object recognition, data fusion, software systems architectures, and use-driven categorization for online reference systems.
- LANL has applied research in the formation, rapid generation, and maintenance of several comprehensive data warehouses. LANL was the birthplace of GenBank, the nucleic acid database that has become the cornerstone for bioinformatics and genomics, and maintains several specialized sequence databases dealing with a variety of pathogens, including:
  - STDGen, Sexually Transmitted Disease Database
  - OralGen, Oral Pathogen Sequence Database

- TPGen, Threat Pathogen Database
- ISD, Influenza Sequence Database
- HIV Databases
- Xbase, Global Gene Expression Database
- AFLP Database
- The LANL's Computational Immunology Research Program bridges the management and analysis of biological data with computer-generated modeling and simulation in a system-to-component approach for immune systems, and has been applied successfully to influenza (basis of EpiSims) and HIV.

Modeling and Simulation

The LANL's modeling and simulation research includes complexity science and agent-based and entity-based modeling ranging from biological to sociological model scales in socio-technical systems (Fig. 3). The LANL's modeling and simulation area research overlaps and complements biofoundations and epidemiology in modeling biological systems and networks, the computational analysis of antibody binding, and the construction of optimal treatment models.

Some of the LANL's distinctive capabilities are:

- The NISAC (National Infrastructure Simulation and Analysis Center) is based on a $150 million software investment supported by the world's largest secure scientific computing environment. It provides vulnerability assessment and policy decision-support system capabilities in the areas of infrastructure policy, education, planning, investment, and crisis response.
- Nicholas C. Metropolis Center for Modeling and Simulation was constructed to model and validate complex code in predictive science.
- EpiSims is a highly resolved epidemic simulation at the level of the individual in a large urban region, taking into account individual differences in immune response and realistic contact patterns of the hosts in order to provide estimates of geographic and demographic distribution of disease as a function of time.
- Urban Transport and Dispersion Modeling rapidly and adaptively models the dispersion of an atmospheric release down to building level detail in a

Figure 3. Modeling and simulation is a vital part of LANL's research effort.

regional context. It also includes "rules-of-thumb" for improving decisions of first responders and local decisions makers.

- Advanced Visualization Environment for Simulation is a cutting edge visualization center that includes two theaters with "PowerWalls" for visual immersion in models.

LANL researchers have also developed several plume dispersion models, available at different levels of fidelity, which model complex terrains, such as urban settings complete with a heterogeneous building set. The model is capable of producing turbulence fluctuations that would be produced by wind patterns flowing between buildings (mechanically produced turbulence) as well as those produced by meteorological effects, such as solar heating of the ground (convectively driven turbulence).

Vulnerability Assessment and Policy Tools

When considering the complexities of ensuring homeland security, LANL scientists place paramount importance on discovering systems weaknesses, opportunities, and enhancements and development of decision support tools. The laboratory develops algorithms for statistical inference and decision making, leading to new techniques for data management (including new methods for classification of biological sequences). The LANL's capabilities include systems analysis and integration, engineering design and analysis, safety and risk assessment, probabilistic risk analysis, threat understanding and preemption, and the development and application of decision-making tools.

GIS science and technology at LANL provides advanced capabilities to manage and utilize spatially referenced data (data tied to map locations). Advanced GIS capabilities include map-based visualization, 3D GIS, enterprise GIS design, integration of GIS with agent-based modeling, spatial decision support systems, and custom applications.

Sensors Technology

The LANL is researching and developing numerous sensor technologies, ranging in scale from handheld units that the first responder might use to a city scale "system of systems" (Fig. 4). Standoff detection technologies include Light Detection and Ranging (LIDAR), hyperspectral imaging, distributed sensor networks, and a Remote Ultra Low Light Level Imager (RULLI). Sensors technology development relies on many of the capabilities at the LANL, such as nanotechnologies (materials science) and fuel cell development, which are crucial for implementation of smaller, fieldable instruments for first responders.

Response, Training, and Deployment

The LANL has several unique capabilities in response, training, and deployment to mitigate chemical and biological threats. For example, its Hazardous Materials Training Facility features a hands-on, situational training ground for Los Alamos' Nuclear Emergency Support Team (NEST), a team of knowledgeable volunteers

Figure 4. LANL field-deployable sensor systems for toxin detection.

qualified to deal with a weapons emergency (Fig. 5). A proposed Biological Emergency Support Team (BEST) will be modeled after NEST's organization and response architecture.

Training and response capabilities are integral to the LANL. Not only do they ensure the success of daily lab operations but they also ensure domestic security. They build on LANL's strengths in incidence planning and response capabilities.

RECENT AND EMERGING ADVANCES IN BIOLOGICAL COUNTERMEASURES

Highlights of the LANL's deployed systems and technology in development include the following:

- The Biological Aerosol Sentry and Information System (BASIS) provides early warning of airborne biological incidents for special events such as large assemblies and high-visibility meetings. The BASIS system was developed in collaboration with LLNL. Deployed at the Salt Lake City Olympics, it had the potential to detect a biological incident within a few hours of an attack, early enough to mount an effective medical response. BASIS has been further refined and deployed as the BioWatch urban monitoring system by DHS, the EPA, and the CDC.
- The need for rapid, quantitative, and facile software for use by urban response planners has driven the development of the fast running, easy to use QUIC (Quick Urban Industrial Complex) sensor

Figure 5. LANL's hazardous materials training facility, the only facility certified outside the state of California by the California Specialized Training Institute (CSTI).

placement tool, which is in routine use for tailored deployment of the BioWatch aerosol monitoring system in urban areas.

- The Reagentless Optical BioSensor (ROB) is a ligand-based optical biosensor developed at the LANL in collaboration with the University of New Mexico that can be used to detect toxins and pathogens in support of counterterrorism efforts, as well as point of care centers for naturally occurring infectious disease. The sensor employs several key innovations, such as exchangeable sensor elements for reuse, biomimetic ligand design for generic agent detection, and signal amplification through fluorescence resonance energy transfer.

- The National Infrastructure Simulation and Analysis Center (NISAC) is a joint effort between the LANL and Sandia National Laboratory founded on decades of agent (single individuals) level modeling with advanced mobility parameters (e.g., TRansportation ANalysis and SIMulation System (TRANSIM)). The model has been used for guiding such critical decisions as vaccination regimes to mitigate terrorist initiated epidemics.

- DNA forensic analysis techniques such as Amplified Fragment Length Polymorphism (AFLP) and Multi Locus Sequence Tandem repeats (MLST) have been developed at LANL with collaboration from academic institutions such as Northern Arizona University, and have rapidly transitioned to operational use in support of the criminal justice community.

- Bio-Surveillance Analysis Feedback Evaluation and Response (B-SAFER), developed in collaboration with the University of New Mexico, is a real-time data collection and integration system for the medical community that both improves daily public health support and rapidly identifies bio-incidences, whether synthetic or natural. The system was demonstrated in the Biodefense Test Bed demonstration program supported by the DoD and NNSA in Albuquerque, New Mexico.

- Bioforensics Demonstration and Application Program (BDAP), designed to expedite development of standardized protocols for bioforensic analysis employing available techniques, makes research lab bioforensic capabilities available to law enforcement and public health entities.

- Biological Defense Initiative (BDI) is a full spectrum metropolitan test bed to develop a predict-to-manage capability, combining BASIS and B-SAFER.

ORGANIZATIONAL TIES

The LANL fosters close working relationships with academia, industry, federal, and other national laboratories to accelerate the discovery process and commercialization of technology for rapid deployment to users. As a member of the University of California community, the LANL is actively engaged in cross-campus activities and a free flow of technical personnel. In addition, investigators are strongly encouraged to collaborate with external colleagues and establish commercial outlets for successful technology development efforts.

THE FUTURE

Continuation of the multidisciplinary, science-based approach to the development of new technologies that improve and transform how national security is thought of and achieved is the LANL strategy. While establishing a biocountermeasures baseline capability by engineering today's technology is critical, fundamental scientific advances are required to develop systems that will meet the diverse user needs and operational constraints. An attention to all aspects of the discovery, innovation, and deployment cycle will ensure the continued role of the LANL in addressing the biological threat to the nation.

MARINE ALGAL TOXINS

Mark A. Poli

Opinions, interpretations, conclusions, and recommendations are those of the author and are not necessarily endorsed by the U.S. Army.

INTRODUCTION

Marine biotoxins are a problem of global distribution, estimated to cause more than 60,000 foodborne intoxications annually. In the United States, over 2500 cases of foodborne illness due to marine toxins were reported to the Centers for Disease Control and Prevention (CDC) from 1993 to 1997. In addition to human morbidity, they cause massive fish kills, such as those occurring during the Florida red tides, and have been implicated in mass mortalities of birds and marine mammals. The long-term environmental and public health effects of chronic exposure are poorly understood.

Ingesting seafood containing marine biotoxins can cause six identifiable syndromes: paralytic shellfish poisoning (PSP), neurotoxic shellfish poisoning (NSP), ciguatera fish poisoning (CFP), diarrhetic shellfish poisoning (DSP), amnesic shellfish poisoning (ASP), and azaspiracid poisoning (AZP). With the exception of CFP, which, as the name implies, is caused by eating contaminated finfish, all are caused by the ingestion of shellfish. And, with the exception of ASP, which is of diatom origin, the causative toxins all originate from marine dinoflagellates. These toxin-producing species are only a small minority of the thousands of known species of phytoplankton. However, under the proper environmental conditions, they can proliferate to high cell densities known as blooms. During these blooms, the toxin-producing species are ingested in large quantities by zooplankton, by filter-feeding shellfish, and by grazing or filter-feeding fishes. Through these intermediates, toxins can be vectored to higher trophic levels, including humans.

In general, marine toxins have not been viewed as important biological warfare (BW) threat agents in recent years. Unlike microorganisms or agricultural crops, marine toxins occur naturally at low concentrations in wild resources that may be difficult to amass in large quantities. Most are nonproteinaceous and therefore not amenable to simple cloning and expression in microbial vectors. Although some toxins can be harvested from laboratory cultures of the producing organism, yields from these cultures are insufficient to support their development as BW agents.

Biological terrorism, especially related to the food supply, is a much different scenario. These toxins occur naturally in seafood products in concentrations sufficient to cause incapacitation or even death. The contaminated foodstuffs appear fresh and wholesome and cannot be differentiated from nontoxic material except by chemical analysis. This negates the requirement for isolation of large quantities of pure toxins and subsequent adulteration of food products. In theory, the toxic seafood need only be harvested, and then inserted into the food supply at the desired location downstream of any regulatory testing.

In some cases, harvesting toxic seafood may be difficult. Such is the case with ciguatoxin, where contaminated fish are typically only a small percentage of the total catch. In other cases, harvesting could be trivial. For instance, the United States and other countries have monitoring programs to ensure seafood safety. On the Gulf coast of the United States, the concentrations of toxin-producing dinoflagellate *Gymnodinium breve* in the water column are watched closely. When they rise above a predetermined concentration, suggesting an imminent bloom, harvesting of shellfish is halted. The shellfish are then monitored by chemical analysis or mouse bioassay until toxin concentrations fall to safe levels, at which point harvesting is resumed. During this period, when the shellfish are toxic, the information is made freely available through the news media and regulatory agencies to discourage recreational harvesting. Thus, any interested party can determine when the local shellfish

Encyclopedia of Bioterrorism Defense,   Edited by Richard F. Pilch and Raymond A. Zilinskas
ISBN 0-471-46717-0   Copyright © 2005 Wiley-Liss

are toxic and surreptitious harvesting, although illegal, could easily ensue.

A brief description of each of the six marine toxin syndromes and the causative toxins are presented here. Some should be of a greater concern for homeland security than others. Issues that may impact or limit their potential use as weapons of bioterrorism are also discussed.

SAXITOXIN AND PARALYTIC SHELLFISH POISONING (PSP)

PSP results from ingesting shellfish contaminated with a family of compounds called *saxitoxins*. The name saxitoxin (STX) is derived from the giant butter clam *Saxidoma giganteus*, from which the toxin was first isolated. While STX was the original toxin isolated, the family of PSP toxins is now known to include over 20 derivatives of varying potency. They are associated with species of marine dinoflagellates belonging to the genera *Alexandrium*, *Pyrodinium*, and *Gymnodinium*, as well as several species of freshwater cyanophytes (blue-green algae). In filter-feeding shellfish such as clams, oysters, mussels, and scallops, toxins accumulate after ingestion either of dinoflagellate cells during bloom conditions or of resting cysts in the bottom sediment. Ingestion of toxic shellfish by humans results in the characteristic signs and symptoms of intoxication. These toxic dinoflagellates occur in both tropical and temperate oceans.

STX and its derivatives elicit their physiological effects by interacting with the voltage-dependent sodium channels in excitable cells of heart, muscle, and nerves. High-affinity binding to the sodium channel blocks ion conductance across the cellular membranes, thereby inhibiting depolarization. While voltage-dependent sodium channels in various tissues are susceptible to saxitoxins, pharmacological considerations make the peripheral nervous system the primary target in seafood intoxications.

Ingestion of PSP toxins results in a rapid onset (minutes to hours) complex of paresthesias, including a prickling, burning, or tingling sensation in the lips and mouth that rapidly progresses to the extremities. At low doses, these sensations can disappear in a matter of hours with no sequelae. At higher doses, numbness can spread from the extremities to the trunk, and then be followed by weakness, ataxia, hypertension, loss of coordination, and impaired speech. At lethal doses, respiratory failure results from paralysis of respiratory musculature. Children appear to be more susceptible than adults. The lethal dose for young children may be as low as 25 µg STX equivalents (Rodrigue et al., 1990), while that for adults is probably about 5 to 10 mg STX equivalents. In adults, clinical symptoms probably occur upon ingestion of 1 to 3 mg equivalents. Shellfish can contain up to 10 to 20 mg equivalents per 100 g of meat, so ingestion of only a few shellfish can cause illness (Rodrigue et al., 1990, Gessner et al., 1997). Fortunately, clearance from the blood via the urine is rapid. After a series of outbreaks of PSP on Kodiak Island, Alaska in 1994, serum half-life was estimated to be less than 10 h. Respiratory failure and hypertension resolved in 4 to 10 h in these victims, and toxin was not detectable in urine 20 h after ingestion (Gessner et al., 1997).

Treatment of PSP consists of removing unabsorbed material from the gastrointestinal tract and providing supportive care. In severe cases, mechanical ventilation may be necessary. Neither antidote nor vaccine is currently available.

Saxitoxin is approximately fivefold more toxic by inhalation than by intraperitoneal administration. The LD_{50} (median lethal dose) in mice for aerosol exposure is about 2 µg/kg, as opposed to about 10 µg/kg by intraperitoneal administration (Franz, 1997). Unlike PSP, which has a relatively slow onset, inhalation of a lethal dose of STX may cause death within minutes. At sublethal doses, symptoms parallel those of PSP, with a more rapid onset. Presumably, this short onset reflects extremely rapid absorption through the pulmonary tissues.

Toxins in clinical samples can be detected by several methods. High-performance liquid chromatography (HPLC) can detect individual toxins but typically requires derivatization reactions (Lawrence and Menard, 1991). While this allows elucidation of toxin profiles that can provide valuable comparative data, it provides no information on toxicity. Receptor-binding assays based upon either rat brain membranes (Poli, 1996, Doucette et al., 1997) or purified STX binding proteins from frogs or snakes (Llewellen and Moczydlowski, 1994) measure total biological activity without regard to toxin profile. All of these assays have been used to detect PSPs in the urine and serum of intoxicated victims (Gessner et al., 1997). Antibody-based assays can detect major toxins, but cross reactivity with other PSPs is highly variable. Commercially available rapid test kits are now attainable from several sources.

In view of its high potency and relative stability, STX is a potential bioterrorist threat agent. Because toxins are easily isolated from laboratory cultures, small-scale aerosol exposure scenarios could be envisioned. However, the more likely threat is via the food supply. Toxins in naturally contaminated shellfish easily reach lethal levels. Blooms of the causative organisms occur annually on both the Atlantic and Pacific coasts of the United States and Canada, as well as many other places around the world. Thus, toxic material is widely available. Contaminated shellfish cannot be differentiated from wholesome shellfish except through chemical analysis. A similar toxin (tetrodotoxin), derived from puffer fish and demonstrating the same mechanism of action, was investigated by the Japanese as a warfare agent in the 1930s and 1940s but was never weaponized.

STX is a much lower threat to the water supply. While small-scale contamination (i.e., water coolers, etc.) would be feasible, dilution effects and lower oral potency preclude contamination of rivers, reservoirs, or even water towers.

BREVETOXINS AND NEUROTOXIC SHELLFISH POISONING (NSP)

NSP results from the consumption of shellfish contaminated with brevetoxins, a group of cyclic polyether neurotoxins produced by the marine dinoflagellate *Karenia brevis* (formerly *Ptychodiscus brevis*). Like PSPs, brevetoxins accumulate in filter-feeding mollusks that are then consumed by humans. Unlike PSP, however, it appears that the causative agents in NSP are actually molluscan metabolites of the parent brevetoxins (Poli et al., 2000). This syndrome has historically been limited to the American states bordering the Gulf of Mexico, although in 1993 an outbreak of shellfish poisoning in New Zealand was identified as NSP. Recently, a bloom of *Chattonella verruculosa* appeared in Rehoboth Bay, Delaware, that was demonstrated to produce significant quantities of brevetoxins. While no cases of human illness were reported, this event opens the possibility of a range extension of NSP into the Delaware and Chesapeake Bays (Bourdelais, et al., 2002).

Brevetoxins and their metabolites bind to voltage-sensitive sodium channels, where they alter the voltage-dependence of activation and inhibit channel inactivation (Huang et al., 1984; Poli et al., 1986). This results in inappropriate and prolonged channel opening.

Symptoms of NSP can manifest within an hour of consumption of contaminated shellfish. These typically include nausea, oral paresthesias, ataxia, myalgia, and fatigue. In severe cases, tachycardia, seizures, and loss of consciousness can occur, but a fatal case of NSP has never been reported. Treatment consists of removing unabsorbed material from the gastrointestinal tract and providing supportive care. Patients typically improve dramatically in 1 to 3 days.

The toxic dose of brevetoxins in humans has not been established, although it is clear that eating only a few shellfish can result in severe intoxication. A severe NSP outbreak occurred in Florida in 1996, when a family ingested whelks collected in Sarasota Bay. Two children were hospitalized with severe symptoms, including seizures. Brevetoxin metabolites were detected in urine samples collected 3 h postingestion, but were undetectable 4 days later (Poli et al., 2000). With supportive medical care, symptoms resolved in 48 to 72 h.

Toxins in clinical samples can be detected either by HPLC coupled to mass spectrometry, receptor-binding assays, or immunoassays (Poli et al., 2000). Because metabolic conversion of the parent toxins occurs in shellfish, and metabolites are less pharmacologically active than the parent toxins, it appears at this time that immunological assays are preferable as screening tools. However, the question of secondary metabolism in humans may impact this issue, and it awaits further study.

Brevetoxins should be considered to have only moderate bioterrorism potential. Blooms of *K. brevis* occur nearly every year on the U.S. Gulf coast, making toxic material available. However, any artificial outbreak of NSP would be of short duration. While victims may be incapacitated, mortality is unlikely. The primary result would be disruption of an ongoing event or perhaps economic disruption of the local seafood industry.

Karenia brevis is easily grown in culture, and toxins can be readily isolated from these cultures. Unpublished animal experiments suggest that brevetoxins will be about 10- to 100-fold more potent by aerosol than by oral exposure. Thus, small-scale aerosol attacks may be possible, although isolation and dissemination of toxins is not trivial.

CIGUATOXINS AND CIGUATERA FISH POISONING (CFP)

CFP is a syndrome caused by exposure to ciguatoxins through the consumption of fresh fish. Like brevetoxins and PSP toxins, ciguatoxins originate with dinoflagellates, in this case the benthic species *Gambierdiscus toxicus*. This organism is an epiphyte, growing in association with filamentous algae on coral reefs and reef lagoons. Specific strains of *G. toxicus* produce precursors of ciguatoxins, which are ingested by grazing herbivorous fish and invertebrates (Holmes et al., 1991). As these precursors move up the food chain to higher trophic levels through predation, they are metabolically modified to form a family of very potent neurotoxins (Lewis, 2001). At present, over 20 members of this family have been identified. The Pacific form of the toxin varies slightly from the Caribbean form, although both are long cyclic polyether compounds reminiscent of the brevetoxins. Large, predatory reef-dwelling carnivores such as grouper, snapper, barracuda, king mackerel, and jacks are especially recognized as frequent carriers of ciguatoxins. However, small reef-dwelling herbivores can also cause ciguatera, especially when consumed whole. This is especially true in the tropical Pacific where these small herbivores are more widely eaten.

Although CFP occurs globally in tropical and subtropical latitudes, approximately paralleling the distribution of reef-building corals, it occurs most frequently in the Pacific Ocean, western Indian Ocean, and the Caribbean. However, modern advances in the shipping of fresh fish have expanded the range of CFP to virtually anywhere in the world. Even within endemic regions, however, it is highly variable and spotty in distribution. In most areas, only a small percentage of the large fish are toxic. Difficulties in predicting toxic areas and detecting toxicity in fish has always been a major impediment to the implementation of control measures.

CFP is estimated to affect more than 25,000 people annually (Lewis, 2001), although substantial underreporting undoubtedly occurs. The symptomatology is complex. In severe cases, symptoms can develop in as little as 30 min; in milder cases, onset can be delayed 24 h or more. The early symptoms are typically gastrointestinal, including nausea, vomiting, diarrhea, and abdominal pain. These generally last only 24 to 48 h, and may co-occur with neurological symptoms such as tingling of the lips and extremities, reversal of or abnormalities in hot/cold temperature sensations, and severe localized itching. These neurological symptoms occur in nearly all cases, and are often accompanied by a wide range of other signs and symptoms (Bagnis et al., 1979). Fatigue, muscle and joint pain, and mood disorders such as anxiety or depression occur in 50 percent or more of cases. Severe cases may also manifest cardiac symptoms such as bradycardia and hypotension. There are also regional differences in symptomatology, probably resulting from regional differences in toxins (Lewis, 2001). Although the gastrointestinal symptoms resolve early, neurological symptoms often persist for weeks or even months. Late in the course of recovery, symptoms may become episodic, recurring during periods of stress or after consumption of certain foods or alcohol.

Ciguatoxins bind to the same receptor site on the voltage-dependent sodium channel as do brevetoxins and cause a hyperpolarizing shift in the voltage dependence of channel activation (Lewis, 2001). In mammals, they are the most potent sodium channel toxins known.

Treatment of CFP consists primarily of symptomatic care and preservation of electrolyte and fluid balances. While not beneficial in all cases, patients diagnosed early may respond to intravenous mannitol treatment (Palafox et al., 1988; Pearn et al., 1989). During the recovery period, avoiding the consumption of fish and alcohol are recommended.

Ciguatoxins in fish are best detected with analytical methods such as liquid chromatography coupled to mass spectrometry. An in vitro competitive receptor-binding assay is also available (Poli, 1996), which is sufficiently sensitive to detect ciguatoxins at levels that are believed to cause human intoxication.

Although rarely fatal, victims of CFP often suffer debilitating symptoms for several weeks, or even months. This alone might make it an attractive agent of bioterror. However, the spotty distribution and relative rarity of ciguatoxic fish among the population make the collection of known toxic material very difficult. Although extremely toxic (LD_{50} in mice is approximately 0.2 µg/kg for the most potent Pacific congener), toxins exist in the fish flesh at ng/g levels, thus making the collection of purified toxins a Herculean task. The causative organism, G. toxicus, has been cultured and found to produce ciguatoxin precursors. These precursors are biologically active in vitro, but their in vivo potency is not well understood. Further, most strains of G. toxicus do not produce ciguatoxin precursors, and those that do produce them at low levels. Thus, the likelihood of terrorists obtaining material from dinoflagellate cultures is extremely low. For these reasons, ciguatoxins must be considered extremely low risk agents at the present time.

OKADAIC ACID AND DIARRHETIC SHELLFISH POISONING (DSP)

Diarrhetic shellfish poisoning occurs after consumption of shellfish containing okadaic acid or its derivatives. Okadaic acid was named for the black sponge Halichondria okadai from which it was first isolated. However, it was later determined that the origin of the toxin was actually dinoflagellates of the genera Prorocentrum and Dinophysis that lived on and within the sponge. These organisms also produce at least seven okadaic acid derivatives denoted dinophysistoxins (DTXs).

Toxic species of Dinophysis and Prorocentrum are distributed worldwide. Consequently, DSP is also widespread. It occurs seasonally and is a major problem to the shellfish industry in Europe and Japan, but has also been documented in South America, South Africa, New Zealand, Australia, Thailand, Mexico, Scandinavia, and Canada (Cohen et al., 1990). The primary vectors to humans are cultured or wild mussels, which accumulate toxins in the digestive glands after filter feeding. Although the toxin-producing species Prorocentrum lima has been associated with cultured mussels in Maine, and low levels of okadaic acid have been measured in shellfish from the Gulf of Mexico (Dickey et al., 1992), no outbreaks of DSP have yet been reported in the United States.

Symptoms of DSP intoxication in humans can occur within 30 min of ingestion and consist entirely of gastroenteritis, including nausea, vomiting, diarrhea, and abdominal pain. Symptoms typically resolve in 2 to 4 days. Treatment is symptomatic, including maintenance of fluid and electrolyte balances, and no deaths have been reported.

DSPs are potent inhibitors of serine/threonine protein phosphatases. Activity is highest for the PP2A class, lower for the PP1 class, and minimal or absent for the PP2B and PP2C classes of phosphatases (Duranas et al., 2001). Diarrhea is thought to be a result of hyperphosphorylation of proteins controlling sodium secretion by intestinal cells (Cohen et al., 1990), causing impaired water balance and fluid loss. In addition, through their effects on diverse protein phosphorylation and dephosphorylation reactions, DTXs can impact such diverse cellular processes as signal transduction, memory, cell division, and apoptosis (Duranas et al., 2001). Whether these effects are important in human intoxications is not fully understood.

Both okadaic acid and at least one of the DTX derivatives are also potent tumor promoters (Suganuma, et al., 1988; Fujiki et al., 1988). The mechanism of this activity is thought to be increased phosphorylation of critical cellular proteins and/or intermediate filaments and changes in DNA gene expression resulting from phosphorylation of suppressor elements. Again, it is not known whether this activity poses a significant public health threat to the seafood consumer. However, the wide distribution of toxic Prorocentrum and Dinophysis species and detection of low levels of DTXs in shellfish in diverse regions raises the question of whether chronic ingestion of subsymptomatic doses of these compounds could pose a health risk.

In most areas of the world where DSP is a problem, shellfish stocks are closely monitored for the presence of toxins and the waters are monitored for toxic dinoflagellates. Since this information is freely available, toxic material can be easily acquired. As with brevetoxins, exposure would likely result in incapacitation rather than lethality, so the terrorist potential is limited to disruption of specific events or economic disruption of the local seafood industry. Thus, these toxins should be considered only moderate threats as agents of terror.

DOMOIC ACID AND AMNESIC SHELLFISH POISONING (ASP)

ASP first came to the attention of public health authorities during an outbreak in Prince Edward Island, Canada, during the winter of 1987. In this event, over 100 people became ill after eating contaminated mussels, and three people died. The causative agent was soon identified as domoic acid (Wright et al., 1989). Domoic acid was not an unknown compound; it had been isolated from red macroalgae in 1958 (Takemoto and Daigo, 1958) and was the active ingredient in an algal extract used as an antihelmitic in fishing villages in rural Japan. It had been evaluated and subsequently rejected as a potential insecticide. Consequently, it was quite surprising to discover a link between domoic acid and an outbreak of human seafood intoxication. Even more surprising was the identification of the diatom *Nitzschia pungens* f. *multiseries* (now known as *Pseudo-nitzschia multiseries*) as the causative organism. This remains the first and only known seafood toxin produced by a diatom.

Domoic acid has since been found to be seasonally widespread along the Pacific coast of the United States (Trainer et al., 2001) as well as the Gulf of Mexico. Around the world, domoic acid has been reported in such diverse locales as New Zealand, Mexico, Denmark, Spain, Portugal, Scotland, Japan, and Korea. Occasionally, levels in shellfish become sufficient to stimulate bans on harvesting. Fortunately, since the initial 1987 Canadian outbreak, no further human cases have been confirmed. This is no doubt attributable to effective survey and monitoring programs.

Domoic acid is a neuroexcitatory amino acid, structurally related to kainic acid. It binds with high affinity to the kainate and α-amino-3-hydroxy-5-methyl-4-isoxazoleproprionic acid (AMPA) subtypes of the glutamate receptor throughout the central nervous system and elicits nonsensitizing or very slowly sensitizing currents. The nonsensitizing nature of these currents causes a protracted influx of cations into the neurons through the receptor channels, stimulating a variety of intracellular biochemical events that lead to cell death in susceptible cells (Hampson and Manolo, 1998). Kainate and AMPA glutamate receptor subtypes are present in high concentrations in the hippocampus, a portion of the brain associated with learning and memory processing.

Symptoms of intoxication occur within hours of ingestion and include vomiting, diarrhea, or abdominal cramps within 24 h and potentially cause confusion, disorientation, memory loss, and in serious cases, seizures, coma, and death. The memory loss involves primarily short-term memory, and in the Canadian outbreak, it was more prevalent in elderly patients (Todd, 1993). Diverse neurological deficits may occur and can persist for months.

On the basis of the levels measured in the Canadian shellfish, it was estimated that mild symptoms can occur after ingesting about 1 mg/kg domoic acid; severe symptoms occur when 2 to 4 mg/kg are ingested. The aerosol toxicity has not been investigated.

The official regulatory testing method for domoic acid in the United States and the European Community is analytical HPLC. However, both immunological methods and a very simple and inexpensive thin-layer chromatographic method have been reported to work very well (Quilliam, 1999).

Because of its seasonally widespread nature and the number of monitoring programs, domoic acid must be considered easily available. The nature of the intoxication, including disorientation, seizures, memory loss, and death is such that this agent has a higher terror potential than some other toxins. Although the reality of an outbreak is probably limited to disruption of ongoing events and economic disruption of the local seafood industry, the emotional response may be greater. Even so, the bioterrorist potential is probably only moderate.

AZASPIRACIDS AND AZASPIRACID POISONING (AZP)

Azaspiracids have only been known since 1995, when at least eight people in the Netherlands became ill after eating cultured mussels from Killary Harbor, Ireland. From these mussels was isolated a new class of cyclic polyether neurotoxins. Further outbreaks elsewhere in Europe were traced back to Irish mussels in 1997 and 1998. Monitoring efforts from 1998 to 2000 in Ireland showed that most of the major shellfish-producing areas suffered periods of contamination by azaspiracids. The causative organism of AZP has recently been shown to be dinoflagellates of the genus *Protoperidinium* (James et al., 2003).

The symptomatology of AZP includes nausea, vomiting, diarrhea, and stomach cramps, and is thus reminiscent of DSP. However, animal studies have demonstrated hepatitis and fatty liver as added complications, and chronic exposure studies demonstrated that stomach and intestinal erosions took many months to heal. In addition, azaspiracids are tumorigenic at much lower doses than required for gastrointestinal (GI) damage, and tumors occur in the absence of added initiators (Ito et al., 2002).

Azaspiracids are not limited to Irish mussels; they have also been detected in mussels from northeastern England and southwestern Norway. The wide distribution of *Protoperidinium* species in coastal oceans suggests that a much wider distribution is likely.

Because so little is known of azaspiracids at this time, and because they are unlikely to cause fatal intoxication, they must be considered minimal threats as agents of bioterror. However, like other polyether marine toxins, these are likely to be more potent by aerosol than by oral administration. Therefore, if the causative organism is found (or can be induced) to produce high levels of toxins in culture, this conclusion should be reassessed.

SUMMARY

Marine foodborne toxins (summarized in Table 1) are a legitimate potential threat to the food supply. Because they occur naturally in fresh and otherwise wholesome seafood products, possess no visual or olfactory clues to their presence, and are impervious to typical cooking temperatures, they can be a difficult problem to identify.

Table 1. Medical Summary of Marine Algal Toxin Syndromes

–	Approx Mouse LD$_{50}$ (μg/kg)	Fatal?	Signs/Symptoms	Onset	Treatment	Duration
PSP	STX: 400 (oral) 10 (ip) 2 (aerosol) (more potent in children than adults)	Yes	Paresthesias, numbness, ataxia, weakness, loss of coordination, respiratory failure	min–h	Supportive, including respiratory support	Days
NSP	PbTx-2: 200 (ip, iv) 6600 (oral) PbTx-3: 100 (iv) 200 (ip) 500 (oral) Active metabolites: ?	No	Nausea, oral paresthesias, ataxia myalgia, fatigue	min–h	Supportive	1–3 days
CFP	Pac: CTX-1: 0.2 (ip) Carib: CTX-1: (~1–2?)	Rarely	Nausea, vomiting, diarrhea, followed by complex neurological presentation, including hot/cold reversal, weakness, tingling; myalgia, fatigue are common	min–h	Supportive; early treatment with iv mannitol may lessen symptoms	Weeks–months
DSP	Okadaic acid: 200 (ip) DSPs: 200–600 (ip)	No	Nausea, vomiting, diarrhea, abdominal pain	min–h	Supportive	2–4 days
ASP	Domoic acid: 1–5 mg/kg orally causes symptoms in humans. (low toxicity in mice)	Yes	Vomiting, diarrhea, abdominal cramps, progressing to confusion, disorientation, memory loss, seizures, coma	h	Supportive	Months (years for memory dysfunction)
AZP	AZA 1–3: 100–200 (ip) AZA4-5: 500–1000 (ip) (20μg/kg AZA1 is tumorigenic in mice)	No	Nausea, vomiting, diarrhea, abdominal cramps	h	Supportive	Days

Note: PSP: paralytic shellfish poisoning; NSP: neurotoxic shellfish poisoning; CFP: ciguatera fish poisoning; DSP: diarrhetic shellfish poisoning; ASP: amnesic shellfish poisoning; AZP azaspiracid poisoning; ip: intraperitoneal; iv: intravenous.

Further, shellfish become toxic naturally, and information on the presence of toxic shellfish is given wide dissemination by the mass media to protect the seafood consumer. Thus, it is very difficult to restrict the availability of toxic material. Should this material be introduced into the food supply, a local outbreak of poisoning is almost certain to occur.

The ramifications of such an outbreak range from possible deaths from STX to merely incapacitation and disruption for most other toxins. While technically not "weapons of mass destruction," these toxins could function as legitimate "weapons of disruption." Certainly, local events could be disrupted, and if the scope of the outbreak was sufficiently great, either actual victims or the "worried well" could overwhelm local health care facilities. In addition, the economic disruption to the seafood industry could be significant.

The potential threat to the water supply or as aerosolized weapons is much less. Dilution effects and the necessity to isolate pure toxins limit the water supply threat to, at best, small-scale events. Although these toxins are significantly more potent by inhalation than by ingestion, the need for large amounts of pure toxin and the technical aspects of efficient dissemination limit their use as aerosol weapons. However, all of these toxins can, either potentially or at present, be grown in laboratory culture. Should significant increases in toxin production

from cultures be accomplished, the threat by these routes will require reassessment.

REFERENCES

Bagnis, R., Kuberski T., and Laugier, S., *Am. J. Trop. Med. Hyg.*, **28**, 1067–1073 (1979).

Bourdelais, A.J., Tomas, C.R., Naar, J., Kubanek, J., and Baden, D.G., *Environ. Health Perspect.*, **110**, 465–470 (2002).

Cohen, P., Holmes, C.F.B., and Tsukitani, Y., *Trends Biochem. Sci.*, **15**, 98–102 (1990).

Dickey, R.W., Fryxell, G.A., Granade, H.R., and Roelke, D., *Toxicon*, **30**, 355–359 (1992).

Doucette, G.J., Logan, M.M., Ramsdell, J.S., and Van Dolah, F.M., *Toxicon*, **35**, 625–636 (1997).

Duranas, A.H., Norte, M., and Fernandez, J.J. *Toxicon*, **39**, 1101–1132 (2001).

Franz, D.R., "Defense Against Toxin Weapons," in R. Zajtchuk, Ed., *Medical Aspects of Chemical and Biological Warfare. Part I: Warfare, Weaponry, and the Casualty*, Office of the Surgeon General, United States Army, Falls Church, Virginia, pp. 603–619.

Fujiki H., Suganuma, M., Suguri, H., Yoshizawa, S., Tagaki, K., Uda, N., Wakamatsu, K., Yamada, K., Murata, M., Yasumoto, T., and Sugimura, T., *Jpn. J. Cancer. Res.*, **79**, 1089–1093 (1988).

Gessner, B.D., Bell, P., Doucette, G.J., Moczydlowski, E., Poli, M.A., Van Dolah, F., and Hall, S., *Toxicon*, **35**, 711–722(1997).

Hampson, D.R. and Manolo, L.J., *Nat. Toxins*, **6**, 153–158 (1998).

Holmes, M.J., Lewis, R.J., Poli, M.A., and Gillespie, N.C., *Toxicon*, **29**, 761–776 (1991).

Huang, J.M.C., Wu, C.H., and Baden, D.G., *J. Pharmacol. Exp. Ther.*, **229**, 615–621 (1984).

Ito, E., Satake, M., Ofuji, K., Higashi, M., Harigaya, K., McMahon, T., and Yasumoto, T., *Toxicon*, **40**, 193–203 (2002).

James, K.J., Moroney, C., Roden, R., Satake, M., Yasumoto, T., Lehane, M., and Furey, A., *Toxicon*, **41**, 145–152 (2003).

Lawrence, J.F. and Menard, C., *J. Assoc. Off. Anal. Chem.*, **74**, 1006–1012 (1991).

Lewis, R.J., *Toxicon*, **39**, 97–106 (2001).

Llewellyn, L.E. and Moczydlowski, E., *Biochemistry*, **33**, 12312–12322 (1994).

Palafox, N.A., Jain, L.G., Pinano, A.Z., Gulick, T.M., Williams, R.K., and Schatz, I.J., *J. Am. Med. Assoc.*, **259**, 2740–2742 (1988).

Pearn, J.H., Lewis, R.J., Ruff, T., Tait, T., Quinn, J., Murtha, W., King, G., Mallet, A., and Gillespie, N.C., *Med. J. Aust.*, **151**, 77–80 (1989).

Poli, M.A., Three-dimensional binding assays for the detection of marine toxins. In: *Proceedings of the Workshop Conference on Seafood Intoxications: Pan American Implications of Natural Toxins in Seafood*, University of Miami, 1996, p. 138.

Poli, M.A., Mende, T.J., and Baden, D.G., *Mol. Pharmacol.*, **30**, 129–135 (1986).

Poli, M.A., Musser, S.M., Dickey, R.W., Eilers, P.P., and Hall, S., *Toxicon*, **38**, 381–389 (2000).

Quilliam, M.A., *J. AOAC Int.*, **82**, 773–781 (1999).

Rodrigue, D.C., Etzel, R.A., Hall, S., de Porras, E., Valesquez, O.H., Tauxe, R.V., Kilbourne, E.M., and Blake, P., *Am. J. Trop. Med. Hyg.*, **42**, 267–271 (1990).

Suganuma, M., Fujiki, H., Suguri, H., Yoshizawa, S., Hirota, M., Nakayasu, M., Ojika, M., Wakamatsu, K., Yamada, K., and Sugamura, T., *Proc. Natl. Acad. Sci. U.S.A.*, **85**, 1768–1771 (1988).

Takemoto, T. and Daigo, K., *Chem. Pharm. Bull.*, **6**, 578 (1958).

Todd, E.C.D., *J. Food Prot.*, **56**, 69–83 (1993).

Trainer, V.L., Adams, N.G., and Wekell, J.C., "Domoic Acid-Producing *Pseudo-nitzschia* Species off the US West Coast Associated with Toxification Events," in G. Hallegraeff, S. Blackburn, C. Bolch, and R. Lewis, Eds., *Harmful Algal Blooms 2000*, Intergovernmental Oceanographic Commission of UNESCO, Paris, 2001, pp. 46–49.

Wright, J.L.C., Boyd, R.K., de Freitas, A.S.W., Falk, M., Foxall, R.A., Jamieson, W.D., Laycock, M.V., McCulloch, A.W., McInnes, A.G., Odense, P., Pathak, V.P., Quilliam, M.A., Ragan, M.A., Sim, P.G., Thibault, P., and Walter, J.A., *Can. J. Chem.*, **67**, 481–490 (1989).

FURTHER READING

Doble, A., "Pharmacology of Domoic Acid," in L. Botana, Ed., *Seafood and Freshwater Toxins: Pharmacology, Physiology, and Detection*, Marcel Dekker, New York, 2000, pp. 359–372.

Lehane, L. and Lewis, R., *Int. J. Food Microbiol.*, **61**, 91–125 (2000).

Poli, M.A., "Brevetoxins: Pharmacology, Toxicokinetics, and Detection," in M. Fingerman and R. Nagabhushanam, Eds., *Recent Advances in Marine Biotechnology, Volume 7: Seafood Safety and Human Health*, Science Publishers, Inc., Enfield, NH, 2002, pp. 1–30.

Van Dolah, F.M., *Env. Health Perspect.*, **108**(Suppl. 1), 133–141 (2000).

See also TOXINS: OVERVIEW AND GENERAL PRINCIPLES.

MARINE CORPS CHEMICAL AND BIOLOGICAL INCIDENT RESPONSE FORCE

JENNIFER RUNYON

INTRODUCTION

The Chemical Biological Incident Response Force (CBIRF) is a unit constituted by approximately 375 marines and sailors whose central mission is to either forward-deploy or respond to a credible threat of chemical, biological, radiological, or nuclear terrorism, or a high-yield explosive incident (Fig. 1). The unit is headquartered at Indian Head, Maryland, which is located approximately 26 miles south of Washington, DC.

BACKGROUND

The 1995 White House Presidential Decision Directive 39, "United States Policy on Counterterrorism," gave priority to developing capabilities for managing the consequences of nuclear, chemical, and biological terrorism. In response, the U.S. Marine Corps established CBIRF in 1996 to assist local, state, and federal agencies in managing the consequences of chemical, biological, radiological, nuclear, and high-yield explosive (CBRNE) terrorism. CBIRF can also be deployed for foreign operations.

CBIRF is composed of two companies: an Initial Response Force (IRF) of approximately 117 marines and sailors and a Follow-on Force (FOF) of approximately 200. The IRF maintains a 24-hour readiness posture, and can deploy via ground transportation within one hour of notification and by air within four hours. If needed, the FOF can be deployed either to assist the IRF or for an independent operation.

The CBIRF's operational capabilities include:

- Detection and identification of chemical, biological, or radiological contaminants;
- Casualty extraction, utilizing an urban search and rescue unit;
- Decontamination of ambulatory and nonambulatory victims;
- Medical triage, stabilization, and regulation;
- Explosive ordnance disposal;
- Command, control, communications, computers, and intelligence (C_4I).

If warranted, CBIRF also has the capability to assist medical personnel in performing vaccinations against smallpox. All members of CBIRF have been appropriately vaccinated.

ACTIVITIES AND EXERCISES

The key activities of CBIRF are training exercises and forward deployment to National Special Security Events (NSSE). CBIRF training exercises include monthly training with Washington, DC, area first responders, training with other first responder communities like those in Seattle and New York City, and overseas deployment in support of exercises and seminars. CBIRF has forward-deployed to several NSSE, including the Atlanta Olympics in 1996, the papal visit to the United States in 1999, and several Presidential State of the Union addresses.

As a result of the anthrax mailings of 2001, several work places had to be decontaminated. CBIRF's deployment for Operation Noble Eagle—the decontamination of the Senate office buildings—began in October 2001 and ended in January 2002. During that time, CBIRF removed 12 tons of unopened, suspicious mail and five tractor-trailer loads of Senate office furniture. The unit also collected over 600 samples from Senate office buildings.

FUTURE OUTLOOK

CBIRF's future plans are to increase its force numbers and to utilize the intellectual capital of its reservists. Between 2004 and 2005, CBIRF will add approximately 100 more

Figure 1. The CBIRF seal. *Source*: http://www.cbirf.usmc.mil/.

Encyclopedia of Bioterrorism Defense, Edited by Richard F. Pilch and Raymond A. Zilinskas
ISBN 0-471-46717-0   Copyright © 2005 Wiley-Liss

marines, bringing its total number of personnel to about 500. CBIRF also plans to enroll reservists possessing advanced degrees and relevant expertise. These reservists will be able to make a commitment to serve 10 to 15 years in the CBIRF, as compared to the 3 to 5 years marines and sailors currently spend with the unit. By employing reservists, the CBIRF brings highly skilled experts on board and decreases its turnover of trained personnel. Currently, CBIRF's budget is $2 million a year.

FURTHER READING

Chemical and Biological Arms Control Institute, *Bioterrorism in the United States: Threat, Preparedness and Response*, November 2000.

Multiple interviews in 2003 with the Commanding Officer of CBIRF, Colonel Thomas X. Hammes.

See also ARMY TECHNICAL ESCORT UNIT.

MAU-MAU (THE LAND OF FREEDOM ARMY)

Monterey WMD-Terrorism Database Staff

OVERVIEW

In 1952, members of the anticolonialist, national liberation movement Mau Mau targeted 33 steer at a British mission in Kikuyu (present-day Kenya) with what British authorities believed to be the toxin of *Synadenium grantii* Hook, the African Milk Bush. Eight steer died in the attack.

BACKGROUND

Mau Mau was formed as a nationalist liberation uprising in protest of British colonial rule during the 1950s. The group attacked white settlers and British outposts in the hopes of gaining Kenyan independence, initially only through acts of sabotage (Plaut, 2003). The movement escalated quickly, however, forcing the British to declare a state of emergency in late 1952. Scholars estimate that by 1956, close to 15,000 people had been killed in the uprising.

THE INCIDENT

When authorities discovered that eight steer from a British station on the Kikuyu tribal land had died under unusual circumstances, the Veterinary Research Laboratory in Kabete opened an inquiry into the incident. All possible etiologies were considered, including poison and infectious disease. Specifically, a number of different toxins were evaluated, including those derived from *Abrus precatorius*, *Ricinis communis*, and *Jatropha curcas*, but all were ruled out as causative agents in the deaths. Finally, the Laboratory considered *S. grantii*, a readily available plant known to grow naturally in the region. When researchers exposed guinea pigs to *S. grantii*, they discovered that the symptoms developed by the infected guinea pigs were similar to those of the steer, indicating that, in view of the other pertinent negatives, the toxin elaborated by *S. grantii* or a related species was the likely cause of death. Authorities soon learned that portions of the plant had been inserted into the skin of the steer via hand-made, superficial incisions (Carus, 1999).

Authorities noted frequent attempts by Mau Mau to kill livestock, some of whom had employed arsenic in the past (commercially available due to its use as an "insect dip" for cattle in the region), and believed that this incident was part of a concerted effort by the group to destroy British livestock in response to a recent crackdown on its movement. No definitive proof implicating Mau Mau was ever obtained.

RELATED INCIDENTS

There is at least one reported case of a member of Mau Mau employing a like agent against a human target. The member fired a "poisoned" arrow at a policeman, but the arrow failed to penetrate the skin (Carus, 1999).

CONCLUSION

The Mau Mau case study suggests that groups may not necessarily require expensive laboratory equipment and complex delivery systems in order to wage a successful campaign of biological or chemical terrorism. Members of Mau Mau had no such capability, nor did they possess significant knowledge regarding production of the agent employed (e.g., the toxin was not extracted from its source), yet they nevertheless succeeded in accomplishing what may well have been the group's primary objective with this attack: to incite fear.

REFERENCES

Carus, S., *Bioterrorism and Biocrimes: The Illicit Use of Biological Agents in the 20th Century*, Working Paper, Center for Counterproliferation Research, National Defense University, July 1999 revision.

Monterey WMD-Terrorism Database, Center for Nonproliferation Studies, 1998–2003, http://cns.miis.edu.

Plaut, M., Kenya Lifts Ban on Mau Mau, *BBC News Online*, 8/31/03, http://news.bbc.co.uk/2/hi/africa/3196245.stm, accessed on 10/24/03.

Encyclopedia of Bioterrorism Defense, Edited by Richard F. Pilch and Raymond A. Zilinskas
ISBN 0-471-46717-0   Copyright © 2005 Wiley-Liss

MEDIA AND BIOTERRORISM

Joby Warrick

INTRODUCTION

Coverage of bioterrorism in the mainstream press was episodic and largely speculative prior to September 11, 2001. But the anthrax attacks of September/October 2001 on public figures triggered months of saturation coverage that highlighted both the best and worst practices of the American mass media in the 24-hour news era. While many news outlets emphasized caution and restraint in reporting the unfolding events in the fall of 2001, the shrill nature of the coverage in some publications and cable news shows very likely contributed to public anxiety over the attacks. The episode underscored the need for improved communications planning by public health agencies and better preparation by news organizations for future attacks. In the aftermath of the attacks, media critics and bioterrorism experts called for a new approach to covering biological terrorism, one that emphasizes planning, patience, and restraint.

HISTORICAL PATTERNS

With his 1971 film "The Andromeda Strain," director Robert Wise introduced moviegoers to the notion of an alien microbe with the power to destroy much of civilization. Over the following three decades, the same fearful image infused articles and television shows about biological terrorism in the popular media—despite a near-absence of actual attacks. Yet, when a real bioterrorism event finally occurred, newspapers and broadcasters found themselves woefully ill-prepared. The anthrax attacks in the fall of 2001 forced newspapers to scramble in search of accurate, reliable information. A near vacuum of official information in the initial days of the crisis was followed by weeks of frenzied, "saturation" coverage, as cable news networks produced an endless stream of commentators and "experts" to speculate about perpetrators, health risks, and future attacks. Rather than easing public anxieties, such coverage is believed to have contributed to a climate of confusion and fear (Gursky et al., 2003).

The handful of bioterrorism-related incidents that occurred in the 1980s and 1990s helped whet the media's interest in the subject but offered little hint of the challenges and pitfalls inherent in reporting a sophisticated, large-scale attack. The second half of the twentieth century witnessed only a single confirmed, successful bioterrorism episode, the 1984 release of *Salmonella typhimurium* bacteria at Oregon restaurants by the Rajneeshee cult. The incident caused hundreds of illnesses but no deaths, and only fleeting interest on the part of the mainstream press. In 1995, in the wake of the sarin attacks in the Tokyo subway by Aum Shinrikyo, major newspapers and broadcasters noted the cult's failed experiments in biological terrorism, including attempts to trigger epidemics by releasing *Bacillus anthracis* and botulinum toxin (Kaplan, 2000). Again, the perpetrators failed to cause even a single death.

Despite the limited impact of these events, major print and broadcast media produced numerous prominent features during the period that stressed the potentially catastrophic nature of a biological attack. ABC News's highly regarded *Nightline* program, for example, built a series of news shows around the dramatization of an anthrax attack on a fictional U.S. city. In the scenario, public health systems break under the strain as the death toll nears 50,000.

Several developments in the 1990s provided inspiration for such reporting. Among them were bioterrorism "war-gaming" exercises such as "Dark Winter," new discoveries about Iraq's biological weapons research, and accounts of previous secret weapons research in the former Soviet Union. One-time Soviet bioweaponeer and defector Ken Alibek, whose book *Biohazard* detailed a massive Soviet campaign to weaponize *B. anthracis* among other select agents, became a prominent voice in news articles and television news features on the "coming plague" of biological terrorism. While such stories increased public awareness of the threat, much of the coverage was sensational in tone and reflected a poor understanding of science (e.g., some reports described *B. anthracis* as a virus, or suggested that anthrax could easily be spread from person to person). Such stories frequently cited high-end estimates for casualties and glossed over the formidable technical obstacles that make it difficult to unleash a truly catastrophic attack (Tucker, 1999). Donald A. Henderson, then the director of the Johns Hopkins Center for Civilian Biodefense Studies, criticized such "misleading" news stories as counterproductive. "By confusing fact and fiction, (such) coverage could cause more harm than good," he warned (Henderson, 1999).

Encyclopedia of Bioterrorism Defense,   Edited by Richard F. Pilch and Raymond A. Zilinskas
ISBN 0-471-46717-0   Copyright © 2005 Wiley-Liss

COVERAGE OF THE 2001 ANTHRAX LETTERS

In early October 2001, reports of a real bioterrorism attack on the United States jarred newsrooms already on edge because of the September 11 tragedies in New York, Pennsylvania, and Washington, DC. Journalists covering the developing story were hobbled from the outset by two factors: a lack of reliable, readily available expertise about anthrax, and confused, conflicting accounts from public health officials and investigators. A 2003 study documenting the latter problem found that reporters had difficulty reaching public health officials during the first few weeks of the crisis; moreover, the experts who made themselves available frequently had limited scientific understanding of the disease. "Every day we got different and conflicting information. The health department press offices didn't know the disease. I can't think of any other public event like this," one reporter said in a confidential interview (Gursky et al., 2003). The study concluded that the public health community did a generally poor job in working with the media during the crisis and failed to heed a well-tested axiom for communicating health-risk information in a crisis: "Authorities should tell the truth as they know it, when they know it."

The response to this information vacuum varied strikingly. Media critics and bioterrorism experts at the height of the attack gave mostly high marks to mainstream newspapers and broadcasters for coverage they described as generally balanced, and not needlessly alarmist (Poynter Online, 2001; Cortes, 2001). "Their approach has been not to increase panic," one biodefense expert said of the media's performance in the early weeks of the crisis. Some, like Nieman Foundation curator Robert Giles, praised the press for not immediately pointing fingers at Usama Bin Ladin, mastermind of the September 11 attacks, as the likely perpetrator (Ricchiardi, 2001).

But with the criminal investigation failing to produce a clear suspect or motive for the attack, speculation and rumor began creeping into reports. This was especially true on cable news shows, which faced intense competition and a grueling 24-hour news cycle. Cable was singled out for criticism by scientists for giving airtime to a parade of nonexperts whose views were often uninformed and even irresponsible. One prominent bioterrorism expert and frequent guest noted that cable shows appeared to favor guests with sensationalist views. "There would be an alarmist, and I would be the one trying to tamp down some of the hysteria. And the alarmist would get 90 percent of the attention," he said (Brookings Institute, 2001). The nonstop nature of cable news contributed to several of the top criticisms of media coverage of the attacks, as identified by a one scholarly survey of journalists and public health officials conducted by researchers at the University of Arkansas at Little Rock and the University of South Florida at St. Petersburg. It concluded that "too-frequent" updates led to public obsession with the unfolding crisis—a "Conditization syndrome." The report criticized TV news reporting as "breathless and sensational," and said the text "crawls" that appeared at the bottom of the screen tended to be alarmist (Swain and Mason, 2003).

NEW MODELS FOR BIOTERRORISM COVERAGE

Despite such problems, the anthrax attacks spurred new thinking about how the news industry covers terrorism in general, and bioterrorism in particular. Special difficulties inherent in reporting biological attacks came into stark relief: Unlike bombs, the microbes used in biological weapons are microscopic and silent, and often go undetected until medical symptoms emerge. The diseases of bioterrorism, such as smallpox and plague, provoke strong, visceral reactions that can lead to panic and hysteria. And the stealthy nature of a biological attack severely limits the ability of public health officials to ascertain facts or make predictions. In fact, it may even be difficult initially to determine whether the attack is real. Journalists may be confronted with hoaxes and false alarms, as well as conflicting information and mixed messages from officials (Swain and Mason, 2003).

New strategies for dealing with these challenges have been put forward by scholars, media ethicists, veteran journalists, and bioterrorism experts. A critical need cited by numerous experts is preparation: News organizations should decide in advance of a crisis to draw up a protocol or road map, spelling out how attacks should be covered and what questions or sensitivities should be considered before a story is printed or aired. Newsroom managers also should invest in training so that key reporters and editors understand the subject matter and are equipped to ask the right questions. And, news organizations should build relationships with agencies and officials who are likely to respond to future bioterrorism events (Steele, 2001).

CONCLUSION

As with any story, journalists covering bioterrorism remain obliged to report the facts fully and completely, and to maintain their independence from government. "Journalists must not be deterred in their pursuit of the truth. Nor should they be unduly influenced by those who might use their power or position to keep essential information from the public," advises Bob Steele, the Nelson Poynter Scholar for Journalism Values at the Poynter Institute. But at the same time, news gatherers must also recognize an obligation to "do no harm" (Steele, 2001). Mindful of the potential for panic during a bioterrorism attack, journalists should err on the side of caution. They should present the facts as they are known without needlessly highlighting drama or extremes. They should avoid giving voice to conspiracy theories, speculation, and doomsday predictions. And they should not allow nonexperts to pose as authorities on technical or medical matters (Swain and Mason, 2003).

A final, crucial element in a reporting about bioterrorism is patience—a virtue not always encouraged in newsrooms. In a future biological attack, the facts are likely to emerge slowly overtime, demanding that journalists resist the temptation to rush to conclusions or to report incomplete information. This is a seemingly simple admonition, yet it reflects one of the most difficult challenges for the press in an age of "24/7 media coverage and expectations of instant answers," concede the authors of a 2003 study of the public health response to the anthrax attacks.

"The media and the public must understand that the need for rapid decisions may be at odds with the desire for complete answers" (Gursky et al., 2003).

REFERENCES

Brookings Institute/Harvard University Shorenstein Center forum, *The Anthrax Scare and Bioterrorism*, (2001).

Cortes, N., *Scientists, Doctors Call Coverage of Anthrax Scare Fair, Balanced*, December 12, 2001, Freedomforum.org, Freedom Forum.

Gursky, E., Inglesby, T.V., and O'Toole, T., *Biosecur. Bioterror. Biodefense Strategy, Pract. Sci.*, **1**(2), 97–110 (2003).

Henderson, D.A., Dangerous Fictions About Bioterrorism, *Washington Post*, November 8, 1999.

Kaplan, D., "Aum Shinrikyo," in *Toxic Terror: Assessing Terrorist Use of Chemical and Biological Weapons*, J.B. Tucker, Ed., MIT Press, Cambridge, Mass., 2003, pp. 207–226.

Poynter Online, *Scientists Rate the Stories: Experts Evaluate the Accuracy of Anthrax Coverage*, October 12, 2001, Poynter Institute.

Ricchiardi, S., *Am. J. Rev.*, **23**, 18–23 (2001).

Steele, B., *Ethics in Journalism and Bioterrorism*, in Poynter Online, posted Jan. 1, 2001.

Swain, K. and Mason, T., *Media Coverage of the 2001 Anthrax Scare: Lessons Learned about Bioterrorism Communications*, from PowerPoint presentation, School of Mass Communication, University of Arkansas at Little Rock; College of Public Health, University of South Florida at St. Petersburg, 2003.

Tucker, J., Bioterrorism is the Least of Our Worries, *New York Times*, Oct. 16, 1999.

FURTHER READING

David, C. and India, L., *Bioterrorism: A Journalist's Guide to Covering Bioterrorism*, Second Edition, Radio and Television News Directors Foundation, Washington, D.C., 2004.

WEB RESOURCES

American Journalism Review, essays on news media coverage of bioterrorism, www.ajr.org, accessed on September, 2003.

Columbia News Video Forum, *Columbia Experts Assess Media Coverage*, http://www.columbia.edu/cu/news/vforum/01/get_it_right/; also *First Amendment Panel Scrutinizes Media Coverage of Anthrax Attack*, http://www.columbia.edu/cu/news/vforum/02/anthrax_media/, accessed on October, 2003.

Freedom Forum, essays on coverage of terrorism and bioterrorism, http://www.freedomforum.org, accessed on October, 2003.

Journalist's Guide to Covering Bioterrorism, published by the Radio and Television News Directors Foundation, http://www.rtnda.org/resources/bioguide.pdf, accessed on September, 2003.

Poynter Online (Poynter Institute); essays on journalism and bioterrorism, http://poynteronline.org, accessed on October, 2003.

METROPOLITAN MEDICAL RESPONSE SYSTEM

John J. Shaw

INTRODUCTION

The Metropolitan Medical Response System (MMRS) Program began in 1996 and currently is funded by the U.S. Department of Homeland Security (Fig. 1). The primary focus of the MMRS program is to enhance existing local and regional emergency preparedness systems to effectively respond to a public health crisis, especially a weapons of mass destruction (WMD) event. Through a process of planning for and implementing specially designed contract deliverables, local law enforcement, fire services, Hazmat teams, Emergency Medical Services (EMS), hospitals, public health agencies, and other "first responder" personnel can more effectively respond during the first 48 hours of a public health crisis.

On March 1, 2003, MMRS joined the Federal Emergency Management Agency (FEMA) and other programs from the Department of Health and Human Services, Department of Energy, and Department of Justice to become the Emergency Preparedness and Response Directorate of the new Department of Homeland Security.

HISTORY OF MMRS IN EMERGENCY PLANNING

The Defense Against Weapons of Mass Destruction Act of 1996 directed the Secretary of Defense to enhance capability and support improvements of local response agencies. The Nunn-Lugar-Domenici Amendment to the National Defense Authorization Act for FY 1997 authorized funding for "Medical Strike Teams (MST)," which subsequently evolved into the MMRS Program. Ongoing Congressional appropriations have funded contracts with 125 MMRS jurisdictions nationally. Through the MMRS, the federal government has provided substantial additional resources to these municipalities in order to support a response to a mass casualty incident, especially an event resulting from the use of a weapon of mass destruction.

ROLE OF MMRS IN EMERGENCY PLANNING

Recognizing the wisdom in the axiom "All emergencies begin and end locally," MMRS enhances local planning efforts to produce a comprehensive response system that includes preparation for a bioterrorism incident. MMRS expands and enhances local planning efforts in four important ways:

- The MMRS contract deliverables ensure that traditional emergency planning by the local fire service, police, and EMS leadership is expanded to include representation from both public health agencies and hospitals, thereby ensuring a public health component in emergency planning.
- The MMRS program encourages, and financially supports, the expansion of emergency planning to the regional level, thereby maximizing the use of resources normally not available to local communities.
- The MMRS pharmaceutical stockpile, an integral part of all MMRS programs, provides a level of protection specifically for the first responder community in the event of a chemical or biological adverse incident.
- The MMRS program ensures that all local and regional planning is compatible with and supports state and federal emergency planning, including the FEMA National Response Plan.

KEY MMRS PROGRAM COMPONENTS

Components of the MMRS program include:

- Activation and notification procedures
- An all-hazards operations plan that includes a bioterrorism component
- Specially trained responders and equipment
- Management of risk communications and public information
- A pharmaceutical plan for first responders
- Enhanced emergency transport and hospital capabilities

Encyclopedia of Bioterrorism Defense,   Edited by Richard F. Pilch and Raymond A. Zilinskas
ISBN 0-471-46717-0   Copyright © 2005 Wiley-Liss

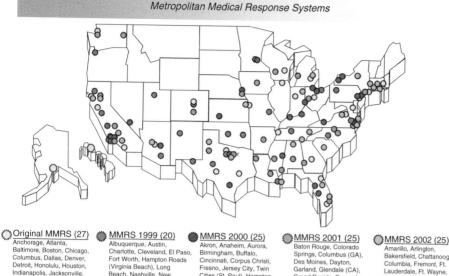

Figure 1. Metropolitan medical response systems.

- Mass fatality management
- Mental health services in support of first responders

MMRS CONTRACT DELIVERABLES

The MMRS contract deliverables ensure that key questions regarding community response to a WMD or mass casualty incident have been asked and answered. For example, MMRS requires that member communities address the following preparedness issues:

- Component MMRS Plan for the rapid forward movement of patients prior to activation of the National Disaster Medical System (NDMS);
- Component MMRS Plan for responding to a chemical, radiological, nuclear, or explosive WMD event;
- Component plan for a comprehensive response to the use of a biological weapon;
- Component plan for the protection of the local hospital and health care system (including procedures for notification, facility protection, mass decontamination, triage and treatment, and the development of a plan for surge capacity);
- Chemical antidote stockpile sufficient to provide care for up to 1000 first responders;
- Biological antidote stockpile sufficient for up to 10,000 victims;
- MMRS training plan (includes initial and refresher training requirements);
- MMRS equipment purchase plan to enhance available regional resources.

UNIQUE FEATURES OF MMRS

The MMRS program is a federal resource made available at the community level in order to ensure a comprehensive and workable response mechanism in the event of a major disaster. By supplying training and education to the first responder community including health and hospitals, MMRS raises the level of awareness at the local level while enhancing the response capabilities of the region.

In most MMRS jurisdictions, volunteers from all emergency response professions come together to plan their joint response to the disasters none of us ever hope to experience. Questions of enormous significance for the safety of the citizens of these communities are asked so that practical answers based on sharing of regional resources can be reached. MMRS is not designed to supplant those long-standing local response plans that have worked so well. Rather, MMRS expands the conversation locally by bringing together new and traditional planning partners to ensure that local and regional plans reflect the latest science, and so that the latest technologies are made available.

Beyond the MMRS role as a facilitator of comprehensive planning that includes a public health component, MMRS uniquely provides protective services specifically targeted at the first responders. The MMRS pharmaceutical program *Project BioShield* provides nerve agent antidotes in the event of a chemical incident and biological antidotes if a biological agent is released, so that first responders can not only survive but also manage the event.

MMRS can also support special projects that enhance the capabilities of its member jurisdictions. MMRS funding frequently has been used to expand the MMRS program to other regions or even throughout an entire state. MMRS dollars have also provided hospitals with special equipment to protect their facilities, including mass decontamination equipment.

But the most important contribution that an MMRS program makes to the local emergency planning process is to bring together planners from all areas of emergency response so that, in the event of a local disaster, skills and leadership are shared with confidence.

THE FUTURE OF MMRS

Since its absorption into the U.S. Department of Homeland Security in March of 2003, the national MMRS program has been thoroughly evaluated for effectiveness and has been found to be remarkably successful in meeting its original goals. By any measure, the 125 MMRS communities have been able to demonstrate superior preparedness and operational readiness in the event of disasters, either natural or man-made. The Department of Homeland Security has recognized the importance of continued expansion of the MMRS model so that more communities can benefit from the expanded capacities and resources that result from completion of the MMRS planning process.

A five-year strategic planning process for MMRS has been authorized at the Department of Homeland Security. The strategies include providing for the sustainability of the MMRS plan in existing communities, thereby ensuring that local emergency planning will remain a vital and dynamic process. Though presently there are no plans to create additional MMRS jurisdictions, existing MMRS entities will be encouraged to share their skills and capacities with neighboring communities and with their states, thus recognizing the importance of uniformity in planning, training, and equipment purchasing in assuring a seamless response to all types of local calamities.

See also DEPARTMENT OF HOMELAND SECURITY.

MINNESOTA PATRIOTS COUNCIL

Jason Pate

INTRODUCTION

In 1995, four Minnesota men were the first to be tried and convicted under the United States Biological Weapons Anti-Terrorism Act of 1989. Douglas Baker, Leroy Wheeler, Dennis Henderson, and Richard Oelrich had acquired the toxin ricin as part of an alleged plot to kill local deputy sheriffs, U.S. marshals, and IRS agents. The four men were members of a radical tax-protesting militia organization called the Minnesota Patriots Council.

BACKGROUND

The Minnesota Patriots Council was founded in 1970 by Colonel (Ret) Frank Nelson of the U.S. Air Force. As a right-wing organization, it opposed the notion of a federal government and refused to recognize any authority above the local county. Its members protested U.S. taxation policies and met periodically in small groups, or cells, to discuss pressing issues of relevance to the group, at times even outlining violent methods (such as blowing up buildings) to combat what was perceived to be tyrannical, illegitimate federal authorities.

AGENT ACQUISITION AND PRODUCTION

In 1991, Oelrich, Henderson, and Wheeler came across a classified notice in the *CBA Bulletin*, a right-wing, open-source publication, advertising a mail order castor bean kit. The castor bean is the source of ricin, which once extracted is extremely toxic to human beings, even in minute quantities. The three ordered the kit in April 1991 with the intention of mixing the extracted ricin with aloe vera hand lotion and dimethylsulfoxide (DMSO) in order to create an effective delivery system that, unlike ricin alone, presumably could penetrate the skin (this in fact is a highly inefficient technique).

After completing this part of the plan, in early 1992, Henderson took the mixture of ricin, aloe vera, and DMSO to his friend Douglas Baker's house, where it was stored in a coffee can along with a cautionary note. Following a marital dispute, however, Baker's wife Colette took the coffee can along with several other weapons to the local sheriff's office, which in turn contacted the FBI. It was determined that the coffee can contained 0.7 g of ricin, theoretically capable of killing "hundreds of people" (according to the FBI report) if appropriately dispersed.

AFTERMATH

Baker and Wheeler were arrested on August 4, 1994, and tried for the possession of a deadly biological substance at the Federal District Court in St. Paul, Minnesota. Each man ultimately received a 33-month prison term and three years of subsequent probation. Henderson and Oelrich were arrested in July and August of 1995, after going into hiding upon the news of their coconspirators' arrests. They stood trial in October and were convicted of the possession and production of ricin. In January 1996, Henderson was sentenced to 48 months in prison followed by three years of probation, and Oelrich received a 37-month prison term and three years of probation.

FURTHER READING

Tucker, J.B. and Pate, J., "Chapter 10: The Minnesota Patriots Council (1991)," *Toxic Terror: Assessing Terrorist Use of Chemical and Biological Weapons*, MIT Press, 2000, pp. 159–183.

See also Ricin and Abrin and United States Legislation and Presidential Directives.

Encyclopedia of Bioterrorism Defense,   Edited by Richard F. Pilch and Raymond A. Zilinskas
ISBN 0-471-46717-0   Copyright © 2005 Wiley-Liss

MINUTEMEN: CASE STUDIES

Monterey WMD-Terrorism Database Staff

OVERVIEW

The 1960s right-wing California militia group the Minutemen strove to defend the United States against what it believed to be inevitable Communist takeover by any means necessary. Its founder, Robert DePugh, made numerous threats to use chemical and biological weapons to further his organization's goals, and apparently developed a plan to spread biological agents throughout the United States. Neither this plan nor any of his other weapons of mass destruction (WMD)-related threats were ever pursued in any active sense.

BACKGROUND

In 1960, Robert Bolivar DePugh, concerned about the growing threat of Communism and wishing to establish a network of citizens dedicated to the defense of the United States in the face of Communist takeover, established a group that he named the Minutemen (Drinkard, 1980). The group acquired funds through membership, with each member paying an initial five dollars to join and a minimum of two dollars a month for membership dues. Members were further responsible for acquiring their own weapons, thus keeping organizational costs low. While DePugh has claimed that his organization was 25,000 members strong at one time, former FBI director J. Edgar Hoover estimated that there were no more that 500 individual "Minutemen." Other estimates of group membership include a report issued by the California Attorney General in 1965, which noted that between 100 and 600 Minutemen members existed in California alone (Jones, 1968).

SELECTED INCIDENTS

In his book *The Minutemen*, author Harry Jones describes an interview with DePugh, in which the Minutemen leader outlined a "hypothetical" plan to disperse a biological weapon throughout the United States in a compressed time frame. The plan entailed developing a virus, immunizing himself against it, then contaminating himself with it and coughing on travelers passing through the Kansas City Municipal Air Terminal. At the time, DePugh was the head of the veterinary drug development company Biolab Corporation in Norbourne, Missouri. DePugh later debated the details in this description of the plan, not debating their validity but rather denying that the plan entailed coughing on people. In his words, "I was talking about how easy it would be to start a national epidemic by going down to the airport and spreading disease germs on the floor so they could be tracked onto the aircraft and carried all over the United States in a short period of time" (Jones, 1968).

The Minutemen organization has also been associated with threats involving chemical agents. According to Jerry Brooks, a former member of the group, in 1962 DePugh sent him on a "dollar ninety-nine bus tour" around the United States with the mission of assassinating suspected Communists. Brooks alleges that DePugh provided him with three vials of strychnine to accomplish the job. DePugh denied Brooks's claim, stating that the purpose of the mission had been to gain access to the offices of the Communists for intimidation purposes in the future. Also according to Brooks, a Minutemen member allegedly discussed plans to introduce hydrogen cyanide into the air-conditioning system of the United Nations building in New York City. Brooks described this plan to prosecutors during a firearms trial brought against DePugh (Jones, 1968).

In 1965, the Attorney General of California released a report on the Minutemen and its antigovernment activities against the state of California and the United States at large. Among other observations, the report emphasized the group's interest in both conventional and unconventional weapons, despite the fact that (according to all available information in the public domain) its activities were limited to plots only with respect to WMD. The report specifically noted that the group's literature contained information on the making of an unspecified chemical nerve agent, a substance that some allege DePugh tested on his family dog in an attempt to determine the minimum lethal dosage.

DePugh is also alleged to have disseminated literature describing methods for building and concealing improvised projectile weapons capable of delivering nerve agents via a modified shotgun. In addition, the Minutemen held a training seminar near Temecula, California, on September 28 and 29, 1963, during which the use of

Encyclopedia of Bioterrorism Defense,   Edited by Richard F. Pilch and Raymond A. Zilinskas
ISBN 0-471-46717-0   Copyright © 2005 Wiley-Liss

methane and nerve agents was discussed. Information provided, as described in the Attorney General's report, was as follows: "Methane gas or nerve gas can be obtained when small slivers of pure teflon plastic are inserted in a cigarette. The results are always fatal, and almost immediate. The only known antidote is atropine, which must be taken immediately" (Jones, 1968).

REFERENCES

Drinkard, J., Ultra-Conservative Says Armed Struggle Inevitable, Associated Press (PM cycle) November 29, 1980.

Jones, J.H., Jr., *The Minutemen*, Doubleday & Company, Inc., Garden City, NY, 1968, pp. 21–22.

Monterey WMD-Terrorism Database, Center for Nonproliferation Studies, 1998–2003, http://cns.miis.edu.

MODELING THE PUBLIC HEALTH RESPONSE TO BIOTERRORISM

Nathaniel Hupert

INTRODUCTION

Over the past three years, the United States has been faced with three separate epidemic outbreaks that have demonstrated the importance of system-wide preparedness for public health emergencies. The 2001 anthrax attacks, 2002–2003 SARS (Severe Acute Respiratory Syndrome) epidemic, and 2003 monkeypox outbreaks, all demonstrated both the nation's vulnerability to infectious diseases of natural and deliberate origin and its limited capacity for hospital-based treatment and patient isolation in response. And though crucial for the public's welfare and for national security, health system preparedness for epidemics, both natural and deliberate, has been studied only minimally in any standardized way. This entry therefore reviews alternative approaches to the modeling and assessment of health system preparedness and response in order to provide a foundation for future standardized study.

BACKGROUND

Modeling is a critical component of the design of optimized health response plans, since accurate models allows planners to "think with numbers" in exploring the consequences of policy, resource allocation, and disease management options. For modeling purposes, the public health response to disease outbreaks can be divided into five main components:

1. Surveillance (either disease-specific or syndromic);
2. Stockpiling of medical materiel;
3. Distribution capacity and transporting that materiel to the site of the outbreak;
4. Dispensing capacity to provide necessary antidotes, antibiotics, or vaccines to affected populations in a timely manner Figure 1;
5. Medical treatment capacity to care for outbreak victims in whom disease cannot be prevented through mass prophylaxis activities (either because of lack of effective remedies or delay in getting such remedies to affected populations while still amenable to therapy).

Of these five activities, the majority of scientific studies have focused on surveillance, while reports describing actual responses to local events such as the anthrax attacks have focused on dispensing. Notable for its near-absence from this body of literature is scientific modeling of treatment capacity for mass casualties from a disease outbreak.

The Centers for Disease Control and Prevention (CDC) website illustrates the depth of scientific investigation focusing on surveillance. There are 60 peer-reviewed papers and technical reports listed in the annotated bibliography posted on the Division of Public Health Informatics and Surveillance section of the CDC site. In contrast, the Public Health Preparedness and Response Capacity Inventory section of the CDC site offers only two 50+ page self-survey forms (one for local and one for state-level planning) but no references to technical or scientific articles. A recent evidence report funded by the Agency for Healthcare Research and Quality (AHRQ) reviewed over 16,000 peer-reviewed articles and 8000 websites for information on detection and management of bioterrorist attacks. In this large sample, the authors identified 217 information technology/decision support systems that met their rigorous quality criteria, of which the majority (168) were surveillance, detection, and diagnostic systems (Bravata et al., 2002). Only 25 of the systems reviewed focused on patient management in the aftermath of an attack and none provided tools for assessing prehospital or hospital-based surge capacity for mass casualties from a bioterrorist event.

Of the limited number of papers and reports that address the stockpiling, distribution, and dispensing components of bioterrorism response, most fall into the category of "lessons learned" from the response to the 2001 "anthrax letters" and earlier exercises and outbreaks of infectious disease. Included in this group are descriptions of the public health response to meningitis outbreaks in Minnesota (1995) and Alberta, Canada (2000), informal reports on the Top Officials (TOPOFF) exercise of 2000, a military exercise using the Neighborhood Emergency Help Center (NEHC) concept, and descriptions of mass antibiotic dispensing clinics in Washington, DC, and New York City as a result of the anthrax letters.

MODELS OF THE HEALTH SYSTEM RESPONSE

Only six papers on the CDC website attempt to model various components or outcomes of health system epidemic response activities. These include four papers that consider smallpox vaccination strategies, one on the efficient design of mass dispensing centers, and

Encyclopedia of Bioterrorism Defense, Edited by Richard F. Pilch and Raymond A. Zilinskas
ISBN 0-471-46717-0 Copyright © 2005 Wiley-Liss

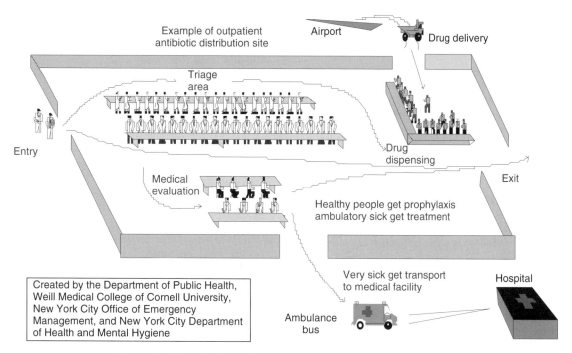

Figure 1. Slide of animated model of mass prophylaxis dispensing site.

one on prehospital and hospital surge capacity for anthrax response. These six papers constitute the entirety of the peer-reviewed science in modeling of health system response to bioterrorism and other public health emergencies.

Meltzer, 2001

The first paper on smallpox vaccination strategies, by Meltzer and colleagues at the CDC, uses a Markov chain model to simulate the effect of targeted ("ring") versus mass vaccination, in combination with quarantine, on the outcome of a hypothetical intentional release of the virus in civilian U.S. communities (Meltzer et al., 2001). On the basis of this model, a minimum stockpile of 40 million doses of smallpox vaccine, a response plan that combined quarantine with targeted vaccination, and a very rapid outbreak response are recommended. The conclusions are driven largely by modeling assumptions about the size and nature of the smallpox outbreak, specifically the number of original infected individuals and the reproductive rate of the outbreak (i.e., the average number of new cases infected by each existing case). As with each of the next three papers reviewed, Meltzer and colleagues do not specifically address the issue of how to accomplish the goal of rapid mass vaccination in the aftermath of a smallpox attack, but rather focus on *whether* and *when* it should occur.

Kaplan, 2002

Next, Kaplan, Wein, and Craft from the Yale and Stanford Schools of Management developed mathematical models that suggest that the optimal public health response to any smallpox outbreak is mass, and not ring, vaccination (Kaplan et al., 2002). Advantages of their

multiple differential equation approach include accounting for queues (or lines) forming at mass vaccination stations and delays caused by epidemiologic investigation of contacts in ring vaccination strategies. The main disadvantage of the employed modeling technique (which may have biased their results in favor of a mass vaccination strategy, according to subsequent analyses) is its mathematical approach, which unrealistically assumes completely homogeneous mixing of cases and susceptibles throughout the 280 million–large population of the United States. Additionally, although Kaplan and colleagues acknowledge certain realistic aspects of mass prophylaxis campaigns (e.g., the fact that queues will form at mass vaccination centers), their conclusion rests on the assumption that rapid nationwide mass smallpox vaccination could in fact be accomplished. This assumption, which the authors never explicitly model, represents the paper's major practical limitation.

Halloran, 2002

Halloran and colleagues from the Emory School of Public Health next addressed what they saw as Kaplan's methodological flaws by developing a discrete event simulation model (what they call a "stochastic heterogeneous simulator") that explores the impact of targeted versus mass vaccination on smallpox outbreaks (Halloran, 2002). Their disease scenarios, which reflect an increased appreciation for the complexity of realistic epidemic response, incorporate factors such as preexisting herd immunity, limited vaccine supply, and vaccine-related side effects. These considerations lead the authors to moderate their model-derived conclusions. For example, while they agree that mass vaccination is generally superior to targeted vaccination, they note that targeted vaccination is a more efficient use of vaccine supplies and may be preferable in particular settings.

Bozzette, 2003

Bozzette and colleagues at RAND Corporation extended this work in stochastic simulation to explore the role of prevaccinating health care workers against smallpox as part of an overall national response strategy (Bozzette et al., 2003). Their model indicates that the choice of an optimal vaccination strategy depended on the type of outbreak (the group considers scenarios ranging from a hoax to infection of thousands of airline passengers simultaneously at multiple airports). Using a harm-benefit analysis that weighs vaccination-related deaths against attack-induced deaths, the authors determine the estimated risk thresholds at which health care worker and general population vaccination should take place in pre- and postattack settings. The authors do not, however, address the development of feasible operational plans for each of the response strategies under consideration.

Hupert, 2002

This task was addressed by Hupert and colleagues at Weill Medical College of Cornell University, who developed a discrete event simulation model of dispensing center staffing and patient flow in order to determine the optimal use of limited staff resources in a high-flow mass prophylaxis setting (Hupert et al., 2002). The model outputs, which concern a scenario involving antibiotic distribution in the aftermath of a hypothetical anthrax attack, provide staffing estimates for clinics located in areas of high-, medium-, and low-level exposure. The paper also discusses methods for budgeting for surges in patient arrivals at mass dispensing centers (similar to the way that Kaplan and colleagues account for queues in their model) by adjusting planned utilization rates of staff workers at patient care stations. Use of stochastic simulation techniques similar to those of Halloran and Bozzette allowed the authors to identify unexpected crowding due to naturalistic variations in patient arrivals. The paper is limited by a lack of experimental or real-life data to quantify patient influx to a mass prophylaxis center in the setting of a bioterrorist event. Additionally, the clinic model presented in the paper was not linked to a hospital treatment model; therefore, queuing backups due to hospital overcapacity were not introduced into the simulated operation of the dispensing clinic.

Wein, 2003

The sixth and last paper on the website is another mathematical model by the research team of Wein, Kaplan, and Craft, this time addressing the question of antibiotic prophylaxis and hospital treatment capacity for a very large-scale anthrax attack (Wein et al., 2003). Simulating the infection of over 1 million individuals in an urban setting, the authors used a system of integropartial differential equations to determine the chief bottlenecks in providing prophylaxis and hospital treatment for victims in arbitrarily defined zones downwind of a release of *Bacillus anthracis* spores. Outputs for this model are uniformly dire: Over 100,000 people die in the baseline scenario, and hospitals are found to be the major system bottleneck leading to increased mortality. The authors conclude from this study that, because of the difficulties involved in postattack mass prophylaxis and treatment, serious consideration should be given to a variety of preattack prophylaxis activities.

ANALYSIS

Wein and Kaplan deserve praise for attempting to model the spectrum of epidemic outbreak response activities in this manner; in fact, theirs is the only original study addressing hospital surge capacity that is identified by a PubMed search for "hospital surge capacity" (although Hupert and colleagues have addressed this subject as well (Hupert and Cuomo, 2003). However, their work has important limitations from both scientific and practical perspectives. The model includes a number of features that are either unrealistic or deserving of sensitivity analyses, such as the extremely high number of exposed patients, a fixed time delay of 48 hours before the start of mass prophylaxis intervention, a fixed rate of prophylaxis once these efforts are underway, fixed (though unreported) numbers of mass prophylaxis sites and hospital beds, and 100 percent mortality among symptomatic patients. Additionally, they assume that civilian mass prophylaxis would be operated according to plans developed by the U.S. military, and mobile military field hospitals would be available within 18 hours of detection of the attack. Taken together, these assumptions seriously weaken the validity of the model outputs reported. Additionally, the study's key conclusions are impractical (such as the finding that a 75fold increase in the number of hospital beds would be needed to overcome bottlenecks at this stage).

All six of these articles suffer from a lack of real patient-level data to drive modeling assumptions. Evaluating the actual health system utilization patterns of different populations will improve future bioterrorism response models by allowing the simulation of more realistic surges in patient arrival following an event and by identifying potential demographic groups for whom a "push" (e.g., delivery) rather than "pull" (e.g., dispensing) method of prophylaxis provision may be needed. Clarifying these initial patient contact rates will have important downstream implications for dynamic hospital capacity. Calculating dynamic prehospital prophylaxis and hospital bed capacity is complex, requiring extensive data collection and the use of discrete event simulation modeling. However, once models of dynamic capacity are complete, they will offer health system planners the opportunity to test, in real time, how well specific facilities and regions are able to manage victims of epidemic outbreaks.

REFERENCES

Bravata, D.M., McDonald, K.M., Owens, D.K., Buckeridge, D., Haberland, C., Rydzak, C., Schleinitz, M., Smith, W., Szeto, H. and Wilkening, D., *Bioterrorism Preparedness and Response:*

Use of Information Technologies and Decision Support Systems, AHRQ Publication No. 02 E028, Agency for Healthcare Research and Quality, Rockville, MD, June 2002.

Bozzette, S.A., Boer, R., Bhatnagar, V., Brower, J.L., Keeler, E.B., Morton, S.C., and Stoto, M.A. *N. Engl. J. Med.*, **348**(5), 416–425 (2003).

Halloran, M.E., Longini, I.M., Jr., Nizam, A., and Yang, Y., *Science*, **298**(5597), 1428–1432 (2002).

Hupert, N. and Cuomo, J., "Markov Chain Modeling of Bioterrorism Response: Limitations and Suggestions for the Future," in R.V. Duncan and A.N. Sobel, Eds., *BTR 2003: Unified Science & Technology for Reducing Biological Threats & Countering Terrorism, Proceedings*, University of New Mexico, Albuquerque, NM, 2003, pp. 142–146.

Hupert, N., Mushlin, A.I., and Callahan, M.A., *Med. Decis. Making*, **22**(Suppl.5), S17–25 (2002).

Kaplan, E.H., Craft, D.L., and Wein, L.M., *Proc. Natl. Acad. Sci. U.S.A.*, **99**(16), 10935–10940 (2002).

Meltzer, M., Damon, I., LeDuc, J., and Millar, J., *Emerg. Infect. Dis.*, **7**(6), 959–969 (2001).

Wein, L.M., Craft, D.L., and Kaplan, E.H., *Proc. Natl. Acad. Sci. U.S.A.*, **100**(7), 4346–4351 (2003).

N

NATIONAL INSTITUTES OF HEALTH AND NATIONAL INSTITUTE OF ALLERGY AND INFECTIOUS DISEASES

LAUREN HARRISON

BACKGROUND

The National Institutes of Health (NIH) is one the largest supporters in the United States of biomedical research and also operates a substantial number of research establishments dedicated to helping prevent, detect, diagnose, and treat diseases. Founded in 1887 as a one-room Laboratory of Hygiene with a budget of $300, today, the NIH is a multi-campus national organization headquartered in Bethesda, Maryland, with more than 18,000 employees and a budget of close to $24 billion. It operates directly under the U.S. Department of Health and Human Services (HHS), and is constituted by 27 different institutes and centers, including the National Cancer Institute, National Eye Institute, National Human Genome Research Institute, National Institute of Biomedical Imaging and Bioengineering, National Institute on Drug Abuse, and National Institute of Allergy and Infectious Diseases (NIAID).

The NIH's mission is to improve worldwide health by uncovering new knowledge to prevent, detect, diagnose, and treat disease. To accomplish this goal, the NIH conducts research in its own facilities, teams with nonfederal researchers in universities, hospitals, and medical schools, trains new researchers, and provides an open means of communication for medical information. Almost 8 percent of the NIH's budget is allocated to support research in NIH facilities, including the Warren Grant Magnuson Clinical Center, the Mark O. Hatfield Clinical Research Center, and the National Library of Medicine. Nearly 84 percent of the NIH's budget is allocated to nonfederal research institutions across the United States, including numerous universities, hospitals, and medical schools, which work privately toward the same goals. The NIH and its associated research institutions are currently working to improve methods of preventing and treating cancer, diabetes, kidney disease, heart disease, Alzheimer's disease, drug abuse and alcoholism, AIDS, and other afflictions.

NIH BIODEFENSE: THE NATIONAL INSTITUTE OF ALLERGY AND INFECTIOUS DISEASES

Created in 1948 as the National Microbiological Institute through the merger of the Rocky Mountain Laboratory, the Biologics Control Laboratory, and NIH's Divisions of Infectious Diseases and Tropical Diseases, the NIAID carries out research to better understand, prevent, and treat infectious, allergic, and immunologic diseases, including those diseases and disease agents considered primary threats with respect to bioterrorism. The NIAID studies the capabilities of these organisms as well as the host response to them in order to create novel interventions and diagnostic tools.

Similar to its parent organization, the NIAID maintains both an internal and external research component. The internal research division is located on the main NIH campus and at its own Rocky Mountain Laboratories in Hamilton, Montana. These facilities focus primarily on research in biochemistry, epidemiology, immunology, immunopathology, molecular biology, parasitology, and virology. In addition, the NIAID provides resources and support for universities and industries throughout the world to conduct research on allergy, immunology, infectious disease, microbiology, and transplantation.

NIAID's Strategic Plan for Biodefense Research

After the "anthrax letters" of fall 2001, the NIAID focused some of its efforts on addressing the need for prevention, detection, and treatment of diseases caused by select bioterrorism agents as defined by the Centers for Disease Control and Prevention (CDC). Of the portion of the Homeland Security budget allocated for terrorism defense, substantial funds have been provided to the NIH to allow it to support scientific research and development efforts that will ultimately enable the United States to respond to a biological event. The NIAID has constructed a Strategic Plan for Biodefense Research for this purpose, which is intended to bolster these efforts. The Strategic Plan is divided into six sections, with each section addressing specific goals:

1. *Biology of the Microbe.* Understanding the essential biology and means by which pathogens cause disease is vital to bioterrorism prevention and intervention. Thus, genomic sequencing, proteomics, and other approaches are taken to better understand a particular pathogen's lifecycle, transmission, invasiveness, and virulence.

2. *Host Response.* A better understanding of human immune defenses against select agents is essential for effective vaccine development and other alternatives to enhancing host response.

3. *Vaccines.* Vaccines that can be applied safely and quickly are essential to establishing an adequate preparedness capability. Because of current concerns, research in this domain focuses primarily on anthrax and smallpox vaccines.

Encyclopedia of Bioterrorism Defense, Edited by Richard F. Pilch and Raymond A. Zilinskas
ISBN 0-471-46717-0 Copyright © 2005 Wiley-Liss

4. *Therapeutics.* Research in this category focuses on the need to examine existing and develop new antiinfectives and immunotherapies against select agents with the goal of expanding the quality and quantity of licensed antimicrobials, immunotherapeutics, and antitoxins available.

5. *Diagnostics.* Some diagnostic tools exist that allow for the correct identification of bioterrorist agents, but tend to be unreliable and expensive. NIAID's objective in this area thus is to develop inexpensive, sensitive, and specific diagnostic tools, and to provide a standard for validation and comparison.

6. *Research Resources.* An area of emphasis for NIAID is the development of centralized sources of expertise in bioterrorism. Thus, scientists are working on ways to develop both general and specific research resources to enable others in the field to work quickly and efficiently.

NIAID Research Agenda for CDC Category A Agents

Supplementing the Strategic Plan for Biodefense Research, the NIAID convened the "Blue Ribbon Panel on Bioterrorism and its Implications for Biomedical Research" to assess current research on bioterrorism agents and make recommendations on how best to move forward in this arena. Through this experience, the NIAID developed the Biodefense Research Agenda, which similar to the Strategic Plan focuses on the need for basic biological research to develop diagnostics, therapeutics, and vaccines in regard to the CDC's Category A agents. This Plan has set research goals for each specific organism in the same categories elucidated by the Strategic Plan for Biodefense: basic biology, immunology and host response, vaccines, drugs, diagnostics, and research resources.

Recent NIAID Biodefense Highlights

- As a result of the fall 2001 "anthrax letters," NIH researchers, in conjunction with the Office of Naval Research and the National Science Foundation, compared the genetic structures of the *Bacillus anthracis* strain recovered from the first victim of the 2001 anthrax letter attacks in Florida, with the well-characterized Ames strain, in order to develop techniques to trace the origin of *B. anthracis* strains and determine whether they have been genetically modified.

- In 2002, NIH scientists conducted research demonstrating that the smallpox vaccine currently stored by the United States can be diluted without losing potency, which means that a much larger number of people can be protected than previously believed.

- In April 2002, NIH researchers uncovered the genetic basis for *Yersinia pestis* transmission.

- In April 2003, NIH scientists sequenced the complete *B. anthracis* genome.

- In August 2003, scientists at the NIH created an experimental Ebola vaccine that tested successfully in preclinical trials (monkeys).

New NIAID Biodefense Initiatives

In 2002 and 2003, NIAID announced more than 50 initiatives related to biodefense, of which 75 percent were new. The following is a sample list of some these initiatives:

- New antibody production facilities will be created in order to increase the production of antibodies specific for diseases caused by bioterrorism agents.

- A Food and Waterborne Diseases Integrated Research Network and a Respiratory Pathogens Research Network will be created to provide means for continued research on these organisms.

- A team of scientists from academia and the biotechnology industry will work to identify possible drug targets against hemorrhagic fever viruses.

- NIAID's vaccine treatment program will be expanded to test vaccines against smallpox, anthrax, and West Nile fever.

- NIAID will construct and renovate biosafety level 3 and 4 (BSL-3 and BSL-4) facilities, thereby providing additional safe research space for the development of bioterrorism countermeasures. These facilities will be located at the NIAID campus in Frederick, Maryland, and at Rocky Mountain Laboratory in Montana.

WEB RESOURCES

The National Institute of Allergy and Infectious Diseases website, http://www.niaid.nih.gov/default.htm, accessed on 10/22/03.

The National Institute of Allergy and Infectious Diseases website, *NIAID Biodefense Research Agenda for Category A Agents*, http://www.niaid.nih.gov/biodefense/research/biotresearch agenda.pdf, accessed on 10/20/03.

The National Institute of Allergy and Infectious Diseases website, *NIAID Strategic Plan for Biodefense Research*, http://www.niaid.nih.gov/biodefense/research/strategic.pdf, accessed on 10/20/03.

The National Institutes of Health website, http://www.nih.gov/, accessed on 10/22/03.

NATO AND BIOTERRORISM DEFENSE

ANDREA CELLINO

NATO IN THE PAST

The North Atlantic Treaty Organization (NATO) is a relative newcomer to the field of bioterrorism defense. Only after September 11, 2001, did the Alliance, for the first time in its more than 50-year history, launch specific initiatives to address the threat of bioterrorism.

Of course, before this, NATO had addressed the potential threat of biological warfare (BW), but this was in the context of the Cold War, when the Alliance's collective defense efforts were focused on deterring military aggression from the Soviet Union. As well as preparing to defend against the Soviet Union's nuclear, biological, and chemical (NBC) weapons, allies regularly consulted on nonproliferation, arms control, and disarmament matters. With specific regard to biological weapons, all NATO members have signed and ratified the 1972 Biological and Toxins Weapons Convention (BWC).

Before 1991, terrorism was scarcely mentioned in NATO's major policy documents. If it was, the language would broadly "urge closer international cooperation" to fight terrorism, and encourage allies "to work together to eradicate this scourge" (North Atlantic Council, 1986).

Things changed slightly after the fall of the Soviet Union. In 1991, the Alliance's Strategic Concept, while reiterating that "any armed attack on the territory of the Allies, from whatever direction, would be covered by Articles 5 and 6 of the Washington Treaty," also indicated that "Alliance security interests can be affected by other risks of a wider nature, including proliferation of weapons of mass destruction, disruption of the flow of vital resources and actions of terrorism and sabotage" (Alliance's Strategic Concept, 1991).

However, despite appearing on the same list of new challenges facing the Alliance, NBC proliferation and terrorism remained quite distinct in NATO's policy. At the 1994 Brussels Summit, the Alliance formally included addressing the threat of NBC weapons among its priorities and decided to intensify efforts against proliferation. At the same meeting, NATO leaders generically condemned terrorism and stressed "the need for the most effective cooperation possible to prevent and suppress" it "in accordance with our national legislation" (North Atlantic Council, 1994).

As a result of the 1994 Summit, two working groups were established under a Joint Committee on Proliferation (JCP): The Senior Politico-Military Group on Proliferation (SGP), charged with addressing the political aspects of proliferation; and the Senior Defence Group on Proliferation (DGP), responsible for identifying the military capabilities that NATO would need to discourage NBC proliferation, deter threats or use of NBC weapons, and protect populations, territory, and forces.

The SGP concentrated on political and diplomatic efforts to prevent proliferation, focusing mainly on the periphery of NATO's territory. It also addressed the Alliance's contribution to implementing and strengthening arms control and nonproliferation regimes. The DGP, after conducting an assessment of proliferation risks, identified a range of capabilities needed to improve NATO's defenses against NBC weapons. According to its findings, alongside prevention efforts and the strengthening of nonproliferation norms, robust military capabilities would "signal to proliferants the utmost seriousness with which NATO approaches proliferation risks." In particular, complementing nuclear forces "with an appropriate mix of conventional response capabilities and passive and active defenses" would reinforce NATO's deterrence posture against NBC threats. The DGP also indicated a number of core capabilities that would strengthen the Alliance's nonproliferation objectives. These included strategic and operational intelligence; automated and deployable command, control, and communications; wide area ground surveillance; standoff/point biological and chemical detection, identification, and warning; extended air defenses for deployed forces; and individual protective equipment for deployed forces (NATO Parliamentary Assembly, 2002).

The DGP recommendations, which included providing the necessary capabilities and adapting NATO's doctrine, plans, and training to counter the NBC threat, were approved by Alliance defense ministers in December 1996. On that occasion, ministers stressed that NATO's defense planning should put particular emphasis on enhancing protection for deployed forces and improving defenses against biological weapons. The endorsement by NATO ministers meant that, from that moment on, force goals addressing NBC risks would be an integral part of collective defense planning.

Building on the work of the two groups on proliferation, NATO ministers indicated in subsequent meetings that they intended to expand the Alliance's efforts to address the evolving proliferation threat. Such efforts resulted in the Alliance WMD (weapons of mass destruction) Initiative adopted by NATO during the April 1999 Washington Summit. The aim of this Initiative was mainly to give coherence to and coordinate the activities of the various NATO bodies involved in proliferation matters. Apart from encouraging "a more vigorous, structured debate" at NATO "leading to strengthened

Encyclopedia of Bioterrorism Defense,   Edited by Richard F. Pilch and Raymond A. Zilinskas
ISBN 0-471-46717-0   Copyright © 2005 Wiley-Liss

common understanding" on proliferation, the Initiative was intended to improve intelligence and information-sharing among allies; increase military readiness to operate in NBC environments; enhance allies' mutual assistance in protecting civil populations; and support a public information strategy (Hain-Cole, 1999).

To implement and coordinate the 1999 Initiative, a WMD centre was created at NATO Headquarters in Brussels. Opened in May 2000, the centre has been headed since its creation by Edward C. Whiteside, a Canadian diplomat, who is assisted by two deputies and about 10 technical staff. This team has expertise in chemical weapons, biological agents, ballistic missiles, force protection, intelligence, and political aspects of arms control and nonproliferation regimes. According to Mr. Whiteside, "the role of the centre is three-fold; (1) to improve intelligence and information-sharing about proliferation issues; (2) to assist Allies in enhancing the military capabilities to work in a WMD environment; and (3) to discuss and bring the Alliance's support to non-proliferation efforts in the world" (http://www.nato.int/multi/video/2003/v030522/v030522a.htm, accessed 11/30/03). More specifically, the centre supports the SGP and the DGP and, through them, the North Atlantic Council in dealing with proliferation.

The WMD centre, in support of the SGP, has also taken up the task of maintaining the Matrix of Bilateral WMD Destruction and Management Assistance Programmes. The purpose of the Matrix is "to highlight the assistance programmes from NATO countries, non-NATO countries, and International Organisations to Russia and NIS countries in the field of decommissioning weapons of mass destruction" and related material, including biological agents. It also aims to avoid overlap between national assistance programmes of allies and to enhance multilateral coordination of assistance (NATO Parliamentary Assembly, 2002).

At the 1999 Washington Summit, NATO also adopted a revised Strategic Concept that, among other things, recognized for the first time in an official Alliance document the possible link between NBC weapons and "non-state actors." Paragraph 22, states that: "The Alliance recognises that proliferation can occur despite efforts to prevent it and can pose a direct military threat to the Allies' populations, territory, and forces. Some states, including on NATO's periphery and in other regions, sell or acquire or try to acquire NBC weapons and delivery means. Commodities and technology that could be used to build these weapons of mass destruction and their delivery means are becoming more common, while detection and prevention of illicit trade in these materials and know-how continues to be difficult. Non-state actors have shown the potential to create and use some of these weapons." Further, at the end of paragraph 53, the document clearly states that "the Alliance's forces and infrastructure must be protected against terrorist attacks" (Alliance's Strategic Concept, 1999).

NATO TODAY

The tragic events of September 11, a little more than two years after the Washington Summit, brought about dramatic changes for NATO, as for the rest of world. The Alliance reacted not only by invoking Article 5 for the first time in its history but also by stepping up its efforts against terrorism and in support of the United States's campaign to fight it. Between the attacks on the World Trade Center and the Prague Summit in November 2002, NATO worked vigorously to expand and reinforce its overall role in the fight against terrorism in both political and military terms. In Prague, NATO approved a new Military Concept for defense against terrorism, which underscored NBC defense as "high priority" (NATO's Military Concept for Defence Against Terrorism, 2002) and therefore launched a comprehensive package of measures to strengthen NATO's "ability to meet the challenges to the security of our forces, populations and territory... including the threat posed by terrorism and by the proliferation of weapons of mass destruction and their means of delivery" (Prague Summit Declaration, 2002). A number of initiatives endorsed by the Prague Summit are extremely relevant in the area of defense against attacks involving biological weapons.

NATO Defense ministers decided in June 2002 to enhance Alliance forces' ability to defend against NBC weapons with five specific capabilities developed by the DGP. These initiatives, adopted in Prague, include:

- A Disease Surveillance System aimed at collecting information on unusual disease outbreaks; alerting NATO Commanders in case of a biological outbreak or incident; and combining and collating data with other information sources, such as the World Health Organization (WHO). NATO is also working on an updated database of terrorist groups and incidents involving NBC and radiological weapons, and is developing a database of global endemic diseases.

- An NBC Event Response Team able to assess the effects of an NBC event, advise NATO Commanders on mitigating these effects, and help them to reach back to national experts for further technical advice. A prototype multinational Team, to which all member countries contribute, has been set up for a one year trial period.

- A Prototype Deployable NBC Analytical Laboratory that is quickly and easily transportable into theatre, able to investigate and collect samples of possible NBC contamination, and capable of conducting highly reliable scientific analysis of samples, has also been set up for one year trial period.

- A NATO Biological and Chemical Defence Stockpile has been developed, enabling allies to identify and share national stockpiles, provide assistance to deployed forces by rapidly moving needed vaccines or defense material into theatre, and improve medical treatment protocols. In this regard, because of the lack of an effective interface between different epidemiological systems, NATO is developing interoperability concepts for medical surveillance that will support deployed forces. Such a network could eventually be linked to the overall WHO system.

- A Virtual Training Centre for NBC Defence has been developed in the WMD Centre, which organizes exercises to enhance NBC education of senior NATO staff, improve operational understanding of NBC defense, and expand NBC defense training (http://www.nato.int/docu/comm/2002/0211-prague/exhibition/index.htm, accessed 11/30/03).

In 2003, Alliance defense ministers also agreed to deploy by July 2004 a multinational Chemical, Biological, Radiological, and Nuclear (CBRN) Defence Battalion (Fig. 1), which will be part of the NATO Response Force (NRF). The NRF, inaugurated on 15 October 2003, is a multinational force, kept on the highest level of alert, ready to be deployed anywhere in the world with just five days' notice to meet threats from terrorists, mount peacekeeping operations, evacuate civilians, or help out in natural disasters. Some capabilities provided in the context of the five Prague initiatives—primarily the prototype Response Team and Deployable Laboratory—may eventually merge into the NBC Battalion and become permanent. Despite the timeliness of all of these initiatives, according to a senior Pentagon official, some member countries are not reacting rapidly enough and very basic capabilities are still lacking. NATO officials, however, are confident that the deadlines will be met (Meeting the Threat of WMD Terrorism, 2003). On 1 December 2003, the new CBRN Defence Battalion achieved its initial operational capability. The Czech Republic leads the unit, which is headquartered in Liberec, in the northern part of the country. Twelve other NATO countries will participate in the formation of the Battalion and will conduct periodic training throughout 2004.

Terrorism and proliferation are also on the agenda of the NATO-Russia Council (NRC), established in May 2002 to improve relations between the Alliance and Moscow. Although working groups on both issues have been created in such context, no substantial decisions have been taken with regard to possible common action.

The Partnership Action Plan against Terrorism, also adopted in Prague and involving NATO and 27 Partner nations in Europe, the Caucasus, and Central Asia, has among its principal objectives "to cooperate in preventing and defending against terrorist attacks and dealing with their consequences." Several specific activities are envisaged under this plan, including measures to control biological and chemical agents, border control assistance, and information exchange on "WMD-related terrorism" (Partnership Action Plan against Terrorism, 2002).

NATO's Civil Emergency Planning (CEP) organization has also recently intensified its efforts to protect civil populations against biological, chemical, or radiological terrorism. Active in this area for a number of years, CEP consists of a number of NATO Committees and the Euro-Atlantic Disaster Response Co-ordination Centre (EADRCC), created in 1998 in cooperation with the UN Office for the Co-ordination of Humanitarian Affairs (UN-OCHA). After September 11, CEP began working on an inventory of national capabilities and assets that could be made available in the event of a biological, chemical, or radiological attack in order to protect

Figure 1. Multinational chemical, biological, radiological, and nuclear (CBRN) defence battalion, November 26, 2003; *Source:* NATO/SHAPE photos.

civilian populations. This project, completed in 2003, has involved NATO members and partners. The inventory includes points of contact; medical experts; lists of medical countermeasures; warning, detection, decontamination and protective equipment; and lists of laboratories and specialized hospitals.

The EADRCC, which is based at NATO headquarters and staffed by experts from several member and partner countries, is ready to react to major incidents involving biological, chemical, or radiological agents across the Euro-Atlantic area and act as a clearing house for international assistance. A CEP Action Plan listing over 50 action items to respond to such incidents was adopted in Prague, with guiding principles aimed to encourage to the maximum extent possible an integrated civil-military response; improve interoperability between partners through the formulation of common minimum standards in equipment, planning, training and procedures; and coordinate efforts with other international organizations. In this regard, interesting developments could occur in the context of the burgeoning security relationship between NATO and the European Union, which is also intensifying its efforts in the area of civil protection against bioterrorism.

While significant progress has been made in the implementation of the CEP Action Plan, according to a NATO official interviewed in September 2003, "preparation and capabilities are generally much better with regard to possible chemical and radiological incidents/attacks than they are for possible biological incidents/attacks... It is reasonable to assume that the adequacy of response capability for certain biological scenarios is somewhat doubtful to say the least" (Interview, 2003).

The CEP, sometimes in coordination with the WMD Centre, organizes workshops on biodefense-related issues. A NATO Advanced Research Workshop on "Preparedness against bio-terrorism and re-emerging infectious diseases—regional capabilities, needs and expectations in Central and Eastern European countries" was held in Warsaw from 15 to 18 January 2003. The objective of the workshop was to provide countries in which a biodefense system is currently under development with the experience of those that are more advanced in this area. Moreover, the CEP, together with the WHO, is exploring possibilities for tabletop exercises focusing on the bioterrorist threat in late 2004.

NATO TOMORROW

In the future, NATO will likely have an important role in the area of bioterrorism defense. Although some of the initiatives and plans outlined above are still in their infancy, the potential for developing robust NBC defenses in the context of the Alliance is substantial. Greater involvement of Russia and other key countries, not only in Europe, the Caucasus, and Central Asia but also in the Mediterranean (through NATO's well-established cooperative initiative), would certainly be a great asset for the Alliance's biodefense strategies. Much depends, however, on the political will of member states and partners to provide essential capabilities and expertise, as well as on the level of cooperation and coordination that the Alliance will establish with other organizations, primarily the European Union.

REFERENCES

Declaration of the Heads of State and Government participating in the Meeting of the North Atlantic Council (*"The Brussels Summit Declaration"*), Brussels, January 11, 1994, paragraphs 1 and 19.

Hain-Cole, C., *NATO Rev.*, **47**(2), 33–34 Summer (1999).

Interview with a NATO CEP senior staff conducted by the NATO Parliamentary Assembly, September 3, 2003.

NATO Parliamentary Assembly in Wilton Park, on *Meeting the Threat of WMD Terrorism*, June 6–8, 2003.

NATO's Military Concept for Defence Against Terrorism, Prague, November 21, 2002.

NATO website, http://www.nato.int/docu/comm/2002/0211-prague/exhibition/index.htm, accessed on 11/30/03.

Partnership Action Plan against Terrorism, Prague, November 22, 2002.

Prague Summit Declaration, issued by the Heads of State and Government participating in the meeting of the North Atlantic Council, November 21, 2002, paragraph 3.

Secretary General of NATO commentary on the Policy Recommendations adopted in 1998 by the NATO Parliamentary Assembly (formerly North Atlantic Assembly), March 1999.

Secretary General of NATO commentary on the Policy Recommendations adopted in 2001 by the NATO Parliamentary Assembly, April 2002.

Statement of the Ministerial Meeting of the North Atlantic Council, Halifax, Canada, May 29–30, 1986.

Ted Whiteside video interview, 22 May 2003, available on the NATO website at http://www.nato.int/multi/video/2003/v030522/v030522a.htm

The Alliance's Strategic Concept agreed by the Heads of State and Government participating in the Meeting of the North Atlantic Council in Rome 7–8 November 1991, paragraph 12.

The Alliance's Strategic Concept approved by the Heads of State and Government participating in the Meeting of the North Atlantic Council, Washington, DC, April 23–24, 1999.

NORTH AMERICAN MILITIA

MONTEREY WMD-TERRORISM DATABASE STAFF

OVERVIEW

In 1998, Randy Graham of Springfield, Bradford Metcalf of Olivet, and Kenneth Carter of Battle Creek, Michigan, all members of the North American Militia, were implicated in a series of plots against government officials. A search of the homes and belongings of these individuals by law enforcement officers uncovered, in addition to conventional weapons materials, a tape with instructions for the manufacture of ricin. The men were ultimately convicted on charges unrelated to biological weapons.

THE INCIDENT

In March 1997, an Alcohol, Tobacco, and Firearms (ATF) agent infiltrated the North American Militia and learned of various plots of three of its members—Graham, Metcalf, and Carter—to assassinate certain government officials and sabotage a number of public facilities (*Detroit News*, 1998; *Wood TV*, 1998). Targets included public bridges, power lines, roads, utility transmitters, television stations, IRS offices in Portage, Michigan, a federal building in Battle Creek, Michigan, and specific government employees (among others, U.S. Senator Carl Levin, Governor John Engler, and Federal Judge Enslen, who later was Graham's sentencing judge) (*AP Online*, 1999; *AP State and Local Wire*, 1999). Weapons of choice included sniper rifles, machine guns, and bombs. When federal law enforcement raided the homes of these men, they discovered an arsenal of such weapons and a videotape entitled "The Poor Man's James Bond." Produced in a cooking-show format, the tape provided instructions on the manufacture of bombs and other assorted militia-type weaponry, including a feature segment on how to extract ricin from castor beans. During the court proceedings, prosecutors drew attention to the ricin segment, stating that the men were "collecting information on the manufacture and use of ricin" (*Detroit News*, 1999). However, other than the videotape, no materials associated with ricin production were found in any of the raids. Graham, Metcalf, and Carter were arrested on weapons and conspiracy charges in March 1998 and indicted one month later.

OUTCOME

During legal proceedings on June 1, 1998, Carter stated that he had "conspired to possess machine guns, to threaten to assault and murder federal employees and to damage and destroy federal buildings by explosives" (*Detroit News*, 1999). He was ultimately sentenced to five years in prison. Metcalf was found guilty on weapons and conspiracy charges, and on May 25, 1999, was sentenced to 40 years in prison with no parole. Graham was found guilty of planning to use explosives to destroy federal property, threatening the lives of various government officials, and cultivating 32 marijuana plants in order to raise money for his organization (*AP Online*, 1999; *WWMT Channel 3*, 1999), and on June 10, 1999, was sentenced to a 55-year prison term to be served in a maximum-security facility (*AP Online*, 1999; Irwin, 1999).

REFERENCES

Associated Press Online, Three Militia Members Sentenced in Mich., June 11, 1999.

Associated Press State & Local Wire, Militia Member Draws Five Years in Prison, Judge's Tongue-Lashing, May 21, 1999.

Detroit News, Militia Member Pleads Guilty in Plot, June 3, 1998, Internet, available from http://detnews.com/1998/metro/9806/03/06030125.htm.

Detroit News, Militia Members' Trial to Begin, January 11, 1999, Internet, available from http://detnews.com/1999/metro/9901/11/01110044.htm.

Irwin, J., Last of Three Militia Members Sentenced in Terrorist Plot, The Associated Press State & Local Wire, June 11, 1999.

Monterey WMD-Terrorism Database, Center for Nonproliferation Studies, 1998–2003.

WOOD-TV, News 8 at Noon, Video Monitoring Services of America, March 24, 1998.

WWMT Channel 3, Militia Member Guilty of Plotting Attack, February 25, 1999 [likely meant to be 26 January 1999]), Internet, available from http://209.41.6.5.News3-1999/WWMT-archive/19990126.htm, accessed on 7/13/99.

WEB SOURCES

Associated Press, North American Militia Members Convicted for Plotting Terrorism, (29 January 1999), Internet, available from http://www.factnet.org/cults/militia/militia.htm, accessed on 7/13/99.

Detroit News, FBI: Militia Threatened Informant, November 12, 1998, Internet, available from http://detnews.com/1998/metro/9811/12/11120087.htm.

NORTHCOM (U.S. NORTHERN COMMAND)

CHARLOTTE SAVIDGE

INTRODUCTION

Headquartered at Peterson Air Force Base in Colorado Springs, Colorado, U.S. Northern Command (NORTHCOM) commenced operations on October 1, 2002, and achieved full operational capability on October 1, 2003 (Miles, 2003). NORTHCOM's mission is homeland defense and civil support (http://www.northcom.mil/index.cfm?fuseaction=s.who_mission, accessed December 17, 2003).

BACKGROUND

Prior to September 11, 2001, the United States believed that the majority of threats to the country originated from outside of North America. The events of September 11 demonstrated that this was not necessarily the case, and pointed to the need for a unified command to protect against threats and aggression launched from within North America (http://www.northcom.mil/index.cfm?fuseaction=s.home_neighbors, accessed December 18, 2003). As a result, President George W. Bush established NORTHCOM in April 2002 with the signing of a new Unified Command Plan. It represents the first U.S. unified command created for homeland defense, a mission that had previously been divided between several commands.

NORTHCOM demonstrated its initial operational capability in September 2002 with the simulation exercise Unified Defense (Fisher, 2003). In August 2003, it conducted Determined Promise (DP) 03, a two-week exercise simulating an outbreak of pneumonic plague in Clark County, Nevada (http://www.esi911.com/esi/news/dp.shtml, accessed December 18, 2003), the successful completion of which demonstrated NORTHCOM's full operational capability. NORTHCOM has also organized several tabletop exercises (Garamone, 2003) and participated in the Department of State and Department of Homeland Security's TOPOFF 2, which simulated the explosion of a radiological dispersal device near downtown Seattle and the release of a biological agent in Chicago (http://www.northcom.mil/index.cfm?fuseaction=news.showstory&storyid=22FD315D-A3E4-E25B-B3758C0521FD3D97, posted May 13, 2003).

CURRENT RESPONSIBILITIES AND EFFORTS

To fulfill its homeland defense and civil support mission, NORTHCOM plans, organizes, and conducts operations to deter, prevent, and defeat external threats and aggression aimed at the United States, its territories, and interests within its specified area of responsibility. This area of responsibility includes the continental United States, Alaska, Puerto Rico, the U.S. Virgin Islands, and the Gulf of Mexico, as well as air, land, and sea approaches such as Canada, Mexico, and surrounding waters to a distance of approximately 500 nautical miles out from U.S. territory. Security cooperation and coordination with Canada and Mexico are also within NORTHCOM's domain. The U.S. Pacific Command is responsible for Hawaii and U.S. territories and possessions in the Pacific (http://www.northcom.mil/index.cfm?fuseaction=news.factsheets, accessed December 17, 2003).

Under the direction of the President or Secretary of Defense, NORTHCOM is also charged with assisting federal agencies with disaster relief, counterdrug operations, and consequence management following incidents of chemical, biological, or nuclear terrorism (http://www.northcom.mil/index.cfm?fuseaction=s.who_civil, accessed December 17, 2003). Such assistance is only provided if it is clear that the seriousness of the event is beyond a state's and local agencies' capacities to manage. Were such an event to take place, NORTHCOM would provide "one-stop shopping" for state and local officials in need of federal support (Carter, 2003).

NORTHCOM's daily activities are implemented through three Joint Task Forces subordinated to NORTHCOM: the Joint Force Headquarters-Homeland Security (JFHQ-HLS), the Joint Task Force-Civil Support (JTF-CS), and the Joint Task Force 6 (JTF-6) (Garamone, 2002). The JFHQ-HLS was established in the weeks immediately following the September 11 terrorist attacks. Under NORTHCOM, it has the responsibility for coordinating homeland defense and civil support activities including prevention, crisis response, and consequence management for federal agencies such as the Federal Emergency Management Agency (FEMA). It also oversees the JTF-CS, which is responsible for consequence management in the event of an incident involving a weapon of mass destruction (http://www.northcom.mil/index.cfm?fuseaction=

Encyclopedia of Bioterrorism Defense, Edited by Richard F. Pilch and Raymond A. Zilinskas
ISBN 0-471-46717-0   Copyright © 2005 Wiley-Liss

news.factsheets&factsheet=4, accessed December 17, 2003). The JTF-6 coordinates the Department of Defense's counter-drug activities in support of federal law enforcement agencies. If needed, the JTF-6 also supports federal law enforcement agencies' counterterrorism efforts (http://www.northcom.mil/index.cfm?fuseaction= news.factsheets&factsheet=2#taskforcesix, accessed December 17, 2003). Other command units would be assigned to NORTHCOM as needed.

NORTHCOM's Combined Intelligence and Fusion Center collects and analyzes intelligence from the U.S. intelligence agencies and almost 50 other government agencies to identify potential threats that fall within NORTHCOM's area of responsibility (Elliot, 2003).

ORGANIZATIONAL TIES

To fulfill its homeland defense and civil support mission, NORTHCOM relies on cooperation with a variety of other organizations. To protect North American air space, NORTHCOM works with the United States-Canadian North American Aerospace Defense Command (NORAD) (Carter, 2003). In its counterterrorism operations, NORTHCOM shares information and coordinates plans and activities with several federal agencies to enhance overall homeland security efforts. In cases where state and local agencies are unable to handle disaster relief or consequence management, NORTHCOM provides such assistance in coordination with lead federal agencies and with the prior approval of lawyers and officials at the Department of Defense (Foster, 2003). In this regard, NORTHCOM acts within the confines of the Posse Comitatus Act, which prevents the military from direct involvement in domestic law enforcement except in cases of crimes involving nuclear materials, emergency situations involving chemical or biological weapons, or national emergencies beyond the capability of domestic law enforcement agencies (http://www.northcom.mil/index.cfm?fuseaction= news.factsheets&factsheet=5#pca, accessed December 18, 2003).

The simulation exercise DP-03 demonstrated the extent of NORTHCOM's intended cooperation with other organizations. For DP-03, NORTHCOM engaged fourteen federal agencies: Assistant Secretary of Defense (Homeland Defense), Department of Homeland Security, FEMA Region IX, Transportation Security Administration, National Security Agency, Defense Intelligence Agency, National Reconnaissance Office, Defense Logistics Agency, National Imagery and Mapping Agency, Department of Veterans Affairs, Department of Health and Human Services, Central Intelligence Agency, U.S. Coast Guard, and the Army Corps of Engineers. In addition, it engaged two nongovernmental organizations (the American Red Cross and American Association of Railroads), seven state and local organizations (State of Nevada Division of Emergency Management, Clark County Office of Emergency Management, Nevada Army National Guard, Nevada Air National Guard, 167th Theater Support Command Alabama Army National Guard, 91st Civil

Support Team, and 103rd Field Artillery Brigade, and Rhode Island Army National Guard), and 28 military units, including the Chemical Biological Rapid Response Team (http://www.jtfcs.northcom.mil/pressreleases/news 08152003NorthcomFactSheet.htm, posted August 15, 2003).

THE FUTURE

NORTHCOM plans to hold a minimum of two major simulation exercises per year through 2008. These exercises will be used to develop skills as well as interagency relationships with federal, state, and local agencies across the country (Miles, 2003). The next DP, focusing on Los Angeles City and the Port of Los Angeles and covering chemical, biological, radiological, and nuclear (CBRN) defense support, will be held in August 2004 (Law and Canfield, 2003).

REFERENCES

Carter, A., One-Stop Shopping for U.S. Defense, Press Release, U.S. Northern Command website, Colorado Springs, U.S. Northern Command, posted June 19, 2003, http://www.northcom.mil/index.cfm?fuseaction=news.show story&storyid=E19C1635-FE83-245E-00B2C107DEC80B00.

Elliott, S., Eberhart Briefs Congress On U.S. Northern Command, Press Release, U.S. Northern Command website, Colorado Springs, U.S. Northern Command, posted March 14, 2003, http://www.northcom.mil/index.cfm?fuseaction=news.show-story&storyid=EE9730BE-E953-A751-135915FCE93DA621.

Fisher, D., Northern Command Announces Full Operational Capability, Press Release, U.S. Northern Command website, Colorado Springs, U.S. Northern Command, posted September 16, 2003, http://www.northcom.mil/index.cfm?fuseaction=news .showstory&storyid=AACAEE8B-F78C-4BAD-391CA9B2384F 586A.

Foster, D., Northcom a growing force to fight terror: Agency providing support, strategies to civilians, military, *Rocky Mountain News*, Colorado Springs, May 12, (2003), http://www.rockymountainnews.com/drmn/state/article/0,1299, DRMN_21_1954997,00.html.

Garamone, J., NORTHCOM Chief Says U.S. Better Prepared Against Terror, American Forces Information Service, Washington, October 2, 2003, http://www.defenselink.mil/news/ Oct2003/n10022003_200310026.html.

Garamone, J., Northern Command to Assume Defense Duties Oct. 1, American Forces Information Service, Washington, September 25, 2002, http://www.defenselink.mil/news/Sep2002/ n09252002_200209254.html.

Law, C. and Canfield, B., Minutes, Meeting of Emergency Operations Board, City of Los Angeles, City of Los Angeles Emergency Preparedness Department, July 21, 2003, http://www.ci.la.ca.us/epd/eobminutes/epdeobminutes 21113780_07212003.pdf.

Miles, D., Teamwork, Planning Behind NORTHCOM Stand-up, American Forces Information Service, Washington, October 15, 2003, http://www.defenselink.mil/news/Oct2003/ n10152003_200310155.html.

WEB RESOURCES

Emergency Services Integrators website, WebEOC Receives High Praise by Users Participating in Determined Promise—03, http://www.esi911.com/esi/news/dp.shtml, accessed on 12/18/03.

Joint Task Force Civil Support News, Top Story: Exercise Determined Promise—03, August 15, 2003, http://www.jtfcs.northcom.mil/pressreleases/news08152003 NorthcomFactSheet.htm.

U.S. Northern Command website, History: Our Neighbors, http://www.northcom.mil/index.cfm?fuseaction=s.home _neighbors, accessed on 12/18/03.

U.S. Northern Command website, Joint Force Headquarters Homeland Security Fact Sheet, http://www.northcom.mil/index. cfm?fuseaction=news.factsheets&factsheet=4, accessed on 12/17/03.

U.S. Northern Command website, Joint Task Force Six Fact Sheet, http://www.northcom.mil/index.cfm?fuseaction=news. fact-sheets&factsheet=2#taskforcesix, 12/17/03.

U.S. Northern Command website, Posse Comitatus Act Fact Sheet, http://www.northcom.mil/index.cfm?fuseaction=news. fact-sheets&factsheet=5#pca, 12/18/03.

U.S. Northern Command website, TOPOFF 2 Another Opportunity to Train, Press Release, posted May 13, 2003, http://www.northcom.mil/index.cfm?fuseaction=news .showstory&storyid=22FD315D-A3E4-E25B-B3758C0521FD3D97.

U.S. Northern Command website, US Northern Command Fact Sheet, http://www.northcom.mil/index.cfm?fuseaction=news .factsheets, accessed on 12/17/03.

U.S. Northern Command website, Who We Are: Civil Support, http://www.northcom.mil/index.cfm?fuseaction=s.who_civil, accessed on 12/17/03.

U.S. Northern Command website, Who We Are: Mission, http://www.northcom.mil/index.cfm?fuseaction=s.who_mission, accessed on 12/17/03.

See also HOMELAND DEFENSE and JOINT TASK FORCE CIVIL SUPPORT.

OFFICE INTERNATIONAL DES EPIZOOTIES: WORLD ORGANIZATION FOR ANIMAL HEALTH

MICHAEL WOODFORD

OVERVIEW

The Office International des Epizooties (OIE) is an intergovernmental organization with 162 Member Countries as of March 2003. Its primary missions are:

- To inform governments on the existence of animal diseases, changes in their distribution worldwide, and means of controlling them;
- To coordinate research at an international level on animal disease surveillance and control; and
- To examine regulations on international trade in animals and animal products, with a view to their harmonization between member countries.

The OIE maintains permanent international relations with numerous organizations, including the Food and Agriculture Organization of the United Nations (FAO), World Health Organization (WHO), World Trade Organization (WTO), Inter-American Institute for Cooperation on Agriculture (IICA), and Pan American Health Organization (PAHO).

ORGANIZATION OF THE OIE

International Committee

The International Committee is the highest authority of the OIE. It comprises all of the delegates, and meets at least once a year during the General Session in Paris in May. Voting by the delegates within the International Committee respects the democratic principle of one country, one vote. The principal functions of the International Committee are:

- To adopt international standards in the field of animal health, especially as pertaining to the international animal trade;
- To adopt resolutions for the control of the major animal diseases;
- To elect members of the OIE's statutory bodies (president and vice president of the committee, members of

the Administrative Commission, Regional Commissions, and Specialist Commissions);
- To appoint the Director General of the OIE; and
- To examine and approve the annual OIE activity report, financial report, and budget presented by the Director General.

During the General Session, two technical items of current interest are presented by rapporteurs who are recognized authorities on these subjects, and closely monitored changes affecting the distribution of the major animal diseases throughout the world are considered.

The work of the International Committee is prepared by the Administrative Commission, comprising nine delegates. This commission meets in February and May of each year.

Administrative Commission

The Administrative Commission, consisting of the president of the International Committee, the vice president, the past president, and six elected delegates, represents the committee in the interval between the General Sessions. The commission meets twice a year to examine, in consultation with the Director General, technical and administrative matters, in particular, the programme of activities and financial documents to be submitted to the International Committee for approval.

Regional Commissions

The five Regional Commissions study specific problems affecting the veterinary services and organize cooperation within each of the following: Africa; Americas; Asia, the Far East, and Oceania; Europe; and the Middle East. Each commission holds a meeting every two years in one of the countries of the region to study technical items and regional cooperation on animal disease control. The Regional Commissions also meet during the General Session of the International Committee. They report to the committee on their activities and submit recommendations.

Encyclopedia of Bioterrorism Defense, Edited by Richard F. Pilch and Raymond A. Zilinskas
ISBN 0-471-46717-0 Copyright © 2005 Wiley-Liss

The Director General and the Central Bureau

The Central Bureau, located in Paris, is managed by the Director General of the OIE. He is appointed by the International Committee. The Central Bureau implements the strategy determined by the International Committee and coordinates the corresponding activities in the fields of information, international cooperation, and scientific dissemination. The Central Bureau also provides the secretariat for the annual General Session of the committee, the various meetings of the commissions, and technical meetings held at the OIE, and contributes to the secretariat for regional and specialized conferences.

With the help of voluntary contributions from some of the member countries, the Central Bureau provides the impetus for activities such as organizing regional training seminars and coordinating control programs.

The Central Bureau has become an international resource centre at the service of animal health officials worldwide.

TOWARDS GREATER TRANSPARENCY IN THE ANIMAL HEALTH SITUATION WORLDWIDE

The OIE is the observatory for animal health. Its key mission is to keep national veterinary services informed on the appearance and course of epizootics representing a threat to animal or public health. This information is classified according to the gravity of the epidemiological events reported, with the various diseases being categorized into two lists based on severity of effect: List A and List B (Fig. 1).

The warning system operated by the OIE Central Bureau allows member countries to react rapidly to animal disease outbreaks if the need arises. A country detecting the first outbreak of a List A disease or any other contagious disease likely to have serious public health implications or economic repercussions for animal production must declare the occurrence to the Central Bureau within 24 hours. This information is then distributed in a number of ways, including:

- Immediately relaying by telex, fax, or e-mail to countries directly threatened;
- Through the weekly publication, *Disease Information*, available on the OIE website or by mail, to all member countries;
- Through the OIE *Bulletin*, published every two months, which allows the course of List A disease outbreaks to be monitored. The *Bulletin* contains a number of other sections, including those on the epidemiology of other contagious diseases (List B) and the activities of the OIE;
- Through the annual compilation entitled *World Animal Health*, which provides a wide variety of information on the animal health situation in member countries and reports on the disease control methods that they apply;
- Via integration into *Handi*STATUS, a regularly updated computerized database available on the OIE website; and

- Through scientific publications, including the *Scientific and Technical Review*, which contains research articles of the very highest standard.

By collecting, processing, and disseminating data on animal diseases throughout the world, the OIE endeavors to ensure transparency in the animal health situation worldwide for the benefit of the member countries. The information thus generated is essential for the success of national and regional disease control programmes and to reduce the health risks arising from international trade in animals and animal products. This warning system also provides a worldwide surveillance network for the early detection and rapid reporting of any suspicious disease occurrence that could have its origin in an act of bioterrorism.

TOWARDS IMPROVED HEALTH SAFEGUARDS IN INTERNATIONAL TRADE

The smooth flow of animals and animal products requires the development and adoption by the international community of animal health regulations aimed at limiting the spread of diseases transmissible to both animals and humans, and the harmonization and greater transparency of sanitary regulations applicable to trade in animals and their products so as to avoid unnecessary obstacles to international trade.

The WTO Agreement on the Application of Sanitary and Phytosanitary Measures advocates the use of standards developed under the auspices of the OIE. In addition, various normative works approved by the OIE International Committee are designed to promote the harmonization of regulations applicable to trade in animals and animal products, for example:

- The *International Animal Health Code* for mammals, birds, and bees, developed by the International Animal Health Code Commission, is an important normative work for international trade. It is regularly updated and is available both as an electronic version on the OIE website and in a printed version.

- The *OIE International Animal Health Code* (the *Code*), which outlines the standards for the member countries and contains guidelines for disease reporting. These standards state that member countries previously considered disease-free should report outbreaks of List A diseases within 24 hours to the OIE. This information is then forwarded immediately to other member countries.

- The *Manual*, developed by the OIE's Standards Commission, presents standard methods for diagnostic tests and vaccine control to be applied notably in the context of international trade. It constitutes the reference work for the international harmonization of diagnosis of animal diseases and vaccine control.

LIST A

- Foot and mouth disease
- Swine vesicular disease
- Peste des petits ruminants
- Lumpy skin disease
- Bluetongue
- African horse sickness
- Classical swine fever
- Newcastle disease
- Vesicular stomatitis
- Rinderpest
- Contagious bovine pleuropneumonia
- Rift Valley fever
- Sheep pox and goat pox
- African swine fever
- Highly pathogenic avian influenza

LIST B

Multiple Species Diseases

- Anthrax
- Aujeszky's disease
- Echinococcosis/hydatidosis
- Heartwater
- Leptospirosis
- New world screwworm (*Cochliomyia hominivorax*)
- Old world screwworm (*Chrysomya bezziana*)
- Paratuberculosis
- Q fever
- Rabies
- Trichinellosis

Cattle Diseases

- Bovine anaplasmosis
- Bovine babesiosis
- Bovine brucellosis
- Bovine cysticercosis
- Bovine genital campylobacteriosis
- Bovine spongiform encephalopathy
- Bovine tuberculosis
- Dermatophilosis
- Enzootic bovine leukosis
- Hemorrhagic septicemia
- Infectious bovine rhinotracheitis/infectious pustular vulvo-vaginitis
- Malignant catarrhal fever
- Theileriosis
- Trichomonosis
- Trypanosomosis (tsetse-transmitted)

Sheep and Goat Diseases

- Caprine and ovine brucellosis (excluding *Brucella ovis*)
- Caprine arthritis/encephalitis
- Contagious agalactia
- Contagious caprine pleuropneumonia
- Enzootic abortion of ewes (ovine chlamydiosis)
- Maedi-visna
- Nairobi sheep disease
- Ovine epididymitis (*Brucella ovis*)
- Ovine pulmonary adenomatosis
- Salmonellosis (*Salmonella abortusovis*)
- Scrapie

Equine Disease

- Contagious equine metritis
- Dourine
- Epizootic lymphangitis
- Equine encephalomyelitis (Eastern and Western)
- Equine infectious anemia
- Equine influenza
- Equine piroplasmosis
- Equine rhinopneumonitis
- Equine viral arteritis
- Glanders
- Horse mange
- Horsepox
- Japanese encephalitis
- Surra (*Trypanosoma evansi*)
- Venezuelan equine encephalomyelitis

Swine Diseases

- Atrophic rhinitis of swine
- Enterovirus encephalomyelitis
- Porcine brucellosis
- Porcine cysticercosis
- Porcine reproductive and respiratory syndrome
- Transmissible gastroenteritis

Avian Diseases

- Avian chlamydiosis
- Avian infectious bronchitis
- Avian infectious laryngotracheitis
- Avian mycoplasmosis (*Mycoplasma gallisepticum*)
- Avian tuberculosis
- Duck virus enteritis
- Duck virus hepatitis
- Fowl cholera
- Fowlpox
- Fowl typhoid
- Infectious bursal disease (Gumboro disease)
- Marek's disease
- Pullorum disease

Lagomorph Diseases

- Myxomatosis
- Rabbit hemorrhagic disease
- Tularemia

Bee Diseases

- Acariosis of bees
- American foulbrood
- European foulbrood
- Nosemosis of bees
- Varroosis

Fish Diseases

- Epizootic haematopoietic necrosis
- Infectious haematopoietic necrosis
- *Oncorhynchus masou* virus disease
- Spring viraemia of carp
- Viral hemorrhagic septicemia

Mollusk Diseases

- Bonamiosis (*Bonamia exitiosus*, *B. ostreae*, *Mikrocytos roughleyi*)
- Marteiliosis (*Marteilia refringens*, *M. sydneyi*)
- Mikrocytosis (*Mikrocytos mackini*)
- MSX disease (*Haplosporidium nelsoni*)
- Perkinsosis (*Perkinsus marinus*, *P. olseni*/*atlanticus*)

Crustacean Diseases

- Taura syndrome
- White spot disease
- Yellowhead disease

Other

- Leishmaniasis

Figure 1. Office International des Epizooties: Classification of Diseases.

- A *Code* and a *Manual* for diseases of aquatic animals has been developed by the OIE Fish Diseases Commission.

- Lastly, the OIE develops and updates lists of countries recognized as being free from the most serious diseases, most notably foot and mouth disease. These lists make a substantial contribution to the health security of international trade.

The OIE now takes a proactive approach to disease reporting and will also report information on a disease outbreak that is provided by the OIE reference laboratories or by unofficial sources such as scientific publications, ProMED, and lay publications (but only after it has been verified by the member country). A recent outbreak of a List A disease that was promptly reported and subjected to effective control has been the appearance of Avian influenza in the Netherlands.

TOWARDS OBJECTIVE AND IMPARTIAL EXPERTISE IN ANIMAL HEALTH

The International Agreement of 25 January 1924, establishing the OIE made it responsible for promoting and coordinating research on the surveillance and control of animal diseases throughout the world. This objective has been attained by the creation of a veritable worldwide animal health network, involving the generation and implementation of Specialist Commissions and working groups, the designation of collaborating centres and reference laboratories, the organization of meetings of experts and the publication of scientific articles.

Specialist Commissions

The Specialist Commissions study both problems of animal disease surveillance and control and questions relating to the harmonization of international regulations. For example,

- The Foot-and-mouth Disease and Other Epizootics Commission contributes to the development of better strategies for animal disease surveillance and control. The commission convenes groups of specialists, particularly in the event of an animal health emergency or to verify the status of member countries in terms of specific animal diseases.

- The Standards Commission harmonizes methods for the diagnosis of animal diseases and the control of biological products, especially vaccines for veterinary purposes. The commission coordinates a program to develop standard reagents aimed at standardizing diagnosis.

- The Fish Diseases Commission collects all available information on diseases of fish, crustaceans, and molluscs, as well as control measures for these diseases. The commission harmonizes rules governing trade in aquacultural products as well as diagnostic methods. It also organizes scientific meetings on these topics.

Working Groups

Four working groups are currently active: Biotechnology, Informatics and Epidemiology, Veterinary Drug Registration, and Wildlife Diseases. These groups meet to review progress made in their subject field and to ensure that the information is rapidly made available to all OIE member countries. They also contribute to the organization of scientific meetings, seminars, workshops, and training courses.

Reference Laboratories and Collaborating Centres

The aim of these laboratories and centres is to provide OIE member countries with support and scientific advice on all matters relating to the surveillance and control of animal diseases. This support can take many forms, including the provision of experts, preparation and supply of diagnostic kits or standard reagents, seminars, courses, organization of scientific meetings, and so on.

WEB RESOURCES

OIE website: http://wwwoie.int

ORANGE OCTOBER: A CASE STUDY

MONTEREY WMD-TERRORISM DATABASE STAFF

INTRODUCTION

In August 1993, a group calling itself "Orange October" mailed envelopes containing syringes allegedly contaminated with an undifferentiated hepatitis virus to a large number of companies in the United Kingdom. While some employees did suffer inadvertent needlesticks as a result, forensic testing concluded that the needles had not been contaminated.

THE INCIDENT

Five weeks prior to this mass mailing, the group had delivered some 400 unstamped envelopes to companies including British Steel and Linpak and Geest, listing 10 organizations (delineated below) and demanding a payment of £100,000 each from 5 of these 10. Recipients were to confirm payment through advertisements in local newspapers. Threats for nonpayment included sending "hit-and-run" drivers after individuals, placing bombs in the cars of top executives, and introducing computer viruses into company networks. In compliance with police orders, no money was paid in response to either group of envelopes.

A second wave of envelopes containing the syringes followed, targeting a total of 700 companies. The envelopes also contained a similar list of organizations and threatened to "poison" food on supermarket shelves if the recipients did not make substantial payments to them. Again, no money was paid.

Under the direction of Scotland Yard, police established a nationwide investigation that failed to identify the perpetrators.

DISCUSSION

While Orange October's primary objective in this incident remains unclear, the group's actions appear to have been politically motivated. The organizations that stood to benefit from the extortion threats included Sinn Fein, Greenpeace, the Animal Liberation Campaign, Help the Homeless, and the Epilepsy Association. Further, as expressed in the group's first declaration signed by "Gerald Butler, Commander Ten Brigade," Orange October claimed to "redistribute power, wealth, and land" by attacking the G7 industrial nations.

Although unsuccessful in the end, the Orange October case serves as a reminder of the potential for inciting fear that even baseless claims regarding biological weapons may hold.

REFERENCES

Marshall, F., Shops Targeted by Blackmailers and Firebombers, *Press Association Newsfile*, December 13, 1995, Home News.

Monterey WMD-Terrorism Database, Center for Nonproliferation Studies, 1998–2003, http://cns.miis.edu.

Stern, C., Hunt for Orange October; Nationwide Terror Alert Gang Blackmailing Top Firms, *Main on Sunday (London)* August 22, 1993, 18.

Encyclopedia of Bioterrorism Defense,   Edited by Richard F. Pilch and Raymond A. Zilinskas
ISBN 0-471-46717-0   Copyright © 2005 Wiley-Liss

P

PALESTINE LIBERATION ORGANIZATION (PALESTINIAN AUTHORITY)

SKYLER J. CRANMER

OVERVIEW

Though the Palestinian Liberation Organization (PLO) draws upon Islam to some extent as a means of inspiring its members, it is basically a secular nationalist organization in that it does not call for the establishment of an Islamic state or the imposition of Islamic law. The PLO is an umbrella organization divided into diverse factions whose stated purpose is to "liberate" Palestine from the Israeli occupation and form a Palestinian state. When the PLO National Council first declared a Palestinian state on November 15, 1988, it listed Jerusalem as its capital; this declaration was then recognized by some 100 nations despite the fact that the PLO had no sovereign territory under its control. However, since February of 1998, the PLO has been considering a second, unilateral declaration of statehood that has yet to be issued. Since the PLO has never been able to control Jerusalem, it has operated from Jordan, southern Lebanon, Tunis, and most recently, the city of Ramallah in the West Bank.

The PLO currently denounces acts of terrorism and therefore does not officially sponsor terrorist operations. Historically, however, armed factions within the PLO have targeted Israeli soldiers, civilians, and interests both in Israel proper and the occupied territories, and its leaders have operated from neighboring states such as Syria, Lebanon, Jordan, Iraq, Egypt, and Tunisia. Today, the PLO has been institutionalized in the governing Palestinian Authority. As a result, it engages primarily in more or less legitimate political activities in the occupied territories, Israel proper, and fora organized by various international organizations, despite the fact that armed terrorist groups have continued to emerge from its ranks.

The exact number of PLO members is unknown, but its active members are estimated in the thousands and its supporters are said to number in the millions. The group considers itself to be the sole representative of the roughly 2.9 million Palestinians living in the occupied territories (Palestinian Central Bureau of Statistics, 1998). Since its inception, the PLO has been aided by a number of its neighbors, including Jordan, Syria, Lebanon, Iraq, and Egypt, and it has received the bulk of its financial support from donations and contributions from Arab governments, taxes collected from Palestinian workers in Arab countries, and the profits of PLO businesses (Mark, 2002).

The PLO is considered currently active despite the fact that it has not been regarded by the U.S. Department of State as a terrorist group since 1988.

HISTORICAL BACKGROUND

The PLO was founded by the Arab League in 1964 as an umbrella organization encompassing three constituent groups: the Popular Front for the Liberation of Palestine, the Popular Democratic Front for the Liberation of Palestine, and al-Fatah (Alexander and Sinai, 1989). Since the organization's inception, Yasir Arafat has been its most prominent figure, though he did not assume the chairmanship of the Executive Committee until 1969 and did not fully consolidate his control until 1974, when al-Fatah shifted the PLO's agenda of terrorism to include political elements for the first time (Federation of American Scientists, Undated). The United Nations (UN) recognized the PLO in December of 1988 when the group renounced terrorism, accepted UN Resolutions 242 and 338 (calling for an exchange of land for peace), and recognized Israel as a state. While the PLO has publicly adhered to its statement before the United Nations, it has often been argued, particularly by Israel, that the group continues to support terrorism.

The PLO has engaged Israel in a long and troubled peace process, the high points of which have been the 1978 Camp David Accords, the 1993 Oslo Accord, and the exchange of letters granting mutual recognition between Arafat and then Israeli Prime Minister Yitzhak Rabin. The low points have been the two orchestrated *Intifadas* ("uprisings"), the first of which ranged from December 1987 to April 1996 and left some 1900 Palestinians and almost 400 Israelis dead (Mark, 2002), and a second which began in September 2000 and is continuing as of this writing (Cordesman, 2002).

SELECTED CHRONOLOGY

A list of terrorist attacks perpetrated and/or supported by the PLO over the years is beyond the scope of this writing, but over the years it has carried out a large number of such attacks. In its early years, the PLO relied heavily on terrorist activities. That posture began to change in 1974 when the PLO entered the political arena, and was

Encyclopedia of Bioterrorism Defense, Edited by Richard F. Pilch and Raymond A. Zilinskas
ISBN 0-471-46717-0 Copyright © 2005 Wiley-Liss

transformed more fully in 1988 when the PLO officially renounced terrorism. Although the PLO is widely believed to have continued indirectly supporting terrorist groups that have emerged from al-Fatah, such as the al-Aqsa Martyrs Brigade, it has not publicly claimed credit for any terrorist attacks since its 1988 declaration before the United Nations.

BIOLOGICAL WEAPONS CAPABILITY

To date, there is no information indicating that the PLO has developed or possessed biological weapons at any time.

MOTIVATION TO PURSUE BIOLOGICAL WEAPONS

In 2001, the PLO made its first, and to date the only, public suggestion that it may be interested in pursuing biological weapons. In the August 13, 2001, edition of *Al-Manar,* a Palestinian weekly periodical published by the PLO-directed Palestinian Center for Information Sources-Gaza, an article entitled "Will We Reach the Option of Biological Deterrence?" appeared. That article, written by Deputy Chairman and manager of the Center Tawfiq Abu Khusa, stated that "serious thinking has began a while ago about obtaining biological weapons." Abu Khusa added that "the Palestinian side is required to use weapons of deterrence that will even the balance of power, at least on the field." He also claimed that "whether biological or chemical," it would be possible to make such weapons "without too much effort," given "the fact that there are hundreds of experts who are capable of handling them and use them as weapons of deterrence, thus creating a balance of horror in the equation of the Palestinian-Israeli conflict" (Abu-Khosa, 2001).

IMPLICATIONS

While it is likely that the Abu Khusa article was intended to exert a psychological effect on the PLO's Israeli adversaries, one cannot automatically dismiss such statements. However, even if it were conceded that the development of biological weapons by the Palestinians would give them a significantly greater leverage in their dealings with Israel, which is debatable, potential Israeli targets lie in close proximity to the population that the PLO purports to defend, making the use of such weapons unappealing given their unpredictable nature. Further, such a serious breach of international norms and sanctions against the use of these weapons would have serious consequences for the PLO, and serious military retaliation from Israel would be almost assured if an attack of this type was carried out.

REFERENCES

Abu-Khosa, T., Will We Reach the Option of Biological Deterrence? *Al-Manar*, August 13, 2001, as translated by the Middle East Media Research Institute, Special Dispatch Series- No. 255 (August 14, 2001).

Alexander, Y. and Sinai, J., *Terrorism: The PLO Connection*, Taylor & Francis, New York, 1989, pp. 29, 34.

Cordesman, A., *Israel versus the Palestinians: The 'Second Intifada' and Asymmetric Warfare*, Working Draft, Center for Strategic and International Studies, July, 2002, p. 9.

Mark, C., *Palestinians and Middle East Peace: Issues for the United States*, CRS Issue Brief for Congress, April 9, 2002.

Palestinian Central Bureau of Statistics (PCBS), President's press conference, Dr. Hassan Abu-Libdeh, November 1998.

Palestine Liberation Organization (PLO), Federation of American Scientists, FAS ONLINE www.fas.org.

FURTHER READING

Bickerton, I. and Klausner, C., *A Concise History of the Arab-Israeli Conflict*, 2nd edition, Prentice-Hall, Englewood Cliffs, 1995.

Livingstone, N. and Halevy, D., *Inside the PLO: Covert Units, Secret Funds, and the War Against Israel and the United States*, William Morrow, New York, 1990.

Mark, C., *Palestinians and Middle East Peace: Issues for the United States*, CRS Issue Brief for Congress, April 9, 2002.

Sela, A. and Ma'oz, M., Eds., *The PLO and Israel: From Armed Conflict to Political Solution, 1964-1994*, St. Martin's Press, New York, 1997.

PALESTINIAN ISLAMIC JIHAD (PIJ, HARAKAT AL-JIHAD AL-ISLAMI AL-FILASTINI, ISLAMIC JIHAD MOVEMENT IN PALESTINE)

JOSHUA SINAI

INTRODUCTION

Palestinian Islamic Jihad (PIJ) is a radical Sunni Muslim group that portrays Palestine as vital to the Muslim world and contends that anti-Zionism is the key to Islam's success. PIJ deems Jews to be the eternal enemies of Islam and excoriates any form of compromise with them, and therefore calls for an Islamic armed struggle against the "Zionist Jewish entity" in order to replace it with an Islamic Palestinian state. PIJ believes that carrying out terrorist attacks against Israel will weaken it in preparation for military confrontation by the "great Islamic Arabic army." It also opposes pro-Western Arab governments.

PIJ is headquartered in Damascus, Syria, and has offices in Beirut, Lebanon, Tehran, and Khartoum. Both its armed and its religious activities are primarily centered in the Gaza Strip and the West Bank.

PIJ has a small support base. At the beginning of the first Intifada in 1987, the PIJ consisted of approximately 250 militants and several hundred sympathizers in universities and mosques. The current size of its militia is unknown. According to opinion polls, it is supported by some 4 to 5 percent of the Palestinian population.

The group is allied with HAMAS, the Lebanese Hizballah, and the Iranian Revolutionary Guards stationed in Lebanon. It is also allied with al-Fatah activists. At the height of the al-Aqsa Intifada, which began in late-September 2000, PIJ, like the other Palestinian terrorist groupings, was granted freedom of action and logistical support by the Palestinian Authority.

The group conducts fundraising through Islamic charities. It receives logistical assistance from Syria and financial and material resources from Iran.

BACKGROUND

Formed by a group of radical Palestinian students at Zaqaziq University in Cairo and, like other radical Islamic groups in Egypt, influenced by the success of the Islamic revolution in Iran, the PIJ emerged in the late 1970s under the initial leadership of Fathi al-Shiqaqi (1951–1995), a physician known for his great charisma and organizational capability. Other early PIJ leaders included Dr. Ramadan 'Abdallah Shalah (a close confidant of Shiqaqi) and Dr. Sami al-'Aryan. Both Shalah and al-'Aryan at various points in their career taught at the University of South Florida at Tampa, and established the World and Islam Studies Enterprise (WISE) there, a research institute promoting PIJ's political beliefs. Later leaders included 'Abd al-Aziz Awda and Bashir Musa, whose current status is unknown.

Unlike other Sunni Islamist terrorist groups, however, PIJ's leaders were also directly inspired by the distinctive ideology of Shi'ite Islamists such as Khomeini.

The PIJ began employing terrorist violence against Israel in 1984. After the outbreaks of the original Intifada of 1987–1992 and the second al-Aqsa Intifada in late September 2000, these efforts were substantially increased. Its terrorist activity is primarily of the suicide variety.

CHRONOLOGY OF SIGNIFICANT EVENTS

1987

In August, a member of PIJ's Shiqaqi faction shot Captain Ron Tal, commander of the Israeli military police in the Gaza Strip, in his car in a main Gaza street.

1993

On October 9, the PIJ's al-Aqsa Squad members murdered Israeli civilians Dror Forer and Aran Machar in Wadi Kelt in the West Bank Judean desert.

On November 17, the PIJ claimed responsibility for the stabbing of Israeli Sgt. 1st class Chaim Darino in a cafeteria at the Nahal Oz roadblock at the entrance to the Gaza Strip.

On December 5, the PIJ's Shiqaqi faction claimed responsibility for the murder of David Mashrati, an Israeli reserve soldier killed as a terrorist squad attempted to board a bus on route 641 at the Holon junction.

1994

On February 9, the PIJ's Shiqaqi faction claimed responsibility for the kidnapping and murder of Israeli taxi driver Ilan Sudri.

On November 11, PIJ claimed responsibility for a suicide bombing at the Netzarim junction in the Gaza Strip. Three Israeli military officers were killed when a bicycle rider detonated a supply of explosives strapped to his body.

Encyclopedia of Bioterrorism Defense,   Edited by Richard F. Pilch and Raymond A. Zilinskas
ISBN 0-471-46717-0   Copyright © 2005 Wiley-Liss

1995

On January 22, the PIJ claimed responsibility for two consecutive bombings at the Beit Lid junction near Netanya, which together killed 18 Israeli soldiers and one civilian.

On April 9, the PIJ claimed responsibility for the deaths of Israeli Sgt. Avraham Arditi and U.S. citizen Alisa Flatow, after a van filled with explosives collided with their bus near Kfar Darom in the Gaza Strip.

1996

On March 4, a PIJ suicide bomber detonated a 20-kg nail bomb outside the Dizengoff Center in Tel Aviv, killing 12 Israeli civilians and one soldier.

2000

On October 2, two Israeli civilians were killed in a PIJ bombing in West Jerusalem.

In mid-December, a PIJ cell in the West Bank town of Nablus was broken up by Israeli security forces.

2001

On April 1, PIJ activist Muhammad Abdalal was killed by Israeli security agents in a helicopter missile attack in the town of Rafiah, at the southern end of the Gaza Strip. Israeli security sources claimed that Abdalal had been involved in numerous terrorist attacks against Israelis.

On April 5, Iyad Hardan, leader of PIJ's military wing in the West Bank, was killed when a bomb detonated while he was standing in a telephone booth opposite the Palestinian police station in the West Bank town of Jenin, where he had been jailed. Though Israel refused to officially comment on the incident, it is believed that Israeli security services were responsible for the explosion.

2002

On June 5, PIJ claimed responsibility for a suicide bus bombing at the Megiddo junction in northern Israel, which killed 17 persons and wounded more than 40. On November 15, PIJ gunmen killed 12 Israeli settlers in the West Bank town of Hebron.

On December 27, two PIJ gunmen infiltrated the Jewish community of Othniel in the West Bank, killing two Israeli civilians and two IDF (Israel Defense Forces) soldiers and wounding 10 others inside a local seminary.

2003

On September 26, PIJ claimed responsibility for a shooting in the West Bank settlement of Negohot, in which two

Israelis (including a baby girl) were killed and two others lightly wounded.

On October 4, a PIJ woman suicide bomber blew herself up in the popular Haifa restaurant Maxim, killing 21 people and wounding more than 60.

PIJ AND BIOLOGICAL WEAPONS

There have been no reports in the open source environment of a biological weapons capability on the part of PIJ. However, in April 1998, PIJ leader Nasir Asad al-Tamimi discussed the possibility of his organization's acquiring biological weapons, at a HAMAS memorial service in Amman, Jordan (Broad and Miller, 1998; Carus, 1998). According to a report by the Center for Israeli Civilian Empowerment, al-Tamimi stated, "Jihad has at last discovered how to win the holy war-lethal germs" (http://www.geocities.com/CapitolHill/Congress/7663/ Quote.html, accessed on 6/28/99).

In 2002, the Israeli press reported that PIJ had planned but never executed an operation to poison the water supply of a Jerusalem hospital with an unidentified agent.

IMPLICATIONS

These reports aside, it is clear that PIJ's weapons' and tactical repertoire consists primarily of shootings, detonation of improvised explosives and car bombs, and its most lethal tactic, suicide bombings. In the future, PIJ can be expected to continue to perpetrate suicide bombings of various degrees of lethality against Israeli soldiers and civilians, whether in the occupied territories of the West Bank and Gaza Strip or in theIsraeli "proper."

REFERENCES

Broad, W.J. and Miller, J., The Threat of Germ Warfare Is Rising. Fear Too, *New York Times*, December 27, 1998, p. 4.

Carus, W.S., *Bioterrorism and Biocrimes, The Illicit Use of Biological Agents in the 20th Century*, Center for Counterproliferation Research, National Defense University, August 1998, http://www.geocities.com/CapitolHill/Congress/7663/Quote .html, accessed on 6/28/99.

Fiamma, N., E Arafat dicharoí guerra al terrorismo; Resa dei conti dopo il attentato all Ingegniere, ma il leader rischia un conflitto algerinoí, *La Stampa*, April 14, 1998, p. 8.

See also ISLAMISM and ISLAM, SHI'A AND SUNNA.

PHARMACEUTICAL INDUSTRY (DRUG INDUSTRY, PHARMA COMPANIES, BIOTECH COMPANIES)

James A. Poupard

INTRODUCTION

The pharmaceutical and biotechnology industries play an important role in providing antiinfective drugs, vaccines, and biologicals (a category of pharmaceutical products consisting not of chemical agents like drugs and not of vaccines, but rather of products such as immunomodulators, interferons, and monoclonal antibodies, which are often produced in facilities similar to vaccine production lines since they are usually derived from tissue cultures or, in some cases, from organisms like modified *Escherichia coli* but are not classic vaccines) for use in responding to a bioterrorist attack. Research, development, and production programs initiated by the pharmaceutical industry will play a key role in providing new therapeutic agents for use against potential bioterrorist threats, and the industry will be an important element in determining future policies relating to bioterrorism defense.

BACKGROUND

In times of war and national catastrophes alike, the pharmaceutical industry has worked closely with governments and international agencies to ensure the greatest possible response capability to those in need. During World War II, for example, the pharmaceutical industry worked with U.S. government and university scientists to rapidly develop critical blood fractions that ultimately saved innumerable lives (Starr, 1998), while a similar joint project resulted in the greatly expedited development of penicillin (Hobby, 1985). When, following the September 11, 2001, attacks on the United States, President Bush declared war on terrorism, the pharmaceutical industry again volunteered to do what was needed in order to aid in the effort. Both the needs and the solutions have since proven to be much more complex than at any other time in history, however, and the industry is still, to some extent, in the process of determining what is expected of it in the post-September 11 era.

Prior to 2001, the distinction between pharmaceutical industry policies relating to biological warfare (BW) and bioterrorism was unclear. Large pharmaceutical companies were interested in such issues as the genetic engineering of microorganisms and the stockpiling of antiinfective drugs and vaccines in anticipation of the use of biological agents as weapons. There was also an interest and concern about the proposed site inspection aspects of the Biological and Toxin Weapons Convention (BWC) verification protocol initiative. Attention to all of these issues was of a general nature except for the brief period prior to the Gulf War. Following September 11 and the subsequent use of *Bacillus anthracis* spores as a bioterrorist agent delivered via the U.S. postal system, however, the pharmaceutical industry's attention became much more focused.

POST-SEPTEMBER 11 PHARMACEUTICAL INDUSTRY RESPONSE

It has been noted that the terrorist attacks of September 2001 using commercial aircraft as weapons produced varied reactions in the United States, including one that was neither directly nor logically connected to the attack: the overwhelming expression by several officials and experts that the next terrorist attack could well involve biological or chemical weapons (Leitenberg, 2002). The subsequent mailing of letters containing *B. anthracis* spores and the infections and deaths associated with this incident enhanced the feeling of vulnerability to this type of weapon, provoking a deluge of inquires from the press, trade groups, and government agencies (both legislative and executive branches) concerning the readiness of the pharmaceutical industry to respond to a bioterrorist attack on a larger scale. Specifically, demands were placed on the industry to define inventories and production and distribution capabilities of relevant antiinfective and wound healing drugs, vaccines, and biologicals. Requests were also submitted for details on drugs and vaccines in development, including time estimates of when these agents would be available for emergency use.

Encyclopedia of Bioterrorism Defense, Edited by Richard F. Pilch and Raymond A. Zilinskas
ISBN 0-471-46717-0 Copyright © 2005 Wiley-Liss

Accurate information on these issues was often difficult to obtain, because no individual or single department within larger pharmaceutical companies possessed the necessary data in the detail that was required. The disclosure of specific information on drugs in development further created legal and patent issues that required consultation and clarification. Regardless, it eventually became apparent that companies did not possess large stockpiles of those agents deemed necessary, and even when they did, distribution directly to the general public in many cases violated current laws and required expertise that did not exist. It soon became apparent that there was a need for a single communication body to speak for the pharmaceutical industry, and thus in October 2001, the Pharmaceutical Research and Manufacturers of America's (PhRMA) Task Force on Emergency Preparedness was formed.

Task Force on Emergency Preparedness

Upon its creation, the Task Force established four initial priorities (PhRMA, 2001):

- Each company was to designate a top scientist to serve on an industry/government working group to identify priority needs, gaps, and capabilities relevant to the pharmaceutical industry.

- Each company with antibiotic capability was to provide data on its production capacity for antibiotics that might be effective against bioterrorism agents.

- Information was to be provided about the ability of the industry to share laboratory capability in order to help analyze suspicious material for the presence of bioterrorism agents.

- It was to be determined whether industry could be of assistance in the storage and distribution of needed therapeutics or vaccines.

The Task Force, today, continues to provide a vehicle to address and better define exactly what is expected of the pharmaceutical industry in addressing public health needs in the face of bioterrorism, and to establish practical policies in this area.

The Impact of Legislation

Legislation such as the House Anti-Terrorism Bill raised significant issues for the pharmaceutical industry. Although many potentially effective antiinfective drugs had demonstrated in vitro activity (low minimum inhibitory concentration) against potential bioterrorism organisms, the Food and Drug Administration's (FDA) approved labeling for these drugs did not include treatment of infection by such organisms. This situation raised complex issues for government agencies charged with protecting public health within the bounds of current laws regulating prescription drugs, because success required promoting their off-label usage. In addition, legislation concerning the security of drug production and research and development (R&D) facilities for global pharmaceutical companies complicated employment and personnel reliability measures, particularly when aliens,

foreign graduate students, and postdoctoral personnel were involved. Another significant issue concerned the cost of stockpiling patented drugs versus the less expensive generic version of these drugs, some of which were not available in the United States at that time because of patent restrictions. The pharmaceutical industry responded to these issues by making the necessary adjustments to ensure that adequate drugs would be available to address the needs of the public.

CURRENT BIOTERRORIST AGENTS OF CONCERN

Anthrax

One of two diseases to receive the majority of attention from pharmaceutical and biotech industries is anthrax, its priority self-evident, given the events of fall 2001. Currently, the pharmaceutical industry is capable of both supplying adequate drugs for limited anthrax outbreaks and filling the needs of stockpiling programs integral to response efforts in a major incident. Distribution of stockpiled drugs remains an issue but falls outside the scope of responsibility of the pharmaceutical industry.

The pharmaceutical industry is fully cooperating with government agencies to perform the necessary animal studies to increase the number of currently marketed drugs and drugs in development against anthrax. Within both the pharmaceutical and biotechnology industries, programs are also in place to develop improved vaccines and biological therapies for the future, and to ensure that adequate supplies of today's drugs are available. Significant attention, particularly by the biotechnology industry, is being directed toward the development of rapid detection and diagnostic procedures using both environmental and human samples as well (Enserink, 2001).

Smallpox

Smallpox has received considerable attention from pharmaceutical and biotechnology industries as well (Enserink, 2002). Initially, priority was given to ensuring that adequate stocks of the current vaccine were available to first vaccinate first responders as a preparedness measure and then vaccinate necessary populations in the event of an outbreak. The ultimate goal was to establish a capability to vaccinate the entire population of the United States if necessary. Several pharmaceutical companies bid on a one-year contract to produce the required amounts of the vaccine, which upon being awarded has led to the manufacture and stockpiling of enough vaccine to meet current public health needs. Research on improved vaccines is ongoing. The pharmaceutical industry is also cooperating with government agencies to expedite the search for antiviral drugs with activity against the smallpox virus. Ultimately, the potential consequences of a bioterrorist threat employing smallpox, although still significant, have in a very short period of time been significantly reduced.

Other Potential Bacterial Threats

Marketed drugs exist that are likely to be effective against those pathogens considered to be potential *bacterial* bioterrorist threats (for more information, please refer to the entry entitled "Centers for Disease Control and Prevention's Bioterrorism Preparedness Program"). However, the potential for use of highly resistant strains of these pathogens necessitates the continued development of new agents with novel mechanisms of action, a need consistent with current pharmaceutical industry drug discovery programs. Thus, the encouragement of the pharmaceutical industry to maintain these programs and develop new antibacterials appears to be a critical element of current approaches to preparedness. The development of new vaccines and biologicals is also critical to addressing the problem of drug-resistant bacterial agents of today and tomorrow.

Potential Viral Threats Other Than Smallpox

The need for the development of drugs against potential *viral* bioterrorism threats remains a serious problem. Most current antiviral drugs are not effective against select viral agents, making pharmaceutical and biotechnology industry research on new vaccines and other biologicals, such as those associated with the enhancement of host defense mechanisms, the best approach for the future (Fox, 2003a).

Toxins

One approach to responding to an anticipated bioterrorist threat is to focus research on neutralizing either a freestanding toxin (such as botulinum toxin) or the toxin produced by an infecting agent (such as lethal factor, or LF, of *B. anthracis*) rather than focusing on an antibacterial drug. This approach appeals to both biotechnology as well as large pharmaceutical companies since such biological products or antitoxin drugs not only fill a definite medical need but also allow for justification of the research investment. Research in this area is of particular interest since it bypasses the issue of drug resistance, which is the most limiting factor to the market value of both novel and classic antimicrobial drugs. Development of drugs directed against toxins, if successful, will yield a product with an extended market value.

PHARMACEUTICAL INDUSTRY DRUG DISCOVERY PROGRAMS

As noted, a vital approach to responding to highly resistant organisms in the future is the continued implementation of drug discovery programs to enhance the pipeline of novel agents. At present, however, a number of large pharmaceutical companies are in the midst of reducing or eliminating these programs because of practical concerns, namely, that the expenses pertaining to maintaining such a program and taking a single drug through development all the way to the market (average cost is about $800 million) are prohibitive (Gwynne and Heebner, 2003). Further, in large pharmaceutical companies, antiinfective programs must compete with other therapeutic areas for funding. When one considers that antiinfective drugs are only required for a short duration (days) whereas drugs from other therapeutic areas are often given to patients for

years or even life, it is difficult to justify their development and production from a financial perspective. Thus, pharmaceutical companies honoring the commitment to their shareholders cannot, in the contemporary situation, pursue antiinfectives as a high priority, creating a void that will most likely be filled by small pharmaceutical and biotechnology companies in the immediate future (though it should be noted that many of the drugs being developed by these companies have come from lead compounds generated by the larger pharmaceutical companies). This is an issue that must receive the attention of policy planners when setting future priorities.

ADDITIONAL PHARMACEUTICAL INDUSTRY INITIATIVES

Most pharmaceutical companies with marketed antiinfectives have offered their drugs either free of charge or at cost to the government for use in an emergency situation. Several pharmaceutical companies have offered teams of scientists and laboratory space to conduct research, leading to solutions against potential microbial weapons as well. Other initiatives include a program coordinated by PhRMA in cooperation with the U.S. government, which involves the delivery by pharmaceutical company representatives of printed summary data on anthrax and smallpox to health care personnel in order to raise their awareness of these diseases and provide an informative source for related questions that these personnel may have. In addition to these programs, some pharmaceutical companies sponsor scientists from foreign countries to work at the Centers for Disease Control and Prevention (CDC) in Atlanta in order to develop skills for addressing bioterrorism in their respective countries.

THE FUTURE

The pharmaceutical industry will continue to play a significant role in providing therapeutic drugs against the most likely bioterrorist threat agents. However, there are significant gaps in the current therapies available for highly resistant bacteria and many viruses, and significant time, research, and money must therefore be invested to address these gaps. This will involve the continuing cooperation of the pharmaceutical and biotech industries with various government agencies, academic researchers, and private foundations. Incentives must be provided to conduct research in identified areas of need (Fox, 2003b), and new paradigms must be developed to evaluate drugs and vaccines that cannot be tested in human clinical trials to determine efficacy. New surrogate markers, such as animal or even in vitro models, will have to be accepted as indicators of efficacy as an incentive to develop some of these needed therapies (Friedlander et al., 1993). The pharmaceutical industry will continue to be an active participant in the dialog necessary to move forward in these areas in the years to come.

REFERENCES

Enserink, M., *Science*, **294**, 1266–1267 (2001).
Enserink, M., *Science*, **296**, 1592–1595 (2002).

Fox, J.L., *ASM News*, **69**, 269–270 (2003a).

Fox, J.L., *ASM News*, **69**, 322–323 (2003b).

Friedlander, A.M., Welkos, S.L., Pitt, M.L.M., et al. *J. Infect. Dis.*, **167**, 1239–1242 (1993).

Gwynne, P. and Heebner, G., *Science*, **299**, 915–921 (2003).

Hobby, G.L., *Penicillin: Meeting the Challenge*, Yale University Press, New Haven, CN, and London, 1985.

Leitenberg, M., *Polit. Life Sci.*, **21**(2), 3–27 (2002).

PhRMA Task Force on Emergency Preparedness (PTF), News Release, October 16, 2001.

Starr, D., *Blood: An Epic History of Medicine and Commerce*, Alfred A. Knopf, Inc., New York, 1998.

WEB RESOURCES

Pharmaceutical Research and Manufacturers of America, http://www.phrma.org.

See also FOOD AND DRUG ADMINISTRATION and PREVENTION AND TREATMENT OF BIOLOGICAL WEAPONS-RELATED INFECTION AND DISEASE.

PINE BLUFF ARSENAL (CHEMICAL WARFARE ARSENAL, PINE BLUFF CHEMICAL ACTIVITY (PBCA))

JOSEPH URREA

INTRODUCTION

The Pine Bluff Arsenal has served as a manufacturing center for the production of both chemical and biological weapons (Fig. 1). Today, the facility remains a storage site for a portion of the U.S. chemical defense stockpile.

BACKGROUND

The Pine Bluff Arsenal was established in November 1941 as the Chemical Warfare Arsenal; it was renamed four months later. Its original mission was to serve as a manufacturing center for magnesium and thermite munitions. After World War II, the Arsenal's capabilities expanded to include the manufacturing, loading, and storing of war gases, and the filling of smoke and white phosphorus munitions. Arsenal-produced conventional munitions were used in both the Korean and Vietnam wars. Millions of pounds of mustard and Lewisite were also produced at the Arsenal, but not nerve agents.

In 1953, a biological weapons mission was added to the Arsenal. The Production Development Laboratories was responsible for the manufacturing and loading of biological munitions until President Nixon banned biological weapons in 1969. As a result of Nixon's ban, the bioweapons production facility at Pine Bluff was abandoned and partly dismantled, and in 1972 was renamed the National Center for Toxicological Research, removed from the jurisdiction of the Arsenal, and placed under the Department of Health, Education, and Welfare.

Located in Southeast Arkansas, the Pine Bluff Arsenal is 8.5 miles long by 2.75 miles wide and covers 14,944 acres. It includes 952 buildings that provide 3.3 million sq ft of floor space, including storage bunkers, and has 42 miles of railroad track, and 2 million square yards of roads and paved surfaces.

CURRENT STATUS

Today, the Pine Bluff Arsenal manufactures chemical, smoke, riot control, incendiary, and pyrotechnic mixes and munitions. Limited production facilities also exist that are used to manufacture chemical defense items such as clothing and protective masks, a pursuit that began in the late 1970s when the Pine Bluff Arsenal was given the mission to produce M24/M25 A1 masks. The Pine Bluff Arsenal is the U.S. Army's sole facility for the repair and rebuilding of the M17 series, M9A1, M24, M25A1, and M40 masks, as well as the M20 breathing apparatus. It is the only active site at which white phosphorous–filled weapons are loaded.

Also located at the complex is the Pine Bluff Chemical Activity (PBCA), which maintains its stockpile on 431 acres of land in the northwestern portion of the Arsenal. The PBCA's specific mission is to stockpile 3850 tons of chemical weapons until it is disposed of through the Chemical Stockpile Disposal Program. The stockpile consists of rockets containing the nerve agents sarin (GB) and VX, and one-ton bulk storage containers with sulfur mustard payloads. The safe storage of the entire stockpile is overseen by the U.S. Army Chemical and Biological Defense Command, which is charged with monitoring the storage.

RECENT DEVELOPMENTS

In the 2001 Defense Authorization Act, Congress included the Arsenal Support Program Initiative to help maintain the viability of Army arsenals, including Pine Bluff. The initiative allows arsenals to enter into cooperative partnerships with private corporations. Companies can work out agreements with arsenals permitting them to use warehouses, office space, and other facilities. In addition, the Justice Department has established a Domestic Preparedness Equipment Technical

Figure 1. Pine Bluff Arsenal chemical demilitarization support facilities; U.S. Army Corps of Engineers, Programs and Project Management Division, Little Rock District: http://www.swl.usace.army.mil/projmgt/milproj.html.

Encyclopedia of Bioterrorism Defense,   Edited by Richard F. Pilch and Raymond A. Zilinskas
ISBN 0-471-46717-0   Copyright © 2005 Wiley-Liss

Assistance Program (DPETAP) at Pine Bluff. DPETAP teaches emergency responders how to choose, operate, and maintain their radiological, chemical, and biological detection and response equipment. These developments indicate that Pine Bluff will continue to play a role in U.S. homeland security in the years to come.

FURTHER READING

Kennedy, H., Army's Pine Bluff Arsenal Seeks New Assignments, *National Defense*, March 2002.

WEB RESOURCES

GlobalSecurity.org, www.globalsecurity.org/index.html.

PLAGUE (*YERSINIA PESTIS*)

Serguei Popov

TAXONOMY

The etiological agent of plague is *Yersinia pestis*, a gram-negative bacterium of the *Enterobacteriaceae* family. The genus *Yersinia* is composed of 11 species, three of which are pathogenic in humans. The three pathogens, *Y. pestis, Y. enterocolitica, and Y. pseudotuberculosis*, cause a broad spectrum of disease ranging from pneumonic plague to acute gastroenteritis.

EVOLUTIONARY ORIGIN

Some evidence suggests that *Y. pestis* might have evolved from *Y. pseudotuberculosis* 1500 to 20,000 years ago. As a result, and unlike *Y. enterocolitica* and *Y. pseudotuberculosis*, *Y. pestis* is unable to survive in the environment outside of an animal host. It also does not infect the host by the enteric route. Some of the genes that are required for invasion by this route are still present in *Y. pestis* but are not expressed as a consequence of mutations.

EPIDEMIOLOGY

Yersinia pestis is found in every continent except Australia and Antarctica. It is transmitted from one animal host to another either directly or via a flea vector (often *Xenopsylla cheopis*). In areas of the world where plague is endemic, the bacterium appears to survive by causing chronic disease in rodent reservoirs. The occasional transfer of the bacteria to other mammalian hosts can result in acute disease, which is recognized as plague. Therefore, outbreaks of plague in humans are often associated with close contact with animal reservoirs. It continues to be a noneradicable threat because of its extreme virulence, vast changing areas of infection in wild rodents and other animals, and residential encroachment on former rural areas that contain enzootic foci of plague.

INCIDENCE OF DISEASE IN HUMANS

It is generally recognized that plague caused three major pandemics of disease in the first, fourteenth through seventeenth, and nineteenth centuries, and resulted in as many as 200 million deaths.

In A.D. 541, the first recorded plague pandemic began in Egypt and swept across Europe, with population losses between 50 and 60 percent in North Africa, Europe, and central and southern Asia. The second plague pandemic, also known as the Black Death or Great Pestilence, began around 1300 and eventually killed 20 to 30 million people in Europe, one-third of the European population. Plague spread slowly and inexorably from village to village by infected rats and humans, or more quickly from country to country by ships. The pandemic lasted more than 130 years and had major political, cultural, and religious ramifications (Gottfried, 1985). The third pandemic began in China in 1855, spread to all inhabited continents, and ultimately killed more than 12 million people in India and China alone.

Although *Y. pestis* no longer causes disease on this scale, there is still a public health problem from plague. Small outbreaks of plague continue to occur throughout the world, especially in Africa, Asia and South America. According to the World Health Organization (WHO), during the period 1967–1993, the average annual incidence of plague was 1666 cases worldwide. However, many cases of plague are not diagnosed, and it is likely that the true incidence of disease is several times the WHO figures. Potential for rapid spread of the disease throughout the world by symptomless infected individuals utilizing air transport systems is of particular concern, especially in the case of the pneumonic form of the disease.

CLINICAL PRESENTATION

The principal clinical forms of plague are bubonic, septicemic, and pneumonic. Unusual manifestations are meningitis and pharyngitis. Both septicemic and pneumonic forms may be primary or secondary to bubonic disease. Plague is characterized by a rapid onset of fever and other systemic manifestations of gram-negative bacterial infection. If not quickly and effectively treated, shock, multiple organ failure, and death can supervene rapidly.

Encyclopedia of Bioterrorism Defense,   Edited by Richard F. Pilch and Raymond A. Zilinskas
ISBN 0-471-46717-0   Copyright © 2005 Wiley-Liss

Bubonic Plague

This most common form is typically caused by the bite of an infected flea that has fed previously on an infected animal, but can also result from direct inoculation of infectious fluids. Most often, the bacteria are disseminated from the initial site of infection to the draining lymph nodes, which rapidly become swollen and tender, forming a bubo. This is accompanied by toxemia and a lack of evident cellulitis or lymphangitis.

Septicemic Plague

Overwhelming sepsis can develop in the absence of a bubo. Gastrointestinal symptoms—nausea, vomiting, diarrhea, and abdominal pain—may be prominent. Early treatment with appropriate antibiotics is vital, as disseminated intravascular coagulation, shock, and organ failure may progress rapidly. Because of the generalized nature of these symptoms, a diagnosis of plague is often delayed, and even with medical intervention, 50 percent of patients die.

Pneumonic Plague

Pneumonic plague, the most likely manifestation of deliberate exposure, is the most rapidly progressive and frequently fatal pneumonia resulting from colonization of the alveolar spaces (Fig. 1). A pneumonic plague outbreak would initially resemble other severe respiratory illnesses, such as legionellosis or severe influenza. Symptoms begin 1 to 6 days following exposure. The first cases could be expected within 1 to 2 days of exposure, with many people dying soon afterwards. Plague pneumonia is invariably fatal if antibiotic treatment is delayed more than 24 h after illness onset. Initial manifestations of fever, chills, headache, myalgia, and weakness would be rapidly followed by cough, dyspnea, and chest pain. Inhalation of airborne droplets resulting from coughing is a major route of disease transmission to other individuals.

DIAGNOSIS

Yersinia pestis is a nonmotile, gram-negative bacillus that can be stained with Wright, Giemsa, or Wayson

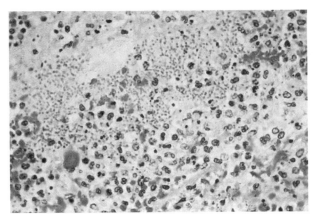

Figure 1. Photomicrograph of lung tissue revealing *Yersinia pestis* organisms, Public Health Image Library (PHIL), id# 1956; *Source*: CDC.

stain. *Yersinia pestis* is a lactose nonfermenter, and is urease and indole negative. It grows optimally at 28 °C on blood agar or MacConkey agar, typically requiring 48 h for observable growth. Blood, sputum, or tracheal aspirates (and, if relevant, cerebrospinal fluid or bubo aspirates) can be used for bacteriological examination. *Yersinia pestis* grows on a wide variety of culture media. Most microbiology laboratories use either automated or semiautomated bacterial identification systems. Some of these systems may misidentify *Y. pestis*. In laboratories without automated bacterial identification, as many as six days may be required for identification, and there is some chance that the diagnosis may be missed entirely. According to the recommendations of the Centers for Disease Control and Prevention (CDC), if a laboratory using automated or nonautomated techniques is notified that plague is suspected, it should carry out two tests in culture: The first culture should be incubated at 28 °C for rapid growth, and the second culture should be incubated at 37 °C for identification of the diagnostic capsular (F1) antigen. Using these methods, up to 72 h may be required following specimen procurement to make the identification. Antibiotic susceptibility testing should be performed at a reference laboratory because of the lack of standardized susceptibility testing procedures for *Y. pestis*.

Chest X ray may show initial segmental or lobar involvement before bronchopneumonic spread to other lobes in both lungs, which may become confluent. If coinciding with an influenza or respiratory epidemic, or if presenting at a time of year in which the onset of such an epidemic might be expected, a case of pneumonic plague is quite likely to be overlooked or misdiagnosed. Bloody sputum, generally watery or mucoid and frothy, typically suggests plague, in contrast to widened mediastinum in inhalation anthrax.

VIRULENCE FACTORS

The pathogenicity of *Y. pestis* results from a number of virulence factors that enable it to survive in humans by facilitating use of host nutrients, causing damage to host cells, and subverting not only the destruction of bacterial cells by the immune system but other host defense mechanisms as well. Multiplication of *Y. pestis* takes place largely outside the cells of the infected lymphatic system and body organs such as the spleen, the lungs, and especially the liver.

The closely related foodborne pathogens, *Y. pseudotuberculosis* and *Y. enterocolitica*, cross the intestinal barrier and multiply in the abdominal lymphoid tissues. All three pathogens require many chromosomally encoded virulence factors, as well as the so-called extra-chromosomal plasmid-encoded factors, for full virulence. The *Yop* virulon, encoded by plasmids, is the core of the *Yersinia* pathogenicity arsenal. It allows extracellular *Yersinia* docked at the surface of a host cell to inject specialized proteins, called *Yops*, across the cellular membrane. The injected Yops disturb the dynamics of the cytoskeleton and block the production of host protective mediators, thereby favoring the survival of the invading *Yersinia*. Similar virulence mechanisms have now been identified in more than a dozen major animal or plant pathogens.

The F1 antigen is a capsular protein, located on the surface of the bacterium, which is thought to have antiphagocytic properties. The V antigen is a protein secreted by *Y. pestis* under low calcium growth conditions. Another important virulence factor, pH6 antigen, appears on the surface of bacteria in certain conditions within infected host cells. Our knowledge of *Yersinia* virulence mechanisms is far from complete (Cornelis, 2000).

MANAGEMENT AND PREVENTION

In general, *Y. pestis* is susceptible to antibiotics. The preferred ones for cases of bubonic plague are streptomycin or gentamicin; alternatives are doxycycline, chloramphenicol, and ciprofloxacin. Chloramphenicol should be included if meningitis is suspected. However, successful treatment of septicemic and pneumonic plague with antibiotics is much less likely because the disease develops so rapidly. If pneumonic plague, treatment must commence as quickly as possible, within 24 h of symptom onset.

Isolation of a multiple antibiotic-resistant strain of *Y. pestis* in 1995 in a 15-year-old boy in Madagascar indicates that the long-term potential for the use of antibiotics to treat plague is uncertain. Given that naturally occurring antibiotic resistance is rare and that information on engineered antibiotic resistance is lacking, some experts believe that initial treatment recommendations should be based on known drug efficacy, drug availability, and ease of administration.

All patients with plague should be isolated to prevent respiratory droplet transmission until either pneumonia has been ruled out or antibiotics have been given for at least 48 h and clinical improvement has occurred. People with close exposure (within 2 m (6 ft)) are at significant risk of infection. Doxycycline, ciprofloxacin, or chloramphenicol should be given for seven days as postexposure prophylaxis for people who have household, health care, or other close contact with a patient. People in close contact with patients should wear gloves, gowns, eye protection, and surgical masks for at least the first 48 h of antibiotic treatment, after which universal precautions are adequate. In addition, patients with pneumonic plague should wear simple masks.

Both live attenuated and killed whole cell vaccines have been used in humans (Titball, 2001). Killed whole cell vaccines are used throughout the Western World, while live attenuated vaccines have been used particularly in the former USSR and in the former French colonies. Although there is circumstantial evidence for the efficacy of the latter vaccines, none have been subjected to controlled and randomized clinical trials. The live attenuated vaccine (EV76 strain) is a pigmentation negative attenuated mutant derived from a fully virulent strain of *Y. pestis*. The vaccine has been in use since 1908 and is given as a single dose of about 6×10^6 cells. Immunization of mice with the EV76 induces an immune response that provides protection against subcutaneous and inhalation challenges, suggesting that immunization with the EV76 vaccine will provide protection against both bubonic and pneumonic plague in humans. However, the EV76 strain is not avirulent, and the safety of this vaccine in humans is of legitimate concern.

Varieties of the killed whole cell vaccine against plague were developed for use in humans in 1940s. Typically, the killed whole cell plague vaccine requires a course of injections over a period of two to six months. Therefore, this vaccine is used mainly in those individuals who might be exposed to the pathogen, for example veterinarians, those engaged in research with the bacterium, and individuals who are traveling to parts of the world where the disease is endemic. Side effects such as malaise, headaches, elevated temperature, and lymphadenopathy occur in approximately 10 percent of those immunized. There are no definitive clinical trials that demonstrate the efficacy of killed whole cell vaccine in humans, however, the vaccine is protective in several animal species against bubonic plague. Circumstantial evidence, such as the low incidence of bubonic plague in immunized U.S. servicemen serving in Vietnam during the period 1961–1971, indicates the efficacy of the vaccine in humans. Evidence for the efficacy of killed whole cell vaccines for the prevention of pneumonic plague is less conclusive. Therefore, in practice, people who have been vaccinated should be managed no differently from those who have not.

In recent years, efforts have focused on the development of a so-called subunit vaccine for plague based on the protective immune response to certain proteins, such as fraction 1 (F1) antigen, virulence (V) antigen, pH6 antigen, plasminogen/coagulase, and other proteins that might be located on the surface of the bacterium. Immunization of animals with all of these components induces high levels of circulating antibodies; however, only the F1 and V antigens induce responses, which consistently provide protection against challenge with *Y. pestis*.

Y. PESTIS AS A WEAPON

Plague bacteria have long attracted attention as biological weapon because they are extremely infectious. In 1346, plague broke out in the ranks of the Tartar army during its siege of Kaffa, and the Tartars hurled the corpses over the city walls using catapults. Some historians believe that infected Kaffans who managed to escape may have started the spread of the Black Death across Europe. The use of dead bodies and excrement as weapons continued in Europe during the Black Plague of the fourteenth and fifteenth centuries, and even as late as the early eighteenth century Russian troops fighting Sweden resorted to catapulting the bodies of plague victims over the city walls of Reval.

During World War II, a secret branch of the Japanese army, Unit 731, investigated the use of plague as a biological weapon and used infected fleas not only against nearby Chinese and Russian territory but as far away as Burma, Thailand, and Indonesia (Williams and Wallace, 1989). The resulting death toll may have run as high as 200,000 people. This weapon, which was dependent of vectors for spread, was cumbersome and unpredictable, however.

In the 1950s and 1960s, the U.S. and Soviet biological warfare (BW) programs developed techniques to directly aerosolize plague particles in order to cause pneumonic plague. Both programs were able to develop methodologies for suspending or dissolving optimal quantities of *Y. pestis* and other pathogens in special solutions containing preservatives, adjuvants, and antistatic chemicals. The final dry powder or liquid suspension is commonly called a *formulation*, which is required for every weaponized pathogen and toxin in order to preserve its virulence or toxicity during shelf storage (up to several years). While U.S. scientists had not succeeded in making quantities of plague organisms sufficient to use as an effective weapon by the time the U.S. offensive program was terminated in 1969, Soviet scientists were able to manufacture large quantities of the agent suitable for placing into weapons in violation of the Biological and Toxin Weapons Convention (BWC), which the Soviet Union ratified in 1972. More than 10 institutes and thousands of scientists were reported to have worked with plague in the former Soviet Union. Plague formulations in both liquid and dry powder forms were developed for effective dissemination of bacteria by explosion or spraying. Further, Soviet research on genetically modified plague resulted in the development of multidrug-resistant strains (Orent, 2004). The plague weapon has been considered one of the most effective weapons ever created by the Soviet Union.

The organism can be manufactured by large-scale fermentation without affecting its properties as a BW agent. *Yersinia pestis* can be stored relatively easily because it can survive at low or freezing temperatures for extended periods as long as there is water available. The organism remains viable in water, moist soil, and grains for several weeks. At near-freezing temperatures, it will remain alive from months to years, but is killed by 15 min of exposure to 55 °C. It can also be freeze-dried and stored for up to several years without loss of viability.

An aerosol made up of plague bacteria–containing droplets of the size best suited for absorption in the lungs (1–5 μm) dispensed over an unprotected population could kill a very large fraction (90–100 percent) of those exposed. The dose of viable bacterial cells in aerosol an individual has to receive to have a 50 percent chance of being infected (ID_{50}) for *Y. pestis* is estimated to be about 1000 (different estimates obtained in laboratory animals and extrapolated to humans range from 100 to 3000 cells). Plague bacteria are, therefore, considered somewhat more effective than anthrax bacteria, but less effective than tularemia bacteria or Q fever rickettsia, in causing infection. The size of a pneumonic plague epidemic following an aerosol attack would depend on a number of factors, including the amount of agent used, the meteorological conditions, and methods of aerosolization and dissemination. A 1970 WHO study concluded that, in a worst-case scenario, 50 kg of *Y. pestis* released as an aerosol over a city of five million could result in 150,000 cases of pneumonic plague, with 80,000 to 100,000 cases requiring hospitalization and 36,000 deaths. The WHO study also found that the organism could remain viable for up to 1 h after dispersal as an aerosol and be carried for a distance of up to 10 km from the point of release. The rate of loss of viability of bacteria in aerosol after dispersal (up to 70 percent per minute) is dependent on temperature, intensity of sunlight, and humidity of the atmosphere. In general, the higher the temperature the more intense the sunlight, and the dryer the atmosphere the quicker the aerosolized pathogen dies.

There are neither effective environmental warning systems to detect an aerosol of plague bacilli nor widely available rapid diagnostic tests for plague. Tests that would be used to confirm a suspected diagnosis—antigen detection, enzyme immunoassay, immunostaining, and polymerase chain reaction—are currently under development. The routinely used passive hemagglutination antibody detection assay is typically only of retrospective value since several days to weeks usually pass after the disease onset before antibodies develop at a detectable concentration in a victim's blood.

The epidemiology of plague following its use as a biological weapon would differ substantially from that of naturally occurring infection. The first indication of a clandestine terrorist attack with plague would most likely be a sudden outbreak of illness presenting as severe pneumonia and sepsis. If there are only small numbers of cases, the possibility of plague as a diagnosis may at first be overlooked given the clinical similarity to other bacterial or viral pneumonias. However, the sudden appearance of a large number of patients with fever, cough, shortness of breath, chest pain, and a fulminant course leading to death should immediately suggest the possibility of pneumonic plague or inhalational anthrax. Indications that plague has been artificially disseminated would be the occurrence of cases in locations not known to have enzootic infection, in persons without known risk factors, and in the absence of prior rodent deaths (Inglesby et al., 2000).

Given the rarity of plague infection and the possibility that early cases are a harbinger of a larger epidemic, the first clinical or laboratory suspicion of plague must lead to immediate notification of the hospital epidemiologist or infection control practitioner, health department, and the local or state health laboratory. Definitive tests can thereby be arranged rapidly, and early interventions instituted.

REFERENCES

Cornelis G.R., *Proc. Nat. Acad. Sci. USA*, **97**, 8778–8783 (2000).

Gottfried, R.S., *The Black Death: Natural and Human Disaster in Medieval Europe*, Free Press, New York, 1985.

Inglesby, T.V., Dennis, D.T., Henderson, D.A., Bartlett, J.G., Ascher, M.S., Eitzen, E., Fine, A.D., Friedlander, A.M., Hauer, J., Koerner, J.F., Layton, M., McDade, J., Osterholm, M.T., O'Toole, T., Parker, G., Perl, T.M., Russell, P.K., Schoch-Spana, M., and Tonat, K., *J. Am. Med. Assoc.*, **283**(17), 2281–2290 (2000).

Orent, W.P., *Plague: The Mysterious Past and Terrifying Future of the World's Most Dangerous Disease*, Free Press, New York, 2004.

Titball R.W. and Williamson E.D., *Vaccine*, **19**, 4175–4184 (2001).

Williams, P. and Wallace, D., *Unit 731: The Japanese Army's Secret of Secrets*, Hodder & Stoughton, London, 1989.

POLISARIO (MOVEMENT FOR THE LIBERATION OF SAGUIA EL HAMRA AND OUED EDDAHAB (MLS) (FORERUNNER OF POLISARIO), SAHRAWI ARAB DEMOCRATIC REPUBLIC (SADR))

GAIL H. NELSON

OVERVIEW

Ideology

The aim of the *Frente Popular para la Liberacion de Saguia el-Hamra y Rio de Oro* (POLISARIO) is the creation of a sovereign Western Saharan state ruled by a democratic government that represents the Sahrawi tribes and Islamic culture (Fig. 1). The means it relies upon to achieve its political ends include international diplomacy, guerrilla warfare, insurgency, and terrorist attacks, all mainly directed against Moroccan interests.

Headquarters

The POLISARIO has avoided fixed positions for guerrilla operations, preferring to rely on mobility and on an intimate knowledge of the Western Sahara to screen its movements. However, the POLISARIO uses the Sahrawi enclaves in southwestern Algeria and Mauritania for refuge and resupply, as well as a base of operations. POLISARIO operations encompass the Western Sahara, western Morocco, and the desert regions within other neighboring states.

Strength

The POLISARIO militants numbered approximately 20,000 during the peak of its conflicts in the 1980s.

Figure 1. Western Sahara as envisioned by the POLISARIO.

Its recruits came primarily from Sahrawi population centers outside the Western Sahara, including Morocco, Mauritania, and Algeria. POLISARIO weapons consist mostly of small arms, but the group also received some tanks and antiaircraft missiles from Algeria, Libya, and the USSR before the collapse of the East Bloc. Libya distanced itself from the POLISARIO in 1984.

Group Ties

At least 30 member states of the Organization of African Unity (OAU) and more than 70 countries worldwide recognize the Sahrawi Arab Democratic Republic (SADR). The OAU admitted the SADR as its fifty-first member in 1982. The extent of its direct and indirect ties to other liberation movements and international terrorist groups are unclear.

Current Status

POLISARIO guerrilla operations are now dormant due to a UN-negotiated ceasefire and the presence of a UN contingent of 2900 peacekeepers (MINURSO—UN Mission for the Referendum in Western Sahara). However, militant factions could resume terrorist attacks without warning because of their frustration with the slowness of UN arbitration and its failure to conduct a referendum over Western Saharan independence.

BACKGROUND

The Western Sahara is an area of 102,703 square miles (265,898 sq km) with an estimated 1997 population of 230,000. Another 196,000 Sahrawi live in refugee camps around Tinduf in southwest Algeria. The Sahrawi are nomadic people with Moorish ancestors, and they are almost entirely Sunni Muslim. Sahrawi society is divided into tribes and castes, and tribal assemblies are composed of family heads under the authority of a shaykh.

Western Sahara History

Spain exercised dominion over most of Morocco and the Western Sahara during the latter nineteenth century, but divided its colonial possessions with France around the turn of the century. Although Morocco obtained independence from France in 1956, Spain retained control over the Western Sahara. The Spanish allowed home rule for the Sahrawi in 1967, but Morocco and Mauritania

Encyclopedia of Bioterrorism Defense,   Edited by Richard F. Pilch and Raymond A. Zilinskas
ISBN 0-471-46717-0   Copyright © 2005 Wiley-Liss

opposed Madrid's decision and made territorial claims on the region. As these neighboring states positioned themselves for territorial possession, Sahrawi nationalists formed their own independence movement. The United Nations called for an independent Western Sahara in 1972, and Spain endorsed the idea of a referendum in 1975 to determine the future status of the region. In contradiction to its support of a referendum, Madrid also declared its intention to transfer sovereignty of Western Sahara as soon as possible, and then signed an accord with Morocco and Mauritania to hand over the territory. Morocco and Mauritania secretly agreed to partition Western Sahara along ancient boundary lines in 1976.

Polisario History

The POLISARIO was founded in May 1973 as a national liberation movement in the Spanish Sahara. The group strongly opposed the secret partition of Western Sahara between Morocco and Mauritania, denouncing it as an act of neocolonialism. It declared the establishment of an independent republic (SADR) in 1976, and thence commenced insurgent operations against both Moroccan and Mauritanian occupying forces. Mauritania withdrew from the conflict in 1979 following POLISARIO successes, leaving the dispute to be resolved between Morocco and the POLISARIO. The protracted conflict raged throughout most of the 1980s, leading to a stalemate but with Morocco in possession of the main lines of commerce. Both sides accepted a UN-sponsored ceasefire formula in 1988, according to which a referendum to determine the wishes of the Sahrawi people would be held. A referendum under UN supervision has yet to be carried out.

POLISARIO AND BIOLOGICAL WEAPONS

Intent

There is no indication that POLISARIO leaders have adopted a policy to use weapons of mass destruction (WMD) to achieve their political ends, despite one unconfirmed incident noted below. However, rogue elements within the movement could conceivably attempt to carry out unsanctioned individual acts of bioterrorism to stoke international fears and thereby focus media attention on their cause.

Capability

The POLISARIO reportedly threatened to contaminate Paris, Madrid, Rabat, and the Nouakchott city water supply with *Vibrio cholerae* in 1975 in an effort to protest the policies of France, Spain, Mauritania, and Morocco concerning West Saharan independence. However, there is no indication that the POLISARIO was in fact seeking a biological weapons capability at that time, nor has any such indication surfaced since.

IMPLICATIONS

The POLISARIO will actively pursue independence for the Western Sahara and may resume protracted guerrilla operations against Morocco. These operations probably will include terrorist attacks against Moroccan interests. Nonetheless, the POLISARIO's modus operandi does not appear to include the deployment of WMD to achieve the group's political objectives. There are no reliable indications that POLISARIO will adopt any methods other than conventional methods to wage their liberation campaign. The UN arbitration may lead to a final settlement following the outcome of a referendum, but Morocco and the POLISARIO are disputing the eligibility of some tribes to register for the vote. Meanwhile, a tenuous ceasefire is in effect, and Morocco has strengthened its hold over most of Western Sahara by means of the emigration of Moroccans, making the Sahrawi a minority in their own lands. The Moroccan offer of autonomy to the Sahrawi may be an alternative outcome to protracted conflict.

SUGGESTED READING

Biger, G., *The Encyclopedia of International Boundaries*, Facts on File, New York, 1995.

van Creveld, M., *The Encyclopedia of Revolutions and Revolutionaries*, Facts on File, New York, 1996.

Economist, An Endless Dance in the Desert, March 25, 1999.

Economist, Fixing the Vote, July 16, 1998.

Economist, He Tried Hard, November 12, 1998.

Economist, In the Sahrawi Camps, November 2, 2000.

Economist, It's A Mirage, January 20, 2000.

Economist, Numbered Days, June 13, 2002.

Economist, Polisario's Sinking Hopes, December 6, 2001.

Economist, Talking at Last, June 19, 1997.

Economist, The UN's U-Turn, June 28, 2001.

Economist, To Be, or not to be, a Saharawi? March 19, 1998.

Economist, Triumph for Procrastination, November 2, 2000.

Longman Current Affairs Series, *Border & Territorial Disputes*, United Kingdom Ltd., 1992.

Longman Current Affairs Series, *Islam & Islamic Groups*, United Kingdom Ltd., 1992.

Longman Current Affairs Series, *Revolutionary & Dissident Movements of the World*, United Kingdom Ltd., 1991.

Longman International Reference, *World Directory of Minorities*, United Kingdom Ltd., 1990.

Pinter Reference, Ewan Anderson, *An Atlas of World Political Flashpoints*, United Kingdom, 1993.

Stockholm International Peace Research Institute (SIPRI), *Yearbook 1999*, Oxford University Press, United Kingdom, 1999.

Stockholm International Peace Research Institute (SIPRI), *Yearbook 2000*, Oxford University Press, United Kingdom, 2000.

Turner, B., *The Statesman's Yearbook 2000*, Macmillan, New York, 2000.

PREVENTION AND TREATMENT OF BIOLOGICAL WEAPONS-RELATED INFECTION AND DISEASE

Paul Setlak

INTRODUCTION

With the fall 2001 dispersal of anthrax spore-laden letters to several individuals in the United States, the use of biological weapons gained recognition as a genuine risk to the common man. Thus, an adequate understanding of current and future approaches to medical prevention and treatment of such weapons is necessary in order to gain an accurate understanding of the threat. When discussing these approaches, one must note that the health care system, including all components that contribute to its effective functioning, will rapidly become strained in the event of a biological attack. If such an attack occurs without warning (as would be expected), the system will undoubtedly be left to accommodate its regular patients as well as the exposed population, including those genuinely affected and those psychologically or psychosomatically affected (the latter comprising the so-called worried well contingent of the patient load). Consequently, health care personnel will have little time to consider what if any medication best suits a patient's indication, a situation that complicates providing efficacious, safe, and timely treatment to a large number of individuals in a relatively short period of time. This chapter provides a blueprint of current and future suitable treatment approaches to various biological warfare (BW) agents in order to ease this burden on the health care professional.

TREATMENT CATEGORIES AND HUMAN IMMUNITY

Three main categories of medications can be used to treat patients with suspected or confirmed disease states brought on by BW agents and natural infections alike: antimicrobials, antivirals, and immunobiologics (Fig. 1). Antimicrobials include medicants such as antibiotics, antifungals, and antiparasitics. Antibiotics, the largest subcategory of antimicrobials, are soluble substances derived from a mold or bacterium or made synthetically. Antibiotic therapies can either be bacteriostatic (inhibiting or retarding the further growth of susceptible bacteria) or bactericidal (causing death in susceptible bacteria) (Carruthers et al., 2000). No compelling clinical data exists that suggests any efficacy in administering antibiotics to treat viral infections, given that the mechanisms by which antibiotics exert their effects relate specifically to bacterial cell structures and mechanisms and not those of viral particles. As a result, it is not advisable that antibiotics be used to treat viral BW casualties. Antibiotics given orally, one of the most common routes of administration, include tablets, capsules, solutions, and suspensions. Many medicants can also be administered intravenously (IV), especially in emergency situations when onset of the drug's action is needed immediately (Greenwood, 2000).

Antiviral medicants comprise a second major category of drug treatment approaches to BW. Antivirals are specifically designed to produce a therapeutic effect in those individuals with a disease state originating from a virus. Although much is known about many viruses, especially those with the potential for weapons utilization, the number of effective antiviral drugs remains relatively small. Further, many drug therapies designed to fight viruses in humans produce adverse effects in patients due to both their high potency and the need to deliver them in large doses in order to eliminate 100 percent of the virus. High potency and dosage, therefore, are key requirements of antivirals, because any viral particle that escapes their effects can potentially induce a recurring pathological state (Flint et al., 2000). Finally, the duration of viral infections is relatively short, such that by the time such an etiology is confirmed the responsible virus may already have ceased to replicate and may in fact have been cleared from the patient altogether. Thus, in order for antivirals to be effective, they must be administered early on in the course of disease or, even better, prophylactically (i.e., as a preexposure treatment).

The third category of medicants used to treat patients in the context of BW, immunobiologics, is composed of vaccines, toxoids, and immune sera (antitoxins and immunoglobulins). A vaccine is a preparation of antigenic material used to induce immunity against pathogenic organisms. Antigenic material refers to any substance, usually foreign, that induces an immune response in an individual when it binds to a T-cell receptor or an antibody. Toxoids are inactivated bacterial toxins that have the potential to form antitoxins and, as a result, are commonly used in vaccines to confer active immunity in an individual. Antitoxins themselves contain antibodies derived from human (immunoglobulin) or equine (horse antitoxin) sources (DiPiro et al., 2002). The routes of

Encyclopedia of Bioterrorism Defense, Edited by Richard F. Pilch and Raymond A. Zilinskas
ISBN 0-471-46717-0 Copyright © 2005 Wiley-Liss

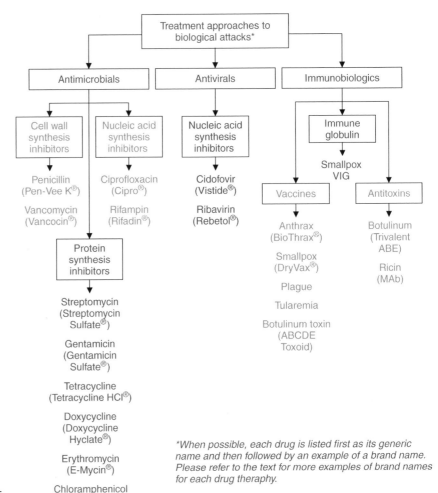

Figure 1. Categorical prevention and treatment approaches to biological attacks.

administration for immunobiologics are intravenous (IV), intradermal (ID), intramuscular (IM), subcutaneous (SC), or through scarification.

Immunobiologics employ one of two methods to help confer immunity in an individual: passive or active immunity. Active immunity is induced in an individual through exposure to a pathogen or by vaccination, while passive immunity refers to an acquired immunity through the transfer of immune products (i.e., antitoxins or immunoglobulin) from an immune to a nonimmune person (Kuby, 1997). Active immunity acquired, for example, from preexposure vaccination aids in preventing an infectious disease, while passive immunity only provides temporary immunity in an individual through antibody mediation (Abbas et al., 1997).

ANTIBIOTICS

Antibiotics are often designated as being either narrow or broad spectrum. Narrow-spectrum antibiotics show efficacy toward either Gram-positive (G+) or Gram-negative (G−) bacteria but not both, while broad-spectrum antibiotics exhibit efficacy toward both G+ and G−bacteria at the same time. The gram stain reaction determining G+ or G−bacteria is therefore extremely useful in

establishing the correct diagnosis and treatment approach in an individual. Both G+ and G−bacteria contain peptidoglycan, a constituent of bacterial cell walls that provides structural support, but G+ bacteria have almost 50 percent more of this "backbone" in their cell wall composition (Madigan et al., 2000). It is this higher ratio of peptidoglycan in the cell wall that allows the bacterial cell to trap the gram stain reaction dye and subsequently present with a G+ reaction (purple color). One structural feature characteristic of G−bacteria (which stain red in color) is the presence of a lipopolysaccharide layer (LPS) in the outer cell wall layer, important in the context of therapy because many antibiotics cannot penetrate this layer and reach their intracellular target. Altogether, there are five recognized classes of antibiotics grouped according to their primary site of action in the bacterial cell: metabolic analogs, cell wall synthesis inhibitors, protein synthesis inhibitors, nucleic acid synthesis inhibitors, and cell membrane function disrupters (Hitner, 1999). The following section focuses on three of these classes in particular—cell wall synthesis inhibitors, protein synthesis inhibitors, and nucleic acid synthesis inhibitors—in order to delineate those antibiotics pertinent to a discussion of biological weapons prevention and management.

Cell Wall Synthesis Inhibitors

The first class, cell wall synthesis inhibitors, includes the Penicillins and Vancomycin.

Penicillins. Penicillins (Penicillin VK®, Pen-Vee K®, Wycillin®) are bactericidal in nature and are most effective when bacteria are actively reproducing (Hutchinson and Shahan, 2003). Penicillin binds to one or more of the penicillin-binding proteins (PBPs) in bacterial cells and subsequently inhibits the final transpeptidation step necessary for the successful cross-linking of peptidoglycan, which gives cell walls their strength. Consequently, the defective cell wall does not allow the bacterium to maintain the osmotic pressure necessary for survival in surrounding bodily fluids, causing it to gain water, swell, lyze (break open), and die. Penicillin has an Food and Drug Administration (FDA)-labeled indication (adult and pediatric cases) for the treatment of anthrax, the disease caused by the G+ spore-forming rod *Bacillus anthracis* (McEvoy, 2002). More specifically, the therapy may be used as a treatment option postexposure (once antibiotic sensitivities have been determined) or prophylactically, though never as a sole treatment because of the potential for β-lactamase positive *B. anthracis* strains (β-lactamase is an enzyme that inactivates Penicillin antibiotics) (Inglesby et al., 2002; CDC, 2001).

Vancomycin. Vancomycin (Vancocin HCl®, Vancomycin HCl®) is another cell wall synthesis inhibitor that acts in a different way than Penicillins, exerting its bactericidal action against a variety of G+ bacteria by preventing the transfer and addition of the building blocks (muramylpentapeptides) that make up peptidoglycan. Although it has not received an FDA-labeled indication for anthrax, Vancomycin has been recommended by the Centers for Disease Control and Prevention (CDC) and the Working Group on Civilian Biodefense as an initial adjunct therapy in a combination regimen against an intentional anthrax release (Inglesby et al., 2002; CDC, 2001).

Protein Synthesis Inhibitors

Antibiotics that are protein synthesis inhibitors (PSIs) include three subdivisions—aminoglycosides, tetracyclines, and macrolides—as well as a stand-alone therapeutic drug, Chloramphenicol.

Aminoglycosides. Aminoglycosides, acting on the bacterial 30S ribosomal subunit, include Streptomycin and Gentamicin. Both of these antibiotics are bactericidal in action and efficacious against many G−aerobes. Streptomycin (Streptomycin Sulfate®) and Gentamicin (Garamycin®, Gentamicin Sulfate®) have similar mechanisms of action: after being taken up by the bacterial cell, both bind to the 30S subunit (and in some cases the 50S subunit) and inhibit protein synthesis by generating errors in the transcription of the genetic code. The exact manner that culminates in bacterial cell death is not entirely understood, although a number of factors, including membrane damage, may contribute. Both Streptomycin and Gentamicin are considered effective therapies for tularemia,

caused by the G−rod *Francisella tularensis*, and plague, caused by the G−rod *Yersinia pestis* (Streptomycin is the antibiotic of choice for plague due to the high level of nephrotoxicity seen with Gentamicin) (Inglesby et al., 2000; Walker et al., 1999).

Tetracyclines. The second subdivision of PSIs is the Tetracyclines. Tetracyclines are largely bacteriostatic and possess a broad spectrum of activity against many aerobic and nonaerobic G+/G−bacteria. Similar to aminoglycosides, Tetracyclines exert their action on the 30S ribosomal subunit, specifically preventing the binding of aminoacyl t-RNA in order to inhibit cell growth. The two main Tetracycline antibiotics relating to BW are Tetracycline itself (Achromycin V®, Bristacycline®, Tetracycline HCl®) and Doxycycline (Vibramycin®, Doxycycline Hyclate®, Vibra-Tabs®). Both Tetracycline and Doxycycline have FDA-labeled indications for anthrax, plague, and tularemia, although Tetracycline therapy is utilized only when Penicillin is contraindicated (e.g., because of a person exhibiting hypersensitivity to it) in the case of anthrax (McEvoy, 2002). Doxycycline is usually chosen over Tetracycline because of its long half-life (permitting less-frequent dosing) and safety when administered to patients with renal insufficiency (Klein and Cunha, 1995). Furthermore, it remains recommended as a "combination-drug" regimen (with Ciprofloxacin, Penicillin, plus 1 or 2 other antibiotics) in cases of anthrax (Brook, 2002; Inglesby et al., 2002).

Macrolides and Chloramphenicol. The last subdivision of protein synthesis inhibitors, macrolides, can for the purpose of this discussion be lumped with the lone drug not belonging to a PSI subdivision, Chloramphenicol, because their mechanisms of action are quite similar. Erythromycin (E-Mycin®, Erythrocin®), a major bacteriostatic macrolide, exhibits its action on G+ bacteria by binding reversibly to the 50S ribosomal subunit, inhibiting protein synthesis and ultimately blocking cell growth. Chloramphenicol (Chloromycetin®, Chloramphenicol Sodium Succinate®) inhibits cell growth of both G+ and G−bacteria in a similar fashion; however, there are more side effects associated with Chloramphenicol than other antibiotics, so it should be used only in serious and life-threatening situations (Inglesby et al., 2000). Both Chloramphenicol and Erythromycin are recommended for the treatment of anthrax. Only Chloramphenicol is advised for plague (Putzker et al., 2001) and tularemia (Dennis et al., 2001). It is important to note, however, that neither therapy has an FDA-labeled indication for any of these three disease states.

Nucleic Acid Synthesis Inhibitors

The third class of antibiotics, nucleic acid synthesis inhibitors (NASIs), includes the anti-tubercular drug Rifampin and the fluoroquinolone Ciprofloxacin.

Rifampin. Rifampin (Rifadin®, Rimactane®), an adjuvant treatment for tuberculosis, has been recommended for postexposure treatment of anthrax (Inglesby et al.,

2002). The drug binds the β-subunit of bacterial RNA-polymerase, inhibiting transcription of DNA to RNA. Depending on the dose, the action of Rifampin can either be bacteriostatic or bactericidal; in terms of limiting BW casualties, healthcare professionals will likely employ bactericidal dosing (i.e., 300 mg every 12 h), and given in conjunction with other antibiotics for the treatment of anthrax.

Ciprofloxacin. Ciprofloxacin (Cipro®), a fluoroquinol-one, has an FDA-labeled indication for both postexposure and prophylactic treatment of cases of anthrax (Inglesby et al., 2002). Ciprofloxacin exerts its broad-spectrum activity by inhibiting DNA gyrase, an enzyme essential for bacterial DNA replication. Ciprofloxacin gained world-wide attention during the anthrax attacks of 2001, when medical personnel consistently administered this drug to patients with confirmed or suspected *B. anthracis* infections (Mayer et al., 2001; Barakat et al., 2002; Mina et al., 2002). The CDC now specifically recommends that Ciprofloxacin be considered initial therapy for postexposure prophylaxis of individuals exposed to inhalational *B. anthracis* (CDC, 2001). In addition, Ciprofloxacin is recommended for the treatment of tularemia and plague, although no FDA-labeled indication has been approved in either case as yet. Nevertheless, a recent report, though based on only one case, illustrates the potential efficacy of administering Ciprofloxacin to patients diagnosed with plague (Kuberski et al., 2003).

ANTIVIRALS

A biological weapon employing a viral pathogen such as Variola virus (smallpox) or one of the various hemorrhagic fever viruses poses a significant challenge to health care personnel because treatment options are severely limited. In this section, the mainstays of antiviral drug therapy (other than vaccines and immunoglobulins), Cidofovir and Ribavirin, are reviewed.

Cidofovir

Cidofovir (Vistide®), a nucleotide analog of cytosine, acts as a competitive inhibitor of viral DNA polymerase. Although it has no FDA-labeled indication for smallpox, cell-based and animal studies have demonstrated some efficacy against this disease (De Clercq, 2002; Smee et al., 2001; Bray et al., 2000). Because of its unusually high level of nephrotoxicity to humans and carcinogenicity in animals (even in low doses), Cidofovir is only available as an Investigational New Drug (IND) protocol sponsored by the CDC for those individuals who do not respond to Vaccinia Immune Globulin (VIG) treatment (Gilead Sciences, 2000). According to a recent report issued by the CDC, Cidofovir would be released to civilian and military populations only if the situation satisfies one or more of three criteria (Cono et al., 2003):

1. The patient fails to adequately respond to VIG treatment.
2. The patient is near death.

3. All stocks of VIG have been exhausted.

Ribavirin

Ribavirin (Rebetol®, Virazole®) is an antiviral agent with an unclear mechanism of action: the interference of multiple pathways may contribute to the drug's overall effect to inhibit viral nucleic acid synthesis (Crotty et al., 2002; Tam et al., 2001). Although Ribavirin may be effective for treating Lassa fever and Rift Valley fever (Andrei and De Clercq, 1993), it fails to demonstrate any significant in vitro efficacy against other hemorrhagic fevers such as the Ebola and Marburg fevers (Bronze and Greenfield, 2003). Unfortunately, no effective treatment has been identified as yet for combating Ebola and Marburg fevers, and as a result, any treatment approach is largely supportive and directed toward the management of symptoms common to these and other hemorrhagic fevers (Sidell, et al., 2002). In contrast to Cidofovir, Ribavirin is available to the public and military through the auspices of a health care provider.

VACCINES AND TOXOIDS

Vaccines and toxoids provide what is called active immunity, meaning that the therapeutic agent stimulates the formation of antibodies in the host to induce immunity and prevent disease. Five diseases warrant discussion in the context of BW prophylaxis with vaccines and toxoids: anthrax, smallpox, plague, tularemia, and botulism.

Anthrax Vaccine

The two most well known vaccines in a BW context target anthrax (BioThrax®) and smallpox (DryVax®), and can be utilized for *both* preexposure prophylaxis and postexposure treatment. The current anthrax vaccine, Anthrax Vaccine Adsorbed, was preceded by attenuated vaccines in the United States (MDPH-PA® and MDPH-AVA®), and has foreign equivalents as well (U.K. Human Anthrax Vaccine® and a Russian vaccine) (Turnbull, 1991). The vaccine is composed of cell-free filtrates of an avirulent, nonencapsulated strain of *B. anthracis*. The antigenic portion, which is identified by antibody and thus confers immunity to the individual upon vaccine administration, is an 83-kDa protective antigen (PA) protein. Importantly, the manufactured vaccine does not contain any live or dead bacteria. The PA protein is utilized in the vaccine because it alone does not create any toxicity or disease in an individual (pathology results only when PA is combined with lethal factor (LF) or edema factor (EF), as occurs during the normal infection process) (Bioport Corporation, 2002). Preexposure prophylaxis for anthrax is provided through three 0.5-mL subcutaneous (SC) injections spaced two weeks apart, followed by three more 0.5-mL SC injections administered at 6, 12, and 18 months. The manufacturer recommends 0.5-mL booster injections annually thereafter. In a postexposure setting, administration guidelines state that 0.5 mL of the vaccine should be injected SC immediately and then repeated at 2 and 4 weeks. Antibiotic treatment (Ciprofloxacin 500 mg twice daily, Doxycycline 100 mg twice daily, or another recommended drug regimen) should be provided concurrently. Although initially available only in the military sector, the anthrax vaccine is now approved for health care personnel and first responders as well (Bioport Corporation, 2002).

Smallpox Vaccine

As noted above, the smallpox vaccine has proven efficacy both pre- and postexposure if administered in a timely fashion. Smallpox vaccine (DryVax®) is a live-virus preparation of Vaccinia virus administered to an individual through scarification (Wyeth Laboratories 2002). This method of administration involves a bifurcated needle that is used to pick up a drop of vaccine from the manufacturer's vial. This drop on the needle is then deposited on a clean dry site of skin on the upper arm. Subsequently, using the same needle, 15 rapid punctures are made in order to trough the drop into the superficial layers. Initial reactions (termed a "take") present in 3 to 5 days with a vesicular or pustular lesion at the site of administration, followed by crusting between the fourteenth and twenty-first days. A successful active vaccination, termed a "major reaction," occurs when, by the end of the third week, a permanent skin-colored scar characterized by induration is present at the site of scarification. If a major reaction does not occur, called an "equivocal reaction," vaccination procedures should be reviewed and the vaccine readministered (Cono et al., 2003). Revaccination every 10 years using the same methodology should be considered for laboratory workers working with smallpox, military personnel, and first responders and physicians having received the vaccine as part of the U.S. bioterrorism preparedness efforts of 2003. Vaccinia can be transmitted from the site of vaccination to another person, but proper and vigorous hand hygiene, along with the covering of the vaccination site, prevents this type of transmission (ACIP, 2001).

A new vaccine based on the DryVax® Vaccinia virus strain is in the process of development through a joint venture between the English company Acambis and the CDC. Acambis hopes to be able to license the vaccine by 2004 (Birmingham and Kenyon, 2001). In addition, a decades-old vaccine made by Aventis-Pasteur, approximately 80 million doses of which were mysteriously discovered in Pennsylvania early in 2002, exists as well. This vaccine is not and will not be approved for licensure but will remain in IND status, to be used only in the event of an overwhelming smallpox outbreak.

Plague Vaccine

Active immunization against plague is possible through a vaccine produced from inactivated Y. pestis cells. The FDA-labeled indication states that this vaccine should only be administered for prevention or symptom relief of plague. The vaccine became unavailable for distribution in 1999. Pending renewed approval by the FDA on the basis of manufacturing changes, plague vaccine will again be available, initially to the military population and those laboratory workers working with Y. pestis. No estimation on the date of approval has been given (Greer

Laboratories, 1995). The vaccine is administered in a series of three injections, the second dose given 1 to 3 months after the first, and the final dose, 6 months after the second. Booster doses are required every six months, demanding a well-organized system for distribution and administration of the vaccine. Whether this vaccine is effective against pneumonic plague is questionable.

Tularemia Vaccine

A live attenuated vaccine to treat tularemia is only available as an IND, and early data suggests that if a person is exposed to a high enough dose of F. tularensis, immunization may not be effective (Dennis et al., 2001). The vaccine itself is indicated for those individuals needing preexposure protection against tularemia. Continued testing of the tularemia vaccine is being conducted at an accelerated pace because of the potential use of this pathogen as a BW agent (Conlan et al., 2002; Elkins et al., 2003; Vogel, 2003).

Botulinum Toxoid

Pentavalent (ABCDE) botulinum toxoid is available from the CDC as an IND to protect certain individuals, namely, those working in high-risk laboratories and, when necessary, selected military personnel, from accidental exposure to botulinum toxin. As the name suggests, the toxoid is composed of five different antigenic types of toxin, A, B, C, D, and E, in order to combat the most common and lethal variants of the toxin.

ANTITOXINS AND IMMUNE GLOBULIN

Antitoxins and immune globulins provide what is called passive immunity, in which preformed antibodies are provided to the host. Three diseases are pertinent to the discussion of such medicants in a BW context: smallpox, botulism, and ricin intoxication.

Smallpox VIG

Vaccinia Immune Globulin (VIG) is a sterile solution that contains antibodies to Vaccinia virus collected from volunteers previously vaccinated against smallpox. VIG is used to treat smallpox vaccine reactions, especially eczema vaccinatum (localized papular or pustular rash), progressive vaccinia (painless progressive necrosis at vaccination site), and severe cases of generalized vaccinia (vesicular rash). VIG may also be administered to treat cases of inadvertent ocular vaccinia (ACIP, 2001). For those individuals who cannot receive the smallpox vaccine because of absolute contraindications such as pregnancy or an immunocompromised state, VIG can be given after or within 24 hours of exposure to confer the highest level of protection (it should be noted, however, that this level of protection is not sustained). Currently, VIG is only available from the CDC through an IND protocol. Furthermore, the CDC has stockpiled enough VIG to treat more than 4000 cases of serious side effects, and production of additional VIG continues.

Botulinum Antitoxin

Trivalent (ABE) botulinum antitoxin is available only from the CDC but is in fact a licensed product. The antitoxin is a solution of horse immune globulin used in the treatment of suspected or confirmed cases of botulinum toxin poisoning. Administration of botulinum antitoxin must be rapid—if delayed, it may be ineffective (Sidell et al., 2002). The antitoxin is kept at several CDC quarantine stations around the country, allowing for rapid distribution as necessary. When a suspected or confirmed case is identified, the treating physician initiates the process of obtaining the antitoxin by notifying local and state public health departments. The final decision for antitoxin release is made after epidemiological staff from the CDC discusses the case with the treating physician. This chain of command helps to control the distribution of a therapy with limited indications and a relatively short expiration time. Furthermore, the necessity of authorization helps promote rapid botulinum intoxication surveillance and outbreak detection.

Ricin Antitoxin

There exists no known treatment licensed to effectively combat ricin intoxication (Sidell et al., 2002). However, specifically constructed monoclonal antibodies (MAbs) have been shown to neutralize the toxin and oral immunization with ricin toxoid has been shown to be highly effective against a ricin aerosol challenge in mice (although this study specifies that the dosing and administration of the vaccine was highly stringent so any deviation from it conferred little or no protection against the toxin; also, one has to be careful about interpolating data from animal studies to humans) (Lemley et al., 1994; Kende et al., 2002).

REFERENCES

Abbas, A.K., Lichtman, A.H., and Pober, J.S., Ed., *Cellular and Molecular Immunology*, W.B. Saunders Company, Philadelphia, PA, 1997, pp. 4–7.

ACIP (CDC), *MMWR*, **50**(RR10), 1–25 (2001).

Andrei, G. and De Clercq, E., *Antivir. Res.*, **22**(1), 45–75 (1993).

Barakat, L.A., Quentzel, H.L., Jernigan, J.A., Kirschke, D.L., Griffith, K., Spear, S.M., Kelley, K., Barden, D., Mayo, D., Stephens, D.S., Popovic, T., Marston, C., Zaki, S.R., Guarner, J., Shieh, W.J., Carver, H.W. 2nd, Meyer, R.F., Swerdlow, D.L., Mast, E.E., and Hadler, J.L.; Anthrax Bioterrorism Investigation Team., *J. Am. Med. Assoc.*, **287**(7), 863–868 (2002).

Bioport Corporation, *BioThrax® [Package insert]*, Bioport Corporation, Lansing, MI, 2002.

Birmingham, K. and Kenyon, G., *Nat. Med.* **7**(11), 1167 (2001).

Bray, M., Martinez, M., Smee, D.F., Kefauver, D., Thompson, E., and Huggins, J.W., *J. Infect. Dis.*, **181**(1), 10–19 (2000).

Bronze, M.S. and Greenfield, R.A., *Curr. Opin. Investig. Drugs*, **4**(2), 172–178 (2003).

Brook, I., *Int. J. Antimicrob. Agents*, **20**(5), 320–325 (2002).

Carruthers, S.G., Hoffman, B., Melmon, K., and Nierenberg, D., *Clinical Pharmacology*, 4th ed., McGraw Hill, New York, 2000, pp. 873–909.

CDC, *MMWR*, **50**(42), 909–919 (2001).

Conlan, J.W., Shen, H., Webb, A., and Perry, M.B., *Vaccine*, **20**(29–30), 3465–3471 (2002).

Cono, J., Casey, C.G., and Bell, D.M., *MMWR*, **52**(RR-4), 1–28 (2003).

Crotty, S., Cameron, C., and Andino, R., *J. Mol. Med.*, **80**(2), 86–95 (2002).

De Clercq, E., *Antivir. Res.*, **55**(1), 1–13 (2002).

Dennis, D.T., Inglesby, T.V., and Henderson, D.A., *J. Am. Med. Assoc.*, **285**(21), 2763–2773 (2001).

DiPiro, J., Talbert, R., Yee, G., Matzke, G., Wells, B., and Posey, L.M., Eds., *Pharmacotherapy—A Pathophysiologic Approach*, 5th ed., McGraw Hill, New York, 2002, pp. 2123–2141.

Elkins, K.L., Cowley, S.C., and Bosio, C.M., *Microbes Infect.*, **5**(2), 135–142 (2003).

Flint, S.J., Enquist, L.W., Krug, R.M., Racaniello, V.R., and Skalka, A.M., Eds., *Virology: Molecular Biology, Pathogenesis, and Control*, ASM Press, Washington, DC, 2000 pp. 673–711.

Gilead Sciences, *Cidofovir [Package insert]*, Gilead Sciences, Inc., Foster City, CA, 2000.

Greenwood, D., Ed., *Antimicrobial Chemotherapy*, Oxford University Press, UK, 2000, pp. 4–28.

Greer Laboratories, Inc., *Plague Vaccine [Package insert]*. Lenoir, NC; 1995.

Hitner, H. and Nagle, B., *Basic Pharmacology*, McGraw Hill, New York, 1999, pp. 551–564.

Hutchinson, T.A. and Shahan, D.R., Eds., *DRUGDEX® System*, Micromedex, Greenwood Village, CO, (Expires 6/2003).

Inglesby, T.V., Dennis, D.T., Henderson, D.A., Bartlett, J.G., Ascher, M.S., Eitzen, E., Fine, A.D., Friedlander, A.M., Hauer, J., Koerner, J.F., Layton, M., McDade, J., Osterholm, M.T., O'Toole, T., Parker, G., Perl, T.M., Russell, P.K., Schoch-Spana, M., Tonat, K., *J. Am. Med. Assoc.*, **283**(17), 2281–2290 (2000).

Inglesby, T.V., O'Toole, T., Henderson, D.A., Bartlett, J.G., Ascher, M.S., Eitzen, E., Friedlander, A.M., Gerberding, J., Hauer, J., Hughes, J., McDade, J., Osterholm, M.T., Parker, G., Perl, T.M., Russell, P.K., and Tonat, K.; Working Group on Civilian Biodefense., *J. Am. Med. Assoc.*, **287**(17), 2236–2252 (2002).

Kende, M., Yan, C., Hewetson, J., Frick, M.A., Rill, W.L., and Tammariello, R., *Vaccine*, **20**(11–12), 1681–1691 (2002).

Klein, N.C. and Cunha, B.A., *Med. Clin. North Am.*, **79**(4), 789–801 (1995).

Kuberski, T., Robinson, L., and Schurgin, A., *Clin. Infect. Dis.*, **36**(4), 521–523 (2003).

Kuby, J., *Immunology*, 3rd ed., W.H. Freedman & Co., New York, 1997, pp. 432–442.

Lemley, P.V., Amanatides, P., and Wright, D.C., *Hybridoma*, **13**(5), 417–421 (1994).

Madigan, M.T., Martinko, J.M., and Parker, J., Eds., *Biology of Microorganisms*, Prentice Hall, Upper Saddle River, NJ, 2000, pp. 69–77.

Mayer, T.A., Bersoff-Matcha, S., Murphy, C., Earls, J., Harper, S., Pauze, D., Nguyen, M., Rosenthal, J., Cerva, D. Jr., Druckenbrod, G., Hanfling, D., Fatteh, N., Napoli, A., Nayyar, A., and Berman, E.L., *J. Am. Med. Assoc.*, **286**(20), 2549–2553 (2001).

McEvoy, G.K., Ed., *AHFS DI Bioterrorism Resource Manual*, American Society of Health-System Pharmacists, Bethesda, MD, 2002, pp. 228–237; 319–356.

Mina, B., Dym, J.P., Kuepper, F., Tso, R., Arrastia, C., Kaplounova, I., Faraj, H., Kwapniewski, A., Krol, C.M., Grosser,

M., Glick, J., Fochios, S., Remolina, A., Vasovic, L., Moses, J., Robin, T., DeVita, M., and Tapper, M.L., *J. Am. Med. Assoc.*, **287**(7), 858–862 (2002).

Putzker, M., Sauer, H., and Sobe, D., *Clin. Lab.*, **47**(9–10), 453–466 (2001).

Sidell, F.R., Patrick, W.C., Dashiell, TR, Alibek, K., and Layne, S., Eds., *Jane's Chem-Bio Handbook*, 2nd ed., Jane's Information Group, Virginia, VA, 2001, pp. 215–249.

Smee, D.F., Bailey, K.W., Wong, M.H., and Sidwell, R.W., *Antivir. Res.*, **52**(1), 55–62 (2001).

Tam, R.C., Lau, J.Y., and Hong, Z., *Antivir. Chem. Chemother.*, **12**(5), 261–272 (2001).

Turnbull, P.C., *Vaccine*, **9**(8), 533–539 (1991).

Vogel, G., *Science*, **302**, 222–223 (2003).

Walker, P.D., Barri, Y., and Shah, S.V., *Ren. Fail.*, **21**(3–4), 433–442 (1999).

Wyeth Laboratories, *DryVax [Package insert]*, Wyeth Laboratories, Marietta, PA, 2002.

WEB RESOURCES

CDC website, http://www.cdc.gov/ncidod/srp/drugs/formulary .html (last accessed July 5, 2004)

See also PHARMACEUTICAL INDUSTRY.

PRION DISEASES

Raymond A. Zilinskas

INTRODUCTION

The term "prion" stands for proteinaceous infectious particle. Prion diseases are infectious diseases of the brain and occur in both animals and people. The disease is given a different name according to the species it affects:

Humans	Creutzfeldt–Jakob Disease (CJD)
	Gerstmann–Sträussler Syndrome (GSS)
Cattle	Bovine Spongiform Encephalopathy (BSE)
Sheep	Scrapie
Deer/Elk	Chronic Wasting Disease (CWD)

"Transmissible Spongiform Encephalopathy" (TSE) is a general term encompassing all prion diseases.

Prion diseases are transmissible between species and bring about the slow degeneration of the central nervous system, which inevitably leads to death. A very long period of time elapses between infection and the appearance of the first clinical symptoms: typically, 2 to 4 years in sheep, 3 to 6 years in cattle, and more than 10 years in humans. As a general rule, once the symptoms have appeared, death will occur within a few months (Prionics, 2004).

THE PRION PATHOGEN

Prions, the pathogens causing prion diseases, differ considerably from better-known pathogens such as bacteria and viruses. Until now, science has not been able to completely characterize prions and their methods of propagation. What is known is that a disease-specific protein can be identified in an infected brain and other organs. This protein, the "scrapie prion protein" (PrP^{Sc}) (called the "BSE prion protein" in the case of BSE), is derived from an unaffected form of the same protein, PrP^C, which occurs normally in the body (Fig. 1). The two proteins, the disease-specific PrP^{Sc} and the normal PrP^C, differ in their spatial structures and the fact that PrP^{Sc} is resistant to destruction by digestive enzymes secreted within the human stomach, whereas PrP^C is completely destroyed by digestive enzymes.

As a result of many studies, scientists now conclude that PrP^{Sc} is a component of the prion pathogen. Some scientists, however, speculate that PrP^{Sc} represents the complete pathogen. According to this theory, an infection brought about by the penetrating PrP^{Sc} causes PrP^C

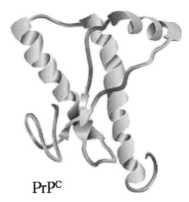

Figure 1. PrP^C.

to be converted into PrP^{Sc}. The newly formed PrP^{Sc} can now, for its part, bring about the conversion of more PrP^C into PrP^{Sc}. This leads to the disastrous chain reaction in which the PrP^{Sc} production increases exponentially and, eventually, causes irreversible damage to the brain. Notably, the increase in PrP^{Sc} during the course of the disease correlates with an increase in infectivity.

Prions are extremely resistant to heat and chemicals. Even heating to 100°C will not completely inactivate prions, and many of the usual disinfectants hardly affect it at all. Prions also resist biological decomposition; they can, for example, survive intact for many years in soil.

DIAGNOSTIC METHODS FOR PRION DISEASES

Diagnostic methods for prion diseases encompass clinical examination, evidence of infectivity, gross examination of the brain and immunohistochemistry, and identification of proteins specific to the infectious agent.

Clinical Symptoms

Early clinical symptoms are often unspecific and include, in the case of the cow, reduced milk production. Later clinical symptoms include timidity and mobility disturbances. Before the introduction of diagnostic tests utilizing specific antibodies, only BSE cases with clinical symptoms were detectable. In the case of classical CJD, electroencephalography (EEG) can be used to

Encyclopedia of Bioterrorism Defense, Edited by Richard F. Pilch and Raymond A. Zilinskas

PRION DISEASES 403

detect abnormal electrical brain activity indicative of the disease.

Evidence of Infectivity

The most accurate method to diagnose prion disease is the detection of infectivity in a test animal. To perform testing, suspected infectious material is injected directly into the brain of laboratory animals such as mice (mouse assay). After a long incubation period (for BSE, approximately 8 to 12 months), animals infected with prions become ill and die. An examination of the brain will show the characteristic spongiform changes and reveal large quantities of the disease-specific prion protein. Because of the length of time that testing takes to complete, transmission testing is not suitable for routine diagnostics but is limited to research experiments.

Gross Examination of the Brain and Immunochemistry

Upon autopsy, an examination of the brain will reveal small holes in various regions of the diseased brain, the number of which increases as the disease progresses, making the brain appear "spongy." Apart from these spongiform changes, there are deposits of the disease-specific prion proteins that, when treated with specific antibodies and suitably stained, are visible under a microscope (so-called immunohistochemistry). Cross-sections of the brain are usually examined for a conclusive BSE diagnosis in animals that have been slaughtered after the appearance of clinical symptoms. Immunohistochemistry is a precise and sensitive method (this is called the "gold" standard); however, it takes from days to weeks to complete and can therefore not be used as a rapid BSE-surveillance tool.

Detection of the BSE-specific Prion Proteins

In Europe, the European Commission has licensed five diagnostic tests for prion diseases. All employ antibodies against the prion protein. Four of them rely on the fact that the abnormal protein is resistant to digestion by proteinase K, while the normal isoform is not. The fifth test utilizes a method without proteinase treatment that compares the antigen/antibody binding affinities of prions in their native and unfolded states (Perkel, 2004). These tests can be used to identify the prion proteins specific to the disease within only a few hours. Currently there is no licensed rapid test in the United States.

PRIONS AND BIOTERRORISM

Under the Bioterrorism Preparedness and Response Act of 2002, BSE prions are considered one of 80 "select agents," which require special security arrangements, including background checks on anyone who may have access to this material in the laboratory. But whether prions have utility for terrorists is questionable. Richard Johnson, a neurologist, is quoted as saying: "I can think of many agents that are far more dangerous than BSE... As a bioterrorism agent, BSE is a loser" (Sabo, 2004). A leading prion researcher in Switzerland, Adriano Aguzzi, has said:

"The malicious introduction of prions into the food chain of humans or of cows would be a terrible scenario... And yet, I continue to believe that a terrorist might probably rather opt for poisons acting with a shorter latency, rather than an agent that may have a hypothetical incubation time of 20 years or longer" (Sabo, 2004).

KEY TERMS

BSE—Bovine Spongiform Encephalopathy, prion disease in cattle.

CJD—Creutzfeldt–Jakob Disease, prion disease in humans. In the new variant (vCJD), suspected to be transmitted to humans by consumption of BSE-contaminated beef, the disease strikes mostly young persons.

CWD—Chronic Wasting Disease, prion disease in certain wild animals.

Luminescence Immuno Assay—Diagnostic method using antibodies and a chemical reaction producing a light signal.

Prion—Pathogen causing prion diseases.

Prion Disease—Neurodegenerative transmissible disease caused by the accumulation of prions in the brain.

Scrapie—Prion disease in sheep.

Western-Blot—Technology for protein analyses.

REFERENCES

Perkel, J.M., Mad cow test options are plentiful, *The Scientist*, January 12, 2004, http://www.biomedcentral.com/news/20040112/02, accessed on 1/16/04.

Prionics, A.G., *Prionic diseases*, http://www.prionics.ch/prionics-e.htm, accessed on 11/21/04.

Sabo, E., Mad cow as bioterrorism? *The Scientist*, January 15, 2004, http://www.biomedcentral.com/news/20040115/04, accessed on 1/16/04.

FURTHER READING AND WEB RESOURCES

Aguzzi, A., http://www.unizh.ch/pathol/neuropathologie/, accessed on 6/28/04.

Blakeslee, S., *Jumble of tests may slow mad cow solution*, *New York Times*, January 4, 2003, http://www.nytimes.com/2004/01/04/national/nationalspecial2/04BEEF.html, accessed on 6/28/04.

Hitt, E., Prion prevention, *The Scientist*, September 27, 2002, http://www.biomedcentral.com/news/20020927/06/, accessed on 6/28/04.

Medical Follow-Up Agency, Institute of Medicine, *Advancing Prion Science: Guidance for the National Prion Research Program*, National Academies Press, Washington, DC, 2003, http://www.nap.edu/books/0309090601/html/, accessed on 6/28/04.

National Agriculture Biosecurity Center, http://www.ksu.edu/nabc/index.html, accessed on 6/28/04.

Shih, J., *Scientists Discover That Enzyme Degrades Mad Cow Disease Prion*, North Carolina State University press release, January 5, 2004, http://www.ncsu.edu/news/press_releases/04_01/001.htm, accessed on 6/28/04.

Stafford, N., Prion researcher awarded, *The Scientist*, October 30, 2003, http://www.biomedcentral.com/news/20031030/03/, accessed on 6/28/04.

PSYCHOLOGICAL AND SOCIAL SEQUELAE OF BIOTERRORISM

Gary Ackerman

INTRODUCTION

By definition, one of the core objectives of terrorism is to impact a broader population than the immediate victims of attack by causing fear, anxiety, and social disruption. Bioterrorism in particular encompasses several features that can serve to amplify these effects; indeed, this is one of the very reasons why terrorists might find this form of attack attractive. Furthermore, the psychological casualties of any terrorism incident involving unconventional weapons are likely to outnumber the physical casualties—some estimates anticipate by as much as a factor of five (Becker, 2001)—which could complicate response efforts. Even a small-scale bioterrorist event with limited direct casualties will likely impact millions, as seen during the 2001 "anthrax mail attacks." Therefore, one cannot overlook the psychological and social consequences of a bioterrorist event in any discussion about preparation for defense against bioterrorism.

DIRECT EFFECTS OF BIOLOGICAL AGENTS

The most basic effects that a biological attack can have on the mental health of victims are those that arise directly from the biological agent's somatic and physiological effects. Most of the primary biological warfare agents (such as *Bacillus anthracis*, *Yersinia pestis*, Variola major or staphylococcal enterotoxin B) induce a high fever, making delirium, with its concomitant cognitive impairment and disorientation, a possibility in almost any biological attack (Kron and Mendlovic, 2002). Q-fever can lead to encephalitis with hallucinations (DiGiovanni, 1999) and the viral encephalitides can produce long-term mood alterations and cognitive impairment (Kron and Mendlovic, 2002). Other possible effects are the depression and irritability sometimes associated with brucellosis and the depression and demoralization that can accompany long periods of ventilator use following exposure to botulinum toxin (Kron and Mendlovic, 2002).

Historical examples exist for such effects. During the 1918 influenza pandemic, for example, there were many recorded cases where the disease directly affected the mind of victims, resulting in hallucinations, amnesia, and even suicide (Iezzoni, 1999). And although no direct

psychiatric symptoms have been recorded from the use of mycotoxins, one can also point historically to the observed bizarre behavior and psychotic symptoms associated with "St. Vitus' Dance" and "tarantism" originating from ergot derivatives of other fungal toxins (Jones, 1995).

THE PSYCHOSOCIAL CHARACTERISTICS OF BIOTERRORISM

To examine the more indirect psychological and social consequences of bioterrorism that arise from the stress and trauma of the incident itself, it is first necessary to examine just why biological weapons carry with them such an aura of horror. A large-scale bioterrorist event would to some degree incorporate elements of a conventional terrorist attack, a national disaster, and a disease epidemic. By its very nature, bioterrorism thus reflects several of the characteristics that have been identified in such events as causing increased levels of concern, and which together magnify the perception of risk (Covello et al., 2001). These stress-inducing characteristics include:

- *Intangibility.* A covert bioterrorist attack leaves no immediate sensory cues (Kron and Mendlovic, 2002; DiGiovanni et al., 2003; Reid, 2002), engendering doubt and obstructing the innate "fight or flight" response (Roth, 1982).
- *Unfamiliarity.* The public at large does not have a copious understanding of the pathogenicity of microorganisms.
- *Human Origin.* Intentional human-precipitated disasters cause more anxiety than technological accidents or natural disasters (Becker, 2003; Kron and Mendlovic, 2002; Siegrist, 1999).
- *Amorality.* Terrorism is viewed by all its potential victims as morally abhorrent.
- *Lack of Personal Control.* Treatment and decontamination are mostly out of the hands of the average individual.
- *Dread.* Human beings almost universally display a deep fear of infection and contamination from communicable disease.

Encyclopedia of Bioterrorism Defense,   Edited by Richard F. Pilch and Raymond A. Zilinskas

- *Catastrophic Potential.* A bioterrorist event with a contagious agent is theoretically boundless.
- *Sudden Nature.* There is likely to be no warning of an attack.
- *Irreversibility of Outcomes.* Many biological agents have high mortality rates.

Sensationalistic media reporting and popular entertainment serve only to heighten these levels of concern. Further, the fact that bioterrorist events will often lack an apparent geographic locus (e.g., in the case of covert or simultaneous attacks) and may be "self-expanding" (Kron and Mendlovic, 2002), either through contagion or the delayed onset of symptoms, means that potential victims will experience an event without a discreet endpoint (such as an earthquake or terrorist bombing). For a protracted period, there may be no opportunity for those fearing exposure to determine that they have escaped harm and to begin to deal with the consequences, resulting in an extended state of stress and high psychological arousal.

It must be emphasized that many of the psychosocial sequelae of bioterrorism stem not from the actual risk of exposure or infection, but rather from the risk as perceived by the public (Reid, 2002; Covello et al., 2001). It is useful in this context to think of the perceived risk during a bioterrorist incident as consisting of both a hazard component (the risk as determined by scientists and other experts) together with an outrage component (the risk as perceived by the public, often including an emotional reaction to the threat) (Sandman, 2001; Covello et al., 2001).

The nature and severity of the psychosocial consequences of a bioterrorist attack will depend on a variety of factors, including the nature of the attack, the agent used, the response of authorities and the media (Smith et al., 2000), and the coping mechanisms of the individual and the community. Actually tracking and predicting these responses is not easy—epidemiological investigations of mental health are particularly difficult in the context of disasters involving terrorism, with findings varying widely between different populations and events (North and Pfefferbaum, 2002). Nonetheless, it is possible to use the existing literature and historical experience of large-scale disasters, terrorism, and infectious disease epidemics to broadly identify a set of possible psychological and social consequences to a bioterrorist event.

Kron and Mendlovic (2002) outline four key phases of victim response to a disaster (the impact phase, the early post-disaster phase, the disillusionment phase, and the recovery-reconstruction phase), but for the purposes of describing the social and psychological effects of bioterrorism, these effects will be divided into acute and long-term sequelae.

ACUTE PSYCHOSOCIAL SEQUELAE

As soon as it becomes apparent that people are falling ill rapidly and on a large scale in a specific locality, presumptive fears of bioterrorism will rise, especially if the initial disease symptoms are similar to those that have been widely mentioned as resulting from bioterrorist attacks (e.g., anthrax or smallpox). Modern communications systems, including television, radio, and the Internet, will almost instantaneously transmit this information beyond the affected area, causing a psychological ripple effect throughout the nation and possibly even worldwide, creating an endless list of witnesses to the event (Tasman, 2002). Factors such as the ambiguity of the threat, the fear of contagion, and the possibility of quarantine will cause a high degree of stress, even in those situated far from the scene of the attack (although, these effects are likely to increase the closer the people are to the incident locus). These high stress levels may result in maladaptive responses amongst the public, but will not necessarily do so.

The first of these responses to consider is that of large-scale panic. There is some disagreement in the literature regarding the potential for panic in such a situation (see, e.g., Final Report, 2001; Siegrist, 1999; Sandman, 2001), much of which can be traced to differing definitions of the term panic. While the stress brought on by a suspected or confirmed bioterrorist event will no doubt elicit strong reactions in the majority of the populace, many researchers consider the probability of panic strictly defined (as overwhelming fear leading to irrational responses such as "freezing" or unreasoned flight) to be fairly small (Sandman, 2001; Siegrist, 1999; Jones, 1995). For example, evacuation can take the form of reasoned withdrawal rather than blind panic, and the run on the antibiotic Ciprofloxacin during the 2001 "anthrax letter" attacks (Graham, 2001) can be regarded as resulting from a rational (if self-interested and individualistic) decision-making process. Indeed, it is estimated that during the initial stage of a disaster (the impact stage) only 10 to 15 percent of individuals display seriously inappropriate symptoms such as screaming, hysterical crying, or freezing (Jones, 1995). Sandman argues that people are most likely to panic only under a limited set of circumstances, namely, when "dire outcomes seem highly probable, but not absolutely certain, and they cannot tell what to do to optimize their chances of survival" (Sandman, 2001). However, it is still possible that a bioterrorist event could result in circumstances (e.g., a lack of sufficient medical supplies) that could lead to the above-mentioned perceptions, and thereby to genuine panic.

A second possible response during a bioterrorist event is that of denial—an under- rather than an overreaction. Denial is a risk management strategy wherein "the risk so threatens the psyche (usually for internal reasons rather than because of actual danger) that it must be psychologically forced into submission" (Reid, 2002). Victims of bioterrorism who engage in denial in a similar fashion to the coping mechanisms displayed by some early AIDS patients can complicate early detection, quarantine, and treatment efforts.

One response that will probably occur to some degree during a bioterrorist event (particularly one involving a contagious pathogen) will be fear and avoidance of those who are infected, even on the part of medical personnel. Stigma and scapegoating resulting in possible complications such as flight from exposed areas and even vigilantism can arise from the desire of the uninfected to distance themselves from those who are (or are perceived to be) ill.

The historical record is colored with many descriptions of intense social and psychological distress during large-scale disasters or outbreaks of infectious disease. While few, if any, of these can be said to describe cases of genuine "panic," many are instructive in terms of illustrating the public reactions described above. Commentators during the Black Death that raged through Europe in the Middle Ages described a situation in which "almost all [the survivors] tended toward one end—to flee from the sick and whatever belonged to them" (Gottfried, 1983, quoting Boccaccio in the *Decameron*). In Venice, for example, nearly all the doctors deserted the town and, on the Iberian Peninsula, local institutions of law and justice broke down (Gottfried, 1983). Matthew Carey's *A Short Account of the 1793 Philadelphia Yellow Fever Epidemic* describes a city descending into anarchy, with family members abandoning one another and masses fleeing Philadelphia under "an universal terror" (Griffith, 1997). At the same time, he emphasizes many cases of selflessness and caring and a heroic recovery brought about by the middle class. The 1918 influenza pandemic, despite killing millions, led to very little flight from the infected—rather, the dominant initial reaction seemed to be denial, with local officials and the press in many U.S. cities focusing their attentions primarily on Liberty Bonds for the First World War (Iezzoni, 1999). There was however no small amount of stigma: City ordinances made it "unlawful to cough and sneeze" in New York (Iezzoni, 1999), and made the wearing of masks mandatory in San Francisco (Iezzoni, 1999). There were even cases in New Mexico of armed vigilantes forcing visitors from more influenza-prone areas to return on the trains that had brought them (Iezzoni, 1999).

In the modern era, during the Three Mile Island disaster in the United States in 1979, 150,000 people took to the highways—45 people evacuated for every person advised to leave (Becker, 2003). In September 1994, when Indian officials announced an outbreak of plague in the city of Surat, an estimated 400,000 to 600,000 people (around 20 percent of the population) fled the city to other parts of India, spreading the disease (Tysmans, 1995). Anxiety-driven purchasing resulted in a shortage of antibiotics. More than half the physicians left as well, many taking large amounts of tetracycline with them. Yet, through all this, the official death toll was only 234 individuals nationwide (Tysmans, 1995).

These cases and the 2001 "anthrax letter" attacks provide some indication of possible reactions to a bioterrorist event. Yet, there have been no large-scale, intentionally caused epidemics with high mortality rates in the developed world since the advent of modern telecommunications media. Consequently, the exact nature of the acute psychosocial reaction to a bioterrorist event remains uncertain, though it is likely to contain some degree of denial and stigmatization and may even evidence cases of panic.

In addition to people attempting to avoid infection, one likely consequence of any bioterrorist attack will be large numbers of "worried well"—people who mistakenly believe they have been infected and whose anxiety leads to psychosomatic symptoms that may in some cases mimic those of the pathogen used in the attack. There have been several recorded cases of mass sociogenic illness associated with terrorism. Two examples will suffice. Following the March 1995 attack using the chemical nerve agent sarin in the Tokyo subway, there were 12 fatalities, 984 injuries, and more than 4000 people who arrived at hospitals seeking treatment with no observable exposure to the lethal chemical (Smithson and Levy, 2000). On October 3, 2001, over 1000 students at several schools in Manila, the Philippines, "deluged local clinics with mundane flu-like symptoms" (Wessely et al., 2001) after hearing rumors of bioterrorism that had been disseminated through text messaging. Large numbers of "worried well" can obstruct an effective response to a bioterrorist incident by unnecessarily demanding treatment and thus diverting medical resources from the treatment of those truly exposed and in need. At the very least, medical staff time will be consumed differentiating between the infected and the psychosomatic.

Fear of becoming or being infected, evacuation, or quarantine; anger; demoralization; feelings of helplessness; grief; and witnessing large-scale suffering and death are all factors causing great stress during or in the immediate wake of a bioterrorist incident. Besides possibly resulting in maladaptive coping responses (such as panic or denial), this stress can also manifest itself in the short term among at least some proportion of the affected population as a clinically diagnosable psychological condition known as Acute Stress Disorder (ASD). ASD involves a constellation of symptoms and consequences, many of which are more extreme forms of behaviors already discussed. These include dissociative symptoms such as emotional numbing, reexperiencing the traumatic event through nightmares etc., anxiety or increased arousal, and avoidance of stimuli that remind the victim of the trauma.

Dr. Robert DeMartino, director of the Program on Trauma and Terrorism at the Department of Health and Human Services (HHS), stated in 2002 that "for every person who gets physically ill from a bioterrorist attack, there will be at least fifty to one hundred who are so distraught that they cannot function normally in their daily lives" (Associated Press, 2002). Almost 60 percent of respondents to a Japanese survey had at least some symptoms of ASD one month after the Tokyo subway attack (Pangi, 2002). If these symptoms persist for more than a month after the attack, a diagnosis of Post Traumatic Stress Disorder (PTSD) which shares many symptoms with ASD, may be required.

In addition to all of the detrimental psychological effects already discussed, first responders are susceptible to added stresses. These include the possibility of claustrophobia from wearing unfamiliar protective equipment, repeated exposure to horrific scenes, being torn between job and family concerns, and the increased risk of becoming victims themselves. During chemical and biological warfare simulations, 10 to 20 percent of participants experienced moderate to severe psychological symptoms, with 4 to 10 percent having to cease their participation because of anxiety, claustrophobia, panic or equipment difficulties (Smith et al., 2000). After the September 11, 2001, terrorist attacks, 350 personnel from the New York Fire Department were placed on stress-related light duty or medical leave, and over 2000 were seen by counseling services for ASD or PTSD-related symptoms (Myers, 2002).

LONG-TERM PSYCHOSOCIAL SEQUELAE

Long-term Psychological Effects

The majority of survivors of a bioterrorist incident will eventually integrate their experiences, and their initial anxieties and other psychological symptoms will dissipate over time. Some people, however, will experience long-term psychological effects, which can include PTSD, anxiety disorders, phobias, depression, substance abuse, sexual dysfunction, and domestic violence. Although the extent of correlation with a bioterrorist attack is unknown, after the bombing of the federal building in Oklahoma City in 1995, 34 percent of 182 adult survivors developed PTSD, another 11 percent developed other disorders including major depression and substance abuse relapses (Smith et al., 2000), and over 80 percent displayed subthreshold PTSD symptoms (Tasman, 2002). Six months after the September 11 attacks, it was estimated in a study conducted by the Mailman School of Public Health at Columbia University that 26.5 percent of the students in New York City schools experienced some form of posttraumatic stress disorder, anxiety disorders, or depression (Hoven, 2002).

The rates of the development of disorders such as PTSD (and ASD) depend on such factors as the nature of the attack, the individual's proximity to the attack location, whether or not a person became infected, previous exposures to trauma, the individual's prior psychological state, the presence or otherwise of a support network, and the amount of exposure to media coverage (Pangi, 2002; Myers, 2002). Children, the elderly, and first responders in particular are at risk for developing PTSD or adverse psychological symptoms (Tasman, 2002; Smith et al., 2000). A relevant statistic in the context of bioterrorism is that women with family responsibilities are twice as likely as men to experience PTSD, anxiety, and depressive disorders—the very group that make up approximately 70 percent of hospital personnel in the United States (Final Report, 2001).

The long-term symptoms of many psychological victims may not meet the diagnostic criteria for PTSD, but these people may nonetheless exhibit what has been termed "subdiagnostic distress" (North and Pfefferbaum, 2002) with several of the symptoms of PTSD. This can be partly explained by the fact that the DSM-IV criteria for PTSD are not usually interpreted as applying in cases of indirect involvement with trauma, for instance through watching television (Schlenger et al., 2002). A study by Silver et al. (2002) found that 17 percent of the U.S. population outside of New York City reported some symptoms of posttraumatic stress two months after the September 11 attacks, and 5.8 percent did so after six months. However, Schlenger et al. (2002) found that signs of probable PTSD one to two months after the attack were more likely among those who lived in New York City, and therefore directly experienced the attack, than among people living elsewhere.

There is also the possibility that those living in the immediate area of a biological attack will experience long-term anxiety owing to ongoing fears of residual contamination or uncertainty surrounding the chronic health effects of some biological weapons agents (Wessely et al., 2001). As has occurred in various cases of toxic exposure or contamination in the past, people may begin to ascribe any symptom whatsoever, for years afterward, to the bioterrorist event (Havenaar et al., 1994).

The persistent presence of stress (whether resulting from PTSD, other mental disorders, or even psychosomatic symptoms) can in turn exert long-term effects on physiological health, by compromising the immune system, exacerbating conditions such as hypertension, and leading the distressed individual to generally neglect his or her health.

Long-term Social Effects

The long-term effects of a bioterrorism incident will not necessarily be limited to chronic physiological or psychological maladies. A bioterrorist incident could have far-reaching effects on society, where, according to Becker (2003), "repercussions from an attack could change the psyche of the American people." This is not unknown in historical records; the Aztecs, for instance, began to convert to Christianity in droves after it appeared that their gods did not fare as well in protecting them from smallpox as did the god of the Spanish. In reality, the dichotomous effect of the disease on the two populations was due to the fact that the Spanish had largely been exposed to smallpox earlier in life and had survived into adulthood, thus displaying immunity in the face of the virgin soil Aztec population (Oldstone, 1998).

A large-scale bioterrorist event with high mortality will likely lead to a prolonged sense of shared grief in which the victims are mourned by the nation, or even much of the world, depending on the scale of the attack. This could have the positive consequence of closer feelings of community as people at least temporarily focus on victims and recovery and put aside traditional social and other disagreements. On the other hand, it could elicit negative attitudes such as jingoism or xenophobia that are fueled by anger and a desire to punish the perpetrators (who in some cases may remain anonymous, inviting scapegoating).

Another likely long-term social consequence of bioterrorism will be an increased sense of vulnerability by the population and a simultaneous loss of faith in the institutions—state, religious, or medical—that failed to protect or save them. The Black Death, for example, is argued by some historians to have exerted a major impact on the development of medieval European society, leading to such varied outcomes as peasant revolts, a backlash against the Church, and greater materialism amongst those who sought to enjoy life to its fullest extent in the face of capricious disease (Gottfried, 1983). Despite the huge advances in medical science over the preceding century, during the 1918 influenza pandemic, people lost faith in the apparently impotent medical profession and regressed to using folk remedies such as necklaces made of chicken feathers (Iezzoni, 1999). While on some occasions the previously faithful may abandon their religious beliefs, a bioterrorist event could also convince some individuals to return to the fold if they are seeking solace or perceive the attack as a form of divine punishment.

Except in the unlikely event of an uncontrolled contagious epidemic precipitated by a bioterrorist attack, it is improbable that a modern, developed society like that of the United States would descend into macro-level social chaos. There may however be areas where local social, economic, and political functions cease to operate for a time, and must be replaced by external assistance, which can be extremely socially disruptive. It is also quite possible that there will be broad-based recriminations specifically against the government, or segments thereof, for not preventing the attack or not responding more effectively.

A large-scale bioterrorist event could very well permanently alter social structures and attitudes, whether as a result of trauma, grief, blame, an increased sense of vulnerability, stricter security measures, or a loss of support for existing institutions. It is almost impossible to predict with any degree of certainty the direction, duration, or extent of these changes since they depend on a complex and dynamic set of variables. However, all these possibilities should be taken into account in any long-term consequence management planning.

MITIGATING MEASURES

In order to address some of the psychological distress associated with bioterrorism, several techniques from the general management of psychological trauma can be applied to the victims of bioterrorism. These include reassurance that symptoms are a normal human response; allowing victims the opportunity to regain control of their lives and express themselves; support from family and other social networks; and encouraging the resumption of a normal routine (Tasman, 2002). Early intervention is considered important in preventing or minimizing PTSD or ASD symptoms (Pangi, 2002), although recently some researchers have stressed the natural ability to recuperate and contend that early intervention such as Critical Incident Stress Management Programs may in fact exacerbate psychological symptoms.

The imposition of quarantine can provide psychological relief for the uninfected outside of the quarantine area, but at the same time, those under quarantine may have their fears compounded by the isolation and dislocation. Here, constant communication (e.g., using telecommunication) with responders and relatives outside the quarantine area may reassure people and identify those most at risk of developing serious psychological symptoms (Kron and Mendlovic, 2002).

In order for authorities to perform any of these ameliorative functions, it is essential that measures to address the mental health and social fallout of a bioterrorist incident are included in any contingency planning for bioterrorism. Response plans should, for example, take into account the possibility of mass panic and the fact that this possibility is fairly small but may still occur. Currently, the Federal Response Plan assigns the duties of various government agencies in responding to a terrorist event, with the Federal Emergency Management Agency (FEMA) designated as the lead for consequence management. FEMA has few resources to deal with psychological issues, save for the Center for Mental Health Services and grants to local communities (Pangi, 2002). Other large-scale psychological assistance can be provided by the Substance Abuse and Mental Health Services Administration within HHS, four specialized Disaster Medical Assistance Teams, and nongovernmental organizations such as the American Red Cross, the American Psychiatric Association, and the American Psychological Association.

Despite these resources, many experts believe that insufficient attention has been paid to psychosocial effects in planning for terrorism. An Institute of Medicine committee found that national preparedness activities for terrorism "are not addressing mental health issues in a manner consistent with the level of risk" (Nuzzo and Schoch-Spana, 2003). Few first responders are trained to deal with psychological problems, psychosocial effects are not incorporated enough into the design of exercises, and what planning does occur in this area tends to emphasize short-term rather than long-term mitigation efforts (Becker, 2001; Pangi, 2002). Even where psychological issues are given attention, social effects are generally ignored. Psychosocial issues did begin to receive more attention after the publication of the National Council on Radiation Protection and Management Report No. 138. Another positive development is the proposed National Resilience Development Act (H.R. 2301), which would include a provision for 1 percent of a state's Department of Homeland Security (DHS) funding to be devoted to building up local psychological resilience. Yet, most weapons of mass destruction response plans still exclude psychosocial aspects (Becker, 2003). Even when psychosocial effects are included, planning for a bioterrorist event must take into account differences between specific populations, distinguished according to factors such as age group, language, and ethnicity, in order to be effective.

One especially vital aspect of dealing with the psychosocial effects of a bioterrorist incident is the quality of communications between officials and the public. Although the public will to a large extent depend on the government for guidance during a crisis, the conditions for communication will be especially difficult during a bioterrorist event. The audience will be anxious and perhaps skeptical, with mental and emotional "noise" potentially interfering with its ability to properly process information (Covello et al., 2001). Officials will have to rely to some degree on the media to disseminate information during a crisis. Yet, media attributes such as a rapid newscycle, competition for the "scoop," and an orientation toward sensationalism, as well as the potential for misinformation over the Internet and other media channels, mean that the facts surrounding the event as well as messages from officials may become distorted before reaching the public (Final Report, 2001). In a simulated bioterrorist attack in 2003, local journalists exhibited some of the highest levels of fear and confusion out of all the population groups studied (DiGiovanni, 2003). In any event, there is also the issue of trust: Many people do not consider the government to be the most credible source of information. During the Three Mile Island accident, multitudes felt as though they were not being told everything they needed to know (Final Report, 2001). And Sandman (2001) argues that during the 2001 "anthrax letter" attacks, the Centers for Disease Control and Prevention (CDC) should have clearly and often acknowledged that they initially made an error regarding whether *B. anthracis* spores could escape a sealed envelope to infect postal workers in order to preserve public trust in the CDC.

Limitations on space preclude a detailed review of crisis and risk communication methods, but some of the most important requirements in the event of a bioterrorist incident can be considered. A successful bioterrorism communication plan includes having a specific and professional communication strategy prepared prior to the event; being honest with the public at all times; refraining from either overreassurance or dwelling on negatives; acknowledging official uncertainty during a crisis; showing empathy and treating the public's concerns as legitimate; and providing medically acceptable actions that individuals can take in order to maintain their sense of control (Covello et al., 2001; Sandman, 2001).

Effective communications may in fact be one of the best avenues to counter much of the psychosocial sequelae of bioterrorism when used proactively. By educating the public beforehand about the risks of and responses to a bioterrorist incident, it may be possible to desensitize the public to some degree and so "psychologically inoculate" the populace prior to an attack. If the public were better prepared mentally to deal with bioterrorism, and the fear and possibility of panic during a future attack thereby reduced, one of the key elements making bioterrorism attractive to terrorists would be rendered nugatory. This may in turn reduce the probability of such an attack in the first place. Together with all the psychosocial sequelae outlined above, this makes acknowledging and preparing for the many psychological and social consequences, in addition to expected physiological consequences, an essential element in defending against the threat of bioterrorism.

REFERENCES

Associated Press, *Experts Gauge Mental Impact of Attack*, *New York Times*, 11/20/2002.

Becker, S.M., *Mil. Med.*, **166**(Suppl. 2), 66–68 (2001).

Becker, S.M., Psychosocial Issues in Radiological Terrorism and Response: NCRP 138 and After, Presented at the *International Workshop on Radiological Sciences and Applications: Issues and Challenges of Weapons of Mass Destruction Proliferation*, Albuquerque, New Mexico, 2003.

Covello, V.T., Peters, R.G., Wojtecki, J.G., and Hyde, R.C., *J. Urban Health: Bull. N. Y. Acad. Med.*, **78**, 382–391 (2001).

DiGiovanni, C., *Am. J. Psychiat.*, **156**, 1500–1505.

DiGiovanni, C., Reynolds, B., Harwell, R., Stonecipher, E.B., and Burkle, F.M., *Emerg. Infect. Dis.*, **9**, 708–712 (2003).

Final Report, *Human Behavior and WMD Crisis/Risk Communication Workshop*, co-sponsored by Defense Threat Reduction Agency, Federal Bureau of Investigation, U.S. Joint Forces Command, 2001.

Gottfried, R., *The Black Death*, Free Press, New York (1983).

Graham, S., *Bayer to Raise Production of Anthrax Antibiotic*, *Associated Press*, New York, 10/10/2001.

Griffith, S.F., "A Total Dissolution of the Bonds of Society": Community Death and Regeneration in Matthew Carey's "*Short Account of the Malignant Fever*," in J.W. Estes and B.G. Smith, Eds., *A Melancholy Scene of Devastation: The Public Response to the 1793 Philadelphia Yellow Fever Epidemic*, Science History Publications, USA, 1997, pp. 45–59.

Havenaar, I.M., Rumiantseva, G.M., and van den Bout, J., *Russ. Soc. Sci. Rev.*, **35**, 87 (1994).

Hoven, C., Children of September 11: The Need for Mental Health Services, *Testimony of Dr. Christine Hoven before the Committee on Health, Education, Labor, and Pensions*, United States Senate, 6/10/2002, p. 43.

Iezzoni, L., *Influenza 1918*, TV Books, New York, 1999.

Jones, F.D., "Neuropsychiatric Casualties of Nuclear, Biological, and Chemical Warfare," *Textbook of Military Medicine: War Psychiatry*, Department of the Army, Office of The Surgeon General, Borden US Government Publication Institute, 1995, pp. 85–111.

Kron, S., and Mendlovic, S., *IMAJ*, **4**, 524–527 (2002).

Myers, D., Psychological Impact of Terrorist Incidents. Presented to Bay Area Terrorism Working Group, 2002.

North, C.S., and Pfefferbaum, B. *J. Am. Med. Assoc.*, **288**, 633–636 (2002).

Nuzzo, J., and Schoch-Spana, M., *Biodef. Q.*, **5.2**(3), 10–11 (2003).

Oldstone, M.B.A., *Viruses, Plagues, and History*, Oxford University Press, Oxford, 1998.

Pangi, R., *Persp. Preparedness*, **7**, 1–20 (2002).

Reid, W.H., "Bioterrorism: Separating Fact, Fiction, and Hysteria," in C.E. Stout, Ed., *The Psychology of Terrorism*, Praeger, Westport, CT, 2002, pp. 159–172.

Roth, W.T., "The Meaning of Stress," in F.M. Ochberg and D.A. Soskis, Eds., *Victims of Terrorism*, Westview Press, Boulder, CO, 1982, pp. 37–57.

Sandman, P.M., *Anthrax, Bioterrorism and Risk Communication: Guidelines for Action*, 2001, accessed at http://www.psandman.com/col/part1.htm on 7/25/2003.

Schlenger, W.E., Caddell, J.M., Ebert, L., Jordan, B.K., Rourke, K.M., Wilson, D., Thalji, L., Dennis, J.M., Fairbank, J.A., and Kulka, R.A., *J. Am. Med. Assoc.*, **288**, 581–588 (2002).

Siegrist, D.W., Behavioral Aspects of a Biological Terrorism Incident, *Presentation to the Association for Politics and the Life Sciences Convention*, Atlanta, GA, 9/2/1999.

Silver, R.C., Holman, E.A., McIntosh, D.N., Poulin, M., and Gil-Rivas, V., *J. Am. Med. Assoc.*, **288**, 1235–1244 (2002).

Smith, C.G., Veenhuis, P.E., and MacCormack, J.N., *NCMJ*, **61**, 150–165 (2000).

Smithson, A., and Levy, L., *Ataxia: The Chemical and Biological Terrorism Threat and the US Response*, Stimson Center Report No. 35, accessed at http://www.stimson.org/cbw/pubs.cfm?ID=12, 2000.

Tasman, A., *Mental Health Issues Related to Terrorism*, Presented at the University of Louisville, accessed at http://www.chse.louisville.edu/bioterrorismTasman.pdf, 11/28/2002.

Tysmans, J.B., *Carolina Pap. Int. Health Dev.*, **2**, 1–12 (1995).

Wessely, S., Hyams, K.C., and Bartholomew, R., *Br. Med. J.*, **323**, 878–879 (2001).

PUBLIC HEALTH PREPAREDNESS IN THE UNITED STATES

REBECCA KATZ

INTRODUCTION

Public health preparedness for bioterrorism refers to the ability of the federal, state, and local governments to organize resources in order to appropriately address the threat of a biological weapons attack on the public. This typically involves close cooperation with private industry, individual clinicians, hospitals, and nongovernmental organizations. The basic responsibilities of the public health system with respect to bioterrorism preparedness are prevention, detection, treatment, containment, communication, and research.

BACKGROUND

The initial indication that a bioterrorism attack has occurred will most likely come from the local or state public health system. A biological weapon is inherently difficult to detect, as an attack becomes evident through the illness and/or death of its victims, in contrast to other weapons systems that can be detected through sight, sound, or smell. When a person becomes ill, he or she enters the medical system, which reports directly to the public health community. The reporting of an unusual cluster of illness or death, or a single anomalous case due to a critical biological agent included in the Centers for Disease Control and Prevention's (CDC) Category A, B, or C list, might be the first indication of a biological weapon attack.

The public health system initially organized to address the threat of bioterrorism in 1951, through the creation of the Epidemic Intelligence Service (EIS). In subsequent years, however, little attention was paid to bioterrorism, and resources were devoted to the fight against naturally occurring diseases. By the late 1970s, funding for the public health infrastructure began to wane, as the ability to fight infectious diseases improved and resources were diverted to address chronic diseases. Twenty years of neglect resulted in a public health infrastructure with a small workforce, poorly maintained facilities, and limited resources to detect and respond to infectious diseases and bioterrorism alike. However, with the resurgence of attention to bioterrorism as a potential public health threat, along with the emergence and reemergence of dangerous and often lethal infectious diseases, resources to improve the public health infrastructure have been appropriated. The limited financial support that the public health system had received prior to this recent surge in funding means that much is needed in the way of repairs and additions in order to adequately address the threat of bioterrorism and protect the public's health.

Today's public health system is responsible for (http://www.phppo.cdc.gov/nphpsp/phdpp/10ES.htm, accessed on 3/17/03):

- Monitoring health status
- Diagnosing and investigating health problems
- Engaging the community to address health problems
- Helping people find appropriate health services
- Enforcing public health laws and regulations
- Assuring a competent workforce
- Evaluating health services
- Communicating to the public
- Disseminating timely information
- Conducting appropriate research

In addition, the CDC recently defined seven focus areas of public health preparedness for the purpose of funding local health agencies to enhance their capacity to respond not only to threats of bioterrorism but also to unintentional cases of infectious disease (CDC, 2002):

- Preparedness, planning, and readiness assessment
- Surveillance and epidemiology capacity
- Laboratory capacity for biological agents
- Laboratory capacity for chemical agents
- Communications and information technology
- Communicating health risks and health information dissemination
- Education and training

In 2004, the CDC expanded thier list of fundamental competencies for bioterrorism preparedness and response to include consequence management, emergency response, partnerships to expand expertise, and supportive management (CDC, 2004).

DISCUSSION

As discussed above, public health preparedness for bioterrorism encompasses prevention, detection, containment, treatment, communication, and research. The following section describes each of these aspects in detail.

Encyclopedia of Bioterrorism Defense, Edited by Richard F. Pilch and Raymond A. Zilinskas
ISBN 0-471-46717-0 Copyright © 2005 Wiley-Liss

Prevention

Several actions can be taken both before and after a bioterrorist attack to prevent infection and protect public health. The primary mode for prevention is the distribution of prophylactic drugs and vaccines. This includes vaccination campaigns in response to credible threats, such as the recent program to begin vaccinating health care workers and other first responders against smallpox. Postattack prevention refers to the distribution of drugs and/or vaccines to unexposed populations in order to protect them from infection.

Detection

Rapid detection of a bioterrorist attack or any other disease epidemic is achieved through a comprehensive surveillance system, encompassing effective computer data monitoring programs, astute clinicians, and efficient laboratorians. The surveillance system must be able to effectively monitor rates of disease, symptoms, and drug use over time, through the cooperation of hospitals, pharmacists, clinicians and public health professionals. Surveillance involves both passive (disease information reported to public health departments) and active (search for data on disease occurrences) action, and must be able to look at information around the clock and from various geographical regions. This surveillance is complemented by epidemiologic skills of public health professionals who are able to use the data to create a temporal and geographical picture of disease events.

Containment

Pathogens arming biological weapons are either communicable or noncommunicable. For noncommunicable diseases, there is very little the public health system needs to do with regard to containment. Those exposed to the agent at the site of an attack are the only ones at risk of developing disease. Containment in this case is the responsibility of law enforcement, intelligence, and security personnel, and involves the discovery and control of future attack sites. Communicable diseases, however, require public health measures to limit exposure and control the spread of disease. Methodologies for disease containment vary by the disease itself and the circumstances for its spread. In some instances, early treatment of infected patients before they are most infectious is the best control. For certain diseases, changing personal or cultural behaviors will limit the spread of disease. This is true for sexually transmitted diseases or Ebola fever, where the cultural practice of preparing a body for a funeral transmits the disease from the corpse to the family members. Containment may involve the physical removal of infected populations from the rest of the public through quarantines, or may involve the forced control of population movements. Additionally, environmental decontamination may assist in controlling the spread of disease.

Treatment

When a population is exposed to and potentially infected by a biological weapon's agent, the medical and public health systems are responsible for its treatment and overall care. Treatment needs to be delivered by medical staff trained to identify and treat diseases caused by biological weapons in a clinical setting that is prepared for large numbers of patients. If a hospital does not have enough beds, alternative sites can be employed. Treatment of mass casualties also involves the provision of medical equipment, such as ventilators, and drugs and vaccines.

Communication

In order for the public health system to be adequately prepared for bioterrorism, both internal and external communication procedures need to exist. Internally, the public health systems at the local, state, and federal levels must be able to communicate with each other to share data, concerns, and instructions. Additionally, the government needs to be able to share information securely with hospitals, clinicians, and laboratories so that all aspects of the public health and medical systems are fully informed and capable of responding to events. External communication refers to the public health system's interaction with the media and general public. Public health officials must be able to inform the public about threats, provide guidance and instructions when appropriate, and ensure calm. This requires a coordinated communications plan so that information passed to the public is correct, consistent, and current.

Research

Basic scientific research is an integral part of public health preparedness in that it leads to a better understanding of the pathogenesis of disease, the genetic makeup of microbes, and the development of methods to diminish the severity of disease in humans. Through the advancement of knowledge of infectious diseases and biological agents in general, it will become easier to rapidly detect new agents, diagnose disease in patients, and develop new methods of treatment.

CURRENT STATUS

In the last few years, many resources have been devoted to improving public health preparedness for bioterrorism. While great improvements have been made to the public health infrastructure, particularly at the local level, this task is by no means complete. This section describes the current status of preparedness in six major focus areas: surveillance, laboratory analysis, pharmaceutical and vaccine stockpiles, hospital readiness, research, and communication.

Surveillance

In order for a surveillance system to be effective, reporting must be complete and record-keeping accurate to allow for the compilation of historical trends. All localities need to participate, and clinicians and hospitals must cooperate. The CDC has over 100 separate surveillance systems, most of which are independent and disease specific. Many local and state agencies have their own surveillance systems that do not communicate with regional or federal systems. Several efforts have been made to coordinate surveillance and design new methodologies for data collection. The CDC created the National Electronic Disease Surveillance System (NEDSS) as a means to integrate disease databases and coordinate local and state surveillance. The $1.1 billion designated by the Bush administration in 2001 for public health infrastructure included the adoption of NEDSS in all states (Lumpkin and Richards, 2002). Additional CDC surveillance projects include the Enhanced Surveillance Project (ESP), the Early Aberration Reporting System (EARS), and BioNet in collaboration with the LRN and the Association of Public Health Laboratories. Individual states and localities, however, have not coordinated their efforts and have created independent systems. Coordination of surveillance systems remains one of the most important challenges for the public health community.

The CDC and the Department of Defense are experimenting with new surveillance systems that could provide a unified approach to disease monitoring. The CDC is building a computer network focused on eight major cities in the United States that will monitor physician reports, emergency room visits, and pharmacy sales (Broad and Miller, 2003). The Defense Advanced Research Project Agency (DARPA) is funding the testing of ESSENCE 2 (Electronic Surveillance System for Early Notification of Community-Based Epidemics), which is based upon a prototype surveillance system developed by the military for monitoring active duty servicemen and their dependents living in the Washington, DC, region. ESSENCE 2 applies the same design to the civilian population of Washington, DC, by tracking drug store sales, school absenteeism, animal health, and hospital and physician records (http://www.geis.ha.osd.mil/GEIS/SurveillanceActivities/ESSENCE/ESSENCE.asp, accessed on 3/22/03).

Laboratory Analysis

Rapid and accurate laboratory analysis is a crucial aspect of public health preparedness. The Association of Public Health Laboratories (APHL) created the National Laboratory Training Network (NLTN) in 1988, and has since trained thousands of scientists in the identification of dangerous pathogens. In response to bioterrorism, the Laboratory Response Network (LRN) was established through Congressional mandate to manage and diagnose biological weapons agents. This system focuses primarily on the capacity of local, state, and federal public health laboratories, and offers swift movement of samples to the appropriate level laboratory. The CDC is also working toward the creation of the National Laboratory System (NLS), designed to network public and private laboratories around the country. From an infrastructure perspective, the National Institutes of Health (NIH) is funding the construction of additional BSL-3 and BSL-4 laboratories around the country, which will improve the overall national capacity to analyze and research biological warfare agents.

Pharmaceutical and Vaccine Stockpiles

The National Pharmaceutical Stockpile Program, now called the Strategic National Stockpile, was created in January 1999 through a Congressional mandate. This program is responsible for stockpiling drugs, vaccines, and medical supplies in locations around the country, which can be delivered on request to any location in the United States within 12 hours. An additional shipment of supplies pertinent to the specific emergency can be delivered within 24 to 36 hours. The program initially was funded at $52 million a year. Funding has dramatically increased in the past years, allowing for the purchase of additional supplies and improved delivery to sites around the country.

Hospital Readiness

A recent survey of rural and urban emergency departments in the mid-Atlantic region of the United States found that only 10 percent were equipped to handle a mass casualty event of 50 to 100 patients (Treat et al., 2001). Hospitals in general have limited excess capacity to handle large numbers of casualties, a limitation that would be further complicated in the event that the casualties are affected by a contagious disease. In FY 2002, the Health Resources and Services Administration allocated $125 million for the development and implementation of regional plans to improve hospital capacity (Treat et al., 2001). Funding for hospital readiness in 2003 was increased to $498 million (http://www.hhs.gov/news/press/2003pres/20030320.html, accessed on 3/22/03). Hospitals will use these funds to create dedicated decontamination facilities, stage disaster training and drills, purchase personal protection equipment, acquire drugs, expand mental health services, and plan for alternative treatment sites if hospital capacity is exceeded. In order to meet these requirements, hospitals are cooperating with regional public health agencies.

Research

Funding for research related to biodefense has dramatically increased in the last few years. Coordinating the research effort is the National Institute of Allergy and Infectious Diseases (NIAID) at the NIH. Since fall 2001, NIAID has begun or expanded 25 research initiatives. Specific disease research and drug development have focused on the Category A and B agents, as defined by the CDC critical agent list. Research related to Category C and chemical agents is currently in the planning process. The research agenda is concentrated on basic biology, immunology and host response, vaccines and drugs, and diagnostics. In addition, NIAID is funding the construction of research facilities around the country capable of biodefense and emerging infectious disease research.

Communication

In 1992, only 45 percent of local health departments possessed the ability to fax health alerts to the community. By 1999, 20 percent of local health departments had no access to email, and by October 2001 more than 30 percent of county agencies still did not have Internet connections (Baker and Koplan, 2002; Frist, 2002). Thus, while the ability of public health agencies to communicate with each other and to the public is a major aspect of preparedness, this need for open lines of communication was grossly neglected for many years. Recent initiatives, such as the Health Alert Network, have aimed to address these shortfalls. Today, approximately 90 percent of the population is served by public health departments with high-speed Internet access (Baker and Koplan, 2002).

In late 2000, the CDC created the Epidemic Information Exchange (Epi-X), a new communication system for secure nationwide transmission of information. Today, it reaches more than 800 public health professionals and provides them with updates on outbreaks and threats (Baker and Koplan, 2002). Although this program may serve as a foundation for future efforts, there is still no single coordinated system utilized by the entire nation for communication of health information.

Future Outlook

In February 2003, President Bush proposed $6 billion to fund a new system for biodefense, Project BioShield, and on May 19, 2004 the Project BioShield Act of 2003 was passed by the Senate by a vote of 99-0. This project aims to develop and produce safer and more effective vaccines and treatments against biological warfare agents. Project BioShield will build capacity to produce vaccines and drugs, hire more staff, research biomedical countermeasures, construct new research resources, enable the Food and Drug Administration (FDA) to process new treatments in a timely fashion, and make funds available to purchase needed supplies in emergency situations.

Conclusion

Public health preparedness is an integral part of the national defense against bioterrorism. It is the most likely means by which the occurrence of an attack will first be detected, and it is the primary actor in treatment and control after the attack. A fully prepared system can significantly limit the morbidity and mortality that might result from a bioterrorist attack. A fully prepared system will also address everyday threats to public health in a way that will limit the morbidity and mortality from naturally occurring diseases.

Much remains to be done before the public health system will be fully prepared to respond to the threat posed by bioterrorism, yet tremendous strides have been made in a short period of time. There is now increased cooperation and communication between public health agencies and the medical, law enforcement, and security communities; regions are planning for mass casualty events and conducting training for medical personnel; the government is stockpiling needed supplies and supporting research to better understand disease; and the ability to monitor the health of the national and international population has greatly improved. It will be important, however, to focus on coordinating surveillance and communication systems around the country and worldwide so that complete information can be assessed and management of victims and response mechanisms can be efficiently coordinated to both contain the spread of disease and adequately treat infected populations.

REFERENCES

Baker, E.L. and Koplan, J.P., *Health Aff.*, **21**(6), 15–27 (2002).

Broad, W. and Miller, J., Health Data Monitored for Bioterror Warning, *New York Times*, January 27, 2003.

Centers for Disease Control and Prevention. Cooperative Agreement Award Notice and Grant Guidance, Guidance for Fiscal Year 2002, Supplemental Funds for Public Health Preparedness and Response for Bioterrorism, Announcement Number 99051-Emergency Supplemental. February 15, 2002, https://www.bt.cdc.gov/planning/coopagreementaward/index.asp, accessed on 3/17/03.

Centers for Disease Control and Prevention, *Agency for Toxic Substances and Disease Registry*. A National Public Health Strategy for Terrorism Preparedness and Response 2003-2008, March 2004.

DoD-GEISS. ESSENCE: Electronic Surveillance System for the Early Notification of Community-based Epidemics: http://www.geis.ha.osd.mil/GEIS/SurveillanceActivities/ESSENCE/ESSENCE.asp, accessed on 3/22/03.

Frist, B., *Health Aff.*, **21**(6), 117–130 (2002).

Harrell, J.A. and Baker, E.L., *The Essential Services of Public Health*, American Public Health Association, http://www.phppo.cdc.gov/nphpsp/phdpp/10ES.htm, accessed on 3/17/03.

Health and Human Services, Press Release: HHS Announces Bioterrorism Aid for States, Including Special Opportunity for Advance Funding, March 20, 2003. http://www.hhs.gov/news/press/2003pres/20030320.html, accessed on 3/22/03.

Lumpkin, J.R. and Richards, M.S., *Health Aff.*, **21**(6), 45–56 (2002).

Treat, N.K., Williams, J.M., Furbee, P.M., Manley, W.G., Russell, F.K, and Stamper, C.D., *Ann. Emerg. Med.*, **38**(5), (2001), as cited in Health Resources and Services Administration, HHS. Bioterrorism Hospital Preparedness Program, Cooperative Agreement Guidance, February 15, 2002, ftp://ftp.hrsa.gov/terrorism/Hospitalpreparednessguidance FINAL2-15.pdf, accessed on 3/18/03.

White House. Press Release: Project BioShield, February 3, 2003, http://www.whitehouse.gov/news/releases/2003/02/print/20030203.html, accessed on 3/1/03.

FURTHER READING

Henderson, D.A., *Emerg. Infect. Dis.*, **4**(3), 488–492 (1998).

Khan, A.S., Morse, S., and Lillibridge, S., *Lancet*, **356**(9236), 179–1182 (2000).

Knobler, S., Mahmoud, A.A.F., and Pray, L., Eds., *Biological Threats and Terrorism: Assessing the Science and Response Capabilities: Workshop Summary*, National Academy Press, Washington, DC, 2002.

Lederberg, J., Ed., *Biological Weapons: Limiting the Threat*, MIT Press, Cambridge, 1999.

WEB RESOURCES

National Institutes of Health, www.niaid.nih.gov/biodefense/.

Centers for Disease Control and Prevention, www.bt.cdc.gov.

See also CENTERS FOR DISEASE CONTROL AND PREVENTION'S BIOTERRORISM PREPAREDNESS PROGRAM; CONSEQUENCE MANAGEMENT; CRISIS MANAGEMENT; DEPARTMENT OF HEALTH AND HUMAN SERVICES; EDUCATION FOR BIODEFENSE; HOMELAND DEFENSE; LABORATORY RESPONSE TO BIOTERRORISM; NATIONAL INSTITUTES OF HEALTH AND NATIONAL INSTITUTE OF ALLERGY AND INFECTIOUS DISEASES; and SYNDROMIC SURVEILLANCE.

R

RAJNEESHEES

W. Seth Carus

INTRODUCTION

In August and September of 1984, the religious cult, the Rajneeshees, employed the biological agent *Salmonella typhimurium* against the inhabitants of The Dalles, a small town in Oregon. Public health officials estimated that 751 people became sick because of the attack. No one died, although nearly four dozen people were hospitalized.

BACKGROUND

The Rajneeshees was founded in Poona, India, during the 1960s by a highly charismatic Indian guru who called himself the Bhagwan Shree Rajneesh. By the 1970s, most of his followers were young Westerners attracted to his eclectic philosophic views, especially his emphasis on sexual practices not encumbered by traditional moralities. As a result of its popularity, the cult gained considerable wealth. By 1981, however, the cult's beliefs had so outraged the local community in Poona that the Bhagwan moved the group to a remote ranch located mostly in Wasco County, Oregon. Wasco County was largely rural, and about half of its 20,000 residents lived in the county seat, the town of The Dalles.

Using the group's substantial resources and the labor of his followers, the Bhagwan built a small town of several thousand people on the ranch. "Rajneeshpuram" was a legally incorporated community with its own police force, trained in part by the state government and—initially at least—having access to law enforcement computer databases. Rajneeshpuram had a formal power structure, including a mayor and other city officials, but its true leader was the Bhagwan.

Day-to-day operation of the cult, however, was the responsibility of Ma Anand Sheela, an Indian woman with a U.S. college education who had convinced the Bhagwan to move to the United States in the first place. Although Sheela brought other senior members of the cult to regular meetings in her living quarters to discuss significant issues, she made all key decisions on her own, or after consultation with the Bhagwan, and brooked no opposition. Members who refused to follow her orders risked being ejected from the community.

The cult quickly alienated the locals and their governance. Although many people in Wasco County were offended by Rajneeshee beliefs and behaviors, the real problems were legal in origin. The building of Rajneeshpuram violated Oregon's stringent land use laws, which were intended to prevent uncontrolled development in rural areas. As conflicts with the local and state authorities grew, the office of Oregon's Attorney General opened an investigation into the legal status of Rajneeshpuram, which threatened the continued existence of the community. Equally problematic, members of the cult had violated U.S. immigration laws, and the U.S. Attorney's office in Portland, Oregon, had opened an investigation that threatened to lead to the deportation of many cult members, possibly including the Bhagwan himself.

By the spring of 1984, Sheela believed that many of the cult's problems resulted from the hostility of the Wasco County Court, the county's governing authority. Two of the three county commissioners were openly hostile to the cult's activities by that time. The Rajneeshees needed numerous county permits to conduct its activities, and the county commissioners were becoming increasingly uncooperative. Although Sheela could contest unfavorable county rulings by calling on the services of a team of skilled lawyers who had joined the cult, this opposition nevertheless posed a significant challenge for the group.

THE PLOT

In early 1984, Sheela and the Bhagwan decided to solve this problem by replacing the two hostile commissioners with their own agents in the upcoming November election. After exploring several options, Sheela adopted a two-pronged approach. First, the local populace would be incapacitated with biological agents just before the election, preventing them from voting. Second, Sheela would bring thousands of homeless people to Rajneeshpuram, register them to vote, and ensure that these new residents voted for candidates favored by the Rajneeshees.

The person most responsible for the Rajneeshees' pursuit of biological weapons was Ma Anand Puja, who had joined the cult in 1979. Although born in the Philippines, Puja was raised in California, where she attended nursing school and became a registered nurse in 1977. Puja joined the cult after hearing the Bhagwan speak at the

Encyclopedia of Bioterrorism Defense, Edited by Richard F. Pilch and Raymond A. Zilinskas
ISBN 0-471-46717-0 Copyright © 2005 Wiley-Liss

ashram in Poona, and in April 1980 was made director of its health clinic. Puja had total authority over Rajneeshpuram's medical facilities. She became close to Sheela, and was considered one of the cult's senior officials.

Even many Rajneeshees saw Puja as a sinister figure. At least some called her "Dr. Mengele," a reference to the notorious Nazi concentration camp doctor. According to one of the cult's former senior officials, Puja "delighted in death, poisons and the idea of carrying out various plots." Among the items discovered among Puja's belongings, after she fled Rajneeshpuram in 1985, were numerous publications discussing poisons and biological agents, including information on *Salmonella* infection.

Sheela and Puja studied reference books looking for suitable agents. According to one Rajneeshee source, Sheela "had talked with Bhagwan about the plot to decrease voter turn out in The Dalles by making people sick. Sheela said that Bhagwan commented that it was best not to hurt people, but if a few died not to worry." Puja considered several biological agents for the contamination scheme, including *S. typhimurium*, *Salmonella typhi*, an unspecified hepatitis virus, HIV, and possibly other organisms. Invoices from the American Type Culture Collection reportedly show that she had purchased cultures of *S. typhi*, *Salmonella paratyphi*, *Francisella tularensis*, and other organisms.

Ultimately, however, Puja focused on *S. typhimurium*, a common cause of self-limited food poisoning, and sometime between October 1, 1983, and February 29, 1984, purchased the organism from a medical supply company, ostensibly for the community's state-licensed medical laboratory (*S. typhimurium* is routinely acquired by clinical laboratories for quality assurance testing).

Puja apparently produced small quantities of *S. typhimurium* in a covert laboratory located in a remote location within Rajneeshpuram. It was dismantled a few days before law enforcement officials had a chance to examine it, but is known to have contained an incubator used to cultivate *S. typhimurium* in Petri dishes. She had learned how to culture the organism from a laboratory technician who belonged to the cult, and had produced it in quantity. One witness remembered receiving two large jars filled with a liquid containing *Salmonella*.

In total, only about 14 people were directly involved in the Rajneeshee efforts to develop and deliver the biological agent. About a dozen people appear to have participated in the planning for the attack, many in the discussions held in Sheela's quarters. Three or four participated in the production of the *Salmonella*, and seven or eight appear to have contributed to the attack.

THE ATTACK

The main efforts by the Rajneeshees to use biological agents took place in September 1984, and focused on using *S. typhimurium* to contaminate food products at restaurants in The Dalles. These attacks were in fact only experiments, intended to validate the tactics to be used immediately prior to the November election. All told, the group appears to have tainted the salad bars of at least ten restaurants in The Dalles. On several occasions,

coffee creamers were also contaminated. According to official Centers for Disease Control and Prevention (CDC) figures, 751 people became ill after eating at these salad bars, but the actual number was probably higher because of the community's location on an interstate highway, which suggests that out-of-state travelers may have been infected as well. At least 45 people were hospitalized during the outbreak.

The attack appears to have occurred in two waves. The CDC data suggest that the first wave, from September 11 to 18, targeted at least two restaurants, and that the second wave, from September 19 to 25, involved at least 10 restaurants. However, Rajneeshee informants subsequently claimed knowledge of attacks involving other restaurants. Although it is possible that the discrepancy reflected faulty memories, it also is possible that nobody became infected after some of the contaminations or that investigators failed to locate the victims of those attacks.

Although Puja appears to have conducted many of the attacks herself, at least four other people also contaminated restaurants on one or more occasions. One of the self-confessed participants recalled that Puja gave him a plastic bag containing a test tube filled with a "mostly clear" light brown liquid. She directed him to pour the contents on food at a restaurant in The Dalles. He arrived after lunchtime and found that most of the salad bar's contents had been removed. The salad dressing remained, however, and he dumped the contents of the test tube into it.

RELATED INCIDENTS

The first clearly documented incident of the Rajneeshees' use of biological agents occurred well before the above-described incident. On August 29, 1984, during a routine fact-finding visit to Rajneeshpuram by Wasco County's three commissioners, the Rajneeshees provided *S. typhimurium*-contaminated water to the two commissioners hostile to the group. Both became sick, and one ultimately required hospitalization.

It is possible, however, that the Rajneeshees began using such agents even earlier. One Rajneeshee told law enforcement authorities about an incident that may have occurred in "late July/August" 1984. Puja gave this man an eyedropper filled with a "brownish clear liquid" and told him to spread the liquid on doorknobs and urinal handles in the Wasco County Courthouse. The nature of the liquid was never determined, but the perpetrator believed that it contained *Salmonella*. To Sheela's great frustration, no one got sick from the contamination.

On another occasion, also reportedly sometime in July or August, a group of Rajneeshees, including Sheela and Puja, stopped at a supermarket in The Dalles and spread liquid thought to contain *S. typhimurium* on produce. The Rajneeshees also targeted schools, nursing homes, and political gatherings for *S. typhimurium* dissemination, although there is no evidence of people becoming sick from the attacks. One Rajneeshee claimed that she threw away the liquid rather than spreading it in a nursing home.

The Rajneeshees tried to contaminate The Dalles' water supply as well. They obtained maps of the water system, and targeted a large, holding tank located on a hill above the community. They apparently were successful in placing something into the tank (a contamination that went unnoticed at the time), but it is unclear exactly what was introduced. One former Rajneeshee thought that it was *S. typhimurium*, while another suggested that it might have been a mixture of raw sewage from Rajneeshpuram and dead rodents. Yet another account suggests that Puja might have used macerated beavers infected with *Giardia lamblia*, a protozoan that causes giardiasis, a diarrheal disease. It is commonly called "beaver fever" in the Rocky Mountains area, since most beavers are infected with the pathogen. There is no evidence to suggest that the water contamination resulted in any illness—it appears that only a small amount of material was put into the tank, and was so thoroughly diluted that it was unlikely to cause disease in the first place.

Despite the success of the restaurant contamination, no follow-on attacks were made. Two factors account for the abandonment of the plot. First, the leadership became increasingly preoccupied with managing the unexpected difficulties of coping with the numerous homeless people that they had brought to Rajneeshpuram. Reportedly, these new residents caused so many problems that Puja took to drugging the food they were served. Second, it became increasingly clear that Oregon officials had no intention of allowing the Rajneeshees to manipulate the political process. The state systematically reviewed new voting registrations and systematically rejected those from people who appeared to have no ties to the community. As a result, there was no longer any prospect of taking over Wasco County through the ballot box.

THE AFTERMATH

Many residents of The Dalles suspected that the *Salmonella* outbreak resulted from the actions of the Rajneeshees, and some actually reported suspicious activity consistent with subsequent accounts of Rajneeshee activity. However, public health officials discounted the theory, and Oregon public health officials declared that the outbreak had probably resulted from unsanitary practices by restaurant workers. The truth about the Rajneeshees' involvement was not established until a year later, when a serious rift developed between Sheela and the Bhagwan: Sheela, Puja, and some other senior officials fled Rajneeshpuram—allegedly absconding with significant funds belonging to the cult—and the Bhagwan retaliated by blaming Sheela and the others for a series of criminal acts, including the biological attacks.

Following an investigation, the U.S. Attorney's office in Oregon charged Sheela and Puja with several criminal violations, including breaking the product tampering laws recently enacted in response to the Tylenol contamination. Sheela and Puja were subsequently extradited from West Germany, and convicted in state and federal courts. Both served two-and-a-half years of four-and-a-half year federal prison sentences before being released. Sheela was deported at that time. Although the two were also given 20-year sentences by the state court for their part in the food contamination, they never spent time in an Oregon prison because of the procedural lapses by the state government.

FURTHER READING

There are three primary sources of information about the Rajneeshee use of biological agents:

REFERENCES

Additional details, not published elsewhere, appear in Miller, J., Engelberg, S., and Broad, W., Eds., *Germs: Biological Weapons and America's Secret War*, Simon and Schuster, New York, 2001, pp. 15–33.

An account that focuses mainly on the Rajneeshee biological weapons activity is W. Seth Carus, "The Rajneeshees (1984)," in J. Tucker, Ed., *Toxic Terror*, MIT Press, Cambridge, MA, 2000, pp. 55–70.

Török, T.J. et al., *J. Am. Med. Assoc.*, **278**, 389–395, (1997), describes the results of the epidemiological investigation into the outbreak.

REPUBLIC OF TEXAS: A CASE STUDY

Monterey WMD-Terrorism Database Staff

OVERVIEW

On July 1, 1998, Johnnie Wise, Jack Abbott Grebe, Jr., and Oliver Dean Emigh, all members of the separatist militia Republic of Texas, were arrested at Wise's mobile home in Olmito, Texas, and charged with conspiracy to use weapons of mass destruction (WMD) (United Press International, 1998; Schiller, 1998). The Republic of Texas believes that the State of Texas was not properly annexed in 1845 and therefore is not part of the United States. The trio in fact belonged to a splinter faction of the group under the leadership of Jesse Enloe (Republic of Texas, 1998). The Republic of Texas organization issued a statement denouncing any connection to the perpetrators of the June 1998 incident and any use of violence.

THE INCIDENT

On June 25 or 26, 1998, the Federal Bureau of Investigation (FBI) received an e-mail, apparently intended for FBI Director Louis Freeh, declaring "war" on the United States (Mclemore and Snyder, 1998; United Press International, 1998). Modified versions of the same e-mail were sent to other federal officials as well. Its message threatened to use biological weapons against President Clinton, the directors of the FBI and Central Intelligence Agency (CIA), U.S. Attorney General Janet Reno, Texas Attorney General Dan Morales, Cameron County Court at Law Judge Migdalia Lopez, their families, and other local, state, and federal agencies and officials (Mclemore and Snyder, 1998; United Press International, 1998; Reuters, 1998; Schiller, 1998).

On two separate occasions earlier that month, Wise and Grebe had met at the house of John Cain, a freelance computer specialist, to send the e-mails (Schiller, 1998). Cain informed the authorities, and the men were arrested (Baro, 1998; *Dallas Morning News*, 1998; Schiller, 1998; Reed, 1998). A search of Wise's mobile home uncovered chemistry and electronics books, other technical materials, photographs, notebooks, a syringe, a jar marked "DMSO" (presumably identifying the jar as containing dimethyl sulfoxide), an unlabeled 25-gallon barrel, approximately four gallons of a clear liquid contained in three jars, and, according to one report, corrosive and noncorrosive materials in rusted 30-gallon metal drums labeled as containing sodium benzoate, sorbic acid, nitric acid, and other undisclosed chemicals (Mclemore and Snyder, 1998; Reed, 1998; Schiller, 1998; Baro, 1998).

Wise claimed that the clear liquid was coconut oil used to make soap and perfume, and that the drums contained orange marmalade (Mclemore and Snyder, 1998; Reed, 1998). However, FBI agents reported that the content of the drums "could include meat, blood infected with the AIDS-causing virus, anthrax spores, and the rabies virus" (Reed, 1998). Further, according to the FBI affidavit, the three men were attempting to produce botulinum toxin with chicken hearts, chicken livers, and "green beans with a little dirt" (Baro, 1998).

Code 3 Inc., a hazardous materials retrieval company, removed the chemicals from Wise's property (Mclemore and Snyder, 1998).

OUTCOME

The group's e-mail to the FBI Director Freeh read as follows:

> Your FBI employees and their families have been targeted for destruction by revenge. We the people are extremely mad and will not accept the inequities any longer. Non-traceable, personal delivery systems have been developed to inject bacteria and/or viruses for the purpose of killing, maiming and causing great suffering. Warn all concerned so that they may protect themselves and be made aware of this threat to themselves and their families. Good Luck! (Schiller, 1998).

According to testimony offered by Cain, the e-mail referred to a plot devised by Wise, Grebe, and Emigh to alter disposable Bic cigarette lighters to propel air rather than propane. A hypodermic needle would then be inserted into the opening of the lighter, followed by a cactus thorn smeared with biological agents ranging from HIV and rabies virus to *Bacillus anthracis* and botulinum toxin (Baro, 1998; Schiller, 1998). A squeeze of the lighter would launch the contaminated thorn at its target, presumably generating a lethal effect in the long term. Cain's testimony suggested that the three were interested in hurting the families of the primary targets because they intended to "show the same 'value of life'" that they believed the government officials had shown to children at Ruby Ridge and the Branch Davidian raid in Waco (Baro, 1998).

On July 21, 1998, the men were indicted by a grand jury on charges including one count of conspiracy and seven counts of threatening government officials with WMD (*Houston Chronicle*, 1998). Facing a life sentence, Wise and Grebe each received 24 years in prison with a subsequent five-year period of supervised release. Emigh was acquitted.

Encyclopedia of Bioterrorism Defense,   Edited by Richard F. Pilch and Raymond A. Zilinskas
ISBN 0-471-46717-0   Copyright © 2005 Wiley-Liss

REFERENCES

Baro, M., FBI: 3 Plotted to Kill Clinton, Associated Press, July 15, 1998, Internet, available from dailynews.yahoo.com, accessed on 7/15/98.

Dallas Morning News, 3 Men Held in Threats on Officials, July 16, 1998, Internet, available from www.dallasnews.com, accessed on 7/16/98.

Houston Chronicle, 3 Men Arrested in Threat to Use Biological Weapons, July 2, 1998, Internet, available from www.chron.com, accessed on 7/2/98.

Mclemore, D. and Snyder, D., Three Face Biological Weapons Charges; Men Accused of Threatening Agencies Remain Jailed, *Dallas Morning News*, July 3, 1998.

Monterey WMD-Terrorism Database, Center for Nonproliferation Studies, 1998–2003, httm://cns.miis.edu.

Reed, C., FBI Foils Plot to Poison Clinton, *Guardian* (London), July 16, 1998.

Republic of Texas Press Release, October 5, 1998, Internet, available from http://www.overland.net/~embthert/em99084.html, accessed on 10/7/98.

Reuters, Men Reportedly Plotted to Kill Clinton, July 14, 1998.

Schiller, D., Three Face Hearing in Assassination Plots; Separatists Suspected in Plan to Kill Clinton, *San Antonio Express-News*, July 15, 1998.

United Press International, Biological Arms Threat Alleged in Texas, July 2, 1998.

RICIN AND ABRIN

Mark A. Poli

Opinions, interpretations, conclusions, and recommendations are those of the author and are not necessarily endorsed by the U.S. Army.

SOURCE AND DISTRIBUTION

Ricin is a protein found in the seed of the castor bean plant, *Ricinus communis* (*Leguminosae*). Abrin is a similar protein found in the seeds of *Abrus precatorius* (*Euphorbiaceae*), known colloquially as the rosary pea or jequirity bean. Both are lectins, members of a group of dichain ribosome-inactivating toxins known to block protein synthesis in eukaryotic ribosomes.

The castor bean plant is native to Africa but has been introduced and cultivated worldwide in tropical and subtropical climates. Although tolerant to a wide range of temperatures, it grows best where temperatures are rather high throughout the year. However, seeds may fail to set if temperatures exceed 38°C for an extended period, and plants are rapidly killed by frost. It can be grown as an annual in the cool temperate zone.

The seeds are cultivated in many regions of the world, but especially in Brazil, Ecuador, Thailand, India, Haiti, and Ethiopia. The beans contain 35 to 55 percent by weight of fast-drying, nonyellowing oil that is an important commercial product used in the manufacture of lubricants and inks, in the dyeing of textiles, and preservation of leather. When dehydrated, castor oil compares favorably with tung oil and is used in paints and varnishes. After the oil is extracted from the seeds, the remaining cake can be detoxified and used as animal feed; the seed hulls have approximately the same fertilizer value as fresh barnyard manure.

Abrus precatorius is also native to tropical regions throughout the world. It occurs naturally in the United States only in Hawaii, Puerto Rico, and the Virgin Islands. However, it currently grows throughout central and southern Florida as an introduced species, and is listed as an invasive weed by the Florida Exotic Pest Plant Council. Lacking oil, its seeds have little commercial value. Because of their bright red and black coloration, however, the seeds are highly attractive and are often made into necklaces, rosaries, and other items for the tourist trade, especially in India and the West Indies.

In India, subcutaneous injections of seed extracts have long been used to kill cattle and as a means of assassination. As with the castor bean, seeds are poisonous by ingestion if the protective seed coat is violated. The roots and seeds of *A. precatorius* also have a long history in ethnobotany. Native peoples have used various preparations of these plant parts for such widely divergent purposes as an aphrodisiac, contraceptive, and treatments for cancer, worms, venereal diseases, convulsions, diarrhea, fever, gingivitis, colic, night blindness, and a host of other conditions. The roots have recently been found to contain potent antiplatelet, anti-inflammatory, and antiallergic properties (Kuo et al., 1995). For this reason, it is of interest to ethnobotanists and others studying traditional medicines.

HISTORY

Because of its potency and ease of production, the U.S. Chemical Warfare Service began considering ricin as a potential biological warfare (BW) agent near the end of World War I. Research during this period included developing methods to adhere ricin to shrapnel bullets and produce aerosol clouds (Smart, 1997). However, the war ended before these technologies evolved into weaponry. During World War II, American and British scientists collaborated in the development of ricin-containing bombs. Although these bombs were tested, they were never used. The United States unilaterally decided to end its offensive BW program in 1969 and 1970, so all offensive research and development were terminated and remaining stocks of ricin-containing munitions were destroyed during 1971–1972. With the entry into force in 1975 of the Biological and Toxin Weapons Convention (BWC), the development, production, and storage of any toxin for offensive purposes were prohibited worldwide. In addition to being covered under the BWC, ricin is one of the only two natural toxins (the other being saxitoxin) that is mentioned by name in the 1993 Chemical Weapons Convention (CWC), which was ratified by the United States in 1997 and entered into force the same year. (Although only two toxins are named in the CWC, all toxins fall under its purview.) In the United States, ricin and abrin are both included in the

Encyclopedia of Bioterrorism Defense,   Edited by Richard F. Pilch and Raymond A. Zilinskas
ISBN 0-471-46717-0   Copyright © 2005 Wiley-Liss

Centers for Disease Control and Prevention's (CDC) select agent list of toxins requiring certification for possession and transfer.

It is generally accepted by the U.S. intelligence community that ricin was a part of the Soviet BW program. It is also believed to have been part of the offensive BW program of Iraq, and possibly other nations as well.

In recent years, ricin has drawn the interest of extremist groups. This interest is likely related to the ready availability of castor beans, ease of toxin extraction, and popularization by the press and on the Internet. Several arrests have been made under the 1989 Biological Weapons Anti-Terrorism Act for possession of ricin, and two tax protestors were convicted in 1995 of possessing ricin as a biological weapon (Tucker and Pate, 2000).

Evidence for the pursuit of abrin by state-sponsored BW programs or terrorist organizations is scant. A brief description, along with crude instructions for its isolation, is included in the al-Qa'ida training manual captured in England by the Manchester Metropolitan Police Force. However, the paucity and inaccuracy of the information suggests that the group's experience with abrin is minimal.

MECHANISM OF ACTION

Ricin and abrin are similar in structure and mechanism of action. Both are proteins of molecular mass approximately 60 to 65 kD and consist of two subunits, an A-chain and a B-chain, linked by a disulfide bond. The B-chains are lectins that bind the toxins to cell-surface receptors. The toxicity lies in the enzymatic activity of the A-chains, which modify the 28S ribosomal subunits of eukaryotic cells to block protein synthesis.

The B-chains of both toxins contain similar amounts of carbohydrates, primarily mannose and glucosamine residues. These chains direct binding to galactose residues on cell-surface glycoproteins and glycolipids, which are widely distributed on eukaryotic cells. This binding triggers endocytic uptake of the toxins. Internalization, primarily via uncoated pits but to a lesser extent through coated pits, occurs within a few hours. Once internalized into endosomes, the toxin may take one of several pathways: It may be degraded in lysosomes, it may cycle back to the cell surface in the intact form, or it may enter the cytosol via the Golgi apparatus. Once in the cytosol, the A- and B-chains dissociate and the A-chain targets the 28S ribosomal subunit. Ricin has a Michaelis constant (K_M) of approximately 0.1 μM for ribosomes and an enzymatic constant (K_{cat}) of 1500/min (K_M and K_{cat} are empirical parameters of enzyme kinetics) (Olsnes and Pihl, 1982). One adenosine residue (A4324) is removed near the 3' end of the 28S RNA, which is sufficient to inhibit binding of elongation factor-2, thereby blocking protein synthesis. This inhibition of protein synthesis is lethal to cells. Prokaryotic cells are insensitive, and plant cells are less sensitive than mammalian cells.

TOXICITY

Ricin

Ricin is extremely toxic to animals, although the potency varies by species. Frogs and chickens appear to be the most resistant, horses the most susceptible (Balint, 1974). The most toxic routes are by inhalation and intravenous injection. In mice, the respective lethal doses for 50 percent of the exposed population (LD_{50}) are 3 to 5 μg/kg by inhalation, 5 μg/kg by intravenous injection, and 22 μg/kg by intraperitoneal injection. By oral ingestion, ricin is several thousand times less potent (LD_{50} is 20 mg/kg in mice), probably due to poor absorbance from the intestinal tract and perhaps some level of enzymatic degradation. Dermal toxicity is negligible (Franz and Jaax, 1997).

Ingestion. Human oral intoxications from ricin have been reviewed (Balint, 1974; Rauber and Heard, 1985), and case studies confirm the relatively low potency by this route. The primary symptoms are gastrointestinal; within hours of ingestion, nausea, vomiting, diarrhea, and localized epigastric pain are common. If untreated, these can lead to severe dehydration and the associated sequelae of tachycardia, oliguria, hypotension, lethargy, confusion, and shock. Direct hepatotoxic effects, which might be predicted from animal studies, are typically absent. Autopsy may show multifocal ulceration and hemorrhage of gastric and small intestinal mucosa as well as diffuse nephritis and splenitis. The number of beans required to elicit symptoms varies widely between individuals and no general prediction can be made.

A phase I trial of ricin as an anticancer drug revealed side effects similar to those observed after ingestion of castor beans (Fodstad et al., 1984). Patients receiving intravenous ricin at 18 to 23 μg/m^2 surface area (approximately 0.5–0.75 μg/kg) suffered mild side effects including flu-like symptoms, fatigue, and myalgia. Neither bone marrow suppression nor hematotoxicity was noted.

Injection. A case of human fatality after intramuscular administration is that of the assassination of Georgi Markov in 1978 (Crompton and Gall, 1980). In this well-known case, Mr. Markov, a Bulgarian dissident, was injected with a lethal dose of what is believed to have been ricin. The injection came from a special umbrella that had been modified so it contained an air rifle that fired a small platinum/iridium pellet. This pellet, which had been drilled and loaded with toxin, penetrated the back of Markov's right thigh and lodged in the superficial musculature. Within hours, he developed symptoms of weakness, fever, nausea, and vomiting. Thirty-six hours later, he entered the hospital with fever, tachycardia, and swollen lymph nodes. On the second day, he developed hypotension, vascular collapse, and shock. Death occurred on the third day post exposure.

Inhalation. Aerosol exposure to ricin causes pathology distinct from those of oral or intramuscular administration noted above. By this route, animal studies on rodents and primates demonstrate necrosis of the airways and interstitial and alveolar edema and inflammation. Death occurs from hypoxia secondary to alveolar flooding.

Abrin

The potency of abrin is slightly higher than that of ricin in animals; the reported LD_{50} in mice via intravenous injection is 0.7 μg/kg. In humans, toxic effects are reported after intravenous administration of doses greater than 5 μm/m^2 (approx. 0.15 μg/kg). The lethal dose in humans is not known with certainty but is likely similar to or slightly lower than that of ricin. Older literature reports suggest that one seed, if masticated, may be toxic, especially in children. The latency period between exposure and manifestation of symptoms may be three days or longer, and death may occur as late as 14 days post exposure. Symptoms are similar to that of ricin: nausea, vomiting, diarrhea, abdominal cramps, and hematemesis. Hemagglutination and hemolysis of red cells can occur, with resulting precipitation of hemoglobin in the renal tubules. Pulmonary edema and hypotension were reported in one recent case report (Fernando, 2001).

Inadvertent poisoning from *Abrus* seeds is rare. No cases of poisoning have been reported to the poison control centers in Florida or Puerto Rico in the past five years. During the same period, the Hawaii center recorded only two calls, one in 1997 and one in 1998. Consequently, recent case reports are scarce in the medical literature. Fernando's case report describes the management of the case of a 13-year-old male in Sri Lanka. Even in Sri Lanka, where *A. precatorius* is native, he could document only nine inquiries to the Poison Center in the preceding decade, none of which were subsequently reported in the medical literature.

TREATMENT AND PROPHYLAXIS

The treatment for ricin and abrin intoxication is supportive and symptomatic. There are no specific chemotherapeutic agents at this time, nor are there commercially available antisera (i.e., antitoxins). Fluid and electrolyte balances should be closely monitored and controlled. Because of the necrotizing action of these toxins, gastric lavage and emetics should be used with caution. In the event of significant hemolysis, alkaline diuresis with sodium bicarbonate can prevent precipitation of hemoglobin in the kidney tubules.

A ricin vaccine is currently under development by the U.S. Army. No vaccine is currently available for abrin. Animal studies have demonstrated that pretreating lungs with specific antibodies protects against aerosol exposure to ricin (Poli et al., 1996). However, these studies have not been extended to humans.

DETECTION

Attempts to detect ricin and abrin typically focus on either their physiological mechanism or their physical structure. Assays have been developed whereby the inhibition of protein synthesis is measured in vitro. These methods use cell-free systems to remove the requirement for cell-surface binding and internalization, thereby decreasing the timeframe required. Antibody-based assays take advantage of the fact that large proteins are immunogenic and typically result in good antibody production when inoculated into animals. Several good antibody-based assays have been reported in the literature, a few able to detect toxin at levels of 1 to 10 ng/ml (reviewed in Wellner et al., 1995). The latest generation of ricin assays, based upon electrochemical luminescence or chemiluminescence ELISA (Enzyme-linked Immunosorbent Assay), can accurately measure ricin at a concentration of 0.1 ng/ml. Rapid, field-deployable assays for ricin are currently being developed by the U.S. military.

Detection in food and/or environmental samples may require extraction from the matrix or elimination of interfering factors before analysis. This sample cleanup, of course, depends upon the matrix being tested. Toxin in tissue samples can be detected by immunohistochemistry techniques (Poli et al., 1996).

REFERENCES

Balint, G.A., *Toxicology*, **2**, 77–102 (1974).

Crompton, R. and Gall, D., *Med. Leg. J.*, **48**, 51–62 (1980).

Fernando, C., *Anaesthesia*, **56**, 1178–1180 (2001).

Fodstad, O., Kvalheim, G., Godal, A., Lotsberg, J., Amdahl, S., Host, H., and Pihl, A., *Cancer Res.*, **44**, 862–865 (1984).

Franz, D.R. and Jaax, N.K., "Ricin Toxin," in R. Zajtchuk, Ed., *Medical Aspects of Chemical and Biological Warfare. Part I: Warfare, Weaponry, and the Casualty*, Office of the Surgeon General, United States Army, Falls Church, VA, 1997, pp. 631–642.

Kuo, S.C., Chen, S.C., Chen, L.H., Wu, J.B., Wang, J.P., and Teng, C.M., *Plant Med.* **61**, 307–312 (1995).

Olsnes, S. and Pihl, A.,"Toxic Lectins and Related Proteins," in P. Cohen and S. van Heyningen, Eds., *Molecular Action of Toxins and Viruses*, Elsevier Biomedical Press, Amsterdam, Netherlands, 1982, pp. 51–105.

Poli, M.A., Rivera, V.R., Pitt, M.L. and Vogel, P., *Toxicon*, **34**, 1037–1044 (1996).

Rauber, A. and Heard, J., *Vet. Human Toxicol.*, **27**, 498–502 (1985).

Smart, J.K. "History of Chemical and Biological Warfare: an American Perspective," in R. Zajtchuk, Ed., *Medical Aspects of Chemical and Biological Warfare. Part I: Warfare, Weaponry, and the Casualty*, Office of the Surgeon General, United States Army, Falls Church, VA, 1997, pp. 9–86.

Tucker, J.B. and Pate, J., "The Minnesota Patriots Council (1991)," in J.B. Tucker, Ed., *Toxic Terror: Assessing Terrorist Use of Chemical and Biological Weapons*, MIT Press, Cambridge, MA, 2000, pp. 159–184.

Wellner, R.B., Hewetson, J.F., and Poli, M.A., *J. Toxicol.—Toxicol. Rev.*, **14**, 483–522 (1995).

See also AL-QA'IDA; ASSASSINATIONS; MINNESOTA PATRIOTS COUNCIL; and TOXINS: OVERVIEW AND GENERAL PRINCIPLES.

RISE: A CASE STUDY

W. Seth Carus

BACKGROUND

On January 18, 1972, the Chicago, Illinois, police arrested two college students on the basis of information that the two were "within a few days" of contaminating the Chicago water system with *Salmonella typhi*, the organism that causes typhoid fever. The two perpetrators, Steven Pera, 18, and Allen C. Schwander, 19, were students at Mayfair College, a community college that subsequently became a component of Chicago City College. The two had vague ecological concerns, and this led them to organize a terrorist group, which they called RISE, in the fall of 1971, possibly in late November.

MOTIVATION

The group's manifesto argued that mankind was destroying the Earth, and that the world would be a better place if it were inhabited only by a small group of like-minded people who shared the same views about how to address the world's problems. Contrary to some subsequent reports, there is no evidence that the group was motivated by neo-Nazi or racist views. Government officials claimed at the time that "Members of 'RISE' were allegedly to be inoculated and immunized, enabling them to survive ... and to form the basis of a new master race. Water filtration plants in the Midwest were allegedly to be infected with typhoid and their deadly bacteria."

TECHNICAL CAPABILITY

Agent Acquisition

Analyses conducted after the arrests confirmed that the group had acquired at least four biological agents: *Corynebacterium diphtheriae*, *Neisseria meningitidis*, *S. typhi*, *Shigella sonnei*, and possibly *Clostridium botulinum*. A hospital microbiology laboratory where Pera worked provided the seed stock for the *S. typhi* and *N. meningitidis*; the source of the other organisms was never determined. In addition, RISE had attempted to acquire *Vibrio cholerae* and *Yersinia pestis*, apparently with no success, and at least discussed obtaining *Bacillus anthracis*. Inside informants told police that they had each received two injections of a purported typhoid vaccine, and were scheduled to receive a third when they decided to report the plot.

Production

The group initially used the hospital microbiology laboratory to culture the biological agents, but was ejected in December 1971 after Pera illegally tried to acquire controlled substances there. The cultures growing at the hospital were destroyed at the time, and the group transferred its activities to the laboratories of Mayfair College.

Delivery

Pera and Schwander discussed three alternative methods for disseminating biological agents: aerosols, food contamination, and water contamination. They were aware of the possibilities of aerial dissemination of biological aerosols, but ultimately focused on water contamination when they suspected that authorities were aware of their activities.

RESOLUTION

Following their arrest, Pera and Schwander were permitted to post bonds and were released. They skipped bail, fled the United States, and ended up in Cuba after hijacking a small plane in Jamaica in early May 1972. Schwander apparently died in Cuba in November 1974. Pera returned to the United States in late 1974, and in June 1975 was sentenced to five years probation. His probation ended in February 1977 at the request of his attorney.

FURTHER READING

The only detailed account of this incident appears in W. Seth Carus, "RISE," in J. Tucker, Ed., *Toxic Terror*, MIT Press, Cambridge, MA, 2000, pp. 54–70.

Encyclopedia of Bioterrorism Defense, Edited by Richard F. Pilch and Raymond A. Zilinskas
ISBN 0-471-46717-0 Copyright © 2005 Wiley-Liss

RISK ASSESSMENT IN BIOTERRORISM (MICROBIAL RISK ASSESSMENT, QUANTITATIVE RISK ASSESSMENT, HUMAN HEALTH RISK ASSESSMENT)

BRUCE K. HOPE

INTRODUCTION

Since 1995, concerns about possible threats posed by terrorist use of biological agents (biological agents) have increased dramatically at state and federal levels and in the popular press (Carus, 2001; Falkenrath et al., 1999; Tucker, 1999). However, while we might envision many scenarios for how such threats could become reality, we also must recognize that some scenarios have a greater potential for being realized than others and that not all potential targets of a bioterrorist attack are equally valuable, equally susceptible to biological agents, or thus require equal protection (GAO, 1998). Implementation of effective security and counterterrorism programs requires a process that allows decision makers to identify and focus on those scenarios that are both the most probable and potentially the most damaging. Risk assessment methodologies developed for human health and the environment can be adapted to move analysts beyond potentially intractable generalizations about who or what is threatened by bioterrorism toward specific scenarios that can be evaluated, anticipated, and managed (Stern and Fineberg, 1996; USEPA, 1989, 1998).

However, because bioterrorist attacks have occurred so infrequently (Carus, 2001), data needed to estimate the probability of a specific attack scenario (e.g., placing pathogens on salad bars), as well as epidemiological data on outcomes of bioterrorist attacks, are extremely limited. This paucity of data suggests that a bioterrorism risk assessment could occur in two stages (Fig. 1): a qualitative or semiqualitative *vulnerability analysis* stage (similar to the "problem formulation" stage of other types of risk assessment (USEPA, 1998)), followed, depending on data availability, by a more quantitative *risk estimation* stage, with the latter involving four components: hazard characterization, hazard identification, exposure assessment, and risk characterization (Fig. 2).

VULNERABILITY ANALYSIS

Vulnerability analysis starts by clearly stating the assessment's scale and scope (e.g., the world, a country, a specific facility?), with the understanding that the risk assessment problem becomes increasingly more tractable as the scope narrows. It then uses currently available technical and intelligence information, along with expert elicitation and best professional judgment, to first conceptualize combinations of targets (i.e., valued assets deemed important to protect), biological agents, exposure pathways (how a biological agent could reach the target), and potential adverse outcomes into "attack scenarios." These scenarios are then comparatively ranked on the basis of their probability of occurrence and the nature and magnitude of consequences should they occur. Such a comparative ranking may be, in and of itself, sufficient to focus and support risk management decision-making.

Humans, as well as agricultural or other natural resources, are valued *assets* that might potentially be attacked by bioterrorists. Every asset has at least one *attribute*, defined as a characteristic potentially at risk and either directly or indirectly measurable (USEPA, 1998). For humans, "health," which implies an absence of illness requiring use of the health care delivery system, could be an attribute requiring protection from bioterrorist activities. This health attribute is measurable in terms of morbidity (number sick to varying degrees) and mortality (number killed). The outcome of vulnerability analysis is one or more attack scenarios describing the pathway by which a specific asset could be put at risk by a specific biological agent. The following are examples of three basic attack scenarios:

1. The bioterrorist's primary objective is to directly cause adverse health outcomes in humans, up to and including mass casualties. This is to be achieved by exposing humans directly, through contact, inhalation, or consumption, to the selected biological agent, with this exposure leading to adverse health outcomes.

2. The bioterrorist's primary objective is to indirectly cause adverse health outcomes in humans. Exposure of resources that humans require (e.g., agricultural commodities) to the biological agent results in a reduction or loss or of these resources for human use. Lack of essential resources (e.g., food, water, fuel, etc.) then leads to adverse health outcomes. Psychological adverse health outcomes are likely if such resources are actually attacked or, simply, if they are perceived by the public as being compromised or limited by bioterrorist activities.

3. The bioterrorist's primary objective is some form of mass disruption, either through market loss or denial of service. Although such disruption may lead to adverse health outcomes, this is not the primary objective. These types of attack may appeal to a bioterrorist morally constrained from directly

Encyclopedia of Bioterrorism Defense,   Edited by Richard F. Pilch and Raymond A. Zilinskas
ISBN 0-471-46717-0   Copyright © 2005 Wiley-Liss

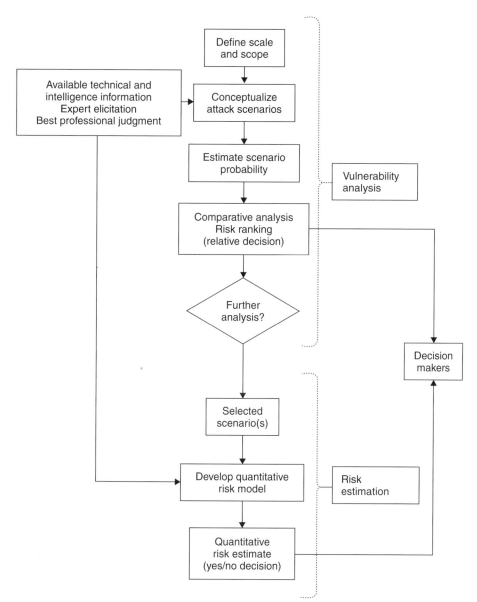

Figure 1. Relationship between vulnerability analysis and quantitative risk assessment as applied to the evaluation of bioterrorist threats.

targeting humans with violence. In a "market loss" attack, nonlethal biological agents are used to adversely affect the marketability of a resource (particularly an agricultural commodity) without destroying it (Ban, 2000). A "denial of service" attack renders infrastructure elements unusable (either temporarily or permanently) due to the reality or perception of contamination with a biological agent. The economic, social, and psychological disruptions associated with either a market loss or denial of service attack could be expected to ultimately have an adverse affect on human health.

If a follow-on, more quantitative assessment is desired, it would be expected to focus on those few attack scenarios judged to be of greatest concern (i.e., those with the highest probability of occurrence and consequence). This focus permits substantive quantitative data gaps, such as limited predictive microbiological data for classic biological

agents, to be overcome with what data are available, aided by expert elicitation and best professional judgment.

RISK ESTIMATION

Developing a quantitative risk estimate for a bioterrorist attack presupposes that a vulnerability analysis to narrow the scope and identify attack scenarios worthy of more detailed analysis has occurred. Risk estimation then gathers what quantitative data is available regarding the attack scenario and proceeds through four steps—hazard characterization, hazard identification, exposure assessment, and risk characterization (Fig. 2)—to provide a numeric estimate of the risk posed by that scenario. Risk estimation generally involves development of a model or models. Human health and environmental risk assessments have become increasingly quantitative and are typically based on models parameterized with empirical data of varying extent and quality. Data can also

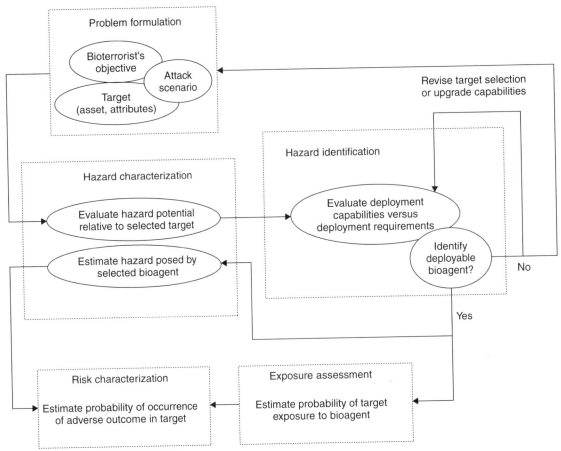

Figure 2. Adaptation of the general risk assessment paradigm to one specifically for the assessment of bioterrorist threats.

be obtained through expert elicitation, by appeals to best professional judgment, and by assuming a bounding range of possible values—rarely is there absolutely no data for risk estimation. Pathogen risks to humans, crops, and livestock have been estimated with a variety of models, for example, quantitative dose-response (Coleman and Marks, 1999; USDA 1998), fault tree (Marks et al., 1998), and epidemiological (Pybus et al., 2001). Similar models have been applied to political terrorism (Koller, 2000) and suggested for application to anticrop bioterrorism (Madden and Scherm, 1999). Risk estimation is typically an iterative process. Information gained in each step is combined to represent a cause-and-effect chain from the prevalence and concentration of an intentionally introduced biological agent to the probability and magnitude of adverse outcomes in the valued asset being targeted.

Hazard Characterization

Hazard characterization evaluates the nature and magnitude of the hazard that could be posed to a specified target by one or more biological agents. Ideally, quantitative dose-response data are available to provide a more precise measure of a biological agent's hazard potential. With biological agents, this step is very difficult because of the shortage of data on pathogen-specific responses and because those responses depend on the immune status of the host (target). However, even limited knowledge of the

shape and boundaries of a dose-response function can be informative in estimating the probability of an adverse response (Haas, 2002). Absent quantitative dose-response data, a number of criteria, which reflect intrinsic features of a biological agent, can be semiquantitatively appraised to characterize the nature and magnitude of the hazard it may pose to a specific target. These include:

- *Virulence (Toxicity).* Relative severity of the adverse outcomes (e.g., illness, incapacitation, mortality) that a biological agent induces in an exposed asset.
- *Morbidity.* Severity and duration of the adverse outcomes.
- *Transmissibility.* Ability of a biological agent to spread from source or reservoir to asset, directly, indirectly (via a vehicle such as food or a vector), or through the air (droplet, nuclei, dust).
- *Infectivity.* Relative ease with which an infectious biological agent can establish itself in a susceptible asset.
- *Persistence (stability, viability).* Length of time a biological agent will remain effective in the environment.
- *Treatment Options.* Availability and effectiveness of all known medical interventions (e.g., supportive measures, isolation, vaccination, pharmaceuticals, or antisera).

Note that the degree of hazard is positively correlated with all of these factors, with the exception of treatment options, with which there is a negative correlation (i.e., the more numerous and/or effective treatment options the less hazardous the biological agent). Similar criteria, in conjunction with expert elicitation, have been used to rank biological threats posed by a number of biological agents (Wade, 2001).

Hazard Identification

Hazard identification estimates which biological agent a bioterrorist is most likely to choose, considering the bioterrorist's deployment capabilities, the biological agent's hazard potential (from Hazard Characterization) relative to that desired by the bioterrorist, and the target's susceptibility to a given agent. Selection of a biological agent emerges from trade-offs between (1) a biological agent's scientific and technical requirements for safe acquisition (i.e., production, isolation, culturing), management (i.e., handling, storage), and deployment (dissemination) versus a bioterrorist's scientific and technical capabilities in these areas; (2) the logistical, financial, and intelligence resource requirements needed to acquire the biological agent and access a target versus a bioterrorist's resources in these areas; and (3) the adverse outcome desired by the bioterrorist (as a function of their objective) versus a biological agent's hazard characteristics. In general, each of these trade-offs must shift in favor of the bioterrorist, before a biological agent, particularly an extremely hazardous one, is likely to be utilized. That this shift will occur with a moderate to high degree of probability is what separates credible threats from hoaxes.

Lack of scientific and technical expertise is likely to limit the probability of a successful deployment (i.e., if a bioterrorist's scientific and technical capabilities are well below those required by the desired biological agent, he will likely be at greater risk of harming himself than the intended target). Pragmatic bioterrorists may thus have to (1) choose a biological agent likely to achieve their desired objective and then increase their technical and resource capabilities as needed or (2) reconsider their ideological motivations and select an objective better suited to a biological agent they can deploy with existing capabilities. Criteria and a scoring system for bioterrorist characteristics that may bear on biological agent selection have been proposed to aid in anticipating a bioterrorist's choice of biological agent (Wilson et al., 2000).

Exposure Assessment

Exposure assessment describes the pathways a biological agent is expected to follow from its release point to the target and attempts to quantify the amount of biological agent reaching the target. Exposure begins when an intentional act releases a biological agent into the environment toward a target. The probability of an intentional release is a function of a bioterrorist's (1) possession of a suitable biological agent, as determined by Hazard Identification, (2) scientific and technical capabilities relative to those required for biological

agent dissemination, and (3) financial, logistical, and intelligence resources relative to those needed for access to the release point.

Ideally, the biological agent would be released as close to the target as is feasible and then travel directly from its release point to the target. The probability of the biological agent reaching the target is dependent on (1) it first being released, (2) suitable environmental conditions (relative to that biological agent's intrinsic environmental requirements for growth or survivability) at that point, and (3) detection and interdiction (by active countermeasures or unfavorable environmental conditions) avoidance after its release. A biological agent's persistence (c.f., Hazard Characterization) is critical to its ability to survive movement from its release point through differing environmental conditions en route to the target. This persistence could also allow discrete time-branching epidemic processes and natural or human-mediated (e.g., trucking, shipping) dispersion to spread a biological agent well beyond its initial release point to (perhaps unintended) targets distant in space and time from the initial release point.

Risk Characterization

Risk characterization integrates information gathered in previous steps to estimate the probability of occurrence of the bioterrorist's desired adverse outcome on the target, where that adverse outcome is a function of its exposure to a biological agent (from Exposure Assessment) and the probable potential for a specified outcome (e.g., morbidity, mortality) due to that exposure (from Hazard Characterization). Although the number of organisms required to establish an infection is potentially greater than one, an initial assumption may be that if one organism reaches the target, that organism alone could be sufficient to induce the desired adverse health outcome. This conservative assumption can be set aside if quantitative dose-response data are available for the biological agent and the exposure assessment is modified to quantify target dose.

RISK MANAGEMENT

The ultimate goal is not a perfectly predictive risk model (any such models for bioterrorism can initially be expected to be crude) but rather a framework within which to accommodate the breadth of available information about a particular scenario or biological agent. Such models are, however, also expected to be constantly updated and improved as each new related study and intelligence update provides it with additional relevant data. Placing all of the information in one consistent framework allows clearer delineation of gaps in knowledge, provides a focus for discussions among workers from diverse disciplines, and best describes what is currently known and unknown. It can also provide estimates of the cost-benefits of proposed research or specific intelligence-gathering efforts. Such models support decision-making not only by providing a risk estimate for a given scenario but also by allowing decision makers to test assumptions and perform "what if" analyses on alternative countermeasures for reducing or eliminating credible threats, as well as to consider and compare countermeasure strategies that would be very difficult to test in a "live" environment (GAO, 1998).

Application of the risk assessment paradigm can provide the decision-maker with insights not typically evident in "piece-meal" considerations of data and may provide the only systematic means to interpret the impact of changes or trends before a significant threat becomes reality. Some are tempted to avoid performing a risk assessment until some level of (perhaps unspecified) certainty is reached in its predictive ability. This is a mistake. Decisions to address bioterrorist risks cannot wait for scientific certainty. While large degrees of uncertainty require that decisions be made with great caution, the best decision must be made on the basis of available information, as captured in a risk assessment model.

REFERENCES

Ban, J., *The Arena*, **9**, 1–8 (2000).

Carus, S.W., *Working Paper—Bioterrorism and Biocrimes: The Illicit Use of Biological Agents Since 1900*, Center for Counterproliferation Research, National Defense University, Washington, DC, 2001.

Coleman, M.E. and Marks, H.M., *Food Control*, **10**, 289–297 (1999).

Falkenrath, R.A., Newman, R.D., and Thayer, B.A., *America's Achilles' Heel: Nuclear, Biological, and Chemical Terrorism and Covert Attack*, BCSIA Studies in International Security, MIT Press, Cambridge, MA, 1999, pp. 167–215.

GAO, *Combating Terrorism: Threat and Risk Assessments Can Help Prioritize and Target Program Investments*, GAO/NSIAD-98-74, Government Accounting Office, National Security and International Affairs Division, Washington DC, 1998.

Haas, C.N., *Risk Anal.*, **22**, 189–193 (2002).

Koller, G.R., "Terrorism Risk Models—Relative and Absolute Risk," *Risk Modeling for Determining Value and Decision Making*, Chapman & Hall/CRC Press, Washington, DC, 2000, pp. 21–65.

Madden, L.V. & Scherm, H., Epidemiology and risk assessment, *A Symposium on Plant Pathology's Role in Anti-Crop Bioterrorism and Food Security*, Joint American and Canadian Phytopathological Society meeting, Montreal, Canada, August 7–11, 1999.

Marks, H.M., Coleman, M.E., Lin, C.-T.J., and Roberts, T., *Risk Anal.*, **18**, 309–328 (1998).

Pybus, O.G., Charleston, M.A., Gupta, S., Rambaut, A., Holmes, E.C., and Harvey, P.H., *Science*, **292**, 2323–2325 (2001).

Stern, P.C. and Fineberg, H.V., *Understanding Risk: Informing Decisions in a Democratic Society*, National Research Council, National Academy Press, Washington, DC, 1996.

Tucker, J.B. "Bioterrorism: Threats and Responses," in J. Lederberg, Ed., *Biological Weapons: Limiting the Threat*, BCSIA Studies in International Security, MIT Press, Cambridge, MA, 1999, pp. 283–320.

USDA, *Salmonella Enteritidis Risk Assessment for Shell Eggs and Egg Products, Final Report*, Food Safety and Inspection Service, U.S. Department of Agriculture, Washington, DC, 1998.

USEPA, *Guidelines for Ecological Risk Assessment, Final*, EPA/630/R-95/002F, Risk Assessment Forum, U.S. Environmental Protection Agency, Washington, DC, 1998.

USEPA, *Risk Assessment Guidance for Superfund, Volume I, Human Health Evaluation Manual, Part A, Interim Final*, EPA/540/1-89/002, Office of Emergency and Remedial Response, U.S. Environmental Protection Agency, Washington, DC, 1989.

Wade, J., *Medical Risk Assessment of the Biological Threat*, Contract No. SPO700-00-D-3180, Chemical Warfare/Chemical and Biological Defense Information Analysis Center, Defense Technical Information Center, Defense Information Systems Agency, Fort Belvoir, VA, 2001.

Wilson, T.M., Logan-Henfrey, L., Weller, R., and Kellman, B., "Agroterrorism, Biological Crimes, and Biological Warfare Targeting Animal Agriculture," in C. Brown and C. Bolin, Eds., *Emerging Diseases of Animals*, American Society of Microbiology Press, Herndon, VA, 2000, pp. 23–57.

WEB RESOURCES

U.S. Department of Agriculture microbial (food safety) risk assessment, www.fsis.usda.gov/OPHS/risk/index.htm, accessed on 10/25/04; last update Jun 12, 03.

U.S. Department of Energy, Oak Ridge National Laboratory human health and ecological risk assessment guidance, (human) risk.lsd.ornl.gov/rap_hp.shtml (ecological) www.esd.ornl.gov/programs/ecorisk/ecorisk.html, accessed on 10/25/04; last updated Feb 26, 03.

U.S. Environmental Protection Agency human and ecological risk assessment guidance, www.epa.gov/superfund/programs/risk/tooltrad.htm, accessed on 10/25/04; last updated Oct 21, 03.

Other risk-related sites: Society for Risk Analysis, (www.sra.org) Risk World, (www.riskworld.com) National Academies First Responder Site, (www.nap.edu/shelves/first/), accessed on 10/25/04; last update 2002.

S

SANDIA NATIONAL LABORATORIES

Lauren Harrison

OVERVIEW

A government-owned, contract-operated facility currently managed by Lockheed Martin, Sandia National Laboratories is involved in research and development projects ranging from biosciences to homeland security and defense. Sandia was officially founded in 1945 as "Z division," the ordnance design, testing, and assembly section of Los Alamos National Laboratory (LANL), and thus was part of the Manhattan Project (the U.S. atomic bomb objective throughout World War II). In 1948, the division became a separate branch of Los Alamos and was relocated to Sandia Base in Albuquerque, New Mexico, and in 1949, President Truman requested that AT&T manage the labs. Thirty years later, in 1979, Sandia was designated as a national laboratory. Current components include a smaller laboratory with some 850 employees in Livermore, California, and the larger facility with over 6800 employees in Albuquerque.

Sandia's mission is to, "[help] our nation secure a peaceful and free world through technology." Much of the work at Sandia supports the National Nuclear Security Organization and the U.S. Department of Energy (DOE). Sandia maintains capabilities in a range of areas, including advanced manufacturing, biological and chemical science, computer information science, electronics, engineering, materials and process science, modeling and simulation, and nanotechnology, along with many other specialties. In addition to these capabilities, Sandia has contributed significantly to homeland security in the nuclear, biological, and chemical domain, including work to ensure the safety of the U.S. nuclear weapons stockpile as well as to monitor energy and critical infrastructure. Sandia's major association with antiterrorism and counterterrorism is a result of the growing need to develop technology to protect nuclear facilities.

STRATEGIC AREAS OF FOCUS

Sandia's four strategic areas of focus include (1) nuclear weapons, (2) energy and infrastructure assurance, (3) proliferation and assessments, and (4) military technologies and applications.

Nuclear Weapons

Within the nuclear weapons arena, Sandia supports efforts such as the enhancement of weapon and surveillance technology, nuclear stockpile evaluation, the development of new defense options, and the creation of safeguards to protect existing weapons. Sandia is actively supporting the verification of related treaties on an international level, and is currently working with nations that once constituted the Soviet Union to manage both the dismantlement of weapons and the safety of nuclear material.

Energy and Infrastructure Assurance

The goal of Sandia's energy and infrastructure program is to secure energy infrastructures and augment their safety. Particular areas of concern include oil and gas distribution, transportation, the electric power grid, telecommunications, and finance and banking.

Proliferation and Assessment, and Military Technologies and Applications

Both the proliferation and assessment program and the military technologies and applications program work on a number of biological warfare (BW) defense efforts (outlined below), as well as many non-BW related initiatives, for example, the development under Sandia's Advanced Concept Group (ACG) of networked sensors capable of detecting and tracking large targets of security concern, such as missile launchers and the movement of hostile forces, and technology to detect the presence of explosives integral to walk-through airport security tools.

The proliferation and assessment and military technologies and applications programs are designed to decrease U.S. vulnerability to weapons of mass destruction (WMD) through the enhancement of existent and creation of new detection technologies, decontamination methods, and attack and response simulations. Specific biological weapons–related efforts include the following:

Microchemlab. Sandia is currently working to develop portable, handheld chemical analysis systems to detect chemical warfare agents as well as protein toxins such as ricin and staphylococcal enterotoxin B. Such units are in the prototype stage, and further research is being conducted to expand the capability and improve the performance of the devices.

Assessing Vulnerability of Facilities to Chemical and Biological Attack. Researchers from Sandia have worked to create simulation and modeling tools that can be used by analysts to assess the vulnerability of various facilities

Encyclopedia of Bioterrorism Defense, Edited by Richard F. Pilch and Raymond A. Zilinskas
ISBN 0-471-46717-0 Copyright © 2005 Wiley-Liss

to chemical and biological agents by replicating how these agents move through a given facility. Using electronic blueprint software, researchers map a particular building and then simulate agent release in order to determine how flow is affected by the infrastructure of the building itself. These simulation programs may someday be effective in creating "smart" buildings and in helping public officials determine how to effectively prevent more widespread contamination after an attack.

Sandia Decon Formulation. Sandia has developed an aqueous, nontoxic solution that can be deployed to decontaminate affected areas in the wake of a chemical or biological release. The solution is effective against both types of agents, and can be deployed as a foam, liquid spray, or fog, allowing delivery via a number of different devices. The Sandia Decon formula was used to decontaminate the Congressional office buildings in Washington, DC, during the 2001 "anthrax letter attacks," and was available for use, if need arose, during the 2002 Winter Olympics in Salt Lake City, Utah.

RECENT HIGHLIGHTS

Numerous projects at Sandia have recently been employed to enhance homeland security. Highlights include the following:

- In February 2002, scientists at Sandia, in conjunction with the Federal Bureau of Investigation (FBI), effectively disabled a bomb attached to the shoe of Richard Reed, arrested in December 2001 after attempting to bring down American Airlines flight 63. For this, the scientists utilized a Percussion-Actuated Nonelectric (PAN) Disrupter, an advanced tool developed at Sandia.

- In July 2002, researchers at Sandia used advanced simulation capabilities to prepare public health officials in Northern California to effectively manage biological and chemical events. This simulation was conducted by the Weapons of Mass Destruction Decision Analysis Center (WMD-DAC), located at Sandia's California laboratory.

- Sandia has recently developed a lightweight device that detects chemical nerve and blister agents. This device, called the SnifferSTAR, was designed for attachment to small aerial drones in order to provide the quick analysis vital to the management of chemical attacks. In addition to aerial use, this sensor can also be used near the ventilation system of a building.

ORGANIZATIONAL TIES

Sandia works in concert with the LANL (so-called cooperative competition) in a number of areas. Sandia has also worked in conjunction with the DOE research and development laboratory, the Department of Defense (DoD), the National Nuclear Security Organization, and the Department of Homeland Security (DHS), as well as universities, small businesses, industry, and other federal groups.

FUTURE INITIATIVES

Each program at Sandia has generated a set of initiatives for the future. Examples include continuing the development of advanced robotic systems program to monitor proliferation activities and of technologies that can provide early detection of such activities. In addition, Sandia is working to establish a simulation system that will model the complete life cycle of a nuclear weapon, and its proliferation and assessment program is in the continuous process of developing more effective technologies to combat chemical and biological threats.

WEB RESOURCES

Sandia National Laboratories website, *Bomber's Shoes Disabled by Sandia Bomb Squad Tool*, 2/21/02, http://www.sandia.gov/media/NewsRel/NR2002/bombtool.htm, accessed on 9/30/01.

Sandia National Laboratories, Press Release, October 7, 2003, *Sandia Seeks Commercialization Partners for Hand-Held Chemical Analysis and Detection Systems*, http://www.sandia.gov/news-center/news-releases/2003/mat-chem/chempartners.html, accessed on 10/29/03.

Sandia National Laboratories website, http://www.sandia.gov/about/index.html, accessed on 10/1/03.

Sandia National Laboratories website, *Flying Snifferstar May Aid Civilians and US Military*, 1/23/03, http://www.sandia.gov/news-center/news-releases/2003/mat-chem/SnifferSTAR.html, accessed on 9/30/03.

Sandia National Laboratories website, *MicroChemLab Fact Sheet*, http://www.ca.sandia.gov/microchem/McCmLab.pdf, accessed on 9/30/03.

Sandia National Laboratories website, *Sandi DeCon Formulation Fact Sheet*, http://www.sandia.gov/SandiaDecon/factsheets/overview_apr2002.pdf, accessed on 9/29/03.

Sandia National Laboratories website, *Sandia Develops Tools for Assessing Vulnerability of Buildings to Chemical and Biological Attacks*, 12/12/01, http://www.sandia.gov/media/NewsRel/NR2001/build.htm, accessed on 10/1/03.

Sandia National Laboratories website, *Sandia Researchers Help Prepare Public Health Officials, Others With Anti-Terror "Decision Analysis" Tool*, 8/20/02, http://www.sandia.gov/news-center/news-releases/2002/gen-science/antiterror.html, accessed on 10/1/03.

SCIENTISTS, SOCIETIES, AND BIOTERRORISM DEFENSE IN THE UNITED STATES

RITA R. COLWELL

INTRODUCTION

Fundamental research in all fields is one of the pillars of United States's strength and prosperity. A hallmark of basic research is that results come from unexpected places and support for basic research contributes to the wellspring from which discoveries and their applications emerge. Years ago, Star Trek completely monopolized sci-fi airwaves. At that time, William Shatner and Leonard Nimoy were leading Americans "to boldly go where no man has gone before." Now, for a nominal fee, any of us can actually book our own flights into space. So have the times changed.

A series of events more specific to the biological sciences and to microbiology has unfolded. Many of us know the landmark year of 1967, when theWorld Heath Organization (WHO) established the smallpox eradication program. At that time, the disease was still prevalent in 33 countries, with estimates of 10 to 15 million cases each year. As a global community and with hubris, we celebrated enthusiastically 10 short years later, in 1977, when the last smallpox patient was successfully treated. Our world was declared smallpox free in 1980. The success of the eradication program was global but fragile, and the international microbiology community played a pivotal role.

Many microbiologists were not converts to the notion of eradication. In fact, many scientists, and microbial ecologists particularly, have doubts about such absolutes when microorganisms are involved. In any case, we now find ourselves facing a most troubling paradox in the return of smallpox as a "number one" priority on the list of potential biological weapons.

This situation symbolizes the obligation to outline the *new* role of scientists and especially that of microbiologists. As scientists, we must deal with the global problem of bioterrorism, and the professional community of microbiologists and the fundamental scientific research they do are critical to the task. For purposes of this chapter, potential actions microbiologists could take to assist federal and state health agencies, first responders, and citizens in the event of a bioterrorist attack or an infectious disease outbreak will be highlighted.

BACKGROUND

First, a brief look back in time. Throughout history, significant events have served as guideposts. In U.S. history, we speak of the Revolution, the Civil War, Pearl Harbor, Sputnik, the Kennedy assassination, the Moonwalk, and, now, September 11. A quote from T.S. Elliot's *Four Quartets* states it well. He wrote, "Time present and time past/Are both perhaps present in time future, And time future contained in time past."

There is much truth in those prophetic words describing the interconnectedness of the past, present, and future. The past holds the seeds of the future, and our future is rooted in the past. Both influence the decisions we make today. In time past, present, and future, are lessons that will stimulate discussion on protecting society from bioterrorism. In time past, valuable lessons portend significant consequences for our present and future.

The impacts of infectious disease and bioterrorism throughout history are not altogether new information today. Bioterrorism existed centuries before the discipline of microbiology, and what is thus an extensive timeline contains a number of both interesting and useful examples relevant to this discussion (Fig. 1). In the sixth century, the Assyrians poisoned the wells of their enemies. Later, the Tartar army catapulted bodies of plague victims over the walls of Kaffa (in what is today Ukraine), leading to a plague epidemic that followed as mercenary forces retreated from the city. And even more recently, smallpox was intentionally spread to Native Americans with contaminated blankets and handkerchiefs during the French and Indian War. Evidence lies in an excerpt from an original letter kept in the archives at the Library of Congress, written by Colonel Henry Bouquet to Lord Jeffrey Amherst in 1763, which suggests the distribution of blankets to "inoculate the Indians" (Fig. 2). It reads, "I will try to inoculate the Indians by means of blankets... taking care however not to get the disease myself" (Poupard and Miller, 1989).

Encyclopedia of Bioterrorism Defense,   Edited by Richard F. Pilch and Raymond A. Zilinskas
ISBN 0-471-46717-0   Copyright © 2005 Wiley-Liss

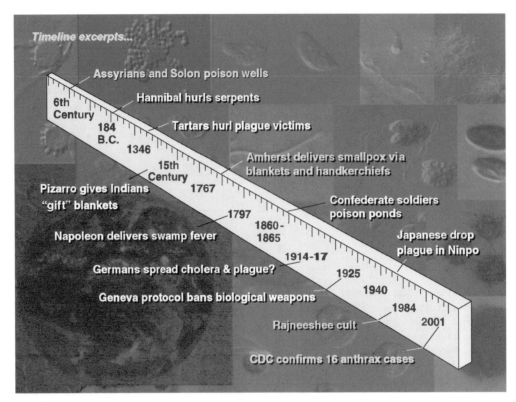

Figure 1. Timeline of bioterrorism.

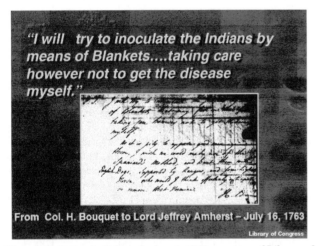

Figure 2. Letter to Jeffrey Amherst, 1763; courtesy of Library of Congress.

anthrax cases. Thus, the nineteenth century experience, in essence, offered a valuable lesson: epidemiology is not just the science of a microorganism, but is also about history and society.

The words of George Santayana, well known philosopher and poet, are nearly biblical: "Those who cannot remember the past are condemned to repeat it." We are fortunate that within the professional society of microbiology, we have the experience, wisdom, and research know-how to heed Santayana's warning. Microbiologists can help guide society by offering both a historical context and solid foundation of scientific research, both of which

Figure 3. Cholera in Paris.

In reading about the first cholera epidemics (Fig. 3), government officials in Paris, France, were quoted as saying, "There is nothing to worry about. We are the most civilized nation in the world. The epidemics that are happening in other places will not happen here." In spite of this statement, in 1848, some 8000 Parisians died of cholera. In Paris at that time, literacy was so low that newspapers were read only by an elite few. News spread by bulletins and posters with updates of who and how many had died. There was rumor, propaganda, and extreme anxiety—much like what was witnessed during the recent are critical to scientific advancement and national and international security in time present.

SCIENTIFIC AND SOCIETAL RESPONSIBILITIES

It is clear that there is a greater need for public understanding of scientific knowledge. Scientists have a responsibility to convey to the public that science is neither inherently good nor bad. Over the centuries, microbiology has been a dual use science. Today, microbiologists have new and important societal responsibilities: as valuable resources for enlightening domestic policy and international collaboration, they must encourage young scientists to study and conduct research abroad in order to gain insight into a wide variety of cultures, pathogens, and predictive and remedial tools. Close international cooperation not only enriches the research but can serve to alert society at large to malevolent microbial traffic. And the expanding diversity of the U.S. domestic workforce is an opportunity to enlist more people of different backgrounds and cultures into the field of microbiology. With the world becoming more of a global community, this must be a goal (Fig. 4).

The scientific community is a global sharing institution, with information readily available to those who need it. Unfortunately, terrorist organizations can use this benign and beneficial research for their own purposes. The will and capability to use biological weapons is increasing among terrorist groups and those states that sponsor such research. In 1997, one analyst observed "eleven countries are pursuing offensive-oriented biological warfare programs, up from just four in the 1960s" (Roberts, 1997).

The Internet and information technologies have drastically improved the commerce factor for scientific advancement. Figure 5 shows global Internet traffic, which depicts the free flow of information. This exchange has led to an accelerated pace of advancement in most areas of research. However, it is also much easier for a sinister-minded individual to usurp this same information to inflict harm.

In addition, biological agents are more suited to terrorist objectives than other potential weapons of mass destruction (WMD). A single individual can manufacture a biological weapon under relatively primitive conditions, as opposed to the large, sophisticated facilities and additional personnel needed to manufacture a nuclear weapon, for example.

This means that knowledge itself has become the critical commodity. Membership rosters of scientific societies list experts (sometimes the only individuals with such expertise in the world) capable of distinguishing whether an individual is conducting purely beneficial research or research that holds the potential for illicit application. Further, with the advent of biotechnology, expertise has progressively become more narrowly focused, such that for the most part only a scientist examining activities in his or her area of expertise is now capable of distinguishing between the two. This makes scientists, notably microbiologists, unique and powerful allies for the international regulatory community.

Scientists as a community should not be reticent. The federal government has become proactive and has responded with new investments in research and technologies to prevent, remediate, and negate biological weapons, and scientists should participate in helping to prioritize such research investments. But as we all now understand, research is only one component critical to the equation. How we organize to prevent and respond to outbreaks is also important.

SPECIAL CONSIDERATIONS: SABOTAGE AND AGROTERRORISM

In addition to human diseases, we must consider indirect threats to our food and water supplies. This is not idle speculation. For instance, al-Qa'ida evidently plotted the landmarks and public water supplies of most major U.S. cities. In addition, many countries considered epicenters of terrorist activity have experimented extensively with agroterrorism. Iraq, for example, developed wheat cover smut as a weapon in the late 1980s, most likely to use against its neighboring rival, Iran. As Bruno Sobral, Director of the Virginia Bioinformatics Institute, said in his recent testimony before the U.S. Senate, "A single agricultural terrorist could launch a pathogen that, spread by wind, water, or soil, could cause an irremediable chain reaction. The food supply and industries involved directly in food production and distribution are especially vulnerable. A terrorist wishing to cause severe and reverberating financial consequences could simply introduce a foreign disease into American livestock or crops. This would set off a chain reaction touching virtually every segment of this nation's economy."

The recent foot-and-mouth disease debacle in the United Kingdom (U.K.) is a case in point. This outbreak occurred as the U.K. cattle industry was already reeling from a $6 billion loss due to the mad cow disease outbreak in 1996, and led to the slaughter of nearly four million animals before control was established. Figure 6 shows that only two areas in the world have remained free of

Figure 4. The global community of infectious disease.

Eick; www.nd.edu/~networks/gallery.htm; Laubenbacher, VBI

Figure 5. Internet traffic across the continents.

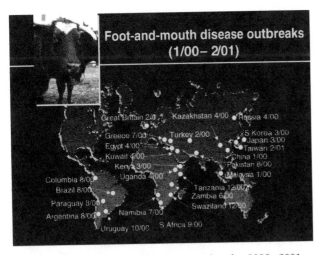

Figure 6. Foot-and-mouth disease outbreaks, 2000–2001.

foot-and-mouth disease over the last few years: Australia and North America.

A disease epidemic or bioterrorist attack on the agricultural system introduces another conundrum: caution must be taken in deploying emergency personnel, because an attack in one sector could conceivably be a feint to divert resources from critical command posts. For example, a major livestock disease outbreak in Texas would shift the primary command and control emphasis, as well as large numbers of military personnel, to that site, potentially leaving other key targets vulnerable.

SCIENTIFIC PROGRESS IN A GLOBAL ENVIRONMENT

There is a flipside to this discussion, however: with any luck, the same basic science that helped create biological weapons will also ultimately provide antidotes to them. Sensors offer a powerful and practical example of science's dual use. The types of tools formally designed for a single purpose can also protect nations, as well as society as a whole, in times of peace. Utilizing nanotechnology, researchers are on the verge of a comprehensive quick-warning system with many practical uses. Nevertheless, the exchange of scientific information and its use will always be a two-way street. We need no longer ponder the chicken or the egg conundrum. Perhaps poet Anne Sexton captured it best when she said, "Even without war, life is dangerous."

Science thrives on open discourse. We cannot limit scientific interaction without limiting scientific progress. Measures that inhibit dialogue will impede advancement. Scientific progress is imperative if a country is ill-prepared to cope with an epidemic. This is the case whether contending with a biological weapon, an accidentally introduced exotic pathogen, or a naturally mutated pathogen. Unfortunately, the United States and developing countries generally have had little experience dealing with epidemics of any proportion, despite the fact that many diseases currently overwhelm preventative and therapeutic measures: AIDS, Ebola fever, West Nile fever, and malaria to name a few. The recent Severe Acute Respiratory Syndrome (SARS) outbreaks have proven beyond any doubt that infectious disease concerns are global in scope. In today's world of rapid travel and large migrant populations, infectious diseases, regardless of introduction mode, pose a growing threat to health, agriculture, and the economy.

BIOTERRORISM: A UNIQUE QUANDARY

Nothing in the realm of a natural outbreak rivals the complex response problems that would follow a bioterrorism attack against a civilian population. Such an attack represents a unique hybrid of a national security crisis and public health emergency. Organizations and agencies involved in bioterrorism response follow different cultural styles, and individuals in leadership positions generally have not worked together before, forcing top officials to rely on the advice of experts they have never met. The officials will further be forced to make decisions about issues with which they are largely unfamiliar. Thus, one of the most important roles microbiologists can play is to fill the information void for decision makers regarding the principles of infectious disease.

Community-wide response plans that incorporate the latest scientific principles of microbiology are critical, and mechanisms for experts to be tapped in the event of an emergency are also essential. In the words of Admiral Stansfield Turner, former CIA Director, only biological weapons and nuclear weapons have the potential to bring the United States "past the point of nonrecovery." Indeed, it was Nikita Krushchev who warned that "after nuclear war, the living would envy the dead," and it appears likely that this comment would hold for bioterrorism as well.

In time future, the potency, diversity, and accessibility of biological weapons will increase as biological science advances and the number of trained biologists rises. The future that will be experienced indeed lies in the past and in the present. Will it be written that we suffered greatly because we were unprepared? Or, will the future be filled with better health care and disease eradication because we took the initiative, and did prepare?

THE WAY AHEAD

There are, in spite of gloomy predictions and forewarnings, some immediate actions that can be taken to counter the concerns set forth above. These are offered here as sketched ideas directed toward the scientific community,

not as absolutes but as a mechanism for future discussion and debate.

- Guides can be developed for certain diseases containing pertinent information—for example, symptomatology, whether or not a given disease is contagious, and what the best initial emergency action is if exposure to such a disease is suspected—and provided to policymakers, frontline physicians, first responders, and most importantly the lay public.

- A format for workshops can be designed for emergency personnel, that is, first responders, providing training on the most effective immediate actions in a bioterrorism crisis.

- Information guides can be developed for public sanitation officials, containing, for example, precautions for food and water supplies.

- Guidelines and workshops to educate the public can be developed and made available in public libraries, public schools, and also via video for public TV and school assemblies.

- An e-mail hotline for questions from public officials can be established such that questions are routed to appropriate experts or rotated through appropriate committees.

Today's society is virtually inundated with rhetoric and general talk about bioterrorism concerns. The above are just a few practical suggestions to make better use of the unique expertise of scientists to meet the current needs of a society reckoning with the spectra of bioterrorism. Ultimately, much can, and should, be done to protect against bioterrorism activities in the future.

REFERENCES

Roberts, B., "Between Panic and Complacency: Calibrating the Chemical and Biological Warfare Problem," in S.E. Johnson, Ed., *The Niche Threat: Deterring the Use of Chemical and Biological Weapons*, Government Printing Office, Washington, DC, 1997, pp. 9–42.

Poupard, J.A., Miller, L.A., and Granshaw, L., *ASM News*, **55**, 122–124 (1989).

WEB RESOURCES

American Society for Microbiology, Biological Weapons Resource Center, http://www.asm.org/Policy/index.asp?bid =520, accessed on 11/22/04.

SMALLPOX

Richard F. Pilch

INTRODUCTION

Pursuant to concerns that Iraq and other states of proliferation concern might possess the smallpox virus and in view of the implications of September 11, 2001, with respect to homeland defense, on December 13, 2002, President George W. Bush announced a multiphase plan for the vaccination of certain subsets of the U.S. population against smallpox. A year later, this plan has essentially stalled, largely because the known risks of the vaccine are thought to outweigh the theoretical risk of a smallpox outbreak. This entry addresses both the threat of the virus and the risk of the vaccine in order to get to the heart of this ongoing debate, and offers both civilians and policymakers the background necessary to make an informed decision about the U.S. vaccination program. Part I outlines the basics of smallpox. Part II covers the smallpox vaccine, including contraindications, side effects, and availability. In Part III, technical hurdles to attaining a smallpox capability are reviewed. Finally, Part IV addresses the U.S. vaccination plan itself.

THE SMALLPOX VIRUS

In the strict sense of the word, "smallpox" is defined as the disease caused by the virus Variola major (or in very rare cases Variola minor), but for the sake of simplicity, the term is used to represent both the virus and its resulting disease in this entry. In the twentieth century, smallpox killed more people than war. Despite its lethality, however, the fact that the virus can only live in humans made it vulnerable.

Viruses differ from bacteria in that, because they generally consist of only a protein shell, a sequence of genetic material (DNA or RNA), and sometimes a lipid membrane, they lack the cellular machinery required for reproduction (termed "replication") and thus survival. As a result, they need to adopt the cellular machinery of a host organism in order to replicate, and are therefore inert until they get into a cell, human or otherwise. The smallpox virus is limited in its choice of hosts: it is only equipped to enter and adopt the machinery of human cells (Fig. 1).

Physicians and scientists in the international community recognized that given this limitation, preventing

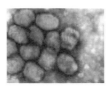

Figure 1. A transmission electron micrograph of smallpox viruses. Public Health Images Library (PHIL) id# 2292 *Source*: CDC/Fred Murphy; Sylvia Whitfield.

human-to-human transmission would halt the spread of the virus and effectively eliminate it from its only reservoir. Thus, in 1967, the World Health Organization (WHO) began its global campaign to eradicate smallpox, utilizing a "surveillance-containment" ring vaccination technique that is now the basis of the Centers for Disease Control and Prevention's (CDC) primary strategy for vaccination to be implemented in the event of a single identified case of the disease (see Part IV below). A decade later, the campaign was deemed successful, and in 1980, WHO recommended the cessation of all vaccinations worldwide. Vaccinations in the United States had been discontinued in 1972. A universal and reportedly successful movement followed to destroy all remaining samples of the virus except for those held in two locations, the CDC in Atlanta and the Institute of Virus Preparations in Moscow. Later, the Moscow samples were moved to the State Research Center for Virology and Biotechnology (Vector) in Siberia. These two laboratories, the CDC and Vector, hold the only official stores of the smallpox virus today.

In the United States, the cessation of vaccinations 30 years ago has created what is effectively a virgin soil population in which previously vaccinated U.S. citizens have little, if any, waning immunity to the disease and those born after 1972 have none at all. Studies have shown that neutralizing antibodies against smallpox, which are believed to signify protection, decrease substantially over 5 to 10 years following vaccination, making protection 30 plus years later negligible at best (Downie and McCarthy, 1958). Historically, this type of unprotected population has not fared well against the virus. For example, when a series of events surrounding the 1519 exploits of Hernando Cortes unwittingly introduced smallpox among the virgin Aztec civilization of Mexico, the impact was so devastating that within a generation the Aztec culture, religion,

Encyclopedia of Bioterrorism Defense,   Edited by Richard F. Pilch and Raymond A. Zilinskas
ISBN 0-471-46717-0   Copyright © 2005 Wiley-Liss

and language were gone (Tucker, 2001). Similarly, the distribution by British forces of blankets that had been used by smallpox patients among virgin populations of American Indians during the French and Indian Wars (1754–1767) led to epidemics of the disease with case-fatality rates (the percentage of deaths among infected individuals) reportedly surpassing 50 percent in affected tribes (Stearn and Stearn, 1954).

Transmission and Infection

How then does the smallpox virus generate this lethal effect? First, it must gain entry into the body, which in nature—and presumably in the event of deliberate dissemination of the virus as well—generally occurs via the respiratory system. Most commonly, the virus spreads from person to person when very small viral particles are expelled from the mouth of an infected individual and inhaled by a person nearby. The virus can also be naturally transmitted via direct contact with the rash (but generally not the scabs) or via fomites such as contaminated clothes or bed linens. Transmission due to the exchange of saliva is a possibility as well.

As a general rule, the distance between individuals must be less than 2 m (approximately 6.5 ft) in order for respiratory transmission to occur (CDC, Undated c), though this proximity limitation is not always the case. In 1970, for example, a hospitalized smallpox patient isolated in a single room managed to infect individuals on three floors of the hospital (Wehrle, 1970). Only a few virions (individual viral particles) need be inhaled to cause infection. For example, in animal studies during the Cold War, Soviet scientists found that fewer than five viral particles were required to infect half of the exposed population (denoted by the value ID_{50}, or infectious dose in 50 percent of those exposed) (Alibek and Handelman, 1999).

Eradication data from the Indian subcontinent showed that when one member of a household became sick with smallpox, 36 to 88 percent of other unvaccinated household contacts would contract the disease (USAMRIID, 2002). However, because individuals suffering from the symptoms of smallpox were usually bedridden by the time the rash formed (essentially signaling contagiousness), such household contacts were often the only ones exposed to the virus. Thus, each smallpox case passed on the disease to only 0 to 3 healthy persons on the average, *the majority of whom were household contacts* (USAMRIID, 2002). In the 49 documented outbreaks of smallpox in Europe between 1950 and 1971, the average case in the general public led to 1.6 second-generation cases (the average case identified in a hospital setting led to 2.4 second-generation cases, reflecting the propensity for transmission of the virus within a given medical establishment) (USAMRIID, 2002). Upon diagnosis, 34 of these 49 outbreaks did not continue beyond the second generation of cases, and 13 demonstrated no transmission of the virus whatsoever beyond the index case (further, 12 of the 49 outbreaks demonstrated transmission to only 1 to 4 individuals) (USAMRIID, 2002).

Pathogenesis and Clinical Course

Upon inhalation, smallpox virions become implanted in the lining of the mouth or lungs. Normal draining of these areas then carries the virions to regional lymph nodes, which exist throughout the body and filter such drain-off in order to ensure that no foreign particles enter the bloodstream. The filtration process works, and the virions are held at bay. However, this affords the viral particles the opportunity to enter the lymph node cells and begin to replicate.

By day 3 or 4 postinfection, the virus enters the bloodstream and is carried to the spleen and bone marrow. In addition to continued replication in the lymph nodes, the virus replicates in these new areas as well, generating virtually limitless copies of itself in order to maximize its own chances of survival as a species. Around day 8 postinfection, the virus reenters the bloodstream in force.

At this point, timing and resultant symptomatology differ depending on what type of form of disease presents. Malignant and hemorrhagic presentations of smallpox demonstrate accelerated disease progression and increased lethality, but these presentations occur only very rarely and are therefore only mentioned in passing here. In the most common presentation of the disease, the infected individual has demonstrated no sign of infection up to this time. Unless testing has been performed, he or she has no idea the infection has taken place. This resurgence of the virus into the bloodstream signals the approach of ill effects, however. The first symptoms usually appear by day 12 to 14 postinfection (with a range of 7 to 17 days), and classically include high fever, generalized discomfort, headache, and backache.

Inside the body, the virus continues to work. It finds its way to small blood vessels within the skin and beneath the lining of the mouth and enters cells in the vicinity, where it again replicates. Because this replication occurs so close to the surfaces of the skin and mouth, a rash begins to form in these areas, and with this rash comes the ability of the virus to spread to other individuals. It is the rash that occurs in the mouth that allows for this. Lesions of this rash break open and distribute the virus into the mouth and throat, after which it is exhaled in minute particles with each breath. At this point, the infected individual is most capable of spreading the virus to others, and the appearance of the rash therefore serves as an indicator that he or she has become "infectious" (Fig. 2). Though most infectious during the 7 to 10 days following onset of the rash, the individual remains capable of transmitting the virus until the rash scabs over and the last scab falls off, a process that takes about three weeks from beginning to end (if the patient survives).

Smallpox infection carries an estimated case-fatality rate of 30 percent among unvaccinated individuals in the absence of specific therapy. Death usually occurs during the second week of the rash (the third to fourth week postinfection), most likely because of the overwhelming presence of toxic components in the blood, namely, viral particles and immune complexes (viral particles bound to antibodies). Even if the victim survives, however, the effects of the virus can last a lifetime. In 1562, for example, Queen Elizabeth I of England was infected with smallpox

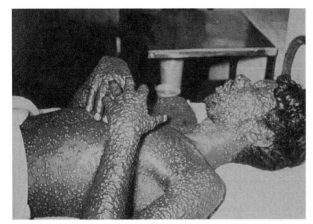

Figure 2. Man with smallpox. Public Health Images Library (PHIL) id# 131. *Source*: CDC/Barbra Rice.

at age 29, and though she survived the disease, she was left without hair and with a permanently scarred face, hence the heavy make-up and red wig with which she is identified today (Tucker, 2001).

THE SMALLPOX VACCINE

Viruses differ from bacteria in that they are not susceptible to antibiotics. Antibiotics specifically target bacterial structures or mechanisms; viruses do not have these features and thus are not targeted. For a given virus, treatment is limited to (1) a vaccine, given pre- or postexposure to induce the formation of protective antibodies by the host; (2) immune globulin, a solution containing preformed antibodies derived from human blood; or (3) specifically designed antiviral agents, which unfortunately have limited application (e.g., cidofovir).

Untold generations ago, the recognition that victims who survived smallpox never contracted the disease again led to the idea of vaccination in its most primitive form: deliberately infecting an individual with a mild form of the illness in order to confer immunity. The first method employed in this regard was known as "variolation," developed sometime before 1000 B.C. This process consisted of inoculating unexposed individuals—through incisions in their skin—with scabs or pus from mildly infected smallpox patients, an action that effectively reduced the fatality rate of the resulting smallpox infection from 30 percent to approximately 1 percent (Tucker, 2001).

In the latter half of the eighteenth century, physician Edward Jenner observed the clear complexion of milkmaids in the English countryside, noting that they exhibited none of the disfiguring lesions associated with a past history of smallpox. Up until this point, variolation remained commonplace as the only known approach to mitigating the effects of the disease. Jenner's observation, however, set him on the path to discovery, and what would be the creation of the first true vaccine. It turned out that the milkmaids, as a result of occupational exposure to the cowpox virus, had developed antibodies to this virus that also served to protect them against the related smallpox virus. In 1796, Jenner demonstrated that this effect could

be recreated via the injection of fluid from a cowpox pustule in order to confer immunity to smallpox. Eventually, he coined the term "vaccine" (from the Latin word *vacca*, meaning "cow") to describe the injected substance. This cowpox vaccine is the predecessor of the vaccinia vaccine used today.

Contraindications of the Vaccine

Certain populations are known to be at high risk for developing severe complications following vaccination with vaccinia vaccine. With respect to these populations, the only scenario in which the vaccine's benefits outweigh its risks is in the event of direct contact (face to face, household, or close, i.e., less than 2 m) with a confirmed smallpox case. Otherwise, the dangers of the vaccine are too great to warrant its use, even if cases of smallpox appear that are remote from the high-risk individual. Preevent voluntary vaccination is not an option. These contraindicated populations include (CDC, Undated a and c):

- Immunocompromised persons, including individuals with HIV/AIDS or cancer, organ transplant recipients, or those receiving chemotherapy, radiation, or high-dose corticosteroids;
- Persons with a history of eczema, whether or not the condition is mild or presently active;
- Pregnant or breast-feeding women;
- Infants less than 1 year old;
- Individuals with life-threatening allergies to the antibiotics neomycin, polymyxin B, streptomycin, or tetracycline, because the major brand of currently available vaccine (DryVax) contains trace amounts of these substances;
- Persons with acute or chronic skin conditions such as atopic dermatitis, burns, impetigo or varicella zoster (shingles) should wait until the condition resolves before being vaccinated;
- Persons with moderate to severe short-term illness should wait until the condition resolves before being vaccinated.

Importantly, any person who lives with a contraindicated individual should not receive the vaccine unless either that person or the contraindicated individual has for certain been exposed to the virus. Also, it is very important to note that the Advisory Committee on Immunization Practices (ACIP) has recommended against the vaccination of any individual under 18 years of age unless exposed to the virus (CDC, Undated a). While not a contraindication per se, this explains why President Bush's voluntary vaccination plan was only offered to general members of the "adult" public.

Side Effects of the Vaccine

Even in healthy individuals, the vaccine can be dangerous. Up to 70 percent of vaccine recipients may experience some sort of self-limited adverse reaction, for example fever or swelling of the lymph nodes (McIntosh et al., 1990).

Further, statistics suggest that for every million people vaccinated approximately 1000 will experience serious complications, for example, the formation of a localized or generalized rash. Some 50 vaccine recipients per million will be expected to develop life-threatening complications, and 1 to 2 healthy individuals per million vaccinated will be expected to die as a result of the vaccination. It is generally accepted that individuals vaccinated for the first time ("primary vaccinees") are at a greater risk for complications and/or death than individuals who have been vaccinated in the past. Serious complications carrying the risk of death include:

- Tissue destruction at the site of inoculation ("progressive vaccinia," a.k.a. "vaccinia necrosum"), frequently fatal in immunocompromised individuals, hence their contraindication as discussed above. See, for example, www.bt.cdc.gov/training/smallpoxvaccine/reactions/prog_vac_path.html: "Death occurred in nearly all individuals with profound [cell-mediated immune (CMI)] defects." The T-cell deficiency seen in untreated HIV/AIDS patients is a classic example of such a CMI defect.

- Widespread skin infection ("eczema vaccinatum") in individuals with preexisting skin conditions, hence their contraindication as discussed above. Of note, Goldstein et al. (1975) reported that vaccinia immune globulin (VIG) has some efficacy against eczema vaccinatum.

- Inflammation of the brain ("postvaccinial encephalitis"), which occurs only in primary vaccinees at a rate of 3 cases per million. Postvaccinial encephalitis carries an estimated lethality of 25 percent. Those who survive often suffer permanent neurological damage, for example paralysis (Henderson et al., 1999).

- Inflammation of the heart (myocarditis) or its surrounding membrane (pericarditis), or a combination of both (myopericarditis). This finding, while noted in the past, only began to gain attention during the recent U.S. vaccination efforts (CDC, Undated d).

An additional point that deserves attention is that the vaccinia vaccine is made from and thus contains a living virus (the vaccinia virus, not the smallpox virus). Following vaccination, this virus can spread from one part of the body to another ("autoinoculation") or from person to person ("contact vaccinia") via contact with the inoculation site.

Autoinoculation occurs when, after touching the vaccination site, an individual touches or scratches another part of the body with the same hand. This action transplants the virus to the new location, where a rash develops. Importantly, transfer of the virus to the eye in this fashion can lead to severe damage and even blindness. Autoinoculation accounts for a substantial portion of the serious but not life-threatening rashes mentioned above, the remainder of which are generally caused by either entry of the virus into the bloodstream with subsequent generalized rash ("generalized vaccinia") or an allergic reaction to the vaccine itself ("erythema multiforme").

Contact vaccinia is a concern because it presents the possibility of spreading vaccinia virus from healthy, voluntary vaccine recipients to immunocompromised—and thus contraindicated—individuals. This person-to-person transmission has been documented in the past at a rate of 27 infections per million (Lane et al., 1970).

Vaccine Availability in the United States

Multiple strains of vaccinia virus exist and are thus available for use in vaccine production. The strain used in all U.S. vaccines is the New York City Board of Health (NYCBH) strain. This strain has been selected for vaccine development because it carries the lowest complication rate of all strains available (Wyeth Smallpox Vaccine [package insert], 1968; Lane et al., 1969).

Two brands of vaccine compose the present U.S. stockpile:

1. Wyeth DryVax (15.4 million doses), which over the past many years has been used to vaccinate laboratory personnel at risk of exposure to the virus. Studies have shown that this vaccine can be diluted up to tenfold and still be highly effective, thus creating a potential supply of 154 million doses (Russell, 2002). However, current plans have only called for fivefold dilution of the stockpile—resulting in 77 million doses—in order to meet the nation's demands. This is the vaccine that was used to implement the early stages of President Bush's current vaccination plan. Only in the event of an emergency requiring the implementation of the Supplemental Strategy and mass vaccination (described in part IV below) will other vaccines be required (ABCNews.com, 2002). On October 15, 2002, DryVax was approved for licensure by the Food and Drug Administration (FDA).

2. Aventis-Pasteur vaccine, the decades old supply mysteriously discovered in Pennsylvania early in 2002 (approximately 80 million doses) (Russell, 2002; Roos, 2002). This vaccine is not and will not be approved for licensure but will remain in what is known as Investigational New Drug (IND) status, only to be used in the event of emergency activation of the Supplemental Strategy.

The CDC has also ordered an additional 209 million doses of vaccine in contracts awarded in September 2000 to Acambis and November 2001 to Acambis/Baxter (Monath, 2002). The first contract calls for the 20-year continuous production of 45 million doses of the vaccine ACAM 1000, now undergoing Phase II clinical trials. The second contract calls for the production of 155 million doses of the vaccine ACAM 2000 in the shortest timeframe possible. ACAM 2000 is now undergoing Phase I clinical trials.

These and other new vaccines are under development, largely because of the dangers associated with the current vaccine. Originally, viruses used in vaccine production ("first generation" vaccines) were grown in embryonated hen eggs; in other words, eggs containing living cells in which the virus can replicate. Embryonated eggs are particularly suited for viral vaccine production because they offer a sterile environment, have no immunologic functions (and thus do not fight off the virus being produced), and are both inexpensive and widely available. Wyeth DryVax and the Aventis-Pasteur vaccine are examples of first generation vaccines.

While this technique continues to be the major approach to such production today, improvements in biotechnology over the past many years have led to the creation of a new growth medium of living cells in which vaccines can be produced ("second generation" vaccines) called "cell culture." However, because the actual product is the same, the second-generation vaccines come with the same complications as their predecessors (Midthun, 2002). Thus, with respect to the smallpox vaccine, these newer products offer little hope for improvement of the traditional side effects. ACAM 1000 and ACAM 2000 are examples of vaccines grown in cell culture (second-generation vaccines).

Because the worst complications of the smallpox vaccine are due to replication of vaccinia virus in the vaccine recipient, research on "third generation" vaccines is focused on genetically removing this characteristic of the virus in order to create nonreplicating vaccinia strains that still retain their immunizing properties (Midthun, 2002; Falkner, 2002). The future, while bright, remains distant.

ATTAINING A SMALLPOX CAPABILITY

A biological weapons capability can most basically be defined as whether a given adversary has the technical ability to carry out an attack. In order to attain such a capability, in general terms, three major hurdles must be overcome: a pathogenic strain of the intended weapon must be acquired, the strain must be used to produce an amount of agent sufficient for use, and the agent must be effectively delivered. Each is discussed in turn below.

Acquisition

As far as is known, U.S. policymakers do not possess any incontrovertible evidence that any of our perceived enemies, whether state or terrorist organization, has acquired the smallpox virus. Potential sources for such acquisition are largely concentrated in two spheres: (1) the former Soviet BW program, which throughout the 1970s is said to have maintained a 20-ton-per-year stockpile of the virus (Alibek and Handelman, 1999); and (2) surviving virus from the preeradication era.

Since the dissolution of the Soviet Union, fears have persisted that seed stocks of certain BW agents, including the smallpox virus, have been illegally removed from one or more of the many facilities of the former Soviet BW program. Two facilities are of primary concern in this regard: Vector, reportedly the central facility of the Soviet civilian smallpox weaponization program from 1987 on; and the former military research and production facility for smallpox under the Ministry of Defense at Sergiyev Posad (Zagorsk), which remains closed to outsiders and the West to this day.

Vector has long been the focus of international threat reduction efforts, and as such has benefited from enhanced physical security and increased

wages for its scientists over the past few years (http://www.dtra.mil/ctr/ctr_russia.html, accessed 9/30/03), though recently the financial situation of this facility has significantly worsened (FBIS CEP 20031030000047, 2003). The security measures essentially serve to guard Vector against outside penetration, while scientist funding is meant to dissuade insiders from stealing and selling strains of the virus, thus addressing the two major concerns with respect to containment at such facilities: poor security and vastly underpaid scientists who might be persuaded to supply agents or expertise to an outsider. This outsider may represent a state of proliferation concern or a terrorist organization, for example al-Jihad, a group associated with Usama Bin Ladin that according to the 1999 testimony of a group member had successfully purchased biological and chemical weapons from countries of the former Soviet bloc (Tucker and Vogel, 2000b). More likely, however, the outsider would be a member of one of Russia's estimated 12 to 15 major mafia groups (Galeotti, 2000a), which via operations in 60 to 65 countries could then distribute the acquired agent on the black market or to interested parties directly (Galeotti, 2000b).

Current research and development efforts behind the closed doors of the Center for Virology (formerly the Scientific Research Institute of Sanitation), commonly known as Sergiyev Posad (formerly Zagorsk) after the city in which it is located, are unclear. The facility's earlier activities, however, suggest the possibility that Sergiyev Posad may retain stores of the smallpox virus. Reportedly, during the Cold War, a specially selected strain of the virus procured by KGB agents in India was produced as a weapon at the facility (Alibek and Handelman, 1999). Weaponization continued until at least the 1980s (Breslow, 1990), such that at any one time stores of the virus there may have measured in the tons (Knyazkov, 1986).

The notion of smallpox virus persisting from the preeradication era generally refers to samples (crusts, pus, or isolates obtained from growth on chick chorioallantoic membrane) taken from smallpox patients that for whatever reason were not destroyed following the disease's eradication. As mentioned, in the wake of WHO's global eradication campaign, the necessity of such destruction was emphasized, and compliance was ultimately reported by all countries without exception. But bits and pieces of evidence to the contrary have come to light, suggesting that this may not in fact have been the case.

A second possible source of virus from the preeradication era is bodies of smallpox victims buried in permafrost ground. The virus is preserved in deep cold, illustrated for example by the fact that it is stored as such at the CDC. Thus, viable virus could theoretically be recovered from smallpox victims buried in Siberia, Alaska, and other permafrost regions. The recovery of viable influenza virus from victims of the 1918 "Spanish flu" pandemic has been attempted in this manner, albeit unsuccessfully (Davies, 1999).

Production

Like vaccinia vaccine production in embryonated cells or cell culture, production of the smallpox virus requires the insertion of a pathogenic seed strain into a living host. For example, throughout the 1970s the former Soviet Union's BW program reportedly collected hundreds of thousands of eggs monthly for its smallpox production process, which took the form of an extended laboratory assembly line (Alibek and Handelman, 1999). Years later, Soviet scientists also developed effective techniques for growing the virus in cell culture (much like the second generation vaccinia vaccines described above), namely, cultured monkey kidney cells (Tucker, 2001). Production on this level does not need to be achieved to attain enough virus for a large-scale attack, however, and in fact contagious agents such as smallpox do not necessarily require any additional production measures upon acquisition of a pathogenic strain whatsoever, a point that is addressed further in the analysis of delivery mechanisms below.

The endpoint of the production process is the puncture and drainage of each egg, the liquid of which contains virtually limitless copies of the virus. This initial product is considered a basic wet preparation. Like with most BW agents, the initial wet preparation of smallpox can be taken a step further in two separate ways. It can either be suspended in a "formulation" of adjuvants, preservatives, and other chemicals, or dried and then milled to attain the proper-sized particles. Either of these processes demands much more technical ability but generally yields a far better agent in terms of ease of dissemination and overall effect. In addition, the resulting agent can usually be stored for much longer periods of time in either case. Dry agents can be taken an additional step further and specially formulated as well, as was the case for example in the "anthrax letters" of 2001, to prevent clumping due to electrostatic forces. This clumping results in large, ineffective particles that are either blocked by the mucociliary response of the respiratory tract or fall harmlessly to the ground. Though largely preferring dry formulations of their BW agents, the Soviets chose only to produce smallpox in a wet formulation because the virus remained virulent for up to a year in such a state and was highly stable when aerosolized (Tucker, 2001). The Soviets had also found the dry formulation to be too dangerous for the workers involved in its production.

Delivery

Little is known about the current delivery capability of any state thought to possess smallpox virus. Conceivably, the virus can be delivered in a number of ways. In the past, for example, the former Soviet Union's Strategic Rocket Forces deployed intercontinental ballistic missiles (ICBMs) armed with single warheads containing smallpox-laden bomblets near the Arctic Circle, where they were maintained in silos on a launch-ready status (Tucker, 2001). Soviet wartime (i.e., World War III) strategy also called for the dropping of cluster bombs by long-range strategic bombers to deliver the virus (Tucker, 2001). But generally speaking, most analysts believe that currently, the most likely delivery system for smallpox would be a simple aerosol device such as an atomizer, inhaler, or handheld sprayer.

One of the most important features of a contagious agent, however, is that a potential perpetrator would not necessarily have to create a delivery system. In theory, humans themselves can be effective delivery devices for contagious diseases, obviating the need for missiles, sprayers, letter "bombs," and so on. Such an approach also eliminates the need for mass production of the contagious agent (only a small amount is needed to initiate the chain of events potentially leading to an epidemic) and specific formulation. From this notion has come the concept of the so-called smallpox suicide bomber, in other words, a knowingly infected individual who serves as a delivery device, spreading the disease by secondary transmission. Numerous news reports have suggested that such an attack could be carried out with relative ease. However, even if the agent were somehow acquired, initiating an epidemic would likely be more complicated than simply injecting it, waiting for a rash, and going to a public place. This process is similar to variolation described above. Although a potential suicide bomber might develop fulminating smallpox from such an injection, he or she would be more likely to develop a mild infection with or without a rash that in most cases would not lead to shedding of the virus and secondary spread. Further, he or she would likely be incapacitated to the extent that mobility would be severely limited. In any event, the virus as noted is not very contagious and requires close contact for person-to-person spread. Thus, while panic or the "worried well" phenomenon might ensue from any indication of such an operation, the overall likelihood of success with this type of attack is believed to be quite low, especially when the difficulty of acquisition is taken into account.

THE MULTIPHASE U.S. VACCINATION PLAN

Phase One

The first phase of President Bush's plan called for (1) the voluntary vaccination, beginning late January 2003, of approximately 500,000 frontline health workers (e.g., emergency room physicians and nurses) and first responders (e.g., paramedics, firemen, and policemen) identified by the Department of Health and Human Services (HHS) and state and local governments as candidates for specialized "Smallpox Response Teams;" and (2) the mandatory vaccination by the Department of Defense (DoD) of some 500,000 military and civilian personnel deployed to locations considered "high threat" by authorities within the U.S. intelligence community (Stevenson and Stolberg, 2002; CDC, Undated b). In addition, the DoD would offer voluntary vaccination to a limited number of additional personnel stationed at certain overseas embassies.

Smallpox Response Teams are meant to consist of any individuals who might be involved in the identification and management of a smallpox outbreak (in view of the disease's eradication, the word "outbreak" here refers to the appearance of one or more confirmed cases of smallpox).

These teams are, as their name implies, meant to initiate response efforts should they be required. However, this is not their only purpose. In fact, ensuring an adequate response may well be only a secondary consideration, because there is a much graver concern that is being addressed with their creation: the need to protect those individuals most likely to be the first ones exposed in the event of an outbreak. Illustrating this concern is the fact that in the 49 documented outbreaks of Variola major in Europe between 1950 and 1971, more than half of the cases occurred within a medical setting and greater than 50 percent of those cases involved medical personnel and hospital staff such as physicians and nurses (USAMRIID, 2002).

In the event of an outbreak, the highest likelihood is that an infectious smallpox patient (or patients) will present to either his or her family practitioner or a hospital emergency room complaining of such prodromal symptoms as fever, headache, and backache, as outlined above. A rash may or may not be present upon initial presentation. Again, because respiratory transmission of the virus is only possible after the lesions of the rash have formed and broken open in the oropharynx, the appearance of the external rash generally signifies contagiousness. However, the internal rash ("enanthem") precedes the external rash, such that a patient may already be exhaling viral particles in respiratory droplets up to 24 hours before a visible rash forms on the skin (USAMRIID, 2002). Thus, anyone who has had face-to-face contact with an infected patient any time after the onset of fever is considered at risk for infection (Henderson et al., 1999). A physician may therefore be unaware of being at risk when he or she first sees the patient. Further, at the earliest phase of the outbreak, before awareness has developed, smallpox may not even be considered in the physician's initial differential diagnosis (a list of possible diagnoses to be systematically ruled out). Without such awareness or a high index of suspicion on the part of the treating physician, identification, management, and isolation of the infected individual may be delayed, increasing the likelihood of transmission to the patient's direct caretakers and other personal, occupational, and incidental contacts in the area.

Risks aside (e.g., the side effects of the vaccine, or the possible spread of vaccinia to patients), preevent vaccination in the above scenario would not only serve to protect these at-risk caretakers from the disease but also would allow for both immediate management of the infected individual and rapid treatment of any contacts. At this stage of the disease, management of the infected individual is limited to supportive care (e.g., IV fluids and pain control), with the lethality rate essentially fixed at approximately 30 percent. Vaccination is of no value once symptoms appear. But supportive care is of inestimable value to an ailing patient and his or her loved ones, and thus the capacity to provide it is an absolute necessity.

Contacts are a different story. Upon making the diagnosis, the public health authority of the city or county where the first case is diagnosed has a narrow window of opportunity in which to identify and vaccinate all of the infected individual's household and close-proximity (again, less than 2 m) contacts, as well as additional response personnel at risk of exposure to the virus, in order to limit further spread of the disease. These additional response personnel include health care workers

responsible for the care or transportation of suspected or confirmed cases; laboratory personnel responsible for collecting or processing clinical specimens of such cases; and laundry or medical waste handlers at the facility where the diagnosis has been made (CDC, Undated c). In such a case, that is, the presentation in a hospital setting of a patient infected with smallpox, the CDC also recommends that "consideration. . . be given to vaccination of all individuals present in the hospital setting during the time a case was present and not isolated in the appropriate manner in a room with ventilation separate from other areas of the hospital" (CDC, Undated c).

The rapid administration of the vaccine to these high-risk groups could mean the difference between life and death: data has indicated that vaccination of a contact within four days of exposure reduces the likelihood of that person becoming sick with smallpox or, should sickness occur, significantly reduces mortality (Dixon, 1962). This process is discussed in Phase Four below, but its implications are critical here. While suspicion of a smallpox case demands notification of the CDC, an action which would lead to the rapid mobilization of teams of specialized CDC personnel, these personnel are only meant to assist and coordinate activities on the local front (CDC, Undated c). Local health care workers must therefore be prepared to act immediately after the initial diagnosis is made in order to save lives, and preevent vaccination allows them to do so.

What happens if these individuals are not preemptively vaccinated? In the event of an outbreak, they would immediately receive the vaccine. A waiting period of approximately 7 days is then necessary to ensure that the inoculation is successful (called a "take") and to allow for the development of protective immunological factors in the vaccinated individual (the appearance of the inoculation site will indicate a "take" in 3 to 4 days and offer a high level of certainty by day 7 to 11) (CDC, Undated c). Additionally, the vaccine, which has been known to cause fever in up to 70 percent of recipients 4 to 14 days after inoculation, may temporarily incapacitate these workers and thus hinder response measures that might otherwise allow for disease containment and the saving of lives.

Phase Two

Following the initial phase, vaccinations were to be extended under President Bush's plan to up to 10 million additional members of the health care and first responder community, again on a voluntary basis. This second phase was meant to be completed as early as summer 2003. Because the threat of smallpox is not quantifiable to any appreciable extent, it is very difficult to assess the need for such widespread vaccination. However, on the basis of historical data it can be estimated that if all 10 million of these individuals were to accept vaccinations, over 10,000 might suffer serious complications and 10 otherwise healthy individuals might die. Of course, not everyone was expected to volunteer. Regardless, these numbers illustrate yet again the very real risks of the current smallpox vaccine.

Phase Three

According to President Bush's plan, building upon this was what could be considered a third phase of vaccinations, those involving the general public, namely, adult civilians insisting on being vaccinated and for whom the vaccine is not contraindicated. At the time of President Bush's announcement, options outlined by the CDC included vaccination in 2003 with the unlicensed vaccine, or in 2004 once the vaccine was licensed by the FDA, or alternatively with next generation vaccines by participating in ongoing clinical trials (for more information, please refer to the U.S. National Institute of Health website, http://www.clinicaltrials.gov/).

Initially, this was presented to the American public as what essentially amounted to a rights issue: In an interview with Barbara Walters, President Bush stated, "I think it ought to be a voluntary plan. In other words, I don't think people ought to be compelled to make the decision which they [don't] think is best for their family" (Stevenson and Altman, 2002). Regardless, the driving logic behind Phase Three seems to have been that its implementation would raise "herd immunity" such that in the event of introduction of the virus into the U.S. population, subsequent spread would be limited. Theoretically, this makes sense, but certain issues persist, namely, whether screening measures would be adequately effective in identifying contraindicated persons, and how an informed decision on the part of those persons electing to receive the vaccine could be ensured. Neither issue has been addressed to date, as the vaccination plan has failed to reach this phase to any appreciable degree.

Phase Four

While the President's plan ends there, additional phases of vaccination exist, as laid out by the CDC (CDC, Undated c). The first of these phases (Phase Four overall), the "Primary Strategy: Contact Identification and Vaccination," is only set in motion in the event of a confirmed smallpox outbreak. Upon identification and isolation of the initial smallpox cases, all individuals having had face-to-face or household contact with those infected will be identified and vaccinated in the now classic "ring" approach, regardless of whether or not they are contraindicated for the vaccine. In addition, all people likely to come into contact with the individuals of this ring (i.e., household members) will also be vaccinated unless contraindicated, creating a second ring of protection around each exposed individual in order to prevent further spread of the disease should that individual later become symptomatic. Additional health care and response personnel deemed to be at high risk for exposure to the virus may receive the vaccine at this time as well.

The key to Phase Four is preparedness. In the event of a documented smallpox outbreak, members of the general public are likely to panic. Individuals will flood hospitals and emergency rooms demanding vaccination, and in doing so will essentially increase their risk of exposure. Misinformation will likely be widespread, as was the case with the early "anthrax letter" reports. The health care community must be prepared to deal with these and any number of other unforeseeable problems, while still mobilizing resources to initiate and sustain the necessary outbreak control measures.

Phase Five

In the unlikely event of an outbreak, historical data suggest that the virus will not spread uncontrollably and ravage the U.S. (or any other) population. That being said, given the fact that the world population is immunologically naive and conceding the point that other variables may favor such spread (e.g., if the virus were simultaneously released in multiple locations nationwide), Phase Five, the "Supplemental Strategy" has been developed as a contingency plan to prevent such an eventuality. In addition to continued contact identification and vaccination, the Supplemental Strategy involves "broader" vaccination of the U.S. population in order to increase herd immunity. Presumably, this refers to nationwide mass vaccination.

The Supplemental Strategy is only to be initiated under one of three circumstances: (1) the size of the first wave of smallpox cases is too great to be effectively managed by contact identification and vaccination alone; (2) no decline is seen in the number of new cases after two or more generations from the index case(s); or (3) no decline is seen in the number of new cases after approximately 30 percent of current vaccine stores has been utilized.

CONCLUSION

Smallpox is a devastating disease, and any unvaccinated population is in fact vulnerable to it. But on the basis of information in the public domain, it is not known with 100 percent confidence that the virus even exists outside the CDC and Vector. Acquisition is the principal hurdle to attaining a biological weapons capability with the virus. Production is difficult but not impossible, as is delivery. Ultimately, strategic vaccination can—and in a limited sense has already—both reduce the vulnerability of a population to the virus and minimize the potential impact of its release.

Acknowledgment

This entry draws extensively from Pilch, R., "Smallpox: Threat, Vaccine, and US Policy," Six-Part Series, Center for Nonproliferation Studies web report, available at www.cns.miis.edu (2003). It also draws in part from Pilch, R., "The Bioterrorist Threat in the United States," in Howard, R.D. and Sawyer, R.L., eds., *Terrorism and Counterterrorism*, Revised ed. (New York: McGraw-Hill, 2003). The author is indebted to the reviewers of *Biosecurity and Bioterrorism*, for which a portion of this manuscript was originally intended. The author would also like to thank Dr. Raymond A. Zilinskas for his extensive editing and review of this entry.

REFERENCES

ABCNews.com, Optional Inoculation: Bush Says Smallpox Vaccine Program Will Be Voluntary, December 11, 2002, www.abcnews.go.com/sections/2020/DailyNews/bush_walters 021211.html.

Alibek, K. and Handelman, S., *Biohazard: The Chilling True Story of the Largest Covert Biological Weapons Program in the World—Told From Inside by the Man Who Ran It*, Random House Inc, New York,, 1999.

Breslow, L., *Annu. Rev. Public Health*, **11**, 1–28 (1990).

CDC (Undated a), *People Who Should NOT Get the Smallpox Vaccine (Unless they are Exposed to the Smallpox Virus)*, www.bt.cdc.gov/agent/smallpox/vaccination/contraindications-public.asp.

CDC (Undated b), *Protecting Americans: Smallpox Vaccination Program*, www.bt.cdc.gov/agent/smallpox/vaccination/vaccination-program-statement.asp.

CDC (Undated c), *Smallpox Response Plan and Guidelines (Version 3.0)*, http://www.bt.cdc.gov/agent/smallpox/response-plan/index.asp.

CDC (Undated d), *Smallpox Vaccine and Heart Problems*, http://www.bt.cdc.gov/agent/smallpox/vaccination/heart problems.asp.

Davies, P., *Catching Cold: 1918's Forgotten Tragedy and the Scientific Hunt for the Virus That Caused It*, Michael Joseph, Canada, 1999.

Dixon, C., *Smallpox*, J&A Churchill Ltd., London, 1962, p. 1460, as cited in Henderson et al., Smallpox as a Biological Weapon, *J. Am. Med. Assoc.*, 1999.

Downie, A. and McCarthy, K., *J. Hyg.*, **56**, 479–487 (1958), as cited in Henderson, D. et al., *J. Am. Med. Assoc.*, **281**, 2127–2137 (1999).

Falkner, F., Highly attenuated vaccinia strains as safe third generation smallpox vaccines, presented at the G7+ Global Health Security Action Group Workshop, *Best Practices in Vaccine Production for Smallpox and other Potential Pathogens*, Paul-Ehrlich-Institut, Langen (Germany), September 5–6, 2002.

FBIS CEP 20031030000047, *Funds Shortage Forces Russian Virology Center to Cut Working Day by Half*, Moscow Radio Rossii, October 30, 2003 (in Russian).

Galeotti, M., *Jane's Intell. Rev.*, **12**(3), 8–9 (2000a), as cited in Zilinskas, R. and Carus, W., *Possible Terrorist Use of Modern Biotechnology Techniques*, Chemical and Biological Defense Information Analysis Center, p. 44 (2002) (For Official Use Only).

Galeotti, M., *Jane's Intell. Rev.*, **12**(3), 10–15 (2000b), as cited in Zilinskas, R. and Carus, W., *Possible Terrorist Use of Modern Biotechnology Techniques*, 44.

Goldstein, V., et al., *Pediatrics*, **55**, 342–347 (1975) as cited in Henderson et al., Smallpox as a Biological Weapon.

Henderson, D. et al., *J. Am. Med. Assoc.*, **281**, 2127–2137 (1999).

Knyazkov, M., Fever as a weapon: Concerning the disappearance of a highly dangerous virus from a Pentagon laboratory (November 2) (in Russian), *Sovetskaya Rossiya* (Moscow), 1986, p. 5.

Lane, J., et al., *New Engl. J. Med.*, **281**, 1201–1208 (1969), as cited in Henderson et al., "Smallpox as a Biological Weapon."

Lane, J. et al., *J. Infect. Dis.*, **122**, 303–309 (1970), as cited in Vaccinia (Smallpox) Vaccine Recommendations of the Advisory Committee on Immunization Practices (ACIP), 2001, www.cdc.gov/mmwr/preview/mmwrhtml/rr5010a1.htm.

McIntosh, K. et al., *J. Infect. Dis.*, **161**, 445–448 (1990), as cited in Henderson et al., Smallpox as a Biological Weapon.

Midthun, K., Regulatory requirements for historical and new smallpox vaccines, presented at the G7+ Global Health Security Action Group Workshop *Best Practices in Vaccine Production for Smallpox and other Potential Pathogens*, September 5–6, 2002, Paul-Ehrlich-Institut, Langen (Germany).

Monath, T., Smallpox vaccine: US manufacturers, presented at the G7+ Global Health Security Action Group Workshop *Best Practices in Vaccine Production for Smallpox and other Potential Pathogens*, September 5–6, 2002, Paul-Ehrlich-Institut, Langen (Germany), See also, http://www.hhs.gov/news/press/2001pres/20011128.html.

Roos, R., *Trials of Aventis Smallpox Vaccine Getting Underway*, Center for Infectious Disease Research & Policy, June 2002, www.umn.edu/cidrap/content/bt/smallpox/news/avent.html.

Russell, P., Smallpox vaccine stockpile for the United States, presented at the G7+ Global Health Security Action Group Workshop Best Practices in Vaccine Production for Smallpox and other Potential Pathogens, September 5–6, 2002, Paul-Ehrlich-Institut, Langen (Germany).

Stearn, E. and Stearn, A., *The Effect of Smallpox on the Destiny of the Amerindian*, Bruce Humphries, Boston, 1954, as cited in Henderson et al., Smallpox as a Biological Weapon.

Stevenson, R. and Altman, L., Smallpox Shots Will Start Soon Under Bush Plan, *New York Times*, December 12, 2002.

Stevenson, R. and Stolberg, S., Bush Lays Out Plan on Smallpox Shots as 'a Precaution', *New York Times*, December 13, 2002.

Tucker, J., *Scourge: The Once and Future Threat of Smallpox*, Atlantic Monthly Press, New York, 2001.

Tucker, J. and Vogel, K., Preventing the Proliferation of Chemical and Biological Weapons Materials and Know-How, *Nonproliferation Review*, 91 (Spring 2000b).

U.S. Army Medical Research Institute of Infectious Diseases (USAMRIID), *Smallpox: Recognition and Response*, satellite broadcast, November 6, 2002.

Wehrle, P. et al., *Bull. World Health Organ.*, **43**, 669–679 (1970), as cited in Henderson et al., Smallpox as a Biological Weapon.

Wyeth Smallpox Vaccine [package insert], Wyeth Laboratories Inc., Lancaster, PA, 1968, as cited in Henderson et al., Smallpox as a Biological Weapon.

FURTHER READING

For a fascinating and frankly alarming survey illustrating the need for public education about smallpox, see Blendon, R. et al., *N. Engl. J. Med.*, **348**, 5 (2003).

WEB RESOURCES

Acambis, www.acambis.com.

Advisory Committee on Immunization Practices (ACIP) Smallpox Vaccine Meeting Briefing, October 17, 2002, http://www.cdc.gov/od/oc/media/transcripts/t021017.htm.

Defense Threat Reduction Agency's (DTRA) Biological Weapons Proliferation Prevention projects (Security Enhancements, Dismantlement, and Cooperative Biodefense Research), http://www.dtra.mil/ctr/ctr_russia.html, accessed on 9/30/03.

U.S. National Institute of Health, http://www.clinicaltrials.gov/.

STAPHYLOCOCCAL ENTEROTOXINS

Mark A. Poli

Opinions, interpretations, conclusions, and recommendations are those of the author and are not necessarily endorsed by the U.S. Army.

INTRODUCTION

Many cases of human foodborne illness throughout the world are caused by enterotoxins produced by various strains of *Staphylococcus aureus*. Along with the pyrogenic exotoxins of group A streptococci, the staphylococcal enterotoxins (SEs) comprise a family of at least 15 structurally related toxins (denoted SEA through SEP, plus TSST-1) that share the ability to stimulate T cells by a common mechanism.

While oral exposure to these toxins commonly results in the traditional food-poisoning symptomatology, nonenteric exposure can cause much more serious, even life-threatening sequelae. From the standpoint of biological warfare (BW) and biological terrorism, staphylococcal enterotoxin B (SEB) is considered the serotype of greatest import. It is a versatile agent, as it can be an effective incapacitating agent when inhaled in very low doses. At slightly higher doses, it can be lethal.

HISTORY OF BW APPLICATION

During the 1960s, SEB was extensively studied as a biological incapacitant (Ulrich et al., 1997). Testing and weaponization by several countries, including the United States, soon followed. These weapons were never used, as the U.S. policy on the use of biological weapons at that time was one of retaliation only (U.S. Dept of Army, 1977). Between 1954 and 1967, SEB was produced at a production plant in Pine Bluff, Arkansas. At this facility, munitions were filled and stored as a deterrent capability. The United States unilaterally renounced all use of biological weapons in 1969. All remaining U.S. military stocks of munitions were subsequently destroyed in accordance with approved demilitarization plans (U.S. Dept of Army, 1977).

The development, production, and storage of toxins, including SEB, for offensive purposes are prohibited by the 1972 Biological and Toxin Weapons Convention (BWC). The international control over the illicit use of toxins was strengthened in 1993, when the Chemical Weapons Convention (CWC) came into existence. While only two toxins are explicitly named in the CWC (ricin and saxitoxin), all toxins come under the authority of this treaty. In the United States, SEB is included in the Centers for Disease Control and Prevention's (CDC) select agent list of toxins requiring certification for possession and transfer.

MECHANISM OF ACTION

Through their ability to simultaneously bind the β chain variable region on the T-cell receptor and the class II major histocompatibility complex (MHC) of antigen-presenting cells (APCs), SEs stimulate the proliferation of specific T cells, including the CD4, CD8, and $\gamma\delta^+$ subsets. Because they interact with the MHC on APCs differently than conventional antigens, causing an abnormally large percentage (up to 20 to 30 percent) of T cells to be activated, they are denoted "superantigens." Activation of T cells by superantigens results in a massive production of cytokines and chemokines, enhanced expression of cellular adhesion molecules, T-cell energy, and apoptosis. Release of cytokines and chemokines results in the fever, gastroenteritis, and toxic shock characteristic of exposure (Krakauer, 1999).

Primates are most sensitive to SEB and related superantigens because their MHC class II receptors have a greater binding affinity than those of other animals. However, there is variability in affinities toward different MHC class II isotypes. This may contribute to individual variability in susceptibility to these toxins.

TOXICITY AND HUMAN EXPOSURE

Toxicity in animals is highly species-specific. Mice are relatively insensitive to SEs, while primates are very sensitive. Even among primates, humans are exquisitely sensitive. By the aerosol route, the human EC_{50} (median effective concentration) of SEB for incapacitation is approximately 0.4 ng/kg. On the basis of extrapolation from primate studies, the human LD_{50} (median lethal dose) has been estimated to be around 20 ng/kg (Ulrich et al., 1997). These numbers are consistent with information derived from accidental human exposures.

Most human exposure to SEB occurs as a result of eating improperly handled food. When the preformed toxin is ingested with food contaminated with *S. aureus*, the result is a common form of food poisoning that presents clinically as emesis with or without diarrhea. This form, known as staphylococcal food poisoning, is typically self-limiting, and signs of systemic intoxication such as fever, hypotension, and shock are absent except in severe cases. It is likely the most common form of food poisoning in the United States, although the true incidence is unknown (Dinges et al., 2000).

Toxic shock syndrome (TSS) is an acute and potentially fatal illness characterized by high fever, rash, hypotension, desquamation, and the involvement of multiple organ systems. In the early 1980s, an epidemic of TSS occurred in the United States among young, menstruating women. This epidemic was eventually traced to the use of highly absorbent tampons and cervical or vaginal colonization by *S. aureus*. Toxic shock syndrome toxin (TSST−1) was identified as the causative toxin, presumably due to its ability, unique among staphylococcal superantigens, to efficiently cross mucosal membranes. TSS cases occasionally occur that are unrelated to tampon use. These cases often involve SEB and, to a lesser extent, SEC.

In addition to these well-studied illnesses, SEs have been associated with several other human diseases, including sudden infant death syndrome (SIDS) and Kawasaki syndrome, the leading cause of acquired heart disease among children in the United States. SEs are also postulated to induce or exacerbate various autoimmune diseases (Krakauer, 1999).

Aerosol exposure to SEB is not a natural route of intoxication but is nevertheless the model for most BW scenarios. Animal studies suggest involvement of some unique mechanisms by this route of exposure. Respiratory involvement is rapid, and the profound hypotension characteristic of TSS is not observed. Fever, which is not observed after SEB ingestion, is a prominent feature after aerosol exposure.

A nonfatal accidental laboratory inhalation exposure (Ulrich et al., 1997) provides a good description of the clinical course of this disease. Fever and muscle aches, beginning 8 to 20 h after exposure, were the first signs of intoxication. Respiratory symptoms and signs included chest pain, nonproductive cough, inspiratory and expiratory rales, and difficulty in breathing. Nausea, vomiting, and anorexia occurred with mild to severe headaches. Patients were normotensive and had normal electrocardiograms. Duration of illness was 3 to 4 days.

DETECTION

A variety of immunological assays have been reported for detecting SEA and SEB (reviewed in Poli et al., 2002). Most have detection limits of 1 ng/ml or less, and some can detect at 0.1 ng/ml. However, most of these were developed primarily to detect toxins in culture medium or food extracts rather than human clinical samples.

It is well documented that most human serum contain antibodies to SEs, presumably from environmental exposures. These antibodies make it extremely difficult to test for human exposure by analyzing serum samples. However, significant amounts of SEs are eliminated in the urine, making this the preferred matrix for evaluating human exposure. Colorimetric ELISAs (Enzyme-linked Immunosorbent Assays) have been reported to evaluate for SEA and SEB in human urine samples (Poli et al., 2002).

For at least 12 to 24 h after inhalation exposure, toxins should be identifiable in nasal swabs. This is probably the best approach to early patient diagnosis after an aerosol attack.

TREATMENT AND PROPHYLAXIS

Treatment of SEB intoxication is symptomatic and supportive. No specific therapy has been identified for managing the respiratory effects. Animal models suggest a role for passive immunotherapy, although no antibody preparations are currently available for human use.

No vaccine for SEB is currently available. A formalin-treated toxoid has shown some success in animal models. A second-generation recombinant vaccine is in advanced stages of development by the U.S. military. In this product, amino acid substitutions are used to eliminate toxicity with minimal effect upon structural determinants, which results in a much higher neutralizing antibody response.

REFERENCES

Dinges, M.M., Orwin, P.M., and Schlievert, P.M., *Clin. Microbiol. Rev.*, **13**, 16−34 (2000).

Krakauer, T., *Immunol. Res.*, **20**, 163−173 (1999).

Krakauer, T. and Stiles, B.G., *Rec Dev. Infect. Immunol.* **1**: 1−27 (2003).

Poli, M.A., Rivera, V., and Neal, D., *Toxicon*, **40**, 1723−1726 (2002).

Ulrich, R.G., Sidell, S., Taylor, T.J., Wilhelmsen, C.L., and Franz, D.R., "Staphylococcal enterotoxins B and related pyrogenic toxins" in R. Zajtchuk, Ed., *Medical Aspects of Chemical and Biological Warfare. Part I: Warfare, Weaponry, and the Casualty*, Office of the Surgeon General, United States Army, Falls Church, Virginia, 1997, pp. 621−630.

U.S. Department of the Army, *US Army Activity in the US Biological Warfare Programs* (Volumes I and II), February 24, 1977, reprinted in *Biological Testing Involving Human Subjects by the Department of Defense, 1977*, Hearings before the Subcommittee on Health and Scientific Research of the Committee on Human Resources, March 8 and May 23, 1977, United States Senate, 95th Congress, First Session (US GPO, 1977), p. 107.

See also TOXINS: OVERVIEW AND GENERAL PRINCIPLES.

SUDAN, REPUBLIC OF

GAIL H. NELSON

INTRODUCTION

With an area of 967,500 square miles, Sudan offers exceptional terrain for insurgent groups in transit and concealment of international terrorist organization in hiding (Fig. 1). The population, which is predominantly Islamic in the north and central regions and Christian in the south, is estimated at almost 30 million. Literacy is 30 percent, and average life expectancy is 52 years. The government has remained an Islamic dictatorship since the 1989 coup, after which all political parties were banned. Its army strength is 75,000, backed by a small air force and navy. Sudan receives most of its equipment from China, Iraq, Libya, and Russia. Sudan is a member of the United Nations (UN), Organization of African Unity (OAU), and the Arab League. Solidarity with other Arab countries is a main feature of Sudan's foreign policy. It sought to steer a nonaligned course during the 1990s, courting Western aid while maintaining cooperative ties with Libya, Syria, North Korea, Iran, and Iraq. However, Sudan's support for regional insurgencies and extremist groups generated international concern.

Sudan broke diplomatic relations with the United States in 1967, following the outbreak of the Arab–Israeli War. Relations improved in 1971, after the Sudanese Communist Party attempted to overthrow the government. However, the March 1, 1973 assassination of the U.S. Ambassador by Palestinian terrorists (Black September), followed by the release of the suspected murderers, led to a decline inrelations. Official U.S. assistance was suspended in 1989, in the wake of the military coup against the elected government.

Sudan's Islamist links with international terrorism led to its designation as a state sponsor of terrorism in 1993. The United States imposed economic sanctions against Sudan in 1997, and on August 20, 1998, launched retaliatory cruise missile strikes against the Shifa Pharmaceutical Complex in Khartoum, which intelligence officials believed to be a possible site for the production of biological-chemical warfare agents. Today, Sudan remains one of the seven state sponsors of terrorism designated by the U.S. State Department. Of those seven, it is one of the two states that is not currently suspected of possessing an active offensive biological warfare (BW) program (the other being Cuba).

BIOLOGICAL WEAPONS HISTORY

Prior to the strike on al-Shifa, the U.S. Central Intelligence Agency (CIA) asserted that Sudan was actively pursuing offensive biological and chemical weapons. The CIA claimed that it had solid evidence of chemical weapons–related activity at the Complex, noting that soil samples had revealed the presence of O-ethylmethylphosphonothioic acid (EMPTA), a key precursor for the deadly nerve agent VX. The CIA further linked the Shifa Pharmaceutical Complex with Usama Bin Ladin and efforts by terrorist organizations to acquire weapons of mass destruction (WMD). The CIA's analysis served as the basis for the 20 August strike. Subsequent litigation and U.S. indemnity to Sudan suggests that the CIA assessment of the Shifa Complex was an error.

CURRENT BIOLOGICAL WEAPONS STATUS

The United States claims that Sudan, a signatory to the Chemical Weapons Convention (CWC) of 29 April 1997 (acceded May 24, 1999), has been building a capability to produce biological and chemical weapons for a number of years (Sudan has been linked with Iraq in this effort). Sudan has not acceded to the Biological and Toxin Weapons Convention (BWC) as of July 2002. The United States is concerned that Sudan is seeking advanced conventional weapons (ACW) and, particularly, ballistic missiles for biological and chemical weapons delivery.

HISTORY OF TERRORIST SUPPORT

Sudan is considered a rogue state by much of the world community because of its support for international terrorism. The United States claimed in 2000 that Sudan served as a central hub for several terrorist groups, including Usama Bin Ladin's al-Qa'ida organization. Moreover, the Sudanese government condoned Iran's assistance to terrorist groups operating in and transiting through Sudan. The United States asserts that Khartoum served as a meeting place, safe haven, and training hub for members of the following terrorist organizations in addition to al-Qa'ida:

Encyclopedia of Bioterrorism Defense, Edited by Richard F. Pilch and Raymond A. Zilinskas
ISBN 0-471-46717-0 Copyright © 2005 Wiley-Liss

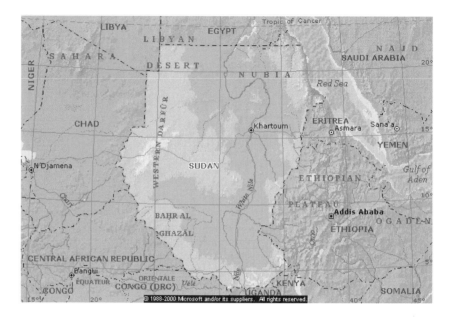

Figure 1. Sudan.

- Lebanese Hizballah (Party of God), a radical Shia group formed in Lebanon and dedicated to the creation of an Iranian-style Islamic republic in Lebanon. It is strongly anti-Western and anti-Israeli.

- Palestinian Islamic Jihad (PIJ), formed among the militant Palestinians in the Gaza Strip during the 1970s. This group is committed to the creation of an Islamic Palestinian state and the destruction of Israel through holy war. The PIJ regards the United States as an enemy because of its support for Israel.

- Abu Nidal Organization (ANO), which split from the Palestinian Liberation Organization (PLO) in 1974. It has carried out terrorist attacks in 20 countries and killed or injured well over 900 people since the 1970s.

- Islamic Resistance Movement (HAMAS), formed in 1987 with the goal of establishing an Islamic Palestinian state in place of Israel.

Sudan's support to these groups included the provision of travel documentation, safe passage, and refuge. Most of the groups maintain offices in Khartoum, using Sudan as a secure base for organizing terrorist operations and assisting compatriots elsewhere. During the early 1990s, Carlos the Jackal, Usama Bin Ladin, Abu Nidal, and other terrorist leaders resided in Khartoum.

CURRENT TERRORIST RELATIONS

The U.S. relations with Sudan focus on counterterrorism, regional stability, conflict resolution, protection of human rights, and humanitarian relief. Since September 11, 2001, Sudan has entered into international and bilateral talks on counterterrorism, providing concrete cooperation that has included unrestricted access to files of suspected terrorists, and has suggested that it would be willing to extradite fugitives to proper authorities. However, Sudan remains ambivalent on international military interventions in the Middle East, including U.S. military strikes in Afghanistan and Iraq.

IMPLICATIONS

There is no firm evidence that the Sudanese government has ever been engaged in the production of chemical or biological weapons. However, Sudan remains a region vulnerable to extremist-terrorist activities, and thus a region requiring close monitoring for potential WMD activity by extremist groups and front enterprises backing terrorist organizations.

REFERENCES

Berger, S., *Special Briefing on Anti-Terrorist Attacks*, USIS Washington File, Washington, DC, November 21, 1998; *Africa South of the Sahara 1998*, Europa Publications, United Kingdom, 1998; *SIPRI Yearbook 1998*, Oxford University Press, Oxford, United Kingdom, 1998; *SIPRI Yearbook 1999*, Oxford University Press, Oxford, United Kingdom, 1999; *SIPRI Yearbook 2000*, Oxford University Press, Oxford, United Kingdom, 2000.

Brown, S. and Congressional Research Service WEB, *Chemical Weapons Convention: Issues Before Congress*, Washington, DC, January 7, 2003; Washington File, Central Intelligence Agency, *Acquisition of Technology Relating to Weapons of Mass Destruction and Advanced Conventional Munitions*, Washington, D.C., January 8, 2003; *Background Note: Sudan*, U.S. Department of State, Washington, D.C., 2003.

Dagne, T., *Sudan: Humanitarian Crisis, Peace Talks, Terrorism, and U.S. Policy*, Congress Issue Brief, Washington, D.C., January 29, 2002; May 8, 2002.

Gannon, J., *Biological-Chemical Weapons Threat*, USIS Washington File, Washington, D.C., November 20, 1998.

Morse, J., *Sudan Home to Lethal Terrorist Groups*, USIS Washington File, Washington, D.C., August 20, 1998; U.S. State

Department, *Fact Sheet: U.S. Strike on Facilities in Afghanistan and Sudan*, USIS Washington File, Washington, D.C., August 24, 1998.

Pickering, J., *Transcript on U.S. Sudan, Afghanistan Strikes*, USIS Washington File, Washington, D.C., August 26, 1998.

Porth, J., *U.S. Chemical Weapons-Related Soil Sample from Sudan Plant*, USIS Washington File, Washington, D.C., August 25, 1998.

Turner, B., *The Statesman's Yearbook*, MacMillan, New York, 2000; U.S. Department of State, *Central Intelligence Agency Report to Congress on the Acquisition of Technology Relating to Weapons of Mass Destruction and Advanced Conventional Munitions, Washington, D.C., (1 July–31 December 2001)*.

See also HAMAS; Hizballah; Palestinian Islamic Jihad; and Al-Qa'ida.

SUICIDE TERRORISM

Adam Dolnik

IDEOLOGICAL SUICIDE IN HISTORY

Despite the heightened focus on suicide terrorism in the last two decades, ideologically and religiously motivated suicide is a much older phenomenon. Perhaps the first recorded suicide operation is the biblical story of Samson, who tore down the pillars of the Temple of Dagon, killing himself along with several thousand Philistine men and women. Throughout history, acts of self-killing have been used as a methodical demonstration of commitment, an effective military tactic, and a persuasive form of protest. The ancient Japanese warrior code *Bushido*, for example, prescribed ritual suicide (*seppuku*) as an honorable way of death for a samurai whose master died or who faced imminent defeat in battle. This tradition was revisited in World War II by the institution of *kamikaze*, in which over 4600 men crashed airplanes into enemy targets (Laqueur, 1990). A different form of suicidal military tactic was adopted during the Iran–Iraq war (1980–1988), in which the Iranians sent young volunteers across minefields to clear the way for their troops (Jaber, 1997). This practice was based on the same religious doctrine of martyrdom that constituted the basis for suicide bombings conducted by the terrorist organization Hizb'allah (Party of God) in the 1980s.

Even more extensive is the use of suicide as a form of protest. In 1936, for example, Stefan Lux shot himself on the floor of the League of Nations to protest Britain's inaction against Nazi Germany. During the Vietnam War, several Buddhist monks set themselves on fire to protest against the regime in South Vietnam (Arad, 2000). This practice was imitated in 1968 in Czechoslovakia, where two students, Jan Palach and Jan Zajic, set themselves on fire to protest the invasion of their country by Warsaw Pact troops. Another form of protest suicide was conducted by 11 members of the Irish Republican Army (IRA), who starved themselves to death in 1981 while incarcerated in prison. This method was imitated on numerous occasions thereafter, mainly in Turkey, Germany, and Spain.

Suicide has also been used as a means of preventing the interrogation of the perpetrators of attacks. For example, several members of the Serbian nationalist movement

Mlada Bosna, which assassinated the Austrian Archduke Franz Ferdinand in Sarajevo in 1914, attempted to swallow cyanide in an attempt to prevent their arrest. The same tactic was adopted by the British-trained Czech commando unit that in 1942 assassinated SS officer Reinhard Heydrich, the Nazi governor of occupied Bohemia and Moravia. More recently, the two North Korean agents responsible for the 1987 bombing of Korean Air Flight 858 swallowed cyanide, though only one died.

This list is far from complete. While these incidents do not fit the classical definition of suicide terrorism, they clearly illustrate the scope of the different motivations and justifications discussed later in this entry. Further, it is crucial to acknowledge that today's practice of suicide terrorism combines all of the above-stated motives for suicide: a demonstration of commitment, a pragmatic military tactic, the prevention of interrogation, and a form of protest. A final implication of the foregoing examples is that contrary to popular belief, an act of ideologically motivated suicide is by no means a unique fingerprint of Islamic fundamentalism.

DEFINITIONS

In order to discuss suicide terrorism, it is extremely important to explain what exactly is meant by this term, because there are several inherent difficulties in defining the phenomenon. The first such difficulty concerns the difference between the *readiness* to die and the *desire* to die. For this reason, it is useful to define a suicide operation on the basis of the perpetrator's death as a precondition of a successful attack (Ganor, 2000). In other words, if the perpetrator does not die, the attack cannot take place. From such a definition, high-risk operations in which the perpetrator has even the smallest chance of survival are excluded. Operations such as Abu Nidal Organization's 1985 machine gun and grenade attacks in Vienna and Rome airports that killed 16 and injured 60, or the Japanese Red Army's 1972 massacre of 26 people at Israel's Lod airport, do not qualify as suicide operations despite the fact that no escape routes were planned and the chances of survival were next to none.

Encyclopedia of Bioterrorism Defense,   Edited by Richard F. Pilch and Raymond A. Zilinskas
ISBN 0-471-46717-0   Copyright © 2005 Wiley-Liss

Another difficulty involves the challenge of distinguishing between cases in which a terrorist dies willingly and cases in which he or she is unaware that his or her death is an integral part of the plan. There have been several cases where the terrorists were tricked into believing that they were only transporting the explosives, which were then remotely detonated while in their possession by the organizers of the attack (Merari, 1998). Even though such cases should not be considered, the difficulty of determining the intentions of suicide bombers after their deaths inevitably results in the misclassification of many such attacks as suicide bombings.

For the purpose of this chapter, a suicide terror attack is defined as a *premeditated act of ideologically or religiously motivated violence, in which the success of the operation is contingent on self-inflicted death by the perpetrator(s) during the attack* (Dolnik, 2003).

WHY TERRORIST ORGANIZATIONS USE SUICIDE BOMBINGS

Various groups have used suicide bombings for various reasons. There are, however, several common characteristics that can be identified. The primary benefit of a suicide operation lies in its tactical advantage over other forms of attack. A suicide bomber has the ability to deliver the payload to places that would be difficult to attack successfully by someone who is trying to stay alive. The fact that the bomber also has the ability to select the exact location, timing, and circumstances of the attack results in the remarkable effectiveness of suicide attacks in terms of delivering a high number of casualties (Sprinzak, 2000). Suicide attacks eliminate the need to plan an escape route, effectively remove the danger of capture and subsequent interrogation, and are highly cost-effective. According to an Al-Aqsa Martyrs Brigades invoice found by Israeli troops at the Palestinian Authority's headquarters during "Operation Defensive Shield" in 2002, the "electrical components and chemical supplies needed to produce a suicide bomb" cost an estimated $150 (Shahar, 2002). The relatively low expenditures involved in the acquisition of explosives make the costs-per-casualty ratio of a suicide operation for a terrorist organization rather favorable. In addition, suicide bombings are extremely difficult to defend against, making the elimination of their indiscriminate use virtually impossible without wide suspensions of civil liberties, which in turn have a secondary effect on the target population. And finally, the universality of the suicide bomber's possible target generates a widespread feeling of vulnerability among the public.

Even if the local population does eventually become used to the idea of being a target, a well-organized terror campaign can adversely affect a country's attractiveness as a tourist destination, resulting in economic strain on the local population. Another intangible benefit of the suicide tactic is the extensive media attention that follows in its wake, which is invaluable for the publicity of the perpetrators' cause. Further, the incomprehensible nature of the suicide bomber's dedication can further serve to legitimize the perpetrator's cause in the eyes of some elements

of the international community—the representative organization can acquire the status of a resolute actor with grievances that cannot be ignored, and for whom death is preferable to life under present conditions.

As is the case with any terrorist attack, the message conveyed by suicide operations is also directed inward. Organizations such as the Kurdistan Workers Party (PKK), the Chechens, or the Liberation Tigers of Tamil Eelam (LTTE), for example, have adopted suicide bombings with the principal goal of solidifying group morale (Schweitzer, 2000). An act of self-sacrifice in the name of the organization's cause creates martyrs, who become a uniting factor within the group. Glorification of the martyr's accomplishment by the organization's prominent members can also increase group members' sense of prestige and serve to inspire future volunteers. If the suicide bomber is seen as a hero and a role model, imitation of the "heroic" accomplishments of the martyr is almost sure to follow. The willingness to die for a cause may also be touted as evidence of the superiority of the group's members over their adversaries, who are portrayed as inherently weak pleasure seekers despite their military dominance (Paz, 2002a). The resulting perception among the terrorist group is that because of superior determination final victory is inevitable. The terror campaign then becomes not a question of winning or losing, but rather of persisting.

The tactic of suicide bombings can, of course, have negative effects on the goals of an organization as well. The lack of understanding of "martyrdom operations" in Western culture can cause the group to be viewed as one composed of irrational fanatics, which tends to shift public opinion into the camp of the adversary. Furthermore, a prolonged suicide terror campaign against civilians is likely to radicalize the target population and make it less willing to compromise. On the other hand, an emotional military overreaction by the attacked government can suit the terrorists' purposes by providing support for their claims of victimization. Regardless of who started the circle of violence, as long as innocents die from the government's counterterrorism efforts, the terrorists have a hope of attracting more popular support against the government.

WHAT MOTIVATES A SUICIDE BOMBER

As in the case of terrorist organizations, the motivations of individual suicide bombers vary considerably. Despite the popular focus on religious justifications used by Islamic fundamentalists, it is important to emphasize that most suicide operations have been carried out by secular rather than religious groups. Furthermore, even though all *religiously motivated* suicide bombings to date have been conducted under the banner of Islam, it is important to note that other religious traditions have also been tapped in an attempt to instigate suicide operations. For instance, Meir Kahane, a fundamentalist Jewish rabbi from Brooklyn, has advocated the use of suicide operations for the sake of victory of Israel: "a man who volunteers for such operations will be called a hero and a martyr" (Hoffman, 1998). It could in fact be argued that almost any religion might be used to justify suicide tactics, because the images of *sacrifice* and *martyrdom* are focal points of all major religious traditions (Juergensmeyer, 2000). A fundamentalist interpretation of these two key spiritual elements during times of uncertainty and perceived external danger can be used to motivate radical believers to die voluntarily for their cause regardless of their religious affiliation. For example, during his 1994 attack on the Tomb of the Patriarchs in Hebron, which killed more than 29 people, Israeli terrorist Baruch Goldstein changed his rifle magazine clip four times and fired 111 bullets before being beaten to death with a fire extinguisher. Even though this attack does not fit the above definition of a suicide operation, it does demonstrate a comparable determination to die for a religious cause.

Differences in the interpretation of suicide bombings are evident based on the very term used to describe this phenomenon among different cultures. While Western cultures apply the term "suicide operations," Islamists refer to the phenomenon as *Istishhad*, or martyrdom and self-sacrifice in the name of Allah, which is significantly different form *Intihar*, or suicide resulting from personal distress (Kushner, 1994). Suicide (*Intihar*) is forbidden by the Qur'an so explicitly that it leaves no room for an alternative interpretation (Qur'an 4:29), which may explain why Islamists do not interpret a suicide bombing as an act of self-killing but rather as an act of killing others while dying in the process. In other words, even though the death of a *shahid* (martyr) is inevitable during his mission, it is not the real goal of the operation and therefore the act is not an act of suicide, at least not the kind of hopeless and impatient suicide that the Qur'an forbids. The Islamic fundamentalist interpretation not only makes the act excusable, it makes it attractive: a system of rewards is in place to encourage martyrs by promising the *shahid* an eternal life in paradise, which is described as a place with rivers of milk and wine, the sexual services of 72 young virgins, and the privilege to "reserve place" in heaven for 70 relatives.

However, even religious *shahidin* are motivated strongly by factors other than religion. In fact, many of the groups that are commonly referred to as "Islamist" do not base their primary rationalization for using violence on religion but rather on the concept of "defensive war." In other words, the struggle is typically seen by terrorists as self-defense, and terrorism is a tool they resort to only because they lack sufficient conventional military capability. Religion provides a useful mobilization tool that is used to overcome differences between different terrorist factions, but the principal justification of violence is more often based on a cultural and historical perception of victimization. In communal cultures, the value placed on individual lives is low, and the sacrifice of an individual for the benefit of the group is a cultural norm. In particular, individuals who have been personally damaged by the enemy and seek revenge, or individuals with nothing to lose in life, are among the most likely candidates for suicide operations.

As indicated above, no single answer to the question concerning motivation of suicide bombers can be given. Individual suicide bombers range from volunteers who participate actively in the planning or the execution of the operation to passive bomb carriers influenced by blackmail, deception, desperation, financial reward, or fanaticism. The often-quoted universal profile of a *shahid* as a young, unmarried, uneducated, unemployed, and heavily indoctrinated male from a refugee camp is largely misleading, as it fails to incorporate information concerning suicide bombers from countries other than Israel. The only true international common denominator seems to be a sense of sacrifice for the "greater good" on part of the bomber. The suicide bomber simply does his or her part to damage the hated enemy in a belief that this act will contribute to the eventual victory of his compatriots. Aside from this common factor, different motivations for suicide bombers can be identified depending on group structure, cultural context, and the nature of the struggle.

SUICIDE BOMBINGS: THE HISTORICAL RECORD

Although the September 11 attacks in the United States and suicide bombings in Israel were and continue to be portrayed by Western media as extraordinary events, suicide operations are much more frequent than is commonly believed. Since the 1980s, at least 25 different terrorist organizations in 26 different countries have resorted to the use of suicide bombings. To date, over 500 suicide bombings have occurred worldwide.

The first modern suicide bombing occurred in 1951 in Indochina, when a young Communist volunteer assassinated a French Brigadier General and the local Governor by exploding a grenade in his pocket (Mickolus, 1980). Subsequently, suicide operations were not recorded for some 30 years, with the exception of several suicide bombings on airliners in the United States in the late 1950s and early 1960s, which were motivated by insurance fraud as opposed to ideological goals.

The systematic modern practice of suicide bombings has its roots in Lebanon. The first major terrorist suicide attack occurred in December 1981, when 61 people died and more than a hundred were wounded in the bombing of the Iraqi embassy in Beirut. The first suicide attack against a Western target occurred on April 18, 1983, when a Hizb'allah bomber drove a van loaded with 2000 pounds of explosives into the American Embassy in Beirut, killing 63 people (Shay, 2000). On October 23, 1983, the same group carried out one of the most devastating events in the history of terrorism, the synchronized suicide bombing of the U.S. Marine camp and the French military headquarters in Beirut, in which 241 U.S. and 58 French troops died. Interestingly, this attack sparked a heated debate within the Hizb'allah about the legitimacy of suicide operations. Even the group's spiritual leader Shaykh Muhammad Husayn Fadlallah raised legal objections to the practice, basing his argument on the Qur'an. The extreme effectiveness of the attack, however, caused the tactic to be continued for the time being. However, following the withdrawal of U.S. and later Israeli troops from Lebanon, the group's leadership reportedly made the decision to end the systematic use of suicide bombings, and since then the tactic has only been used sporadically (Sprinzak 2000).

To date, Lebanon has witnessed more than 50 suicide bombings. About half of these were perpetrated by Hizb'allah and Amal, the two main Shi'ite groups operating in Lebanon, whose success inspired secular groups, namely, the Syrian National Party, the Lebanese Communist Party, and the Socialist-Nasirist Party, to use the same approach (Shay, 2000). Missions of these Syrian-trained groups were generally characterized by a high failure rate, and there are indications that some of the bombers were unaware of their involvement in suicide operations (Merari, 1998). These groups were also the first to introduce the systematic use of women as suicide bombers (in 40 percent of the cases).

Of all the countries in the world, Sri Lanka has by far experienced the most suicide operations. In fact, as of 2003 more suicide attacks have occurred in Sri Lanka than in the rest of the world combined. The LTTE, which has perpetrated over 240 suicide attacks since 1987, has its own suicide volunteer group known as the Black Tigers, to which only the most devoted and able volunteers are admitted (Sprinzak, 2000). The unit's prestige and the glorification of those who deliberately sacrifice their lives for the group's cause play a critical motivational role for the bombers, as was demonstrated by the group's building of a monument in Puthukkuthirippu to celebrate the Black Tigers (Gunaratna, 2001).

In May 1987, the LTTE resorted to suicide operations for the first time, when a volunteer drove a truck full of explosives into an enemy military camp. Previously, the group had unsuccessfully attempted this technique, with the driver jumping out of the truck before the explosion (Gunaratna, 2000). LTTE suicide operations are often highly sophisticated and deadly. Among the most well-known attacks are the 1996 bombing of the central bank, in which 86 people died and 1338 were injured, and the assassinations of two presidents and one minister of defense. The sophistication of LTTE attacks was evident in the assassination of President Premadas, whose suicide assassin infiltrated the president's household and became acquainted with the president's valet many months before the operation was to take place (Gunaratna, 2000).

LTTE suicide attacks have two other distinct features: the use of women, who carry out 30 to 40 percent of the group's suicide attacks, often capitalizing on their ability to appear pregnant as a means of hiding explosives, and the use of capsules containing potassium cyanide, which LTTE members ingest when they are seriously injured or captured (Gunaratna, 2001). For example, during the investigation of the assassination of Rajiv Gandhi, at least 35 LTTE operatives attempted to commit suicide in this manner simply to prevent themselves from being questioned. The majority of LTTE suicide attacks target officials, the security forces, and symbols of power, though civilian targets have likewise been attacked quite extensively. LTTE responsibility for these latter attacks is never officially claimed, however, as the group's code officially prohibits attacks against nonmilitary personnel (Gunaratna, 2001).

In Turkey, the secular neo-Marxist PKK perpetrated 15 suicide attacks between 1996 and 1999. Its apparent goal in adopting this tactic was to strengthen morale within the group after a series of military setbacks in southern Turkey (Schweitzer, 2001). The primary targets were provincial officials, policemen, and military personnel and installations. The total number of casualties in these 15 suicide bombings amounted to six policemen and nine soldiers killed, and 28 policemen and 47 soldiers wounded. Apparently, noncombatants were not deliberately targeted, although four civilians were killed and 63 wounded by collateral damage (Schweitzer, 2001). The bombings were justified on the basis of a doctrine combining Kurdish ethnic nationalism with the religious concept of *shadat*, or martyrdom during war (Ergil, 2000). Interestingly, coercion often played a crucial role during "recruitment," as the suicide bombers were frequently appointees rather than volunteers. Over 65 percent of the attackers were young women, who were largely recruited by exploiting their desire to prove their equal contribution to the organization's cause (Schweitzer, 2001). The group ceased using suicide bombings in 1999, when leader Abdullah Ocalan was arrested, convicted, and sentenced to death.

The country that is most commonly associated with being a target of suicide bombings is Israel. To date, Israel has experienced just over 130 successful suicide bombings, with 59 cases occurring in 2002 alone. Suicide bombings were introduced in Israel in 1993 by HAMAS (Islamic Resistance Movement) and Palestinian Islamic Jihad (PIJ). The principle goal of both of these Palestinian Sunni groups is to disrupt the implementation of the Israeli-Palestinian peace agreement known as the Oslo Accords. It is suspected that the 1992 deportation of 415 Islamic activists from Gaza and the West Bank to Lebanon for a year contributed to the involvement of these groups in suicide operations: while in Lebanon, many of the deportees received training from Hizb'allah (Ganor, 2000). Until recently, HAMAS was responsible for approximately 72 percent and PIJ for the remaining 28 percent of the suicide bombings in Israel. Initially, these operations concentrated primarily on military targets, but following the aforementioned 1994 shooting of Muslim worshippers in Hebron, HAMAS began carrying out attacks on civilian targets, including buses, restaurants, and cafes.

A major shift in suicide operations against Israel occurred in January 2002, when the al-Aqsa Martyrs Brigades, an organization that affiliated with the secular Fatah movement, resorted to the use of suicide bombings. In the first three months of 2002, the al-Aqsa Martyrs Brigades claimed responsibility for nine high-profile suicide bombings, which claimed the lives of 43 people (Jaber, 2002). Unlike the typical HAMAS or PIJ suicide bomber—a religious Palestinian male between 18 to 27 years of age, usually unmarried, uneducated, and often recruited just several days prior to the attack in a refugee camp or an Islamic educational center in Gaza or the West Bank—the al-Aqsa Martyrs Brigades have used both educated men and women to carry out their operations.

In June 2000, following nine months of military setbacks, suicide bombings were also adopted by terrorists operating in Chechnya. The first operation was conducted by two suicide bombers who drove a truck full of explosives into a Russian military checkpoint at Alkhan-Yurt, Chechnya, killing two officers (Paz, 2000). A Russian soldier who converted to Islam delivered another attack several days later, killing two military personnel. The most significant suicide operation in Chechnya was the coordinated attack, in 2000, of five suicide truck bombers against military checkpoints and a police dormitory, which killed 33 people and injured 84 (Weir, 2000). This attack remains one of the most sophisticated suicide operations to date. A similar operation occurred in December 2002, when suicide bombers detonated two truck bombs outside the headquarters of Chechnya's pro-Moscow administration in Grozny, killing more than 70 people. Most recently, a May 2003 suicide truck bomb detonated near a government complex in northern Chechnya resulted in the deaths of at least 52 people. In addition to suicide truck bombings, the Chechens have employed suicide bombers (mainly females) to assassinate Russian military officials and soldiers at checkpoints. The increasing efficiency and sophistication demonstrated by Chechen suicide attacks suggests that this tactic will be used with increasing frequency in the region.

Suicide bombings have also been imitated in other countries of the world, usually in sporadic incidents that do not constitute any type of pattern. On January 30, 1995, the Armed Islamic Group (GIA) perpetrated a suicide attack in Algeria that resulted in 42 fatalities and 286 injuries. A month prior, on December 24, 1994, four GIA terrorists attempted to hijack an Air Algérie airliner with the plan to crash into an unidentified target in Paris, but the plot was thwarted by a rescue operation.

Members of al-Da'wa, a Shi'ite organization that operates mainly in Iraq, were also involved in the 1983 bombing of the U.S. Embassy in Kuwait, in which six people were killed and 80 injured. Although the attack was claimed under the name "Islamic Jihad," the alias used by Lebanese Hizb'allah, the suicide bomber was traced to al-Da'wa (Wright, 2001). Evidence also suggests that the suicide bomber in this case unsuccessfully attempted to prevent his own death by exiting the explosives-packed truck prior to the explosion (Merari, 1998).

In India, 10 soldiers were killed in a suicide attack perpetrated against a military camp by Jaish-e-Mohammed (Mohammed's Army) (Schweitzer, 2001). The same group perpetrated another suicide truck bombing as a part of a larger operation against a state legislature building in Srinagar in 2001, killing at least 29 people. A Sikh group called Babbar Khalsa International (BKI) was responsible for another suicide operation in India in 1995, killing Beant Singh, the chief minister of Punjab (Jurgensmeyer, 2000).

In the past decade, the IRA has on several occasions utilized suicide car bombs to attack Northern Irish police and British troops. In these instances, however, the bombers were suspected informers who were blackmailed into participating by the threat of harm to their abducted families (Sloan and Anderson, 2002).

Suicide bombings have recently been introduced in Pakistan. On June 14, 2002, 11 people were killed and 24 wounded when a suicide bomber drove a truck into the guard post at the U.S. Consulate in Karachi. On May 8, 2002, a suicide car bomber detonated an explosive device next to a Pakistan Navy shuttle bus in Karachi, killing 11 people and injuring 22 others.

A fairly recent development has been the internationalization of suicide bombings not only through the spread of the tactic into new countries but also by the growing number of suicide attacks that are transnational in nature. The first suicide attack outside of the perpetrators' common area of operation took place in 1992, when a Lebanese *shahid* with suspected Hizb'allah links bombed the Israeli Embassy in Buenos Aires, Argentina, in retaliation for the killing of 'Abbas Musawi, the group's secretary general. The attack resulted in 29 fatalities and 250 injuries (Schweitzer, 2001). A similar operation was carried out in Buenos Aires in 1994, in which a suicide bomber detonated a 340-pound truck bomb, destroying a building containing the offices of several Jewish organizations. This attack, which killed 46 people and injured over 200 others, was claimed by Ansar Allah (Partisans of Allah), a group with alleged links to Lebanese Hizb'allah. The LTTE belongs to this list as well: on May 21, 1991, an LTTE suicide bomber killed Rajiv Gandhi, the former Prime Minister of India, during a preelection rally in Tamil Nadu, India (Sloan and Anderson, 2002). This operation was apparently a reprisal for Gandhi's decision to send Indian troops to Sri Lanka in 1987.

Two major Egyptian Islamist organizations, al-Jama'a al-Islamiyya (Islamic Group) and the Egyptian Islamic Jihad (EIJ), have also conducted operations outside of Egypt. In 1995, al-Jama'a al-Islamiyya attacked a police station in Rijeka, Croatia, in retaliation for the disappearance of one of the group's leaders in that country. During the same year, the Jihad Group used two suicide bombers to attack the Egyptian Embassy in Pakistan, killing 15 and injuring many others (Schweitzer, 2001). Factions of both al-Jama'a al-Islamiyya and the Jihad Group are believed to be part of 'Usama Bin Ladin's al-Qa'ida network. This network has been responsible for multiple suicide operations abroad. The 1998 synchronized suicide truck bombings of the U.S. Embassies in Kenya and Tanzania caused 224 fatalities and more than 5000 injuries. The attack on the *USS Cole* in 2000, which involved two suicide bombers on fiberglass boat, resulted in the death of 17 servicemen and the injury of 39 others. The September 2001 assassination of Ahmed Shah Masud, the leader of the Northern Alliance in Afghanistan, was also perpetrated by two suicide bombers who hid their bomb in a television camera (Mashberg, 2001). Finally, the four synchronized suicide bombings of September 11, 2001, which utilized hijacked airplanes to target the World Trade Center and the Pentagon, killed nearly 3000 people. All of al-Qa'ida's suicide operations have been carried out by independent cells, and all have involved the death of the operational planner. Post-9-11 al-Qa'ida suicide bombings include several high-profile operations in Indonesia, Kenya, Yemen, Tunisia, Morocco, Pakistan, Afghanistan, and Saudi Arabia.

FUTURE OUTLOOK: SUICIDE "BOMBERS" AS DELIVERY VEHICLES FOR BIOLOGICAL WEAPONS

Suicide bombings represent the ultimate terrorist tactic. Besides their tactical advantages, they also have the ability to satisfy many terrorist objectives in a single attack, demonstrating dedication and capability, attracting attention and media coverage, producing a high number of casualties, and instigating general feelings of vulnerability. Finding recruits for suicide missions is not difficult once a precedent has been established. Suicide attacks can be justified on any religious or ideological grounds in the appropriate historical and cultural context. Therefore, the use of this tactic will likely become increasingly frequent in areas where it has already been established, and will likely be introduced to many other areas of struggle around the world.

Another source of concern for the future is the possibility of using suicide "bombers" for delivering weapons of mass destruction (WMD) in general, and biological weapons in particular. The successful delivery of a chemical or biological agent would certainly be more feasible if the perpetrator were indifferent to staying alive. Not only would it be possible to eliminate protective gear, which could raise suspicion, but also the delivered substance itself could be used in much higher concentrations if concern for the handler's life was discarded. Furthermore, the attacker's own body could even serve as a delivery system if he or she were infected with a contagious biological agent such as smallpox and sent to a crowded area. Since September 11, 2001, this nightmare scenario has often been cited as a possible example of things to come. There are, however, a number of reasons why the smallpox suicide "bomber" scenario may not be as inevitable as is commonly believed.

The first obstacle is the extremely limited availability of smallpox and other contagious pathogens, which makes the acquisition of these agents by terrorists very unlikely. Second, even if this obstacle were somehow to be overcome, initiating an epidemic would—according to some experts—be more complicated than simply injecting the "bomber" with the agent, waiting for symptoms signaling infectivity to appear, and sending him or her into a crowded public area. With regard to smallpox, for example, although a potential suicide delivery vehicle might develop fulminating smallpox from such an injection, he or she would be more likely to develop a mild infection with or without a rash that in most cases would not lead to shedding of the virus and secondary spread (Pilch, 2003).

Another obstacle to the biological "bomber" lies in the realm of motivational constraints. While recruiting suicide delivery vehicles for attacks involving explosives is in most instances not a problem, the availability of volunteers for a biological suicide operation may not be as widespread as is commonly believed. One of the incentives traditionally used to motivate suicide bombers is the promise of a quick and honorable death. Many such attackers have been observed smiling just prior to detonating their explosives. In the Middle East, this is known as the *bassamat al-farah* (smile of joy); the bombers are experiencing a joyous moment as they are near what they perceive to be the gates of eternity. It may not be possible for an operative in the late stages of a debilitating disease to walk through crowded city areas experiencing such joy, because, in essence, there would be nothing to smile about: the delivery would be taking place but without any immediate impact on the target or perpetrator. The terrorist would not have any sense of success, and the psychological reward associated with a bang would therefore be missing. Furthermore, entering the gates of heaven would not occur in an instant joyful moment, but would rather be a long and painful process in which the suicide delivery vehicle would die under circumstances that are not perceived to be heroic: a decomposing body engenders disgust rather than admiration. The willingness of most terrorists to die in this excruciating manner is highly questionable (Dolnik, 2003).

In addition to these obstacles on an individual level, additional motivational constraints may well be present at the organizational level. On one hand, it is clear that suicide operations are responsible for a large number of fatalities. Of the 30 single most deadly terrorist incidents carried out between 1990 and 2003 19 utilized suicide bombers, and seven of the remaining 11 attacks were perpetrated by groups with a record of using suicide bomb delivery. The fact that groups employing this tactic are responsible for the majority of high-fatality incidents suggests that they have reached an advanced level of enemy dehumanization and are therefore psychologically closer to perpetrating mass-casualty WMD attacks.

On the other hand, despite the common tendency to label suicide attacks as irrational, organizations that have perfected this tactic have demonstrated a great level of rationality, patience, and calculation in their attacks. With the exception of the Algerian GIA, which has only been responsible for one suicide attack, none of the terrorist organizations that have employed suicide bombings on a consistent basis can be safely labeled apocalyptic or entirely indiscriminate. Most have a clearly defined enemy and a well-developed inner logic for their targeting. The potential for the uncontrollable spread of disease by secondary transmission beyond the intended target population may well be inconsistent with the targeting logic of most groups that have utilized suicide bombers in the past. Whether this means that a biological suicide attack will not be carried out in the future, only time will tell.

REFERENCES

Arad, U., "Do Nations Commit Suicide? A Middle Eastern Perspective," *Countering Suicide Terrorism*, International Policy Institute for Counter-Terrorism, Herzliya, 2000, p. 25.

Dolnik, A., *Studies Conflict Terror.*, **26**(1), 17–35 (2003).

Ergil, D., "Suicide Terrorism in Turkey: The Worker's Party of Kurdistan," *Countering Suicide Terrorism*, International Policy Institute for Counter-Terrorism, Herzliya, 2000, p. 120.

Ganor, B., "Suicide Attacks in Israel," *Countering Suicide Terrorism*, International Policy Institute for Counter-Terrorism, Herzliya, 2000.

Gunaratna, R., Suicide Terrorism: Emerging Global Patterns, *Colombo Chronicle*, November 17, 2001.

Gunaratna, R., "Suicide Terrorism in Sri Lanka and India," *Countering Suicide Terrorism*, International Policy Institute for Counter-Terrorism, Herzliya, 2000.

Hoffman, B., *Inside Terrorism*, Orion Publishing Co., New York, 1998.

Jaber, H., *Hizballah*, Columbia University Press, New York, 1997.

Jaber, H., Inside the World of Palestinian Suicide Bomber, *Sunday Times*, Available at www.sunday-times.co.uk/article/0,,178-245592,00.html, accessed on 4/9/02.

Juergensmeyer, M., *Terror in the Mind of God: The Global Rise of Religious Violence*, University of California, London, 2000.

Kushner, H., *Studies in Conflict and Terrorism*, **19**(4), 1996.

Laqueur, W., "Reflections on Eradication of Terrorism," in K. Charles Jr., Ed., *International Terrorism*, St. Martin's Press, New York, 1990.

Mashberg, T., Veil Lifts on New Deadly Breed of Suicide Terrorist, *Boston Herald*, November 30, 2001.

Merari, A., "Readiness to Kill and Die—Suicide Terrorism in the Middle East," in W. Reich, Ed., *Origins of Terrorism*, Woodrow Wilson Center Press, Washington, DC, 1998.

Mickolus, E.F., *Transnational Terrorism: A Chronology of Events, 1968–1979*, Greenwood Press, Westport, 1980.

Paz, R., "The Islamic Legitimacy of Suicide Terrorism," *Countering Suicide Terrorism*, International Policy Institute for Counter-Terrorism, Herzliya, 2000a.

Paz, R., *Middle East Intell. Bull.*, **2**(6) (2000b).

Pilch, R. "The Bioterrorist Threat in the United States," in R.D. Howard and R.L. Sawyer, Eds., *Terrorism and Counterterrorism*, 2nd ed., McGraw-Hill, New York, 2003.

Schweitzer, Y., "Suicide Bombings: The Ultimate Weapon?", International Policy Institute for Counter-Terrorism, Available at http://www.ict.org.il/articles/articledet.cfm?articleid=373, accessed on 04/03/05.

Schweitzer, Y., "Suicide Terrorism: Development & Characteristics," International Policy Institute for Counter-Terrorism, available at http://www.ict.org.il/, accessed on 04/03/05.

Shahar, Y., *"The al-Aqsa Martyrs Brigades: A Political Tool with an Edge*, International Policy Institute for Counter-Terrorism, available at www.ict.org..il, accessed on 3/26/02.

Shay, S., "Suicide Terrorism in Lebanon," *Countering Suicide Terrorism*, International Policy Institute for Counter-Terrorism, Herzliya, 2000.

Sloan, S. and Anderson S.K., *Historical Dictionary of Terrorism*, 2nd ed., The Scarecrow Press, Lanham, 2002.

Sprinzak, E., *Foreign Policy*, October 2000.

Weir, F., Chechen Rebels Go Kamikaze, *Christi an Science Monitor*, June 7, 2000.

Wright, R. *Sacred Rage: The Wrath of Militant Islam*, Simon & Schuster, New York, 2001.

SYNDROMIC SURVEILLANCE

ALAN P. ZELICOFF

INTRODUCTION

This entry examines a vital area of the nation's counterterrorism response: the use of clinicians as sentinel detectors of unusual signs and symptoms in their patient populations, coupled with near-real-time analysis by public health experts of clinical data to provide the earliest possible detection of unusual disease outbreaks and patterns. A relatively novel means for conducting public health via clinical "syndrome-based surveillance" can, at the very least, augment the existing public health reporting infrastructure for classical "reportable diseases" and laboratory-based surveillance, and dramatically increase the likelihood of identification of intentionally introduced disease in humans and animals. In addition, this approach can enhance the robustness of routine public health systems, including the detection of vaccine preventable illness (VPI) and vectorborne disease.

BACKGROUND

A history of biological weapons use might be considered arcane were it not for the posting of letters containing the Ames strain of *Bacillus anthracis* spores to the offices of U.S. Senators Tom Daschle and Patrick Leahy in October 2001. No Senate office employees were affected, but 22 cases of anthrax (11 inhalation anthrax and 11 cutaneous anthrax) occurred in conjunction with the mailings, largely among postal workers. Five people ultimately died, while in excess of 150 people, mostly workers on Capitol Hill, were electively vaccinated against the disease during the incident, and tens of thousands of courses of Ciprofloxacin were prescribed. The Hart Senate office building was closed for months, and multiple decontamination campaigns with ozone and oxidizing foam were required before the building was declared safe for routine use.

The lessons from this event, combined with lessons from the historical development and use of biological weapons for the purposes of sabotage and war, offer a foundation upon which to build a strategy for civilian biodefense:

- Many people become ill (or die) before the source of a biological outbreak can be confidently identified.

- The affected or "exposed" populations may number in the thousands.

- Because of the long delay between use of a biological weapon and onset of illness, the perpetrators may be difficult to find.

- Appropriate prophylaxis may or may not be available, even for those exposed but are not yet ill.

- Uncertainty in the actual number of exposed people, due to an absence of real-time or near-real-time detectors, results in the need for widespread immunization and/or prophylaxis with unclear long-term consequences.

- Costs of a biological event are difficult to calculate, but are probably dominated by disruption of daily activities, resulting in economic losses, and the medical costs associated with diagnostic testing, prophylaxis, and management of postprophylaxis complications.

Several models, usually specific for distinct diseases, suggest that a large scale dissemination of *B. anthracis* or the smallpox virus would create a daunting management problem for public health officials and public safety officers (Kaplan et al., in press). A delay of even a day in identifying a disease caused by an agent of biological warfare (BW) or bioterrorism concern, whether of natural or deliberate etiology, would mean the difference between the loss and salvage of as much as 90 percent of an exposed population. The problem is compounded dramatically by agents such as the smallpox virus or *Yersinia pestis*, which are highly communicable.

There is reason to worry that physicians, nurses, and other health care providers will not recognize the signs and symptoms of a disease caused by a bioterrorism agent. First, these diseases may initially present in the same fashion as any routine flu-like illness, thus requiring a high index of suspicion in order to be identified. Further, few physicians in practice today in the United States have seen cases of plaque, fewer still anthrax, and essentially none a case of smallpox or hemorrhagic fever (Henderson, 1999), and if biological agents are dispersed as a fine aerosol, the atypical route of entry may result in previously

Encyclopedia of Bioterrorism Defense, Edited by Richard F. Pilch and Raymond A. Zilinskas
ISBN 0-471-46717-0 Copyright © 2005 Wiley-Liss

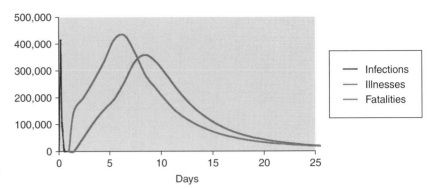

Figure 1. Anthrax dispersal in three large U.S. cities.

undescribed or nonspecific symptoms. In addition, because few physicians report even those diseases required by law to be reported to public health authorities, (Doyle et al., 2002), these authorities will likely be unaware of the scope of the contagion until well into the epidemic, when so many people (or animals or both) become ill that hospitals and clinical offices are overwhelmed. Finally, even when public health officials become aware of disease outbreaks, their ability to reach busy clinicians is limited. The traditional system of peer review and paper-based publication is likely too slow to allow for a meaningful and effective response to a bioterrorism event, and physicians receiving their news on disease outbreaks from the media may be getting incomplete and unreliable information.

However, there are some hopeful signs that medicine and public health are embracing some of the more flexible, rapid communications technologies. For example, during the Severe Acute Respiratory Syndrome (SARS) epidemic of spring 2003, reports were disseminated via the Internet on authoritative websites (such as those of the Centers for Disease Control and Prevention and the World Health Organization) (Tsang et al., 2003). Regardless, a reconsideration of the public health reporting and data dissemination system is necessary for adequate response times in the future.

WHAT STRATEGIES ARE NEEDED?

Almost without exception, the organisms that have been touted (or stockpiled in the past by military programs) for use as biological agents result in diseases that are eminently preventable but usually untreatable once illness has advanced beyond a certain point. Thus, if it is possible to recognize the presence of biological weapons agents in the air (or water) at the time of their dispersal, or to quickly diagnose the earliest cases of disease in humans or animals, medical intervention to protect those exposed is still possible. The technological developments surrounding biological-agent sensor is beyond the scope of this entry; indeed, many of the techniques are at the laboratory bench-top stage, not yet ready for wide-area installation. The technology for improving the responsiveness of public health officials and the medical community to a biological weapons disaster, on the other hand, is in fact available.

Consider, for example, the following bioterrorism event: 100 kg of aerosolized, "weapons-grade" *B. anthracis* spores (not a vaccine strain) are dispersed in the cities of Los Angeles, New York, and Washington, DC. Calculations performed at Sandia National Laboratories (Todd, 2003) show that, undetected until approximately day 5 postdispersal, a time graph of infections and fatalities can be constructed (Fig. 1). By integrating under the Fatalities curve, it is possible to show that approximately 3.5 million people may die within one month of such an attack.

However, if detection can be accomplished by making the diagnosis or at least having a high index of suspicion on day 3 (when individuals who received a large dose or are immunocompromised begin to become symptomatic), the total number of fatalities is reduced to approximately one million, a significant reduction (though still disturbing in its severity).

How might this "early detection" be accomplished in the absence of continuous air sampling to look for unusual pathogens? The answer may lie in the observational powers of physicians. It is well known that it is difficult to make an accurate diagnosis of a presumed acute infectious disease in the outpatient setting. It takes time to culture bacteria (between 24 and 72 h), if they are obtained at all in collected samples, and most physical findings and routine laboratory tests, even if abnormal, are nonspecific. The problem boils down to one of Bayesian statistics: how does a diagnostician achieve adequate "sensitivity" while not sacrificing "specificity"?

The public health community—literally hundreds of individual county, city, regional, and State health departments around the United States—has long depended on physicians to notify it of "reportable diseases," a list of communicable, sexually transmitted, and vectorborne infectious diseases of primary concern to public health. But clinical reporting is poor, both in the United States (Doyle et al., 2002; Schramm et al., 1991) and elsewhere (Allen and Ferson, 2000). Few physicians know what diseases must be reported and how to go about the reporting process, and many (if not most) are dissuaded by the long forms requiring completion (Karim and Dilraj, 1996). Oddly, pediatricians, the most likely specialists to see and recognize reportable diseases, appear to be among the least compliant health care providers in reporting communicable diseases (Camposoutcalt et al., 1991). In addition, there is an implicit requirement for laboratory confirmation in most reporting, meaning that unless laboratory testing (generally culture or serology) is available, it is difficult to meet reporting requirements. And even further, the proper collection and handling (Hurst et al., 1998) of clinical samples in patients with presumed infectious disease is questionable (Nutting et al., 1996). At the same time, there is no question that laboratory-based surveillance and reporting is very useful (and sometimes essential) in identifying a disease outbreak of both natural and deliberate etiology (Skeels, 2000). Regardless, the utility of the clinical laboratory—within or separate from physicians' offices—is limited and probably underestimates the incidence and prevalence of common infectious disease in the community. As an added point, although there is little literature on the capacity for laboratory-based detection of exotic or novel infectious disease agents, especially in the setting of large exposure as might be anticipated in a terrorist attack with a biological agent, the concern is that the sensitivity of the existing clinical laboratory infrastructure for these unusual agents is likely even lower than that for the more common disease pathogens.

WHAT ARE THE CHARACTERISTICS OF A CLINICALLY USEFUL PUBLIC HEALTH REPORTING SYSTEM?

In early 2002, the National Defense University (NDU) sponsored a high-level conference on counterterrorism policy (Center For Counterproliferation Research, 2002), designed to achieve consensus on the specific near-term needs of the defense, public health, and medical communities in preparing for terrorist attacks in which biological weapons may be used. Although some of the details of the meeting are classified, the following observations and broad requirements were openly discussed during sessions with members of the media:

- *All of Medicine and Public Health is Local.* There are approximately 8000 public health agencies in the United States, and these agencies share information poorly both with each other and with physicians in their communities.
- *The Medical Care Delivery System and Public Health System are Clearly Separate Entities.* If there ever is a bioterrorism attack, a hierarchical "Incident Command" response structure will probably fail, as the medical community has few members trained to implement the Incident Command System (ICS) or practice public health.
- *Continuous Surveillance of Human and Animal Health is Essential, but the Risks of Any "Surveillance Solution" Must be Perceived to be Less than the Risks of Disease.* It will be particularly important to remember what the public will accept and what the public will believe. The media will be *the* key conveyor of information, so advice and intervention must be based on accurate, near-real-time data in a bioterrorism incident.
- *Thus, Any Surveillance System Should Make Public Health Part of Medicine, and Medicine Part of Public Health.* The system must be *low cost* and have *low intrusiveness* to individuals, respect the existing culture of medicine and public health, yet foster the rapid sharing of information across jurisdictional boundaries in order to provide information on a "need-to-know" basis to health professionals, public

health officials, emergency response managers, local political leaders, and national-level decision makers.
- *Training for Health Care Providers Should be Inherent in the Surveillance System,* as there is almost no expertise in the medical community on the most likely weapons-related diseases or their causative organisms. If possible, any training should be made part of physicians' Continuing Medical Education (CME).

Although there was no agreement on a specific set of public health and clinical tools that would meet all of these extremely challenging requirements, one thing was clear to conference participants: clinical accuracy (positive and negative predictive values) is only a part of the solution. It is also necessary to get relevant information (as opposed to "all possible" information) to all of the "stakeholders" in the response to a bioterrorist attack. This goes well beyond the physician's traditional sphere of influence—the clinic, office, and hospital—and even beyond the boundaries of traditional public health practice; at some point in a bioterrorism event, national-level response will have to be planned and managed. Since public health is practiced, by design, at a local and regional level, the need for information sharing with flow of data outside of local officials' control may run headlong into bureaucratic obstacles.

Two general approaches to medical surveillance are available: *passive surveillance,* which employs electronic records that accrue during patient care, such as billing records and disease codes on those records, ambulance logs, and even pharmaceutical sales including over-the-counter drugs; and *active surveillance,* which depends on physicians and other health care providers to report unusual symptoms or physical examination findings on acutely ill individuals. Several passive systems have been developed: ESSENCE (by the U.S. Department of Defense) is in use in military installations around the United States and is probably the leading example. Only a few active surveillance systems are in use, one of which is the Rapid Syndrome Validation Program (RSVP^{TM}).

A POSSIBLE SOLUTION: THE RAPID SYNDROME VALIDATION PROGRAM (RSVP)

One of the existing syndrome surveillance systems that meets the NDU requirements is the RSVP, an Internet-based population health surveillance tool designed to facilitate rapid communication between epidemiologists (public health officials in local public health jurisdictions) and health care providers (especially physicians, physician assistants, and nurse practitioners). RSVP was developed with funding from the U.S. Department of Energy's Chemical and Biological Weapons Non-Proliferation (CBNP) program, an open (i.e., unclassified), independently reviewed research effort to develop specific technologies for the early detection and, if necessary, remediation of biological weapons use. Early in the CBNP planning, it was recognized that clinicians in community practice possess a wealth of observational data that is never communicated to public health officials, data that could be critical in understanding the scope of a bioterrorism event simply because people are distributed much more widely than any sensor network ever will be.

RSVP is designed to overcome existing barriers to the reporting of suspicious or unusual symptoms in patients, and capture clinician judgment regarding severity of illness and likely category of disease. For clinicians, RSVP provides immediate feedback with data of relevance to a reported case, as well as continuously updated geographic and temporal characteristics of symptom distribution in the local community. For the epidemiologist in the local public health office, RSVP provides real-time data reporting, the ability to perform a wide array of the Geographic Information System (GIS) analyses, and a fast, convenient way to communicate with all reporting clinicians on the network.

RSVP creates a new environment for the reporting of "reportable" infectious diseases, as well as suspicious or novel symptoms that may or may not be part of a known disease or disease-complex. As its name suggests, RSVP reporting is based on symptom and sign complexes known as *syndromes*. Syndromes can be defined with a high degree of specificity in mind (e.g., hemorrhagic fever syndromes) or can be more generalized, reflecting common parlance in medical care and communication among doctors. Although there are no strict rules for development and promulgation of syndrome definitions, a common approach (and that applied in RSVP) is to organize syndromes around organ systems, taking into account the use of descriptors that characterize severely ill patients. There are six syndrome complexes in RSVP:

- Influenza-like illness
- Fever with skin rash
- Fever with suspected Central Nervous System (CNS) infection (formerly "fever with mental status changes")
- Severe diarrhea (watery or bloody; "severe" is a subjective term to be defined by the treating physician)
- Adult respiratory distress syndrome (ARDS)
- Acute hepatitis

These syndrome complexes were developed as a result of exhaustive discussions among public health officials and practicing clinicians, and are believed to capture the vast majority of diseases of public health importance, including but not limited to diseases that could occur as a result of bioterrorism.

Clinicians and their patients can be the "smart sensors" in any disease outbreak. Without timely reporting, awareness of public health authorities is likely to be outstripped by disease progression, particularly in the setting of communicable infectious disease. Given the critical role that clinicians play in identifying and mitigating the consequences of infectious disease, the RSVP reporting interface has been constructed to meet the following requirements:

- Data entry, including demographic as well as clinical information, must be:

- Fast (less than 30 seconds of clinician time for each case entry);
- Inexpensive, requiring no downloads of special software at any time;
- Simple and without ambiguity;
- Intuitive, so that no training is required;
- Error-free, obviating correction and reentry.
- Feedback to clinicians must be:
- Immediate (less than 10 to 15 seconds);
- Relevant to the patient that the physician is seeing;
- Of importance for the differential diagnosis;
- Accompanied by information on similar cases seen recently by other clinicians in the community.

For RSVP to be of relevance to public health officials, a separate (though sometimes overlapping) set of requirements must be fulfilled:

- The system must automatically notify officials of reported cases that closely match disease entities of *local* prevalence or importance (e.g., hantavirus in New Mexico, botulism in Alaska, or dengue fever in southern Texas).
- GIS tools must be included to facilitate easy analysis.
- Advisory messages to clinicians must be easy to update, perhaps even several times a day.
- Database structure should be compatible with existing medical database standards.
- Specific queries should be easy, using automated interfaces and standard "natural language-like" lexicon.

For such a system to be sustainable on its own merits (particularly in the current cost-containment environment), it must improve the effectiveness of clinicians and public health officials in their day-to-day work.

RSVP for Clinicians

Imagine having a handheld device or computer screen in the doctors' office that continuously updates infectious disease outbreaks in the local region (e.g., city or state) and also provides easy access to up-to-the-minute information on disease outbreaks taking place around the world, so that if a patient presents with a bad skin rash, or fever and confusion after a vacation in the Australian outback, it can quickly be determined whether the differential diagnosis should be expanded to consider exotic illness. This is what RSVP is intended to facilitate: a broader consideration of unusual disease that physicians traditionally do not see. Fig. 2 shows the entry-portal to RSVP.

The functionality of RSVP is contained in each of the buttons and areas of the screen. After signing on to the system, the physician can check reports on disease outbreaks around the world, read updated information of medical importance provided by local public health officials, or enter a case by selecting the appropriate syndrome by which a patient may best be described.

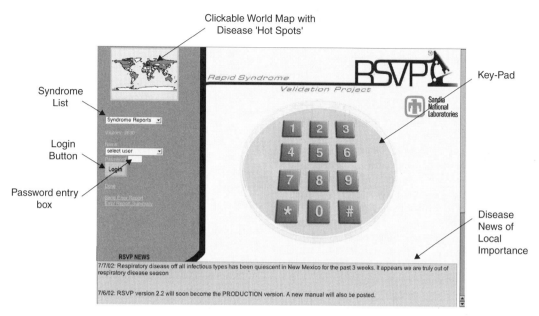

Syndrome
List

Login
Button

Password entry
box

Clickable World Map with
Disease 'Hot Spots'

Key-Pad

Disease
News of
Local
Importance

Figure 2. The RSVP entry-portal.

Reporting takes approximately 30 seconds, and physicians always receive useful information in return after sending in a case report—immediately—via a "clinical feedback" screen (Fig. 3).

The feedback screen is a dynamic document, changing after each report is submitted. It is a combination of automated mapping and graphing functions, advisory messages from local public health authorities (which they can update as frequently as they feel necessary), and culture data for organisms of relevance to the syndrome in question.

In this sample feedback screen, the local advisory message is seen in the upper left-hand corner, representing the net assessment of experts in the public health office of disease patterns and other data of immediate utility to the reporting physician. In the lower left-hand corner is a graph (going back several months or more depending on the locale) of all culture results of major pathogens causative of the syndrome, allowing clinicians to identify patterns in the results or otherwise apply them to "index of suspicion" decisions.

In the lower right-hand corner there is a temporal ("time") graph of all cases of the identified illness (in this case, influenza-like illness, or ILI) reported in the past 30 days. This represents the well-known "epidemiology curve," and may be useful to both clinicians and epidemiologists to determine whether what they are seeing is a true epidemic (based on historical records or experience) or if reports are unrelated and sporadic. Coupled with the temporal graph is a geographic map displayed in the upper right-hand corner of the feedback screen. It is centered on the facility that is doing the

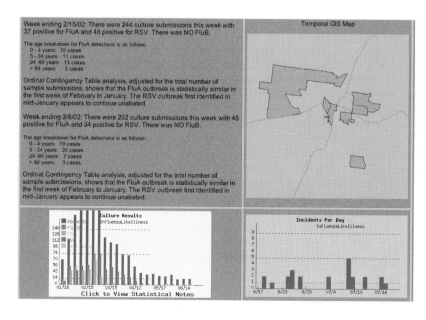

Figure 3. The RSVP "clinical feedback" screen.

reporting (in this example, the crossroads of I-25 and I-40 in Albuquerque, New Mexico); the colored areas are zip codes where there have been reports of similar patients in the past 30 days, color coded by number of cases.

RSVP for Epidemiologists

Public health officials have different skills and different needs than clinicians. Like clinicians, they need a data-reporting system that is fast and easy to use. An epidemiologist, however, must look at large populations (whereas the clinician tends to see patients one at a time) and test hypotheses using statistical correlations to identify outbreaks and their sources. RSVP permits public health practitioners to unify clinical and geographic information, as well as socioeconomic data (census tract information such as median incomes, percentage of black, white, and Hispanic households, median age, etc.), in order to perform robust epidemiological analyses the results of which can be easily communicated to physicians.

As an example, consider a map showing the geographic distribution of respiratory illness in New Mexico, along with the location of forests and habitats for the *Paromyscus* mouse, the carrier of hantavirus (Fig. 4). During early spring, New Mexico public health officials must be on the lookout for this disease, which is difficult to treat but easily preventable. How might they test the hypothesis that respiratory symptoms are occurring in adults because of hantavirus?

On the map, the epidemiologist sees a cluster of flu-like illness in and around Albuquerque. Putting aside sampling error (an important consideration) due to a lack of data from various parts of the state, it is easy to see that the respiratory symptoms are occurring in areas removed from forested areas or areas with heavy vegetation. Thus, it is not very plausible that hantavirus is causing the symptoms.

THE FUTURE OF SURVEILLANCE

As the fall 2001 "anthrax letters" made clear, bioterrorism is no longer a hypothetical risk. Had the perpetrator chosen to employ a (theoretically) more effective means of aerosol dissemination—an agricultural sprayer in the back of a van driving around downtown Washington, DC, for example—tens of thousands of people would have been exposed to potentially lethal doses of infectious and virulent *B. anthracis* spores, with what likely would have been a significant death toll. This catastrophic outcome owes not merely to the lethality of the *B. anthracis* spores used in the attack but also to our inability to recognize, report, and analyze information suggestive of an unusual outbreak and then act on that analysis. Physicians are not trained in statistics, outbreak detection, and management, but they are excellent observers and thoughtful in their questions. Public health officials are not clinicians, but are well trained in population-based data analysis, and in providing advice to physicians, albeit (currently) over an unacceptably long period of time.

Bioterrorism-related disease is a subset of public health problems, one that is particularly loathsome and challenging to be sure, but nonetheless one whose solution is amenable to the analytic skills of well-trained state and local epidemiologists if the data can be gathered and transported in a timely fashion.

It should be possible to put systems such as RSVP in the hands of thousands of expert observers, facilitating reporting of severely ill patients with presumed infectious disease to local public health officials and enabling rapid analysis of this information by experts in infectious disease epidemiology. As an added note, a veterinary version of RSVP is also in development, significant in that almost all BW agents are zoonotic, meaning that they cause disease

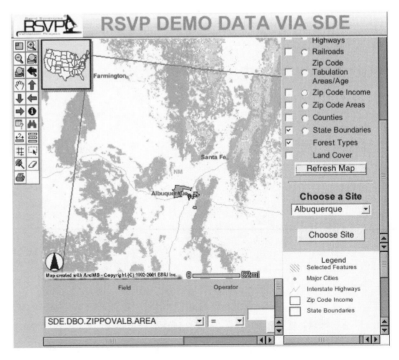

Figure 4. Respiratory symptoms in New Mexico (simulated).

primarily in animals, with humans as accidental, "dead-end" hosts.

In the end, a close partnership among medical, veterinary, and public health practitioners is essential for addressing not only the bioterrorist threat but also the diseases of economic and societal importance that must be faced continuously in the global world of the twenty-first century. The United States has been fortunate in that no major pandemic of an acute infectious disease has occurred here since 1918, but it is folly to bank on this good luck. It is only a matter of time before a large infectious disease outbreak, man-made or natural, visits the United States again. The only question is: will United States be ready?

REFERENCES

Allen, C.J. and Ferson, M.J., *Med. J. Aust.*, **172**(7), 325–328 (2000).

Camposoutcalt, D., England, R., and Porter, B., *Public Health Rep.*, **106**(5), 579–583 (1991).

Center For Counterproliferation Research, National Defense University, Symposium 2002, Toward A National Biodefense Strategy: Challenges And Opportunities, May 29–30, 2002. The Agenda of the Symposium may be seen at: http://www.e-biblus.com/af/ndu/index.cfm, accessed on 4/03/05.

Clarke, K.C., McLafferty, S.L., and Tempalski, B.J., *Emerg. Infect. Dis.*, **2**(2), 85–92 (1996).

Doyle, T.J., Glynn, M.K., and Groseclose, S.L., *Am. J. Epidemiol.*, **155**(9), 866–874 (2002).

Henderson, D.A., *Science*, **283**, 1279–1282 (1999).

Hugh-Jones, M., *Prev. Vet. Med.*, **11**(3–4), 159–161.

Hurst, J., Nickel, K., and Hilborne, L.H., *J. Am. Med. Assoc.*, **279**(6), 468–471 (1998).

Kaplan, E.H., Craft, D.L., and Wein, L.M., *Proc. Natl. Acad. Sci.*, in press.

Karim, S.S.A. and Dilraj, A., *S. Afr. Med. J.*, **86**(7), 834–836 (1996).

Nutting, P.A., Main, D.S., Fischer, P.M., Stull, T.M, Pontious, M., Seifert, M., Boone, D.J., and Holcomb, S., *J. Am. Med. Assoc.*, **275**(8), 635–639 (1996).

Schramm, M.M., Vogt, R.L., and Mamolen, M., *Public Health Rep.*, **106**(1), 95–97 (1991).

Skeels, M.R, *Mil. Med.*, **165**(7/SS), 16–19 (2000).

Timothy, J.D., Kathleen Glynn, M., and Samuel, L.G., *Am. J. Epidemiol.*, **155**(9), 866–874 (2002).

Todd, W., Sandia National Laboratories, Livermore, CA, 2003, personal communication with author.

Tsang, K.W., Ho, P.L, Ooi, G.C., et al., *New Engl. J. Med.*, **348**, 1977–1985 (2003).

SYRIA

Markus Binder

Sundara Vadlamudi

INTRODUCTION

This entry consists of two sections. The first considers Syria's history of terrorist support and examines the state's current status on terrorist relations. The second reviews the literature in order to accurately assess Syria's offensive biological weapons capability in order to appreciate what level of support might be provided to terrorist groups should Syrian policymakers choose to do so. To date, no evidence has been published demonstrating that Syria has transferred biological weapons–related information or material to any of its supported groups.

HISTORY OF SUPPORT OF TERRORISM

Syria was one of the first countries to be included in the list of state sponsors of terrorism released by the U.S. Department of State in 1979 (http://memory.loc.gov/frd/cs/sytoc .html, accessed 06/15/03), and from the mid-1980s to the present day, Syrian intelligence and security agencies have been suspected of supporting terrorism activities throughout the Middle East and Western Europe (Fig. 1). Syria's sponsorship of terrorism is allegedly coordinated by its Air Force Intelligence. According to the *London Times*, Major General Muhammad al Khawli, Air Force Intelligence Chief and late President Hafez al-Assad's adviser on national security, oversaw at least 29 terrorist operations by late 1986 (http://memory.loc.gov/frd/cs/sytoc.html, accessed 06/15/03). Major Khawli's intelligence operatives worked out of Syrian Arab Airline offices abroad or served as military attaches in Syrian embassies.

Military Intelligence Services, the "Mukhabarat," support Palestinian terrorist groups such as the Popular Front for the Liberation of Palestine-General Command (PFLP-GC) and Abu Nidal's terrorist organization, the Fatah-Revolutionary Council. The U.S. Department of State reports that Syria allowed Abu Nidal to utilize facilities in Damascus and also operate training camps in Lebanon's Bekaa valley (http://memory.loc.gov/frd/cs/sytoc.html, accessed 06/15/03). According to a number of news reports, in the 1980s, Syria provided logistical support to multiple Middle Eastern and international terrorists, including Palestinian and Lebanese groups, freelance groups, West European groups, the Japanese Red Army (JRA), Liberation Tigers of Tamil Ealam (LTTE), the Kurdish Labor Party (PKK), and the Moro National Liberation Front (MNLF) of the Philippines (http://memory.loc.gov/frd/cs/sytoc.html, accessed 12/27/03). Syria also provided support for the

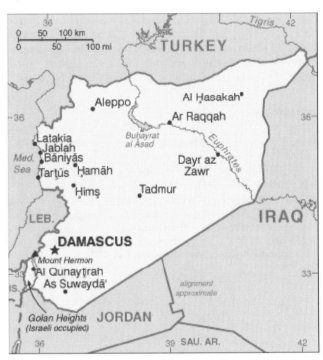

Figure 1. Map of Syria, *CIA World Factbook*, CIA website: http://www.cia.gov/cia/publications/factbook/geos/sy.html.

Iran-backed Hizballah organization. Syria has long been accused of furthering its foreign policy goals through its support for various terrorist organizations.

Current Terrorist Relations

Syria continues to support Palestinian groups such as the Popular Front for the Liberation of Palestine (PFLP), PFLP-GC, Palestine Islamic Jihad (PIJ), and Islamic Resistance Movement (HAMAS). It also acts as a transit point for Iran's supply of arms to Hizballah. The Syrian government, however, has not been accused of direct involvement in a terrorist act since 1986 (U.S. Department of State, 2002). Though Asbat al-Ansar, a group active in Lebanon, is thought to be linked to al-Qa'ida (CRS, 2002), it appears to be operating without active support from Lebanese or Syrian authorities.

Syria has articulated its support for the U.S.-led War on Terrorism and has provided significant cooperation in acting against al-Qa'ida, the Taliban, and other terrorist organizations. During times of crisis, the Syrian government has provided increased security for U.S. interests within its borders, and no terrorist attacks

Encyclopedia of Bioterrorism Defense, Edited by Richard F. Pilch and Raymond A. Zilinskas
ISBN 0-471-46717-0   Copyright © 2005 Wiley-Liss

against U.S. citizens have occurred in Syrian territory in the last five years (U.S. Department of State, 2002). In June 2002, several news reports indicated that Syria had supplied U.S. officials with information gleaned from the interrogation of Muhammad Hayder Zammar, a key suspect in the September 11 attacks (Prados, 2002). And during June 2003, Syrian President Bashar al-Asad stated that Syrian cooperation had prevented an al-Qa'ida attack that could have killed U.S. soldiers, an assertion that U.S. Assistant Secretary of State William Burns confirmed when he told a Congressional panel that "the cooperation the Syrians have provided in their own self-interest on Al-Qaeda has saved American lives" (Prados, 2002).

BIOLOGICAL WEAPONS CAPABILITY

Very little open source information exists on the topic of Syrian biological warfare (BW) activities. The bulk of open source references to Syrian BW activity consists of unclassified statements by U.S. or Israeli government officials, most claiming that there are reasons to believe that an offensive BW program exists in Syria. In contrast to discussions of Syrian chemical warfare (CW) capabilities, no details are provided supporting the BW assertions, either due to concern for protection of sources and methods or, perhaps, due to a simple lack of information. As a consequence, most discussions on this topic are confined to repetition of official assertions or extrapolations based on assessments of Syrian dual-capability industry and political motivations. Such an analysis is limited to educated conjectures of maximum capability. Furthermore, official statements regarding suspicions of the existence of such a program do not in themselves constitute confirmation of a weapons program's existence. It should also be noted that although the existence of a defensive biological weapons research capability would suggest interest—and expertise—in the field of biological weapons, it does not imply or confirm the existence of an offensive biological weapons program. A biotechnological capability does not unequivocally equate to a BW program. Finally, Syria is an authoritarian state surrounded by perceived enemies, so it makes efforts to conceal its military activities and capabilities, especially those associated with its strategic programs.

History

Syria signed the Convention on the Prohibition of the Development, Production and Stockpiling of Bacteriological (Biological) and Toxin Weapons and on their Destruction (BWC) in April 1972. Since that time, Syria has refused to ratify the BWC and has given no indications that it might change its position. For several decades, Syria has expressed a generalized opposition to weapons of mass destruction (WMD). At the same time, Syria has supported the right of any state to adopt those measures that it deems most appropriate to securing itself against outside threats—a position that has in the past been interpreted as representing a political cover for the development of WMD.

Syria appears to have acquired a limited defensive capability against biological weapons in the early 1970s, a result of receiving modern Soviet land warfare systems such as tanks and armored personnel carriers that included NBC (nuclear, biological, and chemical) protective equipment as standard equipment (Finney, 1974). It is unclear whether the Syrian protective capability against biological weapons has improved significantly since then. In the absence of an indigenous production capability, it is likely that there has actually been a decline in the Syrian capability for defense over the last decade because of its inability to obtain replacements for the aging or obsolete Russian equipment.

Since the late 1980s, Syria has undertaken a sustained effort to increase its national capabilities in the pharmaceuticals and biotechnology fields. This has involved the establishment of a number of joint-venture companies and the construction of approximately 12 pharmaceutical factories. These facilities produce for domestic and export markets. Cuba has been mentioned as having supplied Syria with biotechnological products and, possibly, know-how and equipment (see the entry entitled "Cuba"). It is possible that their construction and operation has resulted in the transfer to Syria of skills and technologies relevant to a BW program.

Allegations that Syria possesses an active offensive BW program assert that the primary Syrian biological weapons facility is located at the Damascus-based Scientific Research Council, where pathogens that cause anthrax and cholera, as well as botulinum toxin, supposedly are developed by the Biological Research Facility (Bennett, 2001). In 1992, an additional facility in the Syrian coastal town of Cerin was identified as being responsible for biological weapons production (*Middle East Defense News*, 1992). Since that time, open sources have tended to refer to a probability that Syria is developing biological weapons while avoiding making definitive statements.

Current Status

There are no clear indications that Syria currently possesses an offensive biological-weapons capability. Syria probably possesses the technical and scientific potential to research, develop, produce, and deploy biological weapons, should its leadership so decide. It also is probable, though undemonstrated, that limited research into biological weapons is undertaken by Syrian military scientists. As in many countries, this research may be undertaken to identify defensive needs and, possibly, to explore offensive military applications. Claims that Syria has weaponized botulinum toxin and ricin are dubious, given the difficulties associated with weaponizing either of these toxins (*Jane's Intelligence Digest*, 2003). Research on anthrax may be undertaken in support of efforts to improve the productivity and limit the liability of Syrian animal husbandry. Such research could be used to camouflage a military program, and may be the source of rumors that Syria is attempting to weaponize *Bacillus anthracis*. If *B. anthracis* has in fact been weaponized, it is possible that Syria would employ bomblet technology such as that developed for the dispersal of CW agents (Beaver, 1997). However, on the basis of present knowledge, any conclusions about biological weaponization or deployment modes must be speculative.

Public statements by Western intelligence agencies concur in describing Syria as possessing a limited biotechnical capability that would require significant outside technical assistance before it could undertake weaponization and large-scale production of pathogens or toxins for weapons purposes. Isolated claims that Syria has weaponized and deployed biological agents or toxins are unsupported by facts and probably reflect political goals more than technical analyses. In the absence of new revelations, it is impossible to support or refute allegations that Syria has an active BW program. It is equally difficult to make any claims regarding military or strategic aspects of this alleged program in the absence of reliable information on this subject.

Acknowledgment

The Center for Nonproliferation Studies is grateful to the Nuclear Threat Initiative (nti.org) for its support in the production of a portion of this material.

REFERENCES

Beaver, P., Syria to Make Chemical Bomblets for Scud C's, *Jane's Defence Weekly*, September 3, 1997, p. 3.

Bennett, R.M., *Middle East Intell. Bull.* **3**(8), p. 5, (2001), http://www.meib.org/articles/0108_s1.htm.

CRS Report for Congress, *Terrorism: Near Eastern Groups and State Sponsors, 2002*, Updated February 13, 2002.

Finney, J.W., Abrams Cites Intelligence Gained from Soviet Arms in Mideast, *New York Times*, February 15, 1974, p. 4.

Jane's Intelligence Digest, Syria's Secret Weapons, May 2, 2003, http://www.janes.com.

Library of Congress Country Studies, Syria, available online at http://memory.loc.gov/frd/cs/sytoc.html.

Middle East Defense News (Paris), Investigation: Syrian CW programs, September 28, 1992, pp. 5–6.

Prados, A.B., *Syria: U.S. Relations and Bilateral Issues*, CRS Issue Brief for Congress, updated July 3, 2002.

U.S. Department of State website, *Patterns of Global Terrorism*, 2002, available online at http://www.state.gov/s/ct/rls/pgtrpt/2002/.

WEB RESOURCES

Country profiles and overviews are available online at the Nuclear Threat Initiative website: www.nti.org.

T

TERRORIST GROUP IDENTIFICATION

RICHARD W. HARTZELL

INTRODUCTION

In the United States, the identification of domestic terrorist organizations (DTOs) is coordinated by local and state governments in conjunction with the Department of Homeland Security (DHS). Foreign terrorist organizations (FTOs) are designated on the basis of the criteria of the United States Code (USC); those organizations found to meet all qualifying criteria are included on a master list maintained by the U.S. Department of State. Relevant USC specifications are summarized below, after which U.S. Department of State criteria for designation as an FTO are outlined.

UNITED STATES CODE

[18 USC 2331] Terrorism, Definitions

(1) The term "international terrorism" means activities that (A) involve violent acts or acts dangerous to human life that are a violation of the criminal laws of the United States or of any State, or that would be a criminal violation if committed within the jurisdiction of the United States or of any State; (B) appear to be intended—(i) to intimidate or coerce a civilian population; (ii) to influence the policy of a government by intimidation or coercion; or (iii) to affect the conduct of a government by assassination or kidnapping; and (C) occur primarily outside the territorial jurisdiction of the United States, or transcend national boundaries in terms of the means by which they are accomplished, the persons they appear intended to intimidate or coerce, or the locale in which their perpetrators operate or seek asylum.

Immigration and Nationality (INA) 212 [8 USC 1182] (a)(3)(B), Classes of Aliens Ineligible for Visas

(3) Security and related grounds; (B) Terrorist activities; (iii) TERRORIST ACTIVITY DEFINED. The term "terrorist activity" means any activity which is unlawful under the laws of the place where it is committed (or which, if it had been committed in the United States, would be unlawful under the laws of the United States or any State) and which involves any of the following: (I) The hijacking or sabotage of any conveyance (including an aircraft, vessel, or vehicle); (II) The seizing or detaining, and threatening to kill, injure, or continue to detain, another individual in order to compel a third person (including a governmental organization) to do or abstain from doing any act as an explicit or implicit condition for the release of the individual seized or detained; (III) A violent attack upon an internationally protected person or upon the liberty of such a person; (IV) An assassination; (V) The use of any (a) biological agent, chemical agent, or nuclear weapon or device, or (b) explosive, firearm, or other weapon or dangerous device, with intent to endanger, directly or indirectly, the safety of one or more individuals or to cause substantial damage to property; (VI) A threat, attempt, or conspiracy to do any of the foregoing.

U.S. STATE DEPARTMENT CRITERIA FOR DESIGNATION AS A FOREIGN TERRORIST ORGANIZATION

Former Secretary of State Madeleine K. Albright approved the designation of an initial 30 FTOs in October 1997, on the basis of the following criteria:

1. The organization must be foreign.
2. The organization must engage in terrorist activity as defined in Section 212 (a)(3)(B) of the Immigration and Nationality Act.
3. The organization's activities must threaten the security of U.S. nationals or the national security (national defense, foreign relations, or the economic interests) of the United States.

An updated list is available on the U.S. Department of State website.

WEB RESOURCES

Office of the Coordinator for Counterterrorism, U.S. Department of State, http://www.state.gov/, accessed on 7/4/03.

U.S. Department of Homeland Security, http://www.ready.gov/, accessed on 7/4/03.

U.S. Department of State Foreign, Terrorist Organizations listing, http://www.state.gov/s/ct/rls/fs/2003/17067.htm, accessed on 7/4/03.

Encyclopedia of Bioterrorism Defense,   Edited by Richard F. Pilch and Raymond A. Zilinskas
ISBN 0-471-46717-0   Copyright © 2005 Wiley-Liss

THREAT REDUCTION IN THE FORMER SOVIET UNION

Anne Harrington

INTRODUCTION

Containing the proliferation of former Soviet biological weapons expertise and materials from Russia and Eurasia is a high-priority nonproliferation objective shared by the United States, Russia, and other countries. Various aspects of the U.S. efforts in this area have been reviewed independently by the U.S. General Accounting Office and the National Academy of Sciences and are generally viewed as successful (U.S. General Accounting Office, 2000; National Academy of Sciences, 1997). More than a decade after the dissolution of the Soviet Union, the reasons for launching these efforts continue to be seen as valid. Scientists who were in their early 40 s when the Soviet Union dissolved and had spent 10 to 20 years in weapons work still have a decade or more of productive work time ahead of them and thus continue to pose a proliferation risk. In the meantime, economic conditions do not provide the necessary stability to ensure that long-term, meaningful employment will be available. Although the European Union (both collectively and as individual member states), Japan, South Korea, Norway, and other countries support overall nonproliferation goals and preventing the proliferation of former Soviet biological warfare (BW) expertise and materials, the United States has played the lead role in these efforts.

OVERVIEW OF WEAPONS OF MASS DESTRUCTION ENGAGEMENT

The disintegration of the Soviet Union raised concerns in late 1991 that there could be wholesale proliferation of the former superpower's weapons of mass destruction (WMD) and delivery systems expertise. Many U.S. experts feared that tens if not hundreds of thousands of scientists, engineers, and technicians with expertise in WMD design, production, or weaponization might be vulnerable to offers from states or nonstate entities seeking such expertise. Of greatest initial concern was the risk of nuclear weapons proliferation, with concern

over biological weapons proliferation perceived as a more remote threat.

In addition to U.S. concerns, Germany had its own set of observations based on its experience with the German Democratic Republic. After the Berlin Wall fell in 1989 and the German reunification process began in 1990, officials in the Federal Republic of Germany developed a deeper understanding of the scope and breadth of the Soviet Union's WMD aspirations based on information from former East Germans. In late 1991, U.S. Secretary of State James Baker III and his German counterpart Hans-Dietrich Genscher began a dialogue to look at what might be done to prevent the Soviet Union's WMD experts and expertise from reaching states or nonstate groups seeking to develop WMD programs. Their working plan was to establish a program that could provide peaceful, civilian research opportunities to former Soviet weapons scientists. This concept was explored and developed with the Russian Foreign Minister, Andrei Kozyrev. The three foreign ministers reached an agreement on a general framework, and the three countries issued a joint statement on February 17, 1992, announcing their intent to establish an International Science and Technology Center (ISTC), which many Russian scientists knew by the first initials of the three ministers involved in its creation: "the KGB Center." The European role was ceded by Germany to the European Communities, and after a year of intense preparatory activity, the ISTC made its first grants at the March 1994 inaugural meeting of its Governing Board.

The initial ISTC parties, the United States, Russia, the European Communities (later the European Union), and Japan, funded 23 projects at that meeting. Project funding for the large Russian nuclear weapons facilities easily dominated the funding list, with no projects funded in the biotechnology area. It was not until June 1994 that the first biotechnology project was funded. Funding in this area grew steadily but slowly over the next few years, remaining at relatively low levels well into the 1990s. For a variety of reasons described in more detail below, however, there was a significant expansion

Encyclopedia of Bioterrorism Defense,   Edited by Richard F. Pilch and Raymond A. Zilinskas
ISBN 0-471-46717-0   Copyright © 2005 Wiley-Liss

of BW engagement efforts in the late 1990s. By the end of 1999, ISTC funding for Biotechnology and Life Sciences accounted for 18.6 percent of total program funding. By the end of 2002, that percentage had risen to 25 percent, reflecting a shift in funding priorities led particularly by the United States. Another illustration of the priority shift is drawn from the ISTC 1999 Annual Report, which includes a graph showing that in 1995 fewer than 100 persons at the State Research Institute of Virology and Biotechnology, "Vector," were involved in fewer than five ISTC projects. By the end of 1999, almost 30 projects at Vector engaged nearly 400 people.

In addition to the multilateral ISTC, several bilateral U.S. government–funded nonproliferation programs played an important early role in limiting the proliferation of former Soviet WMD expertise. The U.S. Department of Energy's (DOE) Initiatives for Proliferation Prevention (IPP, initially the Industrial Partnering Program) program also began with a heavy focus on the nuclear weapons community, but included a portfolio of projects in biotechnology, chemistry, materials, and other areas. The U.S. National Academy of Sciences (NAS), with support from the U.S. Department of Defense (DoD), launched a series of institute visits in the mid-1990s, and in 1996 funded eight pilot projects aimed at demonstrating the value of cooperative research on dangerous pathogens. A number of the initial NAS pilot projects were later continued with direct funding from the DoD. The NAS issued a report in 1997 making a number of recommendations to this end, and continues to study the issue actively (National Academy of Sciences, 1997).

THE SLOW EVOLUTION OF BIOENGAGEMENT

There were several reasons for the slow evolution of funding for the Soviet Union's former BW community. First, the specter of "loose nukes" finding their way into unfriendly hands preoccupied decision makers and drove funding in that direction. Second, owing to decades of arms control negotiations, joint verification experiments, and other contacts, there was a reasonable degree of familiarity between the U.S. and Soviet nuclear communities. Scientists at the DOE national laboratories and their former Soviet counterparts generally knew of each other's key personnel, and each side had invested significantly in gathering intelligence on the other. Following initial U.S. visits to Russia by DOE laboratory scientists, cooperation in the nuclear area got off the ground quickly under a series of informal lab-to-lab meetings and contracts. This was followed by project-based work under the ISTC and DOE's IPP program. There was no such avenue for identifying or contacting the former Soviet biological weapons community.

Third, for years there had been suspicions that the Soviet Union had continued an illegal biological weapons program after it signed (April 10, 1972) and ratified (March 26, 1975) the Biological and Toxin Weapons Convention (BWC). A number of events, including a suspicious 1979 anthrax outbreak that killed a still undetermined number of people in and around a military facility in Sverdlovsk

(now Yekaterinburg), fueled concerns. Those suspicions of an ongoing Soviet program were confirmed in 1989 with Dr. Vladimir Pasechnik's defection to the United Kingdom. Dr. Pasechnik, a senior scientist at a Leningrad (now St. Petersburg) institute belonging to an organization called Biopreparat, provided the first major insights into the Soviet Union's use of Biopreparat as a civilian cover for an extensive network of research institutes used, at least in part, to support offensive biological weapons development.

Much more became known about Biopreparat when Dr. Kanatjan Alibekov (now Ken Alibek), a Biopreparat deputy director, defected to the United States in 1992. His book, articles, and interviews provided deeper insight into both the structure and operation of the Soviet BW program as well as the Soviet Union's rationale for continuing this pursuit despite its flagrant violation of international treaty commitments (Alibek and Handelman, 1999). Also in 1992, Russian President Boris Yeltsin made a statement essentially confirming that the Soviet Union had indeed continued an illegal biological weapons program for two decades, adding a pledge to halt all such activities immediately. At that time, an agreement was reached between the three BWC repository states, the United States, United Kingdom, and Russia, to initiate a process to assure each other and the international community that all work inconsistent with the BWC had halted. This trilateral process succeeded in securing visits to Russian and U.S. civilian sites in 1993 and 1994, including major Biopreparat facilities, but stalled when the Russians balked at allowing visits to its military biological sites.

In the late 1990s, further information was revealed in the memoirs of Dr. Igor Domaradskij, who still resides in Russia. The memoirs were expanded and published as a book in 2003 (Domaradskij and Orent, 2003). Concerns were compounded by the lack of a complete and official accounting by the Russian Federation of the Soviet biological weapons program, including the role of Biopreparat. There still is no definitive documentary history of the Soviet Union's offensive biological weapons program.

The Russian Federation's failure to be transparent and forthcoming about the Soviet biological weapons history fueled continued concerns that Russia may not have halted all of its activities in this sphere. The Russian unwillingness to provide a full account in the face of overwhelming evidence was in stark contrast to other Eurasian countries, which inherited smaller, but nonetheless, significant Soviet biological weapons facilities. In Kazakhstan, for example, the government sought and received U.S. government assistance to dismantle a Soviet *Bacillus anthracis* mobilization, production, and weaponization facility at Stepnogorsk. The project provided a firsthand look at the equipment and production technologies developed for the Soviet BW program. The Uzbek government, concerned about potential health and ecological disaster, worked with the U.S. government to dismantle and completely decontaminate the Soviet Union's BW test site on Vozrozhdeniye Island in the Aral Sea. With the Aral Sea shrinking rapidly, a land bridge between the former BW test site and the mainland was formed in 2002. The Uzbeks were concerned that if there were still viable pathogens on the island, they could be transferred to the mainland by animals.

A fourth reason was that a number of individuals who had functioned in the Soviet BW civilian or military leadership continued (and continue to this day) to hold senior official positions, with the apparent approval of the Russian government. The Biopreparat organization also continues to exist and exert influence over a number of facilities, and there has never been any success in gaining transparency into former Soviet (now Russian) military biological facilities. Russia's failure to "clean house" and lack of sensitivity to outside perceptions of the continuing role of individuals and organizations previously involved in the development of offensive biological weapons spurs doubt and suspicion.

A fifth and final point centered on whether or not an engagement plan should include work on dangerous pathogens and to what extent providing support would maintain or even enhance the ability of former Soviet scientists and institutes to maintain or reconstitute an offensive BW capability.

In recent years, the overall atmosphere has been further complicated by the global focus on biodefense research. The very real threat of a bioterrorist attack has spurred research efforts in the United States, Europe, Russia, and elsewhere to better protect populations. Biodefense research is consistent with the BWC, but out of necessity must often look at many of the same issues of pathogenicity, drug resistance, and delivery mechanisms that are studied in the context of offensive application, making it even more difficult to discern the intent of the research and its consistency with the letter and spirit of the BWC.

This atmosphere of distrust was countered by a growing concern that the scientific community involved in the Soviet Union's BW program should be engaged and redirected to legitimate health, pharmaceutical, vaccine, environmental, and other appropriate work before it proliferated beyond the borders of the former Soviet Union. Furthermore, while the United States had abandoned its offensive biological weapons program in 1969, before the greatest advances of molecular biology and genetic engineering, the scientists who had devoted 20 years to developing Soviet biological weapons during that same period were able to incorporate those advances into their work, and thus could potentially provide advanced information to state or nonstate groups seeking biological weapons. Although the decision to engage was compelling, it was still approached with caution because of the dual-use nature of the science, materials, and technologies.

There were some unique psychological barriers to overcome as well before successful engagement was possible. Cooperative work with the Russian nuclear community was aimed at redirecting those portions of the community not needed for weapons work toward civilian scientific and commercial activities. It was assumed that Russia, as a nuclear state under the Nuclear Nonproliferation Treaty, would retain its nuclear weapons capability as well as the technical ability to secure and maintain its nuclear weapons stockpile in a responsible

way. On the other hand, Russia's inherited BW experts could not avoid the stigma of having been involved in treaty violation. Russian scientists who had been part of the Soviet program were tacitly required to admit to contributing to violation of an international treaty in order to qualify for funding under nonproliferation programs. Many felt that despite their isolation and stigmatization, what they had created had been no worse than the weapons their nuclear colleagues had developed, and that they, above all, had been loyal to the Soviet state. The challenge was to find a way to engage the former Soviet biological weapons network, and build trust and transparency, without increasing their ability to sustain or develop further offensive biological weapons in Russia or elsewhere.

THE ENGAGEMENT EFFORT BEGINS

During 1994–1997, a growing number of projects were funded in Russia and other Eurasian countries that had inherited pieces of the Soviet Union's BW program. Most of the largest research and production centers were in Russia, but significant facilities with expertise, technology, and materials of nonproliferation concern were also found in Kazakhstan, Uzbekistan, Georgia, Armenia, and Ukraine. The most notable of these was Biomedpreparat in Stepnogorsk, Kazakhstan, a vast facility intended to weaponize and produce *B. anthracis* during wartime mobilization. The facility has been neutralized under a DoD Cooperative Threat Reduction dismantlement program. The scientists who worked at the facility are now developing an environmental monitoring laboratory with funding from the Department of State's biological and chemical weapons redirection program, which includes collaboration with the U.S. Environmental Protection Agency (EPA) and business support from the International Executive Service Corps (IESC).

By working with research and production facilities in non-Russian states, the United States and other countries hoped to demonstrate that the nonproliferation programs were focused on the present and future, not the past. This was only partially successful. For example, the lag time between the dismantlement of Biomedpreparat in Stepnogorsk and the development of substantial activity at the environmental monitoring laboratory gave some Kazakh and Russian scientists and officials the impression that the United States only intended to tear down facilities without establishing anything in their place.

Early BW redirection projects focused on a relatively few Russian institutes that were considered high priorities for engagement, for example, Vector, the State Research Center for Applied Microbiology (Obolensk), the Institute of Highly Pure Biopreparations (St. Petersburg), and the Institute of Immunological Engineering (Lyubuchany).

The slow pace of engagement changed in the late 1990s, however, when the threat of losing Soviet biological weapons technology to a state with weapons aspirations, Iran, gained public attention in a front page *New York Times* article dated December 8, 1998 (Miller et al., 1998). Judith Miller, the principal author, confirmed in interviews with Russian and Kazakh scientists that Iranians serving in various official capacities had approached them seeking scientific cooperation, training, and, in two reported cases, help in the development of biological weapons. That event, in combination with several other factors delineated below, led to the significantly expanded set of activities that exists today.

The first additional factor was the Russian financial collapse in 1998. Russia's economy had begun to recover in the mid-1990s, and at the time many believed that Russia had finally turned the corner and might even be able to take more responsibility for funding its security requirements, including the cooperative threat reduction programs aimed at redirecting WMD scientists. When instead the situation suddenly and dramatically regressed, the perceived threat to Russia's former BW facilities increased. A second factor that changed the face of engagement was the perceived shift in terrorist activity demonstrated by the attacks of September 11, 2001, and the "anthrax letters" the following month, which together effectively eliminated the theoretical aspect of terrorist deployment of biological weapons on U.S. soil. Seeking ways to foster cooperation between the two states with the most knowledge about biological weapons, the United States and Russia, seemed a logical step toward addressing the threat.

A third factor was the increase of global terrorist activity, including against Russia. As a result of becoming a target of terrorism, Russia's anti- and counterterrorism efforts increased (Russian Ministry of Foreign Affairs, 2001), particularly those directed toward the cycle of violence associated with Chechen elements. Incidents to date primarily have involved conventional explosives, including the October 2002 Moscow theatre hostage crisis and the July 2003 suicide bombing at a Moscow rock concert. However, Chechen terrorists are credited with least one radiological event (Monterey Institute WMD-Terrorism Database, 2003), and there are serious concerns about the possibility of a biological attack based, in part, on threats made by Chechen elements to obtain biological agents from Russian facilities (based on private conversations with institute directors and guidance that they have received). These concerns are reinforced by evidence that certain Chechen factions may have tried to use poisons to contaminate food and have considered contamination of water sources, in addition to the widely accepted ties between Chechen groups and international terrorist networks (http://www.wps.ru/chitalka/media_politic/en/20010816.shtml#7, accessed 12/05/03).

Bioterrorism has become an increasingly common topic at Russian conferences, and Russian government, scientific, health, and academic circles agree on the serious nature of the bioterrorist threat to Russia. The highest-level statements on bioterrorism have come from the president of the Russian Federation, Vladimir Putin. Two of the most significant have been made jointly by President Putin and U.S. President George W. Bush. The first came at the October 2001 Asia-Pacific Economic Conference, when the two presidents "resolved to advance cooperation in combating new terrorist threats: nuclear, chemical and biological, as well as those in cyberspace. They agreed

to enhance bilateral and multilateral action to stem the export and proliferation of nuclear, chemical and biological materials, related technologies, and delivery systems as a critical component of the battle to defeat international terrorism" (United States Department of State, 2001).

The two presidents expanded on this notion on November 13, 2001, at the Washington/Crawford Summit, graduating from an expression of joint concern to a commitment to pursue a joint program to combat bioterrorism (Fig. 1). Since then, however, progress toward implementing that commitment at the official level has been slow.

In addition to President Putin's pronouncements, other members of his government also have made combating bioterrorism a high priority. Former Minister of Health Yuri Shchevchenko highlighted the need to be prepared to respond to a bioterrorism threat on a number of occasions, including in his speeches to the World Health Organization (WHO). He approached the issue with a special perspective, having previously been the director of the S.M. Kirov Military Medical Academy in St. Petersburg, which has a long history of clinical and research activities for military force health and protection. Other Russian officials have focused specifically on biosecurity (FBIS, 2003).

COMBATING BIOTERRORISM AS AN ELEMENT OF ENGAGEMENT

Similar to U.S. and others' efforts to counter bioterrorism, the Russian Government launched a federal biodefense program under Resolution No. 737 of July 2, 1999, aimed at "defending the population and environment against hazardous and extremely hazardous pathogens in natural and manmade emergency situations." The Russian program is similar in concept and structure to those in many other countries where bioterrorism has become a high-priority concern. The program is planned to last from 1999 to 2005, with the Ministry of Health, Ministry of Defense, and Ministry of Food and Agriculture as the implementing agencies. Each agency has a set of mandatory competitive projects, and all work is to be conducted for peaceful purposes with transparency to the Russian and global publics.

The Ministry of Health is responsible for the development of rapid, high-quality diagnostics; clinical treatment plans using immunoglobulins, as well as new antimicrobial and antiviral reagents in combination with natural immune response modifiers; disinfectants; and improved sampling methodologies for bacterial and viral research, and a strategic approach to monitoring pathogens in Russia. The Ministry of Health is also responsible for reconstructing and modernizing the production base needed for combating pathogens.

The Ministry of Defense is responsible for developing mathematical models to track outbreaks and spread of dangerous infectious diseases where military units are deployed during manmade disasters; diagnostic reagents, test systems, devices, and other detection systems for identifying pathogens, including those not endemic to Russia (and ecopathogens that can damage equipment);

13 November 2001

The White House
Office of the Press Secretary
November 13, 2001
Joint Statement By President George W. Bush And President Vladimir V. Putin On
Cooperation Against Bioterrorism

At Shanghai, we resolved to enhance cooperation in combating new terrorist threats, including those involving weapons of mass destruction.

We agree that, as a key element of our cooperation to counter the threat of terrorist use of biological materials, officials and experts of the United States and Russia will work together on means for countering the threat of bioterrorism, now faced by all nations, and on related health measures, including preventive ones, treatment and possible consequence management.

We will continue to work to enhance the security of materials, facilities, expertise,and technologies that can be exploited by bioterrorists. We also confirm our strong commitment to the 1972 Convention on the Prohibition of the Development, Production and Stockpiling of Bacteriological (Biological) and Toxin Weapons and on Their Destruction.

We have directed all of our officials and experts working on these critical matters to expand their cooperation and to consult on strengthening related international efforts.

Credit: http://www.whitehouse.gov/news/releases/2001/11/20011113-7.html

Figure 1. Joint statement by President George W. Bush and President Vladimir V. Putin on cooperation against bioterrorism.

new preventative and treatment reagents and related production technologies against glanders, melioidosis, viral hemorrhagic fever, encephalitis, parenteral hepatitis, and other human and ecological diseases; improved means of collecting air, soil, and water samples for pathogens; and production capability to produce personnel protective gear and protection for equipment and armaments against ecopathogens.

The Ministry of Food and Agriculture is responsible for development and production technologies for a new generation of diagnostics and test systems for agricultural diseases, including foot-and-mouth disease, African and classic swine fever, plague, sheep and goat pox, rabies, anthrax, and others; emergency treatment measures; technical means for express diagnosis and detection of pathogens, mass vaccination, and disinfection; means of protecting food crops; mathematical models for epizootic outbreaks for hazardous and extremely hazardous animal diseases; and reconstructing and modernizing the production base for veterinary reagents.

The tasks laid out for the three Russian ministries bear a striking similarity to the kind of cooperative research and development that has been funded under nonproliferation programs. Diagnostics, new preventative and treatment drugs, research on dangerous pathogens, disinfection and decontamination technologies, mathematical models of biological events, animal health, disease monitoring, and other areas are already the subject of collaborative work under nonproliferation programs and potentially could make significant contributions to combating bioterrorism in Russia, the United States, and elsewhere. In many respects, these projects represent a substantial amount of de facto cooperation to combat bioterrorism.

In 2003, the Russian Ministry of Science, Industry and Technology added its own effort, launching an open competition for a new program to fund biotechnology research and development in a number of biodefense areas.

MAKING ENGAGEMENT WORK

In addition to the motivating factors the led the United States to engage the former Soviet biological weapons community, there were motivations from the Russian side as well. The drastic cut in directed research from Biopreparat and/or the Ministry of Defense left most institutes destitute. If they were to survive, the former Soviet BW community would have to link to external partners who could help them identify meaningful research and development activities for which there was a legitimate scientific or commercial market. However, few of the scientists had substantial contacts outside the Soviet Union, and early project proposals indicated a significant lack of knowledge of the state of the art in many disciplines.

A second motivating factor was the chance to gain professional recognition from international peers. Much of the research during the Soviet period was published in classified form, leaving the scientists performing BW-related research without any means to demonstrate their capabilities to their internal or external peers. Some in the Soviet BW community had the view that engaging themselves in transparent cooperative research was a way to redeem themselves in the eyes of the world.

A related issue concerned the fact that much of the Soviet era research, although interesting on the surface, was not supported by the kind of rigorous methodology and data that is expected by United States and other international-standard research communities. Therefore, to make the leap to the international stage and scientific and/or moral legitimacy, the former Soviet BW community had to learn a new way of doing business, and that required external help.

Despite motivation on both sides, engagement was difficult. Funding for construction and maintenance of most Soviet military-supported facilities ended in the late 1980s. When United States and other visitors made scientific visits to facilities, they found crumbling mortar, common areas overgrown by grass, and largely empty and inactive laboratories. Institutes that tried to continue some level of research often had to resort to using washed food cans in place of standard laboratory glassware. Some early projects involving potentially promising vaccines and drugs quickly encountered the problem of how to properly conduct preclinical trials. Without good data based on well-documented protocols using clean animals, the chances of moving any promising nonproliferation program-sponsored research into health or commercial markets were negligible. Animal facilities at institutes were largely substandard, laboratory animal care and use programs were virtually nonexistent, and no certified animal breeding colonies could be identified in Russia. With growing demands from animal rights groups to reduce or eliminate animal testing, it was not acceptable to U.S., European, Japanese, or other collaborators to export research requiring animal studies to substandard facilities.

The solution to the preclinical trials problem demonstrates how creative program leveraging and coordination not only solved the immediate problem but also introduced a new standard to the emerging Russian biotechnology sector. For example, a facility was identified at the Pushchino Branch of the Shemyakin and Ovchinnikov Institute of Bio-organic Chemistry that was designed for animal breeding. Originally it was financed under an arrangement with the Soviet Ministry of Defense, but its construction was halted in the late 1980s and the building was never used despite the fact that its construction and design were fundamentally sound. With the assistance of a team of U.S. DoD veterinarians, outside consulting services, funding from U.S. contributions to the ISTC, significant amounts of training, and an important collaboration with Charles River Laboratories, the facility is now able to sell clean laboratory mice and rats to research institutes in Russia. In addition, internal animal care and use committees were established in the institutes receiving U.S. nonproliferation funding, and Laboratory Animal Care and Use Guidelines were adopted by the ISTC Governing Board. This example illustrates how complex the task of redirecting former Soviet BW expertise is and the number of obstacles that have to be overcome in order to develop a solid basis for sustainable civilian employment.

Persuading companies to participate in collaborative efforts was an early part of U.S. government strategy, acknowledging that ultimately market forces would determine whether the facilities would survive in a competitive environment. Early contacts showed that Soviet scientists had developed some interesting scientific approaches and technologies that had not been explored in the West, but since there had not been full disclosure of past histories, most companies hesitated to commit to anything beyond supporting basic research. Production was generally out of the question, given the overall poor physical condition of many of the large fermentation facilities. The DOE IPP program had the ability to utilize U.S. industry and the scientific and technical skill of its national laboratory system, and by the late 1990s

had a portfolio of success stories, but even IPP with its strong commercialization focus found it difficult to move biotechnology projects to a sustainable level. It was only in 2002–2003, when the Department of State received Congressional funding for a Bio-Industry Initiative (BII), that more specific focus was given to solving the problems of biotechnology sustainability. Working with a targeted number of facilities, BII focuses on analyzing, then resolving, obstacles to success for production facilities and for accelerated drug and vaccine development, particularly against highly infectious pathogens. The program is driven by its Congressional authority to pursue these activities as an element of the defense against terrorism. Ironically, the same research and technologies that made the Soviet BW program such a threat are now being explored as part of the larger counterterrorism effort.

Another early challenge was the need to expand U.S. participation beyond the traditional security agencies, that is, the Departments of State, Defense, and Energy. To encourage and evaluate projects that redirected scientists to human, animal, and plant health, as well as environmental issues, required much broader U.S. government involvement. Thus, with funding from the Department of State, the Department of Health and Human Services created the Biotechnology Engagement Program (BTEP); the Department of Agriculture engaged the Agricultural Research Service; and the EPA brought its expertise and scientific and technical base to bear. The addition of these agencies was critical to expanding engagement activities into areas of identified research demand. For example, projects to monitor measles and mumps were developed between Vector and the Food and Drug Administration; the Agricultural Research Service became the partner on a project to look at possible aerosol animal vaccines with the Research Center of Toxicology and Hygienic Regulation of Biopreparations; and EPA worked with the same institute to validate a PCB (polychlorinated biphenyl) remediation candidate. Until recently, when more stringent enforcement of visa regulations was implemented, exchanges and travel between former Soviet BW experts and their U.S. collaborators flourished.

As new project potential expanded, concerns still persisted about what technologies should be encouraged. For example, the Soviet Union explored the mechanisms of aerosol delivery, and a number of facilities still maintain both the expertise and technology to do advanced aerosol work. Whereas aerosol delivery is strongly associated with biological weapons, it has now made the transition into the commercial world with the Food and Drug Administration's 2003 approval of Medimmune's FluMist© influenza vaccine. Aerosol vaccination is also one of the potential technologies for quickly vaccinating large human and/or animal populations in the event of either a natural of planned biological event. This provides the potential to link former Soviet scientists and facilities to legitimate, transparent use of their science and technologies.

Promoting a more open dialogue and expanded cooperative research, including that which involves U.S. military experts, could help identify additional areas where former BW expertise can be legitimately employed. At a 2001 international conference on combating bioterrorism sponsored by the ISTC and hosted by the S.M. Kirov Military Medical Academy in St. Petersburg, Russia revealed the extent to which the U.S., Russian, and even European research communities are already working cooperatively and productively in such sensitive areas. A sampling of the papers from that conference follows:

- "Molecular Epidemiology as a Main Approach to Detecting the Terrorist use of Infectious Agents," Vector, Koltsovo

- "Biological Microchips for Identification of Pathogenic Bacteria and Viruses," Engelhardt Institute of Molecular Biology, Moscow

- "Novel Immunoassay Methods in the Fast Detection of Pathogens," Research Center for Molecular Diagnostics and Therapy, Moscow

- "Biosensors and Nanotechnological Methods for Detection and Monitoring of Chemical and Biological Supertoxins: Systems for Destruction of Toxins and Microbes," Moscow State University, Moscow

- "Cytokine Level and Therapy of Infectious Diseases," State Research Center of Highly Pure Preparations, St. Petersburg

- "Antibacterial Resistance Emerging and Dissemination as a Threat to Biological Safety," Institute of Antibiotics, Moscow

- "Clinical Diagnostic Aspects of Bioterrorism," S.M. Kirov Military Medicine Academy, St. Petersburg

- "Risks and Benefits of Natural and Artificial Toxins for the Present and Future," Saclay (France).

PROLIFERATION OF PATHOGEN COLLECTIONS

In the process of engaging the former Soviet BW facilities, it became quickly apparent that multiple pathogen collections existed inside and outside Russia, of which many contained multiple strains of pathogens, including highly dangerous pathogens. Most of them appeared to be in inadequately secured locations. Therefore, a principal thrust of U.S. nonproliferation engagement has since been to identify and secure these collections. Assistance is provided primarily through the DoD Biological Weapons Proliferation Prevention (BWPP) Program. Significant progress has been made at select facilities posing the most immediate concerns for insider or outsider theft, including Vector, Russia's WHO smallpox virus repository. Work in Russia is complicated, however, due to the lack of an implementing arrangement under the DoD Cooperative Threat Reduction Umbrella Agreement that would allow DoD to carry out what are basically engineering tasks at sites reporting to multiple ministries within the Russian government. In the absence of an interlocutor on the Russian side, DoD has worked instead through the ISTC. Although progress has been made, it has been slower than in other Eurasian countries where DoD works directly.

CONCLUSION

The former Soviet BW community in Russia and elsewhere in Eurasia continues to express interest in nonproliferation engagement programs. The ability of some to sustain their activity is influenced significantly by economic developments, including the implementation of bankruptcy laws. Already, the State Research Center for Applied Microbiology, the Biologics Plant at Berdsk, and others are in the process of financial reorganization. In some cases, assets will be stripped and privatized to pay back bills, leaving structures behind that may not be able to support the full scientific complement, and raising the risk again of proliferation of BW expertise. As Russia continues to try to bring its economy into line, the large Soviet BW network, with its sprawling, inefficient design, large utility consumption, and dependency on external funding, will continue to pose a challenge to the nonproliferation community.

REFERENCES

FBIS, *Biological Weapons and Problems of Ensuring Biological Security*, Doc ID CEP20030716000407[0], 07/16/2003.

http://www.wps.ru/chitalka/media_politic/en/20010816.shtml#7, accessed on 12/05/03.

Miller, J. and Broad, W.J., Iranians, Bioweapons in Mind, Lure Needy Ex-Soviet Scientists, *New York Times*, December 8, 1998.

Monterey Institute WMD-Terrorism Database, 2003, http://cns.miis.edu.

National Academy of Sciences News Release, *US-Russian Cooperation in Infectious-Disease Research Could Contribute to National Security, Public Health Goals*, November 7, 1997, http://www4.nationalacademies.org/news.nsf/isbn/11071997?OpenDocument. This entry includes the list of NAS pilot projects. The full report can be found at, http://www.nap.edu/catalog/9471.html?onpi_newsdoc110797. See also a description of the NAS project, *Future Contributions of the Biosciences to Public Health, Agriculture, Basic Research, Counterterrorism, and Non-Proliferation Activities in Russia*, funded by the Nuclear Threat Initiative, http://www4.nas.edu/webcr.nsf/ProjectScopeDisplay/DSCX-N-02-03-A?OpenDocument.

Russian Federation Resolution, No. 737, Concerning the focused federal program for *The creation of methods and means of defending the population and environment against hazardous and extremely hazardous pathogens in natural and manmade emergency situations from 1999 to 2005*, July 2, 1999.

Russian Ministry of Foreign Affairs Daily News Bulletin, 13.11.2001, Speech by Russian President Vladimir Putin at a Meeting with the Commanding Personnel of the Russian Armed Forces, the Ministry of Defense, Moscow, November 12, 2001, http://www.ln.mid.ru/Bl.nsf/arh/CFC5DFF2F9609DD343256B0400336A21?.

United States Department of State, 2001. Joint Statement on Counterterrorism Issued by Bush, Putin: http://usinfo.state.gov/topical/pol/terror/01102204.htm, accessed on 11/23/03.

U.S. General Accounting Office, *Biological Weapons: Efforts to Reduce Former Soviet Threat Offers Benefits, Poses New Risks*, GAO-NSIAD-00-138, GPO, Washington, DC, April 2000, http://www.gao.gov/new.items/ns00138.pdf.

FURTHER READING

Alibek, K. and Handelman, S., *Biohazard: The Chilling True Story of the Largest Covert Biological Weapons Program in the World–Told from Inside by the Man Who Ran It*, Random House, New York, 1999.

Domaradskij, I.V. and Orent, W., *Biowarrior - Inside the Soviet/Russian Biological War Machine*, Prometheus Books, New York, 2003.

Smithson, A.E., *Toxic Archipelago: Preventing Proliferation from the Former Soviet Chemical and Biological Weapons Complexes*, Report No. 32, The Henry L. Stimson Center, December 1999, http://www.stimson.org/cbw.pdf/toxicarch.pdf.

Tucker, J., Bioweapons from Russia: Stemming the Flow, *Issues in Science and Technology Online*, Spring 1999, http://www.nap.edu/issues/15.3/p_tucker.htm.

WEB RESOURCES

Harrington, A.M., *Chemical Weapons Convention Bulletin*, Issue No 29. September 1995, http://www.fas.harvard.edu/hsp/ bulletin/cwcb29.pdf. (This 1995 publication outlines an early argument for engaging former Soviet biological weapons expertise and suggests a number of redirection pathways.)

http://english.pravda.ru/hotspots/2002/07/06/31943_.html, accessed on 11/23/03.

http://www.library.cornell.edu/colldev/mideast/irangrm.htm, accessed on 11/23/03.

http://usinfo.state.gov/topical/pol/terror/01102204.htm, accessed on 11/23/03.

International Science and Technology Center, www.istc.ru, accessed on 11/23/03.

Online Pravda, Chechen terrorists plotting to use Georgia-made poisonous substances in terror acts. http://english.pravda. ru/hotspots/2002/07/06/31943_.html, accessed on 11/23/03.

Russian Ministry of Foreign Affairs Daily News Bulletin, 26.04.2002, Russian President Vladimir Putin Answers Questions During Press Conference on Outcomes of *Conference on Problems of Caspian Region*, Astrakhan, April 25, 2002, http://www.ln.mid.ru/Bl.nsf/arh/D2AD9D877F5E96B243256 BA8003234AE?.

Russian Ministry of Health, http://www.minzdrav-rf.ru/.

Russian Ministry of Science, Industry and Technology, http://www.naukanet.org/friends/naukanet/sponsors/ministry. html(opt,mozilla,unix,english,,MirAquaL.

Russian Ministry of Defense, http://www.mil.ru/index.php?menu_ id=330.

Science and Technology Center in Ukraine, www.stcu.int.

U.S. Department of Defense Cooperative Threat Reduction Program, http://www.dtra.mil/ctr/ctr_index.html.

U.S. Department of State BioIndustry Initiative Fact Sheet, http://www.state.gov/t/np/rls/fs/24242.htm. See also the September 27, 2003, White House fact sheet on HIV/AIDS and BII support, http://www.whitehouse.gov/news/releases/2003/09 /20030927-9.html.

U.S. Department of Health and Human Services Biotechnology Engagement Program, http://www.globalhealth.gov/ europeaffairsdhhs.shtml.

U.S. NNSA Initiatives for Proliferation Prevention, http://www. nnsa.doe.gov/na-20/ipp.shtml.

See also DEFENSE THREAT REDUCTION AGENCY; DEPARTMENT OF DEFENSE; DEPARTMENT OF HEALTH AND HUMAN SERVICES; and DEPARTMENT OF STATE.

TOPOFF

Praveen Abhayaratne

INTRODUCTION

Mandated by the U.S. Congress to evaluate the nation's crisis and consequence management capabilities under duress, in May 2000, the Department of Justice (DOJ) engaged key emergency management personnel in an exercise to prepare senior government officials to respond to an actual terrorist attack involving weapons of mass destruction (WMD), gauge preparedness activities, and identify areas for improvement. The exercise, known as TOPOFF (for "Top Officials"), was a national level, multiagency, multijurisdictional, no notice simulation addressing three fictional events in three major cities: a chemical weapons event in Portsmouth, New Hampshire, a radiological event in the greater Washington, DC, area, and a biological weapons attack in Denver, Colorado (Inglesby et al., 2001). This entry provides an overview of TOPOFF with specific attention paid to the biological weapons scenario.

EXERCISE PARTICIPANTS

TOPOFF was led by the DOJ's Office of Emergency Preparedness, which at the time had been designated as the lead agency to prepare the country for terrorist attacks involving WMD. Participants included the Federal Emergency Management Agency (FEMA), the Attorney General, Secretary of Health and Human Services, the director of the Federal Bureau of Investigation (FBI), and Centers for Disease Control and Prevention (CDC), as well as state, city, and local emergency response and law enforcement officials.

EXERCISE DESIGN

The bioterrorism module in Denver, Colorado, portrayed the covert release of aerosolized *Yersinia pestis* in the Denver Performing Arts Center. In subsequent (simulated) days, emergency plans were set in motion to address the rapidly growing number of people seeking medical attention, who exhibited a similar complex of unusual symptoms. Participants worked around the clock but within the boundaries of their usual official responsibilities. Their decisions and actions affected the development of the exercise. The exercise was guided by simulated "injects" from controllers, which, were intentionally exaggerated events in order to overwhelm local and state emergency management systems. Given the complexity and scale of the scenario, some injects were only notional (e.g., an artificially increased rate of infection).

EXERCISE SCENARIO

On Day 1 of the exercise, the Colorado Department of Public Health and Environment (Department of Public Health) was alerted by local hospitals that increasing numbers of persons were seeking medical attention for cough and fever; by early afternoon, 500 persons had received medical attention and 25 were reported dead. Subsequent laboratory testing by both the CDC and the Department of Public Health identified *Y. pestis* as the causative agent. The state health officer declared a public health emergency, and the governor issued an executive order restricting travel in or out of the Denver metropolitan area. By the end of the day, 123 deaths were reported.

On Day 2, a "push pack" of emergency medical supplies including antibiotics from the National Pharmaceutical Stockpile (NPS, now the Strategic National Stockpile) arrived in Denver, but transportation and distribution problems affected utilization of this resource. Cases of plague were reported from bordering states, as well as from Great Britain and Japan. By the end of the day, several Denver hospitals were filled to capacity, and 1871 cases and 389 deaths of plague had been reported in the United States, London, and Tokyo.

On Day 3, hospitals were overwhelmed by the number of patients and shortages of personnel, medicines and equipment. In addition, distribution problems with the NPS supplies further affected facilities with urgent needs. Despite a need for food and supply lines into the state of Colorado, the governor closed state borders on the advice of the CDC. By the end of the day, the CDC determined that secondary spread of the disease was occurring, and 3060 patients and 795 deaths were reported worldwide.

By the end of the fourth and final day, over 4000 cases were reported with 950 deaths (Inglesby et al., 2001).

OBSERVATIONS

Key Observations and lessons learned from the exercise included:

- Since the governor of Colorado did not actually participate in the exercise, the Emergency Epidemic Response Committee took on a decision-making role, which hampered the decision process. Further delays occurred as a result of participants having to communicate via conference calls.

- Because of indecision and poor coordination among what was a large number of policymakers, observers noted that unrealistic decisions were made in terms of time, data collection, and policy implementation.
- Communication was hampered by the absence of uniform terminology among agencies involved in the crisis management process.
- As the epidemic spread, concerns arose that the manner in which antibiotics were distributed would be driven by politics rather than medical needs.
- Consistent with the design of the exercise, one of the more serious challenges to the public health sector was the influx of large numbers of persons needing medical attention.
- Many public health personnel suffered from emotional and physical exhaustion.

- Observers noted that adequate resources were not allocated to contain the spread of the disease. As a result, one million people were asked to stay indoors without proper explanation, and four million people were stranded inside the state of Colorado without any basic public services or food.

REFERENCES

Hoffman, R.E. and Norton, J.E., *Emerging Infectious Diseases*, **6**, 6 (2000), http://www.cdc.gov/ncidod/eid/vo6no6/hoffman.htm, accessed on 11/21/04.

Inglesby, T.V., Grossman, Rita, and O'Toole, T., *Tara. Clinical Infectious Diseases*, **32**, 436–445 (2001).

TOPOFF, U.S. Department of State, Office of Counterterrorism Fact Sheet, 2002.

See also Consequence Management; Crisis Management; Department of Justice; and TOPOFF 2.

TOPOFF 2

Praveen Abhayaratne

INTRODUCTION

Building on the lessons of the first TOPOFF exercise, which was held in May 2000, TOPOFF 2 was designed to test local, state, and federal agency response capabilities to a combination of natural disasters and terrorism incidents while incorporating new crisis management policies. The exercise was held in King County, Seattle, and Pierce County, Washington; Chicago, Illinois; and Ottawa, Canada. It is the most comprehensive terrorism response simulation conducted in North America to date. Reports and evaluations of TOPOFF 2 had yet to be released at the time of this writing; therefore, information included below is as inclusive as present sources allow.

EXERCISE PARTICIPANTS

TOPOFF 2 sought to incorporate the capabilities of a large and varied community of public service officials from over 100 agencies and organizations in the federal, state, and local sectors, as well as personnel and leaders of U.S. and Canadian law enforcement, emergency response, public works, public health, fire service, and volunteer organizations. The U.S. Department of Homeland Security and the U.S. Department of State sponsored the exercise, which cost an estimated $16 million. Major U.S. actors included the states of Washington, Illinois, Maryland, the District of Columbia, and the cities of Seattle, Chicago. Major Canadian actors included the province of British Columbia and Vancouver city. Resources of the Joint Staff, Federal Bureau of Investigation (FBI), U.S. Northern Command, and National Imagery and Mapping Agency (now the National Geospatial Intelligence Agency) were also utilized. Participants began coordinating, training, and planning activities 18 months prior to the exercise to establish critical networks and relationships that endure to date (City of Seattle, 2003).

EXERCISE DESIGN

Like its predecessor, this full-scale exercise was designed to overwhelm and test the emergency management system of the United States in order to identify areas that need to be strengthened. The exercise was designed to build on the lessons learned in TOPOFF and subsequent preparedness exercises that have been conducted previously in the United States. Participants planned to evaluate local, state, and national "emergency public information, emergency public policy/decision making, communication and connectivity, resource allocation, jurisdiction, anticipating the enemy, and exercise conduct and design" (May, 2003).

EXERCISE SCENARIO

In an effort to challenge the relevant crisis response authorities in a manner similar to what would occur in an actual event, the exercise was not restricted to national borders or a single incident of terrorism involving chemical, biological, radiological, or nuclear weapons (CBRN). On Day 1, participants were informed that a cyber terrorism event had disrupted communication infrastructure one week before, demanding innovative means for communication as three other simulated crises evolved: a radiological dispersal device (RDD, or "dirty bomb") explosion in Seattle followed by a secondary device intended to target emergency responders, a bioterrorist attack in Chicago, and a scenario involving a transit bus and a ferry being taken hostage by terrorists in Washington state (May, 2003).

OBSERVATIONS

Despite the fact that TOPOFF 2 was intended to build on the experiences of earlier exercises, especially the first TOPOFF, this exercise exposed continuing weaknesses that had been identified previously. Observations and lessons from the exercise included:

- A lack of adherence to established procedures for activating and using federal emergency aid supplies, confusion regarding authority, and communication failures prevented resources such as the Strategic National Stockpile from being sufficiently and efficiently utilized.
- Accurate information to the public based on valid medical guidelines in a unified voice was not provided.
- Inadequate information regarding the dispersion of radioactive material from the "dirty bomb" explosion restricted response capability.
- Communications in the Chicago module of the exercise relied heavily on standard modes of communication such as telephone and fax, which were inundated and proved inefficient during the exercise.

Encyclopedia of Bioterrorism Defense,   Edited by Richard F. Pilch and Raymond A. Zilinskas
ISBN 0-471-46717-0   Copyright © 2005 Wiley-Liss

- The Illinois State Health Department had problems adequately staffing emergency centers with trained professionals for the entire duration of the exercise. Despite being assigned 12-hour shifts, staff members often had to stay on duty for longer hours, with the result that they became exhausted, which negatively affected the productiveness and effectiveness of health centers.
- Participants noted that there was no attention paid to long-term recovery plans.

REFERENCES

City of Seattle Brochure, *TOPOFF 2: National Combating Terrorism Exercise*, 2003.

May, S., TOPOFF Exercise Offers Lessons for Preparedness, *Northwest Public Health*, University of Washington School of Public Health and Community Medicine, Fall/Winter 2003.

Kershaw, S., Aftereffects Emergency Preparedness, *New York Times*, May 13, 2003.

Shenon, P., Terrorism Drills Showed Lack of Preparedness, Report Says, *New York Times*, December 19, 2003.

Sweeney, A., Mock Bioliogical Terrorism Attack to Test Chicago's Ability to Respond, *Chicago Sun Times*, May 7, 2003.

WEB RESOURCES

ANSER Institute for Homeland Security, http://www.homeland-security.org/.

Center for Biosecurity at the University of Pittsburgh Medical Center, http://www.upmc-biosecurity.org/post_index.html.

Memorial Institute for the Prevention of Terrorism, http://www.mipt.org/.

Monterey Institute of International Studies, http://cns.miis.edu/research/terror.htm.

RAND, http://rand.org/publications/electronic/terrorism.html.

The Gilmore Commission, http://www.rand.org/nsrd/terrpanel/.

See also CRISIS MANAGEMENT; DEPARTMENT OF HOMELAND SECURITY; DEPARTMENT OF STATE; and TOPOFF.

TOXINS: OVERVIEW AND GENERAL PRINCIPLES

Mark A. Poli

Opinions, interpretations, conclusions, and recommendations are those of the author and are not necessarily endorsed by the U.S. Army.

INTRODUCTION

The world is a much different place than it was even a generation ago. Few people will dispute this point given the great advancements that have been made in recent years, in fields ranging from the natural sciences and medicine to the arts and social sciences. For better or worse, however, the nature of humanity remains the same; we have yet to overcome our propensity for warfare and conflict. Unfortunately, the scientific and engineering advances that enable our constant improvements in health and quality of life also drive our increasing abilities to wreak havoc upon our fellow man. Nowhere is this more evident than in the realm of biological warfare (BW) and terrorism. Both have been with us since the inception of the species. Never have they been so lethal and efficient.

The purpose of this article is to expose the lay reader to an overview of basic concepts regarding toxins as biological weapons or agents of bioterrorism. Specific agents are addressed under separate entries.

DEFINING TOXINS

Toxins are poisons isolated from living organisms. They exert their effects by interrupting normal physiological processes. Unlike infectious agents such as bacteria and viruses, toxins cannot replicate within the body; a lethal dose must be acquired at the time of exposure. Toxins are widely distributed in nature and can be isolated from diverse taxa including bacteria (tetanus and botulinum toxins), algae (saxitoxin, microcystin), vascular plants (ricin from the castor bean), arthropods (insect, scorpion and spider venoms), or animals (snake venoms, batrachotoxin from the poison arrow frog, tetrodotoxin from pufferfish). Treatment for toxin exposure usually consists of drugs, specific antibodies, or supportive medical care.

SCENARIOS FOR USE

Typical scenarios for deploying biological weapons, including toxins, are of three types: open-air delivery, limited-air delivery, and direct application.

Open-Air Delivery

Open-air scenarios may be either line source or point source. A line source would be analogous to opening the valve on a spray tank of agent in the back of a pickup truck while driving through or around a city. A line of agent is dispersed behind the vehicle that is then free to float downwind. A point source would be analogous to opening the same valve in the back of the same pickup truck, but parking the vehicle in the parking lot outside a major sporting event or shopping mall. In this case, the agent emanates from a single point in space after which it is dispersed by the wind. In both cases, dilution in the atmosphere is unlimited, and the agent will eventually disperse or settle out of the air until it reaches nontoxic concentrations.

Limited-Air Delivery

Limited-air delivery describes the dispersal of agent in an enclosed space, such as a building or a single room. In this case, atmospheric dilution is limited, and the concentration inside the enclosure remains elevated until the agent settles out, decays, or is physically removed. Similar situations can be envisioned for contamination of a water supply. Contamination of a single water supply truck, or an isolated building, is conceivable because limited dilution is possible. Conversely, contamination of reservoirs, rivers, or lakes is limited by the huge dilution factors involved, as well as by physical removal of the agents by living organisms, chemical degradation, or binding to sediment. In general, water supplies need to be adulterated close to the end user to minimize dilution effects.

Direct Application

Finally, direct application, or assassination may occur. This is not generally considered to be a major threat in BW or terrorism scenarios, although it can serve a political purpose. In a well-publicized case in 1978, a Bulgarian dissident, Georgi Markov, was assassinated by a Russian intelligence agent by being "shot" in the leg with a pellet containing the plant toxin ricin (Crompton and Gall, 1980).

LIMITATIONS TO TOXINS AS BIOLOGICAL WEAPONS

While literally hundreds of natural toxins occur in nature, most are of little or no value as BW or terrorist weapons. There are several factors, including ease of production and environmental stability, which limit the use of toxins in this way. However, the most important limiting factor is toxicity. Figure 1 demonstrates the relationship between toxicity (expressed in mouse intraperitoneal LD_{50}s, or median lethal dose) and the amount of toxin required to produce a lethal aerosol concentration over a battlefield encompassing $100\ km^2$ under ideal meteorological conditions (Franz, 1997). From these data, one can immediately see that only toxins of the highest potency are viable weapons. At LD_{50}s greater than approximately 0.3 μg/kg, metric tons of toxin are required. For most biological toxins, isolating these amounts from their natural biological matrices is an extremely difficult task and is simply not feasible. In reality, only those compounds designated "most toxic" in Figure 1—those that are toxic in the range of 25 ng/kg and below—are considered viable open-air weapons under ideal conditions (though it should be noted that $100\ km^2$ is a large area and represents a true BW scenario). A realistic bioterrorism event may involve a smaller area, perhaps even a limited-air scenario. In this case, those agents with toxicities up to about 2.5 μg/kg ("highly toxic") might represent a viable threat. Those agents with toxicities above 2.5 μg/kg are not particularly useful in a mass-casualty situation, although they might be used to target a small number of individuals by direct application. Thus, although the number of known natural toxins numbers in the hundreds, only 15 to 20 protein toxins fall into the "most toxic" range and therefore constitute a viable BW threat. Another few dozen, mostly bacterial and marine toxins, can be classified as "highly toxic" and constitute a bioterrorist threat. Further reducing this number are production and stability considerations. A list of comparative toxicities of selected toxins is given in Table 1.

Availability is another important limitation. Traditionally, toxins are extracted from a biological matrix such as plant material, animal tissues, or microbial growth medium. For some toxins, such as marine toxins where organisms must be captured from the wild and naturally occurring tissue concentrations within them are low, accumulating sufficient raw material for extraction can be problematic. For others, such as plant or bacterial toxins, raw materials can be artificially grown in large quantities. However, extracting the toxin from the raw biological matrix may not be trivial, and can be time-consuming and expensive. Of course, advances in molecular genetics have made cloning and expression of both natural protein toxins and new variants a definite possibility. Nonetheless, specialized equipment and some degree of scientific expertise are necessary prerequisites. These issues may keep many toxins out of reach of all but a few well-funded

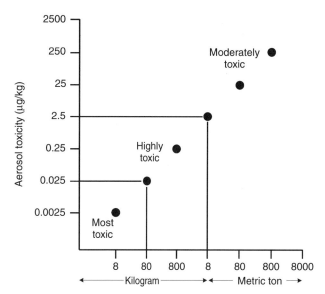

Figure 1. Aerosol toxicity (mouse LD_{50}) versus quantity of toxin providing an effective open-air exposure, under ideal meteorological conditions, over an area encompassing $100\ km^2$. Modified from Franz (1997).

Table 1. Potencies of Selected Toxins. Values are Expressed in Mouse LD_{50} Units (in μg/kg), and Drawn from Various Literature Sources. Route of Administration is Intraperitoneal, Unless Otherwise Indicated

Toxin	LD_{50}	Source
Botulinum toxin A	0.001	Bacterium
Tetanus toxin	0.002	Bacterium
Shiga toxin	0.002	Bacterium
Staphlococcal enterotoxin B	0.02 (est. human, aerosol)	Bacterium
Diptheria toxin	0.1	Bacterium
Maitotoxin	0.1	Marine dinoflagellate
Ciguatoxin (P-CTX-1)	0.2	Marine dinoflagellate
Abrin	0.7	Plant
Batrachotoxin	2	Poison arrow frog
Ricin	3	Plant
Tetrodotoxin	8	Pufferfish
Saxitoxin	10	Marine dinoflagellate
Microcystin	50	Blue-green algae
Aconitine	100	Plant
T-2 toxin	1200	Fungus

laboratories. Thus, while small-scale use of toxins remains possible, the mass casualty threat is probably restricted to governments and state-sponsored terrorist groups that can afford the requisite laboratory infrastructure for mass production and extraction of toxins.

Another important limitation relates to the route of exposure. Because toxins are not volatile and, with rare exceptions, are not dermally active, the routes of exposure are limited. Some toxins, such as marine toxins, are ideally suited to administration via the food supply. Because these toxins occur naturally in seafood in concentrations sufficient to cause incapacitation or death, it is not necessary to isolate pure toxins from natural sources prior to an attack. In theory, toxic seafood can be harvested and inserted into the food supply downstream from any regulatory testing. The food looks and tastes fresh and wholesome, and the toxins cannot be identified without destructive extraction and testing. Because the contamination is natural, it is very difficult to trace. However, although likely to cause disruption of a local event and perhaps economic disruption to the local seafood industry, an attack of this nature is unlikely to result in mass casualties. A true mass-casualty attack with toxins is likely to result only from an aerosol attack.

Aerosol dissemination of toxins in a respirable form is not a trivial matter. Although aerosol dissemination equipment is readily available from commercial sources, most are not designed to produce particles in the respirable size range of approximately 1 to 5 μm in aerodynamic diameter. Only particles within this size range have a high probability of reaching and depositing in the alveolar regions of the lung. Particles smaller than 1 μm have a high probability of being exhaled, and particles larger than 5 μm typically impact the nasopharyngeal region and never reach the deep lung. In addition, larger particles (10 to 20 μm and larger) quickly settle out of the air and fall to the ground, where they become unavailable for inhalation. Although a single individual with the correct scientific background could manage an attack of this nature, the equipment and expertise necessary to efficiently produce respirable aerosols is more likely to be found only in state-supported or other well-funded terrorist groups.

Finally, stability in the environment can be an important limiting factor for BW agents. This is a complex subject and cannot be sufficiently addressed here. Suffice it to say that molecules vary widely in their structural stability, which in turn affects their toxicity. They may be degraded during the dissemination process, or may be environmentally labile. Some toxins are degraded quite quickly by sunlight. Others are tightly bound to soils or sediments and can be quickly degraded by microorganisms.

In short, a good toxin weapon needs to be easily produced, highly toxic, nonlabile, and environmentally stable. Very few natural toxins fit this description sufficiently well to be considered a viable threat.

THE POTENTIAL IMPACT OF BIOTECHNOLOGY AND GENETIC ENGINEERING

Recent advances in biotechnology and genetic engineering have and will continue to fundamentally change the fields of science and medicine in the twenty-first century. However, genetic engineering, for all of its tremendous power to enhance and improve the human condition, also has the power to do great harm. The creation of novel or "improved" biological weapons is no longer theoretical. Potentially, we may someday have to face:

- benign organisms genetically altered to produce a toxin, venom, or endogenous regulator;
- toxin-producing microorganisms resistant to standard antibiotics, vaccines, or therapeutic drugs;
- toxins or microorganisms with enhanced aerosol and/or environmental stability;
- toxin-producing microorganisms altered to defeat an immunocompetent host;
- toxins altered to enhance potency.

Although this type of work requires significant scientific expertise and financial support, it is not out of the range of state-supported programs, and well-funded terrorist groups could succeed in acquiring the products of this research.

CONCLUSION

The threat of biological weapons use by terrorist groups and rogue states is real. While the threat needs to be continuously evaluated, the description of biological agents as "the ultimate terrorist weapon" is likely overstated, at least for toxins. The realistic number of potential toxin weapons is relatively low due to limitations stemming from potency, stability, and availability. Effective weaponization and dissemination are not trivial. Medical countermeasures, such as vaccines, diagnostics, and therapeutics confer significant protection. Although gaps exist in our knowledge base and in the availability of a full range of countermeasures for each toxin, research programs are in place to address these needs. Federal, state, and local governments are beginning to address critical needs in education, communication, and infrastructure. While a great deal has been accomplished, more needs to be done. Difficult funding choices may need to be made, but with good communication, integration, and teamwork, the goal is achievable.

REFERENCES

Crompton, R. and Gall, D., *Med. Leg. J.*, **48**, 51 (1980).

Franz, D.R., "Defense Against Toxin Weapons," in R. Zajtchuk, Ed., *Medical Aspects of Chemical and Biological Warfare. Part I: Warfare, Weaponry, and the Casualty*, Office of the Surgeon General, United States Army, Falls Church, VA, 1997, pp. 603–619.

TULAREMIA (*FRANCISELLA TULARENSIS*) (RABBIT FEVER, HARE FEVER, DEERFLY FEVER, O'HARA'S DISEASE, LEMMING FEVER, *FRANCISELLA TULARENSIS, BACTERIUM TULARENSE, PASTEURELLA TULARENSIS, F. TULARENSIS* SUBSPECIES *TULARENSIS, F. TULARENSIS* SUBSPECIES *HOLARCTICA, F. TULARENSIS* SUBSPECIES *MEDIASIATICA, F. TULARENSIS* SUBSPECIES *NOVICIDA, F. TULARENSIS BIOVAR TULARENSIS, F. TULARENSIS BIOVAR PALEARCTICA, F. TULARENSIS* TYPE A, *F. TULARENSIS* TYPE B.)

Mats Forsman

Anders Johansson

DISEASE

Tularemia is a zoonosis, a disease of animals transmissible to humans, caused by the facultative intracellular bacterium *Francisella tularensis*. The disease first attracted scientific attention in 1911 as a plague-like disease of rodents in Tulare County, California, and successful cultivation of the causative agent was reported soon after (McCoy, 1911; McCoy and Chapin, 1912). Subsequently, this small, gram-negative bacterium has been associated with disease in a wider range of animal species than any other zoonotic disease (Hopla and Hopla, 1994). Natural transmission to humans usually occurs through the bite of a blood-feeding arthropod such as a tick, biting fly, or mosquito, but ingestion of infected food or water or inhalation of the bacterium may also cause infection. *Francisella tularensis* is considered a potential agent of biological warfare (BW) and bioterrorism, and as such has been listed among the top six Category A agents of the Centers for Disease Control and Prevention (CDC) (Dennis et al., 2001; Rotz et al., 2002). In this context, respiratory tularemia (acquired through inhalation) is judged to be the greatest threat. Tularemia in humans with natural etiology has been reported from most countries in the Northern Hemisphere (Fig. 1). The reservoir of *F. tularensis* in nature is still unknown.

HISTORY OF BW APPLICATION

During the period 1932 to 1945, Japanese research units examined the utility of *F. tularensis* as a biological weapon. *Francisella tularensis* has since been an organism of concern in defense plans against biological attack (Dennis et al., 2001). During World War II, tularemia outbreaks affecting tens of thousands of Soviet and German soldiers were reported. After the war, there were continuous military studies of tularemia. The U.S. military, in a large state-funded BW program spanning from 1943 to 1969, developed weapons that could disseminate *F. tularensis* aerosols. The USSR, as part of the civilian component of its offensive BW program (Biopreparat), also incorporated *F. tularensis* into weapons (Davis, 1999). In 1970, the World Health Organization (WHO) published a report estimating that an aerosol dispersal of 50 kg of virulent *F. tularensis* over a metropolitan area with 5 million inhabitants would result in 250,000 incapacitating causalities, including 19,000 deaths (WHO, 1970). Today, the emphasis has shifted toward defending against biological terrorism, but the same features of this organism that attracted the attention of the superpowers decades ago are still pertinent: a potential to cause fatal disease, but no widely available vaccine for prophylaxis (Vogel, 2003).

CRITICAL ASPECTS OF ACQUISITION, PRODUCTION, AND DELIVERY

The natural reservoirs of *F. tularensis* still await complete delineation. *Francisella tularensis* may survive in water and mud for months, but recovery from such sources is difficult due to overgrowth of other bacterial species on most growth media and because *F. tularensis* is a fastidious organism (Parker, 1951). Hence, successful cultivation of the agent from the environment has often been performed by inoculation of samples into laboratory animals. More easily, *F. tularensis* can be recovered from infected mammals or humans by inoculation of samples such as infected tissues or liquids onto rich agar media. Generally, poorer growth is obtained in liquid media. Importantly, there is a considerable risk for laboratory-acquired infection when handling *F. tularensis*, so work with this organism should be performed under BSL-3 (BSL-3) laboratory conditions (U.S. Department of Health and Human Services, 1999). Aerosol delivery or, less likely, delivery via water or food is considered to be the greatest threat if *F. tularensis* were to be used in BW or bioterrorism (Dennis et al., 2001).

PROPERTIES

At present, there are four recognized subspecies of *F. tularensis: tularensis, holarctica, mediasiatica*, and

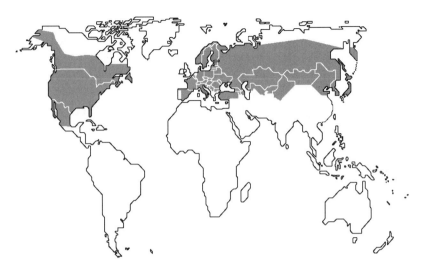

Figure 1. The geographical distribution of tularemia is indicated in gray. Locally, the disease is known to have an uneven distribution with recurrent epidemic outbreaks in geographically restricted natural foci. *Francisella tularensis* ssp. *holarctica* (type B) is found worldwide, whereas *F. tularensis* ssp. *tularensis* (type A) is almost exclusively confined to North America (CDC, 2002; Helvaci et al., 2000; Jellison, 1974; Jusatz, 1952; Kudelina and Olsufjev, 1980; Ohara et al., 1996; Olsufjev, 1966; Pang, 1987; Perez-Castrillon et al., 2001; Reintjes et al., 2002).

novicida (Sjöstedt, in press). Each is predominantly associated with a specific geographical distribution, but all four share most biochemical characteristics. Although strains of the four different subspecies are antigenically similar and show a close phylogenetic relationship (Titball et al., 2003), they have marked variations in their virulence for mammals (Table 1). There are two types of tularemia (type A and type B) that can be distinguished with regard to mortality and virulence in humans and animals (Jellison, 1974). The two types reflect the virulence of the two clinically dominant subspecies, *F. tularensis* subspecies (ssp.) *tularensis* (type A) and *F. tularensis* ssp. *holarctica* (type B). Type A isolates, in contrast to type B isolates, cause life-threatening disease in humans and therefore are considered more likely as BW or bioterrorism agents.

Little is known about the virulence mechanisms of *F. tularensis* (Sjöstedt, 2003). *Francisella* is not known to produce any toxins. Although the bacterium replicates successfully in a wide range of cell types, the macrophage is generally believed to be the main site of replication in mammalian hosts (Tärnvik, 1989). Several candidate genes of importance in the infection of macrophages have

been reported, but it seems clear that many virulence mechanisms await discovery, (Anthony et al., 1994; Baron and Nano, 1998; Golovliov et al., 1997; Gray et al., 2002; Telepnev et al., 2003).

TRANSMISSION AND EPIDEMIOLOGY

Tularemia occurs endemically, although with a patchy distribution, in most countries of the Northern Hemisphere (Fig. 1). Humans can be infected by various modes, including arthropod bites (ticks, deerflies, mosquitoes, etc.), handling of infectious animal tissues or fluids, direct contact with or ingestion of contaminated water, food, or soil and by inhalation of infective aerosols. In fact, *F. tularensis* is one of the most infectious pathogenic bacteria known, requiring inoculation or inhalation of as few as 10 organisms to cause human infection (Table 1).

In the 1960s, Jellison and coworkers introduced the concept of the two distinct types of tularemia, type A and type B, based on biochemical, epidemiological, and virulence data (Jellison, 1974). This concept recognizes the association of the highly virulent organism *F. tularensis* ssp. *tularensis* (type A) with rabbits, ticks, and sheep

Table 1. Virulence of *Francisella tularensis* Subspecies in Animals and Humans

| | Lethal Dose in Animals | | | | | Infectious Dose in Humans | |
| | Subcutaneous or Intracutaneous Inoculation | | | | Inhalation | Intracutaneous Inoculation | Inhalation |
Biotype	Mice	Guinea Pigs	Rabbits	Nonhuman Primates	Nonhuman Primates	Humans	Humans
Francisella tularensis ssp. *tularensis* (type A)	<10 cfu	<10 cfu	1–10 cfu	About 10 cfu	30–50 cfu	<10 cfu	About 10 cfu
Francisella tularensis ssp. *holarctica* (type B)	<10 cfu (s.c.)	<10 cfu	>10^6 cfu	Self-limiting disease	Self-limiting disease	<10 cfu	<200 cfu
Francisella tularensis ssp. *mediasiatica*	<10 cfu	<10 cfu	>10^6 cfu	NR	NR	NR	NR
Francisella tularensis ssp. *novicida*	1–10^2 cfu	10–10^3 cfu	Resistant	NR	NR	NR	NR

Note: NR, Not reported. (Eigelsbach and Hornick, 1972; Eigelsbach and McGann, 1984; Eigelsbach et al., 1968; Gurycova, 1998; Hall et al., 1973; McCrumb, 1961; Olsufjev and Meshcheryakova, 1982; Owen et al., 1964; Saslaw et al., 1961a; Saslaw et al., 1961b; Schricker et al., 1972a; White et al., 1964)

Table 2. Clinical Characteristics of Human *Francisella tularensis* Infection

Clinical Form of Tularemia	Route of Infection	Symptoms, Signs, and General Features[a]
Respiratory	Inhalation of an infectious aerosol, e.g., from farming activities creating dust of *F. tularensis*-contaminated hay or from deliberate spread of infectious aerosol (bioterrorism, biological warfare).	Dry cough, chest pain, or tightness. Sometimes, objective signs of pneumonia such as tachypnea, purulent sputum, and pleuritic pain. X-ray examination may reveal enlarged mediastinal lymph nodes and less frequently also parenchyma consolidation.
Ulceroglandular	Inoculation into minute skin breaks by an arthropod bite or handling of infectious materials (e.g., infected animal tissue or fluids).	Skin ulcer and enlarged tender regional lymph node(s). Most common clinical form of tularemia.
Glandular	By the skin but no ulcer detected.	Regional enlarged tender lymph node(s).
Oculoglandular	Self-inoculation of the eye after handling infectious material (e.g., animal tissue) or possibly from an infectious aerosol.	Conjunctivitis, regional enlarged tender lymph node(s).
Oropharyngeal	Through the mucosal membranes of the mouth and pharynx, by ingestion of infectious foodstuffs (e.g., undercooked infected rabbit meat, contaminated water, fruit juice, or diary product).	Ulceration in the oropharynx and regional enlarged tender lymph node(s).
Typhoidal tularemia	Unknown infection route, probably often from inhalation of an infectious aerosol.	A typhoid fever-like illness. In old literature, it is used to describe a severe (septic) tularemia infection, where the infectious route was unknown. When the infectious route is known, the use of the term typhoidal tularemia is discouraged.
Enteral tularemia	Ingestion of infectious foodstuffs (e.g., contaminated water, fruit juice, or diary product).	Abdominal pains, vomiting, and diarrheas. Uncommon clinical form has only been reported from Eurasia.

[a]The severity of disease is highly dependent on the causative *F. tularensis* subspecies. Human mortality (5–30 percent if not treated with antibiotics) may result from infection with *F. tularensis* ssp. *tularensis* (type A). Other subspecies rarely cause life-threatening human disease. Pneumonia can result in all clinical forms of tularemia, probably from the spread of *F. tularensis* into the blood stream. The onset of all clinical forms of tularemia is usually flu-like with fever (38–41 °C), generalized body aches, headache, rigors, and chills.

in comparatively dry environments in contrast to the association of the less virulent organism *F. tularensis* ssp. *holarctica* (type B) with streams, ponds, lakes, rivers, and semiaquatic animals such as muskrats and beaver. It has, however, been recognized that the epizootology of tularemia is highly complex, and that the above general associations are not always true (Hopla, 1974). Whereas *F. tularensis* ssp. *tularensis* (type A) is predominantly found in North America, *F. tularensis* ssp. *holarctica* (type B) is found over much of the Northern Hemisphere. The association of type B tularemia with water is supported by the recent discovery that the bacteria can survive and replicate in water-associated nonmammalian cells such as protozoa (Abd et al., 2003).

PATHOGENESIS AND CLINICAL COURSE

The clinical expression of tularemia largely depends on the route of entrance of the infectious agent (Table 2). The incubation period is usually 3 to 5 days, but may range from 1 to 21 days (Cross and Penn, 2000). The onset of disease is typically sudden and influenza-like, with symptoms including high fever, chills, fatigue, general body aches, headache, and nausea. Respiratory tularemia may present like pneumonia, with symptoms such as dry cough, difficulty in breathing, and chest pain. However, nearly as frequent in respiratory tularemia is the development of fever and general illness with no respiratory symptoms (Tärnvik and Berglund,

2003). Thus, respiratory tularemia without any specific symptoms or signs makes the clinical diagnosis a great challenge. Studies of aerosol-induced tularemia in nonhuman primates indicate that the terminal respiratory bronchioles and adjacent alveoli are the primary sites of bacterial invasion (Hall et al., 1973; Schricker et al., 1972a; White et al., 1964). Macrophages contain bacteria as early as 20 min after exposure. Later, at 24 to 72 h, lesions are seen in nearby structures, including bronchioles, lung, and lymphatic tissue. The studies reveal no clear differences in pathology between infections caused by the *holarctica* ssp. and *tularensis* ssp., but do emphasize the difference in disease severity. Tularemia from the highly virulent *tularensis* ssp. progresses rapidly by severe focal necrosis in lungs, spleen, and liver to a fatal outcome, while infection from the *holarctica* ssp. spontaneously resolves in nonhuman primates.

Natural outbreaks of human respiratory type A or type B tularemia have repeatedly been recorded in farmers and landscape workers (Dahlstrand et al., 1971; Feldman et al., 2001; Syrjälä et al., 1985). For example, a large outbreak of type B tularemia, including nearly 700 cases of respiratory tularemia, occurred in Sweden in 1966–1967 (Dahlstrand et al., 1971). Inhalation of *F. tularensis*-contaminated hay dust was implicated in 83 percent of the cases, illustrating the potential of effective *F. tularensis* transmission by dry aerosols.

The prognosis in tularemia is highly dependent on the causative subspecies of *F. tularensis*. For the highly virulent *F. tularensis* ssp. *tularensis*, a mortality rate of 5 to 30 percent was recorded prior to the advent of effective antibiotic therapy, and even today fatal cases occur (Dennis et al., 2001; Dienst, 1963; Stuart and Pullen, 1945). In contrast, isolates belonging to *F. tularensis* ssp. *holarctica* or *mediasiatica* are of lower virulence, causing a disease in humans similar to that caused by the *tularensis* ssp. but with no significant mortality (Cross and Penn, 2000; Olsufjev and Meshcheryakova, 1983). Isolates of *F. tularensis* ssp. *novicida* are reported to cause disease predominantly in immunocompromised individuals (Clarridge et al., 1996; Hollis et al., 1989).

DETECTION AND DIAGNOSIS

An early tularemia diagnosis is strongly dependent on clinical suspicion by a health professional. The microbiological diagnosis of tularemia has traditionally relied mainly on serology. The tube agglutination test or Enzyme-linked Immunosorbent Assay (ELISA) shows high sensitivity and specificity. However, antibodies do not appear until 10 to 14 days after onset of disease, making serology a less suitable tool in acute decision-making. Owing to the risk of laboratory-acquired infection, culturing of the agent is often avoided and performed only at designated reference laboratories. Moreover, the bacterium may require 5 to 7 days of incubation to yield visible colonies. Routine culturing of human specimens that possibly contain *F. tularensis* can be performed as biosafety level 2 (BSL-2), whereas subculturing of primary colonies and subsequent manipulations requires BSL-3 conditions (Dennis et al., 2001). *Francisella tularensis* is frequently recovered from wound specimens in the ulceroglandular form of disease and may be recovered from respiratory specimens as well as gastric aspirates in the respiratory form of tularemia (Johansson et al., 2000a; Overholt et al., 1961). With tularemia, growth of *F. tularensis* is diagnostic, and cultures are indispensable for analysis of antibiotic susceptibility. At designated laboratories, *F. tularensis* may be identified by direct flourescent antibody or immunohistochemical stains in human specimens such as secretions or biopsies (Dennis et al., 2001). The use of the polymerase chain reaction (PCR) for direct detection of *F. tularensis* DNA in clinical wound specimens has emerged as a rapid and reliable diagnostic tool in human ulceroglandular tularemia (Johansson et al., 2000a). Although theoretically attractive, there is, however, still a lack of clinical evaluation for detection of *F. tularensis* DNA in respiratory tularemia.

In addition to direct detection in clinical samples, DNA-based technology allows rapid typing down to the level of individual *F. tularensis* isolates. Real-time PCR protocols that distinguish the highly virulent *tularensis* ssp. from the less virulent *holarctica* ssp. have been developed, as well as a single PCR assay that discriminates all four subspecies. For typing of individual isolates, variable-number of tandem repeats (VNTR) analysis of *F. tularensis* has been successfully used (Farlow et al., 2001; Johansson et al., 2001b).

PREVENTION AND TREATMENT

There is currently no licensed and widely available tularemia vaccine. More than 50 years ago, attenuated strains were extensively used for immunization of humans in the Soviet Union (Olsufjev et al., 1958). Attenuated vaccine strains were developed by repeated passage of virulent strains on serum-supplemented media or by drying the strains. Such a vaccine was transferred to the United States in 1956, and further work by U.S. researchers resulted in the live vaccine strain (LVS) shown to be effective in inducing protection against tularemia in both animal models and humans (Tigertt, 1962). LVS vaccine affords excellent protection against human respiratory tularemia and mitigates the course of ulceroglandular tularemia (Burke, 1977). Until recently, immunization of certain risk groups with LVS vaccine was performed in the United States. Owing to the uncharacterized nature of the vaccine strain and the difficulties of standardizing the vaccine itself, the LVS vaccine is no longer licensed. Currently, no effective vaccine is available in the Western countries, though vaccination is still performed in Russia.

Tularemia warrants antibiotic treatment. Aminoglycosides, which are bactericidal, are the primary drugs for treatment of severe disease (Enderlin et al., 1994). The alternative drug, doxycycline, is bacteriostatic and associated with a relatively high risk of relapse. Nevertheless, doxycycline is the preferred drug in most cases of type B tularemia because it can be administered orally. Betalactams, macrolides, lincosamides, and trimoxazole are not reliable for treatment of tularemia. Recently, quinolones, which are characterized by bactericidal action, intracellular accumulation, and high bioavailability after oral administration, have been introduced for treatment of tularemia and are attractive alternatives (Limaye and Hooper, 1999). The experience with quinolones in the treatment of tularemia in Europe is excellent but remains lacking for severely ill patients in North America (Gilligan, 2002; Johansson et al., 2001a; Perez-Castrillon et al., 2001). Antibiotic susceptibility analysis, however, supports the idea that quinolones would be equally effective for treatment of type A tularemia (Johansson et al., 2002).

INFECTION CONTROL AND DECONTAMINATION

Tularemia is not transmitted from person to person, but standard hygiene precautions are recommended for hospital care of tularemia patients. If tularemia is suspected, the clinical microbiologist must be notified by those sending samples for culture in order to avoid laboratory-acquired infection. Primary cultures can be handled under BSL-2 laboratory conditions, but manipulation creating a risk of infectious aerosols needs BSL-3-level precautions. If an autopsy is performed in a suspected fatal case of tularemia, avoiding aerosol generation is recommended. Decontamination of physical surfaces can be achieved by spraying standard disinfectants and allowing 10 minutes contact time.

FUTURE OUTLOOK

Although knowledge of *F. tularensis* is limited, there are ongoing research activities that will undoubtedly result in rapid progress of our understanding of the pathogenesis and virulence mechanisms of *F. tularensis* (Vogel, 2003). For example, genome sequencing of two strains of *F. tularensis* are completed and genome sequencing of several strains are ongoing. These efforts have already resulted in a substantial increase of knowledge about *F. tularensis* relevant to future work on virulence and epidemiology. Also, large-scale proteomics analyses have been performed and are increasingly used to identify differences between strains. Efforts are needed to unravel the natural reservoir(s) and gain more insight into the epidemiology and the different modes of transmission of the organism. Overall, the reawakened interest in tularemia holds great promise for the development of new diagnostics, prophylactics, and therapies for *F. tularensis* in the near future.

REFERENCES

Abd, H., Johansson, T., Golovliov, I., Sandström, G., and Forsman, M., *Appl Environ. Microbiol.*, **69**, 600–606 (2003).

Anthony, L.S., Cowley, S.C., Mdluli, K.E., and Nano, F.E., *FEMS Microbiol. Lett.*, **124**, 157–165 (1994).

Baron, G.S. and Nano, F.E., *Mol. Microbiol.*, **29**, 247–259 (1998).

Burke, D.S., *J. Infect. Dis.*, **135**, 55–60 (1977).

CDC, *MMWR Morb. Mortal Wkly. Rep.*, **51**, 181–184 (2002).

Clarridge, J.E., III, Raich, T.J. Sjosted, A., Sandström, G., Darouidche, R.O., Shawar, R.M., Georghiou, P.R., Osting, C., and Vo, L., *J. Clin. Microbiol.*, **34**, 1995–2000 (1996).

Cross, J.T. and Penn, R.L., "*Francisella tularensis* (Tularemia)," in G.L. Mandell, J.E. Bennet, and R. Dolin, Eds., *Mandell, Douglas and Bennet's Principles and Practice of Infectious Diseases*, Churchill Livingstone, Philadelphia, PA, 2000, pp. 2393–2402.

Dahlstrand, S., Ringertz, O., and Zetterberg, B., *Scand. J. Infect. Dis.*, **3**, 7–16 (1971).

Davis, C.J., *Emerg. Infect. Dis.*, **5**, 509–512 (1999).

Dennis, D.T., Inglesby, T.V., Henderson, D.A., Bartlett, J.G. Ascher, M.S., Eitzen, E., Fine, A.D., Friedlander, A.M., Hauer, J., Layton, M., Lillibridge, S.R., McDade, J.E., Osterholm, M.T., O'Toole, T., Parker, G., Perl, T.M., Russell, P.K., and Tonat, K., *J. Am. Med. Assoc.*, **285**, 2763–2773 (2001).

Dienst, F.T., *J. La State Med. Soc.*, **115**, 114–127 (1963).

Eigelsbach, H.T. and Hornick, R.B., *Continued Studies on Aerogenic Immunization of Man with Live Tularemia Vaccine*, New York, 1972.

Eigelsbach, H.T. and McGann, V.G., "Genus Francisella Dorofe'ev 1947, 176^AL," in N.R. Krieg and J.G. Holt, Eds., *Bergey's Manual of Systematic Bacteriology*, Williams & Wilkins, Baltimore, pp. 394–399.

Eigelsbach, H.T., Saslaw, S., Tulis, J.J., and Hornick, R.B., "Tularemia: The Monkey as a Model for Man," in H. Vagtborg, Ed., *Use of Nonhuman Primates in Drug Evaluation, a Symposium*," San Antonio, Texas, 1968, pp. 230–248.

Enderlin, G., Morales, L., Jacobs, R.F., and Cross, J.T., *Clin. Infect. Dis.*, **19**, 42–47 (1994).

Farlow, J., Smith, K.L., Wong, J., Abrams, M., Lytle, M., and Keim, P., *J. Clin. Microbiol.*, **39**, 3186–3192 (2001).

Feldman, K.A., Enscore, R.E., Lathrop, S.L., Matyas, B.T., McGuill, M., Schriefer, M.E., Stiles-Enos, D., Dennis, D.T., Petersen, L.R., and Hayes, E.B., *N. Engl. J. Med.*, **345**, 1601–1606 (2001).

Gilligan, P.H., *Curr. Opin. Microbiol.*, **5**, 489–495 (2002).

Golovliov, I., Ericsson, M., Sandström, G., Tärnvik, A., and Sjöstedt, A., *Infect. Immun.*, **65**, 2183–2189 (1997).

Gray, C.G., Cowley, S.C., Cheung, K.K., and Nano, F.E., *FEMS Microbiol. Lett.*, **215**, 53–56 (2002).

Gurycova, D., *Eur. J. Epidemiol.*, **14**, 797–802 (1998).

Hall, W., Kovatch, R.M., and Schricker, R.L., *J. Pathol.*, **110**, 193–201 (1973).

Helvaci, S., Gedikoglu, S., Akalin, H., and Oral, H.B., *Eur. J. Epidemiol.*, **16**, 271–276 (2000).

Hollis, D.G., Weaver, R.E., Steigerwalt, A.G., Wenger, J.D., Moss, C.W., and Brenner, D.J., *J. Clin. Microbiol.*, **27**, 1601–1608 (1989).

Hopla, C.E., *Adv. Vet. Sci. Comp. Med.*, **18**, 25–53 (1974).

Hopla, C.E. and Hopla, A.K., "Tularemia," in G.W. Beran, Ed., *Handbook of Zoonosis*, CRC Press, Boca Raton, FL, 1994, pp. 113–126.

Jellison, W.L., *Tularemia in North America, 1930–1974*, University of Montana Foundation, Missoula, MT, 1974.

Johansson, A., Berglund, L., Eriksson, U., Göransson, I., Wollin, R., Forsman, M., Tärnvik, A., and Sjöstedt, A., *J. Clin. Microbiol.*, **38**, 22–26 (2000a).

Johansson, A., Berglund, L., Sjöstedt, A., and Tärnvik, A., *Clin. Infect. Dis.*, **33**, 267–268 (2001a).

Johansson, A., Göransson, I., Larsson, P., and Sjöstedt, A., *J. Clin. Microbiol.*, **39**, 3140–3146 (2001b).

Johansson, A., Urich, S.K., Chu, M.C., Sjöstedt, A. and A. Tärnvik, *Scand. J. Infect. Dis.*, **34**, 327–330 (2002).

Jusatz, H.J., "Tularemia in Europe, 1926–1951," in E. Rodenwaldt, Ed., *Welt-Suchen Atlas*, Falk-Verlag, Hamburg, 1952, pp. 7–16.

Kudelina, R.I. and Olsufjev, N.G., *J. Hyg. Epidemiol. Microbiol. Immunol.*, **24**, 84–91, (1980).

Limaye, A.P. and Hooper, C.J., *Clin. Infect. Dis.*, **29**, 922–924 (1999).

McCoy, G.W., *Publ. Hlth. Bull.*, **43**, 53–71 (1911).

McCoy, G.W. and Chapin, C.W., *Publ. Hlth. Bull.*, **53**, 17–23 (1912).

McCrumb, F.R., *Bacteriol. Rev.*, **25**, 262–267 (1961).

Ohara, Y., Sato, T., and Homma, M., *FEMS Immunol. Med. Microbiol.*, **13**, 185–189 (1996).

Olsufjev, N.G., Emelyanova, O.S., Uglovoi, G.P., Silchenko, V.S., Borodin, V.P., Samsonova, A.P., Konkina, N.S., Shelanova, G.M., Lebacheva, Z.A., Tsareva, M.I., Zykina, N.A., and Lebedeva, T.F., *Pergamon*, **29**, 386–391 (1958).

Olsufjev, N.G., "Tularemia," in Y.N. Pavlovsky, Ed., *Human Diseases with Natural Foci*, Foreign Languages Publishing House, Moscow, 1966, pp. 219–281.

Olsufjev, N.G. and Meshcheryakova, I.S., *J. Hyg. Epidemiol. Microbiol. Immunol.*, **26**, 291–299 (1982).

Olsufjev, N.G. and Meshcheryakova, I.S., *Int. J. Syst. Bacteriol.*, **33**, 872–874 (1983).

Owen, C.R., Buker, E.O., Jellison, W.L., Lackman, D.B., and Bell, J.F., *J. Bacteriol.*, **87**, 676–683 (1964).

Overholt, E.L., Tigertt, W.D., Kadull, P.J., Ward, M.K., Charkes, N.D., Rene, R.M., Salzman, T.E., and Stephens, M., *Am. J. Med.*, **30**, 785–806 (1961).

Pang, Z.C., *Zhonghua Liu Xing Bing Xue Za Zhi*, **8**, 261–263 (1987).

Parker, R.R., *Contamination of Natural Waters and Mud with Pasteurella tularensis*, U. S. Govt. Print Off., Washington, DC, 1951.

Perez-Castrillon, J.L., Bachiller-Luque, P., Martin-Luquero, M., Mena-Martin, F.J., and Herreros, V., *Clin. Infect. Dis.*, **33**, 573–576 (2001).

Reintjes, R., Dedushaj, I., Gjini, A., Jorgensen, T.R., Cotter, A., Lieftucht, B., D'Ancona, F., Dennis, D.T., Kosoy, M.A., Mulliqi-Osmani, G., Grunow, R., Kalaveshi, A., Gashi, L., and Humolli, I., *Emerg. Infect. Dis.*, **8**, 69–73 (2002).

Rotz, L.D., Khan, A.S., Lillibridge, S.R., Ostroff, S.M., and Hughes, J.M., *Emerg. Infect. Dis.*, **8**, 225–230, (2002).

Saslaw, S., Eigelsbach, H.T., Prior, J.A., Wilson, H.E., and Carhart, S., *Arch. Intern. Med.*, **107**, 689–701 (1961a).

Saslaw, S., Eigelsbach, H.T., Prior, J.A., Wilson, H.E., and Carhart, S., *Arch. Intern. Med.*, **107**, 702–714 (1961b).

Schricker, R.L., Eigelsbach, H.T., Mitten, J.Q., and Hall, W.C., *Infect. Immun.*, **5**, 734–744 (1972a).

Sjöstedt, A., *Curr. Opin. Microbiol.*, **6**, 66–71 (2003).

Sjöstedt, A., "Family XVII. *FRANCISELLACEAE*, Genus I. *Francisella*," in *Bergey's Manual of Systematic Bacteriology*, 2nd ed. Vol. 2, D.J. Brenner, Eds., Springer-Verlag, New York (In press).

Stuart, B.M. and Pullen, R.I., *Am. J. Med. Sci.*, **210**, 223–236 (1945).

Syrjälä, H., Kujala, P., Myllyla, V., and Salminen, A., *Scand. J. Infect. Dis.*, **17**, 371–375 (1985).

Tärnvik, A. and Berglund, L., *Eur. Respir. J.*, **21**, 361–373 (2003).

Telepnev, M., Golovliov, I., Grundström, T., Tärnvik, A., and Sjöstedt, A., *Cell Microbiol.*, **5**, 41–51 (2003).

Tigertt, W.D., *Bacteriol. Rev.*, **26**, 354–373 (1962).

Titball, R.W., Johansson, A., and Forsman, M., *Trends Microbiol.*, **11**, 118–123 (2003).

Tärnvik, A., *Rev. Infect. Dis.*, **11**, 440–451 (1989).

U.S. Department of Health and Human Services, *Biosafety in Microbiological and Biomedical Laboratories (BMBL)*, 4th ed., U.S. Government Printing Office, Washington, DC, http://www.cdc.gov/od/ohs/ page reviewed on September 9, 2003.

Vogel, G., *Science*, **302**, 222–223 (2003).

White, J.D., Rooney, J.R., Prickett, P.A., Derrenbacher, E.B., Beard, C.W., and Griffith, W.R., *J. Inf. Dis.*, **114**, 277–283 (1964).

WHO, *Health Aspects of Chemical and Biological Weapons*, World Health Organization, Geneva, 1970, pp. 105–107.

WEB RESOURCES

CDC Public Health Emergency Preparedness & Response Site, www.bt.cdc.gov, accessed on 4/3/05.

TYPHUS, EPIDEMIC (*RICKETTSIA PROWAZEKII*)

Marina E. Eremeeva

Gregory A. Dasch

TAXONOMY

Rickettsia prowazekii, the etiological agent of epidemic typhus, is a small, obligate intracellular gram-negative bacterium in the Order Rickettsiales, which comprises a major division of the Class Alphaproteobacteria. Bacteria in other Orders in this class can be cultivated axenically (in pure culture). The Order Rickettsiales contains a very diverse group of bacteria with properties ranging from commensal to pathogenic in relation to eukaryotic cells and organisms (Eremeeva and Dasch, 2000).

The name *Rickettsia* honors Dr. Howard T. Ricketts, who discovered the etiological agent of Rocky Mountain spotted fever (RMSF) in 1906 and established the role of ticks in its transmission. Ricketts was the first person who recognized that the microbial agents that cause RMSF and epidemic typhus are similar, but distinct, microorganisms. In 1909, Charles Nicolle established that the human body louse transmits epidemic typhus to humans and is the crucial target for control, findings for which he won the Nobel Prize (Gross, 1996). The name *R. prowazekii* also honors Stanislav von Prowazek, who greatly contributed to the understanding of this disease with his studies of typhus epidemics in Central Europe (Andersson and Andersson, 2000).

DISEASE

Rickettsia prowazekii is extremely infectious to humans by the parenteral route, requiring only one viable organism to cause infection. The median infectious dosage by aerosol route is less well defined but may be less than 1000 organisms. Uncertainty in this case arises from the fact that the relative importance of conjunctival, intranasal, and intratracheal routes of infection following aerosol exposure is unclear.

Onset of disease is abrupt following a 1- to 2-week incubation period (length of the incubation period varies depending on the dosage and route of exposure) (Zdrodovskii and Golinevich, 1960). Early clinical manifestations include intense headache, chills, fever and myalgia. A characteristic rash develops on the third to fifth day of disease (absent in only 6–8 percent of cases) and may last from 12 to 14 days. It first appears on the upper trunk but then becomes more generalized, involving the whole body except the face, palms, and soles. As the disease progresses, particularly in untreated patients, significant alterations in mental status from stupor to coma are observed. In patients with severe disease, hypotension and renal failure are common. Epidemic typhus may be a life-threatening illness even for young, previously healthy persons, but the severity of disease increases greatly with age over 40 years as well as with poor nutrition. Mortality may exceed 40 percent in the older age groups.

Patients who have suffered from acute epidemic typhus and recovered may develop an asymptomatic persistent infection with *R. prowazekii*. Ten to twenty percent of these people may suffer a relapse known as Brill-Zinsser disease, which can be as severe as primary infection. Immunologically naïve populations infested with body lice may suffer epidemics initiated by louse acquisition of *R. prowazekii* from a single individual suffering such relapses more than 40 years post primary infection. Although the relative importance of factors contributing to such relapses are poorly understood, recrudescence may occur in response to stress, malnutrition, decreased immune function, or co-occurrence of another infectious disease.

BIOLOGICAL WARFARE HISTORY

The devastation of epidemic typhus in warfare has been elegantly summarized by Hans Zinsser in *Rats, Lice, and History* (Zinsser, 1934). Over 30 million cases of epidemic typhus with 3 million fatalities are estimated to have occurred in World War I in Russia and Eastern Europe. As recently as 1996, more than 30,000 cases of epidemic typhus occurred in Burundi following a catastrophic breakdown in social conditions. The detrimental effect of typhus on both military and civilian populations during World War I and the postrevolution Civil War in Russia stimulated the Soviet Union to evaluate *R. prowazekii* as a weapon. Similarly, the Japanese Army conducted human and field testing with *R. prowazekii* as a biologic weapon in Manchuria from the late 1930s until the end of World War II. According to Ken Alibek and others, advanced efforts with *R. prowazekii* within the former Soviet Union's offensive biological warfare (BW) program included genetic engineering, large-scale cultivation and manufacturing of stable aerosolizable suspensions, and directed selection of bacterial strains resistant to standard therapeutic antibiotics (Alibek and Handelman, 1999). The feasibility of these approaches with *R. prowazekii* and other rickettsiae has been clearly demonstrated in the United States and elsewhere.

IMPORTANT ASPECTS OF WEAPONIZATION

Because of their high infectivity, fastidious nature, 8-hour generation time, and metabolic dependence on host cells for replication and survival, typhus and spotted fever rickettsiae are technically challenging to grow safely and quickly and in large quantities, and require precise and efficient procedures for preserving their viability. However, an intrinsic potential for fulfilling these basic objectives and other requirements for successful BW use was apparent very early on with rickettsiae. Natural experiences indicated that highly infectious dried louse, tick, and flea feces contain copious amounts of rickettsial organisms (*R. prowazekii*, *R. rickettsii* or *R. conorii*, and *R. typhi*, respectively) that can be readily aerosolized. Further consideration of *R. prowazekii* as a potential weapon is warranted due to the very high morbidity and mortality of epidemic typhus infections; nonspecific clinical features upon onset of the illness, rendering it difficult to diagnose; the general absence of clinical diagnostic experience among today's practicing physicians; and the similar absence of specialized reagents and expertise required for laboratory confirmation of disease. Furthermore, there is a very low level of immunity in most developed countries to all rickettsiae, and effective rickettsiacidal antibiotics are not known.

PROPERTIES

Like other rickettsiae (please see "Rickettsiae and Rickettsial Diseases: Overview and General Principles" for more details), *R. prowazekii* primarily targets endothelial cells *in vivo* (Wlaker, 1988). Once inside the cell, the bacterium multiplies to large numbers. Unlike other rickettsial species, which undergo active cell-to-cell spread, *R. prowazekii* cells are only released after the host cell has burst (Eremeeva and Dasch, 2000).

R. prowazekii is a typical gram-negative bacterium with a cell wall containing a lipopolysaccharide (LPS) and protein-rich outer membrane layer, a thin peptidoglycan sacculus in the periplasmic space, and an innermost cytoplasmic membrane containing numerous respiratory, transport, and sensor proteins. Most rickettsiae also contain an external microcapsular paracrystalline protein layer; in *R. prowazekii*, this layer consists of an acidic sulfhydryl-linked methylated autotransporter protein of 120 kDa, known as rOmpB or the species-specific protein antigen (SPA). SPA is the primary surface antigen of *R. prowazekii* that is recognized by antibodies and T-cell mediated immune responses, and is thus the protective vaccine antigen. LPS of rickettsiae is chemically and structurally related to LPS in most gram-negative organisms, but it is not highly toxic compared to endotoxins of enteric bacteria. LPS of *R. prowazekii*

contains several epitopes that are also found in LPS of *Proteus* OX19 and several species of *Legionella*. Rickettsial LPS elicits human antibodies recognized in the classic Weil-Felix agglutination tests.

Although the complete genome sequences of seven species of *Rickettsia* have been obtained, the relative importance of different virulence factors is not established because genetic systems for manipulation of rickettsiae are still cumbersome (Andersson and Andersson, 2000). It is commonly accepted that rickettsial virulence is due to the pathogen's ability to (1) invade and colonize human endothelial cells and be subsequently disseminated into surrounding and distant tissues; and (2) interfere with and withstand host defense systems, which allows for persistent infections. Rickettsiae induce a cascade of host responses, including reactions of proinflammatory, procoagulant, and fibrinolytic systems, and release of prostaglandins that control tone and permeability of vascular vessels. *R. prowazekii* hemolyzes the erythrocytes of vertebrates *in vitro*, but the specific roles of hemolysins, proteases, and phospholipases in the pathogenesis of epidemic typhus are not understood. A number of other putative rickettsial virulence factors have been detected that may be required for intracellular life, including a Type IV secretion system.

TRANSMISSION AND EPIDEMIOLOGY

Epidemic typhus is a disease of humans. The human body louse *Pediculus humanus corporis* is responsible for transmission of the agent from human to human; the head louse *P. humanus capitis* may also become infected. A louse acquires rickettsiae while feeding on an infected person, then leave when the person becomes febrile. The infected louse then carries the infection to others, who upon infestation autoinoculate the rickettsiae by scratching (which leads to contamination of the bite wound or skin abrasion with crushed lice or louse feces). Epidemic typhus commonly occurs in cold climates where people live in overcrowded unsanitary conditions and have few opportunities to change their clothes or bathe. Such conditions often occur during war and natural disasters, and typically facilitate louse infestation. Epidemic typhus is currently prevalent in the mountainous regions of Africa, South America, and Asia. Recovery from epidemic typhus results in nonsterile immunity, permitting persistence of *R. prowazekii* between epidemics and subsequent recrudescent Brill-Zinsser disease to initiate new epidemics.

In the Eastern United States, *R. prowazekii* exists in populations of the southern flying squirrel, *Glaucomys volans volans*, where it appears to be transmitted by fleas and lice (Reynold et al. 2003). Persons exposed to infected squirrel fleas or lice may acquire an infection with *R. prowazekii*, referred to as "sylvatic typhus," which may be milder than classic epidemic typhus.

DIAGNOSIS AND TREATMENT

Epidemic typhus can be difficult to diagnose clinically because of its nonspecific signs and symptoms, particularly at the onset of the disease. Owing to the severity and malignancy of untreated disease, treatment of suspected cases should be initiated on the basis of clinical criteria without waiting for specific laboratory diagnostic testing. Diagnosis of epidemic typhus is based on detection of specific antibodies in patient sera and is most accurate when rises in titers are demonstrated. The indirect immunofluorescence test is the most widely used antibody method, although Enzyme-Linked Immunosorbent Assay (ELISA), latex agglutination, and Dip-S-tick assays are available. Most tests are group reactive so that adsorption is required to accurately differentiate epidemic and murine typhus infections. *R. prowazekii* may also be identified in infected tissues by immunohistochemical staining with specific antisera. Polymerase chain reaction assays provide sensitive and specific detection of *R. prowazekii* in patient samples and body lice, and provide rapid and accurate means for differentiation from *R. typhi*. The use of clinical and epidemiological data is necessary to distinguish among classic typhus, Brill-Zinsser disease, and sylvatic typhus, but these clues would likely be absent if the agent were used as a biological weapon. In addition to other rickettsioses, differential diagnosis should include but not be limited to viral hemorrhagic fevers, meningitis, malaria, and dengue.

Doxycycline is highly effective for treatment of typhus: 100 mg of doxycycline taken orally twice for 15 to 21 days is a frequently prescribed therapy. Single-dose therapy has been shown to be effective in controlling epidemic situations when hospitalization and prolonged therapy are not possible. When doxycycline is not available, epidemic typhus infections also typically respond well to treatment with other tetracycline class antibiotics or chloramphenicol.

PREVENTION AND CONTROL

When washing of clothes in hot water is not possible to kill body lice, the use of insecticides to kill body lice and disinfect louse-infested clothing is the major measure employed to prevent the spread of epidemic typhus. Control requires significant efforts to improve sanitary conditions and living standards, as well as health education. Respiratory protection of health care workers may be necessary when substantial aerosolization is expected from handling of infected louse-feces-contaminated clothing. Environments contaminated by release of aerosolized suspensions of *R. prowazekii* may be rapidly disinfected with 10 percent Chlorox or other common bacterial disinfectants in water or 70 percent ethanol. The pathogen is not stable to elevated temperatures, formaldehyde gas or peroxide, acids or bases, or to humidification, so persistent exposure is not a problem. Commercial killed and attenuated live vaccines for epidemic typhus have been demonstrated to reduce morbidity and mortality in humans, but they are not available in the United States. Several excellent subunit vaccine candidates have been protective and nonreactogenic in animal model studies.

REFERENCES

Alibek, K. and Handelman, S., *Biohazard: The Chilling True Story of the Largest Covert Biological Weapons Program in the World—Told from Inside by the Man who Ran It*, Dell, New York, 1999.

Andersson, J.O. and Andersson, S.G.E., *Res. Microbiol.*, **151**, 143–150 (2000).

Eremeeva, M.E. and Dasch, G.A., "Rickettsiae," in *Encyclopedia of Microbiology*, Vol. 4. 2nd ed., Academic Press, San Diego, CA, 2000, pp. 140–180.

Eremeeva, M.E., Dasch, G.A., and Silverman, D.J., *Subcell. Biochem.*, **33**, 479–516 (2000).

Gross, L. *Proc. Natl. Acad. Sci. U.S.A.*, **93**, 10539–10540 (1996).

Reynolds, M.G., Krebs, J.W., Comer, J.A., Sumner, J.W., Rushton, T.C., Lopez, C.E., Nicholson, W.L., Rooney, J.A., Lance-Parker, S.E., McQuiston, J.H., Paddock, C.D., and Childs, J.E., *Emerg. Infect. Dis.*, **9**, 1341–1343 (2003).

Walker, D.H., Chapter 9, "Pathology and Pathogenesis of the Vasculotropic Rickettsioses," in D.H. Walker, Ed., *Biology of Rickettsial Diseases*, Vol. 1, CRC Press, Boca Raton, Florida, 1988, pp. 116–138.

Zdrodovskii, P.F. and Golinevich, H.M., *The Rickettsial Diseases*, Pergamon Press, New York, 1960.

Zinsser, H., *Rats, Lice and History*, Little, Brown and Company, Boston, Massachusetts, 1934.

U

UNITED KINGDOM: BIOTERRORISM DEFENSE

Graham S. Pearson

INTRODUCTION

The U.K. government has long been prepared to deal with disaster in whatever form that may take—natural, accidental, or deliberate. The guidance "Dealing with Disaster," in its revised third edition (Cabinet Office, 2003a), is designed to deal with a major emergency, defined as:

> Any event or circumstance (happening with or without warning) that causes or threatens death or injury, disruption to the community, or damage to property or to the environment on such a scale that the effects cannot be dealt with by the emergency services, local authorities and other organizations as part of their normal day to day activities.

Its foreword by the Home Secretary points out that the truth of the statement "disasters can strike suddenly, unexpectedly and anywhere" has been witnessed during the past two years, both at home and abroad, noting that in that timeframe "many of our perceptions of both risk and the nature of threats and crises have been considerably altered, not least by the events of 11 September 2001." He goes on to say that:

> However, some things have not been changed. One that has certainly not changed is our continued dependence on the dedication and professionalism of all those people and organizations involved in the response to and resolution of major emergencies and crises. However, just as our dependence upon the various response organizations, the emergency services, local and central government is understood, we acknowledge now, more than ever, the importance of coherent strategies and systems for the harmonisation of contingency plans and procedures.

The guidance provided in "Dealing with Disaster" offered a generic framework for civil protection, defined as "the application of knowledge, measures and practices to anticipate, guard against, prevent, reduce, or overcome any hazard, harm or loss that may be associated with natural, technological or man-made crises and disasters in peacetime." It is stated that the guidance is intended to establish good practice based on lessons learned from planning for and dealing with major peacetime emergencies at all levels during recent years. As its covering note states, the guidance has been revised to reflect the changes to the machinery of government in 2001, which has seen the transfer of responsibilities for civil protection to a new Civil Contingencies Secretariat in the Cabinet Office, along with other changes that have come about since the third edition was issued in 1998.

The United Kingdom is currently considering enactment of a new Civil Contingencies Bill, as it is recognized that existing legislation such as the Emergency Powers Act 1920 and the Civil Defence Act 1948 are outdated and a new legal framework is required to ensure a wide range of coordinated and capable responses to disasters. The consultative document (Cabinet Office, 2003c) on the new Bill notes that the United Kingdom's resilience to disruptive challenge is already high and points out that there is a long standing tradition of effective planning and response at the local level, with emergency services generally working well with local authorities and others to deliver multiagency planning and response. Furthermore, it notes that 30 years of Irish terrorism have led to a capability within government and anawareness among both businesses and the public, which puts the United Kingdom in a comparatively strong position. Nevertheless, the flooding and fuel crisis in 2000 and the foot and mouth disease (FMD) outbreak in 2001 exposed weaknesses in existing arrangements, and September 11 changed the frame of reference for counter terrorism.

The new Bill would include a definition that would enable the use of special legislative powers where appropriate in the event of:

- natural disasters, the effects of severe weather, and epidemics in animals or plants;
- major accidents (including nuclear accidents);
- major health crises, such as a flu pandemic;
- serious economic crises (both financial and nonfinancial in origin);
- attacks on or disruption to infrastructure, both traditional and electronic;
- disruption to the essentials of life (food, water, energy, fuel, communications);
- disruption to the proper functioning of government, public, and other vital services;
- the effect of major acts of terrorism;
- warlike situations or threat thereof;
- contamination of air, water, or land such as to threaten human or animal health or the natural environment;
- disruption to and/or overloading of services and infrastructure, or elements of it, such as to threaten or cause its collapse.

Encyclopedia of Bioterrorism Defense, Edited by Richard F. Pilch and Raymond A. Zilinskas
ISBN 0-471-46717-0 Copyright © 2005 Wiley-Liss

Against this background, this entry examines the U.K. approach to bioterrorism. It recognizes that there has long been a necessity to be able to respond to outbreaks of disease that may affect humans, animals, or plants. The response to bioterrorism is primarily based upon the capability to respond to terrorist incidents and upon the capability to respond to outbreaks of disease. The entry therefore first sets out the integrated approach in the United Kingdom to a major emergency, then examines how that approach is effective in preparedness to counter deliberate releases of biological materials. The approaches are described as those of the United Kingdom in that while there may be differences between the details relating to England and Wales and those in other parts of the United Kingdom, the overall approach and strategy is the same.

RESPONSES TO MAJOR EMERGENCIES

The broad approach that is adopted in the United Kingdom in providing an integrated response to a major emergency is made up of the following elements:

1. Assessment
2. Prevention
3. Preparation
4. Response
5. Recovery management.

In the United Kingdom, the above five activities have collectively been labeled an Integrated Emergency Management (IEM) approach. They are critical activities in responding to any major emergencies. However, events over the last few years have prompted some significant extensions in thinking about major emergencies. Firstly, there is now greater emphasis on assessment and prevention. Secondly, integrated arrangements for responding to and recovering from major emergencies must not only consider the sudden impact disaster with identifiable scenes (transport accidents, flooding, etc.) but also the "creeping crisis" where a specific scene is less apparent (epidemics, widespread protest, etc.). The point is made in the United Kingdom that this wider concept of IEM is geared toward the idea of building greater overall resilience in the face of a broad range of disruptive challenges.

The principal emphasis in developing response and recovery plans needs to be on responding effectively to the common consequences of incidents or events rather than focusing primarily on the different causes. The generic planning arrangements for responding to a range of emergencies must be cohesive and consistent, whether the emergency arises from natural causes, human error, technical failure, or through malicious acts. A plan has to be flexible; it has to work on a public holiday, on the weekend, or in freezing weather conditions, and at any location. It needs to be tested against different specific scenarios to ensure that it is both appropriate and flexible enough to deliver the required functions in varied circumstances.

Major emergencies do not respect boundaries, whether physical or organizational, and their consequences often have widespread ramifications. If the response is to be truly effective in meeting the needs of everyone affected, then all leaders of the community, industry, and commerce have to be aware of the contributions of different organizations.

The need for mutual aid agreements with parallel organizations should be considered within the planning process. In a major emergency, the added amount of work may be overwhelming, while everyday work will also need to continue. Organizations must explore all options for maintaining critical services not only during the response but also throughout the recovery, which may be lengthy.

There is no one-model response to an emergency. The response will need to vary just as the nature and effects of the individual emergency will vary. Nevertheless, any response has to be a combined and coordinated operation, and certain features will be common to a variety of different eventualities:

- The basic objectives of the combined response will be similar on each occasion.
- The same basic management structure should apply when responding to most eventualities.
- Accurate records will be required for briefings, debriefings, formal inquiries, and disseminating information about the lessons learned.

For sudden events, the emergency services—that is, the police, fire, ambulance, and coastguard services—maintain a state of readiness so that they can provide a rapid initial response and early alerting of local authorities and other services. All organizations that need to respond quickly to an emergency will have arrangements that can be activated at short notice. These arrangements should be clearly established and promulgated to all who may be involved in the response.

The emergency services, local authorities, government departments, and other organizations such as the utilities, voluntary organizations, and faith communities have produced single service (and sometimes single issue) planning documents for this purpose. The aforementioned U.K. government publication "Dealing with Disaster" draws on such planning documents to offer guidance on how the procedures and operations of each of the organizations involved can be combined and coordinated to provide an efficient and effective response.

Lead Government Departments

In the United Kingdom, there are designated lead government departments in charge of handling major emergencies. Should the lead not be clear, then the Cabinet Office Civil Contingencies Secretariat is responsible for taking the immediate lead and then for ensuring that one government department is confirmed as the lead government department. The lead departments are designated as follows (Cabinet Office, 2004):

- *Terrorism: Conventional/Siege/Hostage.* The Home Office leads during the counter terrorist phase. If matters move to the stage of managing the consequences of a terrorism incident, then the lead is transferred to the Civil Contingencies Secretariat.

- *Terrorism: Chemical, Biological, Radiological, Nuclear (CBRN).* The Home Office leads in dealing with the effects of the emergency. The Home Office would be supported by other departments including the Department for Environment, Food & Rural Affairs (DEFRA), which also has responsibility for coordinating the government's contribution to the decontamination and recovery phase of such incidents or emergencies in the open environment. At some point, to be determined on a case-by-case basis once the crisis management phase is concluded, the lead department responsibility would be transferred to DEFRA, although depending on the nature and location of the incident—for example, where releases of CBRN materials occur primarily within buildings and infrastructure—other departments might take the lead.

- *CBRN Incidents Arising from Nonterrorist Causes.* The Civil Contingencies Secretariat would ensure that, dependent upon the cause of the incident, a lead department was identified for the emergency phase. If matters move to the stage of managing the consequences, then the lead is transferred to DEFRA.

- *Serious Industrial Accidents.* The Civil Contingencies Secretariat is responsible for confirming the lead department in good time to support the response to an industrial accident.

- *Animal Disease and Welfare.* DEFRA and its State Veterinary Service have the lead, in conjunction with the Department of Health and its Food Standards Agency if there is a threat to human health (e.g., in the case of a zoonotic disease).

- *Infectious Diseases.* Department of Health has the lead, with assistance from the Public Health Laboratories Service (Health Protection Agency, 2003a).

The role of the Ministry of Defence in responding to major emergencies is to provide support to civil authorities (Cabinet Office, 2003b). There are three categories of Military Aid to the Civil Community (MACC):

Category A: Assistance to the civil authorities in dealing with an emergency such as a natural disaster or major incident.

Category B: Short-term routine assistance on special projects of significant social value to the civil community.

Category C: The full-time attachment of volunteers to social service (or similar) organizations for specific periods.

MACC is one of the three strands of Military Aid to the Civil Authorities (MACA). Other strands are Military Aid to the Civil Power (MACP), used only in the maintenance of law and order, and Military Aid to Other Government Departments (MAGD), used for work of national importance and in maintaining services essential to the life, health, and safety of the community.

With respect to terrorist incidents that result in major emergencies, MACC Category A will generally be invoked in obtaining military assistance.

BIOTERRORISM

There has long been recognition in the United Kingdom that terrorists might seek to threaten or actually use nuclear, chemical, or biological materials. Contingency plans have therefore been in existence for many years to respond to such incidents. These plans are regularly reviewed, tested through exercises, and updated in the light of changing circumstances in order to safeguard the interests of national security and ensure the protection of the public. The broad approach to be followed in the United Kingdom in responding to bioterrorism is that offered above for responding to a major emergency (i.e., Assessment, Prevention, Preparation, Response, and Recovery management). Some of the specific aspects relating to responding to the deliberate release of biological agents and toxins are outlined in subsequent sections of this entry, using the same headings as in the broad approach.

Assessment: Biological Agents and Toxins

In considering which biological agents and toxins present a threat, it is useful to consider both chemical and biological terrorism, as there is a spectrum of chemical and biological materials that might be used in a terrorist attack (Fig. 1). Considering such a spectrum is comprehensive and avoids any possible omission of midspectrum materials that can be regarded as either biological terrorism or chemical terrorism.

In considering chemical and biological terrorism, it is misleading to focus attention solely on classical chemical or biological warfare (BW) agents or weapons, as there is no necessity for such warfare agents or weapons to be used by a would-be terrorist group given that many other toxic or infectious materials that may cause harm to humans, animals, or plants exist. It also needs to be appreciated that in selecting biological agents or toxins for use as weapons in military programs, there may well be other criteria such as shelf life and storability that may influence the selection of particular agents, and such criteria can be irrelevant when a terrorist group is considering what agent to obtain. Consequently, preparedness for chemical and biological terrorism must be broad-based and capable of countering the use by terrorists of any toxic chemical or infectious biological material, regardless of whether the target population consists of humans, animals, or plants.

An indication of the biological agents and toxins that are considered to present a risk in the United Kingdom can be gained from Department of Health guidance (Department of Health, 2002a), first published in March 2000, which in a section entitled "Substances likely to be used" states that "the organisms considered by security experts as those most likely to be used in a deliberate attack with some of their characteristics are listed in Annex 2." This Annex lists the following (as written):

- Bacteria
  - *Bacillus anthracis*

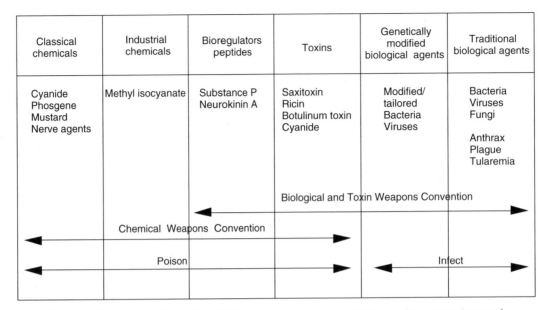

Figure 1. Spectrum of chemical and biological materials that might be used in a terrorist attack.

- *Brucella* species
- *Burkholderia*
- *Chlamydia psittaci*
- *Vibrio cholerae*
- *E. coli* V Tec positive
- *Salmonella*
- *Shigella*
- Rickettsiae, fungi, and viruses
  - *Rickettsia prowazekii*
  - *Rickettsia rickettsii*
  - *Rickettsia tsutsugamushi*
  - Influenza virus
  - *Coxiella burnetii Wood*
  - *Coccidioides immitis*
  - *Histoplasma capsulatum*
  - Ebola virus
  - Lassa virus
  - CCHF virus
  - Smallpox virus
  - Other viruses
- Toxins
  - Botulinum toxin
  - Staphylococcal enterotoxin B
  - Other toxins.

A more comprehensive listing is provided in the Schedule of Pathogens and Toxins attached to the 2001 Anti-Terrorism, Crime and Security Act (Her Majesty's Stationery Office, 2001), which lists the following (as written):

- Viruses
  - Chikungunya virus
  - Crimean-Congo haemorrhagic fever virus
  - Dengue fever virus

- Eastern equine encephalitis virus
- Ebola virus
- Hantaan virus
- Japanese encephalitis virus
- Junin virus
- Lassa fever virus
- Lymphocytic choriomeningitis virus
- Machupo virus
- Marburg virus
- Monkeypox virus
- Rift Valley fever virus
- Tick-borne encephalitis virus (Russian Spring-Summer encephalitis virus)
- Variola virus
- Venezuelan equine encephalitis virus
- Western equine encephalitis virus
- Whitepox
- Yellow fever virus
- Rickettsiae
  - *Coxiella burnetii*
  - *Bartonella quintana (Rochalimea quintana, Rickettsia quintana)*
  - *Rickettsia prowazeki*
  - *Rickettsia rickettsii*
- Bacteria
  - *Bacillus anthracis*
  - *Brucella abortus*
  - *Brucella melitensis*
  - *Brucella suis*
  - *Chlamydophila psittaci*
  - *Clostridium botulinum*
  - *Francisella tularensis*
  - *Burkholderia mallei (Pseudomonas mallei)*

- *Burkholderia pseudomallei (Pseudomonas pseudo-mallei)*
- *Salmonella typhi*
- *Shigella dysenteriae*
- *Vibrio cholerae*
- *Yersinia pestis*
- Toxins
  - Botulinum toxins
  - *Clostridium perfringens* toxins
  - Conotoxin
  - Ricin
  - Saxitoxin
  - Shiga toxin
  - *Staphylococcus aureus* toxins
  - Tetrodotoxin
  - Verotoxin
  - Microcystin (Cyanginosin)
  - Aflatoxins.

It should, however, be noted that as of August 2003, the United Kingdom listing of such agents does not currently include animal or plant pathogens, although the 2001 Anti-Terrorism, Crime and Security Act does include the power to extend the requirements to animal or plant pathogens, pests, or toxic chemicals. This extension "may be exercised in relation to any pathogen or pest only if the Secretary of State is satisfied that there is a risk that the pathogen or pest is of a description that could be used to cause:

a. widespread damage to property;
b. significant disruption to the public; or
c. significant alarm to the public."

With respect to chemicals, this extension "may be exercised in relation to any chemical only if the Secretary of State is satisfied that the chemical could be used as an act of terrorism to endanger life or cause serious harm to human health."

Prevention

In considering the various activities that contribute to the prevention of bioterrorism, there are several elements of importance: first, national legislation that prohibits the development and production of biological weapons; second, national legislation that addresses safe and secure storage of biological agents and toxins; and third, national legislation that controls access to biological agents and toxins.

National Legislation that Prohibits Biological Weapons. Article IV of the Biological and Toxin Weapons Convention (BWC), signed by the United Kingdom on April 10, 1972, and ratified on March 26, 1975, obliges each State Party to "take any necessary measures to prohibit and prevent" the actions delineated in Article I. The basic prohibition in Article I requires that:

Each State Party to this Convention undertakes never in any circumstances to develop, produce, stockpile or otherwise acquire or retain:

(1) Microbial or other biological agents, or toxins whatever their origin or method of production, of types and in quantities that have no justification for prophylactic, protective or other peaceful purposes;
(2) Weapons, equipment or means of delivery designed to use such agents or toxins for hostile purposes or in armed conflict.

Reaffirmations by successive Review Conferences have emphasized the comprehensiveness of the applicability of the Convention. Thus, at the Fourth Review Conference, the Article I section of the Final Declaration (United Nations, 1996) stated that "[t]he Conference also reaffirms that the Convention unequivocally covers all microbial or other biological agents or toxins, naturally or artificially created or altered, as well as their components, whatever their origin or method of production, of types and in quantities that have no justification for prophylactic, protective or other peaceful purposes," and went on to add that "[t]he Conference, conscious of apprehensions arising from relevant scientific and technological developments, *inter alia*, in the fields of microbiology, biotechnology, molecular biology, genetic engineering, and any applications resulting from genome studies, and the possibilities of their use for purposes inconsistent with the objectives and the provisions of the Convention, reaffirms that the undertaking given by the States Parties in Article I applies to all such developments."

In the United Kingdom, the requirement to prevent and prohibit biological weapons was achieved by the Biological Weapons Act 1974 (Her Majesty's Stationery Office, 1974a), which sets out that:

No person shall develop, produce, stockpile, acquire or retain—

(a) any biological agent or toxin of a type and in a quantity that has no justification for prophylactic, protective or other peaceful purposes; or
(b) any weapon, equipment, or means of delivery designed to use biological agents or toxins for hostile purposes or in armed conflict.

This Act goes on to define biological agent and toxin as follows: "'biological agent' means any microbial or other biological agent; and 'toxin' means any toxin, whatever its origin or method of production." It then sets out the penalty that "[a]ny person contravening this section shall be guilty of an offence and shall, on conviction on indictment, be liable to imprisonment for life." It should be noted that the language used in the U.K. Biological Weapons Act very closely parallels that in Article I of the Convention.

National Legislation that Addresses Safe and Secure Storage of Biological Agents and Toxins. States have long had measures to protect the health and safety of people as well as measures to protect the health of animals and plants. In the United Kingdom, the basic provisions for health and safety of people are provided by the Health and Safety at Work Act 1974 (Her Majesty's Stationery Office, 1974b): "to make further provision for securing the health, safety and welfare of persons at work, for protecting others against risks to health or safety in connection with the activities of persons at work, for controlling the keeping and use and preventing the unlawful acquisition, possession and use of dangerous substances..."

It sets out the principal provisions of the Act as being:

(A) Securing the health, safety and welfare of persons at work;

(B) Protecting persons other than persons at work against risks to health or safety arising out of or in connection with the activities of persons at work;

(C) Controlling the keeping and use of explosive or highly flammable or otherwise dangerous substances, and generally preventing the unlawful acquisition, possession and use of such substances.

A key element in the implementation of the Health and Safety at Work Act is the Control of Substances Hazardous to Health (COSHH) Regulations 2002 (Her Majesty's Stationery Office, 2002a), intended to protect both workers and others who may be exposed from work activities to the risks of hazardous substances. In the context of the regulations, a hazardous substance is anything that can harm health if it is not adequately controlled. Consequently, all pathogens and toxins are covered by these regulations. The 2002 COSHH regulations define a "substance hazardous to health" as meaning a substance:

a. which is listed in Part I of the approved supply list as dangerous for supply ... and for which an indication of danger specified for the substance is very toxic, toxic, harmful, corrosive or irritant;

b. for which the Health and Safety Commission has approved a maximum exposure limit or an occupational exposure standard;

c. which is a biological agent;

d. which is a dust of any kind...;

e. which, not being a substance falling within subparagraphs a. to d., because of its chemical or toxicological properties and the way it is used or is present at the workplace creates a risk to health.

It defines biological agent as "a micro-organism, cell culture or human endo-parasite, whether or not genetically modified, which may cause infection, allergy, toxicity or otherwise create a hazard to human health."

The basic approach with regard to substances hazardous to health, including biological agents, adopted in the COSHH regulations consists of the following elements:

- Assessment of risk to health created by work involving substances hazardous to health

- Prevention or control of exposure to substances hazardous to health

- Use of control measures, and so on

- Maintenance, examination, and testing of control measures

- Monitoring of exposure at the workplace

- Health surveillance

- Information, instruction, and training for persons who may be exposed to substances hazardous to health

- Arrangements to deal with accidents, incidents, and emergencies.

Additional requirements apply to work with biological agents, including:

- Classification of biological agents

- Special control measures for laboratories, animal rooms, and industrial processes

- List of employees exposed to certain biological agents (in Group 3 and Group 4)

- Notification of the use of biological agents (in Groups 2, 3, and 4)

- Notification of the consignment of biological agents (in Group 4).

Biological agents are categorized into four hazard Groups, as alluded to above:

Group 1: unlikely to cause human disease.

Group 2: can cause human disease and may be a hazard to employees; unlikely to spread to the community, and there is usually effective prophylaxis or treatment available.

Group 3: can cause severe human disease and may be a serious hazard to employees; may spread to the community, but there is usually effective prophylaxis or treatment available.

Group 4: causes severe human disease and is a serious hazard to employees; likely to spread to the community, and there is usually no effective prophylaxis or treatment available.

An approved list of the categorization to be applied to biological agents is issued by the Health and Safety Executive (Health and Safety Executive, 2004).

The COSHH regulations also set out the minimum requirements for containment measures in facilities and laboratories handling such biological agents as follows:

(a) level 2 for activities that involve working with a Group 2 biological agent;

(b) level 3 for activities that involve working with a Group 3 biological agent;

(c) level 4 for activities that involve working with a Group 4 biological agent;

(d) level 2 for laboratories that do not intentionally propagate, concentrate or otherwise increase the risk of exposure to a biological agent but work with materials in respect of that it is unlikely that a Group 3 or Group 4 biological agent is present;

(e) level 3 or 4, where appropriate, for laboratories that do not intentionally propagate, concentrate or otherwise increase the risk of exposure to a Group 3 or Group 4 biological agent but where the employer knows, or it is likely, that such a containment level is necessary; and

(f) level 3 for activities where it has not been possible to carry out a conclusive assessment but where there is concern that the activity might involve a serious health risk for employees.

The basic requirement is thus that the containment level must at the minimum match the hazard grouping of the agent. COSHH 2002 requirements for containment levels include requirements regarding access and safe storage of biological agents as listed in Table 1.

The requirement for keeping a list of employees exposed to agents in Group 3 and Group 4 is detailed in the COSHH regulations as follows:

List of employees exposed to certain biological agents

4.-(1) Subject to sub-paragraph (2), every employer shall keep a list of employees exposed to a Group 3 or Group 4 biological agent, indicating the type of work done and, where known, the biological agent to which they have been exposed, and records of exposures, accidents and incidents, as appropriate.

(2) Sub-paragraph (1) shall not apply where the results of the risk assessment indicate that

(a) the activity does not involve a deliberate intention to work with or use that biological agent; and

(b) there is no significant risk to the health of employees associated with that biological agent.

(3) The employer shall ensure that the list or a copy thereof is kept available in a suitable form for at least 40 years from the date of the last entry made in it.

(4) The relevant doctor referred to in regulation 11, and any employee of that employer with specific responsibility for the health and safety of his fellow employees, shall have access to the list.

(5) Each employee shall have access to the information on the list which relates to him personally.

The requirement for notification of first use is set out in the regulations as follows:

Notification of the use of biological agents

5.-(1) Subject to subparagraphs (7) and (8), an employer shall not use for the first time one or more biological agents in Group 2, 3 or 4 at particular premises for any of the activities listed in paragraph 3(3) unless he has

(a) notified the Executive in writing of his intention to do so at least 20 working days in advance, or such shorter period as the Executive may allow;

Table 1. COSHH 2002 Requirements for Containment Levels

Containment Measure	2	3	4
Access to be restricted to authorized persons only	Yes	Yes	Yes, via air-lock key procedure
Safe storage of biological agents	Yes	Yes	Yes, secure storage

(b) furnished with that notification the particulars specified in sub-paragraph (5); and

(c) received the acknowledgement required by sub-paragraph (4).

(2) Subject to sub-paragraphs (7) and (9), an employer shall not use a biological agent which is specified in Part V of this Schedule, except where the use of that agent has been notified to the Executive in accordance with sub-paragraph (1), for any of the activities listed in paragraph 3(3) unless he has

(a) notified the Executive in writing of his intention to do so at least 20 working days in advance, or such shorter period as the Executive may allow;

(b) furnished with that notification the particulars specified in sub-paragraph (5); and

(c) received the acknowledgement required by sub-paragraph (4).

(3) The Executive may accept a single notification under sub-paragraph (2) in respect of the use of more than one biological agent by the same person.

(4) Upon receipt of the notification required by sub-paragraph (1) or (2), the Executive shall, within 20 working days

(a) send to the notifier an acknowledgement of receipt; or

(b) if the notification does not contain all of the particulars specified in sub-paragraph (5)

(i) inform the notifier in writing of the further particulars required, and

(ii) within 10 working days of receipt of those further particulars, send to the notifier an acknowledgement of receipt.

(5) The particulars to be included in the notification referred to in sub-paragraphs (1) and (2) shall be

(a) the name and address of the employer and the address of the premises where the biological agent will be stored or used;

(b) the name, qualifications and relevant experience of any employee of that employer with specific responsibility for the health and safety of his fellow employees;

(c) the results of the risk assessment;

(d) the identity of the biological agent and, if the agent does not have an approved classification, the Group to which the agent has been assigned; and

(e) the preventive and protective measures that are to be taken.

(6) Where there are changes to processes, procedures or the biological agent which are of importance to health or safety at work and which render the original notification invalid the employer shall notify the Executive forthwith in writing of those changes.

(7) Sub-paragraphs (1) and (2) shall not apply in relation to a biological agent where an intention to use that biological agent has been previously notified to the Executive in accordance with the Genetically Modified Organisms (Contained Use) Regulations 2000.

(8) The requirement in sub-paragraph (1) to notify first use of a biological agent in Group 2 or 3 shall not apply to an employer whose only use of that agent

is in relation to the provision of a diagnostic service provided that use will not involve a process likely to propagate, concentrate or otherwise increase the risk of exposure to that agent.

(9) The requirement in sub-paragraph (2) to notify use of a biological agent specified in Part V of this Schedule shall not apply to an employer whose only use of that agent is in relation to the provision of a diagnostic service provided that use will not involve a process likely to propagate, concentrate or otherwise increase the risk of exposure to that agent.

The requirement for notification of the consignment of biological agents is set out in the regulations as:

Notification of the consignment of biological agents

6.-(1) An employer shall not consign a Group 4 biological agent or anything containing, or suspected of containing, such an agent to any other premises, whether or not those premises are under his ownership or control, unless he has notified the Executive in writing of his intention to do so at least 30 days in advance or before such shorter time as the Executive may approve and with that notification has furnished the particulars specified in sub-paragraph (4).

(2) Sub-paragraph (1) shall not apply where

(a) the biological agent or material containing or suspected of containing such an agent is being consigned solely for the purpose of diagnosis;

(b) material containing or suspected of containing the biological agent is being consigned solely for the purpose of disposal; or

(c) the biological agent is or is suspected of being present in a human patient or animal which is being transported for the purpose of medical treatment.

(3) Where a Group 4 biological agent is imported into Great Britain, the consignee shall give the notice required by sub-paragraph (1).

(4) The particulars to be included in the notification referred to in sub-paragraph (1) shall be

(a) the identity of the biological agent and the volume of the consignment;

(b) the name of the consignor;

(c) the address of the premises from which it will be transported;

(d) the name of the consignee;

(e) the address of the premises to which it shall be transported;

(f) the name of the transport operator responsible for the transportation;

(g) the name of any individual who will accompany the consignment;

(h) the method of transportation;

(i) the packaging and any containment precautions which will be taken;

(j) the route which will be taken; and

(k) the proposed date of transportation.

It is thus evident that U.K. regulations require that (1) biological agents be categorized; (2) containment

provisions, including provisions for access and safe storage, be appropriately implemented; (3) lists be kept of employees exposed to agents in Group 3 or 4; (4) notification of first use of agents in Groups 2, 3, and 4 be provided to appropriate authorities; and (5) notification of consignments of agents in Group 4 be provided to appropriate authorities.

National Legislation that Controls Access to Biological Agents and Toxins. In the United Kingdom, additional requirements addressing the security of biological agents and toxins were enacted in the above-described Anti-Terrorism, Crime and Security Act 2001 (Her Majesty's Stationery Office, 2001). In "Part 7: Security of Pathogens and Toxins," the following elements are outlined: duty to notify Secretary of State before keeping or using any dangerous substance; power to require information about security of dangerous substances; power to require information about persons with access to dangerous substances; duty to comply with security directions; duty to dispose of dangerous substances; and denial of access to dangerous substances.

Part 7 sets out the pathogens and toxins to which these requirements apply as being those listed in Schedule 5 to the Act (and reproduced earlier in this entry), and as noted includes provision for the Secretary of State to add any pathogen or toxin to that Schedule if he is satisfied that the pathogen or toxin could be used in an act of terrorism to cause widespread damage to property, significant disruption to the public, or significant alarm to the public. The term "dangerous substance" is defined to include, in addition to the substances listed in Schedule 5 themselves, anything such as a plant or animal that is infected by or is the carrier of a pathogen listed in Schedule 5, unless it satisfies prescribed conditions or is kept or used under prescribed conditions.

Specifically, the Act highlights the following points.

- The notification before keeping or using any dangerous substance must:
  (a) identify the premises in which the substance is kept or used;
  (b) identify any building or site of which the premises form part;
  (c) contain such other particulars (if any) as may be prescribed.

- The information about security of dangerous substances includes both:
  (a) measures taken to ensure the security of any building or site of which the premises form part; and
  (b) measures taken for the purpose of ensuring access to the substance is given only to those whose activities require access and only in circumstances that ensure the security of the substance.

- The information required about persons with access to dangerous substances includes a list of:
  (a) each person who has access to any dangerous substance kept or used there;
  (b) each person who, in such circumstances as are specified or described in the notice, has access to such part of the premises as is so specified or described;

(c) each person who, in such circumstances as are specified or described in the notice, has access to the premises; or

(d) each person who, in such circumstances as are specified or described in the notice, has access to any building or site of which the premises form part.

- Regarding denial of access to dangerous substances, the provisions set out that "the Secretary of State may give directions to the occupier of any relevant premises requiring him to secure that the person identified in the directions:

    (a) is not to have access to any dangerous substance kept or used there;

    (b) is not to have, in such circumstances (if any) as may be specified or described in the directions, access to such part of the premises as is so specified or described;

    (c) is not to have, in such circumstances (if any) as may be specified or described in the directions, access to the premises; or

    (d) is not to have, in such circumstances (if any) as may be specified or described in the directions, access to any building or site of which the premises form part."

Preparation

As noted above, the new Civil Contingencies Bill under consideration would include provisions to deal with both the effect of major acts of terrorism and the contamination of air, water, or land such as to threaten human or animal health or the natural environment. Clearly, biological terrorism against humans, animals, and plants would therefore be covered by the Bill.

Preparation for biological terrorism is clearly addressed in guidance provided by the Cabinet Office Civil Contingencies Secretariat in October 2001 to local authorities on the response to deliberate release of chemicals and biological agents (Cabinet Office, 2001). This guidance makes it clear that the measures required to manage the consequences of any crisis featuring the accidental release of toxic substances or a major infectious disease outbreak would be similar to those required for a deliberate release. Consequently, it is appropriate to have a generic plan that can be adapted to either a deliberate or an accidental or natural outbreak. The guidance affirms that although tried and tested contingency measures were already in place in the United Kingdom beforehand, following September 11, 2001, the scrutiny of those plans has been stepped up and the guidance note revised accordingly.

The guidance note sets out as its purpose to give local authorities an overview of the multiagency response to deliberate release of chemical and biological materials in the United Kingdom by terrorists; draw local authorities' attention to particular problems on the effective consequence management of such an incident; highlight key areas of preplanning activity and resource management which should be considered by local authorities; and indicate where local authorities can obtain further technical or specialist information and advice. It then goes on to make a number of planning assumptions:

- There will be strong public interest and concern. Accompanying this will be the inevitable demand for press and media statements indicating what actions the general public should be taking to protect individual health and property.

- There will be a requirement for special coordination and control arrangements to be established (as national and international political interest and pressure is applied) in order to ensure a speedy and successful resolution of the incident.

- With an incident of this nature, it is possible but by no means certain that advance warning will be given, and arrangements must therefore take into account the possibility of a no-notice attack.

- What is evident is that a deliberate release may have the potential to cause serious harm and disruption over a wide geographical area, with no regard for administrative boundaries of local authorities or other agencies. In contrast, a deliberate release of chemicals or biological material would be particularly dangerous where a large number of people were assembled in an enclosed area.

The guidance then goes on to give an appreciation of the characteristics of such a release. A biological release is described as being a major public health emergency that may take days to first become apparent and weeks to evolve. For this reason, biological attacks can be hard to detect and/or identify. Furthermore, the effects of an airborne release of a biological agent will depend upon the nature of the disease that the agent causes and the effectiveness of preventative measures and treatment. A crucial determinant of the potential number of casualties is the ability of the disease to spread from person to person via aerosol. Unless the release is announced, detection and minimization of casualties will depend on early identification of unusual patterns of illness by doctors and laboratories. Water or food supplies could also be used to spread biological agents. Chlorination and/or boiling may eliminate or minimize the danger in most, but not all, cases.

An outline is then given of the multiagency coordination that would be implemented in the event of a deliberate release. It is noted that the arrangements for responding to a credible threat of a release of chemicals or biological material are unusual because (1) the response may be initiated by the strategic level; (2) it involves central government as a key player; and (3) the military are likely to be involved. It is critical therefore that the response is genuinely multiagency and that communication between all agencies is of the highest order. Ministers and senior officials representing the appropriate departments lead the central government involvement from the Cabinet Office Briefing Room (COBR).

Notification and Confirmation. The guidance notes that any terrorist incident is a crime, and as such the Police Incident Commander remains in operational command. As soon as it appears that a terrorist threat or action has taken place, the police will notify central government and the national plan will be brought into operation to augment local resources. This will involve the mobilization of a range of national assets, some of which focus on the management of the crisis (preventing or mitigating the incident itself) while others concentrate on consequence management. A Government Liaison Officer (GLO) from the Home Office leads a Government Liaison Team (GLT) based at the Police Main Base Station (PMBS), which will include a central government official acting as Consequence Management Liaison Officer (CMLO).

The role of the GLO is to keep COBR fully informed of the development of the incident; to ensure that the police interest is taken fully into account at COBR, and conversely to ensure that the Government's views are kept in mind at the scene; and to see that the flow of communications between the scene and COBR works smoothly. In this way, the GLO takes some of the pressure off the Police Incident Commander.

The CMLO is part of the GLT and provides the link between the center and the incident, concentrating on consequence management issues. The CMLO ensures that the appropriate local agencies are fully engaged in the response, and later ensures that the recovery phase is being considered. Normally, the local authorities will be informed directly by the police, but the CMLO serves to ensure that this has been done.

Command and Control. The strategic command for an incident of this nature will be set up in a Police Main Base Station. The Police Main Base Station will normally be at the Police Headquarters of the force in whose area the incident has occurred. However, the Police Incident Commander may decide, taking into account the location of the incident, toxicity of the material being dispersed, and prevailing meteorological conditions, that the location is unsafe, and thus may move strategic command to an alternative site.

In the GLT at the Police Main Base Station will be representatives of the Home Office, the Cabinet Office Civil Contingencies Secretariat, Foreign and Commonwealth Office, Department of Health/National Health Service (NHS), Crown Prosecution Service (CPS), and the military. There may also be representatives from the Department of the Environment, Food and Rural Affairs (including the Drinking Water Inspectorate), Food Standards Agency, Environment Agency, Department of Trade and Industry, and others, dependent upon the nature of the incident.

Also at the Police Main Base Station will be representatives of the Ministry of Defence, whose specialist resources may be required by the Police Incident Commander. These include personnel trained to operate in protective equipment, to gain access to controlled space, to carry out specialist searches, and to render safe any devices that may be found. The military may also be able to provide assistance of a more general nature falling outside the counter terrorism contingency, under the above-described terms of MACC. Requests for such assistance should be made either through the normal channels at the local level or through the local Divisional Headquarters Liaison Officer at the Police Main Base Station.

The local authority Chief Executive (possibly with the Emergency Planning Officer in support) would normally be expected to attend multiagency strategic meetings at the Police Main Base Station. Whoever represents the local authority must be able to make strategic decisions on behalf of that authority. If more than one local authority is involved or affected by the incident, a swift decision will be required as to how their interests are to be represented at the Police Main Base Station to prevent unnecessary duplication and use of resources. Issues to be discussed at multiagency strategic meetings include the decontamination strategy, media issues and public information strategy, the response to central and local government political figures and VIPs, and the strategy for remediation, reoccupation, and recovery. Multiagency tactical level issues that may be referred for a strategic view include evacuation/stay-put decisions, effective road closures, and temporary accommodation for displaced persons.

Joint Health Advisory Cell. A key element of the multiagency response to an incident of this nature, in which there is an identified threat to public health, will be the establishment of a Joint Health Advisory Cell (JHAC) at the Police Main Base Station. This is a strategic group set up at the request of the Police Incident Commander by the Director of Public Health (DPH) for the area affected by the incident (who has the legal responsibility for public health issues). The DPH in turn will be supported by a Consultant in Communicable Disease Control and other medical experts, a press officer, an environmental health officer (EHO) from the local authority, and representatives from DEFRA, the Environment Agency, and the water company. The choice of EHO will depend on the nature of the incident and the expertise of the EHO. Additional members would be alerted by central government and would normally include a senior scientific adviser from the Chemical and Biological Defence (CBD) Sector, Porton Down, a military medical adviser, and a police liaison officer. Although the DPH is responsible for servicing meetings of the JHAC, individual members will wish to ensure that they have their own admin/secretarial support.

The main purposes of the JHAC are to take advice on the health aspects of the incident from a range of experts; provide advice to the Police Incident Commander on the health consequences of the incident, including those relating to evacuation or containment; agree with the Police Incident Commander regarding advice to give to the public on the health aspects of the incident; and maintain a written record of decisions made and the reasons for those decisions. It will also, as necessary, liaise with the Department of Health; liaise with other health authorities; formulate advice to health professionals in hospitals, ambulance services and general practice; and formulate advice on the strategic management of the health service response. The local authority representative will be part of the information gathering process of the JHAC and will help the public health officials to assess how best to inform the general public.

The Role of the Local Authority. On being informed of a deliberate release or credible threat, the appropriate local authority (or authorities) would be expected to invoke its generic major incident procedure. As with all major incidents that actually do, or have the potential to, affect large numbers of people across a widespread area, there is a key role for local authorities in terms of practical measures and logistical support.

For an incident of this type, an extensive program of health care and advice will be set up by the Health Authority (or its successor bodies following reorganization), and there may be requests to local authorities for the use of buildings and assistance with travel, distribution of medicines, provision of equipment, and a range of other services. Local authority emergency plans should clearly identify the nature and extent of existing links to partners, in particular, relationships with health authorities and NHS Trusts. This is especially important in terms of services provided to the general public, for example, counseling, welfare, access to general practitioners, and the establishment of temporary medical centers and/or body holding areas and mortuaries.

Assessment and monitoring for a deliberate release of chemical or biological material would be done by specialist agencies, for example, CBD Porton Down, but it may be some time before national resources arrive on the scene. Local authorities may wish to consider whether their limited equipment could be useful in the interim. It is worth noting that some local authorities also have limited supplies of environmental monitoring and sampling equipment. However, careful consideration needs to be given to the deployment of local authority officers to any incident scene, bearing in mind the availability of protective clothing and equipment and the dangers for staff of unnecessary or prolonged exposure to a harmful agent. The advice of CBD Porton Down should be considered.

Information about the properties of any chemical and biological materials, their behavior, and how they react with other substances is vital to local authorities in formulating an appropriate response to the situation. In addition to their normal sources of advice, environmental health officers can seek advice from CBD Porton Down personnel, who have particular expertise in warfare agents; this should normally be done through the Joint Health Advisory Cell or Consequence Management Liaison Officer on the Government Liaison Team.

The extreme toxicity of materials that could be used in a deliberate attack arguably make health and safety issues as or even more important than after an accidental release or naturally occurring outbreak. It should not normally be necessary for local authority staff to enter a known heavily contaminated area. As soon as a release of chemical or biological material is known or suspected, every effort must be made to protect local authority staff. Immediate advice should be obtained from the health service on medical surveillance and the recording of staff who may have been exposed to the hazardous material.

Local authorities will have a major role to play following an incident of this type. The potential scale and nature mean that large numbers of people are likely to be affected and thus will require some form of help during the recovery and rehabilitation phase. These will include not only casualties and survivors but also relatives, friends, colleagues, and members of the local and wider community. Frontline staff who have been involved in responding to the incident may also be affected.

Local authorities will need to consider with the health authorities the provision of help-lines and also drop-in centers for those in need of counseling. Social service staff may have to visit clients in their own homes to offer comfort and reassurance. They will also provide the gateway to other care and support services, on the basis of local knowledge and accurate information about individual requirements.

Health Service Response. Further insight into the role of the JHAC is provided in a Department of Health guidance (Department of Health, 2002a), initially issued in March 2000 and reissued in August 2002, to help plan the health service response to the deliberate release of biological and chemical agents. In the foreword, the then Chief Medical Officer states that "[m]uch of the preparation [for a deliberate release of biological or chemical agents] is the same as that which is needed to deal with a chemical accident or a naturally occurring disease outbreak." He goes on to note that "[a] deliberate attack does however have some special features. Agents used could be extremely toxic and unfamiliar to NHS staff. There will be special arrangements established by the police to manage the incident." He concludes by pointing out that "the Directors of Public Health and their teams will have a lead role in providing advice to the police, NHS staff and the public and coordinating the health service response. This guidance is intended to assist them understand the nature of the threat and prepare accordingly."

The guidance makes clear that efforts should complement existing arrangements for dealing with communicable disease surveillance and control. The guidance deals primarily with the management of the threat and possible consequences of a deliberate airborne release of those chemical or biological agents currently considered most likely to be used in a terrorist attack. It goes on to note that food or water contamination, which could have significant health consequences and cause massive public alarm, would be dealt with largely through the existing mechanisms for responding to chemical contamination of water or food and therefore are not given detailed consideration in the guidance. The guidance further comments that arrangements are described in greatest detail for a situation in which authorities are given warning of an impending release, as it is in this situation where the command, control, and advisory structures will differ most from an accidental release or naturally occurring disease outbreak (if there is no warning of release, in the first instance the incident will be managed in the same way as an accident or naturally occurring disease outbreak, and only once a deliberate release is suspected will the arrangements in the guidance be applied).

The guidance outlines the following differences between dealing with a terrorist incident and dealing with a naturally occurring disease outbreak or chemical accident.

- The numbers of actual or potential casualties: normal NHS resources could rapidly be overwhelmed.

Abnormal solutions to exceptional problems will have to be considered.

- The scale of public concern: the police will be looking to the JHAC for advice on public statements.
- Central government and national political interest: special arrangements for coordination and control will be established at the Cabinet Office.
- Advance warning: may offer insight into the location and/or type of release.

The guidance then goes on to set out some planning considerations relating to the differences between an accidental and a deliberate release of biological agents. It sets the scene for this by referring back to the anthrax outbreak in Sverdlovsk (the former Soviet Union) in April and May 1979, where most of the victims worked or lived in a narrow zone extending from a military microbiological facility to the southern city limit. The zone paralleled the prevailing northerly wind. Most observers now believe the epidemic, which caused over 60 deaths, was caused by inhalation of spores accidentally released from a military facility. The guidance then states that biological agents, especially those with a capacity for person-to-person spread, are potentially even more devastating for a civilian population than chemicals. It also notes that terrorists might seek to obtain biological agents from the governments of countries sympathetic to their cause (though it may be observed that no such transfer has yet been documented).

The guidance differentiates overt and covert release of a biological agent. If overt, however, the details given may be vague, inaccurate, or deliberately misleading. The NHS has well-established procedures for dealing with outbreaks of infectious diseases, and these procedures will be used in the event of a deliberate release. However, a deliberate release may present special difficulties, for example, regarding the agents used, the concentration of the agent, and the method of dispersal, all of which may be unfamiliar to those responsible for preparing plans or responding to incidents. Moreover, surveillance systems are not well tuned to detecting a deliberate covert release, which may result in unusual and misleading disease presentations.

Unlike a chemical release, where the effects are immediate and are likely to involve the NHS from the outset, the effects of a biological release are likely to be delayed and prolonged for the following reasons:

- People exposed are unlikely to know immediately that they have been affected.
- Incubation periods between infection and development of symptoms may vary (commonly from 1 day to 2 weeks, and exceptionally longer).
- A single aerosol release may continue to be effective for some time after it is discharged.
- Material dispersed, particularly if dispersed in dry form, will be deposited on clothing, equipment, the ground, and other surfaces. When these surfaces are subsequently disturbed, secondary dispersal will occur.

- Alternative and more insidious methods of dispersal can be used, for example, contamination of the food chain.
- Secondary infection of contacts may lead to epidemics if the disease is contagious (e.g., plague and smallpox).

Although the biological agents likely to be released are ones that cause disease in humans through natural processes, the location of release, route of infection (e.g., inhalation, ingestion, injection), and concentration of agent may give rise to presenting symptoms and disease distribution that would not normally be seen in the United Kingdom. The organisms considered by security experts as those most likely to be used in a deliberate attack are listed earlier in this entry. The guidance says that if exposure to one of these agents is suspected, the DPH and his team will require expert clinical advice and laboratory support. This will come in the first instance from the local infectious disease consultant and the local director of the Public Health Laboratory Service (now subsumed into the Health Protection Agency). In addition, a national expert in the suspected disease may be consulted, and the reference facility that will be able to provide laboratory support notified. The guidance also points out that newly emerging organisms and the possibility of genetic manipulation make it impossible to predict with complete confidence what could be used.

As noted, the guidance deals primarily with the management of the consequences of a deliberate airborne release of a biological agent, although some advice is included on substances that could be released into water supplies or the food chain (any substance released into the air has the potential to contaminate food and water supplies as well). The most practical method of disseminating the agent is as minute airborne particles in aerosol form. It is recognized that sophisticated means of droplet release or the use of explosives could lead to (but by no means guarantees) more widespread dispersion, and that release from an aircraft could be highly effective. Dispersal could be geographically widespread and also spread over time, and infections that normally spread only through the food chain or in water may be transmitted in unusual ways, for example, through the respiratory tract or direct contact with mucous membrane or abraded skin.

The guidance notes that the collection, transport, and processing of specimens from human or environmental sources following a release of a BW agent is potentially extremely hazardous. Although accurate identification of the agent is urgently required, this process should not endanger staff. The plans of health authorities and laboratories in their district should therefore include protocols for obtaining samples and for their subsequent handling. Many NHS facilities are not equipped to deal with the potentially very dangerous biological agents (Advisory Committee on Dangerous Pathogens categories 3 and 4, categories into which most potential BW agents fall). During an incident, specialist advice on environmental sampling may be available through the JHAC.

The health authority incident control team, established on suspicion or recognition of a release of a biological agent, should determine what sampling is required, who should collect and transport specimens, and which laboratory should receive them. The team should also discover if any samples have already been taken by other agencies. Health Authorities and Trusts should ensure that their sampling arrangements account for the biological hazards identified in this guidance.

The guidance notes that the following occurrences may be an indication of a covert release:

- Reports of illnesses not endemic to the area
- Reports of unusual presentations of diseases
- A sudden outbreak involving large numbers of victims
- Illness reported with an unusual geographical distribution (this may be downwind of a release site)
- Illness reported at unusual times of year
- Unexplained illness in animals
- Information from security services that a terrorist release is suspected (if a deliberate release is suspected, the police should be informed immediately).

In preparing for such a deliberate release, as noted, health authorities already have frequently used plans for the control of naturally occurring outbreaks of infectious disease. However, the following points are either unique to a deliberate release or require greater consideration than during a naturally occurring outbreak:

- Confirmation of scale and nature of threat
- Restriction of access to potentially contaminated areas
- Immunization desirability and availability
- Public information
- Guidance for health professionals
- Availability of suitable laboratory facilities to deal with very large numbers of specimens
- Ability of the NHS to deal with large numbers of patients with infectious disease.

Health Protection Agency. The Health Protection Agency established on April 1, 2003, is a new national organization, which is dedicated to protecting people's health and reducing the impact of infectious diseases, chemical hazards, poisons, and radiation hazards. The Agency in its Corporate Plan (Health Protection Agency, 2003b) for 2003 to 2008 includes as Strategic Goal 6 the requirement "to improve preparedness of responses to health protection emergencies, including those caused by deliberate release." The plan proposes that the Health Protection Agency:

- build upon existing major incident plans;
- develop the infrastructure for surveillance and early recognition of events;
- continue to provide guidance for health protection for these new hazards;
- identify specific countermeasures and make sure they are available quickly;
- provide training and particularly testing of new plans by exercise scenarios;
- coordinate its assets in emergency situations.

First year milestones are identified, which include to complete and implement standard operating procedures for CBRN emergencies for the Health Protection Agency and coordinate operations centers; to deliver training to general practitioners, health care staff, medical microbiologists, lead Primary Care Trusts (PCTs), and Strategic Health Authorities (SHAs); and to implement the Department of Health's short-term strategy for CBRN training across the health service and to develop the medium and long-term strategies.

In addition, the Emergency Response Division of the Health Protection Agency is charged with improving the speed and effectiveness of overall response, both locally and nationally, in the event of any future incident or threat (this includes providing positive and authoritative messages about health protection measures in order to reduce public anxiety), and providing a central source of authoritative scientific and medical information and other specialist advice on both the planning and operational responses to major incidents and wider public health or other emergencies. Recently, the Health Protection Agency has issued guidance for the initial investigation and management of outbreaks and incidents of unusual disease with particular reference to events that may be due to biological and other causes, including deliberate releases (Health Protection Agency, 2004).

Response

Indications of the U.K. preparedness to respond to biological terrorism are provided in a listing of recent achievements (Cabinet Office, 2003c), which as of June 2003 includes the following:

- The Department of Health has provided 360 mobile decontamination units and 7250 national specification Personal Protection Equipment (PPE) suits around the country, which will enable the Ambulance Service and Accident & Emergency Departments to treat people contaminated with CBRN material.
- The CBRN Police Training Centre has been established at Winterbourne Gunner and has already delivered command training to at least four Commanders from each force.
- The police now have over 2350 officers trained and equipped in CBRN response, and this training roll-out is continuing.
- Her Majesty's Fire Service Inspectorate has signed a Memorandum of Understanding with the Department of Health, such that firefighters will support ambulance services by decontaminating people in a CBRN incident.
- Fire Brigades have been involved in work to prepare for this role, and an interim decontamination methodology has been disseminated to all brigades.
- £57 million has been provided for the Fire Service to provide a national mass decontamination capability. Procurement of equipment (response vehicles,

portable decontamination facilities, and specialist protective clothing) is underway, supported by development of training.

- The Department of Health is funding measures to counter bioterrorism, which include:

  - A U.K. reserve national stockpile of vaccines and antibiotics suitable for the treatment of infectious diseases as well as specialist equipment has been built up over the past year and now stands ready for deployment. Guidance on handling infectious diseases was disseminated throughout the NHS in October 2001.

  - Twelve regional smallpox response groups are being established around the United Kingdom. Vaccine will be offered to volunteer health care personnel, who upon being vaccinated will be able to react quickly and work safely with patients of actual or suspected smallpox. A similar group of specialist military personnel will also receive the smallpox vaccine. In addition, the government has identified reference laboratory centers capable of rapid diagnosis of the disease.

  - £16 million was allocated by the Department of Health in 2001–2002 to provide health care countermeasures against CBRN agents, and a further £80 million has been allocated for 2002–2003, including spending on extra vaccines and antibiotics.

- All the emergency services have specific protocols dealing with chemical or biological attack, which are regularly practiced and refined. On February 3, 2003, the Home Office published Strategic National Guidance (Home Office, 2003) on the decontamination of people exposed to CBRN materials, for use by emergency services and other responders.

The foreword to the Home Office Strategic National Guidance (Home Office, 2003) notes that releases of CBRN material can occur without warning as a result of a wide range of events, including industrial accidents, terrorism, and natural outbreaks of disease, and recognizes that the emergency services and other responding agencies have always worked closely together to deal with the scene of a disaster. They have carried out their roles in accordance with a range of guidance issued by government departments, specialist agencies, and the emergency services themselves. The strategic guidance is therefore not intended to replace the existing guidance but instead is meant to bring the main areas together. It is designed to enable the emergency services, the military, local authorities, health professionals, and government departments and agencies to work together more effectively at an incident. The strategic guidance provides an agreed set of principles, common terminology, and a shared understanding of each organization's roles and responsibilities.

The introduction to the guidance makes it clear that it is not confined to the deliberate release of CBRN material by terrorists, as accidental releases, outbreaks of serious communicable diseases, contamination from overseas incidents, and even domestic spillages also represent real threats. It goes on to make the point that accidental releases of CBRN material or naturally occurring disease outbreaks may be on a more manageable scale than terrorist incidents, because of the lack of intent, the limited nature of sites at risk, and the safety systems already in place. It is likely that a terrorist attack would involve a specific target such as a VIP, critical or iconic location, or a high-profile event. Consequently, victim management requires careful consideration by responding agencies. As with other types of terrorism, the multiagency response to a CBRN incident will be coordinated by the police.

Insofar as the types of incident are concerned, the guidance identifies two types, deliberate and unintentional (e.g., at industrial and commercial sites, laboratories, hospitals, materials in transit, nuclear sites (at home or abroad), and domestic spillages). The existence of well-rehearsed emergency plans and associated safety measures will usually assist responders dealing with accidental releases at industrial sites. Where there is a suspicion of terrorism, the very real possibility that the terrorists may still be at the scene or that secondary devices may be present should always be considered. The immediate and surrounding area must be checked for the presence of secondary devices (either CBRN or explosive) before any decontamination point is set up. While this will normally be a police responsibility, all responders must remain vigilant to this operational threat.

The guidance goes on to set out the Strategic Objectives for a Combined Response to a CBRN incident. It points out that irrespective of the particular responsibilities of organizations and agencies responding to the incident, the strategic intention is to coordinate effective multiagency activity in order to preserve and protect lives; mitigate and minimize the impact of an incident; inform the public and maintain public confidence; prevent, deter, and detect crime; and assist an early return to normality (or as near to it as can be reasonably achieved). Other important common objectives flowing from these principles are outlined in the introduction to this entry.

Regarding decontamination, the guidance points out that decontamination is not an automatic or inevitable response to CBRN incidents. Whether or not to initiate decontamination procedures will depend on the assessment of the nature of the incident by first responders. It goes on to note that once the decision to decontaminate has been made, the principle is that all casualties (injured or not) that are suspected of being contaminated will receive decontamination at the scene. If decontamination procedures are initiated, the first objective is to remove the contaminated person from the area of greatest contamination. Usually this will be to the open air and upwind of the incident. If the CBRN release is still in progress and airborne, a risk assessment should be carried out to determine if removing people to an enclosed area is more appropriate.

An Appendix to the guidance sets out information on the U.K. National Reserve Stock ("DoH pods") for use following the release of CBRN material. This notes that the U.K. National Reserve Stock has been established by the Department of Health, and that the stock's use is not limited to terrorist events but can be called upon for use in major accidental releases as well. The stock is for rapid deployment in major incidents, including mass-casualty situations. Part of the stock has been packed into pods (containers) designed to treat 100 people. Part has been distributed to strategic sites throughout the United Kingdom, and part is held centrally. The stocks include Nerve Agent Antidote Pods, Biological "Cipro" Pods, Biological "Doxy" Pods, and Equipment Pods (for dealing with resuscitation, ventilation etc.), as well as Modesty Pods for use by Ambulance and Acute Trusts following decontamination (each modesty pod contains sufficient paper towels, paper suits, and space blankets for 100 people).

Recovery Management

As noted earlier, lead responsibility in the recovery management phase moves from the Home Office (which leads during the counter terrorist phase). At some point, to be determined on a case-by-case basis and once the crisis management phase is concluded, the lead department responsibility would be transferred to DEFRA, although depending on the nature and location of the incident—for example, when releases of CBRN materials occur primarily within buildings and infrastructure—other departments might take the lead. Strategic national guidance has been issued recently regarding the decontamination of buildings and infrastructure (Office of Deputy Prime Minister, 2004) and of the open environment (Department for the Environment, Food and Rural Affairs, 2004) exposed to CBRN substances and material.

CONCLUSIONS

The overall approach adopted in the United Kingdom to countering bioterrorism can be described as being a web of reassurance—to reassure the State and its public that such bioterrorism is totally prohibited, and that biological materials that might be misused are kept secure and, if used, will have minimal effect. The elements of such a web are:

1. Strong international and national prohibition regimes reinforcing the norm that biological weapons are totally prohibited.
2. Broad international and national controls on the handling, storage, use, and transfer of biological agents and toxins.
3. Preparedness, including both active and passive protective measures and response plans that have been exercised.
4. Determined national and international response to any use or threat of use of biological weapons, ranging from diplomatic sanctions through to armed intervention.

Together, these elements are mutually reinforcing and lead those contemplating acquisition or use of biological weapons to judge that such a capability would not be valuable, would be detected, and would incur an unacceptable penalty. In the United Kingdom, the situation in regard to the elements of the web of reassurance can be summarized as follows:

International and National Prohibition Regimes: The United Kingdom is a codepository to the BWC, which is implemented nationally through the Biological Weapons Act of 1974 (Her Majesty's Stationery Office, 1974a). Insofar as toxins are concerned, these are also prohibited nationally through the Chemical Weapons Act of 1996 (Her Majesty's Stationery Office, 1996).

Broad International and National Controls on the Handling, Storage, Use, and Transfer of Biological Materials and Toxins: The United Kingdom has controls on the handling, storage, use and transfer of biological materials nationally through the Health & Safety at Work Act 1974 (Her Majesty's Stationery Office, 1974b) and the COSHH 2002 (Her Majesty's Stationery Office, 2002a) regulations. Security of and access to dangerous pathogens and toxins are addressed through the Anti-Terrorism, Crime and Security Act 2001 (Her Majesty's Stationery Office, 2001). Transfers of biological materials internationally are controlled by the Export Control Act 2002 (Her Majesty's Stationery Office, 2002b) and related regulations.

Preparedness: The U.K. approach is based on the national guidance "Dealing with Disaster" issued in June 2003 in the interim third edition (Cabinet Office, 2003a). Guidance addressing deliberate releases of chemical and biological materials have been issued by the Cabinet Office in October 2001 (Cabinet Office, 2001) and by the Department of Health in 2000, republished in 2002 (Department of Health, 2002a and 2002b). Strategic National Guidance on decontamination has been issued in February 2003 (Home Office, 2003). It is also evident from this guidance that the U.K. approach is a multiagency one, which draws upon the expertise available from the Ministry of Defence at CBD Porton Down as well as from the Health Protection Agency of the Department of Health established in April 2003.

Determined National and International Response: The U.K. approach is to ensure that breaches of national laws and regulations are prosecuted in the national courts. The relevant government authorities and agencies provide oversight of the implementation of the national laws and regulations, and any noncompliance is addressed.

REFERENCES

Cabinet Office, Civil Contingencies Secretariat, *Response to Deliberate Release of Chemicals and Biological Agents, Guidance for Local Authorities*, October 2001. Available at http://www.lga.gov.uk/lga/publicprotection/biological.pdf, accessed on 8/10/03. A revised version, Cabinet Office, *The Release of Chemical, Biological, Radiological or Nuclear (CBRN) Substances or Material, Guidance for the Local Authorities*, August 2003 has recently been sent to local authorities and to the emergency services, available at http://www.ukresilience.info/cbrn/lacbrnintro.htm, accessed on 7/9/04.

Cabinet Office, *Dealing with Disaster*, Interim 3rd ed., June 19, 2003a. Available at http://www.ukresilience.info/contingencies/dwd/dwdrevised.pdf, accessed on 8/8/03.

Cabinet Office, *Dealing with Disaster*, Interim 3rd ed., June 19, 2003b, pp. 13–16. Available at http://www.ukresilience.info/contingencies/dwd/dwdrevised.pdf, accessed on 8/8/03.

Cabinet Office, *Draft Civil Contingencies Bill*, Consultation Document, June 2003c. Available at http://www.ukresilience.info/ccbill/draftbill/consultation_doc.pdf, accessed on 8/8/03.

Cabinet Office, *Handling a Crisis: Who Does What?* Available at http://www. ukresilience.info/thandling2.htm, accessed on 7/9/04. Civil Contingencies Secretariat, The Lead Government Department and its role: Guidance and Best Practice, March 2004. Available at http://www. ukresilience.info/thandling2.htm, accessed on 7/9/04.

Department of Health, *Deliberate Release of Biological and Chemical Agents, Guidance to Help Plan the Health Service Response*, March 2000, republished August 2002a. Available at http://www.doh.gov.uk, accessed on 8/1403.

Department of Health, UK, Department of Health, Emergency Planning Coordination Unit, *Public Health Response to Deliberate Release of Biological and Chemical Agents*, August 18, 2002b. Available at http://www.doh.gov.uk/epcu/cbr/response/introph.htm, accessed on 8/14/03.

Department for the Environment, Food and Rural Affairs, *The Decontamination of the Open Environment Exposed to Chemical, Biological, Radiological or Nuclear (CBRN) Substances or Material—Strategic National Guidance*, March 2004. Available at http://www.ukresilience.info/cbrn/index.htm, accessed on 7/9/04.

Health and Safety Executive, Advisory Committee on Dangerous Pathogens, *The Approved List of Biological Agents*, March 2, 2004. Available at http://www.hse.gov.uk/hthdir/noframes/biolhaz.htm, accessed on 7/9/04.

Health Protection Agency, As the Public Health Laboratories Service has now been subsumed into the Health Protection Agency established on April 1, the Health Protection Agency would now provide such assistance to the Department of Health. Information on the Health Protection Agency is available at http://www.hpa.org.uk, 2003a.

Health Protection Agency, *Health Protection Agency Corporate Plan 2003–2008*, 2003b. Available at http://www.hpa.gov.uk/publications/corporateplan2003_8.pdf.

Health Protection Agency, *Initial Investigation and Management of Outbreaks and Incidents of Unusual Illnesses with Particular Reference to Events that may be Due to Chemical, Biological or Radiological Causes, including Deliberate Releases*, Version 3, March 2004. Available at http://www.ukresilience.info/cbrn/index.htm, accessed on 7/9/04.

Her Majesty's Stationery Office, *Anti-Terrorism, Crime and Security Act 2001*, 2001. Available at http://www.opbw.org and at http://www.legislation.hmso.gov.uk/acts/acts2001/10024, accessed on 2/16/03.

Her Majesty's Stationery Office, *Biological Weapons Act 1974*, 1974a. Available at http://www.opbw.org, accessed on 8/14/03.

Her Majesty's Stationery Office, *Control of Substances Hazardous to Health Regulations, 2002*, 2002a, Statutory Instrument S.I. 2002/2677. Available at http://www.opbw.org, accessed on 8/14/03.

Her Majesty's Stationery Office, *Export Control Act 2002*, 2002b. Available at http://www.hmso.gov.uk, accessed on 2/16/03.

Her Majesty's Stationery Office, *Health and Safety at Work etc. Act 1974*, Ch. 37, 1974b. Available at http://www.healthandsafety.co.uk/haswa.htm, accessed on 7/9/04.

Her Majesty's Stationery Office, *The Chemical Weapons Act 1996*, Ch. 6, 1996. Available at http://www. hmso.gov.uk/acts/acts1996/1996006.html, accessed on 7/9/04.

Home Office, *The Decontamination of People Exposed to Chemical, Biological, Radiological or Nuclear (CBRN) Substances or Material, Strategic National Guidance*, 1st ed., February 2003. Available at http://www.homeoffice.gov.uk/docs/decontamination.pdf, accessed on 8/8/03.

Office of Deputy Prime Minister, *Strategic National Guidance: the decontamination of buildings and infrastructure exposed to Chemical, Biological, Radiological or Nuclear (CBRN) substances or material*, First version, May 2004. Available at http://www.ukresilience.info/cbrn/index.htm, accessed on 7/9/04.

United Nations, *The Fourth Review Conference of the States Parties to the Convention on the Prohibition of the Development, Production and Stockpiling of Bacteriological (Biological) and Toxin Weapons and on their Destruction*, Final Document, Geneva, November 25—December 6, 1996, BWC/CONF.IV/9, Part II. p. 15. Available at http://www.opbw.org, accessed on 8/8/03.

WEB RESOURCES

United Kingdom Cabinet Office Civil Contingencies Secretariat, http://www.ukresilience.info, accessed on 7/9/04.

United Kingdom Department for Environment, Food and Rural Affairs, http://www.defra.gov.uk, accessed on 7/9/04.

United Kingdom Department of Health, http://www.dh.gov.uk, accessed on 7/9/04.

United Kingdom Health Protection Agency, http://www.hpa.org.uk, accessed on 7/9/04.

United Kingdom Home Office, http://www.homeoffice.gov.uk, accessed on 7/9/04.

United Kingdom Office of Deputy Prime Minister, http://www.opdm.gov.uk, accessed on 7/9/04.

UNITED STATES DEPARTMENT OF AGRICULTURE

Jennifer Mitchell

INTRODUCTION

The United States Department of Agriculture (USDA) was founded in 1862 by President Abraham Lincoln (Welcome to USDA, 2003). The "people's Department," as Lincoln called it, was meant to improve America's living conditions by assisting agricultural production; safeguarding drinking water, meat, poultry, and egg products; and researching human nutrition, crop technologies, and related technologies (Mission of USDA Agencies and Offices, 2003; Welcome to USDA, 2003). Today, services provided by the USDA include the Farm and Foreign Agricultural Service; Food Safety; Natural Resources and Environment; Rural Development; and Food, Nutrition, and Consumer Services (Fig. 1) (Agencies, Services, and Programs, 2003).

SAFEGUARDING AGAINST BIOLOGICAL WEAPONS

The USDA has a number of responsibilities with respect to bioterrorism defense. Its Animal and Plant Health Inspection Service (APHIS) is accountable for protecting agricultural crops and plants against diseases, whether naturally or deliberately introduced (McIntire Peters, 2003). Its Food Safety and Inspection Service (FSIS) is responsible for inspecting meat and poultry products for contamination and improving food safety (Enhancing Public Health: Strategies for the Future, 2003). And USDA scientists at Plum Island Animal Disease Center, research and create countermeasures against foreign animal disease (FAD) agents capable of damaging animal industries and exports (USDA Mission at Plum Island, 2003).

SAFEGUARDING THE UNITED STATES: USDA'S NEW DUTIES

In response to the events of September 11, 2001, the USDA has taken several steps to protect the United States from terrorism involving the nation's food supply and agricultural products including:

- The Homeland Security Council has been formed to coordinate the USDA's efforts.
- The Bioterrorism Preparedness and Response Act of 2002 made the USDA responsible for improving

controls on dangerous biological agents and toxins. The Act also increased the capacity of APHIS to make inspections at points of origin, upgrade surveillance, and respond to outbreaks of plant and animal diseases; allowed FSIS to increase its number of food and meat inspections; and supported upgrades of the USDA's laboratories and facilities at Plum Island (The Public Health Security and Bioterrorism Preparedness and Response Act of 2002, 2002).

- FSIS developed "FSIS Safety and Security Guidelines for the Transportation and Distribution of Meat, Poultry, and Egg Products," which are designed to improve the safety and security plans for shipper meat, egg, and poultry safety and security plans by providing security guidance regulations and safety plans as well as biosecurity information guidelines for processors and constituents (USDA Homeland Security Efforts, 2003; FSIS Announces New Food Safety and Security Guidelines to Assist Transporters and Distributors of Meat, Poultry and Egg Products, 2003). The new guidelines call for shippers, transporters, distributors, and receivers to create and employ controls that guard against the contamination of products through all phases of distribution. In addition, FSIS calls for the implementation of standardized response plans in the event of accidental or deliberate contamination (FSIS Announces New Food Safety and Security Guidelines to Assist Transporters and Distributors of Meat, Poultry and Egg Products, 2003). FSIS has also undertaken simulation exercises to prepare for food safety emergencies, for example, "Crimson Winter" was held in 2003 for the purpose of training managers and staff on the management of foodborne-disease outbreaks, whether intentional or unintentional (Enhancing Public Health: Strategies for the Future, 2003).

- The USDA's Office of Food Security and Emergency Preparedness has been founded to lead prevention and response efforts against attacks on the food supply (USDA Homeland Security Efforts, 2003).

Encyclopedia of Bioterrorism Defense, Edited by Richard F. Pilch and Raymond A. Zilinskas
ISBN 0-471-46717-0 Copyright © 2005 Wiley-Liss

Headquarters organization

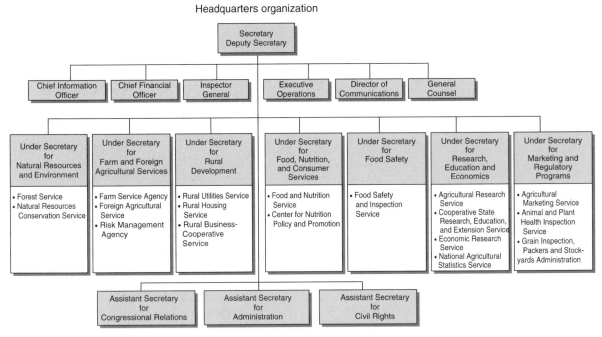

Figure 1. USDA organizational chart, USDA website: http://www.usda.gov/agencies/agchart.htm.

- Twenty "import surveillance liaison" inspectors have been hired to inspect meat and poultry in the United States.

- Identification and diagnostic abilities at federal and state laboratories are being upgraded (McIntire Peters, 2003).

- Border patrol efforts to prevent the entry of foreign agricultural pests and diseases into the United States have been increased.

- The USDA has distributed $43 million to states, universities, and tribal lands as a part of its security, prevention, detection, and response efforts.

- And $18 million has been allocated toward the development of tests for agents that threaten U.S. agriculture, including the causative agents of foot-and-mouth disease, rinderpest, and wheat rust (USDA Homeland Security Efforts, 2003).

COLLABORATION WITH THE DEPARTMENT OF HOMELAND SECURITY

Over the past year, USDA has transferred a number of its subordinate agencies and responsibilities related to the prevention of and response to agricultural terrorism to the newly established Department of Homeland Security (DHS). Thus, the DHS has taken over from APHIS's Agricultural Quarantine Inspection Programs the task of inspecting food and agricultural products entering the United States for foreign plant and animal pests and diseases (McIntire Peters, 2003). The USDA-controlled section of APHIS will continue to determine policies outlining the necessary levels of import restrictions on agricultural products, while the

newfound DHS side of APHIS will carry out the USDA-dictated policy (The Animal and Plant Health Inspection Service and Department of Homeland Security: Working Together to Protect Agriculture, 2003). USDA has also reallocated some of the functions of the Plum Island Animal Disease Center to DHS. Plum Island will serve as the Science and Technology directorate of DHS in order to assist with the biological countermeasures program, while USDA research staff at Plum Island will retain accountability for maintaining research, development, and diagnostics programs, in addition to examining agricultural terrorism concerns (United States Department of Homeland Security Fact Sheet: Plum Island Animal Disease Center Transition, 2003). The effectiveness of this collaborative effort remains to be seen.

REFERENCES

McIntire Peters, K., Growing Threat, *GovExec*, June 15, 2003, www.govexec.com.

United States Department of Agriculture, *Agencies, Services, and Programs*, 2003, www.usda.gov/services.html.

United States Department of Agriculture, *The Animal and Plant Health Inspection Service and Department of Homeland Security: Working Together to Protect Agriculture*, May 2003, www.aphis.usda.gov.

United States Department of Agriculture, *Enhancing Public Health: Strategies for the Future*, 2003, www.fsis.usda.gov.

United States Department of Agriculture, *FSIS Announces New Food Safety and Security Guidelines to Assist Transporters and Distributors of Meat, Poultry, and Egg Products*, August 4, 2003, www.fsis.usda.gov.

United States Department of Agriculture, *Mission of USDA Agencies and Offices*, 2003, www.usda.gov/mission/miss-toc.htm.

United States Department of Agriculture, *USDA Homeland Security Efforts*, June 2003, www.usda.gov.

United States Department of Agriculture, *USDA Mission at Plum Island*, 2003, www.ars.usda.gov/plum/USDAmission. htm.

United States Department of Agriculture, *Welcome to USDA*, 2003, www.usda.gov/welcome/.

United States Department of Homeland Security, *Fact Sheet: Plum Island Animal Disease Center Transition*, 2003, www.dhs.gov.

U.S. Food and Drug Administration, *The Public Health Security and Bioterrorism Preparedness and Response Act of 2002*, 2002, http://www.fda.gov/oc/bioterrorism/bioact.html.

WEB RESOURCES

Bioterrorism Preparedness and Response Act of 2002, http://www. fda.gov/oc/bioterrorism/bioact.html.

United States Department of Agriculture website, http://www. usda.gov/.

United States Department of Homeland Security website, http://www.dhs.gov.

See also AGRICULTURAL BIOTERRORISM and DEPARTMENT OF HOMELAND SECURITY.

UNITED STATES LEGISLATION AND PRESIDENTIAL DIRECTIVES

Leslie Gielow Jacobs

Elizabeth Rindskopf Parker

INTRODUCTION

Lawmaking is event-driven. Lawmakers constructed the complicated mosaic that is the current bioterrorism defense policy over the past 10 to 20 years to address aspects of the biological weapons threat as events revealed those aspects to them. What began as a single idea—eliminating the biological weapons threat—has grown into a broad and complex body of laws that make up the U.S. bioterrorism defense legal regime. At its heart are an international agreement that bans the entire class of weapons and, domestically, statutes that define and punish biological weapons crimes. Beyond these prohibitions exist a wide range of policy measures aimed at preventing and responding to a bioterrorism attack.

The process of creating the many components of bioterrorism defense policy is ongoing. Bioterrorism defense measures are mushrooming in the federal statute books and in the code of federal agency regulations. Presidential directives, programs, and proposals also address multiple aspects of bioterrorism defense. The president's executive department, led by the Attorney General, is actively investigating and prosecuting a wide range of terrorism and terrorism related offenses. Numerous federal courts at different levels across the nation are engaged in reviewing these federal government actions. States, localities, and private parties are participating with the federal government to achieve its bioterrorism defense goals, the first two adding to their books of statutes and regulations as well.

It is a large undertaking, creating a bioterrorism defense policy almost from scratch. What follows must therefore be a snapshot of the major legislation, presidential directives, and agency actions that form the core of the current U.S. bioterrorism defense.

BACKGROUND

The U.S. bioterrorism defense policy has evolved according to policymakers' view of the national security interest. There was a time, from the beginning of World War II until the late 1960s, when the United States actively developed offensive biological weapons (Committee on the Research Standards, 2003). In 1969, the United States switched course. President Nixon renounced the use of chemical and biological weapons, called for their destruction, and pledged that U.S. war-related biological research would be confined to defensive measures only (Nixon, 1969). This policy change reflected a determination that biological weapons developed by the United States could not be contained, but instead would spur proliferation and emulation by other nations, such that an offensive program would threaten rather than enhance national security (Meselson, 1989). The U.S. policy focus with respect to biological weapons became to eliminate them from the world.

The United States's first steps in this new policy direction were to ratify the 1925 Geneva Protocol, which prohibits the use of bacteriological and chemical weapons in war (26 U.S.T. 571, 1972), and to promote the adoption by nations of the Convention on the Prohibition of the Development, Production and Stockpiling of Bacteriological (Biological) and Toxin Weapons and on their Destruction, commonly referred to as the Biological Weapons Convention of 1972 (BWC) (26 U.S.T. 583, 1975). The United States ratified the Geneva Protocol and the BWC in 1975.

Although the biological weapons elimination goal has remained consistent since that time, U.S. policy as to the means to achieve this goal has shifted in response to changing circumstances. One change is the dissolution of the Soviet Union and the current status of the United States as the sole superpower. This means that security threats increasingly come asymmetrically in the form of groups or individuals, rather than nations, and that the United States is increasingly the focus of their attention (National Strategy to Combat Weapons of Mass Destruction, 2002).

The United States's status as the sole superpower has also made possible, and arguably prompted, the shift from international to unilateral action in its foreign and counterterrorism policy. A representative example in the case of conventional weapons is the United States's withdrawal from The Treaty on the Limitation of Anti-Ballistic Missile Systems (23 UST 3435, 1972). Other recent actions suggest this policy with respect to bioterrorism defense. As one example, the United States explained its 2001 refusal to endorse a protocol to strengthen the widely violated and currently ineffective BWC as in part reflecting its own skepticism that an international organization could effectively verify compliance with the treaty (Lacey, 2001). The United

Encyclopedia of Bioterrorism Defense,   Edited by Richard F. Pilch and Raymond A. Zilinskas

ISBN 0-471-46717-0   Copyright © 2005 Wiley-Liss

States articulated a preference for relying on national intelligence to detect BWC violations (Lacey, 2001). Another example is Operation Iraqi Freedom, the spring 2003 war against Iraq. The United States rejected the efforts of UN weapons inspectors, bypassed the UN Security Council, and, along with the United Kingdom, conducted a war against Iraq that it justified on the basis of intelligence that allegedly demonstrated that Saddam Husayn possessed an ongoing weapons of mass destruction (WMD) and, more specifically, biological and chemical weapons programs that threatened the United States and the entire world (Powell, 2003; Bush State of Union, 2003). This intelligence has since been determined to have been faulty (Senate, 2004).

Another change in circumstances that affects U.S. bioterrorism defense policy is the accelerating pace of developments in molecular biology and recombinant technologies, as well as the ever-expanding means and increasing speed of global transportation and communication (National Strategy for Combating Terrorism, 2003). These changes create the "dual-use" dilemma, which characterizes efforts to identify and eliminate biological weapons. The underlying issue confirms the fact that the scientific and technological advances that may lead to great good may also create great harm in the hands of those who want to create a biological weapon. That the materials and scientific knowledge necessary to create biological weapons are widely available for legitimate uses makes it difficult to determine when weapons related activities are occurring. The policy challenge is to eliminate the weapons without stifling beneficial scientific and technological advances in the process.

The dual-use dilemma is evident in the recent U.S. bioterrorism defense policy. Along with its skepticism as to the abilities of international bodies, the United States relied upon the legitimate uses of biological agents, equipment, and expertise by its booming biotechnology industry, and that industry's need for trade secrecy, as another primary reason to reject the proposed verification protocol to strengthen the BWC (Lacey, 2001). Disagreement among experts in the United States about whether mobile units found after the war in Iraq were designed to produce biological weapons and not to fill weather balloons with helium demonstrates the difficulty in detecting violations (Jehl, 2003).

A final change that affects bioterrorism defense policy is a combination of reality and perception. Biological weapons have existed for thousands of years (Mayor, 2003). Thus, the threat or even use of such weapons is really nothing new. Nevertheless, the September 11 terrorism attacks and the anthrax letter attacks that followed within the United States have made the bioterrorism threat "more real," perhaps in truth and certainly as a matter of public and policymakers' perception. Even before 2001, other events had highlighted the bioterrorism threat for U.S. policymakers and the public that they represent. These include the dissolution of the Soviet Union and the revelation of its massive biological weapons program, Iraq's use of chemical weapons against Iran and dissident Kurds during the 1980s and subsequent revelations of its own offensive biological weapons program, the Aum Shinrikyo cult's sarin attack in the Tokyo subway and failed attacks with biological weapons previously, and, even earlier, the Rajneesh cult's contamination of salad bars in Oregon that sickened 751 of its inhabitants. Regardless of the accuracy of this perception, the perceive reality of the bioterrorism threat means that government policy must address not only prevention but also response.

THE STRUCTURE OF U.S. BIOTERRORISM DEFENSE

The powers granted in the United States Constitution determine the broad structure of bioterrorism defense. Most fundamentally, the federal government is one of enumerated powers. What authority it is not granted in the Constitution is left to the states, which, unlike the federal government, possess a general police power to take action to promote the health, safety, and welfare of the population. Federal government powers that may authorize bioterrorism defense actions include Congress's power to regulate interstate commerce (U.S. Const. Art. I, §8, cl. 3) and immigration (U.S. Const. Art. I, §8, cl. 4) and the powers of Congress and the president to protect national security (U.S. Const. Art. I, §8, cl. 11–16; U.S. Const. Art. II, §1, cl. 1, §2, cl. 1, §3). These powers authorize a wide range of federal government actions to address bioterrorism threats that may impact more states than one, or which threaten national security. Nevertheless, that the federal government must act pursuant to enumerated powers means that there may be some limit to its ability to address all aspects of bioterrorism defense through the direct commands of legislation.

The federal government's power to spend money to achieve its policy objectives is broader than its power to legislate, and includes the power to spend money to protect and promote public health and safety (U.S. Const. Art. I, §8, cl.1). Thus, many federal bioterrorism defense actions, especially in areas of traditional state responsibility such as monitoring and responding to disease outbreaks, are structured to provide financial or other types of assistance rather than to impose legal mandates.

Within the federal government's constitutionally granted powers, congressional legislation, executive orders and directives, and agency rules and regulations determine the more specific structure of bioterrorism defense. The president possesses the power and responsibility to execute laws enacted by Congress, and so may issue orders and directives under his own granted authority or when authorized to do so by congressional legislation. Within the president's executive branch also exist a wide range of administrative agencies, which Congress has created by legislation and to which the Constitution allows Congress to grant broad policymaking authority (Whitman, 2001). Because administrative agencies have the expertise and the infrastructure to deal with complex issues in a flexible manner, bioterrorism defense policy will often take the structure of a broad direction from either Congress or the president that is followed by detailed agency action in the form of rules and regulations.

A structural change important to bioterrorism defense policy is the recent creation of the Department of Homeland Security (DHS) (Pub. L. No. 107–296, 116 Stat. 2135 (2002)). The statute that creates DHS directs it to absorb a number of administrative agencies, and restructures both their alignment and a number of their duties. Ongoing DHS action may alter some of the agency responsibilities described in this entry. DHS also has substantial new powers and responsibilities designed to enhance homeland defense.

THE SUBSTANCE OF U.S. BIOTERRORISM DEFENSE

The substance of U.S. bioterrorism defense policy has been rapidly adapting to reflect new national and homeland security priorities. The components of the current U.S. bioterrorism defense policy can be grouped into three general categories: defining and punishing biological weapons offenses and expanding law enforcement tools and methods; preventing access to biological weapons materials, equipment, and expertise, along with key terrorism targets; and preparing for response to a bioterrorism event.

Defining and Punishing Biological Weapons Offenses and Expanding Law Enforcement Tools and Methods

Traditional criminal law objectives are at the core of U.S. bioterrorism defense policy. These objectives are to deter bioterrorism, detect it before it happens, or, if those fail, to punish it after it occurs. The fundamental criminal components of current U.S. bioterrorism defense policy are a series of defined biological weapons and related crimes and their punishment; enhanced investigatory tools to aid detection; and streamlined enforcement provisions that make prosecution more certain and efficient.

Against Nations. Although the national focus of biological weapons elimination efforts is no longer exclusive, it remains a tool in the U.S. arsenal of bioterrorism defense. A number of pieces of legislation direct the president or his subordinates to monitor the activities of foreign nations and to impose sanctions on those that engage in terrorism or assist other nations that do so (28 U.S.C. §1605 (2000), 50 App. U.S.C. §2410 (2000)). Several of these specifically address biological weapons.

Chemical and Biological Weapons Control and Warfare Elimination Act of 1991. The Chemical and Biological Weapons Control and Warfare Elimination Act of 1991 directs the president to use existing export authorities to "control the export of those goods and technology that the President determines would assist the government of any foreign country in acquiring the capability to develop, produce, stockpile, deliver, or use chemical or biological weapons" (22 U.S.C. §5603 (2000)); to determine, when prompted by persuasive information, whether a foreign country has made substantial preparation to use or has used chemical or biological weapons and to report such determination to Congress (22 U.S.C. §5604 (2000)); and to impose sanctions if the determination

is that the foreign country has used such weapons (22 U.S.C. §5605 (2000)). Sanctions that the president must impose are (1) withdrawal of foreign assistance (save for an urgent humanitarian situation), (2) cessation of arms sales, (3) cessation of financing of the countries' armies, (4) denial of U.S. government credit or other financial assistance, and (5) a block on exports of national security-sensitive goods and technology (22 U.S.C. §5605(a) (2000)). The Act directs the president to impose further sanctions if, after three months, he has not determined that the country has ceased its biological weapons program (22 U.S.C. §5605(b) (2000)).

Export Statutes. The Arms Export Control Act (AECA) (22 U.S.C. §§2751–2799 (2000)) is the principal law governing procedures on sales of military equipment and related services. The AECA directs the president to impose sanctions against "foreign persons" (including corporations or other foreign entities) that knowingly contribute to the chemical or biological weapons acquisition efforts of certain countries (22 U.S.C. §2798 (2000)).

The Export Administration Act (EAA) applies to items that are not "munitions" within the meaning of AECA, but which may have "dual-use" (i.e., both peaceful and nefarious) applications (50 App. U.S.C. §2401 et seq. (2000)). The EAA has currently lapsed, and Congress has not reenacted its provisions. Instead, the International Economic Emergency Powers Act (IEEPA) now authorizes export administration, imposing the EAA system but with slightly different penalties and enforcement (50 U.S.C. §1702(a)(1)(B) (2000)). Like the AECA, the EAA directs the president to impose sanctions against foreign countries or entities that assist certain nations' chemical or biological weapons acquisition efforts. (50 App. U.S.C. §2410c (2000)).

Against Individuals. Individually perpetrated domestic terrorist events such as the first World Trade Center bombing, the Oklahoma federal building bombing, and the September 11, 2001, attacks prompted a series of pieces of legislation to address the threat of terrorism carried out by nonstate actors. These statutes define and establish the punishment for individual terrorism crimes, and, more specifically, biological weapons crimes, and expand the investigatory tools and methods that may be used to discover and prosecute the violations.

Biological Weapons Anti-terrorism Act of 1989. The United States first implemented the BWC domestically through the Biological Weapons Anti-Terrorism Act of 1989 (BWATA) (18 U.S.C. §§175–178 (1994)). The BWATA makes it a federal crime, punishable by up to life imprisonment, to knowingly create, keep, or transfer any biological agent, toxin, vector, or delivery system "for use as a weapon," or to knowingly assist a foreign state or any organization to do so (18 U.S.C. §175(a) (1994)). Restricted agents are defined as those generally capable of producing harm or causing disease, but were not in this statute defined to include the new or more virulent agents that biotechnology can produce (18 U.S.C. §178(1) (1994)). "For use as a weapon" does not include the listed activities if undertaken "for prophylactic, protective, or other peaceful purposes" (18 U.S.C. §175(b) (1994)). The BWATA grants extraterritorial jurisdiction over an offense committed by or against a U.S. national (18 U.S.C. §175(a) (1994)).

The BWATA also provides additional investigative and injunctive powers to the government to enforce its provisions. The BWATA allows the government to apply for a warrant to seize (which may lead to forfeiture or destruction), or to obtain an injunction to prohibit the development of, biological agents or delivery systems based upon a showing either of the "for use as a weapon" activities prohibited in Section 175 or upon the lesser showing that a biological agent, toxin, or delivery system is "of a type or in a quantity that under the circumstances has no apparent justification for prophylactic, protective, or other peaceful purposes" (18 U.S.C. §§176(a) & 177(a) (1994)). That the activities are justified by one of the listed purposes is an affirmative defense (18 U.S.C. §§176(c) & 177(b) (1994)).

Antiterrorism Act of 1990 and Violent Crime Control and Law Enforcement Act of 1994.

The Antiterrorism Act of 1990 establishes the offense of "international terrorism," which means violent or dangerous activities with impact outside the United States that are U.S. crimes or would be if committed within the United States, and that appear to be intended to intimidate or coerce a civilian population; to influence the policy of a government by intimidation or coercion; or to affect the conduct of a government by assassination or kidnapping (18 U.S.C. §2331(1) (2000)). Punishments apply that, where a killing occurs that meets the murder definition, may include death (18 U.S.C. §2332 (2000)), and federal district courts have exclusive jurisdiction (18 U.S.C. §2338 (2000)).

The Violent Crime Control and Law Enforcement Act of 1994 (VCCLEA) amends the Antiterrorism Act of 1990 to add the crime of using certain weapons of mass destruction, including one involving a "disease organism" (18 U.S.C. §2332a (2000)). "Disease organism" was later changed to "biological agent, toxin, or vector" as defined by the BWATA (18 U.S.C.A. §2332a(c)(2)(c) (West 2000 & Supp. 2003)). The VCCLEA also criminalizes providing material support to terrorists (18 U.S.C. §2339A (2000)).

Anti-terrorism and Effective Death Penalty Act of 1996.

The Anti-Terrorism and Effective Death Penalty Act of 1996 (AEDPA) amends the definition of the biological weapons crime to include anyone who "attempts, threatens, or conspires" to do the activities prohibited in the BWATA (18 U.S.C. §175 (1994 & Supp. II 1996)). It also amends the definitions of "biological agent," "toxin," and "vector" to include products of biotechnology (18 U.S.C. §178 (1994 & Supp. II 1996)).

In numerous provisions that relate to terrorism more generally, the AEDPA narrows the immunity of certain foreign governments with respect to engaging in or providing support for terrorism (28 U.S.C. §1605 (1994 & Supp. II 1996)), provides compensation and other assistance to terrorism victims (42 U.S.C. §1603b (1994 & Supp. II 1996)), prohibits international terrorist fundraising (18 U.S.C. §2339B (Supp. II 1996)), allows the

Secretary of the Treasury to freeze the assets of foreign organizations engaging in terrorist activities that threaten the security of the United States (8 U.S.C. §1189 (1994 & Supp. II 1996)), facilitates the removal or exclusion of alien terrorists from the United States (8 U.S.C. §§1533-37 (1994 & Supp. II 1996)), modifies procedures that apply to aliens in numerous respects, broadens the scope of nuclear (18 U.S.C. §831 (1994 & Supp. II 1996)) and chemical weapons (18 U.S.C. §2332c (Supp. II 1996)) crimes, expands the definitions of and increases the penalties for certain federal crimes in order to effectively outlaw international terrorism (18 U.S.C. §§2332-2332b (1994 & Supp. II 1996)), modifies the overseas applicability of various federal criminal laws, and provides appropriations for law enforcement.

USA Patriot Act.

The Uniting and Strengthening America by Providing Appropriate Tools Required to Intercept and Obstruct Terrorism Act of 2001 (USA Patriot Act) amends the biological weapons crime definition of "for use as a weapon" to exclude "bona fide research" (18 U.S.C. §175(c) (Supp. I 2001)). The USA Patriot Act defines as an additional offense, punishable by fine or imprisonment of up to 10 years, possession of a biological agent, toxin, or delivery system of a type or in a quantity that is not justified by one of the above-listed beneficent purposes, that is, for "prophylactic, protective, or other peaceful purposes" (18 U.S.C. §175(b) (Supp. I 2001)). The USA Patriot Act also creates several new crimes related to use of biological weapons against mass transportation systems (18 U.S.C. §1993 (Supp. I 2001)).

With respect to terrorism generally, the USA Patriot Act amends 18 U.S.C. §2331, which previously defined the crime of "international terrorism," to include "domestic terrorism," defined to include the same activities when occurring primarily within the United States (18 U.S.C. §2331 (2000 & Supp. I 2001)). The USA Patriot Act also adds new crimes involving harboring or concealing terrorists (18 U.S.C. §2339 (Supp. I 2001)), amends the crime of providing material support to terrorists to include providing "expert advice or assistance" (18 U.S.C. §§2339A-2339C (Supp I 2001)), and criminalizes additional terrorist activities with respect to airplanes (18 U.S.C. §2332b (2000 & Supp. I 2001)).

The USA Patriot Act broadens the government's investigative powers with respect to terrorism in numerous ways. These provisions include expanding the grounds for searches that do not require prior notice (18 U.S.C. §3103a (Supp. I 2001)), applying the low standard for obtaining a telephone pen register to Internet dialing information (18 U.S.C. §3123 (2000 & Supp. I 2001)), providing for nationwide applicability of certain surveillance orders (18 U.S.C. §§2703 & 2711 (2000 & Supp. I 2001)), extending access in an intelligence investigation to a broad range of business records (50 U.S.C. §1861 (2000 & Supp. I 2001)), extending foreign intelligence surveillance powers to a range of criminal investigations, and authorizing the sharing of information between law enforcement and intelligence agencies (18 U.S.C. §2517 (2000 & Supp. I 2001)).

The USA Patriot Act provisions were controversial when enacted and remain so. Some of the provisions expire on December 31, 2005 (Pub. L. No. 107-56, §224, 115 Stat. 272, 295 (2001)).

The government is currently seeking further expansion of the USA Patriot Act investigative authorities, arguing that new powers are necessary to effectively combat terrorism (Patriot Act, Part II, 2003).

Agricultural Bioterrorism Act of 2002. The Agricultural Bioterrorism Act of 2002 amends the law to provide enhanced animal enterprise terrorism provisions (18 U.S.C.A §43 (West 2000 & Supp. 2003)). Persons who travel in interstate or foreign commerce and intentionally cause the loss of property used by an animal enterprise face harsher punishments if economic damage, bodily damage to an individual, or death to an animal results from their action. The punishments range from a fine to life imprisonment (18 U.S.C.A §43 (West 2000 & Supp. 2003)).

Preventing Access to Biological Weapons Materials and Key Terrorism Targets

Beyond deterring and punishing biological weapons crimes, an important component of U.S. bioterrorism defense policy is prevention, in the form of restricting access to the expertise, materials and equipment that may be used to produce biological weapons, and key targets for bioterrorism attacks.

Controlling Access by Foreign Governments or Entities. Exposure of the offensive biological weapons programs in the former Soviet Union and other nations such as Iraq prompted enhanced export controls directed toward these nations and, with respect to the former Soviet Union, revealed the need to identify and destroy dangerous agents and to relocate former weapons scientists in a safe environment (22 U.S.C. §5902(b) (2000)).

Export Controls

Arms Export Control Act. The AECA establishes procedures for both government-to-government and commercial sales of military equipment (22 U.S.C. §§2751–2799 (2000)). The AECA stipulates the purposes for which weapons may be transferred and establishes a process by which the executive branch must give Congress advance notice of major sales over a specific dollar value (22 U.S.C. §2776(b) (2000)). It further requires that foreign governments or entities gain the U.S. government's approval before retransferring arms acquired from the United States to a third party (22 U.S.C. §2778(b) (2000)). The AECA prohibits the export of defense articles or services to any country the Secretary of State determines "has repeatedly provided support for acts of international terrorism" (22 U.S.C. §2780(d) (2000)) or which the president determines are not fully cooperating with U.S. antiterrorism efforts (22 U.S.C. §2781 (2000)).

The AECA requires private parties to seek approval before exporting "munitions." The International Traffic in Arms Regulations (ITAR), as overseen by the Department of State's Office of Defense (Trade) Controls, implements the AECA in this capacity (22 C.F.R. §§120.1–130.17

(2003)). The ITAR provide specific definitions of the things that require a license for export from the Department of State (22 C.F.R §124.1 (2003)). The ITAR also lists countries and destinations that are unable to import arms from the United States (22 C.F.R §126.1 (2003)). The items regulated under ITAR are not necessarily tangible "things." For example, encryption technologies are "munitions," and technical data may be a "defense article" for which approval is required. Criminal penalties apply to AECA violations (22 U.S.C. §2778(c) (2000)).

Export Administration Act. The EAA, as implemented under the IEEPA, requires U.S. exporters to obtain a license from the Department of Commerce to export listed dual-use items (50 App. U.S.C. §2403 (2000)). In addition to physical "exports," the "deemed export rule" in the regulations made pursuant to the EAA restricts "release" of items and information within the United States to a foreign national (15 C.F.R. §§734.2(b)(2)-(3) (2003)). "Fundamental research" that is "ordinarily published and shared broadly within the scientific community" is exempt (15 C.F.R. §734.8 (2003)). Amendments to the EAA address the need to license exportation of goods and technology that might lead to deployment of a biological weapon (50 App. U.S.C. §2405(m) (2000)). The Department of Commerce recently amended its list of dual-use items and made other changes in response to recommendations of the Australia Group, a group of nations formed to monitor and restrict WMD proliferation through exports (Implementation of the Understandings, 2003). The EAA imposes both civil and criminal penalties for violations (50 App. U.S.C. §2410 (2000)).

Nunn-Lugar Cooperative Threat Reduction Program. The Soviet Nuclear Threat Reduction Act of 1991 is part of an amendment to the implementing legislation for the Conventional Armed Forces in Europe (CFE) Treaty (Pub. L. No. 102–228, 105 Stat. 1691 (1991)) sponsored by Senators Sam Nunn and Richard Lugar. It establishes the Nunn-Lugar Cooperative Threat Reduction (CTR) Program to reduce the threat posed by the weapons and weapons development programs of the former Soviet Union in light of its disintegration (22 U.S.C. §§5951–5959 (2000)). The CTR goals are to ensure safe transportation and destruction of existing weapons, to prevent their proliferation, and to prevent the diversion of scientific expertise that could contribute to weapons programs in other nations (22 U.S.C. §5951 (2000)).

Although the original and continuing focus of the CTR Program is nuclear weapons, other WMD threats, including biological weapons, are addressed as well (22 U.S.C. §5951 (2000)). In the 1997 National Defense Authorization Act, Congress directed that more attention be paid to security at facilities with materials that could be used in chemical or biological weapons (Pub. L. No. 104–201, §1501, 110 Stat. 2422, 2731 (1996)). In the 2003 National Defense Authorization Act, Congress granted a three-year waiver of the preexisting condition that the president must certify that certain weapons-control benchmarks had been met before providing additional CTR funding (Pub. L. No. 107–314, §1306, 116 Stat. 2458,

2673 (2002)). In June 2002, the leaders of the countries that form the "Group of Eight" (G8) industrial democracies reached an agreement under which they will match U.S. expenditures under the CTR Program to identify and neutralize elements of the former Soviet Union's WMD programs (Larson, 2002).

Controlling Access Within the United States. Consistent with the recognition of the dangers posed by biological weapons development and proliferation, the United States and other nations have, since World War II, increasingly recognized the human threat that dangerous biological agents pose, even if not weaponized, and have crafted safety codes to protect both those who work with or transport dangerous pathogens and members of the public who might, through mishandling of dangerous agents, accidentally become exposed. Particular events, such as the fraudulent acquisition of *Yersinia pestis* from a domestic laboratory by a potential terrorist (Henry, 1998) and the October 2001 anthrax letter attacks in which the strain employed was found to exist in domestic laboratory culture collections (Johnston and Broad, 2001), awakened lawmakers to the need to tighten access controls to dangerous agents as part of bioterrorism defense policy.

Laboratory and Transportation Safety Controls. A combination of voluntary guidelines and mandatory regulations govern the acquisition, possession, and transfer of dangerous biological agents, organisms, and toxins in the United States (Committee on the Research Standards, 2003). The Centers for Disease Control and Prevention (CDC) and National Institutes of Health (NIH) publication *Biosafety in Microbiological and Biomedical Laboratories* (*BMBL*) applies to work with live biological agents, categorizing agents and laboratory activities from biosafety level 1 (BSL-1) (minimally dangerous) to BSL-4 (agents or activities highly dangerous to humans and with no known antidote), and establishes safety requirements for each level. The "NIH Guidelines for Research Involving Recombinant DNA Molecules" similarly address genetic and recombinant organism research. Federal regulations and BMBL appendix I apply to the handling of toxins (29 C.F.R. §1910.1450 (2003); 29 C.F.R. §1910.1200 (2003)).

The Public Health Service Foreign Quarantine Regulations govern the importation of dangerous biological agents (42 C.F.R. Part 71 (2002)). Foreign imports of such agents cannot take place until a permit has been issued by the CDC (42 C.F.R. §71.54 (2002)). CDC issues, and is currently in the process of revising, the Interstate Shipment of Etiological Agents regulations, which control transportation of such agents within the United States (42 C.F.R. pt. 72 (2002)). Both regulations generally impose packaging, labeling, and safe shipping requirements. Department of Transportation (DOT) regulations on the transportation of etiologic agents (49 C.F.R. §173.196 (2002)) and the Dangerous Goods Regulations Manual of the International Air Transport Association also control shipment of biological agents.

Select Agent and Activity Access Controls

Anti-terrorism and Effective Death Penalty Act of 1996. The AEDPA directs the Department of Health and Human Services (DHHS) Secretary to identify dangerous biological agents and to regulate their transfer and use (Pub. L. No. 104–132 §511(d) & (e), 110 Stat. 1214, 1284-85 (1996)). The DHHS Secretary has delegated this responsibility to CDC, which has established a "select agent" list of the most dangerous biological agents (42 C.F.R. pt. 72, app. A (2002)) and promulgated other regulations. Under the AEDPA, these regulations require that laboratories register to transfer select agents, and require that laboratories receiving transfers be registered as well (42 C.F.R. §72.6 (2002)). They establish a system for tracking transfers of select agents and impose additional safe handling and shipping requirements. The regulations impose criminal penalties of a fine or up to five years imprisonment for violations (43 C.F.R. §72.7 (2002)).

USA Patriot Act. The USA Patriot Act prohibits transfer or possession of a select agent by a "restricted person" (18 U.S.C. §175b(a) (Supp. I 2001)). A restricted person includes most felons, fugitives from justice, unlawful drug users, illegal aliens, persons with mental problems, aliens from nations designated by the Secretary of State as supporters of terrorism, and dishonorably discharged armed forces members (18 U.S.C. §175b(d) (Supp. I 2001)). The USA Patriot Act also requires background checks of persons carrying hazardous substances (49 U.S.C. §5103a (Supp. I 2001)).

Public Health Security and Bioterrorism Preparedness and Response Act of 2002. The Public Health Security and Bioterrorism Preparedness and Response Act of 2002 (PHSBPRA) continues the tightening of dangerous biological agent access restrictions. It expands the laboratory registration requirement beyond transfer to those that possess select agents (42 U.S.C.A. §262a (West 2003)), and requires registered laboratories to limit access to select agents on a "need to use" basis and to submit the names of users to the Attorney General for background checks (42 U.S.C.A. §262a(e)(2) (West 2003)). Users must report any loss or theft of select agents to the DHHS Secretary (42 U.S.C.A. §262a(e)(8) (West 2003)). The Act establishes a national database of information on the nature and location of select agents (42 U.S.C.A. §262a(d)(2) (West 2003)). Criminal penalties apply to transfer to or possession by unregistered persons of select agents (18 U.S.C.A. §175b (West Supp. 2003)).

The DHHS Secretary has designated CDC as the agency responsible for implementing the provisions of the Act. Interim regulations that implement the provisions of the Act are currently in force (Possession, Use & Transfer, 2003). These provide an updated select agent list and identify "overlap" select agents and toxins, which pose a danger both to humans and to plants or animals (67 *Fed. Reg.* 76886, 76898-99). The regulations also contain detailed registration, safety, security, and training provisions (67 *Fed. Reg.* 76900-03). Civil and/or criminal penalties apply to violations (67 *Fed. Reg.* 76886, 76904-05).

Agricultural Bioterrorism Act of 2002. The Agricultural Bioterrorism Act of 2002 is a part of the PHSBRA. It establishes parallel authority for the United States Department of Agriculture (USDA) to regulate possession and transfer of biological agents that pose a threat to plant or animal health (7 U.S.C.A. §8401 (West Supp. 2003)). The Secretary of Agriculture has designated the USDA's Animal and Plant Health Inspection Service (APHIS) as the agency to implement the Act's provisions (7 C.F.R. pt. 331 (2003)). Similarly to the CDC's select agents list, the USDA has developed a list of biological agents and toxins, including plant and animal pathogens, that was published in early 2003 (Agricultural Bioterrorism Protection Act, 2002). Entities that possess listed plant and animal pathogens must register with APHIS, while entities that possess "overlap agents," may register with either APHIS or the CDC (Agricultural Bioterrorism Protection Act, 2002). Security requirements and penalties similar to those that apply to entities that possess human pathogens apply to those regulated by the USDA as well (7 U.S.C.A §8401 (West Supp. 2003)).

Visa and Tracking Programs. Other programs tighten the access of aliens to select agents and, more broadly, to biological research activities in the United States. The Visas Condor program checks a visa applicant's name against a number of government databases and denies visas to aliens suspected of having terrorist connections (Boucher, 2002). The Department of State has created a Technology Alert List that identifies categories of specialized research activities designated on an individual's visa application that will provoke further intelligence review. Included on the list are "technologies associated with the development or production of biological and toxin agents, pathogenics, biological weapons research" (U.S. Department of State, 2003). Both types of review require affirmative authorization by the reviewing agency before a visa can be issued. Implementation of these systems has led to laments by research universities and institutions that the cumbersome process and its delay have led to the loss of access to important and talented foreign scientists (Stroh, 2003).

Once aliens are in the United States, the new Student and Exchange Visitor Information System tracks their activities, including their areas of study. The October 2001 Presidential Decision Directive "Combating Terrorism through Immigration Policies" directs federal agencies to develop student immigration policies to "prohibit certain international students from receiving education and training in sensitive areas... " (Homeland Security Presidential Directive-2, 2001). The White House Office of Science and Technology Policy, along with its Homeland Security Council, are to create an Interagency Panel for Advanced Science and Security (IPASS) to implement this objective. In May 2002, the White House unveiled its IPASS proposal, which would screen foreign graduate students, postdoctoral fellows, and scientists who apply for visas to study "sensitive topics... uniquely available" on U.S. campuses. The final policy on IPASS, which will be implemented by the DHS, is pending further study.

Information Controls. The federal government can restrict for national security purposes dissemination of the information that it generates, and does so according to a classification system that designates both the level of security (e.g., "Secret" or "Top Secret") and which persons cleared at that level of security may have access to the information (Exec. Order No. 12,958; Exec. Order No. 12,968; Exec. Order No. 13,292). Classified information is not subject to disclosure under the Freedom of Information Act, a statute which otherwise makes most government information available upon an individual's request (5 U.S.C. §552 (2000)).

Beyond information that it creates, the government often tries to restrict dissemination of information generated by contracted private parties. A government agency generally designates a project as classified from the onset when entering into a contract with a university or research institute. The research partner can then decide whether to accept the conditions imposed on release of the information generated, can segregate classified research to a particular physical location, or can negotiate the terms of the contract to allow for greater public access to the information created. Recently, government agencies have extended the scope of secrecy applied to research partners by imposing prepublication review requirements on contract research (Borrego, 2002; Shea, 2003) and seeking in other ways to limit dissemination of "sensitive" information that is not formally classified. These restriction of private scientific speech raise constitute concern. In response to concern expressed by research institutes (Broad, 2002), the Bush administration reaffirmed that research information that is not classified will be publicly available (Abraham, 2003).

The Constitution severely limits the government's power to restrict dissemination of privately generated information for national security purposes (U.S. Const. Amend. I). The Invention Secrecy Act of 1951, on the other hand, authorizes the government to impose secrecy orders on patent applications, the release of which may threaten national security (35 U.S.C. §§181–188 (2000)). Included on the 1971 Patent Security Category Review List, once classified but recently made public, are biological warfare agents and methods of manufacture, biological agent detectors, and antibiological warfare agents (Armed Services Patent Advisory Board, 1971). Because the patents are secret, it is not possible to know how many biological weapons related inventions are covered by the orders. The *Washington Post* reports that one man holds five secret patents for methods of weaponizing *B. anthracis* (Thompson, 2003).

The Atomic Energy Act prohibits dissemination of nuclear research information even if privately created (42 U.S.C. §2014(y) (2000)). No such legislation currently limits dissemination of biological research information. Publication of benignly directed but potentially dangerous scientific research, such as contraception research on mice that revealed how to make smallpox more deadly, has led policymakers to consider imposing restrictions on dissemination of biological research that could assist in creating a biological weapon (Card, 2002).

Scientists have taken the lead in proposing information controls in order to avert a government imposed regulatory structure. In February 2003, editors of the country's leading scientific journals announced a policy to conduct reviews of publications to identify those that pose national security concerns (Atlas et al., 2003). In October 2003, an ad hoc committee of the National Academy of Sciences recommended creating an agency within the federal government to review scientific information in order to identify national security risks (Committee on Research Standards, 2003). It is unclear to what extent the voluntary controls will be successful in avoiding more restrictive government action (Wade, 2003).

Critical Target Protection. Recent terrorism attacks caused both the president and federal legislators to recognize the vulnerability of key targets, such as infrastructure and supplies, and to implement policy measures to protect them.

Presidential Decision Directive 63 (PPD-63). Presidential Decision Directive 63: Critical Infrastructure Protection, May 1998, seeks to "swiftly eliminate any significant vulnerability to both physical and cyber attacks on our critical infrastructures" by revising and expanding the federal system of infrastructure protection. The National Coordinator established by Presidential Decision Directive 39 is generally responsible for coordinating implementation of the directive, and within the directive receives a number of new roles and responsibilities. The new organization directed by PDD-63 includes designated lead federal agencies for liaison with targets and for performing certain functions related to critical infrastructure protection, creation of a Critical Infrastructure Coordination Group to coordinate activities among federal agencies, and creation of a National Infrastructure Assurance Council composed of infrastructure provider representatives and state and local government officials to enhance coordination of efforts at various levels of government and with the private sector.

PDD-63 authorizes the Federal Bureau of Investigation (FBI) to create within its organization a National Infrastructure Protection Center composed of various department and agency representatives as a focal point for gathering and disseminating information on threats to infrastructures and for coordinating the government's response to threats or attacks. In addition, the National Coordinator is to strongly encourage owners and operators of critical infrastructures to create a private sector Information Sharing and Analysis Center (ISAC). Executive Order 13, 231 adds further details to the PDD-63 critical infrastructure plan (Exec. Order No. 13, 231). With respect to ISACs, it divides the critical infrastructure sectors into eight specific entities for which a lead agency is assigned. These are (1) Department of Commerce, for information and communication; (2) Department of Treasury, for banking and finance; (3) Environmental Protection Agency (EPA), for water; (4) DOT, for transportation; (5) DOJ, for emergency law enforcement; (6) Federal Emergency Management Agency (FEMA), for emergency service; (7) DHHS, for emergency medicine; and (8) Department

of Energy (DOE), for electric power, gas, and oil. Efforts to form ISACs, and to otherwise improve information-sharing and coordination between the private sector and government agencies, is ongoing (Government Accounting Office, 2003).

Public Health Security and Bioterrorism Preparedness and Response Act of 2002. A substantial portion of the PHSBPRA mandates efforts to protect key supplies from contamination by terrorists. Title III addresses protecting the food and drug supply from adulterated imports. The provisions direct the Food and Drug Administration (FDA) Commissioner to take numerous actions to enhance food and drug safety, including requiring notice of shipments (21 U.S.C.A. §381(m) (West Supp. 2003)), imposing registration and record keeping requirements (21 U.S.C.A. §§350c & 350d (West Supp. 2003), providing for detention and enhanced inspection of imports (21 U.S.C.A. §334(h) (West Supp. 2003)), and providing grants to states for inspections and improved surveillance (21 U.S.C.A. §399 (West Supp. 2003)), 42 U.S.C.A. §247b-20 (West Supp. 2003)). It is estimated that 400,000 companies worldwide will be affected.

Title III also directs the USDA Secretary to take numerous actions to upgrade agricultural security. These include enhancing the inspection capabilities of APHIS (7 U.S.C.A. §8320 (West Supp. 2003)), conducting biosecurity upgrades at Agricultural Research Service facilities, awarding grants to research institutes and universities to conduct biosecurity upgrades (7 U.S.C.A. §3353 (West Supp. 2003)), and generally conducting research to protect the United States food supply (7 U.S.C.A. §3354 (West Supp. 2003)). USDA has taken numerous actions to protect the food supply against a bioterrorism attack. These activities include establishing an Office of Food Security and Emergency Preparedness; appointing a national surveillance system coordinator who will, among other responsibilities, oversee creation of a National Animal Health Laboratory system; developing training materials and conducting education events, including ongoing Foreign Animal Disease Awareness Training Seminars; developing the National Animal Health Reserve Corps, which will mobilize private veterinarians to respond to a terrorist event; and funding research and awarding grants to upgrade security and prepare for events that threaten food safety (USDA, 2003).

Title IV addresses drinking water security and safety by mandating water system vulnerability assessments and updates of water system emergency response plans, including a review of "methods and means" employed to detect and deal with a contaminated water supply (42 U.S.C.A. §300i-2 (West 2003)). It amends portions of the Public Health Services Act of 1944 to require specific actions of community water systems and EPA. To implement Title IV's mandates, EPA has coordinated with numerous government and private groups in order to provide training, vulnerability assessment and emergency response tools, and technical and financial assistance to water delivery entities. EPA provided support to the Association of Metropolitan Water Agencies to create the WaterISAC, which provides secure Web-based information about water system security and safety threats. EPA is also in the process of developing the Water Security Research and Technical Support Action Plan (Research Action Plan) to define a continuing program of research and technical support to protect water supplies and systems from terrorist attacks (U.S. EPA, 2003).

Preparing for Response to a Bioterrorism Event

States and localities have traditionally had the authority for preparing for and responding to public health crises. States, counties, and cities have developed integrated systems for disease reporting and prevention, such as vaccination, and for responding to disease outbreaks, including contact tracing, treatment, isolation, and quarantine. State governors generally have the authority to declare a state of emergency when responding to a public health crisis, including a bioterrorism attack. Governors' emergency powers include such abilities as issuing executive orders, including an order for quarantine, suspending regulatory statutes and state agency rules and regulations, commandeering property or personnel, and spending funds available for such emergencies.

In recognition of these traditional state responsibilities with respect to disease response, and in light of possible constitutional obstacles to federal legislation that directs disease control measures at the local level, federal policy to promote preparedness for disease emergencies and bioterrorism response is generally structured to provide financial and other assistance only when the event exceeds a state's ability to respond.

Presidential Decision Directives 39 & 62. Several presidential directives establish the general structure of agency action to achieve bioterrorism defense. National Security Decision Directive 207 (NSDD-207), issued as a secret document in 1986, created the National Program for Combating Terrorism and established the initial structure of U.S. terrorism response. NSDD-207 designated lead agencies for terrorism incident response: Department of State for international events; DOJ, through the FBI, for domestic events; and the Federal Aviation Administration (FAA) for airplane hijackings within U.S. jurisdiction. NSDD-207 also established several interagency groups to assist lead agencies to respond to terrorism events.

Two directives issued by President Clinton established a more specific structure for domestic WMD defense. Presidential Decision Directive 39: U.S. Policy on Counter-Terrorism, June 1995 (PDD-39), divides event response into two categories, "crisis management" and "consequence management," and assigns lead agencies to perform each function. "Crisis management" is primarily law enforcement activities to prevent or respond to terrorism. The lead agency for domestic crisis management is the FBI. "Consequence management" refers to activities necessary in the aftermath of an attack to protect the public, restore basic government services, and provide emergency relief. FEMA is the lead agency to fulfill the "consequence management" functions.

Presidential Decision Directive 62, signed May 1998, reaffirms and builds upon PDD-39. While the complete PDD-62 remains classified, in its unclassified portions, PDD-62 creates the Office of the National Coordinator for Security, Infrastructure Protection and Counter-Terrorism. The National Coordinator oversees the wide range of policies and programs that relate to counterterrorism efforts and produces for the president an annual Security Preparedness Report.

The DOJ's Guidelines for the Mobilization, Deployment, and Employment of U.S. Government Agencies in Response to a Domestic Terrorist Threat or Incidence in Accordance with Presidential Decision Directive 39 (Domestic Guidelines) establish the roles of various agencies and the use of the military to respond to a terrorist event (Blitzer, 1997; Reno, 1998). The Posse Comitatus Act prohibits the president from using the military to enforce civil law (18 U.S.C. §1385 (2000)). Recently, this law raised issues with respect to federal military involvement in pursuing the sniper targeting Virginia, Maryland, and Washington, DC, residents (Clymer, 2002; Vogel, 2002). An exception to the Posse Comitatus Act allows the Attorney General to request military assistance to respond to a biological weapons emergency (18 U.S.C. §175a (2000)). Regulations establish the types of assistance that the military may provide, but such assistance may not include investigatory or law enforcement activities unless such actions are considered necessary to protect human life and only in the event that civilian authorities are not capable of taking action (10 U.S.C. §382 (2000)).

"CONPLAN: United States Government Interagency Domestic Response Concept of Operations Plan" implements the structure and interagency coordination mandated by PDD-39 and PPD-62. In addition to DOJ/FBI and FEMA, primary federal participating agencies are the Department of Defense (DoD), DOE, EPA, and DHHS. Several of the subordinate organs of these agencies that deal with terrorism, or the agencies as a whole (e.g., DOJ and FEMA), have now been incorporated into the DHS, making the DHS a primary participant as well.

Each primary participating agency has designated responsibilities. The FBI will manage the federal response to a terrorist event until the Attorney General transfers the lead federal agency role to FEMA. Its crisis management responsibilities include designating an on-scene commander to ensure coordination of federal agency and state and local responses, and deploying a Domestic Emergency Support Team composed of representatives from various agencies with expertise in responding to the particular terrorist event (Blitzer, 1997).

Although FEMA is the lead agency for consequence response, CONPLAN confirms the primary role of the states to respond to the consequences of terrorism and notes that the role of the federal government is to provide assistance as required. This primary state authority for bioterrorism event response is consistent with the primary state authority for disaster relief more generally established by the Robert T. Stafford Disaster Relief and Emergency Assistance Act (42 U.S.C. §5121 et seq. (2000)), and the Federal Response Plan (FRP) issued by FEMA. As with other disasters, FEMA's responsibility with respect to bioterrorism consequence management is to coordinate the federal relief and assistance given to states and localities by various government agencies. An interagency annex to the FRP specifically addresses consequence management in the wake of a terrorist event.

DoD is to provide support generally to both crisis and consequence management activities (50 U.S.C. §2314 (2000)). Specific DoD resources that would respond to a bioterrorism event include the U.S. Army Medical Research Institute for Infectious Diseases (USAMRIID), which is the DoD's primary laboratory for researching the medical aspects of biological warfare defense and for formulating strategies and training for medical defense against biological weapons (USAMRIID, 2003) and the chemical and biological rapid response forces of the various military branches, such as the U.S. Marine Corps' Chemical Biological Incident Response Force and the U.S. Army's Chemical and Biological Defense Command.

DOE participation is most relevant to a nuclear or radiological WMD event. DHHS participation is most relevant to a bioterrorism event. DHHS assistance includes threat assessment through surveillance and other devices; epidemiological investigation and postevent support, which may include mass immunization and/or prophylaxis; management of persons and records; and pharmaceutical and other material support operations (U.S. DOJ, 2001). PDD-62 specifically designates DHHS as the lead federal agency to plan and prepare for a national response to medical emergencies in the event of a WMD attack. Within the DHHS, the Office of Emergency Preparedness (OEP) and the CDC play key roles in implementing the mandates of PDD-39 and PDD-62. The OEP manages the National Disaster Medical System (NDMS), which is a cooperative asset-sharing partnership among DHHS, DoD, Department of Veterans Affairs (VA), FEMA, state and local governments, and volunteers. The NDMS's response capacity is organized into National Medical Response Teams, which will provide treatment after a WMD event.

The EPA assists during both crisis and consequence management. Its activities include threat assessment, sample collection and analysis, identification of contaminants, and environmental cleanup and other safety and decontamination functions (U.S. DOJ, 2001).

Public Health Services Act of 1944, as Amended. The Public Health Services Act of 1944 (PHSA) sets out a wide range of public health responsibilities (42 U.S.C. §201 et. seq. (2000)). Most of these are the primary responsibility of the DHHS Secretary. The DHHS Secretary has, in turn, delegated many of the responsibilities that relate to infectious disease prevention and response to CDC.

Most fundamentally, the PHSA directs the DHHS Secretary to cooperate with and assist states in their disease-control efforts (42 U.S.C. §243 (2000)). These efforts now include assisting states to prepare for responding to a bioterrorism event (42 U.S.C.A. §247d-6 (West 2003)). In 2001, CDC commissioned the drafting of the Model State Emergency Health Powers Act (Model Act) to provide a comprehensive template of provisions and emergency powers necessary for a complete and effective state response to a bioterrorism event. The Model Act is drafters report that it has been introduced through bills or resolutions in 43 state legislatures, and that 32 states

and Washington, DC, have passed bills or resolutions that include provisions from or closely related to it (Center for Law and the Public's Health, 2003).

The PHSBPRA amends the PHSA to direct a number of specific bioterrorism preparedness activities. These include requiring the DHHS Secretary to develop a National Preparedness Plan coordinating federal and state bioterrorism responses (42 U.S.C.A. §247d-6 (West 2003)); creating a National Disaster Medical System to coordinate federal and state agency first responders (42 U.S.C.A. §300hh-11(b) (West 2003)); funding revitalization of the CDC; funding a number of specific preparedness programs; and establishing an emergency system for the advance registration of health professions volunteers (42 U.S.C.A. §247d-7b (West 2003)). The PHSBPRA further directs the DHHS Secretary to create a strategic national stockpile of drugs, vaccines, and other supplies (42 U.S.C.A. §300hh-12 (West 2003)), to fund and accelerate countermeasure production (21 U.S.C.A. §356-1 (West Supp. 2003)), and to arrange for smallpox vaccine production and the provision of potassium iodide to protect against radiation exposure (42 U.S.C.A. §§300-12 (West 2003)).

The PHSBPRA funding has allowed the CDC to establish a broad range of new programs to address the bioterrorism threat, including sponsoring multiple types of first responder training programs, providing grants to states and localities to improve their bioterrorism surveillance and response capabilities, and expanding the CDC's capabilities to gather and disseminate disease information (CDC, 2003). It manages the National Pharmaceutical Stockpile Program, which contains packages of vaccines, drugs, and other supplies that can be shipped anywhere in the United States within 12 hours. It has also developed and is in the process of implementing a smallpox preparedness program, which includes vaccinating members of the military and civilian medical and first responder volunteers (Bush, 2002).

Project BioShield is, a 10-year $6 billion plan that amends the PHSA to further address bioterrorism preparedness. The three prime components of the plan are (1) permanent, indefinite funding authority for the federal government to promote development and purchase of large amounts of vaccines and drugs necessary for bioterrorism defense; (2) enhanced flexibility and authority within the NIH to quickly award research and development contracts and grants, hire experts, and obtain items necessary for bioterrorism research; and (3) empowerment of the FDA to authorize treatments in a bioterrorism emergency that are not otherwise formally approved (Bush Bio Shield, 2003).

The PHSA also grants direct federal government authority over infectious disease response. It authorizes the DHHS Secretary to declare a public health emergency in response to a disease outbreak (whether natural or the result of a bioterrorism attack) and to provide assistance, and establishes a Public Health Emergency Fund for this purpose (42 U.S.C.A. §247d (West 2003)). It also establishes direct federal government quarantine authority. Consistent with the scope of the federal government's enumerated constitutional powers, this authority may be enacted to contain the spread of a communicable disease when the disease threatens to spread from foreign countries into the United States or from one state to another (42 U.S.C.A. §264(a) (West 2003)). To this end, CDC, specifically its Division of Global Migration and Quarantine, has the authority to isolate or quarantine travelers seeking to enter the United States (42 U.S.C. §264(b) (West 2003); 42 C.F.R pt. 71 (2002)). This authority extends only to certain diseases specified by executive order, including most recently SARS (Exec. Order No. 13295, 2003). In addition, the CDC has authority to impose a quarantine when an infectious disease threatens to cross state lines (42 C.F.R. §70.2 (2002)). These domestic quarantine powers are also limited to certain listed diseases (42 C.F.R. §70.6 (2002)).

The PHSBPRA amended several federal quarantine and emergency power provisions. The PHSBPRA eliminated the National Advisory Health Council's involvement in a quarantine decision (42 U.S.C.A §264(b) (West 2003)). The president now makes the decision upon recommendation of the DHHS Secretary, in consultation with the Surgeon General. The PHSBPRA also broadened the quarantine authority beyond contagious persons to persons in a precommunicative stage of a dangerous infectious disease (42 U.S.C.A. §264d (West 2003)). It allows the DHHS Secretary to waive the requirements of various federal medical assistance programs during national emergencies (42 U.S.C.A. §1301 et. seq. (West 2003)). The DHHS Secretary has the authority to declare a public health emergency and to "take such action as may be appropriate to respond" (42 U.S.C.A. §247d(a) (West 2003)). The PHSBPRA establishes a 90-day limit on the Secretary's declaration, subject to extension, and requires a report to Congress of a declaration within 48 hours (42 U.S.C.A. §247d(a) (West 2003)).

Defense Against Weapons of Mass Destruction Act of 1996. The Defense Against Weapons of Mass Destruction Act of 1996 is an amendment to the 1997 Defense Authorization Act. It directs DoD to create a Domestic Preparedness Program, commonly referred to as the Nunn-Lugar-Domenici Program after the sponsors of the legislation (50 U.S.C.A. §2311 et. seq. (West 2002)). The Nunn-Lugar-Domenici Program provides training, exercises, and equipment money to major U.S. cities to enhance their ability to respond to WMD incidents. DOJ, within DHS, now oversees the program. The Act directs the president to take action to enhance domestic preparedness both at the federal and the state and local levels (50 U.S.C.A. §2311 (West 2002)), and authorizes funding to achieve these objectives (50 U.S.C.A. §2312(h) (West 2002)). This law also addresses theft and diversion of nuclear materials in the former Soviet Union, as had the previous Nunn-Lugar legislation, and establishes a National Coordinator for Nonproliferation Matters (50 U.S.C.A. §2351 (West 2002)).

CONCLUSION

Although biological weapons have existed for a long time, U.S. bioterrorism defense policy has not. It is a new and rapidly growing body of law, constructed to respond to threats that have changed and that will continue to

do so. The U.S. bioterrorism defense policy is directed toward achieving its fundamental objective—protecting the nation from a biological weapons attack—by means that more specifically address the goals of deterrence, prevention, and response. These measures in turn address threats posed both by nations and by individuals and groups. The broad outlines of a U.S. bioterrorism defense policy are now apparent. It is, however, a sketch that must become more detailed over time.

REFERENCES

Abraham, S., Secretary of Energy, Memorandum for Heads of all Departmental Elements on National Security Decision Directive-189, May 12, 2003.

Agricultural Bioterrorism Protection Act of 2002, Possession, Use and Transfer of Biological Agents and Toxins, 67 Fed. Reg. 76,908, (December 13, 2002) (to be codified at 7 C.F.R. pt. 331 & 9 C.F.R. pt. 121).

Armed Services Patent Advisory Board, *Patent Security Category Review List, at* http://www.fas.org/sgp/othergov/invention/pscrl.pdf, accessed on January 1971.

Atlas, R.M. et al., Uncensored Exchange of Scientific Results, *100 PNAS 1464*, www.pnas.org/cgi/doi/10.1073/pnas.0630491100, accessed on 2/18/03.

Blitzer, R.M., Chief, Domestic Terrorism/Counterterrorism Planning Section, National Security Division, FBI, *Federal Response to Domestic Terrorism Involving Weapons of Mass Destruction and the Status of the Department of Defense Support Program*, Testimony Before the House Committee on National Security, *at* http://www.fas.org/spp/starwars/congress/1997_h/h971104b.htm, accessed on 11/4/97.

Borrego, A.M., Colleges See More Federal Limits on Research, *Chron. Higher Educ.*, November 1, 2002, p. 24.

Boucher, R., Daily Press Briefing of State Department (transcript available at http://www.state.gov/r/pa/prs/dpb/2002/13645.htm), accessed on 9/24/02.

Broad, W., Threats and Responses: Security Measures; Researchers Say Science is Hurt by Secrecy Policy Set Up by the White House, *New York Times*, October 19, 2002, p. A8.

Bush, George W., President, President Delivers Remarks on Smallpox, *at* http://www.whitehouse.gov/news/releases/2002/12/20021213-7.html, accessed on 12/13/02.

Bush, George W., President, President Details Project BioShield, *at* http://www.whitehouse.gov/news/releases/2003/02/print/20030203.html, accessed on 2/3/03.

Bush, George W., President, State of the Union, *at* http://www.whitehouse.gov/news/releases/2003/01/20030128-19.html, accessed on 1/28/03.

California Emergency Services Act, Cal. Gov't Code §8550 et. seq. (Deering 1997).

Cantigny Conference on State Emergency Health Powers and the Bioterrorism Threat (2001).

Card, J.A.H., Assistant to the President and Chief of Staff, Memorandum for the Heads of the Executive Departments and Agencies on Action to Safeguard Information Regarding Weapons of Mass Destruction and Other Sensitive Documents Related to Homeland Security, *at* http://www.fas.org/sgp/bush/wh031902.html, accessed on 3/19/02.

CDC, *Public Health Emergency Response: The CDC Role, at* http://www.bt.cdc.gov/DocumentsApp/ImprovingBioDefense/ImprovingBioDefense.asp, accessed on 9/24/03.

Committee on the Research Standards and Practices to Prevent the Destructive Applications of Biotechnology, National Research Council of the National Academies, *Biotechnology Research in an Age of Terrorism: Confronting the Dual Use Dilemma*, (2003).

Chyba, C.F., *Foreign Affairs*, May-June 2002, at 122.

Clymer, A., Big Brother Joins the Hunt for the Sniper, *New York Times*, October 20, 2002, at D3.

Exec. Order No. 12,958, 60 *Fed. Reg.* 19,825 (April 20, 1995).

Exec. Order No. 12,968, 60 *Fed. Reg.* 40,245 (August 7, 1995).

Exec. Order No. 13,231, 66 *Fed. Reg.* 53,063 (October 18, 2001)

Exec. Order No. 13,292, 68 *Fed. Reg.* 15,315 (March 28, 2003).

Exec. Order No. 13,295, 68 *Fed. Reg.* 17,255 (April 9, 2003).

Government Accounting Office, *Critical Infrastructure Protection: Challenges for Selected Agencies and Industry Sectors*, GAO-03-233, *at* http://energycommerce.house.gov/108/pubs/GAO-03-233.pdf, accessed on 2/1/03.

Henry, L., *SUN Profile: Harris' Troubled Past Includes Mail Fraud, White Supremacy*, Las Vegas Sun, *at* http://www.lasvegassun.com/dossier/crime/bio/harris.html, accessed on 2/23/98.

Homeland Security Presidential Directive-2, *Combating Terrorism Through Immigration Policies*, *at* http://www.whitehouse. gov/news/releases/2001/10/print/2001-1030-2.html, accessed on 10/30/01.

Implementation of the Understandings Reached at the June 2002 Australia Group (AG) Plenary Meeting and the AG Intersessional decision on Cross Flow Filtration Equipment—Chemical and Biological Weapons Controls in the Export Administration Regulations, 68 Fed. Reg. 34526 (to be codified at 15 C.F.R. pts. 742, 745, and 774) (June 10, 2003).

Jehl, D., After the War: Arms; Iraqi Trailers Said to Make Hydrogen, not Biological Arms, *New York Times*, August 9, 2003, at A1.

Johnston, D. and Broad, W.J., A Nation Challenged: The Investigation; Link Suspected in Anthrax and Hijackings, *New York Times*, October 19, 2001, at B5.

Lacey, E.J., Principal Deputy Assistant Secretary of State for Verification and Compliance, *The Biological Weapons Convention (BWC)*, Testimony before the Subcommittee on National Security, Veterans Affairs, and International Relations of the House Comm. on Gov't Reform, *at* http://www.state.gov/t/vc/rls/rm/2001/5168.htm, accessed on 7/10/01.

Larson, A.P., Under Secretary for Economic, Business and Agricultural Affairs, U.S. Dept. of State, *G-8 Global Partnership Against the Spread of Weapons and Materials of Mass Destruction*, Testimony Before the House International Relations Committee, *at* http://www.state.gov/e/rls/rm/2002/12190 pf.htm, accessed on 7/25/02.

Mayor, A., *Greek Fire, Poison Arrows and Scorpion Bombs: Biological and Chemical Warfare in the Ancient World* (2003).

Meselson, M., *Germ Wars: Biological Weapons Proliferation and the New Genetics, Global Spread of Chemical and Biological Weapons*, Testimony before the Senate Committee on Governmental Affairs and its Permanent Subcommittee on Investigations, May 17, 1989.

National Strategy for Combating Terrorism at http://www. whitehouse.gov/news/releases/2003/02/counter_terrorism/ counter_terrorism_strategy.pdf, 2/20/03.

National Strategy to Combat Weapons of Mass Destruction at http://www.whitehouse.gov/news/releases/2002/12/ WMDStrategy.pdf, accessed on 12/1/02.

Nixon, R., President, *Remarks Announcing Decisions on Chemical and Biological Defense Policies and Programs at* http://www.nixonfoundation.org/Research_Center/1969_pdf_ files/1969_0462.pdf, accessed on 11/25/69.

Patriot Act, Part II, *New York Times*, September 22, 2003, at A18.

Possession, Use, and Transfer of Select Agents and Toxins, 60 *Fed. Reg.* 76886 (December 13, 2002) (to be codified at 42 C.F.R. pt. 73).

Powell, C.L., Secretary of State, Remarks before the United Nations Security Council, transcript *at* http://www.state.gov/ secretary/rm/2003/17300.htm, accessed on 2/5/03.

Reno, J., Attorney General, *The Threat of Chemical and Biological Weapons*, Statement of testimony before the Subcommittee on Technology, Terrorism and Government Information, the Committee on the Judiciary and the Select Committee on Intelligence, Senate, *at* http://www.fas.org/irp/congress/1998_hr/ s980422-jr.htm, accessed on 4/22/98.

Shea, D.A., *Report for Congress: Balancing Scientific Publication and National Security Concerns: Issues for Congress at* www.fas.org/irp/crs/RL31695.pdf, accessed on 1/10/03.

Stroh, M., Tougher Visa Rules Aim to Stop Terrorism, But End Up Hampering Research, *Baltimore Sun*, July 14, 2003, at 1A.

The Center for Law and the Public's Health at Georgetown & Johns Hopkins Universities, *Model Public Health Laws: The Model State Emergency Health Powers Act*, *at* http://www.publichealthlaw.net/Resources/Modellaws.htm, accessed on 8/11/03.

The Convention on the Prohibition of the Development, Production and Stockpiling of Bacteriological (Biological) and Toxin Weapons and on their Destruction, 26 U.S.T. 583 (1975).

The Prohibition of the Use in War of Asphyxiating, Poisonous or Other Gases, and of Bacteriological Methods of Warfare (Geneva Protocol), 26 U.S.T. 571 (1972).

The Treaty on the Limitation of Anti-Ballistic Missile Systems, 23 UST 3435 (1972).

The White House, *Progress Report on the Global War on Terrorism at* https://www.whitehouse.gov/homeland/progress/progress_ report_0903.pdf, accessed on 9/1/03.

Thompson, M.W., The Persuit (sic) of Steven Hatfill, *Washington Post*, September 14, 2003, at W06.

U.S. Army Medical Research Institute of Infectious Diseases, *Mission*, at http://www.usamriid.army.mil/aboutpage.htm, accessed on 4/15/04.

U.S. Department of Agriculture, USDA Homeland Security Efforts, *at* http://www.usda.gov/homelandsecurity/hs-efforts.pdf, accessed on June 2003.

U.S. Department of Justice, *CONPLAN: United States Government Interagency Domestic Response Concept of Operations Plan*, at http://www.fbi.gov/publications/conplan/conplan.pdf, accessed on January 2001.

U.S. Environmental Protection Agency, *Water Infrastructure Security*, at http://www.epa.gov/safewater/security/index.html, accessed on 5/27/04.

U.S. State Department, *Technology Alert List*, at http://foia.state. gov/masterdocs/09fam/0940031X1.pdf, accessed on 5/22/00.

Vogel, S., Military Aircraft with Detection Gear to Augment Police, *Washington Post*, October 16, 2002, A01.

Wade, N., Panel of Scientists Supports Review of Biomedical Research that Terrorists Could Use, *New York Times*, October 9, 2003, at A20.

Whitman v. Am. Trucking Assn., 531 U.S. 457 (2001).

WEB RESOURCES

National Security Decision Directive 207, *The National Program for Combatting Terrorism at* http://www.fas.org/irp/offdocs/nsdd/nsdd-207.htm, accessed on 9/24/03.

Presidential Decision Directive 39, *United States Policy on Counterterrorism*, Synopsis, *at* http://www.ojp.usdoj.gov/odp/docs/pdd39.htm, accessed on 9/24/03.

Presidential Decision Directive 62, *Protection against Unconventional Threats to the Homeland and Americans Overseas*, Abstract *at* http://www.ojp.usdoj.gov/odp/docs/pdd62.htm, accessed on 5/22/98.

Presidential Decision Directive 63, *at* http://www.ojp.usdoj.gov/odp/docs/pdd63.htm, accessed on 9/24/03.

Senate Select Committee on Intelligence, 108th Cong., Report on the U.S. Intelligence Community's Prewar Intelligence Assessment's on Iraq (2004), *at* http://intelligence.senate.gov/iraq-report2.pdf.

WATER SUPPLY: VULNERABILITY AND ATTACK SPECIFICS

W. Dickinson Burrows
Sara Renner Birkmire

INTRODUCTION

The wartime use of corpses of men and animals to pollute—and thereby deny—wells and other water sources has been common practice for centuries, and many epidemics of gastrointestinal illness could have resulted from this practice. In terms of terrorism, however, water supplies are generally unfavorable targets. Most biological warfare (BW) agents, whether replicating (live) agents or toxins, have been weaponized as aerosols, and the threat of these agents is thus primarily to the respiratory tract, not the digestive system. Furthermore, the amount of material needed to achieve any useful level of morbidity by the drinking water route makes it necessary to apply the agent as close in the distribution process to the consumer as possible. For example, for any toxic or infectious material applied to a reservoir with 100-days holding capacity, assuming perfect mixing, more than 30,000 times the effective dose for each individual placed at risk would be required to generate an effect, even neglecting natural attenuation and ordinary treatment efficacy. Even for an agent as deadly as botulinum toxin, this would likely be beyond the capabilities of a domestic terrorist group and barely achievable by a hostile nation. For this reason, the successful terrorist will think small and direct his attack locally. Where a lesser objective than mass casualties is intended, biological contamination of water supplies can be a useful tactic, and further obviates the need for sophisticated weaponry.

AGENTS

There are many biological agents potentially useful to terrorists and other hostile elements. In terms of agent selection, important considerations include availability, infectivity or toxic dose, environmental stability, and resistance to disinfection. Among the likely candidates are bacterial agents known (or strongly believed) to have been weaponized in the past, and which are thus available in bulk: *Bacillus anthracis* (anthrax), *Yersinia pestis* (plague), *Francisella tularensis* (tularemia), *Brucella suis* (brucellosis), and *Coxiella burnetii* (Q fever). Accurate knowledge of the infective dose by the oral route is lacking for these agents. Weaponized viruses include Variola major (smallpox), alphaviruses (encephalomyelitis), and the viral hemorrhagic fever viruses (Ebola, Marburg, and others). While all are extremely infectious, no evidence exists that any of these weaponized viruses

are transmittable by the oral route. All the weaponized agents appear to be adequately stable in water; however, most municipal water supplies in the United States are chlorinated, and little firm data exists on chemical disinfection of BW agents by this or any other means.

The weaponized toxins are attractive as waterborne agents because they are about as toxic when ingested as by inhalation. Botulinum toxin, elaborated by *Clostridium botulinum*, is the most toxic natural substance known; ingestion of 1 µg/kg body weight will kill most people, but it is largely inactivated by chlorine under standard disinfection conditions and is also heat sensitive. Ricin is a thousand times less toxic but is readily available due to the widespread existence and use of castor beans, from which it is extracted. Aflatoxin, a mold metabolite, causes massive gastrointestinal bleeding and is a potent carcinogen but does not appear to be particularly suited for weapons use, though it should be noted that Iraq weaponized this agent in quantity. The trichothecene mycotoxin T-2 is another unlikely water threat. Staphylococcal enterotoxin B (SEB) is an incapacitating agent (rather than a lethal agent) that causes severe gastrointestinal illness. Saxitoxin (paralytic shellfish poison) and tetrodotoxin (pufferfish poison) are typical marine toxins, which, like botulinum toxin, cause death by progressive paralysis. The marine toxins, as potent as they are, are not available in bulk at this time, though the simplicity of their molecular structures makes them amenable to mass production by any major pharmaceutical manufacturer. Microcystin, also known as "fast death factor," is a hepatotoxin from certain blue-green algae that may in fact be common in U.S. water supplies. Like ricin and aflatoxin, microcystin causes death by organ necrosis and massive internal hemorrhaging.

Regardless of which agents are perceived to pose the greatest public health threat, truly perceptive adversaries with limited resources will probably ignore exotic BW agents in favor of the more familiar waterborne pathogens. These pathogens are easily obtained, highly infectious, and relatively stable in water, and all cause acute (in some cases terminal) gastrointestinal disease. There are already credible reports of contamination of food and water with *Salmonella* subspecies (spp.) (salmonellosis, typhoid fever) by terrorists. *Shigella* spp. (shigellosis), which have also been implicated in intentional food contamination, are much more infectious, and *Vibrio cholerae* (cholera), for which the outcome can be even more serious, has been implicated in World War II BW attacks on food and

Encyclopedia of Bioterrorism Defense, Edited by Richard F. Pilch and Raymond A. Zilinskas
ISBN 0-471-46717-0 Copyright © 2005 Wiley-Liss

water supplies, as well as a postwar terrorist incident of drinking water contamination. Emerging pathogens such as *Escherichia coli* O157:H7 have caused some recent outbreaks of waterborne disease in North America and may pose an illicit threat as well. The enteric viruses are also strong candidates, and rotavirus and enterovirus 70 have been researched as potential BW agents by Iraq. In addition, *Cryptosporidium parvum*, a protozoan parasite of many warm-blooded animals, is highly infectious and has the added advantage to the terrorist of being resistant to chlorine disinfection.

There are a few other waterborne pathogens that warrant consideration. Among the protozoan parasites are *Cyclospora* and the microsporidia, which cause profuse diarrhea and (like *Cryptosporidium*) are most serious in immunocompromised individuals, and *Toxoplasma*, a parasite of cats that causes flu-like symptoms in humans. *Clostridium perfringens*, a common organism in sewage, causes generalized gastrointestinal disorders and is relatively resistant to chlorine disinfection. Among the emerging bacterial pathogens are *Mycobacterium avium* complex (MAC), which is extremely resistant to conventional disinfection, and *Helicobacter pylori*. Both of these agents cause various gastrointestinal symptoms. In addition, there are a few truly exotic agents that might warrant consideration, such as prions and HIV, but even if these agents could be shown to be waterborne, they are unattractive from a terrorist's point of view in that progression of the disease takes years.

VULNERABILITY

As noted, the huge volumes of water involved make rivers and reservoirs impractical targets. Thus, it is highly unlikely that collateral fallout to a water supply reservoir from an aerosol attack would result in significant contamination, and it seems equally unlikely that a declared adversary would intentionally target a source that gives such poor prospects of success. A rational adversary with limited resources will attempt to contaminate the "finished" water, that is, water downstream from the treatment plant. In order of increasing likelihood of success, this would involve water storage areas, distribution lines, isolated communities or complexes, a single building, and in an extreme example could even involve contaminating bottled water. Contamination of finished water generally requires an agent resistant to residual chlorine used as water disinfectant, such as *B. anthracis* spores, *Cryptosporidium parvum* oocysts, or most toxins. The adversary may seek access to a finished water reservoir; failing that, he can still inject an agent into the distribution mains through a hydrant or valve used for testing. Other options are to gain access through a manhole to water lines servicing a targeted building and install a sampling valve, or to inject an agent back through the plumbing of a rented building where backflow preventers are not in use.

There are circumstances whereby a raw water source could represent a vulnerability. Many farms in the United States serve tens of thousands of animals each. Animal wastes are stored in lagoons prior to land application, and on occasion, these lagoons have failed and discharged tons of animal wastes into water supply sources, typically resulting in the need to close water intakes to treatment plants and apply elevated levels of chlorine. There is no record of any attempt by vandals or terrorists to blow up critically located waste lagoons in the United States, but if this happens, there could be a heavy price to pay in interrupted water service and public dismay. Similarly, where wells are used for water supply, an unsecured wellhead could present a significant vulnerability: a single human stool could be used to infect the well and everyone consuming the water with calicivirus (Norwalk-like virus).

MONITORING

There are for the most part no rapid means of detecting a waterborne BW event. Indeed, the event may have passed before it is recognizable through epidemiology as a terrorist incident. At this time, confirmation of a bioterrorist attack would likely require that the capabilities of the Centers for Disease Control and Prevention (CDC) and/or the U.S. Army Medical Research Institute of Infectious Diseases (USAMRIID) are brought to bear on a suspicious outbreak. There are presently no assays for detecting reproducing agents or toxins that are suitable for use in the field or at most municipal water treatment plants, although there are promising technologies under investigation. The U.S. Army Soldier and Biological Chemical Command (SBCCOM) is developing rapid, handheld, reagent-less test kits for many of the major BW agents described above. For on-line detection of a BW event, environmental sentinel biomonitors or other utility-based methods may provide the most useful indication. An approach developed by the U.S. Army Center for Environmental Health Research (USACEHR) employs the bluegill sunfish as sentinel species. Fish are confined to a flow-through monitoring chamber, and a pair of electrodes record ventilatory signals that are then electronically processed and matched against calibrated standards. With this system (presently deployed at several sites), about 10 µg/L of brevetoxin (a dinoflagellate toxin) is detectable in 40 min. Other biosensors are based on bacteria, microcrustaceans, and bivalves. On-line chemical monitors, individually or in combination, can also provide useful information. For example, a loss of free chlorine residual in the distribution system could warn that a highly impure pathogen preparation has been injected or that a technically sophisticated terrorist, in order to protect a chlorine-intolerant BW agent, has added a chemical to destroy residual chlorine.

DEFENSE

Security is the most important first step in reducing vulnerability. Every effort should be made to protect well-heads, treatment facilities, and distribution and storage systems from vandals as well as terrorists. Vigilance is essential; most communities lack the resources to man routine patrols of their water supply infrastructure, but municipal personnel, particularly personnel who work late shifts, should be imbued with the importance of noting suspicious or abnormal conditions when performing routine duties. For example, an adversary, realizing that he could not hope to contaminate a whole reservoir, might still succeed in incorporating the essential concentration of a threat agent at the intake of a municipal water treatment plant by operating at night from a small boat. Floodlights and obvious vigilance should discourage this.

A well-designed and well-operated municipal water treatment plant can be expected to achieve 99 to 99.9 percent removal of bacteria before disinfection and 99 to 99.99 percent removal of protozoan cysts. For both proteinaceous toxins (e.g., botulinum toxin, ricin, and SEB) and viruses, as much as 90 percent removal may be achievable under optimal conditions of coagulation, flocculation, and filtration. Chlorine disinfection inactivates most common waterborne pathogens, and maintaining an adequate disinfection residual provides significant protection in the distribution system. As noted above, some agents, including *B. anthracis* spores, *Cryptosporidium* oocysts, and most toxins, are nearly impervious to chlorine disinfection; for others, there is no useful information to this end. Ozone, mixed oxidants, ultraviolet radiation and combinations thereof may be more effective, but again data is incomplete.

At the end of the distribution system, there are point-of-entry and point-of-use (POE/POU) water treatment devices that could in principle provide protection from virtually any contaminant, chemical or biological. Membrane microfilters will remove all protozoan cysts and almost all bacteria, ultrafilters will remove viruses as well, and reverse osmosis systems or nanofilters will remove all toxins, as will many granular activated carbon filters. Finally, most replicating biological agents and some toxins are heat labile and thus would not survive a boil-water advisory.

POE/POU treatment devices will likely find application in protection of attractive targets, not to make up for deficiencies in municipal water treatment (although that could be a consideration) but to intercept contaminants specifically directed at that facility. Once a BW incident has been confirmed, the recommended course of action is to seek an alternative source of water, but the POE/POU device can provide interim protection. Ultimately, the dilemma as it currently stands is that the ability to treat BW agent-contaminated water far exceeds the present ability to detect or monitor for their presence.

FURTHER READING

Burrows, W.D. and Renner, S.E., *Environ. Health Perspect.*, **107**(12), 975–984 (1999).

Christopher, G.W., Cieslak, T.J., Pavlin, J.A., and Eitzen, E.M., Jr., *J. Am. Med. Assoc.*, **278**(5), 412–417 (1997).

Khan, A.S., Swerdlow, D.L., and Djuranek, D.D., *Public Health Rep.*, **116**(1), 3–14 (2001).

Rose, J., *Environ. Sci. Technol.*, **36**(11), 247A–250A (2002).

States, S., Scheuring, M., Kuchta, J., Newberry, J., and Casson, L. *J. Am. Water Works Assoc.*, **95**(4), 103–114 (2003).

WEB RESOURCES

Health Canada, Population and Public Health Branch, Office of Laboratory Security: Material Safety Data Sheets, http://www.hc-sc.gc.ca/pphb-dgspsp/msds-ftss/index.html.

U.S. Food and Drug Administration, Center for Food Safety and Applied Nutrition, *The Bad Bug Book: Foodborne Pathogenic Microorganisms and Natural Toxins Handbook*, http://vm.cfsan.fda.gov/mow/intro.html.

See also FOOD AND BEVERAGE SABOTAGE and FOOD AND WATERBORNE PATHOGENS.

WEAPONS OF MASS DESTRUCTION CIVIL SUPPORT TEAMS

LAUREN HARRISON

INTRODUCTION

In response to the 1995 Oklahoma City bombing and the sarin attack in the Tokyo subway, Congress commenced discussions on the creation of a group of teams that could respond quickly to weapons of mass destruction (WMD) attacks in the United States. President Clinton announced a plan to establish the Civil Support Team (CST) program, overseen by the Department of Defense (DoD), in May 1998. Congress subsequently approved a total of 32 teams to be positioned throughout the United States, the first 10 of which were authorized in the National Defense Appropriations Act for fiscal year 1999. As of February 5, 2003, all 32 teams have been authorized, and 27 of the teams have been certified as operational under DoD requirements.

BACKGROUND

The CST program was created to detect, assess, and respond to WMD threats and incidents. Each CST team is trained, equipped, and funded by the federal government, but operates under the jurisdiction of the governor in its home state. Each team provides specialized response and advice in assisting local civil authorities at WMD disaster sites, and essentially serves as a bridge between military and civil support and a link between local response and individual federal agencies.

Each CST team is composed of 22 full-time Army and/or Air National Guard personnel under the command of a Lieutenant Colonel. Members of each unit are trained extensively and then appointed to the following divisions to provide assistance during a crisis involving WMD: command, operations, communications, administration/logistics, medical, and survey. Team members must complete an additional 600 hours of specialized training beyond standard military requirements, addressing such topics as emergency response situations, emergency management systems, and technical issues related to a WMD attack. Other federal agencies, including the Federal Emergency Management Agency (FEMA), Department of Energy (DOE), and Environmental Protection Agency (EPA) also provide assistance in training CST members.

Each team maintains the ability to assess a chemical or biological threat or event, provide logistical analysis of the situation, and refer the situation to a higher authority as necessary. Teams maintain specific equipment to respond to WMD situations, including specialized detection equipment for chemical and biological agents, organic substances, and toxic industrial chemicals, and are prepared to operate insituations involving unknown contamination. Satellite and cellular phone communications and personal protective equipment at a standard higher than typical military equipment are also maintained by each team. CSTs are able to respond to a WMD disaster situation within 3 to 5 hours of departure from their home station.

CSTs IN ACTION

Early in the program, CSTs were for the most part called upon to manage hazardous materials releases such as chemical spills (Fig. 1). The Alaskan CST unit, for example, was summoned to identify unknown material that turned out to be remnants of World War II and

Photo by PH1(AW) William R. Goodwin

Survey team members attached to 93rd Weapons of Mass Destruction Civil Support Team (WMD CST) perform a presumptive analysis on a container and compressed gas bottle with a Hazardous Air Pollutants On-Site Detection Device for chemical weapons, during a weapons of mass destruction exercise held on Ford Island. The scenario was an aerial chemical attack by terrorists against Pearl Harbor ships. The WMD-CST is an active-duty Hawai'i Army and Air National Guard unit composed of command, operations, communications, administration, medical and survey teams. Within four hours of notification, WMD-CST can deploy to a suspected incident in support of a local incident commander to assess the situation, advise civilian authorities on appropriate action, and facilitate requests for assistance to expedite arrival of additional state and federal assets to help save lives, prevent human suffering and mitigate great property damage during a crisis.

Figure 1. A CST in action. Hawai'i Navy News: http://www.hnn.navy.mil/Archives/030530/weapons_030530.htm.

Encyclopedia of Bioterrorism Defense, Edited by Richard F. Pilch and Raymond A. Zilinskas
ISBN 0-471-46717-0 Copyright © 2005 Wiley-Liss

Korean War munitions and material. Anthrax hoaxes are also commonly investigated by CST units.

The most noteworthy example of a CST in action is the response by the 2nd CST, New York National Guard, on September 11, 2001. After the September 11 attacks, the 2nd CST was able to activate within 90 minutes to provide key technical service to other emergency groups. In addition to assessing the hazardous conditions surrounding ground zero, the unit supplied situational awareness by conducting reconnaissance, and became a significant information conduit between the military and involved emergency agencies.

THE FUTURE

In the past, critics of the CST program have suggested that money allocated to it could be better utilized by training local first responders to respond to WMD situations. Some critics have further questioned whether the 3- to 5-hour response time is realistic, let alone sufficient, to respond to a WMD situation. Supporters of the CST program, however, explain that with experience, each CST unit will prove most effective in dealing with WMD situations. Currently, CST units continue to participate in terrorism response exercises across the country in order to improve cooperation and emergency preparedness at the local, state, and federal levels.

WEB RESOURCES

Bogard, A.T., and Borel, W.L., Jr., Civil Support Team in Action, *Center for Army Lessons Learned*, http://call.army.mil/products/trngqtr/tq3-02/borel.htm, accessed on 9/26/03.

Koehler, D., Weapons of Mass Destruction, *Mobile Radio Technology*, November 2002, http://iwce-mrt.com/ar/radio_weapons_mass_destruction/, accessed on 9/26/03.

National Guard website, *WMD Fact Sheet*, http://www.ngb.army.mil/downloads/fact_sheets/wmd.asp, accessed on 9/25/03.

Patterson, K., Civil Scrutiny, *National Guard Magazine*, December 2000, http://www.ngaus.org/ngmagazine/main1200.asp, accessed on 9/25/03.

See also CRISIS MANAGEMENT; DEPARTMENT OF DEFENSE; and JOINT TASK FORCE CIVIL SUPPORT.

WEATHER UNDERGROUND: A CASE STUDY (STUDENTS FOR A DEMOCRATIC SOCIETY (SDS), WEATHERMEN)

Monterey WMD-Terrorism Database Staff

OVERVIEW

In November 1970, Jack Anderson reported in the *Washington Post* that the radical leftist group, the Weather Underground, known to be responsible for numerous bombings in protest of the Vietnam War, had attempted to obtain an unknown biological agent from the United States Army Medical Research Institute of Infectious Diseases (USAMRIID) at Fort Detrick, near Frederick, Maryland. No evidence has been found to support this accusation to date (Parachini, 2000).

BACKGROUND

Originally named the "Weathermen," the Weather Underground was formed in the late 1960s as a radical off-shoot of the antiwar student movement, Students for a Democratic Society (SDS), and was originally connected with Abbie Hoffman's 1967–1968 Youth International Party, or "Yippie" movement. After the 1968 Democratic National Convention, which was protested by a number of antiwar student groups, the Weathermen evolved into an independent, comparatively extreme antiwar movement (Terrorism on the Left). The group's slogan derived from a lyric in Bob Dylan's *Subterranean Homesick Blues*, "you don't need a weatherman to know which way the wind blows." At the time, the group was primarily involved in protests against the Vietnam War. In 1969, members closed the national SDS office and began conducting their "revolution" via underground cells known collectively as the "Weathermen," which they believed would be better protected against outsider threats than public meetings and large conventions. In December 1970, the group officially changed its name to the "Weather Underground" (Parachini, 2000). By 1975, the group had more than 200 members, most of whom were white women connected with the feminist movement (including Bernardine Dohrn, one of the group's leaders). The Weather Underground disbanded in 1976 (Parachini, 2000).

THE INCIDENT

On November 20, 1970, an informant reportedly warned the U.S. Customs Bureau that a revolutionary group called the "SDS Weathermen" was attempting to blackmail a homosexual Army lieutenant stationed at Fort Detrick to steal biological weapons agents from USAMRIID. According to the report, the group planned to contaminate water supplies of a major city to "incapacitate a population by infection for 7–10 days" (*New York Times*, 1970; Clark, 1980). A U.S. Army spokesperson acknowledged receipt of a report describing these allegations, but stated that he was unaware of any such blackmail attempt and that no homosexual officer at Detrick had access to the agents in question. In addition, the spokesperson stated that there was insufficient material at the location to carry out such a threat.

There is no evidence to suggest that the Weathermen ever actually attempted to acquire weapons of mass destruction (WMD), biological or otherwise. Five Freedom of Information Act requests of different relevant federal entities yielded no documents on the case, and aside from news articles on November 20 and 21, 1970, and a series of scholarly journal articles that cite these articles, there has been no further information reported in the public domain (Parachini, 2000).

RELATED INCIDENTS

In addition to this allegation, founder of the Weatherman-Yippie antiwar movement Abbie Hoffman threatened in 1968 to dump lysergic acid diethylamide (LSD) into Lake Michigan, the main water source for Chicago, to "space-out" delegates attending the Democratic National Convention in protest of President Johnson and his administration's role in the Vietnam War. While officials stated at the time that the filtration system would have been able to counter any effects the LSD may have had (Clark, 1980), it is far more likely that the dilutional effect of the Lake would have been the instrumental factor in preventing any appreciable effect from such sabotage. Followers of Abbie Hoffman claimed that his announcement was a hoax, saying he was "just teasing America" (Banko, 1989; Drugs and Hoffman, 1989).

ASSESSMENT

A thorough academic research study conducted in 1997–1998 uncovered no new evidence that this event ever occurred, or that the Weather Underground ever attempted to acquire WMD. Other evidence that calls to question the credibility of the report includes the reliability of the source of information (he was a known drug dealer) and the fact that both the current historian at Fort Detrick and the former chief of product development William C.

Encyclopedia of Bioterrorism Defense, Edited by Richard F. Pilch and Raymond A. Zilinskas
ISBN 0-471-46717-0 Copyright © 2005 Wiley-Liss

Patrick have no recollection of such an event (Parachini, 2000).

Although the Weather Underground did participate in numerous conventional bombings around the United States in protest of the Vietnam War, the idea of inflicting mass casualty is not reflected in their doctrine and public statements. Nearly all of the targets of the bombings for which they were responsible were symbolic in nature, and in its treatise *Prairie Fire*, the Weather Underground clearly contested the idea of nuclear war (Parachini, 2000).

After examining the evidence surrounding this case, most scholars therefore conclude that it is not an example of attempted acquisition of WMD by a terrorist organization. However, this case highlights both the role of the media in such incidents and the ability of groups to use the media to incite fear in the general public. Although the Weather Underground appears to have employed only conventional weapons in its "protests," one cannot rule out the idea that similar groups will attempt to acquire mass casualty weapons to establish better bargaining power in the future.

REFERENCES

Banko, S.T., III, Strange How We Pick Our Heroes, *St. Louis Post-Dispatch*, 3B July 4, (1989).

Clark, R.C., *Technological Terrorism*, Devin-Adain, Old Greenwich, CT, 1980, pp. 110–111.

Drugs and Hoffman, *Newsday*, 14 April 28, (1989).

Monterey WMD-Terrorism Database, Center for Nonproliferation Studies, 1998–2003.

New York Times, Army Tells of Plot to Steal Bacteria From Ft. Detrick, 34 November 21, (1970).

Parachini, J., "The Weather Underground," in J.B. Tucker, Ed., *Incidents of Terrorism Involving Weapons of Mass Destruction*, MIT Press, 2000.

Terrorism on the Left, *Newsweek*, 26–31 (March 23, 1970).

FURTHER READING

Anderson, J., Weatherman Seeking BW Germs, *Washington Post*, November 20, 1970, p. D15.

Carus, S., *Bioterrorism and Biocrimes: The Illicit Use of Biological Agents in the 20th Century*, National Defense University, Washington, DC, August 1998.

INDEX

Abbas, Abu, 283
Aberdeen Proving Ground, 191–192
Abrin, 420–422
   detection of, 422
   mechanism of action, 421
   source and distribution, 420
   toxicity of and prophylaxis for, 422
Abrus precatorius, 346, 420, 422
Abu Khusa, Tawfiq, 380
Abu Nidal Organization (ANO), 328, 450. *See also* Nidal, Abu
   Sudan and, 448
Abu Sayyaf Group (ASG), 1–3, 328
   terrorist operations of, 2–3
Accomodationist political tactics, 297
Acquired Immunodeficiency Syndrome (AIDS), 255
Activatable toxins, 66
Active immunity, 396
Active surveillance, 199, 459
Active syndromic surveillance systems, 173
Acute psychosocial sequelae, of bioterrorism, 405–407
Acute Stress Disorder (ASD), 406–407
Adenoviruses, 65
Administrative Commission (OIE), 373
Advanced Chemical and Biological Incident Response Course, 189
Advanced Concept Group (ACG), Sandia National Laboratories, 429
Advanced Nucleic Acid Analyzer, 321
Advisory Committee on Immunization Practices (ACIP), 438
Aerial applicators, 148
Aerobiology, 4
Aerosol delivery/dispersal/ transmission, 167. *See also* Aerosols
   of biological weapons, 63, 76–77
   of bioterrorism agents, 457–458

of *Burkholderia mallei*, 239
of smallpox virus, 441
of toxins, 481, 482
of VEE virus, 204
of *Yersinia pestis*, 392
Aerosol exposure
   to *Burkholderia mallei*, 236
   to ricin, 421–422
   to staphylococcal enterotoxin B, 446
Aerosol particle sizers, 168
Aerosols, 4–8. *See also* Bioaerosols
   with animal excreta, 248
   influenza, 260
   Serratia marcescens, 53
Aerosol vaccination, 473
Aflatoxin, 525
Africa
   Libyan role in, 328
   Mau Mau in, 346
Agency for Healthcare Research and Quality (AHRQ), 155, 356
Agent dispersal. *See* Delivery methodologies
Agreements, international. *See* International regulations and agreements
Agricultural bioterrorism, 9–15
   future of, 14–15
   history of, 9–10
   U.S. vulnerability to, 10–12
Agricultural Bioterrorism Act of 2002, 516, 518
Agricultural disease agents, 14
Agricultural disease outbreaks, federal response to, 10–11
Agricultural pathogens, as weapons, 12–13
Agriculture. *See also* Agroterrorism; U.S. agriculture
   motives to attack, 14
   vulnerabilities of, 13–14
Agroterrorism, 9–15, 254, 433–434
Ahmad, Abu, 243
Airborne biological agent dissemination, 147–148.

See also Aerosol delivery/ dispersal/ transmission
al-A'la Mawdudi, Sayyid Abu, 297
Al-Aqsa Martyrs Brigade, 380, 381, 451
Albright, Madeleine K., 467
Aleph, 46
Alexandrium, 338
Al-Fatah, 379, 380, 381
Algae. *See* Marine algal toxins
Algeria. *See* Armed Islamic Group (GIA)
Al Harakatul Islamiya, 1
Ali (first *Shi'i* Imam), 287–288
Alibekov, Kanatjan, 469
Aliens, classes of, 467
Aliens of America, 16
Al-Jama'a al-Islamiyya, suicide bombings by, 454
Al-Jami'at al-Salafiyya al-Jihadyya, 284
"All-hazards" approach, 128
Alliance WMD Initiative, NATO, 363
All source analysis, 264
Al-Manar weekly, 380
Alphabet Bomber, 16
AlphaVax, 204
Alphaviruses, 203–205
Al-Qa'ida, 1, 17–21, 283. *See also* Al-Qa'ida network
   background, 17
   biological weapons and WMD capability, 19–20
   capabilities, 20–21
   Chechen separatists and, 112
   chronology, 19
   current status, 18
   history, 17–18
   ideology, 18
   intent to acquire and use WMD, 20
   Sudan and, 447
   suicide attacks, 290
   WMD use, 298

Encyclopedia of Bioterrorism Defense, Edited by Richard F. Pilch and Raymond A. Zilinskas
ISBN 0-471-46717-0 Copyright © 2005 Wiley-Liss